Advances in Experimental Medicine and Biology

Volume 1066

More information about this series at http://www.springer.com/series/5584

Tilman Borggrefe
Benedetto Daniele Giaimo
Editors

Molecular Mechanisms of Notch Signaling

Springer

Editors
Tilman Borggrefe
Institute of Biochemistry
University of Giessen
Giessen, Germany

Benedetto Daniele Giaimo
Institute of Biochemistry
University of Giessen
Giessen, Germany

ISSN 0065-2598 ISSN 2214-8019 (electronic)
Advances in Experimental Medicine and Biology
ISBN 978-3-030-07789-1 ISBN 978-3-319-89512-3 (eBook)
https://doi.org/10.1007/978-3-319-89512-3

Printed on acid-free paper

This Springer imprint is published by the registered company Springer International Publishing AG part of Springer Nature.
The registered company address is: Gewerbestrasse 11, 6330 Cham, Switzerland

Preface

Dear reader,

It is our pleasure to present this book *Molecular Mechanisms of Notch Signaling* to you. Over the last few years, we noticed that the latest book on Notch signaling represented an excellent starting point for scientists when entering the Notch field. This book, edited by Raphael Kopan, was published in 2010 and it was followed in 2014 by a Notch book, edited by Shinya Yamamoto and Hugo Bellen, focused on current methods and protocols. By now, we believe it is time for an update. To our pleasant surprise, it was not so difficult to find many dear colleagues in the field willing to contribute with their relevant expertise. This book can be subdivided into three sections: (1) molecular mechanisms of receptor/ligand interactions, (2) intracellular signaling mechanisms, and (3) disease links and therapeutics. Receptor ligand interactions are covered by summarizing structural aspects, mechanotransduction, regulation by glycosylation, endocytic trafficking but also by modeling the Notch response. Intracellular signaling covers a detailed discussion of the Notch interactome, comparing *Drosophila melanogaster* and human genetics to understand Notch-related pathologies. The second section also covers noncanonical aspects of Notch signaling and oscillatory mechanisms particularly relevant during development. Other developmental aspects of Notch in neurogenesis and stem cell biology are also discussed in this section. Transcriptional regulation, with a focus on canonical transcription factor RBPJ (also known as CSL), is described in depth. In the third section, the recently established link between Notch and senescence is elucidated and advances in our understanding of physiological and tumor angiogenesis are discussed. In regard to the immunological role of Notch, its function in T-cell development and activation is discussed and this is nicely complemented with the well-known role of Notch in leukemia.

This book summarizes molecular aspects of Notch signaling, and it is not only intended for experts, but it should also be a useful resource for young, sprouting scientists or interested scientists from other research areas, who may use this book as a stimulating starting point for further discoveries and developments. Thinking translational, we hope that this will help to encourage the development of better diagnostic tools and/or therapeutic applications for Notch-related diseases.

Tilman Borggrefe
Benedetto Daniele Giaimo

Contents

Contributors

Victor Alfred IFOM, Istituto FIRC di Oncologia Molecolare at IFOM-IEO Campus, Milan, Italy

Spyros Artavanis-Tsakonas Neurology Research, Biogen Inc., Cambridge, MA, USA

Department of Cell Biology, Harvard Medical School, Boston, MA, USA

Diana Bellavia Department of Molecular Medicine, Sapienza University of Rome, Rome, Italy

Anna Bigas Program in Cancer Research, Institut Hospital del Mar d'Investigacions Mèdiques (IMIM), Barcelona, Spain

Centro de Investigación Biomédica en Red de Cáncer (CIBERONC), Barcelona, Spain

Udi Binshtok Department of Biochemistry and Molecular Biology, George S. Wise Faculty of Life Science, Tel Aviv University, Tel Aviv, Israel

Stephen C. Blacklow Department of Biological Chemistry and Molecular Pharmacology, Harvard Medical School, Boston, MA, USA

Tilman Borggrefe Institute of Biochemistry, University of Giessen, Giessen, Germany

Saula Checquolo Department of Medico-Surgical Sciences and Biotechnologies, Sapienza University of Rome, Rome, Italy

Mark Chiang Division of Hematology-Oncology, Department of Internal Medicine, University of Michigan, Ann Arbor, MI, USA

Anna Engler Embryology and Stem Cell Biology, Department of Biomedicine, University of Basel, Basel, Switzerland

Andreas Fischer Vascular Signaling and Cancer (A270), German Cancer Research Center (DKFZ), Heidelberg, Germany

Vascular Biology, CBTM, Medical Faculty Mannheim, Heidelberg University, Mannheim, Germany

Department of Medicine I and Clinical Chemistry, Heidelberg University Hospital, Heidelberg, Germany

Benedetto Daniele Giaimo Institute of Biochemistry, University of Giessen, Giessen, Germany

Wendy R. Gordon Department of Biochemistry, Molecular Biology and Biophysics, University of Minnesota, Minneapolis, MN, USA

Anna C. McCarter Cell and Molecular Biology Program, University of Michigan, Ann Arbor, MI, USA

K. G. Guruharsha Neurology Research, Biogen Inc., Cambridge, MA, USA

Robert S. Haltiwanger Department of Biochemistry and Cell Biology, Stony Brook University, Stony Brook, NY, USA

Complex Carbohydrate Research Center, University of Georgia, Athens, GA, USA

Penny A. Handford Department of Biochemistry, University of Oxford, Oxford OX1 3RE, UK

Beth M. Harvey Department of Biochemistry and Cell Biology, Stony Brook University, Stony Brook, NY, USA

Present Address: Department of Cell and Developmental Biology, University of Pennsylvania, Philadelphia, PA, USA

Matthew Hoare Cancer Research UK Cambridge Institute, University of Cambridge, Cambridge, UK

Department of Medicine, Addenbrooke's Hospital, University of Cambridge, Cambridge, UK

Diana M. Ho Department of Cell Biology, Harvard Medical School, Boston, MA, USA

Akihiro Isomura Institute for Frontier Life and Medical Sciences, Kyoto University, Kyoto, Japan

Japan Science and Technology Agency, PRESTO, Saitama, Japan

Ryoichiro Kageyama Institute for Frontier Life and Medical Sciences, Kyoto University, Kyoto, Japan

World Premier International Research Initiative–Institute for Integrated Cell-Material Sciences (WPI-iCeMS), Kyoto University, Kyoto, Japan

Kyoto University Graduate School of Medicine, Kyoto, Japan

Kyoto University Graduate School of Biostudies, Kyoto, Japan

Thomas Klein Institute of Genetics, Heinrich-Heine-University, Duesseldorf, Germany

Boguslawa Korona Department of Biochemistry, University of Oxford, Oxford OX1 3RE, UK

Rhett A. Kovall Department of Molecular Genetics, Biochemistry and Microbiology, University of Cincinnati College of Medicine, Cincinnati, OH, USA

Susan M. Lea Sir William Dunn School of Pathology, University of Oxford, Oxford OX1 3RE, UK

Klaus N. Lovendahl Department of Biochemistry, Molecular Biology and Biophysics, University of Minnesota, Minneapolis, MN, USA

Masashi Narita Cancer Research UK Cambridge Institute, University of Cambridge, Cambridge, UK

Barbara A. Osborne Department of Veterinary & Animal Sciences, Program in Animal Biotechnology and Biomedical Sciences, University of Massachusetts, Amherst, MA, USA

Program in Molecular & Cellular Biology, University of Massachusetts, Amherst, MA, USA

Franz Oswald University Medical Center Ulm, Center for Internal Medicine, Department of Internal Medicine I, University of Ulm, Ulm, Germany

Rocco Palermo Center for Life Nano Science@Sapienza, Istituto Italiano di Tecnologia, Rome, Italy

Cristina Porcheri Program in Cancer Research, Institut Hospital del Mar d'Investigacions Mèdiques (IMIM), Barcelona, Spain

Christina Redfield Department of Biochemistry, University of Oxford, Oxford OX1 3RE, UK

Jose L. Salazar Department of Molecular and Human Genetics, Baylor College of Medicine (BCM), Houston, TX, USA

Björn Schnute Institute of Genetics, Heinrich-Heine-University, Duesseldorf, Germany

Isabella Screpanti Department of Molecular Medicine, Sapienza University of Rome, Rome, Italy

Hiromi Shimojo Institute for Frontier Life and Medical Sciences, Kyoto University, Kyoto, Japan

David Sprinzak Department of Biochemistry and Molecular Biology, George S. Wise Faculty of Life Science, Tel Aviv University, Tel Aviv, Israel

Richard Suckling Sir William Dunn School of Pathology, University of Oxford, Oxford OX1 3RE, UK

Verdon Taylor Embryology and Stem Cell Biology, Department of Biomedicine, University of Basel, Basel, Switzerland

Fabian Tetzlaff Vascular Signaling and Cancer (A270), German Cancer Research Center (DKFZ), Heidelberg, Germany

Vascular Biology, CBTM, Medical Faculty Mannheim, Heidelberg University, Mannheim, Germany

Tobias Troost Institute of Genetics, Heinrich-Heine-University, Duesseldorf, Germany

Thomas Vaccari IFOM, Istituto FIRC di Oncologia Molecolare at IFOM-IEO Campus, Milan, Italy

Dipartimento di Bioscienze, Università degli Studi di Milano, Milan, Italy

Jyothi Vijayaraghavan Department of Veterinary & Animal Sciences, Program in Animal Biotechnology and Biomedical Sciences, University of Massachusetts, Amherst, MA, USA

Qing Wang Division of Hematology-Oncology, Department of Internal Medicine, University of Michigan, Ann Arbor, MI, USA

Shinya Yamamoto Department of Molecular and Human Genetics, Baylor College of Medicine (BCM), Houston, TX, USA

Program in Developmental Biology, BCM, Houston, TX, USA

Department of Neuroscience, BCM, Houston, TX, USA

Jan and Dan Duncan Neurological Research Institute, Texas Children's Hospital, Houston, TX, USA

Runrui Zhang Embryology and Stem Cell Biology, Department of Biomedicine, University of Basel, Basel, Switzerland

List of Abbreviations

A	Alanine
AAA-ATPase	ATPase Associated with diverse cellular Activities
ACC	Adenoid Cystic Carcinoma
AD	Alzheimer's Disease
ADAM	A Disintegrin And Metalloprotease
AES	Amino-terminal Enhancer of Split
AF4p12	ALL1-Fused Gene From Chromosome 4p12 Protein
AFM	Atomic Force Microscopy
AGM	Aorta-Gonad-Mesonephros
ago	*archipelago*
AIP4	Atrophin-1 Interacting Protein 4
AITL	Angioimmunoblastic T-Cell Lymphomas
Aki1	Akt Kinase-Interacting Protein 1
AKT	AK Mouse Transforming
ALK	Activin receptor-Like Kinase
AMKL	Acute MegaKaryocytic Leukemia
AML	Acute Myeloid Leukemia
AML1	Acute Myeloid Leukemia 1
amx	*almondex*
ANK	Ankyrin
AOS	Adams-Oliver Syndrome
AP	Adaptor Protein
AP-3	Adaptor Protein-3
AP-MS	Affinity Purification and Mass Spectrometry
APC	Adenomatous Polyposis Coli protein
APC	Antigen Presenting Cell
APL	Acute Promyelocytic Leukemia
APH1A	Anterior Pharynx Defective 1 Homolog A
Aph	Anterior Pharynx Defective
aPKC	atypical Protein Kinase C
APOE	*APOlipoprotein E*
APP	Amyloid Precursor Protein
AhR	Aryl hydrocarbon Receptor
ARHGAP31	*Rho GTPase-Activating Protein 31*
Arp2/3	Actin-related protein 2/3

AS	Alagille Syndrome
AS-C	Achaete-Scute Complex
ASCL1	Achaete-Scute homolog 1
ATM	Ataxia Telangiectasia Mutated
AVD	Aortic Valve Disease
AVS	Aortic Valve Stenosis
β4GalT-1	β4-GalactosylTransferase-1
B-CLL	Chronic Lymphocytic Leukemia of the B-cell lineage
BAC	Bacterial Artificial Chromosome
BAV	Bicuspid Aortic Valve
BBC3	BCL-2 Binding Component 3
BCC	Basal Cell Carcinoma
BCL2	B-Cell CLL/Lymphoma 2
BCL6	B-Cell CLL/Lymphoma 6
BCL11B	B-Cell CLL/Lymphoma 11B
BCR	B-Cell Receptor
BDSC	Bloomington *Drosophila* Stock Center
BFP	Biomembrane Force Probe
bHLH	basic Helix-Loop-Helix transcription factor
bib	*big brain*
BioPlex	Biophysical interactions of ORFeome-based complexes
BLK	B Lymphocyte Kinase
BLNK	B-cell Linker Protein
BM	Bone Marrow
BMDC	Bone Marrow-derived Dendritic Cells
BMDMs	Bone Marrow Derived Macrophages
BMI1	B lymphoma Mo-MLV Insertion region 1 homolog
BMP	Bone Morphogenic Protein
BORIS/CTCFL	Brother Of Regulator of Imprinted Sites/CTCF-like protein
BPTES	Bis-2-(5-Phenylacetamido-1,2,4-Thiadiazol-2-Yl)Ethyl Sulfide
BRAF	BRaf proto-oncogene
BRD4	BromoDomain containing 4
BRG1	Brahma-Related Gene-1
C	Cysteine
C. elegans	*Caenorhabditis elegans*
CADASIL	Cerebral Autosomal Dominant Arteriopathy with Subcortical Infarcts and Leukoencephalopathy
CARM1	Coactivator-associated Arginine Methyltransferase 1
CAR-T	Chimeric Antigen T-cell Receptor
CAS9	CRISPR Associated Protein 9
CBF1	C-promoter Binding Factor 1
CBFβ	Core Binding Factor β
Cbl	Casistas B-lineage lymphoma
CBM	CARMA1, BCL10 and MALT1

CBP/CREBBP	C-Adenosine Mono Phosphate Responsive Element (cAMP-RE)-Binding protein (CREB)-Binding Protein; KAT3A
cbEGF	calcium binding Epidermal Growth Factor
CC2D1A	Coiled-Coil And C2 Domain Containing 1A
CC2D1B	Coiled-Coil And C2 Domain Containing 1B
CCL	C-C motif Chemokine Ligand
CCNC	Cyclin C
CCND1	Cyclin D1
CCND3	Cyclin D3
CCNE1	Cyclin E1
CCR7	C-C Chemokine Receptor Type 7
CCZ1	Calcium-Caffeine-Zinc sensititivity protein
CDK	Cyclin-Dependent Kinase
CDK3	Cyclin-Dependent Kinase 3
CDK4	Cyclin-Dependent Kinase 4
CDK6	Cyclin-Dependent Kinase 6
CDK8	*Cyclin-Dependent Kinase 8*
CDK9	Cyclin-Dependent Kinase 9
CDK19	Cyclin-Dependent Kinase 19
CDKN1B	Cyclin Dependent Kinase Inhibitor 1B
CDKN2	Cyclin Dependent Kinase Inhibitor 2D
cDNA	complementary DeoxyriboNucleic Acid
C/EBPβ	CCAAT/Enhancer Binding Protein beta
CHARGE	Cohorts for Heart and Aging Research in Genomic Epidemiology
CHD4	Chromodomain Helicase DNA Binding Protein 4
ChIP	Chromatin ImmunoPrecipitation
CHMP	CHarged Multivesicular body Protein
CIR	Corepressor Interacting with RBPJ
CK1	Casein Kinase 1
CLL	Chronic Lymphocytic Leukemia
CLP	Common Lymphoid Progenitor
CNS	Central Nervous System
CNS-2	Conserved Noncoding Sequence-2
COA	Coarctation of the Aorta
CORVET	class C core Vacuole/Endosome Tethering
COMMD9	COMM (Copper metabolism) Domain containing protein 9
CR	Cysteine-Rich
CR2	Complement C3d Receptor 2
crb	crumbs
CREB	cAMP Response Element Binding protein
CRISPR	Clustered Regularly Interspaced Short Palindromic Repeats
CSCs	Cancer Stem Cells
cSCC	cutaneous Squamous Cell Carcinoma
CSL	CBF1/Su(H)/Lag-1

CTBP	C-terminal Binding Protein
CTIP	CTBP Interacting Protein
CTL	Cytotoxic T lymphocytes
CXCL	chemokine (C-X-C motif) Ligand
CXCR	C-X-C motif chemokine Receptor
CXCR4	C-X-C chemokine Receptor type 4
CYLD	Ubiquitin Carboxyl-Terminal Hydrolase
DDD	Dowling-Degos Disease
DDR	DNA Damage Response
DDX5	DEAD-box RNA helicase 5
DEPTOR	DEP domain containing MTOR interacting protein
DFS	Dominant Female Sterile
DG	Dentate Gyrus
DGGR	*Drosophila* Genomics and Genetic Resources
DGRC	*Drosophila* Genomics Resource Center
DIOPT	*Drosophila* RNAi screening center Integrative Ortholog Prediction Tool
Dl	*Delta*
Dll	Delta-like
Dll1	Delta-like 1
Dll3	Delta-like 3
DLL4	Delta-like 4
DLX5	Distal-Less Homeobox 5
DM14	*Drosophila melanogaster* 14
DN	Double Negative
DN1/DN3	Double Negative (thymocytes)
DN-MAML	Dominant-Negative Mastermind
DNA	DeoxyriboNucleic Acid
dNedd4	Neural precursor cell expressed developmentally down-regulated protein 4
dNotch	*Drosophila* Notch
DOCK6	*Dedicator Of CytoKinesis 6*
DP	Double Positive
DPiM	*Drosophila* Protein interaction Map
DSBs	Double Strand Breaks
DSHB	Developmental Studies Hybridoma Bank
DSL	Delta, Serrate, LAG-2 ligands
DTX1	Deltex E3 ubiquitin ligase 1
DTX2	Deltex E3 ubiquitin ligase 2
DTX3	Deltex E3 ubiquitin ligase 3
Dvl	Disheveled
dx	*deltex*
E	Glutamic acid
E	Embryonic day
E2A	E2A immunoglobulin enhancer-binding factor E12/E47
E8	Embryonic day 8
EAE	Experimental Autoimmune Encephalomyelitis
EB	Enteroblasts

EBNA2	Epstein–Barr Nuclear Antigen 2
EBV	Epstein–Barr Virus
EBV-LCL	Epstein-Barr Virus lymphoblastoid B cells
EC	Endothelial Cell
ECD	Extracellular Domain
ECM	ExtraCellular Matrix
EE	Early Endosome
EE2A	Early Endosomal Entigene 2
EEVs	Early Endosomal Vesicles
EGF	Epidermal Growth Factor
EGFR	Epidermal Growth Factor Receptor
EHBA	Biliary Atresia, ExtraHepatic
EHBP-1	*EH (Eps15 Homology) domain Binding Protein-1*
EIF2A	Eukaryotic Translation Initiation Factor 2A
elav	*embryonic lethal abnormal vision*
EM	Electron Microscopy
EMPs	Erythroid-Myeloid Progenitors
EMS	Ethyl MethaneSulfonate
EMT	Epithelial-Mesenchymal Transition
Eogt	*EGF-domain O-GlcNAc transferase*
EP300	E1A Binding Protein P300, KAT3B
ER	Endoplasmic Reticulum
ER	Estrogen Receptor
ERK	Extracellular signal-Regulated Kinase
ES	Embryonic Stem cells
ESCRT	Endosomal Sorting Complex Required for Transport
E(spl)-C	*Enhancer of split-Complex*
ETO	Eight-Twenty One
ETS1	E26 avian leukemia oncogene 1
ETP-ALL	Early T-cell Precursor Acute Lymphoblastic Leukemia
F	Phenylalanine
FAD	Familial Alzheimer's Disease
FBXW7	F-box and WD repeat domain containing 7
FDA	Food and Drug Administration
FGF	Fibroblast Growth Factor
FHL1	*Four and a Half LIM domains 1*
FL	Follicular Lymphoma
FLP	FLiPpase
Fng	*Fringe*
FOX	Forkhead box protein
FOXA2	Forkhead box A2
FRET	Förster Resonance Energy Transfer
Freud-1	FRE under Dual Repression-Binding-Protein 1
Freud-2	FRE under Dual Repression-Binding-Protein 2
Fringe	Beta-1,3-N-Acetylglucosaminyltransferase
FRT	Flippase Recognition Target
Fuc	Fucose
FYVE	Fab1 YOTB VAC1 EEA1

G	Glycine
G-CSF	Granulocyte Colony Stimulating Factor
GABPA	GA Binding Protein Transcription Factor Alpha Subunit
Gal	Galactose
GAP	GTPase-Activating Protein
GATA3	GATA Binding Protein 3
GATAD2B	GATA Zinc Finger Domain Containing 2B
GBM	Glioblastoma
GCN5	General Control Of AmiNo Acid Synthesis Protein 5-Like 2; KAT2A
GDE2	Glycerophosphodiester phosphodiesterase 2
GDF	GTPase Dissociation Factor
GDI	GDP-Dissociation Inhibitor
GDP	Guanosine DiPhosphate
GEF	Guanine nucleotide Exchange Factor
GEMM	Genetically Engineered Mouse Model of T-ALL
GFAP	Glial Fibrillary Acidic Protein
GFI1	Growth Factor Independent 1 transcriptional repressor
GFP	Green Fluorescent Protein
GI	GastroIntestinal
Glc	Glucose
glcNAc	N-Acetylglucosamine
glp-1	*germ line proliferation defective-1*
GOF	Gain-Of-Function
GOM	Granular Osmophilic Material
GPI	GlycosylPhosphatidylInositol
Gro	*Groucho*
GSI	γ-Secretase Inhibitor
GSK3β	Glycogen Synthase Kinase 3β
GTP	Guanosine TriPhosphate
GVHD	Graft Versus Host Disease
GWAS	Genome-Wide Association Studies
GXYLT	Glucoside α3-Xylosyltransferase
H	*Hairless*
H3K27ac	Histone 3 acetylated at lysine 27
H&N	Head and Neck
hAPF	hours After Puparium Formation
HCC	HepatoCellular Carcinoma
HCOP	Human genome organization gene nomenclature committee Comparison of Orthology Predictions search
HCS	Hadju-Cheney Syndrome
HD	Heterodimerization Domain
HDACs	Histone Deacetylases
HDAC1	Histone Deacetylase 1
HECT	Homologous to the E6-AP Carboxyl Terminus
hDll4	Human delta-like 4
HDR	Homology Directed Repair
HEB	E2A/Hela E Box-Binding

HER2	Human Epidermal growth factor Eeceptor 2
Hes	Hairy and enhancer of split
Hey	Hairy/enhancer-of-split related with YRPW motif
HF	Hair Follicles
hGAVPO	humanized and optimized factor consisting of Gal4 DNA-binding domain, *Neurospora crassa* photoreceptor Vivid, and p65 activation domain
HIF-1α	Hypoxia Inducible Factor 1 alpha
hJ-1	Human Jagged-1
HLH	Helix-Loop-Helix
HLHS	Hypoplastic Left Heart Syndrome
hN1	Human Notch1
HOPS	HOmotypic fusion and Protein Sorting
HPV	Human Papilloma Virus
Hrs	Hepatocyte growth factor-regulated tyrosine kinase substrate
HSC	Hematopoietic Stem Cell
Hsc70	Heat shock cognate70
HSP90	Heat Shock Protein 90
HUD	RNA-binding protein D of the ELAV/Hu family
I	Isoleucine
IκB	Inhibitor of kappa B
ICD	IntraCellular Domain
ICN1	IntraCellular Notch1
ID3	Inhibitor of DNA binding 3 HLH protein
IFE	InterFollicular Epidermis
IFNγ	Interferon gamma
IGF1R	Insulin like Growth Factor 1 Receptor
IHC	ImmunoHistoChemistry
IK	Ikaros
IK-DN	Ikaros Dominant Negative isoforms
IKK	Inhibitor of Kappa-B Kinase
IKKα	Inhibitor of Kappa-B Kinase subunit alpha
IKKβ	Inhibitor of Kappa-B Kinase subunit beta
IKZF1	Ikaros Family Zinc Finger 1
IL	Interleukin
IL1	Interleukin 1
IL1R	Interleukin 1 Receptor
IL2	Interleukin 2
IL4	Interleukin 4
IL6	Interleukin 6
IL7	Interleukin 7
IL7R	Interleukin 7 Receptor
IL9	Interleukin 9
IL10	Interleukin 10
IL12	Interleukin 12
IL13	Interleukin 13
IL17	Interleukin 17

IL22	Interleukin 22
ILV	Intraluminal Vesicle
IMF	Infantile MyoFibromatosis
INK4	Inhibitor of cyclin-dependent Kinase 4
IP	Intermediate Progenitor
IPC	Intermediate Progenitor Cell
IRAK2	Interleukin-1 Receptor-Associated Kinase-like 2
IRF4	Interferon Regulatory Factor 4
IRF8	Interferon-Regulatory Factor 8
ISCs	Intestinal Stem Cells
Itch	Itchy E3 ubiquitin protein ligase
JAG	JAGged
JAK	Janus Kinase
JME	JuxtaMembrane Extracellular
JNK	c-Jun N-terminal Kinase
JUN	Ju-nana (Japanese number 17)
K	Lysine
KDM5A	Lysine(K) Demethylase 5A
KO	Knockout
KMT2D	Lysine MethylTransferase 2D
KRAS	Kirsten Rat Sarcoma viral oncogene homolog
krz	kurtz
kuz	*kuzbanian*
KYOT2/FHL1	Four and a Half LIM domains 1
L	Leucine
l(2)gd1	*lethal (2) giant discs 1*
L3MBTL3	Lethal(3)Malignant Brain Tumor-Like Protein 3
LAG-1	Abnormal cell LINeage-12 (Lin-12) And abnormal Germ line proliferation phenotype-1 (Glp-1)
Lamp	Lysosome-associated membrane glycoprotein
LBR	Ligand-Binding Region
LE	Late Endosome
LEF1	Lymphoid Enhancer binding Factor 1
LFNG	*Lunatic FriNGe*
lgd	lethal (2) giant discs
LGR5	Leucine rich repeat containing G protein-coupled Receptor 5
LIC	Leukemia-Initiating Cells
LID	Little Imaginal Disks
LIF	Leukemia Inhibitory Factor
LIM	*Lin11, Isl-1 and Mec-3*
lin-12	*cell LINeage defective-12*
LMNS	Lateral MeNingocele Syndrome
LMO2	Lim domain Only 2
LN	Lymph Node
LNR	Lin-12/Notch Repeat
LOAD	Late-Onset Alzheimer's Disease
LOF	Loss-Of-function

LPS	Lipopolysaccharide
LRRK2	Leucine Rich Repeat Kinase 2
LSC	Leukemia Stem Cells
lSCC	lung Squamous Cell Carcinoma
LSD1	Lysine (K)-Specific Demethylase 1A
LUNAR1	Leukemia-associated Non-coding IGF1R Activator RNA 1
LV	Lateral Ventricle
LVNC	Left Ventricular NonCompaction
LW	Lateral Wall
LYN	Lck/Yes-related Novel protein tyrosine kinase
MAL	Megakaryocytic Acute Leukemia
mam	*mastermind*
MAML	Mastermind-Like
MAPK	Mitogen-Activated Protein Kinase
MARRVEL	Model organism Aggregated Resources for Rare Variant ExpLoration
Mash1	Mammalian achaete scute homolog-1
MCL	Mantle Cell Lymphoma
MCL1	Myeloid leukemia cell differentiation protein 1
ME	Maturing Endosome
MEF	Mouse Embryonic Fibroblast
MESP2	*Mesoderm posterior bHLH transcription factor 2*
MFNG	*Manic FriNGe*
mib	mindbomb
MiMIC	*Minos*-Mediated Integration Cassette
MINT	Msx2-interating protein
miR	microRNA
Mon1	Monensin sensitivity protein 1
MPC	Memory Precursor Cell
MPECs	Memory Precursor Effector Cells
MPN	Magnetic nanoparticle
mRNA	messenger RiboNucleic Acid
MST	Mammalian Sterile 20-like kinase
MT1-MMP	Membrane-Tethered-Matrix Metallo-Protease 1
mTOR	Mechanistic (or Mammalian) Target Of Rapamycin
MVB	MultiVesicular Body
MYC	Myelocytomatosis proto-oncogene
MyoD	Myogenic differentiation antigen
N-EGF3	N-terminus-EGF3
N-ME	Notch-MYC Enhancer
N1ICD	NOTCH1 Intracellular Domain
N2ICD	Notch2 Intracellular Domain
N3ICD	Notch3 Intracellular Domain
nCC	nonCutaneous Carcinoma
NCoR	Nuclear receptor CoRepressor
Nct/NCSTN	*Nicastrin*
Ndfip	Nedd4 family interacting protein

NDME	Notch-Dependent MYC Enhancer
NECD	Notch ExtraCellular Domain
Nedd4	Neural precursor cell expressed developmentally down-regulated protein 4
nej	*nejire*
NEMO	NF-κB Essential Modulator
NEPs	Neuroepithelial cells
neur/NEURL	*neuralized*
NEXT	Notch EXtracellular Truncation
NFAT	Nuclear Factor of Activated T-cells
NF-κB	Nuclear Factor-κB
NFκB2	Nuclear Factor of Kappa light polypeptide gene enhancer in B cells 2
Ngn2	Neurogenin2
NICD	Notch IntraCellular Domain
NK	Natural Killer
NKT	Natural Killer T
NLK	Nemo-Like Kinase
NLS	Nuclear Localisation Sequences
NOS	Not Otherwise Specified
$Notch^{MCD}$	Notch *microchaetae defective*
NRARP	Notch-Regulated Ankyrin Repeat-containing Protein
Nrp1	Neuropilin-1
NRP	Neuropilin
NRR	Negative Regulatory Region
NSCLC	Non-Small Cell Lung Cancer
NSCs	Neural Stem Cells
NTM	Transmembrane subunit
Numbl	Numb-like
NURD	Nucleosome Remodeling Deacetylase
OB	Olfactory Bulb
OGT	*O*-GlcNAc Transferase
O-fut1	*O-fucosyltransferase-1*
OLIG2	Oligodendrocyte transcription factor
OMIM	Online Mendelian Inheritance in Man
OTT	One-Twenty Two
P	Proline
P	Postnatal day
P14	Postnatal day 14
p70S6K	Ribosomal protein S6 Kinase beta-1
PanIN	Pancreatic Intraepithelial Neoplasia
PAX3/7	Paired homeobox transcription factors
PB1	Polybromo 1
PBAF	Polybromo-Associated BRG1- or HBRM-associated Factors
PBMCs	Peripheral Blood Mononuclear Cells
PC	PhosphatidylCholine
PCAF	P300/CBP-Associated Factor; KAT2B

PCR	Polymerase Chain Reaction
PD1	Programmed Death 1
PDAC	Pancreatic Ductal AdenoCarcinoma
PDGFRB	*Platelet Derived Growth Factor Receptor Beta*
PDX	Patient-Derived Xenograft
PDZ	PSD-95/Dlg/ZO-1
PE	PhosphatidylEthanolamine
pen/PSENEN	*presenilin enhancer*
PEST	proline (P), glutamic acid (E), serine (S) and threonine (T)
PFKFB3	6-Phosphofructo-2-kinase
PHF8	PHD Finger Protein 8
PI(3,5)P2	PhosphatidylInositol 3,5-bisphoshphate
PI3K	PhosphatidylInositol 4,5-bisphosphate 3-Kinase
PI3P	PhosphatidylInositol 3-Phosphate
PIAS	Protein Inhibitor of Activated STAT
PIK3CD	PhosphatidylInositol-4,5-bisphosphate 3-Kinase Catalytic subunit Delta
PIM	Proto-Oncogene, Serine/Threonine Kinase
PIN1	Peptidylprolyl Cis/Trans Isomerase, NIMA-Interacting 1
PIP3	PhosphatidylInositol (3,4,5)-trisphosphate P_3
PLCγ	PhosphoLipase Cγ
PlGF	Placenta Growth Factor
PML	Promyelocytic Leukemia locus gene
pN	picoNewton
POFUT1	*Protein O-fucosyltransferase 1*
POGLUT1	*Protein O-glucosyltransferase 1*
PPI	Protein-Protein Interaction
PR	Progesterone Receptor
PRMT4	Protein Arginine(R) N-MethylTransferase 4
PRL2	Phosphatase of Regenerating Liver
PS	PhosphatidylSerine
PS1/2	*Presenilin 1/2*
PSM	PreSomitic Mesoderm
Psn/PSEN	*Presenilin*
Ptcra	invariant preTα chain of the pre-T cell receptor
PTEN	Phosphatase and tensin homolog
PTM	Post-Translational Modification
QRT-PCR	Quantitative Real Time PCR
R	Arginine
RA	Retinoic Acid
Rab	ras-related in brain
RAM	RBPJ-Associated Molecule
RARA	Retinoic Acid Receptor-Alpha
RAS	Rat sarcoma virus oncogene
RB	Retinoblastoma Protein
RBBP4	RB Binding Protein 4, chromatin remodeling factor
RBC	Red Blood Cell

RBPJ	Recombination signal Binding Protein for immunoglobulin kappa J region
RCC	Renal Cell Cancer
RE	Recycling Endosome
RECK	Reversion-inducing Cysteine-rich protein with Kazal motifs
RelA	v-rel reticuloendotheliosis viral oncogene homolog A
RELB	Relaxed B proto-oncogene
RFNG	*Radical FriNGe*
RGC	Radial Glia Cell
RHD	Rel Homology Domain
RIP	Regulated Intramembrane Proteolysis
RIPPLY2	*RIPPLY transcriptional repressor 2*
RITA11	RBPJ Interacting and Tubulin Associated 1
RMCE	Recombinase Mediated Cassette Exchange
RME8	Receptor Mediated Endocytosis 8
RMS	Rostral Migratory Stream
RNA	RiboNucleic Acid
RNAi	RNA interference
RNAPII	RNA Polymerase II
RNF40	Ring Finger protein 40
ROCK2	Rho-associated protein kinase 2
RORγt	RAR-related Orphan Receptor gamma t
RUNX	Runt related transcription factor
S	Serine
S2 site	Metalloprotease cleavage site
S3 site	Gamma-secretase cleavage site
SA	Splice Acceptor
SA β-GAL	Senescence-Associated beta-galactosidase
SAHM	MAM-like Stapled Peptides
SASP	Senescence-Associated Secretory Phenotype
SAXS	Small Angle X-ray Scattering
SCC	Squamous Cell Carcinoma
SCDO	SpondyloCostal DysOstosis
SCF	S-Phase Kinase Associated Protein1/Cullin/F-Box Protein
SCLC	Small Cell Lung Cancer
Sec15	*Secretory 15*
SEL10	Suppressor and/or Enhancer of abnormal cell LINeage-12 (Lin-12)-10
Ser	*Serrate*
Ser	Serine
SERCA	Sarco/Endoplasmic Reticulum Calcium ATPase
SG	Sebaceous Glands
SGZ	SubGranular Zone
SHARP	SMRT/HDAC1 Associated Repressor Protein
SHH	Sonic hedgehog
shi	*shibire*
Shrb	Shrub

Sia	Sialic Acid
SIRT-1	Sirtuin-1
SKIP	Sloan-KetterIng-retroviral oncogene (SKI) -Interacting Protein
SKP2	S-phase Kinase associated Protein 2
SLE	Systemic Lupus Erythematosus
SLECs	Short Lived Effector Cells
SLL	Small Lymphocytic Lymphoma
SMAD	Mothers Against Decapentaplegic
SMRT	Silencing Mediator for Retinoid or Thyroid-hormone receptors
SMZL	Splenic Marginal Zone Lymphoma
SNARE	Soluble *N*-ethylmaleimide-sensitive-factor Attachment R*e*ceptor
SNX	Sorting NeXin
SOP	Sensory Organ Precursor
SP	Single Positive
spdo	*sanpodo*
SPEN	*Split ENds family transcriptional repressor*
spl	*split*
SPOC	Spen Paralog and Ortholog C-terminal domain
SPR	Surface Plasmon Resonance
SPS	Su(H) Paired Sites or Sequence Paired Sites
SSB	Single-Strand Binding protein
Stam	Signal transducing adaptor molecule
STAT	Signal Transducer and Activator of Transcription
STAT4	Signal Transducer and Activator of Transcription 4
STAT5	Signal Transducer and Activator of Transcription 5
STUB1	STIP1 Homology And U-Box Containing Protein 1
Su(dx)	*Suppressor of deltex*
Su(H)	*Suppressor of Hairless*
SYK	Spleen tyrosine Kinase
Synj2bp	Synaptojanin-2-binding protein
SVZ	SubVentricular Zone
T	Threonine
T-ALL	T-cell Acute Lymphoblastic Leukemia
TACE	Tumor Necrosis Factor (TNF)-Alpha-Converting Enzyme
TAD	Trans-Activation Domain
TAL1	T-cell Acute Lymphoblastic Leukemia 1
TAP	Transient Amplifying Progenitor
TAPE	TBK1-Associated Protein in Endolysosomes
TAV	Tricuspid Aortic Valve
TBX6	*T-box 6*
TCF1	Transcription Factor 1
TCR	T-Cell Receptor
TCRB	T-Cell Receptor β
TEC	Terminal Effector Cells
Temp	*Tempura*

TGF	Transforming Growth Factor
TGFβ	Transforming Growth Factor beta
TGT	Tension Guage Tether
Th1	Lymphocyte T-helper 1
Th2	Lymphocyte T-helper 2
Thr	Threonine
TKO	Triple KnockOut
TLR	Toll-Like Receptor
TLE	Transducin Like Enhancer protein
TM	Transmembrane domain
TM2D3	*TM2 Domain containing 3*
TMPs	Transmembrane proteins
TMZ	Temozolomide
TNBC	Triple-Negative Breast Cancer
TNF	Tumor Necrosis Factor
TNFα	Tumor Necrosis Factor alpha
TOF	Tetralogy Of Fallot
TORC1/2	mTOR signaling complex 1/2
TP53	Tumor protein P53
TPR	TetratricoPeptide Repeat
TRAF6	TNF-Receptor-Associated Factor 6
Tregs	T regulatory cells
TRIB2	Tribbles pseudokinase 2
TSC1/2	Tuberous Sclerosis 1/2
Tsg101	Tumor susceptibility gene 101
TSS	Transcriptional Start Site
UAS	Upstream Activation Sequence
Ub	Ubiquitin
UIM	ubiquitin interacting motif
UTR	Untranslated Region
V	Valine
V-ATPase	Vacuolar-ATPase
VDRC	Vienna *Drosophila* Resource Center
VEGF	Vascular Endothelial Growth Factor
VEGFR	Vascular Endothelial Growth Factor Receptor
Vegfr1	Vascular endothelial growth factor receptor 1
Vps	Vacoular protein sorting
VUS	Variant of Unknown Significance
VZ	Ventricular Zone
WASp	Wiskott-Aldrich Syndrome protein
wg	*wingless*
Wnt	Wingless
wry	weary
WT	Wild Type
WWP2	WW domain containing E3 ubiquitin Protein ligase 2
Xyl	Xylose
XXYLT	Xyloside α3-Xylosyltransferase
Y2H	Yeast two-Hybrid

YAP/TAZ	Yes-associated protein and WW domain containing transcription regulator 1
ZFP36L1	Zinc Finger Protein C3H type 36-Like 1
ZFP36L2	Zinc Finger Protein C3H type 36-Like 2
ZMIZ1	Zinc finger MIZ-type containing 1
ZNF143	Zinc finger protein 143

Part I

Introduction

Introduction to Molecular Mechanisms in Notch Signal Transduction and Disease Pathogenesis

Benedetto Daniele Giaimo and Tilman Borggrefe

Abstract

The Notch signaling pathway plays a pivotal role in development, physiology and diseases such as cancer. In this chapter, we first give an overview of the different molecular mechanisms that regulate Notch signaling. Each subject is covered in more depth in the subsequent chapters of this book. Next, we will use the inflammatory system as an example to discuss the physiological function of Notch signaling. This is followed by a discussion of recent advances in the different pathophysiological roles of Notch signaling in leukemia as well as a wide range of solid cancers. Finally, we discuss how information about pathogenic mutations in Notch pathway components, combined with structural biological data, are beginning to provide important biological and mechanistic insights about the pathway.

Keywords

Notch · Transcription · Cancer · Inflammation

B. D. Giaimo (✉) · T. Borggrefe (✉)
Institute of Biochemistry, University of Giessen, Giessen, Germany
e-mail: Benedetto.Giaimo@biochemie.med.uni-giessen.de; Tilman.Borggrefe@biochemie.med.uni-giessen.de

Abbreviations

A	Alanine
ACC	Adenoid cystic carcinoma
ADAM	A Disintegrin And Metalloprotease
ANKs	Ankyrin
AML	Acute Myeloid Leukemia
AOS	Adams-Oliver Syndrome
ASCL1	Achaete-scute homolog 1
AVD	Aortic valve disease
AVS	Aortic valve stenosis
BAV	Bicuspid aortic valve
BCC	Basal cell carcinoma
BMDM	Bone marrow derived macrophages
BMI1	B lymphoma Mo-MLV insertion region 1 homolog
BRD4	BromoDomain-containing 4
CADASIL	Cerebral Autosomal Dominant Arteriopathy with Subcortical Infarcts and Leukoencephalopathy
CDK9	Cyclin-dependent kinase 9
CLL	Chronic lymphocytic leukemia
COA	Coarctation of the aorta
CR	Cysteine-rich
CSCs	Cancer stem cells
cSCC	Cutaneous squamous cell carcinoma
DKO	Double knockout
DLL	DELTA-LIKE
DN-MAML	Dominant-Negative Mastermind
DSL	DELTA, SERRATE, LAG-2
E	Glutamic acid

T. Borggrefe, B. D. Giaimo (eds.), *Molecular Mechanisms of Notch Signaling*, Advances in Experimental Medicine and Biology 1066, https://doi.org/10.1007/978-3-319-89512-3_1

EGF	Epidermal growth factor
EGFR	Epidermal growth factor receptor
EMT	Epithelial-Mesenchymal Transition
EP300	E1A Binding Protein P300
ER	Endoplasmic reticulum or estrogen receptor
ERK	Extracellular signal-Regulated Kinase
ETS1	E26 avian leukemia oncogene 1
F	Phenylalanine
FBXW7	F-box and WD repeat domain-containing 7
FOXA2	Forkhead box A2
G	Glycine
GABPA	GA Binding Protein Transcription Factor Alpha Subunit
GBM	Glioblastoma
GOF	Gain-Of-Function
GSI	γ-Secretase Inhibitor
HCC	Hepatocellular carcinoma
HCS	Hadju-Cheney syndrome
HD	Heterodimerization Domain
HDACs	Histone deacetylases
HER2	Human epidermal growth factor receptor 2
Hes1	Hairy and Enhancer of Split 1
HEY1	Hairy/enhancer-of-split related with YRPW motif 1
HLH	Helix-loop-helix
HLHS	Hypoplastic left heart syndrome
H&NHLHS	head and neck
I	Isoleucine
IκB	Inhibitor of kappa B
IKKα	Inhibitor of Kappa-B Kinase subunit alpha
IL1R	Interleukin 1 receptor
IL4	Interleukin 4
IL6	Interleukin 6
IL10	Interleukin 10
IL12	Interleukin 12
IL13	Interleukin 13
IFNγ	Interferon γ
IRAK2	Interleukin 1 receptor-associated kinase-like 2
IRF8	Interferon-regulatory factor 8
JAG	JAGGED
K	Lysine
KO	Knockout
KMT2D	Lysine methyltransferase 2D
L	Leucine
LNR	Lin-12/Notch Repeat
LOF	Loss-Of-function
LPS	Lipopolysaccharide
lSCC	Lung squamous cell carcinoma
MAML	MASTERMIND-LIKE
MAPK	Mitogen-activated protein kinase
MCL	Mantle cell lymphoma
MINT	Msx2-interating protein
miR	microRNA
MST	Mammalian sterile 20-like kinase
mTOR	mammalian target of rapamycin
Myc	myelocytomatosis proto-oncogene
nCC	Noncutaneous carcinoma
NCoR	Nuclear receptor corepressor
NECD	Notch extracellular domain
NEXT	Notch EXtracellular Truncation
NF-κB	Nuclear Factor-κB
NICD	Notch intracellular domain
NRR	Negative regulatory region
NSCLC	Non-small cell lung cancer
OLIG2	Oligodendrocyte transcription factor
P	Proline
PanIN	Pancreatic intraepithelial neoplasia
PDAC	Pancreatic ductal adenocarcinoma
PDZ	PSD-95/Dlg/ZO-1
PEST	proline (P), glutamic acid (E), serine (S) and threonine (T)
PI3K	PhosphatidylInositol 4,5-bisphosphate 3-Kinase
PIM	Proto-Oncogene, Serine/Threonine Kinase
PIN1	Peptidylprolyl Cis/Trans Isomerase, NIMA-Interacting 1
POFUT1	Protein O-fucosyltransferase 1
PR	Progesterone receptor
Ptcra	invariant preTα chain of the pre-T cell receptor
PTEN	Phosphatase and tensin homolog

PTM	Post-translational modification
R	Arginine
RAM	RBPJ-associated module
RAS	Rat sarcoma virus oncogene
RBPJ	Recombination signal binding protein for immunoglobulin kappa J region
RCC	Renal cell cancer
RUNX1	Runt related transcription factor 1
RUNX3	Runt related transcription factor 3
S	Serine
SCC	Squamous cell carcinoma
SCLC	Small cell lung cancer
SHARP	SMRT/HDAC1 Associated Repressor Protein
SLE	Systemic lupus erythematosus
SMRT	Silencing mediator for retinoid or thyroid-hormone receptors
SMZL	Splenic marginal zone lymphoma
SPEN	Split ENds family transcriptional repressor
SPOC	Spen paralog and ortholog C-terminal domain
T	Threonine
T-ALL	T-cells acute lymphoblastc leukemia
TAD	Trans-activation domain
TAV	Tricuspid aortic valve
TCR	T-cell receptor
TGFβ	Transforming growth factor beta
TKO	Triple knockout
TLR	Toll-like receptor
TLR4	Toll-like receptor 4
TM	Transmembrane
TMZ	Temozolomide
TNBC	Triple-negative breast cancer
TNFα	Tumor necrosis factor alpha
TORC1/2	mTOR signaling complex 1/2
V	Valine
Vegfr1	Vascular endothelial growth factor receptor 1
YAP/TAZ	Yes-associated protein and WW domain containing transcription regulator 1
ZNF143	Zinc finger protein 143

1 Overview on Notch signaling

The Notch mutant phenotype was first described over a hundred years ago by John Dexter, who noticed the appearance of notches at the wing margins of fruit flies *Drosophila melanogaster*. Thomas Hunt Morgan identified the alleles of the corresponding genes (Morgan 1917). Several decades later, the *Notch* gene, encoding a transmembrane receptor controlling *Drosophila* neurogenesis, was identified (Artavanis-Tsakonas et al. 1983; Wharton et al. 1985; Kidd et al. 1986; del Amo et al. 1993). Soon after that, it became apparent that the *Notch* gene is evolutionary conserved and controls a plethora of developmental decisions, regulating homeostasis as well as development and differentiation of several different tissues and cell types during both embryonic and postnatal life. Thus, it is one of a few signaling pathways, like Wnt, transforming growth factor beta (TGFβ) and Hedgehog that is repeatedly used in multicellular organisms throughout embryonal and adult development. In Integration of *Drosophila* and Human Genetics to Understand Notch Signaling Related Diseases, Yamamoto and colleagues introduce how biological and genetic experiments in *Drosophila* contributed to the identification of key players in Notch signaling, and further discuss how mechanistic information obtained in flies can be translated to understand Notch signaling related genetic disorders in human. Notch signaling has also been implicated in carcinogenesis, of which we will highlight in this chapter and further dedicate several chapters in this book (The Notch3 Receptor and Its Intracellular Signaling-Dependent Oncogenic Mechanisms and Notch in Leukemia).

The mechanisms of how Notch signaling pathway regulates a wide range of functions can be grouped in three main categories: lateral inhibition, lateral induction and lineage decisions. During lateral inhibition, equipotent cells establish a hierarchy mediated by NOTCH receptors and ligands. During these signaling events, one cell "A" signals more to the adjacent ones preventing them to adopt the same "A" cell fate. In the lateral induction model, non-equipotent cells are involved. In particular, one group of cells signals to another group determining the acquisition of different cell fates. Finally, in the lineage decision model, asymmetrical cell division allows daughter cells to adopt different cell fates by the differential expression and/or segregation of NOTCH receptors or modulators of the Notch pathway. These models are described in depth in "Modeling the Notch Response", "Integration of *Drosophila* and Human Genetics to Understand Notch Signaling Related Diseases", "Notch and Stem Cells" and "Notch and Senescence" of this book. T-cell differentiation is a well-characterized example of the lineage decision model that was investigated in depth. In particular, loss-of-function (LOF) of Notch leads to a complete block in T-cell development (Radtke et al. 1999), whereas gain-of-function (GOF) of Notch, by introducing a constitutive-active form of Notch into hematopoietic progenitor cells, leads to T-cells acute lymphoblastic leukemia (T-ALL) in mice (Pear et al. 1996). In fact, the human *NOTCH1* gene was identified in T-ALL patients as a hot spot of chromosomal translocations (Ellisen et al. 1991). The role of Notch in the early stages of T-cell development is discussed by Osborne and colleagues in "Notch and T Cell Function – A Complex Tale" of this book. Regarding pathogenesis, Chiang and colleagues discuss the aspects of Notch signaling in leukemogenesis (Notch in Leukemia) and Screpanti and colleagues focus on NOTCH3 related functions (The Notch3 Receptor and Its Intracellular Signaling-Dependent Oncogenic Mechanisms).

2 Molecular Mechanisms Controlling the Notch Signal Transduction Pathway

At the molecular level, the Notch signaling pathway is a seemingly simple pathway that does not involve any second messengers. Ligand-triggered activation of the NOTCH receptor leads to the release of the cleaved NOTCH intracellular domain (NICD) that drives the signaling response (Fig. 1). NOTCH receptors are single-pass transmembrane proteins that are synthesized in the endoplasmic reticulum (ER) and processed in the Golgi apparatus. During their maturation, NOTCH receptors are proteolytically processed by cleavage at the S1 site (Blaumueller et al. 1997; Logeat et al. 1998; Lake et al. 2009) and further post-translationally modified (discussed in detail in "Regulation of Notch Function by *O*-Glycosylation" of this book), producing the mature heterodimeric NOTCH receptor that is exposed on the plasma membrane. In mammals, four NOTCH receptors (NOTCH1-4) are expressed in a tissue- and cell-type specific manner. Mature NOTCH receptors consist of a NOTCH extracellular domain (NECD) and an intracellular portion (NICD) which are connected by a transmembrane (TM) domain. The NECD is characterized by epidermal growth factor (EGF)-like repeats that vary in number among the different isoforms, followed by three Lin-12/Notch repeats (LNR) and finally by a hydrophobic region required for the heterodimerization of the receptor. The LNR and heterodimerization (HD) domains form a negative regulatory region (NRR) that prevents ligand-independent cleavage of the receptor at the S2 cleavage site (Sanchez-Irizarry et al. 2004). The NOTCH-TM domain contains the S3 cleavage site which is the target of the γ-secretase complex that releases the NICD (Fortini 2002). The NICD is characterized by an N-terminal RBPJ (recombination signal binding protein for immunoglobulin kappa J region)-associated module (RAM) followed by ankyrin repeats (ANKs) that together form the RBPJ-interacting region (Tamura et al. 1995). These domains are followed by a transactivation domain (TAD) required for transcriptional activation and by a proline (P), glutamic acid (E), serine (S) and threonine (T) (PEST)-rich domain involved in regulating the turnover of the NICD protein. It must be noted that the TAD is not conserved in all NOTCH proteins but it is specifically found within NOTCH1 and NOTCH2, suggesting different mechanisms of transcriptional activation used by the different NOTCH proteins.

Similar to NOTCH receptors, the NOTCH ligands are single-pass transmembrane proteins. They are members of two different families: the DELTA/DELTA-LIKE and the

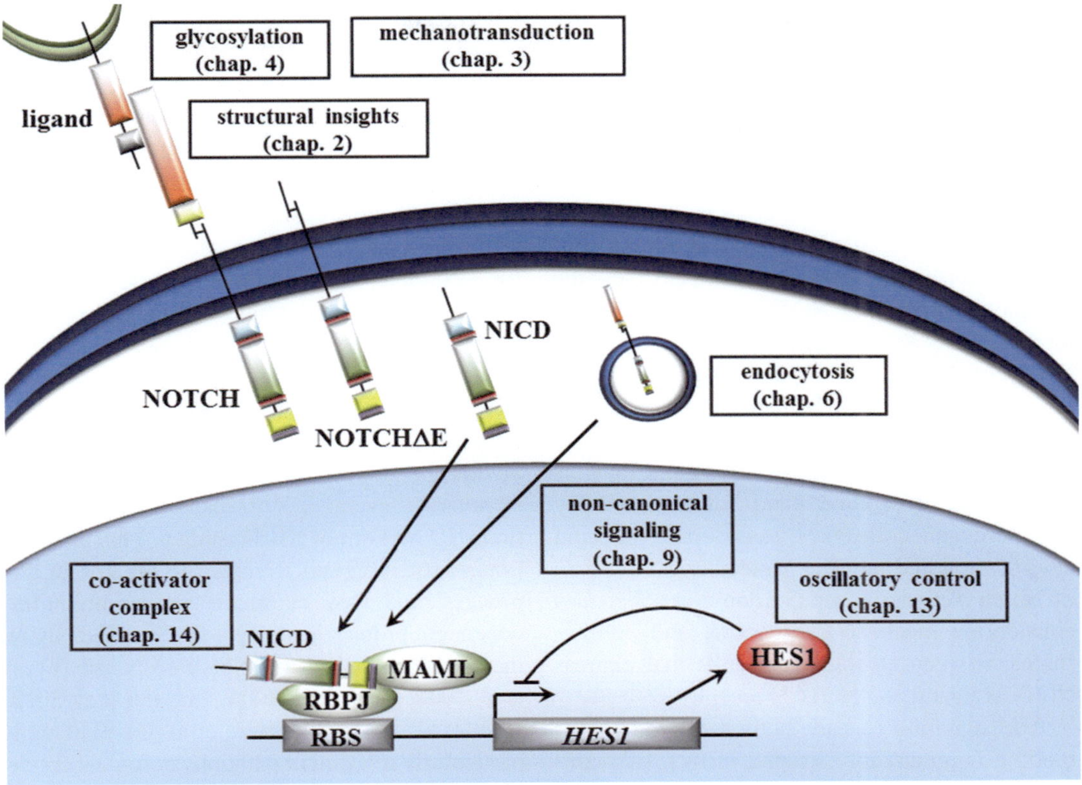

Fig. 1 Overview of the Notch signaling cascade. Ligand binding to NOTCH receptor leads to proteolysis-dependent release of the NOTCH intracellular domain (NICD). Structural aspects of the ligand/receptor interaction are discussed in "Structural Insights into Notch Receptor-Ligand Interactions" whereas mechanotransduction of the signal and glycosylation of the NOTCH receptor are discussed in "The Molecular Mechanism of Notch Activation" and "Regulation of Notch Function by *O*-Glycosylation", respectively. The first cleavage, mediated by ADAM metalloproteases, generates an intermediate proteolytic product called NEXT (Notch EXtracellular Truncation) which is substrate for a γ-secretase complex that releases the NICD. The NICD subqsequently translocates into the nucleus where it interacts with the transcription factor RBPJ and cofactor MAML leading to activation of Notch target genes (see "CSL-Associated Corepressor and Coactivator Complexes" in this book). Several Notch target genes are involved in feedback regulation of the Notch pathway, as in the case of *HES1* which also regulates its own expression leading to an oscillatory control (see Oscillatory Control of Notch Signaling in Development" in this book). The Notch pathway is also regulated by endocytosis and vesicle trafficking of the NOTCH receptor (see "Endocytic Trafficking of the Notch Receptor" in this book) which can lead to degradation or ligand independent activation of the pathway (see "Mechanisms of Non-canonical Signaling in Health and Disease: Diversity to Take Therapy up a Notch?" in this book)

SERRATE/JAGGED families. The *Drosophila* genome encodes one member of each family (*Delta* and *Serrate*) while mammalian ligands are more complex as three members of the DELTA family [DELTA-LIKE (DLL) 1, 3 and 4] and two members of the JAGGED family (JAG1 and 2) are encoded. All NOTCH ligands present with a DSL (DELTA, SERRATE, LAG-2) domain that contains the NOTCH receptor-interacting region followed by EGF repeats which vary in number among the different members of the families (Parks et al. 2006; D'Souza et al. 2008). Compared to the DELTA family, only the JAGGED family presents a cysteine-rich (CR) region proximal to the TM domain. Additionally, the intracellular domain of some Notch ligands is characterized by a PDZ (PSD-95/Dlg/ZO-1) domain that supports interactions with proteins of the adherens junctions (Mizuhara et al. 2005).

A major breakthrough in the Notch field was the recent elucidation of the molecular structure of the NOTCH/ligand complex (see "Structural Insights into Notch Receptor-Ligand Interactions" of this book). Genetic and biochemical studies already revealed that Notch receptor glycosylation is pivotal for its function. Reassuringly, the structures showed that sugars are in the midst of the receptor/ligand structure. This aspect and the complex regulation by NOTCH glycosylation are discussed in "Regulation of Notch Function by *O*-Glycosylation". In addition to glycosylation, the exact molecular mechanisms of receptor/ligand interactions and the signal triggering mechanisms are discussed in "Structural Insights into Notch Receptor-Ligand Interactions" by Handford and colleagues considering the lipid environment and in "The Molecular Mechanism of Notch Activation" by Gordon and colleagues considering mechano-transduction and pulling-forces between two adjacent cells that express NOTCH ligand and NOTCH receptor.

The signaling cascade activated upon ligand binding is remarkably simple; in fact, two consecutive proteolytic cleavages of the NOTCH receptor release the NICD from the membrane. First, ADAM (a disintegrin and metalloprotease) metalloproteases (Brou et al. 2000; Mumm et al. 2000) cleave off the majority of the NECD; this is known as the S2 cleavage. Subsequently, the intracellular domain of the remaining Notch receptor (NICD) is liberated by an intramembrane cleavage mediated by the γ-secretase complex, a process known as S3 cleavage. The intricate regulation of receptor cleavage and endocytic trafficking as part of this process is discussed in detail by Klein and colleagues in "Endocytic Trafficking of the Notch Receptor" of this book. Upon activation, the NICD translocates into the nucleus, associates with transcription factor RBPJ and activates the expression of Notch target genes (Fig. 1). Pivotal cofactors within the RBPJ/NICD complex are MASTERMIND-LIKE (MAML) proteins which are required for the complex to be fully functional (Wu et al. 2000; Lin et al. 2002; Wu et al. 2002; Nam et al. 2003; Nam et al. 2006; Wilson and Kovall 2006); this trimeric complex recruits several additional coactivators such as acetyltransferase EP300 [E1A Binding Protein P300, (Oswald et al. 2001; Hansson et al. 2009; Jung et al. 2013)]. This is known as the canonical pathway of Notch activation and these nuclear events are discussed in "CSL-Associated Corepressor and Coactivator Complexes". Regarding non-canonical Notch signaling, which is represented for example by RBPJ-independent events, Vaccari and colleagues elucidate these aspects of Notch signaling in "Mechanisms of Non-canonical Signaling in Health and Disease: Diversity to Take Therapy up a Notch?". Interestingly, the protein half-life of the NICD is pivotal for amplitude and duration of the Notch response. Several post-translational modifications (PTMs) of the NICD, such as phosphorylation, acetylation and methylation are key in this process, and they culminate in the ubiquitin-dependent proteasomal degradation of the NICD, thereby terminating the Notch response (Fryer et al. 2002; Fryer et al. 2004; Palermo et al. 2012; Hein et al. 2015; Borggrefe et al. 2016). This is particularly relevant in pathophysiological conditions such as leukemogenesis. Here, stabilizing NOTCH mutations are found in several leukemias, such as T-ALL and chronic lymphocytic leukemia (CLL). Additionally, the NOTCH ubiquitin-ligase FBXW7 (F-box and WD repeat domain-containing 7) is frequently mutated in leukemia patients. The interested reader is referred to our recent review (Borggrefe et al. 2016) as well as "The Notch3 Receptor and Its Intracellular Signaling-Dependent Oncogenic Mechanisms" and "Notch in Leukemia" of this book.

In the absence of a Notch signal, the central transcription factor RBPJ remains in the nucleus bound to its target nucleotide sequence and recruits corepressors to prevent the expression of Notch target genes. In the last few years several groups including our have set out to characterize the composition of the RBPJ corepressor and coactivator complexes (Oswald et al. 2001; Hansson et al. 2009; Jung et al. 2013; Kao et al. 1998; Oswald et al. 2002; Oswald et al. 2005; Salat et al. 2008; Borggrefe and Oswald 2009; Moshkin et al. 2009; Liefke et al. 2010; Mulligan et al. 2011; Yatim et al. 2012; Oswald et al. 2016; Xu et al. 2017 and "The Notch Interactome: Complexity in Signaling Circuitry" and "Oscillatory Control of Notch Signaling in Development" of this book) and to unveil their structure (Nam et al. 2003; Nam et al. 2006; Wilson and Kovall 2006; Kovall and Hendrickson 2004; Kovall 2007; VanderWielen et al. 2011; Collins et al. 2014; Contreras et al. 2015; Yuan et al. 2016 and "Structural Insights into Notch Receptor-Ligand Interactions" of this book). These studies resulted in two important findings: First, the Notch signaling pathway is not based on a simple ON/OFF-state concerning Notch target gene expression; second, the individual RBPJ/NICD complex does not operate alone but functions as homodimer and may collaborate with other DNA binding proteins. The first observation is supported by the characterization of the protein interaction network of SHARP [SMRT (silencing mediator for retinoid or thyroid-hormone receptors)/HDAC1 (histone deacetylase)-associated repressor protein; also known as mouse MINT (Msx2-interating protein) or SPEN (Split ENds family transcriptional repressor)] which, focusing on its SPOC (Spen paralog and ortholog C-terminal) domain, unveiled an interesting and surprising scenario (Oswald et al. 2016). In fact, while previously SHARP was exclusively considered as a transcriptional repressor (Oswald et al. 2002; Oswald et al. 2005; Salat et al. 2008), proteomics studies revealed that SHARP does not exclusively interact with the corepressor NCoR (nuclear receptor corepressor) complex but also with the coactivator KMT2D (lysine methyltransferase 2D) complex (Oswald et al. 2016). These observations identified SHARP as a key regulator of the Notch pathway where NCoR and KMT2D compete for the same binding site of SHARP (Oswald et al. 2016). Thus, it is likely that SHARP is a central chromatin regulator tuning the output of the Notch response by balancing histone methylation and deacetylation.

The second observation is based on the identification of NICD homodimers that are required to specifically induce a subset of Notch target genes such as *Hes1* (hairy and enhancer of split 1), *Myc* (myelocytomatosis proto-oncogene) and *Ptcra* [invariant preTα chain of the pre-TCR (T-cell receptor)] that are characterized by paired RBPJ binding sites oriented and spaced in a specific manner (Nam et al. 2007; Liu et al. 2010; Hass et al. 2015). Additionally, genome-wide studies unveiled that NOTCH1 and RBPJ binding occurs at sites that are also bound by additional transcription factors such as AML1 [acute myeloid leukemia 1, also known as RUNX1 (Runt related transcription factor 1)], ETS1 (E26 avian leukemia oncogene 1), GABPA (GA binding protein transcription factor alpha subunit) and ZNF143 [Zinc finger protein 143, (Wang et al. 2011a; Ngondo-Mbongo et al. 2013; Wang et al. 2014)], suggesting that several transcriptional factors synergize to fine-tune the expression of Notch target genes. Alternatively, competitive binding may have different transcriptional outputs in regard to the expression of Notch target genes.

Apart from chromatin regulation prior to the Notch response and combinatorial activities of several transcription factors, differential gene regulation is achieved by different promoter structures and feedback loops, which can result in oscillatory mechanisms that play key roles in development (Fig. 1). One particularly well-studied example is the basic helix-loop-helix (HLH) transcription factor HES1, encoded by a prototypic Notch target gene. Kageyama and colleagues discuss in depth the function of HES1 in "Oscillatory Control of Notch Signaling in Development" of this book.

3 Notch in Inflammation

Notch signaling has been shown to play important roles in both innate and adaptive immunity. In innate immunity, Notch signaling promotes the differentiation of specific cell types as well as supports the activation of specific cells. Macrophages are key mediators of innate immunity but are also involved in supporting specific aspects of the adaptive immunity. Based on the activating stimulus, macrophages polarize into so called M1 or M2 states: while M1 macrophages are involved in supporting inflammatory responses by producing inflammatory molecules such as interleukin 12 (IL12), IL6 or tumor necrosis factor alpha (TNFα), M2 macrophages regulate the resolution of inflammation by producing anti-inflammatory molecules such as IL10 or TGFβ (Porta et al. 2015; Kapellos and Iqbal 2016; Patel et al. 2017). Polarized macrophages can be further distinguished in M2a, M2b or M2c based on the different gene expression profile and the activating stimulus, for example IL4 and IL13 induce the M2a phenotype, the M2b is induced by exposure to immune complexes with Toll-like receptors (TLRs) or interleukin 1 receptor (IL1R) while M2c is induced by IL10 (Mantovani et al. 2004).

In bone marrow derived macrophages (BMDMs) from *Rbpj* conditional knockout (KO) mice (*Rbpj$^{flox/flox}$;Mx1-Cre*) the expression of lipopolysaccharide (LPS)-induced genes is inhibited (Xu et al. 2012). RBPJ positively regulates LPS-mediated transcription via the canonical Notch signaling pathway as treatment with inhibitors of the γ-secretase complex (GSIs), that block the activation of the Notch pathway, *Adam10* or *Notch1* deficiencies impair gene expression of LPS targets (Xu et al. 2012). Mechanistically RBPJ controls the expression of IRAK2 (interleukin-1 receptor-associated kinase-like 2) protein that supports a cascade that culminates with the synthesis of IRF8 (interferon-regulatory factor 8) (Xu et al. 2012), a key transcription factor of the inflammatory gene expression program (Mancino et al. 2015). The control of this program in macrophages is more complex as LPS treatment also leads to upregulation of Notch target genes, such as *HES1* and *HEY1* (*Hairy/enhancer-of-split related with YRPW motif 1*), which are involved in a negative feedback loop that controls the expression of pro-inflammatory cytokines (Hu et al. 2008). Importantly, treatment with interferon γ (IFNγ) leads to downregulation of *HES1* and *HEY1* gene expression. This suggests a mechanism how IFNγ may augment the production of pro-inflammatory cytokines (Hu et al. 2008). As these studies pointed out to the RBPJ-dependent induction of *Il12* gene upon LPS stimulation (Xu et al. 2012; Hu et al. 2008), another study could demonstrate that this effect does not involve the transcriptional activity of the NICD/RBPJ complex as overexpression of a dominant negative form of MAML (DN-MAML) does not influence the expression of *Il12* in BMDMs (Boonyatecha et al. 2012). The reasons for this contrasting results are still not clear but it must be noted that another study could show that the pro-inflammatory cytokine IL6 is positively and directly regulated by the Notch signaling pathway upon treatment of BMDMs with LPS and IFNγ. In fact, *Il6* expression is downregulated by GSIs and upregulated by overexpression of NICD1 upon LPS and IFNγ treatment and finally the *Il6* locus is bound by NOTCH1 (Wongchana and Palaga 2012). Fung and colleagues observed that *NOTCH3* expression increases during differentiation of human monocytes into macrophages in culture, while *DLL4* expression increases upon pro-inflammatory stimulation of human macrophages (Fung et al. 2007). Of note, the LPS-mediated *DLL4* induction is dependent on TLR4 (Toll-like receptor 4) and NF-κB (nuclear factor-κB) pathways and triggers the Notch signaling cascade that finally increases the pro-inflammatory properties of human macrophages (Fung et al. 2007). Similarly, also *JAG1* is induced upon LPS stimulation of human macrophages in an NF-κB-dependent manner (Foldi et al. 2010) as well as *Notch1* induction is observed upon macrophages activation and GSIs pretreatment leads to reduced expression of pro-inflammatory genes upon stimulation with LPS and IFNγ (Palaga et al. 2008), suggesting Notch signaling as an important determinant of macrophages-mediated inflammatory responses. Myeloid-specific LOF of *Notch1*, obtained from *LysMCre;Notch1$^{flox/flox}$* mice, leads to decreased macrophages recruitment at wounds as well as GSIs treatment results in failure of *Vegfr1* (vascular endothelial growth factor receptor 1) induction upon macrophages stimulation with LPS and INFγ (Outtz et al. 2010).

In peritoneal macrophages, Notch signaling determines a switch from pro-inflammatory cytokines (TNFα and IL6) to anti-inflammatory cytokines (IL10) production upon stimulation with LPS in a way that is dependent on the PEST domain of NICD proteins (Zhang et al. 2012). This pro-inflammatory inhibitory effect of Notch signaling is based on the inhibition of the MAPK (mitogen-activated protein kinase) pathway leading to reduced transcriptional activity of NF-κB (Zhang et al. 2012). In contrast, another study observed that Notch signaling increases pro-inflammatory properties of macrophage derived Raw 264.7 cells upon LPS stimulation by promoting nuclear translocation of NF-κB (Monsalve et al. 2009). The reasons for the differences observed in these studies are not clear and more work is required to better dissect the role of Notch signaling upon LPS stimulation in these cells.

RBPJ controls also the M2 polarization of macrophages as RBPJ KO macrophages from *Rbpj$^{flox/flox}$;Lyz2-Cre* mice treated with chitin, a major structural component of fungi and helminthes that induce the M2 polarization, present impaired expression of genes associated with the M2 phenotype (Foldi et al. 2016). It must also be noted that *Rbpj* KO results in M2 polarization of BMDM upon LPS stimulation (Wang et al. 2010a), suggesting that RBPJ may play different roles in the M1 vs M2 polarization based on the activating stimulus. Additionally, stimulation of macrophages with IL4, an interleukin that drives the M2 polarization, leads to upregulation of *Jag1* (Outtz et al. 2010).

Interestingly, in a mouse model of systemic lupus erythematosus (SLE), Notch signaling is required to induce macrophage polarization versus the M2b phenotype through PI3K (phosphatidylInositol 4,5-bisphosphate 3-kinase)/AKT-ERK (Extracellular signal-regulated kinase)-1/2 and p38 MAPK signaling pathways (Zhang et al. 2010).

In summary, an important role for Notch in inflammation is evident, but further studies are required to differentiate between direct and indirect effects and to clarify how the Notch pathway orchestrates different polarization of macrophages.

4 Dysregulation of Notch Signaling in Diseases

Accurate regulation of the Notch signaling pathway is required for development, differentiation and homeostasis of a wide variety of tissues during both adult and embryonic life (see "The Notch3 Receptor and Its Intracellular Signaling-Dependent Oncogenic Mechanisms", "Notch and Neurogenesis", "Notch and Stem Cells", "Notch and Senescence", "Control of Blood Vessel Formation by Notch Signaling" and "Notch and T Cell Function – A Complex Tale" in this book) and dysregulation of Notch signaling is associated with many diseases (see "Integration of *Drosophila* and Human Genetics to Understand Notch Signaling Related Diseases", "Mechanisms of Non-canonical Signaling in Health and Disease: Diversity to Take Therapy up a Notch?", "The Notch3 Receptor and Its Intracellular Signaling-Dependent Oncogenic Mechanisms", "Notch and Senescence", "Control of Blood Vessel Formation by Notch Signaling" and "Notch in Leukemia" in this book).

Notch signaling has been associated with several congenital disorders, for example Notch LOF has been linked to Alagille and Adams-Oliver syndromes (AOS) whereas Notch GOF results in Hadju-Cheney syndrome [HCS, see "Integration of *Drosophila* and Human Genetics to Understand Notch Signaling Related Diseases" in this book and Masek and Andersson 2017]. In addition, missense mutations that affect the structure of NOTCH receptors have been found in genetic diseases. For example, *NOTCH3* mutations that affect specific domains of the NECD have been linked to CADASIL (Cerebral Autosomal Dominant Arteriopathy with Subcortical Infarcts and Leukoencephalopathy). There are several instances where somatic mutations of *NOTCH* or Notch pathway components or modulators lead to cancer. We will briefly discuss the current knowledge about Notch signaling in cancer and the interested reader is referred to "The Notch3 Receptor and Its Intracellular Signaling-Dependent Oncogenic Mechanisms" and "Notch in Leukemia" of this book and other recent reviews (Aster et al. 2017).

4.1 Notch in Leukemia

In 1991, recurring mutations in the *NOTCH1* gene were first described in patients with T-ALL, thus implicating Notch signaling directly in leukemogenesis (Weng et al. 2004). Those mutations lead to a C-terminal truncation of the intracellular NOTCH1 protein, thereby removing the destabilizing PEST domain and leading to increased NICD1 half-life (Weng et al. 2004). Other *NOTCH1* activating mutations in T-ALL have also been identified in the NECD leading to constitutive cleavage of the NOTCH receptor (Weng et al. 2004). Similar activating *NOTCH1* mutations were also identified in CLL (Puente et al. 2011), in line with previous data showing activation of the Notch pathway in CLL (Rosati et al. 2009). These data suggest Notch signaling as a potential therapeutic target in the treatment of T-ALL and CLL and lead to some clinical trials in the last years.

GSIs can be used to prevent the activation of the Notch pathway by blocking the release of the NICD from the membrane. However, this approach is unfortunately limited due to two reasons: 1) GSIs cause severe gastrointestinal side effects due to the important role of Notch signaling in differentiation of the highly proliferating gut epithelium; 2) Drug resistance to GSIs also fairly frequently arises and it is associated with mutational loss of *PTEN* [phosphatase and tensin homolog, (Palomero et al. 2007)] or *FBXW7* (O'Neil et al. 2007) and dependent on BRD4 [bromodomain-containing protein 4, (Knoechel et al. 2014)] as well as on miR (microRNA)-223 (Kumar et al. 2014). The problems encountered with the clinical use of GSIs pointed out the need for a better dissection of the molecular mechanisms that define the Notch signaling response with the final goal to identify additional potential therapeutic targets to block Notch signaling or its oncogenic target genes. This will be of benefit not exclusively for T-ALL and CLL as aberrant Notch signaling is also observed in acute myeloid leukemia [AML, (Thiel et al. 2017)], mantle cell lymphoma [MCL, (Kridel et al. 2012)] and splenic marginal zone lymphoma [SMZL, (Rossi et al. 2012)].

4.2 Notch in Solid Tumors

NOTCH1 was originally identified as an oncogene in leukemia but surprisingly *NOTCH* genes have also been found to have tumor suppressive roles in other contexts (Table 1). In this section we will discuss the different functions of Notch signaling in different types of solid tumors.

4.2.1 Notch in Glioblastoma

Glioblastoma (GBM) represents one of the most aggressive forms of brain tumor and the Notch signaling pathway has been implicated in the molecular pathogenesis of gliomas. NOTCH1 receptor as well as JAG1 and DLL1 ligands are upregulated in GBM cell lines and in primary human gliomas and their knockdown results in decreased luciferase activity, using a Notch-dependent reporter assay (Purow et al. 2005). When human cell lines, transfected with *NOTCH1* siRNAs, were intracranially injected into recipient mice, an increased survival was observed compared to controls (Purow et al. 2005). In line with these observations, GSIs treatment of GBM neurospheres reduces their proliferation while overexpression of an active form of NOTCH2 has the opposite effect (Fan et al. 2010). This phenotype is linked to cancer stem cells (CSCs), as GSIs treatment downregulates the expression of CSCs markers such as CD133, NESTIN, BMI1 (B lymphoma Mo-MLV insertion region 1 homolog) and OLIG2 (oligodendrocyte transcription factor 2). The most striking observation is that GSIs treatment reduces the mortality in mouse models (Fan et al. 2010), suggesting Notch signaling as a good candidate for therapeutic intervention. Even if GSIs lead to increased apoptosis of GBM neurosphere cells, as revealed by increased cleaved CASPASE-3 (Fan et al. 2010), the molecular mechanisms behind are poorly defined. Similarly, expression of DN-MAML reduces the proliferation of GBM cells but the same study pointed out to a cell type-specific dependence on different NOTCH receptors (Chen et al. 2010). Given the poor outcomes of GSIs in clinical applications, it will be important to identify additional targets that may be used to modulate the Notch pathway. One of this targets is potentially RBPJ which is upregulated in brain CSCs (Xie et al. 2016). Knockdown of RBPJ in CSCs has a stronger effect compared to GSIs in term of proliferation and it significantly increases

Table 1 Mutations in FBXW7, NOTCH1 or RBPJ associated with tumors and/or genetic diseases. The list includes insertions, deletions, missense and nonsense mutations. ACC adenoid cystic carcinoma; AOS: Adams-Oliver syndrome; AVD: aortic valve disease; AVS: aortic valve stenosis; BAV: bicuspid aortic valve; CLL: B-cells chronic lymphocytic leukemia; COA: coarctation of the aorta; cSCC: cutaneous squamous cell carcinoma; HLHS: hypoplastic left heart syndrome; H&N: head and neck; lSCC: lung squamous cell carcinoma; MCL: mantle cell lymphoma; nCC: noncutaneous carcinoma; RCC: renal cell cancer; SCLC: small cell lung cancer; SMZL: splenic marginal zone lymphoma; T-ALL: T-cells acute lymhoblastic leukemia; TAV: tricuspid aortic valve; TNBC: triple-negative breast cancer

Gene	Domain	Disease	Reference
FBXW7	F-box	Melanoma, SCLC	George et al. (2015) and Aydin et al. (2014)
	WD40 repeats	Melanoma, SCLC, T-ALL	George et al. (2015), Aydin et al. (2014) and Larson-Gedman et al. (2009)
NOTCH1	EGF repeats	ACC, AOS, AVD, AVS, BAV, bladder cancer, breast cancer (TNBC), COA, cSCC, HLHS, lSCC, MCL, nCC, SCLC, TAV	George et al. (2015), Iascone et al. (2012), Wang et al. (2011b), Foffa et al. (2013), Mohamed et al. (2006), McBride et al. (2008), Kridel et al. (2012), Kent et al. (2013), Ducharme et al. (2013), Garg et al. (2005), Wang et al. (2015), McKellar et al. (2007), Stittrich et al. (2014), Stoeck et al. (2014), Southgate et al. (2015) and Rampias et al. (2014)
	LNR repeats	AOS, AVD, bladder cancer, breast cancer (luminal B), breast cancer (TNBC), cSCC, endometrial cancer, HLHS, RCC	Iascone et al. (2012), Wang et al. (2011b), Garg et al. (2005),Wang et al. (2015), Stittrich et al. (2014), Stoeck et al. (2014), Southgate et al. (2015) and Rampias et al. (2014)
	HD domain	ACC, AOS, AVS, BAV, bladder cancer, breast cancer (TNBC), cervical adenocarcinoma, colon adenocarcinoma, cSCC, glioblastoma, H&N, MCL, melanoma, neuroendocrine carcinoma, pancreatic cancer, SCLC, T-ALL	George et al. (2015), Larson-Gedman et al. (2009), Wang et al. (2011b), Foffa et al. (2013), McBride et al. (2008), Kridel et al. (2012), Wang et al. (2015), Stoeck et al. (2014), Southgate et al. (2015), Rampias et al. (2014), Weng et al. (2004), Breit et al. (2006), Malecki et al. (2006), Zhu et al. (2006), Mansour et al. (2006), Mansour et al. (2007) and De Keersmaecker et al. (2008)
	RAM domain	BAV, cSCC	Wang et al. (2011b) and Mohamed et al. (2006)
	ANK repeats	AOS, bladder cancer, cSCC, SCLC, T-ALL	George et al. (2015), Wang et al. (2011b), Stittrich et al. (2014) , Southgate et al. (2015), Rampias et al. (2014) and Zhu et al. (2006)
	TAD/PEST domain	BAV, breast cancer (ER$^+$, PR$^+$, HER2$^+$), breast cancer (TNBC), CLL, COA, MCL, SCLC, SMZL, T-ALL	George et al. (2015), Larson-Gedman et al. (2009), McBride et al. (2008), Kridel et al. (2012), Wang et al. (2015), Stoeck et al. (2014), Weng et al. (2004), Breit et al. (2006), Zhu et al. (2006), Mansour et al. (2006), De Keersmaecker et al. (2008), Rossi et al. (2012), Puente et al. (2011), Bea et al. (2013), Bittolo et al. (2017), D'Agaro et al. (2017), Fabbri et al. (2011), Pozzo et al. (2017) and Pozzo et al. (2016)
RBPJ	NTD	AOS	Hassed et al. (2012)

the life-span of tumor-bearing hosts (Xie et al. 2016). The differences between GSIs treatment and RBPJ knockdown depend on the fact that RBPJ regulates also a Notch-independent transcriptional program and the effect of RBPJ is based on its interaction with CDK9 (cyclin-dependent kinase 9) to support transcriptional elongation (Xie et al. 2016). It must be also noted that a difference in regard to Notch activity in different GBMs cannot be excluded and that this difference is likely dependent on the *P53* status; in fact, cells with a mutated *P53* background seem to be more sensitive to Notch inhibition compared to cells with a wild type *P53* background (Chen et al. 2010). In line with that, *P53* wild type GBM cells present with low Notch activity as revealed by GSI treatment and DN-MAML overexpression (Xu et al. 2017).

Gliomas are usually treated by surgical intervention aimed to remove the tumor mass followed by radiotherapy and chemotherapy but, while an initial response to radiotherapy is visible, gliomas are refractory (Grossman and Batara 2004; Furnari et al. 2007), probably associated to radiation resistance of CSCs (Bao et al. 2006). GSIs treatment increases the sensitivity of glioma stem cells to clinical doses of radiation while GOF of active forms of NOTCH1 or NOTCH2 protects them from apoptosis upon radiation (Wang et al. 2010b). Importantly, when CSCs were subjected to NOTCH1 or NOTCH2 knockdown before radiation, they showed a reduced tumorigenic activity in mouse models (Wang et al. 2010b), suggesting that a combined therapy, based on radiotherapy and GSIs may be used as a therapeutic approach. In line with these data, Gilbert and colleagues could show that GSIs treatment significantly reduces the recovery of neurospheres treated with Temozolomide (TMZ), a chemotherapeutic agent used to treat gliomas (Gilbert et al. 2010). Given that the neurospheres number was reduced only when TMZ was added before GSIs, one can imagine that Notch signaling, in gliomas, is a mechanism that is activated as part of a resistance upon chemotherapy. Additionally, Gilbert and colleagues could show that combined TMZ and GSIs treatment reduces tumorigenicity in mouse models (Gilbert et al. 2010).

In summary, although the oncogenic role of Notch is clear, Notch inhibition alone remains ineffective in therapeutic terms. Thus, a combination therapy seems to be highly desirable and targeting the CSCs or preventing the tumor plasticity may lead the way.

4.2.2 Notch in Breast Cancer

Interestingly, Notch signaling has been linked to the triple-negative breast cancer (TNBC), negative for estrogen receptor (ER), progesterone receptor (PR) and human epidermal growth factor receptor 2 (HER2) (Foulkes et al. 2010). Similar to the subset of mutations identified in leukemia, Notch activating mutations are found in TNBC at the C-terminal PEST domain of NOTCH1, NOTCH2 and NOTCH3 (Wang et al. 2015). The prolyl-isomerase PIN1 (Peptidylprolyl Cis/Trans Isomerase, NIMA-Interacting 1) is a positive regulator of the Notch signaling pathway (Rustighi et al. 2009) and it supports Notch signaling in TNBC cells by antagonizing the FBXW7-dependent degradation of NICD1 and NICD4 (Rustighi et al. 2014). Overexpression of NUMB, a negative regulator of the Notch signaling pathway, in TNBC cells reduces the epithelial-mesenchymal transition (EMT), a process associated with cancer progression and metastasis and suppresses tumor growth in xenografts mouse models (Zhang et al. 2016a). In line with these observations, NUMB expression is lost in several breast cancer cell lines including lines established from TNBC (Stylianou et al. 2006), as well as in primary samples, leading to increased Notch signaling (Pece et al. 2004).

Mechanistically, Notch signaling regulates cell proliferation in TNBC by directly modulating the expression of CYCLIN D1 (encoded by the *CCND1* gene). In fact, NOTCH1 binds to the *CCND1* locus and LOF of the Notch ligand JAG1 leads to downregulation of *CCND1* associated with cell cycle defects (Cohen et al. 2010).

TNBC frequently presents with alterations in the PI3K/AKT/mTOR (mammalian target of rapamycin) pathway (Lehmann et al. 2011; Banerji et al. 2012; Cancer Genome Atlas 2012) but pharmacological inhibition of this pathway proved to be ineffective. Recently, Bhola and colleagues showed that inhibition of PI3K/mTOR or TORC1/2 (mTOR signaling complex 1/2) in TNBC cells enriches for CSCs and leads to increased expression of NICD1 and JAG1 as well as increased Notch activity (Bhola et al. 2016). Importantly, inhibition of Notch signaling decreases the induction of CSCs upon PI3K/mTOR or TORC1/2 inhibition (Bhola et al. 2016), suggesting a possible combined therapy. In line with this, monoclonal antibodies that prevent Notch signaling activation can reduce tumor growth of TNBC xenografts and increase the efficacy of the chemotherapeutic agent docetaxel in mice (Qiu et al. 2013).

In MCF7 cells (ER^+ PR^+ $HER2^-$), Notch controls a metabolic switch involved in tumorigenesis (Landor et al. 2011). Mechanistically, this process is controlled by PIM (Proto-Oncogene, Serine/Threonine Kinase) kinases that phosphorylate NOTCH1 increasing both its nuclear localization and activity (Santio et al. 2016). Notch signaling is also upregulated upon anti-estrogen treatment of ER^+ patient derived samples and xenografts (Simoes et al. 2015). Additionally, MCF7 cells that undergo EMT upon irradiation, present increased expression of Notch pathway components, namely NOTCH2, DLL4 and JAG1 (Kim et al. 2016). Interestingly, pharmacological inhibition of Notch signaling with GSIs or knockdown of NOTCH2, DLL4 or JAG1 leads to reduced EMT upon radiation of MCF7 cells (Kim et al. 2016), supporting the idea that Notch signaling may contribute to radiation resistance.

In summary, Notch might be a valuable lead target for future therapeutic approaches in TNBC, possibly making use of combined therapies.

4.2.3 Notch in Pancreatic Cancer

Pancreatic ductal adenocarcinoma (PDAC) is one of the leading cause of cancer death and it is believed that it develops from pancreatic intraepithelial neoplasia (PanIN). Notch signaling plays a dual role in pancreatic cancer: on one hand it is oncogenic in PDAC whereas it acts as a tumor suppressor in PanIN.

In PDAC, several Notch pathway components are upregulated including *NOTCH2*, *NOTCH3* and *JAGGED1* (Miyamoto et al. 2003) whereas, using a conditional pancreatic mouse model based on the expression of the RAS (Rat sarcoma virus oncogene) mutant K-RAS^{G12D} (*Pdx1-Cre;LSL-Kras*^{G12D}), Hanlon and colleagues observed an increase of PanIN upon inactivation of *Notch1* (*Pdx1-Cre;LSL-Kras*^{G12D}*;Notch1*$^{flox/flox}$, (Hanlon et al. 2010)), supporting the tumor suppressive role of Notch signaling in PanIN. This conclusion is further supported by the observation that conditional inactivation of *Notch1*, in the *Ptf1a*$^{+/Cre}$*;LSL-Kras*$^{+/G12D}$ (*Ptf1a*$^{+/Cre}$*;LSL-Kras*$^{+/G12D}$*;Notch1*$^{flox/flox}$) background, slightly reduces survival (Mazur et al. 2010a). In contrast, the same study pointed out that *Notch2* might play an entirely different role. In fact, its conditional inactivation (*Ptf1a*$^{+/Cre}$*;LSL-Kras*$^{+/G12D}$*;Notch2*$^{flox/flox}$) leads to decreased PanIN and PDAC development associated with increased survival (Mazur et al. 2010a), suggesting a different and opposing role for the different NOTCH receptors in pancreatic cancer. However, De La O and colleagues observed the opposite in regard to the role of *Notch1*. Its conditional GOF (*Pdx1-CreERT;Rosa26-NICD1*) in the *Kras*^{G12D} background leads to increased PanIN (De La O et al. 2008). These discrepancies are potentially explained by the different genetic approaches used (loss- versus gain-of-function). Thus, it is possible that different NOTCH receptors are involved in different steps of pancreatic tumorigenesis. In line with this hypothesis, conditional LOF of *Lunatic Fringe* in the *Pdx1-Cre;LSL-Kras*^{G12D} background (*Lfng*$^{flox/flox}$*;Pdx1-Cre;LSL-Kras*$^{G12D/+}$), that encodes for an O-fucosylpeptide 3-β-N-acetylglucosaminyltransferase known to modify the epidermal growth factor repeats of NOTCH proteins, caused increased NOTCH3 activation during PDAC initiation and progression but activation of NOTCH1 only at a later time point, suggesting that *Lunatic Fringe* is a tumor suppressor (Zhang et al. 2016b). It must also be noted that conditional expression of DN-MAML in *Kras*^{G12D} background (*p48-Cre;LSL-Kras*^{G12D}*;Rosa26*$^{dn-MAML/+}$), that blocks the canonical activity of all NOTCH receptors, delays PanIN development (Thomas et al. 2014). In agreement with the above, GSIs treatment efficiently blocks Notch signaling and reduces proliferation of both PanIN and PDAC cell lines. GSIs also attenuate PDAC development in mouse models (Plentz et al. 2009). Surprisingly, GSIs treatment of the PDAC mouse model *LSL-Kras*$^{G12D/+}$*;p53*$^{R172H/+}$*;Pdx-Cre*$^{tg/+}$ only modestly increases survival but, when used in combination with the chemotherapeutic agent gemcitabine, a significant increase in survival is observed (Cook et al. 2012). Similarly, GSIs treatment enhances radiosensitivity in xenografts (Bi et al. 2016). A significant reduction in tumor volume was also observed when anti-DLL4 antibodies, in combination with gemcitabine, were used in pancreatic xenografts models (Yen et al. 2012). Furthermore, genetic inactivation of FBXW7, the E3-ubiquitin ligase that supports the degradation of the NICD, in the *p48-Cre;LSL-Kras*^{G12D} mouse model increases pancreatic tumorigenesis (Zhang et al. 2016c).

Finally, both JAG2 and NOTCH1 have been linked to cell migration of pancreatic cancer cells but this mechanism does not seem to require Notch downstream signaling as GSIs treatment has no effect on PDAC cell migration (Hu et al. 2015).

In summary, Notch signaling plays a key role in pancreatic cancer and a better dissection of the molecular mechanisms involved in this context may lead to develop more effective therapies.

4.2.4 Notch in Hepatocellular Carcinoma

Notch plays an oncogenic role in hepatocellular carcinoma (HCC). *NOTCH1* (Cantarini et al. 2006; Zhu et al. 2017) and *NOTCH3* (Hu et al. 2013) are upregulated and inhibition of Notch signaling with antibodies directed against NOTCH2 or JAG1 in a mouse model of liver cancer has a tumor suppressive effect (Huntzicker et al. 2015) while liver specific GOF of NICD2 leads to HCC (Dill et al. 2013). Similar results were observed in mice upon liver specific overexpression of NICD1 and Notch pathway activation is observed in human HCC (Villanueva et al. 2012). Knockdown of *NOTCH1* reduces the migration and invasion of HCC cells (Hu et al. 2014) without influencing cell viability (Zhou et al. 2013) and, in line with these data, GSIs treatment reduces invasion of HCC cells but surprisingly also their viability (Zhou et al. 2012), suggesting that cell viability may be regulated by a different member of the NOTCH family.

POFUT1 (protein O-fucosyltransferase 1), a glycosyltransferase that modifies the EGF repeats of NOTCH receptors promoting ligand interaction, is upregulated in HCC and its expression correlates with poor prognosis (Ma et al. 2016). *POFUT1* knockdown reduces cell growth, proliferation and migration of HCC cells, associated with reduced activation of the Notch pathway (Ma et al. 2016), suggesting POFUT1 as a possible therapeutic target in HCC. Hyperactivation of the Notch pathway in HCC is also mediated by the upregulation of JAG1, caused by the loss of the transcriptional repressor *RUNX3* (Nishina et al. 2011). In addition, RUNX3 also physically interacts with the NICD1/RBPJ complex and decreases its transactivating capacity in HCC cells (Gao et al. 2010). Another study pointed out to a link between IKKα [IκB (inhibitor of kappa B) kinase subunit alpha] and Notch signaling in HCC (Liu et al. 2012). IKKα is upregulated in HCC tumor samples and inactivates the transcription factor FOXA2 (forkhead box A2) by phosphorylation leading to downregulation of NUMB (Liu et al. 2012). Recently, a crosstalk between the Notch and Hippo pathways was described as a mechanism involved in HCC pathogenesis (Kim et al. 2017). Double KO (DKO) of mammalian sterile 20-like kinase 1 and 2 (MST1/2), involved in inhibition of the Hippo pathway by phosphorylation of the transcription factors YAP/TAZ (Yes-associated protein and WW domain containing transcription regulator 1), results in HCC (Song et al. 2010) associated with activation of Notch signaling which forms a positive feedback loop with YAP/TAZ (Kim et al. 2017). GSI treatment leads to reduced HCC in the MST1/2 DKO mouse model and while these data suggest an oncogenic role for Notch signaling in HCC, Wnt pathway plays the opposing role having a tumor suppressive function in HCC (Kim et al. 2017), suggesting the involvement of several different signaling pathways in HCC pathogenesis.

Notch signaling plays a positive role in HCC CSCs as its inhibition reduces their invasion and migration (Luo et al. 2016) and it may also be important in radio-resistance of HCC CSCs. In fact, CD133^{+} HCC CSCs exhibit upregulation of ADAM17, associated with increased Notch signaling, upon irradiation (Hong et al. 2016).

Of note, some reports also provide evidence for a tumor suppressive role of Notch signaling in HCC. Liver specific deletion of all the three members of the *Retinoblastoma protein* family [*Rb*, *p107* and *p130*; triple knockout (TKO) mice] leads to HCC associated with increased expression of Notch pathway components due to upregulation of E2F transcription factors with transactivation capacity (Viatour et al. 2011). Although this suggests that Notch signaling may be an oncogenic driver, GSIs treatment of TKO mice enhances HCC development, revealing a tumor suppressive role of Notch signaling (Viatour et al. 2011). The Sage laboratory could also show that TKO liver progenitors do not show increased expression of *Notch1*, *Hes1*, *Hey1* or *Nrarp* Notch target genes, suggesting that deregulation of Notch signaling by *Rb* family members is cell type-specific and occurs during tumor progression (Viatour et al. 2011).

As consequence, Notch signaling may be an important player in HCC and a better comprehension of its function in this disease may lead to significant improvement of the current therapies.

4.2.5 Notch in Lung Cancer

Lung cancer is the leading cause of cancer-associated mortality worldwide. Based on histopathology and molecular characteristics two main subtypes can be distinguished: non-small cell lung cancer (NSCLC) and small cell lung cancer (SCLC).

SCLC is distinguished from NSCLC by its characteristic small-cell phenotype that reflects its origin from the neuroendocrine lineage. SCLC is highly refractory to chemotherapy. Recent whole-genome sequencing studies of SCLC have identified recurrent mutations in the *NOTCH1-4* genes (George et al. 2015), suggesting that Notch needs to be inactivated for SCLC development. As the Notch signaling pathway is a physiological regulator of neuronal and neuroendocrine differentiation, mutations in *NOTCH* genes are likely responsible for the characteristic neuroendocrine phenotype of SCLC.

In cancer, lineage specification genes often provide survival advantages of which cancer cells become dependent on - similar as they become addicted to - activated oncogenes (Garraway and Sellers 2006). In line with an addiction of SCLC to the neuroendocrine lineage, the Notch target gene *ASCL1* (*achaete-scute homolog 1*), encoding for a transcription factor which is physiologically required to establish the lineage of neuroendocrine cells in the lung (Borges et al. 1997), was previously shown to be required for the continued survival of SCLC cells (Osada et al. 2005; Jiang et al. 2009). Thus, in this setting, Notch signaling most likely plays a tumor suppressive role and it would be attractive to reactivate Notch target genes to induce the cancer initiating cells to differentiate into a different lineage to block its malignancy. Only then, conventional chemotherapeutic agents could eliminate this devastating cancer cells.

4.2.6 Notch in Skin Cancer and Melanoma

Counterintuitively, Notch may also play the role of a tumor suppressor in other contexts. Notch signaling has a tumor suppressive function in the skin as conditional inactivation of *Notch1* leads to epidermal and corneal hyperplasia followed by the development of skin tumors (Nicolas et al. 2003). Similar results were obtained by skin specific deletion of *Notch1*, mediated by *Pdx1-Cre*, using the RAS mutant $Kras^{+/LSL\text{-}G12D}$ mouse model (Mazur et al. 2010b). The same study also pointed out a specific tumor suppressive role for *Notch1* as genetic depletion of *Notch2* does not support carcinogenesis (Mazur et al. 2010b). Demehri and colleagues showed that *Notch1* depletion in epidermal keratinocytes induces tumorigenesis in a non-cell autonomous manner (Demehri et al. 2009). Similarly to *Notch1* LOF, conditional expression of *dn-Maml* driven by *SM22-Cre* in the skin leads to development of cutaneous squamous cell carcinoma [SCC, (Proweller et al. 2006)]. In line with these data, mesenchymal deletion of the

Notch signaling effector *Rbpj* results in skin tumor (Hu et al. 2012). Notch signaling may also play a tumor suppressive role in human skin cancer as several Notch pathway components are downregulated in human basal cell carcinoma [BCC, (Thelu et al. 2002)]. This hypothesis is further supported by the identification of mutations in human *NOTCH1* in cutaneous SCC that impair the Notch function (Wang et al. 2011b).

At molecular level, data from keratinocytes and SCC cell lines suggest that *NOTCH1* is under the positive control of P53 (Lefort et al. 2007), which is frequently mutated in skin SCC (Backvall et al. 2004). This positive function of P53 is counteracted by EGFR (epidermal growth factor receptor) signaling as its inhibition promotes *P53* expression and, consequently, *NOTCH1* expression with increased Notch signaling (Kolev et al. 2008). Of note, EGFR inhibition in SCC cells induces differentiation and, when EGFR inhibition is combined with inhibition of the Notch signaling pathway, increased apoptosis is observed (Kolev et al. 2008).

Recently, the involvement of Notch signaling in melanoma has gained attention. NOTCH receptors and ligands as well as Notch effectors are upregulated in melanomas (Balint et al. 2005; Massi et al. 2006) and Notch signaling inhibition, via GSIs or expression of DN-MAML, suppresses melanoma cell growth (Balint et al. 2005). In line with this, GOF of the active form of *NOTCH1* increases melanoma cell growth as well as enhances primary melanoma and lung metastasis in adult mice (Balint et al. 2005; Liu et al. 2006). In addition, *FBXW7* was found to be mutated in melanoma patients and these mutations compromise the function of FBXW7 protein leading to accumulation of the active form of NOTCH1 (Aydin et al. 2014). At mechanistic level, active NOTCH1 stabilizes the Wnt signaling effector protein β-CATENIN rather than acting through RBPJ. Indeed, LOF of *β-CATENIN* in melanoma cells mirrors the proliferative defects observed upon LOF of Notch signaling (Balint et al. 2005). Such non-canonical functions of the intracellular active form of Notch affecting other conserved signaling pathways have been recently reviewed (Borggrefe et al. 2016) and are also discussed by Vaccari and colleagues in "Mechanisms of Non-canonical Signaling in Health and Disease: Diversity to Take Therapy up a Notch?" of this book. There is another study by Liu and colleagues showing that NOTCH1 increases melanocyte growth by activating the MAPK and PI3K/AKT signaling pathways (Liu et al. 2006), suggesting that Notch signaling is involved in melanoma by regulating crosstalk with even more signaling pathways.

In conclusion, Notch signaling may be a valuable target also for the treatment of melanoma and skin cancer. However, in the case of skin cancer, this will be particularly challenging because of the tumor-suppressive function of Notch.

5 Mutational Spectrum of Notch Pathway Components

Several mutations involving Notch pathway components have been identified in cancer and genetic disorders as discussed in the previous sections and selectively summarized in Table 1. One striking observation is that the same protein domains are mutated in different diseases (see also "Integration of *Drosophila* and Human Genetics to Understand Notch Signaling Related Diseases" of this book), suggesting that common molecular mechanisms are probably used to confer pathogenicity.

A number of structure biological studies have fully or partially solved the molecular structure of key Notch signaling components, allowing us to understand the effect of disease-linked mutations in the context of 3D protein structure. For example, mutations occurring in the *FBXW7* gene (Fig. 2a and Table 1), encoding for the E3 ubiquitin ligase involved in the degradation of the NICD, are frequently found in melanoma, SCLC and T-ALL. These mutations can compromise the activity of FBXW7, leading to increased protein stability of the NICD and of the other FBXW7 substrates (Aydin et al. 2014). *RBPJ* has been reported to be mutated in AOS, a genetic disease characterized in most of the patients by terminal limb malformations (Hassed

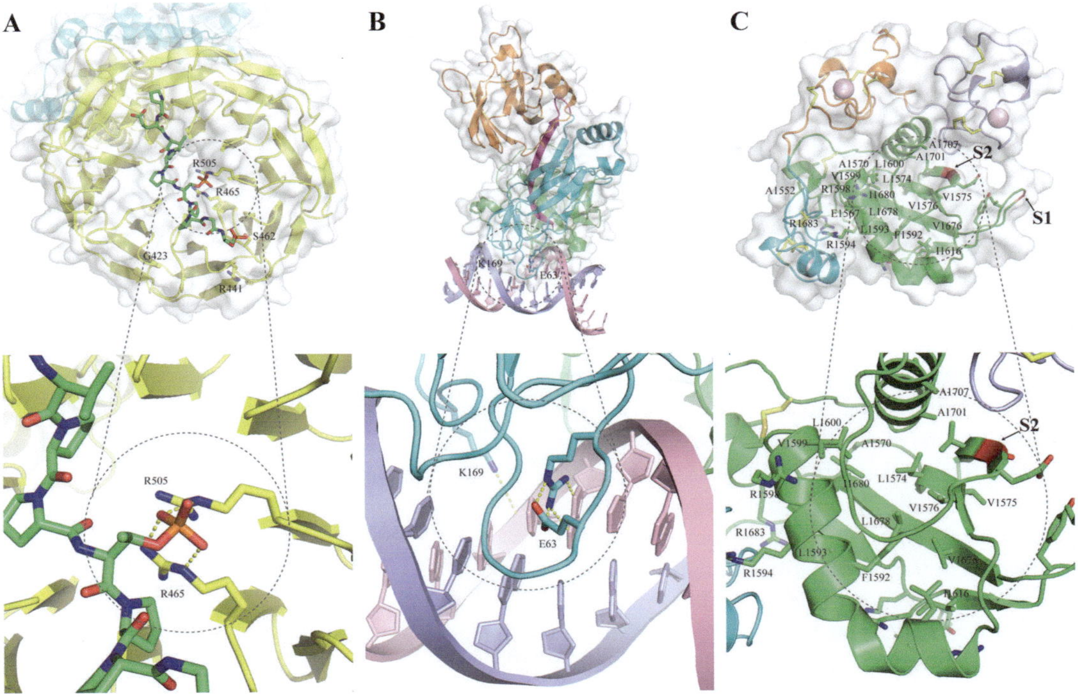

Fig. 2 Examples of pathogenic mutations in key Notch pathway components in the context of protein structure. (**a**) Structure of the WD40 repeats of FBXW7 (PDB ID, 5V4B). Residues for which mutations have been identified in diseases are indicated. (**b**) Structure of the transcription factor RBPJ in complex with the DNA (PDB ID, 3IAG). Indicated are residues mutated in AOS. (**c**) Structure of the NRR of NOTCH1 (PDB ID, 3ETO). Indicated are residues for which mutations have been identified in diseases and that have been functionally analyzed. A: alanine; E: glutamic acid; F: phenylalanine; G: glycine; I: isoleucine; K: lysine; L: leucine; R: arginine; S: serine; V: valine

et al. 2012). The AOS-associated missense mutations identified in the *RBPJ* gene (Fig. 2b) compromise its DNA binding ability (Hassed et al. 2012) and mutations in *NOTCH1* and *DLL4* have been also identified in AOS patients (Meester et al. 2015; Stittrich et al. 2014). The reader is also referred to "Integration of *Drosophila* and Human Genetics to Understand Notch Signaling Related Diseases" for an in-depth review of genetic mutations of Notch pathway components. Chromosomal translocations and aberrations involving *FBXW7* and *RBPJ* are also linked to diseases. *FBXW7* is translocated in renal cell cancer [RCC; (Kuiper et al. 2009)] while *RBPJ* in the proximal 4p deletion syndrome (Nakayama et al. 2014).

Mutations occurring in the *NOTCH1* gene are clustered in different regions (Table 1). Among them, mutations that occur in the LNR repeats, HD and PEST domains are seen in many types of diseases as well as genetic disorders. Typically, mutations involving the LNR repeats and HD domain lead to disruption of the negative regulatory region (Fig. 2c) and promote ligand-independent cleavage of the receptor, leading to increased Notch signaling. Similarly, mutations that influence the structure of the PEST domain lead to increased half-life of the NICD resulting in aberrant transcriptional activity. Similarly to *FBXW7* and *RBPJ*, also the *NOTCH1* gene is subjected to chromosomal translocations that impair its activity (Ellisen et al. 1991).

Thus, there are indeed viable genetic mutations of the Notch receptor or Notch signaling components, that could in future provide even more insights in Notch-related pathologies, not only in the context of development but also in the cancer context.

6 Perspectives

Given the important function of the Notch signaling pathway in cancer as well as in genetic diseases, it will be important to deeper understand its regulation focusing on the molecular basis that characterize this signaling cascade. This approach will allow in the future the development of new and more efficient therapies that can overcome the limitations of the current approaches, primarily the side effects and resistance observed by using GSIs. New cancer therapies might be based on small molecule inhibitors of Notch modulators or Notch pathway components to reactivate the tumor suppressive function or to block the oncogenic activities of the pathway depending on the different pathological contexts. Another fascinating alternative would be the use of antibodies aimed to stimulate or block the activation of NOTCH receptors, an approach that seems to be promising. This can be achieved by using antibodies directed against NOTCH receptors (Aste-Amezaga et al. 2010; Wu et al. 2010; Canalis et al. 2017), ligands (Billiard et al. 2011; Lafkas et al. 2015; Xu et al. 2016; Wang et al. 2017) or the γ-secretase complex (Hayashi et al. 2012). Similar approaches can be employed to modulate the Notch function in macrophages as inflam-mation is one the key processes that drive tumorigenesis. In conclusion, more work is needed to deeply understand the regulation of the Notch signaling pathway and modulate its activity for clinical use.

Acknowledgements We are grateful to Dr. Rhett A. Kovall (University of Cincinnati, USA) for providing us with Fig. 2 and to Drs. Shinya Yamamoto and Jose L. Salazar (Baylor College of Medicine, Houston, USA) and Francesca Ferrante (University of Giessen, Germany) for critical reading. This work was supported by the collaborative research grant TRR81, the Excellence Cluster for Cardio Pulmonary System (ECCPS) and the University of Giessen (Germany).

References

Artavanis-Tsakonas S, Muskavitch MA, Yedvobnick B (1983) Molecular cloning of Notch, a locus affecting neurogenesis in Drosophila melanogaster. Proc Natl Acad Sci U S A 80(7):1977–1981

Aste-Amezaga M, Zhang N, Lineberger JE, Arnold BA, Toner TJ, Gu M, Huang L, Vitelli S, Vo KT, Haytko P, Zhao JZ, Baleydier F, L'Heureux S, Wang H, Gordon WR, Thoryk E, Andrawes MB, Tiyanont K, Stegmaier K, Roti G, Ross KN, Franlin LL, Wang H, Wang F, Chastain M, Bett AJ, Audoly LP, Aster JC, Blacklow SC, Huber HE (2010) Characterization of Notch1 antibodies that inhibit signaling of both normal and mutated Notch1 receptors. PLoS One 5(2):e9094. https://doi.org/10.1371/journal.pone.0009094

Aster JC, Pear WS, Blacklow SC (2017) The varied roles of Notch in cancer. Annu Rev Pathol 12:245–275. https://doi.org/10.1146/annurev-pathol-052016-100127

Aydin IT, Melamed RD, Adams SJ, Castillo-Martin M, Demir A, Bryk D, Brunner G, Cordon-Cardo C, Osman I, Rabadan R, Celebi JT (2014) FBXW7 mutations in melanoma and a new therapeutic paradigm. J Natl Cancer Inst 106(6):dju107. https://doi.org/10.1093/jnci/dju107

Backvall H, Stromberg S, Gustafsson A, Asplund A, Sivertsson A, Lundeberg J, Ponten F (2004) Mutation spectra of epidermal p53 clones adjacent to basal cell carcinoma and squamous cell carcinoma. Exp Dermatol 13(10):643–650. https://doi.org/10.1111/j.0906-6705.2004.00211.x

Balint K, Xiao M, Pinnix CC, Soma A, Veres I, Juhasz I, Brown EJ, Capobianco AJ, Herlyn M, Liu ZJ (2005) Activation of Notch1 signaling is required for beta-catenin-mediated human primary melanoma progression. J Clin Invest 115(11):3166–3176. https://doi.org/10.1172/JCI25001

Banerji S, Cibulskis K, Rangel-Escareno C, Brown KK, Carter SL, Frederick AM, Lawrence MS, Sivachenko AY, Sougnez C, Zou L, Cortes ML, Fernandez-Lopez JC, Peng S, Ardlie KG, Auclair D, Bautista-Pina V, Duke F, Francis J, Jung J, Maffuz-Aziz A, Onofrio RC, Parkin M, Pho NH, Quintanar-Jurado V, Ramos AH, Rebollar-Vega R, Rodriguez-Cuevas S, Romero-Cordoba SL, Schumacher SE, Stransky N, Thompson KM, Uribe-Figueroa L, Baselga J, Beroukhim R, Polyak K, Sgroi DC, Richardson AL, Jimenez-Sanchez G, Lander ES, Gabriel SB, Garraway LA, Golub TR, Melendez-Zajgla J, Toker A, Getz G, Hidalgo-Miranda A, Meyerson M (2012) Sequence analysis of mutations and translocations across breast cancer subtypes. Nature 486(7403):405–409. https://doi.org/10.1038/nature11154

Bao S, Wu Q, McLendon RE, Hao Y, Shi Q, Hjelmeland AB, Dewhirst MW, Bigner DD, Rich JN (2006) Glioma stem cells promote radioresistance by preferential activation of the DNA damage response. Nature 444(7120):756–760. https://doi.org/10.1038/nature05236

Bea S, Valdes-Mas R, Navarro A, Salaverria I, Martin-Garcia D, Jares P, Gine E, Pinyol M, Royo C, Nadeu F, Conde L, Juan M, Clot G, Vizan P, Di Croce L, Puente DA, Lopez-Guerra M, Moros A, Roue G, Aymerich M, Villamor N, Colomo L, Martinez A, Valera A, Martin-Subero JI, Amador V, Hernandez L, Rozman M, Enjuanes A, Forcada P, Muntanola A, Hartmann

EM, Calasanz MJ, Rosenwald A, Ott G, Hernandez-Rivas JM, Klapper W, Siebert R, Wiestner A, Wilson WH, Colomer D, Lopez-Guillermo A, Lopez-Otin C, Puente XS, Campo E (2013) Landscape of somatic mutations and clonal evolution in mantle cell lymphoma. Proc Natl Acad Sci U S A 110(45):18250–18255. https://doi.org/10.1073/pnas.1314608110

Bhola NE, Jansen VM, Koch JP, Li H, Formisano L, Williams JA, Grandis JR, Arteaga CL (2016) Treatment of triple-negative breast cancer with TORC1/2 inhibitors sustains a drug-resistant and Notch-dependent cancer stem cell population. Cancer Res 76(2):440–452. https://doi.org/10.1158/0008-5472.CAN-15-1640-T

Bi YL, Min M, Shen W, Liu Y (2016) Numb/Notch signaling pathway modulation enhances human pancreatic cancer cell radiosensitivity. Tumour Biol J Int Soc Oncodev Biol Med 37(11):15145–15155. https://doi.org/10.1007/s13277-016-5311-8

Billiard F, Kirshner JR, Tait M, Danave A, Taheri S, Zhang W, Waite JC, Olson K, Chen G, Coetzee S, Hylton D, Murphy AJ, Yancopoulos GD, Thurston G, Skokos D (2011) Ongoing Dll4-Notch signaling is required for T-cell homeostasis in the adult thymus. Eur J Immunol 41(8):2207–2216. https://doi.org/10.1002/eji.201041343

Bittolo T, Pozzo F, Bomben R, D'Agaro T, Bravin V, Bulian P, Rossi FM, Zucchetto A, Degan M, Macor P, D'Arena G, Chiarenza A, Zaja F, Pozzato G, Di Raimondo F, Rossi D, Gaidano G, Del Poeta G, Gattei V, Dal Bo M (2017) Mutations in the 3' untranslated region of NOTCH1 are associated with low CD20 expression levels chronic lymphocytic leukemia. Haematologica 102(8):e305–e309. https://doi.org/10.3324/haematol.2016.162594

Blaumueller CM, Qi H, Zagouras P, Artavanis-Tsakonas S (1997) Intracellular cleavage of Notch leads to a heterodimeric receptor on the plasma membrane. Cell 90(2):281–291

Boonyatecha N, Sangphech N, Wongchana W, Kueanjinda P, Palaga T (2012) Involvement of Notch signaling pathway in regulating IL-12 expression via c-Rel in activated macrophages. Mol Immunol 51(3–4):255–262. https://doi.org/10.1016/j.molimm.2012.03.017

Borges M, Linnoila RI, van de Velde HJ, Chen H, Nelkin BD, Mabry M, Baylin SB, Ball DW (1997) An achaete-scute homologue essential for neuroendocrine differentiation in the lung. Nature 386(6627):852–855. https://doi.org/10.1038/386852a0

Borggrefe T, Oswald F (2009) The Notch signaling pathway: transcriptional regulation at Notch target genes. Cell Mol Life Sci 66(10):1631–1646. https://doi.org/10.1007/s00018-009-8668-7

Borggrefe T, Lauth M, Zwijsen A, Huylebroeck D, Oswald F, Giaimo BD (2016) The Notch intracellular domain integrates signals from Wnt, Hedgehog, TGFbeta/BMP and hypoxia pathways. Biochim Biophys Acta 1863(2):303–313. https://doi.org/10.1016/j.bbamcr.2015.11.020

Breit S, Stanulla M, Flohr T, Schrappe M, Ludwig WD, Tolle G, Happich M, Muckenthaler MU, Kulozik AE (2006) Activating NOTCH1 mutations predict favorable early treatment response and long-term outcome in childhood precursor T-cell lymphoblastic leukemia. Blood 108(4):1151–1157. https://doi.org/10.1182/blood-2005-12-4956

Brou C, Logeat F, Gupta N, Bessia C, LeBail O, Doedens JR, Cumano A, Roux P, Black RA, Israel A (2000) A novel proteolytic cleavage involved in Notch signaling : the role of the disintegrin-metalloprotease TACE. Mol Cell 5(2):207–216

Canalis E, Sanjay A, Yu J, Zanotti S (2017) An antibody to Notch2 reverses the osteopenic phenotype of Hajdu-Cheney mutant male mice. Endocrinology 158(4):730–742. https://doi.org/10.1210/en.2016-1787

Cancer Genome Atlas N (2012) Comprehensive molecular portraits of human breast tumours. Nature 490(7418):61–70. https://doi.org/10.1038/nature11412

Cantarini MC, de la Monte SM, Pang M, Tong M, D'Errico A, Trevisani F, Wands JR (2006) Aspartyl-asparagyl beta hydroxylase over-expression in human hepatoma is linked to activation of insulin-like growth factor and notch signaling mechanisms. Hepatology 44(2):446–457. https://doi.org/10.1002/hep.21272

Chen J, Kesari S, Rooney C, Strack PR, Chen J, Shen H, Wu L, Griffin JD (2010) Inhibition of notch signaling blocks growth of glioblastoma cell lines and tumor neurospheres. Genes Cancer 1(8):822–835. https://doi.org/10.1177/1947601910383564

Cohen B, Shimizu M, Izrailit J, Ng NF, Buchman Y, Pan JG, Dering J, Reedijk M (2010) Cyclin D1 is a direct target of JAG1-mediated Notch signaling in breast cancer. Breast Cancer Res Treat 123(1):113–124. https://doi.org/10.1007/s10549-009-0621-9

Collins KJ, Yuan Z, Kovall RA (2014) Structure and function of the CSL-KyoT2 corepressor complex: a negative regulator of Notch signaling. Structure 22(1):70–81. https://doi.org/10.1016/j.str.2013.10.010

Contreras AN, Yuan Z, Kovall RA (2015) Thermodynamic binding analysis of Notch transcription complexes from Drosophila melanogaster. Protein Sci 24(5):812–822. https://doi.org/10.1002/pro.2652

Cook N, Frese KK, Bapiro TE, Jacobetz MA, Gopinathan A, Miller JL, Rao SS, Demuth T, Howat WJ, Jodrell DI, Tuveson DA (2012) Gamma secretase inhibition promotes hypoxic necrosis in mouse pancreatic ductal adenocarcinoma. J Exp Med 209(3):437–444. https://doi.org/10.1084/jem.20111923

D'Agaro T, Bittolo T, Bravin V, Dal Bo M, Pozzo F, Bulian P, Rossi FM, Zucchetto A, Degan M, D'Arena G, Chiarenza A, Zaja F, Pozzato G, Di Raimondo F, Rossi D, Gaidano G, Del Poeta G, Gattei V, Bomben R (2017) NOTCH1 mutational status in chronic lymphocytic leukaemia: clinical relevance of subclonal mutations and mutation types. Br J Haematol. https://doi.org/10.1111/bjh.14843

D'Souza B, Miyamoto A, Weinmaster G (2008) The many facets of Notch ligands. Oncogene 27(38):5148–5167. https://doi.org/10.1038/onc.2008.229

De Keersmaecker K, Lahortiga I, Mentens N, Folens C, Van Neste L, Bekaert S, Vandenberghe P, Odero MD, Marynen P, Cools J (2008) In vitro validation of gamma-secretase inhibitors alone or in combination with other anti-cancer drugs for the treatment of T-cell acute lymphoblastic leukemia. Haematologica 93(4):533–542. https://doi.org/10.3324/haematol.11894

De La OJ, Emerson LL, Goodman JL, Froebe SC, Illum BE, Curtis AB, Murtaugh LC (2008) Notch and Kras reprogram pancreatic acinar cells to ductal intraepithelial neoplasia. Proc Natl Acad Sci U S A 105(48):18907–18912. https://doi.org/10.1073/pnas.0810111105

del Amo FF, Gendron-Maguire M, Swiatek PJ, Jenkins NA, Copeland NG, Gridley T (1993) Cloning, analysis, and chromosomal localization of Notch-1, a mouse homolog of Drosophila Notch. Genomics 15(2):259–264. https://doi.org/10.1006/geno.1993.1055

Demehri S, Turkoz A, Kopan R (2009) Epidermal Notch1 loss promotes skin tumorigenesis by impacting the stromal microenvironment. Cancer Cell 16(1):55–66. https://doi.org/10.1016/j.ccr.2009.05.016

Dill MT, Tornillo L, Fritzius T, Terracciano L, Semela D, Bettler B, Heim MH, Tchorz JS (2013) Constitutive Notch2 signaling induces hepatic tumors in mice. Hepatology 57(4):1607–1619. https://doi.org/10.1002/hep.26165

Ducharme V, Guauque-Olarte S, Gaudreault N, Pibarot P, Mathieu P, Bosse Y (2013) NOTCH1 genetic variants in patients with tricuspid calcific aortic valve stenosis. J Heart Valve Dis 22(2):142–149

Ellisen LW, Bird J, West DC, Soreng AL, Reynolds TC, Smith SD, Sklar J (1991) TAN-1, the human homolog of the Drosophila notch gene, is broken by chromosomal translocations in T lymphoblastic neoplasms. Cell 66(4):649–661

Fabbri G, Rasi S, Rossi D, Trifonov V, Khiabanian H, Ma J, Grunn A, Fangazio M, Capello D, Monti S, Cresta S, Gargiulo E, Forconi F, Guarini A, Arcaini L, Paulli M, Laurenti L, Larocca LM, Marasca R, Gattei V, Oscier D, Bertoni F, Mullighan CG, Foa R, Pasqualucci L, Rabadan R, Dalla-Favera R, Gaidano G (2011) Analysis of the chronic lymphocytic leukemia coding genome: role of NOTCH1 mutational activation. J Exp Med 208(7):1389–1401. https://doi.org/10.1084/jem.20110921

Foffa I, Ait Ali L, Panesi P, Mariani M, Festa P, Botto N, Vecoli C, Andreassi MG (2013) Sequencing of NOTCH1, GATA5, TGFBR1 and TGFBR2 genes in familial cases of bicuspid aortic valve. BMC Med Genet 14:44. https://doi.org/10.1186/1471-2350-14-44

Fan X, Khaki L, Zhu TS, Soules ME, Talsma CE, Gul N, Koh C, Zhang J, Li YM, Maciaczyk J, Nikkhah G, Dimeco F, Piccirillo S, Vescovi AL, Eberhart CG (2010) NOTCH pathway blockade depletes CD133-positive glioblastoma cells and inhibits growth of tumor neurospheres and xenografts. Stem Cells 28(1):5–16. https://doi.org/10.1002/stem.254

Foldi J, Chung AY, Xu H, Zhu J, Outtz HH, Kitajewski J, Li Y, Hu X, Ivashkiv LB (2010) Autoamplification of Notch signaling in macrophages by TLR-induced and RBP-J-dependent induction of Jagged1. J Immunol 185(9):5023–5031. https://doi.org/10.4049/jimmunol.1001544

Foldi J, Shang Y, Zhao B, Ivashkiv LB, Hu X (2016) RBP-J is required for M2 macrophage polarization in response to chitin and mediates expression of a subset of M2 genes. Protein & cell 7(3):201–209. https://doi.org/10.1007/s13238-016-0248-7

Fortini ME (2002) Gamma-secretase-mediated proteolysis in cell-surface-receptor signalling. Nat Rev Mol Cell Biol 3(9):673–684. https://doi.org/10.1038/nrm910

Foulkes WD, Smith IE, Reis-Filho JS (2010) Triple-negative breast cancer. N Engl J Med 363(20):1938–1948. https://doi.org/10.1056/NEJMra1001389

Fryer CJ, Lamar E, Turbachova I, Kintner C, Jones KA (2002) Mastermind mediates chromatin-specific transcription and turnover of the Notch enhancer complex. Genes Dev 16(11):1397–1411. https://doi.org/10.1101/gad.991602

Fryer CJ, White JB, Jones KA (2004) Mastermind recruits CycC:CDK8 to phosphorylate the Notch ICD and coordinate activation with turnover. Mol Cell 16(4):509–520. https://doi.org/10.1016/j.molcel.2004.10.014

Fung E, Tang SM, Canner JP, Morishige K, Arboleda-Velasquez JF, Cardoso AA, Carlesso N, Aster JC, Aikawa M (2007) Delta-like 4 induces notch signaling in macrophages: implications for inflammation. Circulation 115(23):2948–2956. https://doi.org/10.1161/CIRCULATIONAHA.106.675462

Furnari FB, Fenton T, Bachoo RM, Mukasa A, Stommel JM, Stegh A, Hahn WC, Ligon KL, Louis DN, Brennan C, Chin L, DePinho RA, Cavenee WK (2007) Malignant astrocytic glioma: genetics, biology, and paths to treatment. Genes Dev 21(21):2683–2710. https://doi.org/10.1101/gad.1596707

Gao J, Chen Y, Wu KC, Liu J, Zhao YQ, Pan YL, Du R, Zheng GR, Xiong YM, Xu HL, Fan DM (2010) RUNX3 directly interacts with intracellular domain of Notch1 and suppresses Notch signaling in hepatocellular carcinoma cells. Exp Cell Res 316(2):149–157. https://doi.org/10.1016/j.yexcr.2009.09.025

Garg V, Muth AN, Ransom JF, Schluterman MK, Barnes R, King IN, Grossfeld PD, Srivastava D (2005) Mutations in NOTCH1 cause aortic valve disease. Nature 437(7056):270–274. https://doi.org/10.1038/nature03940

Garraway LA, Sellers WR (2006) Lineage dependency and lineage-survival oncogenes in human cancer. Nat Rev Cancer 6(8):593–602. https://doi.org/10.1038/nrc1947

George J, Lim JS, Jang SJ, Cun Y, Ozretic L, Kong G, Leenders F, Lu X, Fernandez-Cuesta L, Bosco G, Muller C, Dahmen I, Jahchan NS, Park KS, Yang D, Karnezis AN, Vaka D, Torres A, Wang MS, Korbel JO, Menon R, Chun SM, Kim D, Wilkerson M, Hayes N, Engelmann D, Putzer B, Bos M, Michels S, Vlasic I, Seidel D, Pinther B, Schaub P, Becker C, Altmuller J, Yokota J, Kohno T, Iwakawa R, Tsuta K, Noguchi M,

Muley T, Hoffmann H, Schnabel PA, Petersen I, Chen Y, Soltermann A, Tischler V, Choi CM, Kim YH, Massion PP, Zou Y, Jovanovic D, Kontic M, Wright GM, Russell PA, Solomon B, Koch I, Lindner M, Muscarella LA, la Torre A, Field JK, Jakopovic M, Knezevic J, Castanos-Velez E, Roz L, Pastorino U, Brustugun OT, Lund-Iversen M, Thunnissen E, Kohler J, Schuler M, Botling J, Sandelin M, Sanchez-Cespedes M, Salvesen HB, Achter V, Lang U, Bogus M, Schneider PM, Zander T, Ansen S, Hallek M, Wolf J, Vingron M, Yatabe Y, Travis WD, Nurnberg P, Reinhardt C, Perner S, Heukamp L, Buttner R, Haas SA, Brambilla E, Peifer M, Sage J, Thomas RK (2015) Comprehensive genomic profiles of small cell lung cancer. Nature 524(7563):47–53. https://doi.org/10.1038/nature14664

Gilbert CA, Daou MC, Moser RP, Ross AH (2010) Gamma-secretase inhibitors enhance temozolomide treatment of human gliomas by inhibiting neurosphere repopulation and xenograft recurrence. Cancer Res 70(17):6870–6879. https://doi.org/10.1158/0008-5472.CAN-10-1378

Grossman SA, Batara JF (2004) Current management of glioblastoma multiforme. Semin Oncol 31(5):635–644

Hanlon L, Avila JL, Demarest RM, Troutman S, Allen M, Ratti F, Rustgi AK, Stanger BZ, Radtke F, Adsay V, Long F, Capobianco AJ, Kissil JL (2010) Notch1 functions as a tumor suppressor in a model of K-ras-induced pancreatic ductal adenocarcinoma. Cancer Res 70(11):4280–4286. https://doi.org/10.1158/0008-5472.CAN-09-4645

Hansson ML, Popko-Scibor AE, Saint Just Ribeiro M, Dancy BM, Lindberg MJ, Cole PA, Wallberg AE (2009) The transcriptional coactivator MAML1 regulates p300 autoacetylation and HAT activity. Nucleic Acids Res 37(9):2996–3006. https://doi.org/10.1093/nar/gkp163

Hass MR, Liow HH, Chen X, Sharma A, Inoue YU, Inoue T, Reeb A, Martens A, Fulbright M, Raju S, Stevens M, Boyle S, Park JS, Weirauch MT, Brent MR, Kopan R (2015) SpDamID: marking DNA bound by protein complexes identifies Notch-Dimer responsive enhancers. Mol Cell 59(4):685–697. https://doi.org/10.1016/j.molcel.2015.07.008

Hassed SJ, Wiley GB, Wang S, Lee JY, Li S, Xu W, Zhao ZJ, Mulvihill JJ, Robertson J, Warner J, Gaffney PM (2012) RBPJ mutations identified in two families affected by Adams-Oliver syndrome. Am J Hum Genet 91(2):391–395. https://doi.org/10.1016/j.ajhg.2012.07.005

Hayashi I, Takatori S, Urano Y, Miyake Y, Takagi J, Sakata-Yanagimoto M, Iwanari H, Osawa S, Morohashi Y, Li T, Wong PC, Chiba S, Kodama T, Hamakubo T, Tomita T, Iwatsubo T (2012) Neutralization of the gamma-secretase activity by monoclonal antibody against extracellular domain of nicastrin. Oncogene 31(6):787–798. https://doi.org/10.1038/onc.2011.265

Hein K, Mittler G, Cizelsky W, Kuhl M, Ferrante F, Liefke R, Berger IM, Just S, Strang JE, Kestler HA, Oswald F, Borggrefe T (2015) Site-specific methylation of Notch1 controls the amplitude and duration of the Notch1 response. Sci Signal 8(369):ra30. https://doi.org/10.1126/scisignal.2005892

Hong SW, Hur W, Choi JE, Kim JH, Hwang D, Yoon SK (2016) Role of ADAM17 in invasion and migration of CD133-expressing liver cancer stem cells after irradiation. Oncotarget 7(17):23482–23497. https://doi.org/10.18632/oncotarget.8112

Hu X, Chung AY, Wu I, Foldi J, Chen J, Ji JD, Tateya T, Kang YJ, Han J, Gessler M, Kageyama R, Ivashkiv LB (2008) Integrated regulation of Toll-like receptor responses by Notch and interferon-gamma pathways. Immunity 29(5):691–703. https://doi.org/10.1016/j.immuni.2008.08.016

Hu B, Castillo E, Harewood L, Ostano P, Reymond A, Dummer R, Raffoul W, Hoetzenecker W, Hofbauer GF, Dotto GP (2012) Multifocal epithelial tumors and field cancerization from loss of mesenchymal CSL signaling. Cell 149(6):1207–1220. https://doi.org/10.1016/j.cell.2012.03.048

Hu L, Xue F, Shao M, Deng A, Wei G (2013) Aberrant expression of Notch3 predicts poor survival for hepatocellular carcinomas. Biosci Trends 7(3):152–156

Hu YJ, Li HY, Qiu KJ, Li DC, Zhou JH, Hu YH, Zhang FM (2014) Downregulation of Notch1 inhibits the invasion of human hepatocellular carcinoma HepG2 and MHCC97H cells through the regulation of PTEN and FAK. Int J Mol Med 34(4):1081–1086. https://doi.org/10.3892/ijmm.2014.1889

Hu Y, Su H, Li X, Guo G, Cheng L, Qin R, Qing G, Liu H (2015) The NOTCH ligand JAGGED2 promotes pancreatic cancer metastasis independent of NOTCH signaling activation. Mol Cancer Ther 14(1):289–297. https://doi.org/10.1158/1535-7163.MCT-14-0501

Huntzicker EG, Hotzel K, Choy L, Che L, Ross J, Pau G, Sharma N, Siebel CW, Chen X, French DM (2015) Differential effects of targeting Notch receptors in a mouse model of liver cancer. Hepatology 61(3):942–952. https://doi.org/10.1002/hep.27566

Iascone M, Ciccone R, Galletti L, Marchetti D, Seddio F, Lincesso AR, Pezzoli L, Vetro A, Barachetti D, Boni L, Federici D, Soto AM, Comas JV, Ferrazzi P, Zuffardi O (2012) Identification of de novo mutations and rare variants in hypoplastic left heart syndrome. Clin Genet 81(6):542–554. https://doi.org/10.1111/j.1399-0004.2011.01674.x

Jiang T, Collins BJ, Jin N, Watkins DN, Brock MV, Matsui W, Nelkin BD, Ball DW (2009) Achaete-scute complex homologue 1 regulates tumor-initiating capacity in human small cell lung cancer. Cancer Res 69(3):845–854. https://doi.org/10.1158/0008-5472.CAN-08-2762

Jung C, Mittler G, Oswald F, Borggrefe T (2013) RNA helicase Ddx5 and the noncoding RNA SRA act as coactivators in the Notch signaling pathway. Biochim Biophys Acta 1833(5):1180–1189. https://doi.org/10.1016/j.bbamcr.2013.01.032

Kao HY, Ordentlich P, Koyano-Nakagawa N, Tang Z, Downes M, Kintner CR, Evans RM, Kadesch T (1998) A histone deacetylase corepressor complex regulates the Notch signal transduction pathway. Genes Dev 12(15):2269–2277

Kapellos TS, Iqbal AJ (2016) Epigenetic control of macrophage polarisation and soluble mediator gene expression during Inflammation. Mediat Inflamm 2016:6591703. https://doi.org/10.1155/2016/6591703

Kent KC, Crenshaw ML, Goh DL, Dietz HC (2013) Genotype-phenotype correlation in patients with bicuspid aortic valve and aneurysm. J Thorac Cardiovasc Surg 146(1):158–165 e151. https://doi.org/10.1016/j.jtcvs.2012.09.060

Kidd S, Kelley MR, Young MW (1986) Sequence of the notch locus of Drosophila melanogaster: relationship of the encoded protein to mammalian clotting and growth factors. Mol Cell Biol 6(9):3094–3108

Kim RK, Kaushik N, Suh Y, Yoo KC, Cui YH, Kim MJ, Lee HJ, Kim IG, Lee SJ (2016) Radiation driven epithelial-mesenchymal transition is mediated by Notch signaling in breast cancer. Oncotarget 7(33):53430–53442. https://doi.org/10.18632/oncotarget.10802

Kim W, Khan SK, Gvozdenovic-Jeremic J, Kim Y, Dahlman J, Kim H, Park O, Ishitani T, Jho EH, Gao B, Yang Y (2017) Hippo signaling interactions with Wnt/beta-catenin and Notch signaling repress liver tumorigenesis. J Clin Invest 127(1):137–152. https://doi.org/10.1172/JCI88486

Knoechel B, Roderick JE, Williamson KE, Zhu J, Lohr JG, Cotton MJ, Gillespie SM, Fernandez D, Ku M, Wang H, Piccioni F, Silver SJ, Jain M, Pearson D, Kluk MJ, Ott CJ, Shultz LD, Brehm MA, Greiner DL, Gutierrez A, Stegmaier K, Kung AL, Root DE, Bradner JE, Aster JC, Kelliher MA, Bernstein BE (2014) An epigenetic mechanism of resistance to targeted therapy in T cell acute lymphoblastic leukemia. Nat Genet 46(4):364–370. https://doi.org/10.1038/ng.2913

Kolev V, Mandinova A, Guinea-Viniegra J, Hu B, Lefort K, Lambertini C, Neel V, Dummer R, Wagner EF, Dotto GP (2008) EGFR signalling as a negative regulator of Notch1 gene transcription and function in proliferating keratinocytes and cancer. Nat Cell Biol 10(8):902–911. https://doi.org/10.1038/ncb1750

Kovall RA (2007) Structures of CSL, Notch and Mastermind proteins: piecing together an active transcription complex. Curr Opin Struct Biol 17(1):117–127. https://doi.org/10.1016/j.sbi.2006.11.004

Kovall RA, Hendrickson WA (2004) Crystal structure of the nuclear effector of Notch signaling, CSL, bound to DNA. EMBO J 23(17):3441–3451. https://doi.org/10.1038/sj.emboj.7600349

Kridel R, Meissner B, Rogic S, Boyle M, Telenius A, Woolcock B, Gunawardana J, Jenkins C, Cochrane C, Ben-Neriah S, Tan K, Morin RD, Opat S, Sehn LH, Connors JM, Marra MA, Weng AP, Steidl C, Gascoyne RD (2012) Whole transcriptome sequencing reveals recurrent NOTCH1 mutations in mantle cell lymphoma. Blood 119(9):1963–1971. https://doi.org/10.1182/blood-2011-11-391474

Kuiper RP, Vreede L, Venkatachalam R, Ricketts C, Kamping E, Verwiel E, Govaerts L, Debiec-Rychter M, Lerut E, van Erp F, Hoogerbrugge N, van Kempen L, Schoenmakers EF, Bonne A, Maher ER, Geurts van Kessel A (2009) The tumor suppressor gene FBXW7 is disrupted by a constitutional t(3;4)(q21;q31) in a patient with renal cell cancer. Cancer Genet Cytogenet 195(2):105–111. https://doi.org/10.1016/j.cancergencyto.2009.07.001

Kumar V, Palermo R, Talora C, Campese AF, Checquolo S, Bellavia D, Tottone L, Testa G, Miele E, Indraccolo S, Amadori A, Ferretti E, Gulino A, Vacca A, Screpanti I (2014) Notch and NF-kB signaling pathways regulate miR-223/FBXW7 axis in T-cell acute lymphoblastic leukemia. Leukemia 28(12):2324–2335. https://doi.org/10.1038/leu.2014.133

Lafkas D, Shelton A, Chiu C, de Leon Boenig G, Chen Y, Stawicki SS, Siltanen C, Reichelt M, Zhou M, Wu X, Eastham-Anderson J, Moore H, Roose-Girma M, Chinn Y, Hang JQ, Warming S, Egen J, Lee WP, Austin C, Wu Y, Payandeh J, Lowe JB, Siebel CW (2015) Therapeutic antibodies reveal Notch control of transdifferentiation in the adult lung. Nature 528 (7580):127-131. doi:https://doi.org/10.1038/nature15715

Lake RJ, Grimm LM, Veraksa A, Banos A, Artavanis-Tsakonas S (2009) In vivo analysis of the Notch receptor S1 cleavage. PLoS One 4(8):e6728. https://doi.org/10.1371/journal.pone.0006728

Landor SK, Mutvei AP, Mamaeva V, Jin S, Busk M, Borra R, Gronroos TJ, Kronqvist P, Lendahl U, Sahlgren CM (2011) Hypo- and hyperactivated Notch signaling induce a glycolytic switch through distinct mechanisms. Proc Natl Acad Sci U S A 108(46):18814–18819. https://doi.org/10.1073/pnas.1104943108

Larson Gedman A, Chen Q, Kugel Desmoulin S, Ge Y, LaFiura K, Haska CL, Cherian C, Devidas M, Linda SB, Taub JW, Matherly LH (2009) The impact of NOTCH1, FBW7 and PTEN mutations on prognosis and downstream signaling in pediatric T-cell acute lymphoblastic leukemia: a report from the children's oncology group. Leukemia 23(8):1417–1425. https://doi.org/10.1038/leu.2009.64

Lefort K, Mandinova A, Ostano P, Kolev V, Calpini V, Kolfschoten I, Devgan V, Lieb J, Raffoul W, Hohl D, Neel V, Garlick J, Chiorino G, Dotto GP (2007) Notch1 is a p53 target gene involved in human keratinocyte tumor suppression through negative regulation of ROCK1/2 and MRCKalpha kinases. Genes Dev 21(5):562–577. https://doi.org/10.1101/gad.1484707

Lehmann BD, Bauer JA, Chen X, Sanders ME, Chakravarthy AB, Shyr Y, Pietenpol JA (2011) Identification of human triple-negative breast cancer subtypes and preclinical models for selection of targeted therapies. J Clin Invest 121(7):2750–2767. https://doi.org/10.1172/JCI45014

Liefke R, Oswald F, Alvarado C, Ferres-Marco D, Mittler G, Rodriguez P, Dominguez M, Borggrefe T (2010) Histone demethylase KDM5A is an integral part of the core Notch-RBP-J repressor complex. Genes Dev 24(6):590–601. https://doi.org/10.1101/gad.563210

Lin SE, Oyama T, Nagase T, Harigaya K, Kitagawa M (2002) Identification of new human mastermind

proteins defines a family that consists of positive regulators for notch signaling. J Biol Chem 277(52):50612–50620. https://doi.org/10.1074/jbc.M209529200
Liu ZJ, Xiao M, Balint K, Smalley KS, Brafford P, Qiu R, Pinnix CC, Li X, Herlyn M (2006) Notch1 signaling promotes primary melanoma progression by activating mitogen-activated protein kinase/phosphatidylinositol 3-kinase-Akt pathways and up-regulating N-cadherin expression. Cancer Res 66(8):4182–4190. https://doi.org/10.1158/0008-5472.CAN-05-3589
Liu H, Chi AW, Arnett KL, Chiang MY, Xu L, Shestova O, Wang H, Li YM, Bhandoola A, Aster JC, Blacklow SC, Pear WS (2010) Notch dimerization is required for leukemogenesis and T-cell development. Genes Dev 24(21):2395–2407. https://doi.org/10.1101/gad.1975210
Liu M, Lee DF, Chen CT, Yen CJ, Li LY, Lee HJ, Chang CJ, Chang WC, Hsu JM, Kuo HP, Xia W, Wei Y, Chiu PC, Chou CK, Du Y, Dhar D, Karin M, Chen CH, Hung MC (2012) IKKalpha activation of NOTCH links tumorigenesis via FOXA2 suppression. Mol Cell 45(2):171–184. https://doi.org/10.1016/j.molcel.2011.11.018
Logeat F, Bessia C, Brou C, LeBail O, Jarriault S, Seidah NG, Israel A (1998) The Notch1 receptor is cleaved constitutively by a furin-like convertase. Proc Natl Acad Sci U S A 95(14):8108–8112
Luo J, Wang P, Wang R, Wang J, Liu M, Xiong S, Li Y, Cheng B (2016) The Notch pathway promotes the cancer stem cell characteristics of CD90+ cells in hepatocellular carcinoma. Oncotarget 7(8):9525–9537. https://doi.org/10.18632/oncotarget.6672
Ma L, Dong P, Liu L, Gao Q, Duan M, Zhang S, Chen S, Xue R, Wang X (2016) Overexpression of protein O-fucosyltransferase 1 accelerates hepatocellular carcinoma progression via the Notch signaling pathway. Biochem Biophys Res Commun 473(2):503–510. https://doi.org/10.1016/j.bbrc.2016.03.062
Malecki MJ, Sanchez-Irizarry C, Mitchell JL, Histen G, Xu ML, Aster JC, Blacklow SC (2006) Leukemia-associated mutations within the NOTCH1 heterodimerization domain fall into at least two distinct mechanistic classes. Mol Cell Biol 26(12):4642–4651. https://doi.org/10.1128/MCB.01655-05
Mancino A, Termanini A, Barozzi I, Ghisletti S, Ostuni R, Prosperini E, Ozato K, Natoli G (2015) A dual cis-regulatory code links IRF8 to constitutive and inducible gene expression in macrophages. Genes Dev 29(4):394–408. https://doi.org/10.1101/gad.257592.114
Mansour MR, Linch DC, Foroni L, Goldstone AH, Gale RE (2006) High incidence of Notch-1 mutations in adult patients with T-cell acute lymphoblastic leukemia. Leukemia 20(3):537–539. https://doi.org/10.1038/sj.leu.2404101
Mansour MR, Duke V, Foroni L, Patel B, Allen CG, Ancliff PJ, Gale RE, Linch DC (2007) Notch-1 mutations are secondary events in some patients with T-cell acute lymphoblastic leukemia. Clin Cancer Res 13(23):6964–6969. https://doi.org/10.1158/1078-0432.CCR-07-1474
Mantovani A, Sica A, Sozzani S, Allavena P, Vecchi A, Locati M (2004) The chemokine system in diverse forms of macrophage activation and polarization. Trends Immunol 25(12):677–686. https://doi.org/10.1016/j.it.2004.09.015
Masek J, Andersson ER (2017) The developmental biology of genetic Notch disorders. Development 144(10):1743–1763. https://doi.org/10.1242/dev.148007
Massi D, Tarantini F, Franchi A, Paglierani M, Di Serio C, Pellerito S, Leoncini G, Cirino G, Geppetti P, Santucci M (2006) Evidence for differential expression of Notch receptors and their ligands in melanocytic nevi and cutaneous malignant melanoma. Mod Path : an official Journal of the United States and Canadian Academy of Pathology, Inc 19(2):246–254. https://doi.org/10.1038/modpathol.3800526
Mazur PK, Einwachter H, Lee M, Sipos B, Nakhai H, Rad R, Zimber-Strobl U, Strobl LJ, Radtke F, Kloppel G, Schmid RM, Siveke JT (2010a) Notch2 is required for progression of pancreatic intraepithelial neoplasia and development of pancreatic ductal adenocarcinoma. Proc Natl Acad Sci U S A 107(30):13438–13443. https://doi.org/10.1073/pnas.1002423107
Mazur PK, Gruner BM, Nakhai H, Sipos B, Zimber-Strobl U, Strobl LJ, Radtke F, Schmid RM, Siveke JT (2010b) Identification of epidermal Pdx1 expression discloses different roles of Notch1 and Notch2 in murine Kras(G12D)-induced skin carcinogenesis in vivo. PLoS One 5(10):e13578. https://doi.org/10.1371/journal.pone.0013578
McBride KL, Riley MF, Zender GA, Fitzgerald-Butt SM, Towbin JA, Belmont JW, Cole SE (2008) NOTCH1 mutations in individuals with left ventricular outflow tract malformations reduce ligand-induced signaling. Hum Mol Genet 17(18):2886–2893. https://doi.org/10.1093/hmg/ddn187
McKellar SH, Tester DJ, Yagubyan M, Majumdar R, Ackerman MJ, Sundt TM 3rd (2007) Novel NOTCH1 mutations in patients with bicuspid aortic valve disease and thoracic aortic aneurysms. J Thorac Cardiovasc Surg 134(2):290–296. https://doi.org/10.1016/j.jtcvs.2007.02.041
Meester JA, Southgate L, Stittrich AB, Venselaar H, Beekmans SJ, den Hollander N, Bijlsma EK, Helderman-van den Enden A, Verheij JB, Glusman G, Roach JC, Lehman A, Patel MS, de Vries BB, Ruivenkamp C, Itin P, Prescott K, Clarke S, Trembath R, Zenker M, Sukalo M, Van Laer L, Loeys B, Wuyts W (2015) Heterozygous Loss-of-Function Mutations in DLL4 Cause Adams-Oliver Syndrome. Am J Hum Genet 97(3):475–482. https://doi.org/10.1016/j.ajhg.2015.07.015
Miyamoto Y, Maitra A, Ghosh B, Zechner U, Argani P, Iacobuzio-Donahue CA, Sriuranpong V, Iso T, Meszoely IM, Wolfe MS, Hruban RH, Ball DW, Schmid RM, Leach SD (2003) Notch mediates TGF alpha-induced changes in epithelial differentiation during pancreatic tumorigenesis. Cancer Cell 3(6):565–576

Mizuhara E, Nakatani T, Minaki Y, Sakamoto Y, Ono Y, Takai Y (2005) MAGI1 recruits Dll1 to cadherin-based adherens junctions and stabilizes it on the cell surface. J Biol Chem 280(28):26499–26507. https://doi.org/10.1074/jbc.M500375200

Mohamed SA, Aherrahrou Z, Liptau H, Erasmi AW, Hagemann C, Wrobel S, Borzym K, Schunkert H, Sievers HH, Erdmann J (2006) Novel missense mutations (p.T596M and p.P1797H) in NOTCH1 in patients with bicuspid aortic valve. Biochem Biophys Res Commun 345(4):1460–1465. https://doi.org/10.1016/j.bbrc.2006.05.046

Monsalve E, Ruiz-Garcia A, Baladron V, Ruiz-Hidalgo MJ, Sanchez-Solana B, Rivero S, Garcia-Ramirez JJ, Rubio A, Laborda J, Diaz-Guerra MJ (2009) Notch1 upregulates LPS-induced macrophage activation by increasing NF-kappaB activity. Eur J Immunol 39(9):2556–2570. https://doi.org/10.1002/eji.200838722

Morgan TH (1917) The theory of the gene. Am Nat 51:513–544

Moshkin YM, Kan TW, Goodfellow H, Bezstarosti K, Maeda RK, Pilyugin M, Karch F, Bray SJ, Demmers JA, Verrijzer CP (2009) Histone chaperones ASF1 and NAP1 differentially modulate removal of active histone marks by LID-RPD3 complexes during NOTCH silencing. Mol Cell 35(6):782–793. https://doi.org/10.1016/j.molcel.2009.07.020

Mulligan P, Yang F, Di Stefano L, Ji JY, Ouyang J, Nishikawa JL, Toiber D, Kulkarni M, Wang Q, Najafi-Shoushtari SH, Mostoslavsky R, Gygi SP, Gill G, Dyson NJ, Naar AM (2011) A SIRT1-LSD1 corepressor complex regulates Notch target gene expression and development. Mol Cell 42(5):689–699. https://doi.org/10.1016/j.molcel.2011.04.020

Mumm JS, Schroeter EH, Saxena MT, Griesemer A, Tian X, Pan DJ, Ray WJ, Kopan R (2000) A ligand-induced extracellular cleavage regulates gamma-secretase-like proteolytic activation of Notch1. Mol Cell 5(2):197–206

Nakayama T, Saitsu H, Endo W, Kikuchi A, Uematsu M, Haginoya K, Hino-fukuyo N, Kobayashi T, Iwasaki M, Tominaga T, Kure S, Matsumoto N (2014) RBPJ is disrupted in a case of proximal 4p deletion syndrome with epilepsy. Brain Dev 36(6):532–536. https://doi.org/10.1016/j.braindev.2013.07.009

Nam Y, Weng AP, Aster JC, Blacklow SC (2003) Structural requirements for assembly of the CSL. intracellular Notch1.Mastermind-like 1 transcriptional activation complex. J Biol Chem 278(23):21232–21239. https://doi.org/10.1074/jbc.M301567200

Nam Y, Sliz P, Song L, Aster JC, Blacklow SC (2006) Structural basis for cooperativity in recruitment of MAML coactivators to Notch transcription complexes. Cell 124(5):973–983. https://doi.org/10.1016/j.cell.2005.12.037

Nam Y, Sliz P, Pear WS, Aster JC, Blacklow SC (2007) Cooperative assembly of higher-order Notch complexes functions as a switch to induce transcription. Proc Natl Acad Sci U S A 104(7):2103–2108. https://doi.org/10.1073/pnas.0611092104

Ngondo-Mbongo RP, Myslinski E, Aster JC, Carbon P (2013) Modulation of gene expression via overlapping binding sites exerted by ZNF143, Notch1 and THAP11. Nucleic Acids Res 41(7):4000–4014. https://doi.org/10.1093/nar/gkt088

Nicolas M, Wolfer A, Raj K, Kummer JA, Mill P, van Noort M, Hui CC, Clevers H, Dotto GP, Radtke F (2003) Notch1 functions as a tumor suppressor in mouse skin. Nat Genet 33(3):416–421. https://doi.org/10.1038/ng1099

Nishina S, Shiraha H, Nakanishi Y, Tanaka S, Matsubara M, Takaoka N, Uemura M, Horiguchi S, Kataoka J, Iwamuro M, Yagi T, Yamamoto K (2011) Restored expression of the tumor suppressor gene RUNX3 reduces cancer stem cells in hepatocellular carcinoma by suppressing Jagged1-Notch signaling. Oncol Rep 26(3):523–531. https://doi.org/10.3892/or.2011.1336

O'Neil J, Grim J, Strack P, Rao S, Tibbitts D, Winter C, Hardwick J, Welcker M, Meijerink JP, Pieters R, Draetta G, Sears R, Clurman BE, Look AT (2007) FBW7 mutations in leukemic cells mediate NOTCH pathway activation and resistance to gamma-secretase inhibitors. J Exp Med 204(8):1813–1824. https://doi.org/10.1084/jem.20070876

Osada H, Tatematsu Y, Yatabe Y, Horio Y, Takahashi T (2005) ASH1 gene is a specific therapeutic target for lung cancers with neuroendocrine features. Cancer Res 65(23):10680–10685. https://doi.org/10.1158/0008-5472.CAN-05-1404

Oswald F, Tauber B, Dobner T, Bourteele S, Kostezka U, Adler G, Liptay S, Schmid RM (2001) p300 acts as a transcriptional coactivator for mammalian Notch-1. Mol Cell Biol 21(22):7761–7774. https://doi.org/10.1128/MCB.21.22.7761-7774.2001

Oswald F, Kostezka U, Astrahantseff K, Bourteele S, Dillinger K, Zechner U, Ludwig L, Wilda M, Hameister H, Knochel W, Liptay S, Schmid RM (2002) SHARP is a novel component of the Notch/RBP-Jkappa signalling pathway. EMBO J 21(20):5417–5426

Oswald F, Winkler M, Cao Y, Astrahantseff K, Bourteele S, Knochel W, Borggrefe T (2005) RBP-Jkappa/SHARP recruits CtIP/CtBP corepressors to silence Notch target genes. Mol Cell Biol 25(23):10379–10390. https://doi.org/10.1128/MCB.25.23.10379-10390.2005

Oswald F, Rodriguez P, Giaimo BD, Antonello ZA, Mira L, Mittler G, Thiel VN, Collins KJ, Tabaja N, Cizelsky W, Rothe M, Kuhl SJ, Kuhl M, Ferrante F, Hein K, Kovall RA, Dominguez M, Borggrefe T (2016) A phospho-dependent mechanism involving NCoR and KMT2D controls a permissive chromatin state at Notch target genes. Nucleic Acids Res 44(10):4703–4720. https://doi.org/10.1093/nar/gkw105

Outtz HH, Wu JK, Wang X, Kitajewski J (2010) Notch1 deficiency results in decreased inflammation during wound healing and regulates vascular endothelial growth factor receptor-1 and inflammatory cytokine expression in macrophages. J Immunol 185(7):4363–4373. https://doi.org/10.4049/jimmunol.1000720

Palaga T, Buranaruk C, Rengpipat S, Fauq AH, Golde TE, Kaufmann SH, Osborne BA (2008) Notch signaling is activated by TLR stimulation and regulates macrophage functions. Eur J Immunol 38(1):174–183. https://doi.org/10.1002/eji.200636999

Palermo R, Checquolo S, Giovenco A, Grazioli P, Kumar V, Campese AF, Giorgi A, Napolitano M, Canettieri G, Ferrara G, Schinina ME, Maroder M, Frati L, Gulino A, Vacca A, Screpanti I (2012) Acetylation controls Notch3 stability and function in T-cell leukemia. Oncogene 31(33):3807–3817. https://doi.org/10.1038/onc.2011.533

Palomero T, Sulis ML, Cortina M, Real PJ, Barnes K, Ciofani M, Caparros E, Buteau J, Brown K, Perkins SL, Bhagat G, Agarwal AM, Basso G, Castillo M, Nagase S, Cordon-Cardo C, Parsons R, Zuniga-Pflucker JC, Dominguez M, Ferrando AA (2007) Mutational loss of PTEN induces resistance to NOTCH1 inhibition in T-cell leukemia. Nat Med 13(10):1203–1210. https://doi.org/10.1038/nm1636

Parks AL, Stout JR, Shepard SB, Klueg KM, Dos Santos AA, Parody TR, Vaskova M, Muskavitch MA (2006) Structure-function analysis of delta trafficking, receptor binding and signaling in Drosophila. Genetics 174(4):1947–1961. https://doi.org/10.1534/genetics.106.061630

Patel U, Rajasingh S, Samanta S, Cao T, Dawn B, Rajasingh J (2017) Macrophage polarization in response to epigenetic modifiers during infection and inflammation. Drug Discov Today 22(1):186–193. https://doi.org/10.1016/j.drudis.2016.08.006

Pear WS, Aster JC, Scott ML, Hasserjian RP, Soffer B, Sklar J, Baltimore D (1996) Exclusive development of T cell neoplasms in mice transplanted with bone marrow expressing activated Notch alleles. J Exp Med 183(5):2283–2291

Pece S, Serresi M, Santolini E, Capra M, Hulleman E, Galimberti V, Zurrida S, Maisonneuve P, Viale G, Di Fiore PP (2004) Loss of negative regulation by Numb over Notch is relevant to human breast carcinogenesis. J Cell Biol 167(2):215–221. https://doi.org/10.1083/jcb.200406140

Plentz R, Park JS, Rhim AD, Abravanel D, Hezel AF, Sharma SV, Gurumurthy S, Deshpande V, Kenific C, Settleman J, Majumder PK, Stanger BZ, Bardeesy N (2009) Inhibition of gamma-secretase activity inhibits tumor progression in a mouse model of pancreatic ductal adenocarcinoma. Gastroenterology 136(5):1741–1749. e1746. https://doi.org/10.1053/j.gastro.2009.01.008

Porta C, Riboldi E, Ippolito A, Sica A (2015) Molecular and epigenetic basis of macrophage polarized activation. Semin Immunol 27(4):237–248. https://doi.org/10.1016/j.smim.2015.10.003

Pozzo F, Bittolo T, Arruga F, Bulian P, Macor P, Tissino E, Gizdic B, Rossi FM, Bomben R, Zucchetto A, Benedetti D, Degan M, D'Arena G, Chiarenza A, Zaja F, Pozzato G, Rossi D, Gaidano G, Del Poeta G, Deaglio S, Gattei V, Dal Bo M (2016) NOTCH1 mutations associate with low CD20 level in chronic lymphocytic leukemia: evidence for a NOTCH1 mutation-driven epigenetic dysregulation. Leukemia 30(1):182–189. https://doi.org/10.1038/leu.2015.182

Pozzo F, Bittolo T, Vendramini E, Bomben R, Bulian P, Rossi FM, Zucchetto A, Tissino E, Degan M, D'Arena G, Di Raimondo F, Zaja F, Pozzato G, Rossi D, Gaidano G, Del Poeta G, Gattei V, Dal Bo M (2017) NOTCH1-mutated chronic lymphocytic leukemia cells are characterized by a MYC-related overexpression of nucleophosmin 1 and ribosome-associated components. Leukemia 31(11):2407–2415. https://doi.org/10.1038/leu.2017.90

Proweller A, Tu L, Lepore JJ, Cheng L, Lu MM, Seykora J, Millar SE, Pear WS, Parmacek MS (2006) Impaired notch signaling promotes de novo squamous cell carcinoma formation. Cancer Res 66(15):7438–7444. https://doi.org/10.1158/0008-5472.CAN-06-0793

Puente XS, Pinyol M, Quesada V, Conde L, Ordonez GR, Villamor N, Escaramis G, Jares P, Bea S, Gonzalez-Diaz M, Bassaganyas L, Baumann T, Juan M, Lopez-Guerra M, Colomer D, Tubio JM, Lopez C, Navarro A, Tornador C, Aymerich M, Rozman M, Hernandez JM, Puente DA, Freije JM, Velasco G, Gutierrez-Fernandez A, Costa D, Carrio A, Guijarro S, Enjuanes A, Hernandez L, Yague J, Nicolas P, Romeo-Casabona CM, Himmelbauer H, Castillo E, Dohm JC, de Sanjose S, Piris MA, de Alava E, San Miguel J, Royo R, Gelpi JL, Torrents D, Orozco M, Pisano DG, Valencia A, Guigo R, Bayes M, Heath S, Gut M, Klatt P, Marshall J, Raine K, Stebbings LA, Futreal PA, Stratton MR, Campbell PJ, Gut I, Lopez-Guillermo A, Estivill X, Montserrat E, Lopez-Otin C, Campo E (2011) Whole-genome sequencing identifies recurrent mutations in chronic lymphocytic leukaemia. Nature 475(7354):101–105. https://doi.org/10.1038/nature10113

Purow BW, Haque RM, Noel MW, Su Q, Burdick MJ, Lee J, Sundaresan T, Pastorino S, Park JK, Mikolaenko I, Maric D, Eberhart CG, Fine HA (2005) Expression of Notch-1 and its ligands, Delta-like-1 and Jagged-1, is critical for glioma cell survival and proliferation. Cancer Res 65(6):2353–2363. https://doi.org/10.1158/0008-5472.CAN-04-1890

Qiu M, Peng Q, Jiang I, Carroll C, Han G, Rymer I, Lippincott J, Zachwieja J, Gajiwala K, Kraynov E, Thibault S, Stone D, Gao Y, Sofia S, Gallo J, Li G, Yang J, Li K, Wei P (2013) Specific inhibition of Notch1 signaling enhances the antitumor efficacy of chemotherapy in triple negative breast cancer through reduction of cancer stem cells. Cancer Lett 328(2):261–270. https://doi.org/10.1016/j.canlet.2012.09.023

Radtke F, Wilson A, Stark G, Bauer M, van Meerwijk J, MacDonald HR, Aguet M (1999) Deficient T cell fate specification in mice with an induced inactivation of Notch1. Immunity 10(5):547–558

Rampias T, Vgenopoulou P, Avgeris M, Polyzos A, Stravodimos K, Valavanis C, Scorilas A, Klinakis A (2014) A new tumor suppressor role for the Notch pathway in bladder cancer. Nat Med 20(10):1199–1205. https://doi.org/10.1038/nm.3678

Rosati E, Sabatini R, Rampino G, Tabilio A, Di Ianni M, Fettucciari K, Bartoli A, Coaccioli S, Screpanti I, Marconi P (2009) Constitutively activated Notch signaling is involved in survival and apoptosis resistance of B-CLL cells. Blood 113(4):856–865. https://doi.org/10.1182/blood-2008-02-139725

Rossi D, Trifonov V, Fangazio M, Bruscaggin A, Rasi S, Spina V, Monti S, Vaisitti T, Arruga F, Fama R, Ciardullo C, Greco M, Cresta S, Piranda D, Holmes A, Fabbri G, Messina M, Rinaldi A, Wang J, Agostinelli C, Piccaluga PP, Lucioni M, Tabbo F, Serra R, Franceschetti S, Deambrogi C, Daniele G, Gattei V, Marasca R, Facchetti F, Arcaini L, Inghirami G, Bertoni F, Pileri SA, Deaglio S, Foa R, Dalla-Favera R, Pasqualucci L, Rabadan R, Gaidano G (2012) The coding genome of splenic marginal zone lymphoma: activation of NOTCH2 and other pathways regulating marginal zone development. J Exp Med 209(9):1537–1551. https://doi.org/10.1084/jem.20120904

Rustighi A, Tiberi L, Soldano A, Napoli M, Nuciforo P, Rosato A, Kaplan F, Capobianco A, Pece S, Di Fiore PP, Del Sal G (2009) The prolyl-isomerase Pin1 is a Notch1 target that enhances Notch1 activation in cancer. Nat Cell Biol 11(2):133–142. https://doi.org/10.1038/ncb1822

Rustighi A, Zannini A, Tiberi L, Sommaggio R, Piazza S, Sorrentino G, Nuzzo S, Tuscano A, Eterno V, Benvenuti F, Santarpia L, Aifantis I, Rosato A, Bicciato S, Zambelli A, Del Sal G (2014) Prolyl-isomerase Pin1 controls normal and cancer stem cells of the breast. EMBO Mol Med 6(1):99–119. https://doi.org/10.1002/emmm.201302909

Salat D, Liefke R, Wiedenmann J, Borggrefe T, Oswald F (2008) ETO, but not leukemogenic fusion protein AML1/ETO, augments RBP-Jkappa/SHARP-mediated repression of notch target genes. Mol Cell Biol 28(10):3502–3512. https://doi.org/10.1128/MCB.01966-07

Sanchez-Irizarry C, Carpenter AC, Weng AP, Pear WS, Aster JC, Blacklow SC (2004) Notch subunit heterodimerization and prevention of ligand-independent proteolytic activation depend, respectively, on a novel domain and the LNR repeats. Mol Cell Biol 24(21):9265–9273. https://doi.org/10.1128/MCB.24.21.9265-9273.2004

Santio NM, Landor SK, Vahtera L, Yla-Pelto J, Paloniemi E, Imanishi SY, Corthals G, Varjosalo M, Manoharan GB, Uri A, Lendahl U, Sahlgren C, Koskinen PJ (2016) Phos phorylation of Notch1 by Pim kinases promotes oncogenic signaling in breast and prostate cancer cells. Oncotarget 7(28):43220–43238. https://doi.org/10.18632/oncotarget.9215

Simoes BM, O'Brien CS, Eyre R, Silva A, Yu L, Sarmiento-Castro A, Alferez DG, Spence K, Santiago-Gomez A, Chemi F, Acar A, Gandhi A, Howell A, Brennan K, Ryden L, Catalano S, Ando S, Gee J, Ucar A, Sims AH, Marangoni E, Farnie G, Landberg G, Howell SJ, Clarke RB (2015) Anti-estrogen resistance in human breast tumors is driven by JAG1-NOTCH4-dependent cancer stem cell activity. Cell Rep 12(12):1968–1977. https://doi.org/10.1016/j.celrep.2015.08.050

Song H, Mak KK, Topol L, Yun K, Hu J, Garrett L, Chen Y, Park O, Chang J, Simpson RM, Wang CY, Gao B, Jiang J, Yang Y (2010) Mammalian Mst1 and Mst2 kinases play essential roles in organ size control and tumor suppression. Proc Natl Acad Sci U S A 107(4):1431–1436. https://doi.org/10.1073/pnas.0911409107

Southgate L, Sukalo M, Karountzos ASV, Taylor EJ, Collinson CS, Ruddy D, Snape KM, Dallapiccola B, Tolmie JL, Joss S, Brancati F, Digilio MC, Graul-Neumann LM, Salviati L, Coerdt W, Jacquemin E, Wuyts W, Zenker M, Machado RD, Trembath RC (2015) Haploinsufficiency of the NOTCH1 receptor as a cause of adams-oliver syndrome with variable cardiac anomalies. Circ Cardiovasc Genet 8(4):572–581. https://doi.org/10.1161/CIRCGENETICS.115.001086

Stittrich AB, Lehman A, Bodian DL, Ashworth J, Zong Z, Li H, Lam P, Khromykh A, Iyer RK, Vockley JG, Baveja R, Silva ES, Dixon J, Leon EL, Solomon BD, Glusman G, Niederhuber JE, Roach JC, Patel MS (2014) Mutations in NOTCH1 cause Adams-Oliver syndrome. Am J Hum Genet 95(3):275–284. https://doi.org/10.1016/j.ajhg.2014.07.011

Stoeck A, Lejnine S, Truong A, Pan L, Wang H, Zang C, Yuan J, Ware C, MacLean J, Garrett-Engele PW, Kluk M, Laskey J, Haines BB, Moskaluk C, Zawel L, Fawell S, Gilliland G, Zhang T, Kremer BE, Knoechel B, Bernstein BE, Pear WS, Liu XS, Aster JC, Sathyanarayanan S (2014) Discovery of biomarkers predictive of GSI response in triple-negative breast cancer and adenoid cystic carcinoma. Cancer Discov 4(10):1154–1167. https://doi.org/10.1158/2159-8290.CD-13-0830

Stylianou S, Clarke RB, Brennan K (2006) Aberrant activation of notch signaling in human breast cancer. Cancer Res 66(3):1517–1525. https://doi.org/10.1158/0008-5472.CAN-05-3054

Tamura K, Taniguchi Y, Minoguchi S, Sakai T, Tun T, Furukawa T, Honjo T (1995) Physical interaction between a novel domain of the receptor Notch and the transcription factor RBP-J kappa/Su(H). Curr Biol 5(12):1416–1423

Thelu J, Rossio P, Favier B (2002) Notch signalling is linked to epidermal cell differentiation level in basal cell carcinoma, psoriasis and wound healing. BMC Dermatol 2:7

Thiel V, Giaimo BD, Schwarz P, Soller K, Vas V, Bartkuhn M, Blatte TJ, Dohner K, Bullinger L, Borggrefe T, Geiger H, Oswald F (2017) Heterodimerization of AML1/ETO with CBFbeta is required for leukemogenesis but not for myeloproliferation. Leukemia 31(11):2491–2502. https://doi.org/10.1038/leu.2017.105

Thomas MM, Zhang Y, Mathew E, Kane KT, Maillard I, Pasca di Magliano M (2014) Epithelial Notch signaling is a limiting step for pancreatic carcinogenesis. BMC Cancer 14:862. https://doi.org/10.1186/1471-2407-14-862

VanderWielen BD, Yuan Z, Friedmann DR, Kovall RA (2011) Transcriptional repression in the Notch pathway: thermodynamic characterization of CSL-MINT

(Msx2-interacting nuclear target protein) complexes. J Biol Chem 286(17):14892–14902. https://doi.org/10.1074/jbc.M110.181156

Viatour P, Ehmer U, Saddic LA, Dorrell C, Andersen JB, Lin C, Zmoos AF, Mazur PK, Schaffer BE, Ostermeier A, Vogel H, Sylvester KG, Thorgeirsson SS, Grompe M, Sage J (2011) Notch signaling inhibits hepatocellular carcinoma following inactivation of the RB pathway. J Exp Med 208(10):1963–1976. https://doi.org/10.1084/jem.20110198

Villanueva A, Alsinet C, Yanger K, Hoshida Y, Zong Y, Toffanin S, Rodriguez-Carunchio L, Sole M, Thung S, Stanger BZ, Llovet JM (2012) Notch signaling is activated in human hepatocellular carcinoma and induces tumor formation in mice. Gastroenterology 143(6):1660–1669. e1667. https://doi.org/10.1053/j.gastro.2012.09.002

Wang YC, He F, Feng F, Liu XW, Dong GY, Qin HY, Hu XB, Zheng MH, Liang L, Feng L, Liang YM, Han H (2010a) Notch signaling determines the M1 versus M2 polarization of macrophages in antitumor immune responses. Cancer Res 70(12):4840–4849. https://doi.org/10.1158/0008-5472.CAN-10-0269

Wang J, Wakeman TP, Lathia JD, Hjelmeland AB, Wang XF, White RR, Rich JN, Sullenger BA (2010b) Notch promotes radioresistance of glioma stem cells. Stem Cells 28(1):17–28. https://doi.org/10.1002/stem.261

Wang H, Zou J, Zhao B, Johannsen E, Ashworth T, Wong H, Pear WS, Schug J, Blacklow SC, Arnett KL, Bernstein BE, Kieff E, Aster JC (2011a) Genome-wide analysis reveals conserved and divergent features of Notch1/RBPJ binding in human and murine T-lymphoblastic leukemia cells. Proc Natl Acad Sci U S A 108(36):14908–14913. https://doi.org/10.1073/pnas.1109023108

Wang NJ, Sanborn Z, Arnett KL, Bayston LJ, Liao W, Proby CM, Leigh IM, Collisson EA, Gordon PB, Jakkula L, Pennypacker S, Zou Y, Sharma M, North JP, Vemula SS, Mauro TM, Neuhaus IM, Leboit PE, Hur JS, Park K, Huh N, Kwok PY, Arron ST, Massion PP, Bale AE, Haussler D, Cleaver JE, Gray JW, Spellman PT, South AP, Aster JC, Blacklow SC, Cho RJ (2011b) Loss-of-function mutations in Notch receptors in cutaneous and lung squamous cell carcinoma. Proc Natl Acad Sci U S A 108(43):17761–17766. https://doi.org/10.1073/pnas.1114669108

Wang H, Zang C, Taing L, Arnett KL, Wong YJ, Pear WS, Blacklow SC, Liu XS, Aster JC (2014) NOTCH1-RBPJ complexes drive target gene expression through dynamic interactions with superenhancers. Proc Natl Acad Sci U S A 111(2):705–710. https://doi.org/10.1073/pnas.1315023111

Wang K, Zhang Q, Li D, Ching K, Zhang C, Zheng X, Ozeck M, Shi S, Li X, Wang H, Rejto P, Christensen J, Olson P (2015) PEST domain mutations in Notch receptors comprise an oncogenic driver segment in triple-negative breast cancer sensitive to a gamma-secretase inhibitor. Clin Cancer Res Off J Am Assoc Can Res 21(6):1487–1496. https://doi.org/10.1158/1078-0432.CCR-14-1348

Wang S, Zhou R, Sun F, Li R, Wang M, Wu M (2017) The two novel DLL4-targeting antibody-drug conjugates MvM03 and MGD03 show potent anti-tumour activity in breast cancer xenograft models. Cancer Lett. https://doi.org/10.1016/j.canlet.2017.09.004

Weng AP, Ferrando AA, Lee W, JPt M, Silverman LB, Sanchez-Irizarry C, Blacklow SC, Look AT, Aster JC (2004) Activating mutations of NOTCH1 in human T cell acute lymphoblastic leukemia. Science 306(5694):269–271. https://doi.org/10.1126/science.1102160

Wharton KA, Johansen KM, Xu T, Artavanis-Tsakonas S (1985) Nucleotide sequence from the neurogenic locus notch implies a gene product that shares homology with proteins containing EGF-like repeats. Cell 43(3 Pt 2):567–581

Wilson JJ, Kovall RA (2006) Crystal structure of the CSL-Notch-Mastermind ternary complex bound to DNA. Cell 124(5):985–996. https://doi.org/10.1016/j.cell.2006.01.035

Wongchana W, Palaga T (2012) Direct regulation of interleukin-6 expression by Notch signaling in macrophages. Cell Mol Immunol 9(2):155–162. https://doi.org/10.1038/cmi.2011.36

Wu L, Aster JC, Blacklow SC, Lake R, Artavanis-Tsakonas S, Griffin JD (2000) MAML1, a human homologue of Drosophila mastermind, is a transcriptional co-activator for NOTCH receptors. Nat Genet 26(4):484–489. https://doi.org/10.1038/82644

Wu L, Sun T, Kobayashi K, Gao P, Griffin JD (2002) Identification of a family of mastermind-like transcriptional coactivators for mammalian notch receptors. Mol Cell Biol 22(21):7688–7700

Wu Y, Cain-Hom C, Choy L, Hagenbeek TJ, de Leon GP, Chen Y, Finkle D, Venook R, Wu X, Ridgway J, Schahin-Reed D, Dow GJ, Shelton A, Stawicki S, Watts RJ, Zhang J, Choy R, Howard P, Kadyk L, Yan M, Zha J, Callahan CA, Hymowitz SG, Siebel CW (2010) Therapeutic antibody targeting of individual Notch receptors. Nature 464(7291):1052–1057. https://doi.org/10.1038/nature08878

Xie Q, Wu Q, Kim L, Miller TE, Liau BB, Mack SC, Yang K, Factor DC, Fang X, Huang Z, Zhou W, Alazem K, Wang X, Bernstein BE, Bao S, Rich JN (2016) RBPJ maintains brain tumor-initiating cells through CDK9-mediated transcriptional elongation. J Clin Invest 126(7):2757–2772. https://doi.org/10.1172/JCI86114

Xu H, Zhu J, Smith S, Foldi J, Zhao B, Chung AY, Outtz H, Kitajewski J, Shi C, Weber S, Saftig P, Li Y, Ozato K, Blobel CP, Ivashkiv LB, Hu X (2012) Notch-RBP-J signaling regulates the transcription factor IRF8 to promote inflammatory macrophage polarization. Nat Immunol 13(7):642–650. https://doi.org/10.1038/ni.2304

Xu Z, Wang Z, Jia X, Wang L, Chen Z, Wang S, Wang M, Zhang J, Wu M (2016) MMGZ01, an anti-DLL4 monoclonal antibody, promotes nonfunctional vessels and inhibits breast tumor growth. Cancer Lett 372(1):118–127. https://doi.org/10.1016/j.canlet.2015.12.025

Xu T, Park SS, Giaimo BD, Hall D, Ferrante F, Ho DM, Hori K, Anhezini L, Ertl I, Bartkuhn M, Zhang H, Milon E, Ha K, Conlon KP, Kuick R, Govindarajoo B, Zhang Y, Sun Y, Dou Y, Basrur V, Elenitoba-Johnson KS, Nesvizhskii AI, Ceron J, Lee CY, Borggrefe T, Kovall RA, Rual JF (2017) RBPJ/CBF1 interacts with L3MBTL3/MBT1 to promote repression of Notch signaling via histone demethylase KDM1A/LSD1. EMBO J. https://doi.org/10.15252/embj.201796525

Yatim A, Benne C, Sobhian B, Laurent-Chabalier S, Deas O, Judde JG, Lelievre JD, Levy Y, Benkirane M (2012) NOTCH1 nuclear interactome reveals key regulators of its transcriptional activity and oncogenic function. Mol Cell 48(3):445–458. https://doi.org/10.1016/j.molcel.2012.08.022

Yen WC, Fischer MM, Hynes M, Wu J, Kim E, Beviglia L, Yeung VP, Song X, Kapoun AM, Lewicki J, Gurney A, Simeone DM, Hoey T (2012) Anti-DLL4 has broad spectrum activity in pancreatic cancer dependent on targeting DLL4-Notch signaling in both tumor and vasculature cells. Clin Cancer Res Off J Am Assoc Can Res 18(19):5374–5386. https://doi.org/10.1158/1078-0432.CCR-12-0736

Yuan Z, Praxenthaler H, Tabaja N, Torella R, Preiss A, Maier D, Kovall RA (2016) Structure and function of the Su(H)-Hairless repressor complex, the major Antagonist of Notch signaling in Drosophila melanogaster. PLoS Biol 14(7):e1002509. https://doi.org/10.1371/journal.pbio.1002509

Zhang W, Xu W, Xiong S (2010) Blockade of Notch1 signaling alleviates murine lupus via blunting macrophage activation and M2b polarization. J Immunol 184(11):6465–6478. https://doi.org/10.4049/jimmunol.0904016

Zhang Q, Wang C, Liu Z, Liu X, Han C, Cao X, Li N (2012) Notch signal suppresses Toll-like receptor-triggered inflammatory responses in macrophages by inhibiting extracellular signal-regulated kinase 1/2-mediated nuclear factor kappaB activation. J Biol Chem 287(9):6208–6217. https://doi.org/10.1074/jbc.M111.310375

Zhang J, Shao X, Sun H, Liu K, Ding Z, Chen J, Fang L, Su W, Hong Y, Li H, Li H (2016a) NUMB negatively regulates the epithelial-mesenchymal transition of triple-negative breast cancer by antagonizing Notch signaling. Oncotarget 7(38):61036–61053. https://doi.org/10.18632/oncotarget.11062

Zhang S, Chung WC, Xu K (2016b) Lunatic fringe is a potent tumor suppressor in Kras-initiated pancreatic cancer. Oncogene 35(19):2485–2495. https://doi.org/10.1038/onc.2015.306

Zhang Q, Zhang Y, Parsels JD, Lohse I, Lawrence TS, Pasca di Magliano M, Sun Y, Morgan MA (2016c) Fbxw7 deletion accelerates KrasG12D-Driven pancreatic tumorigenesis via Yap accumulation. Neoplasia 18(11):666–673. https://doi.org/10.1016/j.neo.2016.08.009

Zhou L, Wang DS, Li QJ, Sun W, Zhang Y, Dou KF (2012) Downregulation of the Notch signaling pathway inhibits hepatocellular carcinoma cell invasion by inactivation of matrix metalloproteinase-2 and -9 and vascular endothelial growth factor. Oncol Rep 28(3):874–882. https://doi.org/10.3892/or.2012.1880

Zhou L, Wang DS, Li QJ, Sun W, Zhang Y, Dou KF (2013) The down-regulation of Notch1 inhibits the invasion and migration of hepatocellular carcinoma cells by inactivating the cyclooxygenase-2/Snail/E-cadherin pathway in vitro. Dig Dis Sci 58(4):1016–1025. https://doi.org/10.1007/s10620-012-2434-7

Zhu B, Sun L, Luo W, Li M, Coy DH, Yu L, Yu W (2017) Activated Notch signaling augments cell growth in hepatocellular carcinoma via up-regulating the nuclear receptor NR4A2. Oncotarget 8(14):23289–23302. https://doi.org/10.18632/oncotarget.15576

Zhu YM, Zhao WL, Fu JF, Shi JY, Pan Q, Hu J, Gao XD, Chen B, Li JM, Xiong SM, Gu LJ, Tang JY, Liang H, Jiang H, Xue YQ, Shen ZX, Chen Z, Chen SJ (2006) NOTCH1 mutations in T-cell acute lymphoblastic leukemia: prognostic significance and implication in multifactorial leukemogenesis. Clin Cancer Res 12(10):3043–3049. https://doi.org/10.1158/1078-0432.CCR-05-2832

Part II

Molecular Mechanisms of Receptor/Ligand Interactions

Structural Insights into Notch Receptor-Ligand Interactions

Penny A. Handford, Boguslawa Korona, Richard Suckling, Christina Redfield, and Susan M. Lea

Abstract

Pioneering cell aggregation experiments from the Artavanis-Tsakonas group in the late 1980's localized the core ligand recognition sequence in the *Drosophila* Notch receptor to epidermal growth factor-like (EGF) domains 11 and 12. Since then, advances in protein expression, structure determination methods and functional assays have enabled us to define the molecular basis of the core receptor/ligand interaction and given new insights into the architecture of the Notch complex at the cell surface. We now know that Notch EGF11 and 12 interact with the Delta/Serrate/LAG-2 (DSL) and C2 domains of ligand and that membrane-binding, together with additional protein-protein interactions outside the core recognition domains, are likely to fine-tune generation of the Notch signal. Furthermore, structure determination of *O*-glycosylated variants of Notch alone or in complex with receptor fragments, has shown that these sugars contribute directly to the binding interface, as well as to stabilizing intra-molecular domain structure, providing some mechanistic insights into the observed modulatory effects of *O*-glycosylation on Notch activity.

Future challenges lie in determining the complete extracellular architecture of ligand and receptor in order to understand (i) how Notch/ligand complexes may form at the cell surface in response to physiological cues, (ii) the role of lipid binding in stabilizing the Notch/ligand complex, (iii) the impact of *O*-glycosylation on binding and signalling and (iv) to dissect the different pathologies that arise as a consequence of mutations that affect proteins involved in the Notch pathway.

Keywords

EGF12 · Calcium binding · Fringe · C2 domain · Lipid binding

Abbreviations

DSL Delta Serrate LAG-2
EGF epidermal growth factor-like

1 Notch Receptor Ligand-Binding Region (LBR)

The Notch receptor is part of a short-range cell-cell signaling system in metazoans and comprises a large extracellular domain and a short intracel-

P. A. Handford (✉) · B. Korona · C. Redfield
Department of Biochemistry, University of Oxford, Oxford OX1 3RE, UK
e-mail: penny.handford@bioch.ox.ac.uk

R. Suckling · S. M. Lea
Sir William Dunn School of Pathology, University of Oxford, Oxford, Oxford OX1 3RE, UK

T. Borggrefe, B. D. Giaimo (eds.), *Molecular Mechanisms of Notch Signaling*, Advances in Experimental Medicine and Biology 1066, https://doi.org/10.1007/978-3-319-89512-3_2

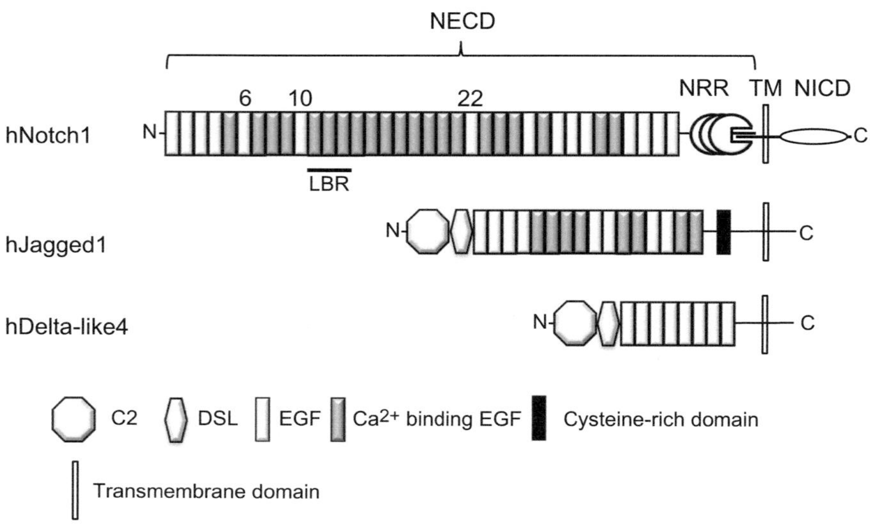

Fig. 1 Modular organisation of the extracellular domains of human Notch1 (hN1) receptor and the Jagged1 (hJ-1) and Delta-like 4 (hDll4) ligands. The negative regulatory region (NRR) of Notch1 and the transmembrane domain (TM) of Notch1, hJ-1 and hDll4 are indicated, as is the complete extracellular domain (NECD). Individual domains belonging to the Notch intracellular domain (NICD) are not indicated separately. Ca^{2+}-binding and non-Ca^{2+}-binding EGF domains and the cysteine-rich domain are indicated as grey, white and black rectangles, respectively. The C2 and DSL domains are indicated as white octagons and hexagons, respectively. The ligand-binding region (LBR, corresponding to EGF11–13) of hN1 is shown

lular domain (Fig. 1). Ligand binding to the Notch extracellular domain (Rebay et al. 1991) initiates a process known as intra-membrane regulated proteolysis which releases the Notch intracellular domain (NICD); this then translocates to the nucleus where it assembles into a transcriptional activation complex to initiate expression of target genes (Bray 2016). The extracellular domain comprises the membrane-proximal negative regulatory region (NRR) and a contiguous set of EGF domains. The number of EGF domains varies from species to species, as does the number of Notch receptors but the molecular architecture (represented in Fig. 1) is similar. Mammals have 4 Notch paralogues (Notch1–4), while *Drosophila melanogaster* has 1 (dNotch) and *C.elegans* 2 (LIN-12, GLP-1). Both dNotch and human Notch1 (hN1) have 36 disulphide-rich EGF domains. These domains are extensively *O*-glycosylated and many of them contain an additional consensus sequence which confers the ability to bind Ca^{2+} (cbEGF, calcium binding EGF).

1.1 Structure of Unmodified Human Notch1 (hN1) EGF11–13

The solution structure of unmodified (unglycosylated) hN1 EGF11–13, containing the ligand-binding region, was determined using NMR in 2004 (Hambleton et al. 2004). All three EGF domains contain a calcium-binding consensus sequence and binding of this fragment to ligand-expressing cells was shown to be calcium dependent. The high-resolution structure, together with a dynamic analysis, showed that EGF11 and 12 adopted a near linear rod-like orientation, stabilized by calcium binding at the domain-domain interface and a conserved hydrophobic packing interaction between Y444 of EGF11 and I471 of EGF12 (Fig. 2). The tilt angle was similar to that observed for other calcium-binding EGF domain pairs from the extracellular matrix protein fibrillin-1 (reflecting the near-linear structure) but the twist angle was very different (119° compared with ~155°) (Downing et al. 1996; Smallridge et al. 2003). These features were also observed in

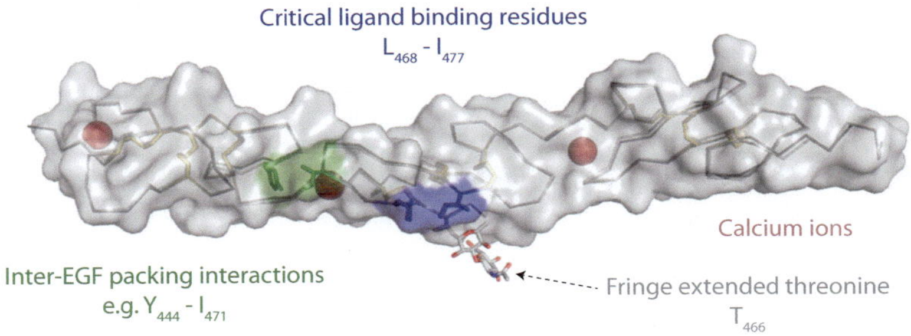

Fig. 2 Structure of ligand-binding hN1 EGF11–13. Key features are highlighted including calcium ions, hydrophobic packing interaction of EGF11–12 between Y444 and I471, key ligand-binding residues in EGF12, L468 and I477, and the GlcNAc fucose disaccharide added by POFUT1 and Fringe, respectively onto T466. Note the overall rod-shape of the triple domain fragment stabilized by calcium binding to each EGF domain and hydrophobic packing interactions and the proximity of the disaccharide to the ligand-binding site

a crystal structure (at 2.6 Å resolution) of a C-terminal tagged form of unmodified EGF11–13 which, in addition, revealed a very similar pairwise organization of EGF12 and 13 [(Cordle et al. 2008a), Fig. 2]. From these data, it was possible to identify other regions of Notch which were likely to show a similar calcium-stabilized architecture.

1.2 Defining the Ligand-Binding Site in hN1 11–13

Site-directed mutagenesis of each of the calcium-binding sites within EGF11–13 indicated that only loss of calcium binding to EGF12 abrogated binding of this fragment to ligand Delta-like1 (Dll1) in flow cytometry experiments (Cordle et al. 2008b). This suggested that the Notch EGF11–12 interface and/or Notch EGF12 contained the main ligand-binding site within this fragment. Further amino acid substitutions within EGF12 of hN1 and dN EGF11–13 triple domain fragments were made to probe ligand recognition, whilst retaining calcium binding. Using flow cytometry and cell aggregation assays, L468 and I477 in EGF12 of hN1 and equivalent residues L504 and V513 in dN were identified as being essential for binding (Whiteman et al. 2013). In the crystal structure of EGF11–13, the side chains of these residues are located opposite each other on the central β hairpin of EGF12 and are involved in a hydrophobic interaction (Cordle et al. 2008a; Whiteman et al. 2013). Collectively, these data suggested that these residues, together with contributions from β strand residues in close proximity, provide a stable ligand-binding platform which is conserved from *Drosophila* to human (Fig. 2). Interestingly, this site is adjacent to the amino acid residue T466 which forms part of the *O*-fucosylation consensus (C^2-X_4-(S/T)-C^3) and is subjected to further GlcNAc modification by Fringe enzymes (Rampal et al. 2005; Moloney et al. 2000a; Moloney et al. 2000b; Moloney et al. 1997; Wang et al. 2001).

1.3 *O*-Glycosylation of the Ligand-Binding Region and Effect on Structure

Since *O*-glycosylation of the extracellular domain was known to regulate Notch signaling (Rana and Haltiwanger 2011) and the ligand-binding site in EGF12 was found to map adjacent to an *O*-fucosylation consensus site, a series of *in vitro* modified forms of hN1 EGF11–13 was prepared and utilized in molecular and cell binding assays (Taylor et al. 2014). Stoichiometric addition of *O*-glucose at S458 (EGF12) and S496 (EGF13) catalyzed by protein O-glucosyltransferase (POGLUT1), and subsequent extension with xylose by glucoside α3-xylosyltransferase 2 (GXYLT2), showed no effect on the binding of EGF11–13 to cells expressing either Notch ligand Jagged-1 or Delta-like 4 (Dll4), (Taylor et al. 2014). In contrast, *O*-fucose modification of T466 by protein O-fucosyltransferase (POFUT1) within EGF12 and extension with GlcNAc by the enzyme Lunatic Fringe (Lfng) showed effects on cell binding. *O*-fucose monosaccharide addition had a minor effect on the binding of EGF11–13 to Jagged-1 and Dll1 but the subsequent enzymatic addition of GlcNAc by Fringe to form the disaccharide conferred substantial enhancement of binding to Jagged-1 and Dll1. Any further enzymatic extension to produce tri- and tetra-saccharide forms failed to show any additional effect. No effect of the mono- or disaccharide was observed on binding to Dll4 in these flow cytometry assays. The molecular basis of these interactions was further investigated by surface plasmon resonance (SPR, a biophysical method by which biomolecular interactions can be evaluated in real time) which showed that the disaccharide form of EGF11–13 caused a 9-fold and 18-fold increase in binding to Jagged-1 and Dll1 respectively compared to the monosaccharide, in agreement with the flow cytometry data. Dll4 by comparison had a high inherent affinity for unmodified EGF11–13 in the absence of *O*-glycans, explaining why *O*-glycosylation had no effect on binding in flow cytometry experiments and only minor effects when binding was quantified by SPR (Taylor et al. 2014). The high inherent affinity of Dll4 for Notch was also observed in an independent study (Andrawes et al. 2013).

Crystal structures of hN1 EGF11–13 modified with either *O*-fucose or GlcNAc-fucose were subsequently determined in the presence of Ca^{2+} and compared to the previously determined structure for the unmodified protein (Taylor et al. 2014). The *O*-fucose sugar added to T466 was found to point away from the central β-sheet region. Both the side chain and sugar modification were well ordered in the structure and were seen to make intramolecular contacts with EGF12 (specifically the C6 methyl group of the *O*-fucose ring was packed between residues I477 and M479). Fringe-catalyzed addition of GlcNAc to the *O*-fucose group was shown to extend the sugar further away from the central β hairpin, thus increasing the potential ligand-binding surface (Fig. 2). Similar to the *O*-fucose sugar, GlcNAc was observed to make intra-molecular contacts with neighboring residues D464 and M479. Although both the monosaccharide and disaccharide made extensive contacts with the protein, no conformational change was observed in EGF11–13 and the backbone structure, tilt and twist angles were unaffected (Fig. 2). Importantly, these data indicate that the increase in binding seen on Fringe modification is most likely due to the increased affinity of hN1 EGF11–13 for Jagged-1 and Dll1 ligands. Since other EGF domains in the Notch extracellular domain contain the *O*-fucose consensus and are modified (Shao et al. 2003; Harvey et al. 2016), it is possible that Fringe modification may modulate the receptor/ligand interaction at other sites. Very recently, this has been confirmed using a mass spectrometry approach (Kakuda and Haltiwanger 2017). Mammalian Lunatic, Manic, and Radical Fringe proteins leave distinctive enzymatic signatures on hN1 (resulting in patterns of disaccharide modifications to Notch EGF domains which are specific to each enzyme) which influence whether or not the different ligands activate or inhibit signaling.

1.4 Structure of Unmodified Human Notch1 (hN1) EGF4–13

Although many of the EGF domains of the EGF5–25 region of the human Notch extracellular domain are of the calcium-binding type and predicted to form rod-like structures similar to that observed for EGF11–13, they are interspersed with non-calcium-binding domains EGF6, EGF10 and EGF22 (Downing et al. 1996; Hambleton et al. 2004; Handford et al. 1991). These have the potential to introduce sites of flexibility and impart dynamic properties to the extracellular domain or allow it to adopt a non-linear structure which could impose a "jack-knife" conformation or stabilise a linear conformation which extends the extracellular domain away from the cell surface. To address this, $\{^1H\}$-^{15}N nuclear Overhauser effects, residual dipolar couplings and X-ray crystallography were used to identify a non-linear organization for the EGF4–13 region comprising a rigid bent conformation for EGF4–7, a single flexible linkage between EGF9–10 connected to the calcium-stabilized ligand-binding region EGF11–13 (Weisshuhn et al. 2016). Overall, these data suggested a non-linear but not jack-knifed organization. The near-linear calcium-stabilised section of EGF6–9, N-terminal to the flexible EGF9–10 linkage, suggested that the receptor may align with ligand along its longitudinal axis, with additional weak interactions outside the core recognition region contributing to the overall affinity of receptor for ligand. These data highlight the need for a careful study of pairwise domain interfaces involving a C-terminal non-calcium binding EGF domain, since their properties are not easily predictable from sequence.

1.5 Calcium Binding to Notch EGF Domains

The free Ca^{2+} concentration in the extracellular milieu is ~1.4 mM (Breitwieser 2008). Ca^{2+} affinities of EGF domains containing the calcium-binding consensus sequence D/N-x-D/N-E/Q-x_m-D*/N*-x_n-Y/F (where* denotes possible β-hydroxylation) in the EGF4–13 region have been measured to identify which sites would be saturated under physiological conditions thus conferring rigidity to interdomain interfaces (Handford et al. 1990; Handford et al. 1991; Weisshuhn et al. 2016). Two methods have been useful for obtaining these data; chromophoric chelation for high affinity Ca^{2+} binding sites (Linse et al. 1991), NMR titrations for low and medium affinity sites and to assign high affinity sites to specific domains [Fig. 3, (Suk et al. 2004; Whiteman et al. 2014)]. Almost all the calcium-binding domains contain the canonical calcium-binding consensus sequence and the aromatic packing residue in the preceding domain but EGF7 and 9 replace the E/Q consensus residue with a D residue. All calcium-binding EGF domains in a native context (i.e. with a covalently linked N-terminal domain) have K_d values in the range 1–60 μM, with the exception of EGF5 which has an affinity of 170 μM [Fig. 3, (Weisshuhn et al. 2016)]. Since all measurements were collected at pH 7.5 and at physiological ionic strength ($I = 0.15$), these data suggested that in the extracellular milieu all sites measured so far would be occupied in >90% of molecules, conferring rigidity to domain interfaces (Fig. 3).

2 Ligand Structure

All Notch ligands contain a variable number of EGF domains, a Delta/Serrate/Lag-2 (DSL) domain and an N-terminal domain which until recently was known as the MNNL (Module at the N terminus of Notch Ligand). There are two ligand families which may be distinguished by the presence (Serrate/Jagged) or absence (Delta/Delta-like) of a cysteine–rich region [(Chillakuri et al. 2012), Fig. 1]. A number of mutagenesis and deletion studies previously showed that the DSL domain conferred specificity of binding to Notch and that covalent linkage of EGF1 and 2 to the DSL domain facilitated binding (Shimizu et al. 1999; Henderson et al. 1997).

2.1 Structure of hJagged-1 DSL-EGF3

Based on these data a DSL-EGF3 four-domain fragment from human Jagged-1 was expressed in bacteria, *in vitro* refolded to form the native disulphide-stabilised fold and purified to homogeneity for structure determination (Cordle et al. 2008a). The X-ray structure of hJagged-1 DSL-EGF3 was determined at a resolution of 2.5 Å and showed an extended linear arrangement of domains (Cordle et al. 2008a). All EGF domains in this fragment were of the non-calcium-binding type but EGF1 and 2 had much shorter loops between the cysteine residues involved in disulphide bonding, whilst EGF3 adopted a more classical fold. The DSL domain was shown to have a distinct fold but with some similarity to EGF domain structure, suggesting it may have evolved from a pair of EGF domains. Sequence alignments of DSL domains from both Jagged/Serrate

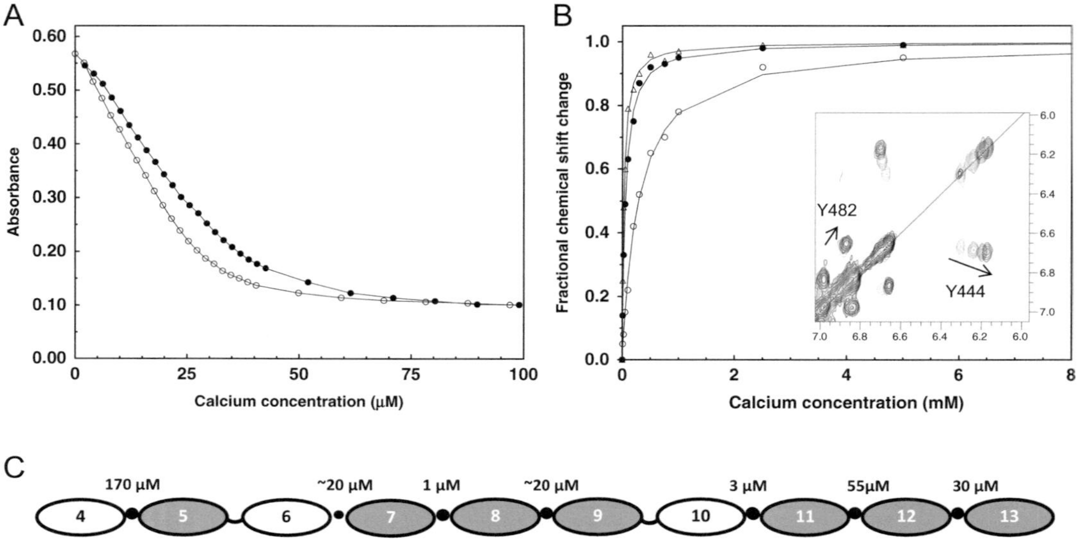

Fig. 3 Measurement of calcium affinities of Notch EGF domains. (**a**) High-affinity Ca^{2+} binding can be determined using the chromophoric chelator 5,5'-Br_2BAPTA. The open circles show the decrease in absorbance at 263 nm when Ca^{2+} is added to a 30 μM solution of 5,5'-Br_2BAPTA alone. This response is altered (filled circles) when Ca^{2+} is added to a mixture of 30 μM 5,5'-Br_2BAPTA and 30 μM hN1 EGF9–11, due to the competition of the high-affinity EGF11 site with the chelator for Ca^{2+}. A K_d value of 3 μM can be determined for EGF11 in the EGF9–11 construct. (**b**) Lower-affinity Ca^{2+} binding can be determined using NMR. The fractional chemical shift change for aromatic residues in EGF11 (open circles), EGF12 (filled circles) and EGF13 (open triangles) in the EGF11–13 construct are plotted as a function of the free Ca^{2+} concentration. Overlaid ^{1}H-^{1}H NOESY spectra collected with 0, 0.4, 1.2 and 15 mM Ca^{2+} are shown in the inset. K_d values of 310, 55 and 30 μM are fitted for EGF11, EGF12 and EGF13; weak binding is observed for EGF11 because this domain is not in a native context in the EGF11–13 construct. (**c**) The measured Ca^{2+} dissociation constants (K_d) at pH 7.5 and I = 0.15 for the EGF4–13 region of human Notch-1 receptor. K_d values in the 1–20 μM range were determined by chromophoric chelation. K_d values in the 20 μM to mM range were determined by NMR. Ca^{2+}-binding and non-Ca^{2+}-binding EGF domains are indicated in grey and white, respectively. Ca^{2+} is indicated by a black sphere at the N-terminus of each Ca^{2+}-binding EGF domain

and Delta ligand families identified a series of highly conserved residues which were mapped onto the Jagged-1 structure. A subset of these clearly performed a structural role but others (F199, R201, R203, D205, F207) were surface exposed on one face of the DSL domain suggesting these may form a Notch-binding site (Fig. 4a). To confirm these observations, a series of alanine substitutions was generated at equivalent residues within Serrate, the Jagged homologue in *Drosophila*, for testing in an *in vivo* wing imaginal disc model of Notch activity. Each construct was ectopically expressed along the anterior/posterior compartment boundary and *wingless* expression measured as a downstream marker of Notch activity. All Serrate variants gave functional effects on Notch signaling, consistent with a role in Notch binding. F207 was found to have a particularly crucial role, confirmed by alanine substitution of this residue in either *Drosophila* Serrate (F257A) or hJagged-1 (F207A) which abrogated Notch interaction in binding assays. It was also possible to show that the same face of the DSL domain was involved in mediating both Notch trans-activation and cis-inhibition, since both induction of an ectopic wing margin (trans) and suppression of endogenous Notch activity at the dorsal-ventral boundary (cis) could be observed in this system. Subsequent to this, it was demonstrated that a region containing the ligand binding portion of the receptor (dNotch EGF10–12) was required for cis-inhibition medi-

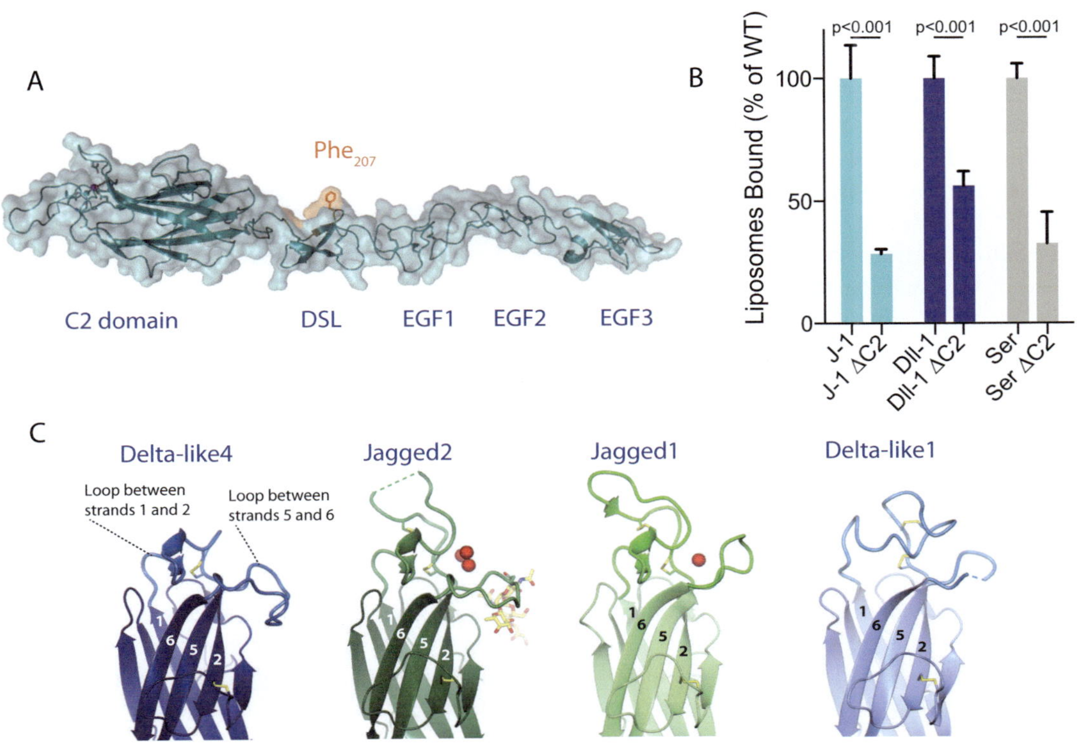

Fig. 4 Notch ligand structure and liposome binding. (**a**) Structure of hJagged-1 N-EGF3 fragment showing the C2 domain at the N-terminus, the DSL domain (with the Notch binding residue F207 in the Notch-binding loop is indicated) and three contiguous EGF domains of the non-calcium binding type. (**b**) The presence of the C2 domain confers liposome binding properties to three diverse Notch ligands (J-1, Dll1, Serrate), Conversely, each ligand with the C2 domain deleted (ΔC2) shows greatly reduced binding to phosphatidylcholine/phosphatidylserine/phosphatidylethanolamine (PC/PS/PE) liposomes. (**c**) Comparison of C2 domains of different human Notch ligands Delta-like-4, Jagged-2, Jagged-1 and Delta-like-1 showing that the Jagged family bind calcium ions but the Delta family does not. Note the diversity in loop structures at the apex of each domain suggesting that each ligand may show different lipid preferences

ated by Serrate (Becam et al. 2010), suggesting that interactions between DSL and EGF11–12 underlie both cis- and trans-Notch complexes.

2.2 Structure of N-Terminus-EGF3 (N-EGF3) of Human Jagged-1

The importance of the region N-terminal to the DSL domain of Notch ligands was suggested by an early experiment performed by Henderson and colleagues (Henderson et al. 1997), since its deletion in the *C.elegans* ligand LAG-2 abolished function. Expression of an N-terminal fragment of human Jagged-1, comprising the complete N-terminus, DSL domain and three contiguous EGF domains (N-EGF3), in HEK cells enabled purification of sufficient material for structure determination (Chillakuri et al. 2013). The N terminal region was found to adopt the fold of a common phospholipid-recognition C2 domain which packed on top of the DSL domain, thereby extending the linear organization of the ligand identified in the DSL-EGF3 structure (Fig. 4a). The Jagged-1 C2 domain had strong structural homology to the calcium-binding C2B domain of Munc13 (an intracellular protein involved in priming synaptic vesicles) with a typical β sandwich fold comprising two four-stranded β sheets at its centre (Cho and Stahelin 2006; Shin et al. 2010). Although crystallization conditions for Jagged-1 N-EGF3 did not originally contain

Ca^{2+}, the strong homology to Munc13 prompted us to assess the Jagged C2 domain for its calcium-binding properties using limited proteolysis performed in the presence and absence of Ca^{2+}. Addition of Ca^{2+}, but not Mg^{2+}, conferred protection against proteolysis, suggesting the presence of Ca^{2+} binding sites within the protein. Further crystallization trials were performed in the presence of Ca^{2+} and the structure of the holo form was solved. Comparison of both apo- and holo-crystal forms showed that Ca^{2+} occupancy increased the degree of structure in the loop regions responsible for co-ordinating Ca^{2+}, regions that are located at the apex of the C2 domain.

In parallel with our structural data, the biological significance of the C2 domain was investigated (Chillakuri et al. 2013). Liposome binding assays confirmed that the C2 domain conferred a phospholipid-binding capability to Jagged-1 N-EGF3 and other Notch ligands which was abrogated by a C2 domain deletion (Fig. 4b). In a quantitative split luciferase Notch activation assay, a Jagged-1 variant containing a double D140A/D144A substitution, designed to prevent Ca^{2+} coordination, was found to substantially reduce activation compared to its wild type (WT) counterpart, despite still being able to bind to the Notch receptor (Chillakuri et al. 2013). Collectively, these data suggested that phospholipid binding, in addition to the core receptor/Jagged-1 interaction, facilitates generation of the Notch signal.

2.3 Additional Ligand Structures

Other crystal structures of the N-terminal region of Notch ligands have been determined since the Jagged-1 structure was published. An eight-domain fragment of hDll1 showed an extended, near-linear conformation for C2, DSL and EGF1–4 (Kershaw et al. 2015), confirming the presence of a C2 domain and a similar arrangement of domains in both ligand families. Interestingly, a 90° bend was then observed between EGF4 and 5 of Dll1, with EGF5 and 6 forming a near linear arrangement. Since the EGF4–5 interface was the site of a crystal contact, it is unclear whether or not the bent conformation observed is present in solution or is more dynamic. However, this structure, together with the identification of EGF domain interfaces with very different properties (bent, flexible, rigid) in the Notch EGF4–13 region, underscores the importance of a detailed study of these regions when elucidating the architecture of both ligand and receptor. With the availability of crystal structures for a member of each ligand family, the authors further identified by comparative sequence analysis a conserved patch in EGF2 which may represent an additional receptor-ligand interaction site. More recently, the structures of the N-terminal regions of hJagged-2 and hDll4 have been determined (Suckling et al. 2017). The C2 domains of Jagged-1 and -2 bind calcium ions while the Delta family do not and, in each case, the loop regions of the C2 domains are highly variable suggesting different lipid preferences (Fig. 4c).

3 Structures of Notch/Ligand Complexes

3.1 Structure of Notch/Dll4

A milestone in Notch structural biology was reached in 2015 when the crystal structure of a Notch1/Dll4 complex, comprising the core recognition regions, was solved at 2.3 Å [(Luca et al. 2015), Fig. 5)]. The technical challenge of crystallising the relatively low affinity Notch complex was circumvented by targeting the rat Notch1/Dll4 interaction for affinity maturation. The N-terminal region-EGF5 region from Dll4 was expressed as a fusion protein on the surface of yeast and higher affinity variants (generated using error-prone PCR) were identified using Notch1 EGF1–14, immobilized to magnetic beads, as bait. Any ligand construct selected was subjected to a second round of mutagenesis/selection to further enhance affinity. Subsequent DNA sequencing revealed a series of missense mutations in the C2 and DSL domains of the ligand construct. Affinity-matured variants were

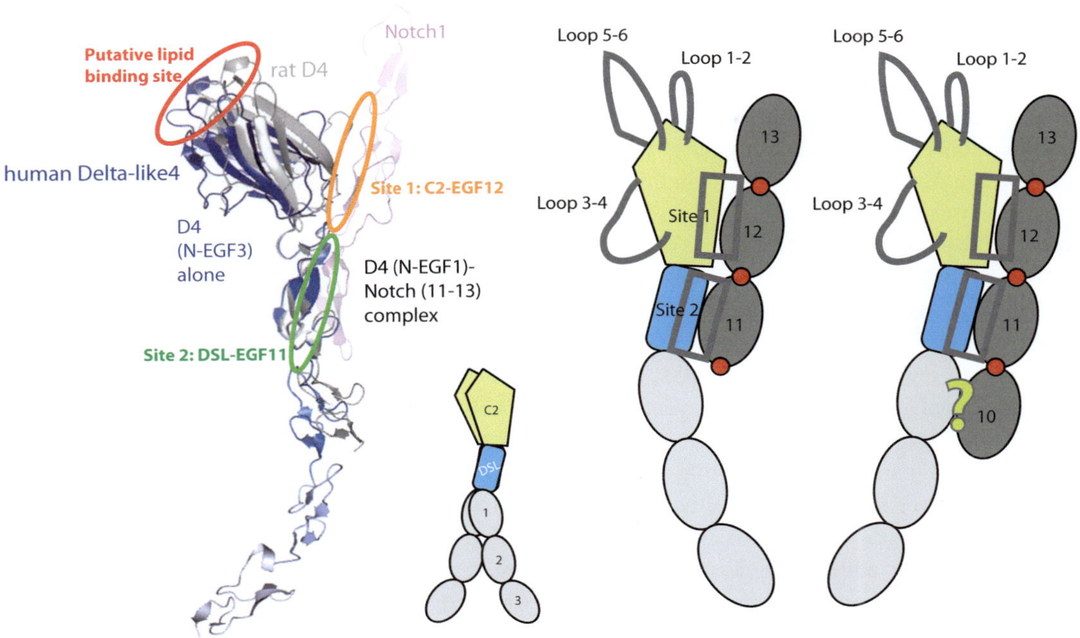

Fig. 5 The Delta-like-4/Notch1 complex. Key structural features are indicated including the putative lipid-binding site of the C2 domain (CHILLAKURI et al. 2013) and receptor/ligand interaction Site 1 (C2-EGF12) and Site 2 (DSL-EGF11) (LUCA et al. 2015). Shown, adjacent in cartoon form, are two different ligand conformations seen in crystal structures of various ligands (SUCKLING et al. 2017) which suggest that i) the straighter form is required to accommodate longer constructs of Notch and ii) that additional contacts may occur along the longitudinal axis (WEISSHUHN et al. 2016). C2 domain loops 1–2, 3–4, 5–6 are indicated

subsequently expressed in insect cells. The N-EGF2 form of Dll4$_{SLP}$ (containing G28S, F107 L and L206P substitutions) showed a > 70-fold enhancement in affinity relative to the WT construct, attributed to the slower dissociation rate of the complex. In activity assays using a luciferase reporter construct, N-EGF2 Dll4$_{SLP}$ showed no reduction in activity compared to the WT construct and the dose response curve was shifted to the left consistent with the increased affinity measured. Crystal forms of glycan-trimmed Dll4$_{SLP}$ (N-EGF2 or N-EGF1) bound to EGF11–13, also purified from insect cells, were obtained and their structures solved by molecular replacement.

The structures of the individual components of the complex confirmed the elongated structure of ligand and receptor identified previously as well as the lack of calcium binding to the C2 domain of Delta ligands first observed by Kershaw and colleagues (Kershaw et al. 2015). However, striking new observations included an antiparallel arrangement of receptor relative to ligand in the complex and two sites of interaction observed between C2 and EGF12 (Site 1) and DSL and EGF11 [Site 2, (Fig. 5)]. The antiparallel arrangement, coupled with the observed sites of flexible/bent EGF interfaces in hNotch1 (Weisshuhn et al. 2016), may suggest that a single receptor/ligand complex underlies trans-activation and cis-inhibition modes of Notch activity, since flexible regions either side of the core recognition element Notch EGF11–12 would allow a single complex to form in cis or trans.

The direct binding role of the *O*-glycan modifications on T466 in EGF12, proposed by Taylor and colleagues (Taylor et al. 2014), was confirmed in this structure with the *O*-fucose making a network of glycan-amino acid contacts within the C2 domain. Modelling of the Fringe extension based on this complex suggested that, as in the crystal structure of Fringe-modified hN1 EGF11–13 (Fig. 2), the disaccharide extends away from the *O*-fucose and makes further contacts with ligand and receptor. *O*-glucose modifications present within EGF11–13 in the complex were not involved in direct contacts with ligand, in agreement with the lack of effect on ligand binding observed by Taylor and colleagues (Taylor et al. 2014). In contrast, DSL-EGF11 contacts at Site 2 were mainly mediated by protein-protein interactions but an *O*-glycan at S435 in EGF11 was observed to interact directly with DSL residues. The functional significance of this is yet to be elucidated but Ser 435 is not absolutely conserved across Notch paralogues suggesting it plays a lesser regulatory role in Notch activity than the Fringe modification. On the basis of sequence comparisons, the authors suggested that Site 2 (EGF11-DSL) is the main common determinant of ligand binding, while Site 1 (EGF12-C2) is less specific, allowing different receptor/ligand combinations to form. It should be noted, however, that other regions of contact may exist away from the core recognition surfaces that also modulate different receptor ligand pairings. A model for hN1 EGF10–13 (Weisshuhn et al. 2016) has been superimposed on EGF11–13 of the complex; small rearrangements were required in order to accommodate the longer receptor fragment and suggested that additional interactions may occur between EGF10 of Notch and EGF1 of ligand. Since possible interface residues are not conserved between ligand families, this and other similar sites such as in EGF 2 of Dll1 which are outside the core recognition sites may contribute to ligand-specific differences in binding. A comparison of new ligand structures from both Jagged and Delta families showed that two different conformations existed in the crystals. The straighter conformation is compatible with the binding of longer fragments of Notch [Fig. 5, (Suckling et al. 2017)].

3.2 Structure of Notch/Jagged-1

Very recently a 2.5 Å crystal structure of a Notch/Jagged-1 complex has been obtained using affinity maturation to select for high affinity ligand variants (Luca et al. 2017). It confirms the Site 1 and Site 2 core interaction sites, as observed in the Notch/Dll4 complex, but utilizing longer five-domain constructs it shows an extensive interface that forms along the whole length of the longitudinal axis. Specifically, Notch EGF12, EGF11, EGF10, EGF9 and EGF8 interact with Jagged-1 C2, DSL, EGF1, EGF2 and EGF3. This complex thus corroborates the observation made by Weisshuhn and colleagues, that domain rearrangements were necessary to accommodate longer Notch constructs in complex with ligand (Weisshuhn et al. 2016). The Notch EGF8/Jagged-1 EGF3 interaction provides a structural explanation for the *jigsaw* mutation identified in a *Drosophila* screen which mapped to V324 in dNotch EGF8 (Yamamoto et al. 2012). Interestingly, the Notch/Jagged-1 complex showed that *O*-fucose modification of T311 in Notch EGF8 plays a direct role in the interaction with Jagged-1 EGF3, highlighting the importance of *O*-glycans in Notch domains other than EGF12 for selective ligand binding. It was also observed in this study that Jagged-1 altered its conformation on Notch binding and exhibited catch bond behavior (where the lifetime of the receptor/ligand interaction is increased on application of a tensile mechanical force), providing an explanation of how relatively weak interactions between receptor and ligand are able to survive cellular forces and result in NRR cleavage.

4 Implications for Disease

These structural data for the extracellular portions of the receptor and ligands have provided insight into molecular mechanisms underlying genetic disease associated with the Notch pathway (Table 1). Many missense mutations affecting Jagged-1 have been identified which cause Alagille syndrome, a disease affecting liver, heart and kidney development (Penton et al. 2012) and related disorders such as extrahepatic biliary atresia and Tetralogy of Fallot. Prior to structural work identifying the C2 domain, a number of *JAG1* mutations associated with nonsense-mediated decay of RNA were known to

Table 1 Molecular consequences of disease-causing missense mutations based on structural information for extracellular domains of ligand and receptor discussed in this review

Protein affected	Domain affected	Missense mutation	Disease	Predicted effect of mutation
J1	C2	L20P	AS	defective secretion
J1	C2	C22R	AS	defective secretion
J1	C2	A31V	AS	disruption of signal peptide cleavage
J1	C2	G33D/S/V	AS	stability/ folding
J1	C2	L37S	AS	stability/folding
J1	C2	I39S	AS	stability/folding
J1	C2	L40P	AS	stability/folding
J1	C2	V45L	EHBA	lipid binding affected
J1	C2	N53D	EHBA	lipid binding affected
J1	C2	K65M	EHBA	lipid binding affected
J1	C2	D69G	AS	lipid binding affected
J1	C2	F75S/L	AS	stability/folding
J1	C2	C78R/G/Y/S	AS	Notch binding site1 C2-EGF12 perturbed
J1	C2	L79H/F	AS	stability/folding
J1	C2	K80E	AS	Notch binding site 1 C2-EGF12 perturbed
J1	C2	C92R/Y	AS	Notch binding site 1 C2-EGF12 perturbed
J1	C2	I120N	AS	stability/folding
J1	C2	L122P	AS	stability/folding
J1	C2	P123S	AS	exposed residue (effect unclear)
J1	C2	A127T	AS	Notch binding site 1 C2-EGF12 perturbed
J1	C2	P129R	AS	Notch binding site 1 C2-EGF12 perturbed
J1	C2	L134F	AS	stability/folding
J1	C2	V136G	AS	stability/folding
J1	C2	I152T	AS	stability/folding
J1				
J1	C2	A155P	AS	stability/folding
J1	C2	P163L/R	AS	misfolded protein
J1				
J1	C2	F179S	AS	stability/folding
J1	C2	Y181N	AS	stability/folding
J1	C2	R184H/G/L/C	AS	stability/folding
J1	DSL	C187Y/S	AS	misfolded protein
J1	DSL	R203K	EHBA	Notch binding site 2 DSL-EGF11 perturbed
J1	DSL	C220F	AS	misfolded protein
J1	DSL	W224C	AS	stability/folding
J1	DSL	C229G/Y	AS	misfolded protein
J1	EGF1	R252K/G	AS	exposed residue (effect unclear)
J1	EGF1	G256S/C	AS	disrupt loop conformation
J1	EGF1	G259V	AS	disrupt loop conformation
J1	EGF2	C265F	AS	misfolded protein
J1	EGF2	P269L	AS	disrupt loop conformation
J1	EGF2	C271R	AS	misfolded protein
J1	EGF2	V272F	TOF	exposed residue (effect unclear)
J1	EGF2	G274D	TOF	stability/folding
J1	EGF2	E278D	TOF	exposed residue (effect unclear)
N2	EGF9	C373R	AS	misfolded protein

(continued)

Table 1 (continued)

N2	EGF10	P383S	AS	disrupt loop conformation (based on homology to EGF11)
N2	EGF10	P394S	AS	disrupt loop conformation (based on homology to EGF11)
N2	EGF11	C444Y	AS	misfolded protein
N2	EGF12	C480R	AS	misfolded protein
Dll4	C2	A121P	AOS	stability/folding
Dll4	DSL	R186C	AOS	Notch binding site 2 DSL-EGF11 perturbed
Dll4	DSL	F195L	AOS	Notch binding site 2 DSL-EGF11 perturbed (confirmed experimentally)
Dll4	EGF2	P267T	AOS	disrupt loop conformation
Dll4	EGF5	C390R/Y	AOS	no structure available, misfolded protein predicted
Dll4	EGF7	C455W	AOS	no structure available, misfolded protein predicted
N1	EGF10	P407R	AOS	disrupt loop conformation (based on homology to EGF11)
N1	EGF11	C429R	AOS	misfolded protein
N1	EGF11	R488Q	AOS	Notch binding site 2 DSL-EGF11 perturbed (confirmed experimentally)
N1	EGF11	C449R	AOS	misfolded protein
N1	EGF12	C456Y	AOS	misfolded protein
N1	EGF12	A465T	AOS	stability/folding
N1	EGF35	C1374R	AOS	no structure available, misfolded protein predicted

AS (Alagille syndrome), AOS (Adams-Oliver syndrome), EHBA (Biliary atresia, extrahepatic), TOF (Tetralogy of Fallot). Mutation data obtained from the Human Gene Mutation Database (STENSON et al. 2014).

cause Jagged-1 haploinsufficiency. Mapping Alagille missense mutations onto the structure of Jagged-1 C2 suggested they destabilized the hydrophobic core and prevented native folding (Chillakuri et al. 2013; Luca et al. 2015). This was corroborated by expression of Jagged-1 N-EGF3 disease-causing variants, which mainly resulted in little or no protein secretion, thus underscoring haploinsufficiency as the main dominant mechanism underlying Alagille syndrome (Chillakuri et al. 2013). However, a few missense mutations were observed which might act through disruption of ligand domain interfaces, a direct effect on Notch binding (Luca et al. 2015) or an effect on lipid binding (Suckling et al. 2017).

Adams Oliver syndrome, a developmental disorder affecting the scalp and cranium, and limb development is associated with a subset of mutations in the *NOTCH1* and *DLL4* genes. The missense mutations identified affected C2, DSL, EGF6, 8, 9 domains of Dll4 (Meester et al. 2015). Many are implicated in domain misfolding by altering the highly conserved cysteine residues which stabilize the fold, suggesting that the main autosomal dominant mechanism underlying *DLL4* mutations is haploinsufficiency. Missense mutations causing Adams Oliver disease have also been identified in *NOTCH1* (Stittrich et al. 2014); a C429R change in EGF11 is also likely to cause domain misfolding of which one can speculate may lead to cellular retention of this variant.

In summary, there have been substantial exciting advances made recently in structure determination of the extracellular domains of Notch and its ligands. The future challenge will be to determine the remaining architecture to build plausible models of the receptor/ligand complex at the cell surface and combine these data with cell biology experiments. Currently, we have been unable to harness the advances made in electron microscopy (EM) due to the narrow dimensions of both receptor and ligand. However, improvements in protein expression systems will allow us to use small angle X-ray scattering (SAXS) to complement the high-resolution structure data and we may be able to apply EM methods in future when we will have a better understanding of individual structures and their interacting regions. Structure-informed mutagenesis, combined with model organism studies, will be required to determine the physiological role of lipid binding by the C2 domain. These multidisciplinary data will then help to optimise the design of novel therapeutic agents aimed at modulating the Notch signal.

Acknowledgements This work was funded by the MRC (MR/L001187/1), a Wellcome Investigator award to SML (100298) and by the EPA Cephalosporin Trust. We would like to thank Dr. Pat Whiteman for proof-reading and helpful comments.

References

Andrawes MB, Xu X, Liu H, Ficarro SB, Marto JA, Aster JC, Blacklow SC (2013) Intrinsic selectivity of Notch 1 for Delta-like 4 over Delta-like 1. J Biol Chem 288:25477–25489

Becam I, Fiuza UM, Arias AM, Milan M (2010) A role of receptor Notch in ligand cis-inhibition in Drosophila. Curr Biol 20:554–560

Bray SJ (2016) Notch signalling in context. Nat Rev Mol Cell Biol 17:722–735

Breitwieser GE (2008) Extracellular calcium as an integrator of tissue function. Int J Biochem Cell Biol 40:1467–1480

Chillakuri CR, Sheppard D, Lea SM, Handford PA (2012) Notch receptor-ligand binding and activation: insights from molecular studies. Semin Cell Dev Biol 23:421–428

Chillakuri CR, Sheppard D, Ilagan MXG, Holt LR, Abbott F, Liang SY, Kopan R, Handford PA, Lea SM (2013) Structural analysis uncovers lipid-binding properties of Notch ligands. Cell Rep 5:861–867

Cho W, Stahelin RV (2006) Membrane binding and subcellular targeting of C2 domains. BBA-Mol Cell Biol L 1761:838–849

Cordle J, Johnson S, Tay JZY, Roversi P, Wilkin MB, de Madrid BH, Shimizu H, Jensen S, Whiteman P, Jin B, Redfield C, Baron M, Lea SM, Handford PA (2008a) A conserved face of the Jagged/Serrate DSL domain is involved in Notch trans-activation and cis-inhibition. Nat Struct Mol Biol 15:849–857

Cordle J, Redfield C, Stacey M, van der Merwe PA, Willis AC, Champion BR, Hambleton S, Handford PA (2008b) Localization of the delta-like-1-binding site in human notch-1 and its modulation by calcium affinity. J Biol Chem 283:11785–11793

Downing AK, Knott V, Werner JM, Cardy CM, Campbell ID, Handford PA (1996) Solution structure of a pair of calcium-binding epidermal growth factor-like domains: implications for the Marfan syndrome and other genetic disorders. Cell 85:597–605

Hambleton S, Valeyev NV, Muranyi A, Knott V, Werner JM, Mcmichael AJ, Handford PA, Downing AK (2004) Structural and functional properties of the human Notch-1 ligand binding region. Structure 12:2173–2183

Handford PA, Baron M, Mayhew M, Willis A, Beesley T, Brownlee GG, Campbell ID (1990) The 1st Egf-like domain from human factor-IX contains a high-affinity calcium-binding site. EMBO J 9:475–480

Handford PA, Mayhew M, Baron M, Winship PR, Campbell ID, Brownlee GG (1991) Key residues involved in calcium-binding motifs in EGF-like domains. Nature 351:164–167

Harvey BM, Rana NA, Moss H, Leonardi J, Jafar-Nejad H, Haltiwanger RS (2016) Mapping sites of O-glycosylation and fringe elongation on Drosophila Notch. J Biol Chem 291:16348–16360

Henderson ST, Gao D, Christensen S, Kimble J (1997) Functional domains of LAG-2, a putative signaling ligand for LIN-12 and GLP-1 receptors in Caenorhabditis elegans. Mol Biol Cell 8:1751–1762

Kakuda S, Haltiwanger RS (2017) Deciphering the fringe-mediated Notch code: identification of activating and inhibiting sites allowing discrimination between ligands. Dev Cell 40:193–201

Kershaw NJ, Church NL, Griffin MDW, Luo CS, Adams TE, Burgess AW (2015) Notch ligand delta-like1: X-ray crystal structure and binding affinity. Biochem J 468:159–166

Linse S, Helmersson A, Forsen S (1991) Calcium-binding to calmodulin and its globular domains. J Biol Chem 266:8050–8054

Luca VC, Jude KM, Pierce NW, Nachury MV, Fischer S, Garcia KC (2015) Structural basis for Notch1 engagement of Delta-like 4. Science 347:847–853

Luca VC, Kim BC, Ge C, Kakuda S, Wu D, Roein-Peikar M, Haltiwanger RS, Zhu C, Ha T, Garcia KC (2017) Notch-Jagged complex structure implicates a catch bond in tuning ligand sensitivity. Science 355:1320–1324

Meester JAN, Southgate L, Stittrich AB, Venselaar H, Beekmans SJA, den Hollander N, Bijlsma EK, den Enden AHV, Verheij JBGM, Glusman G, Roach JC, Lehman A, Patel MS, de Vries BBA, Ruivenkamp C, Itin P, Prescott K, Clarke S, Trembath R, Zenker M, Sukalo M, van Laer L, Loeys B, Wuyts W (2015) Heterozygous loss-of-function mutations in DLL4 cause Adams-Oliver syndrome. Am J Hum Genet 97:475–482

Moloney DJ, Lin AI, Haltiwanger RS (1997) The O-linked fucose glycosylation pathway–evidence for protein-specific elongation of O-linked fucose in Chinese hamster ovary cells. J Biol Chem 272:19046–19050

Moloney DJ, Panin VM, Johnston SH, Chen JH, Shao L, Wilson R, Wang Y, Stanley P, Irvine KD, Haltiwanger RS, Vogt TF (2000a) Fringe is a glycosyltransferase that modifies Notch. Nature 406:369–375

Moloney DJ, Shair LH, Lu FM, Xia J, Locke R, Matta KL, Haltiwanger RS (2000b) Mammalian Notch1 is modified with two unusual forms of O-linked glycosylation found on epidermal growth factor-like modules. J Biol Chem 275:9604–9611

Penton AL, Leonard LD, Spinner NB (2012) Notch signaling in human development and disease. Semin Cell Dev Biol 23:450–457

Rampal R, Li ASY, Moloney DJ, Georgiou SA, Luther KB, Nita-Lazar A, Haltiwanger RS (2005) Lunatic fringe, Manic fringe, and Radical fringe recognize similar specificity determinants in O-fucosylated epidermal growth factor-like repeats. J Biol Chem 280:42454–42463

Rana NA, Haltiwanger RS (2011) Fringe benefits: functional and structural impacts of O-glycosylation on the extracellular domain of Notch receptors. Curr Opin Struct Biol 21:583–589

Rebay I, Fleming RJ, Fehon RG, Cherbas L, Cherbas P, Artavanis Tsakonas S (1991) Specific EGF repeats of notch mediate interactions with Delta and serrate–implications for notch as a multifunctional receptor. Cell 67:687–699

Shao L, Moloney DJ, Haltiwanger R (2003) Fringe modifies O-fucose on mouse Notch1 at epidermal growth factor-like repeats within the ligand-binding site and the abruptex region. J Biol Chem 278:7775–7782

Shimizu K, Chiba S, Kumano K, Hosoya N, Takahashi T, Kanda Y, Hamada Y, Yazaki Y, Hirai H (1999) Mouse Jagged1 physically interacts with Notch2 and other Notch receptors–assessment by quantitative methods. J Biol Chem 274:32961–32969

Shin OH, Lu J, Rhee JS, Tomchick DR, Pang ZPP, Wojcik SM, Cmacho-Perez M, Brose N, Machius M, Rizo J, Rosenmund C, Sudhof TC (2010) Munc13 C2B domain is an activity-dependent Ca2+ regulator of synaptic exocytosis. Nat Struct Mol Biol 17:280–288

Smallridge RS, Whiteman P, Werner JM, Campbell ID, Handford PA, Downing AK (2003) Solution structure and dynamics of a calcium binding epidermal growth factor-like domain pair from the neonatal region of human fibrillin-1. J Biol Chem 278:12199–12206

Stenson PD, Mort M, Ball EV, Shaw K, Phillips AD, Cooper DN (2014) The human gene mutation database: building a comprehensive mutation repository for clinical and molecular genetics, diagnostic testing and personalized genomic medicine. Hum Genet 133:1–9

Stittrich AB, Lehman A, Bodian DL, Ashworth J, Zong ZY, Li H, Lam P, Khromykh A, Iyer RK, Vockley JG, Baveja R, Silva ES, Dixon J, Leon EL, Solomon BD, Glusman G, Niederhuber JE, Roach JC, Patel MS (2014) Mutations in NOTCH1 cause Adams-Oliver syndrome. Am J Hum Genet 95:275–284

Suckling R, Korona B, Whiteman P, Chillakuri C, Holt L, Handford PA, Lea SM (2017) Structural and functional dissection of the interplay between lipid and Notch binding by human Notch ligands. EMBO J:e201796632

Suk JY, Jensen S, Mcgettrick A, Willis AC, Whiteman P, Redfield C, Handford PA (2004) Structural consequences of cysteine substitutions C1977Y and C1977R in calcium-binding epidermal growth factor-like domain 30 of human fibrillin-1. J Biol Chem 279:51258–51265

Taylor P, Takeuchi H, Sheppard D, Chillakuri C, Lea SM, Haltiwanger RS, Handford PA (2014) Fringe-mediated extension of O-linked fucose in the ligand-binding region of Notch1 increases binding to mammalian Notch ligands. Proc Natl Acad Sci U S A 111:7290–7295

Wang Y, Shao L, Shi SL, Harris RJ, Spellman MW, Stanley P, Haltiwanger RS (2001) Modification of epidermal growth factor-like repeats with O-fucose–molecular cloning and expression of a novel GDP-fucose protein O-fucosyltransferase. J Biol Chem 276:40338–40345

Weisshuhn PC, Sheppard D, Taylor P, Whiteman P, Lea SM, Handford PA, Redfield C (2016) Non-linear and flexible regions of the human Notch1 extracellular domain revealed by high-resolution structural studies. Structure 24:555–566

Whiteman P, de Madrid BH, Taylor P, Li DM, Heslop R, Viticheep N, Tan JZ, Shimizu H, Callaghan J, Masiero M, Li JL, Banham AH, Harris AL, Lea SM, Redfield C, Baron M, Handford PA (2013) Molecular basis for Jagged-1/serrate ligand recognition by the Notch receptor. J Biol Chem 288:7305–7312

Whiteman P, Redfield C, Handford PA (2014) Bacterial expression and in vitro refolding of limited fragments of the Notch receptor and its ligands. Methods Mol Biol (Clifton, N J) 1187:193–208

Yamamoto S, Charng WL, Rana NA, Kakuda S, Jaiswal M, Bayat V, Xiong B, Zhang K, Sandoval H, David G, Wang H, Haltiwanger RS, Bellen HJ (2012) A mutation in EGF Repeat-8 of Notch discriminates between serrate/jagged and Delta family ligands. Science 338:1229–1232

The Molecular Mechanism of Notch Activation

Klaus N. Lovendahl, Stephen C. Blacklow, and Wendy R. Gordon

Abstract

Research in the last several years has shown that Notch proteolysis, and thus Notch activation, is conformationally controlled by the extracellular juxtamembrane NRR of Notch, which sterically occludes the S2 protease site until ligand binds. The question of how conformational exposure of the protease site is achieved during physiologic activation, and thus how normal activation is bypassed in disease pathogenesis, has been the subject of intense study in the last several years, and is the subject of this chapter. Here, we summarize the structural features of the NRR domains of Notch receptors that establish the autoinhibited state and then review a number of recent studies aimed at testing the mechanotransduction model for Notch signaling using force spectroscopy and molecular tension sensors.

Keywords

Mechanotransduction · Single molecule force spectroscopy · Molecular tension sensors · Notch signaling · X-ray structure

Abbreviations

NRR	Negative Regulatory Region
EGF	Epidermal Growth Factor
LNR	Lin12 Notch Repeat
HD	Heterodimerization
DSL	Delta-Serrate-Lag-2 ligands
ADAM	A Disintegrin and Metalloprotease
BFP	Biomembrane Force Probe
TGT	Tension guage tether
pN	PicoNewton
MPN	Magnetic nanoparticle
NICD	Notch Intracellular domain
NTM	Transmembrane subunit
CAR-T	Chimeric antigen T-cell receptor
GFP	Green fluorescent protein
AFM	Atomic force microscopy
FRET	Förster Resonance Energy Transfer
S2 site	Metalloprotease cleavage site
S3 site	Gamma-secretase cleavage site
RIP	Regulated intramembrane proteolysis
EGFR	Epidermal Growth Factor Receptor
SSB	Single-strand binding protein
RBC	Red blood cell
NTM	Transmembrane subunit

K. N. Lovendahl · W. R. Gordon (✉)
Department of Biochemistry, Molecular Biology and Biophysics, University of Minnesota, Minneapolis, MN, USA
e-mail: wrgordon@umn.edu

S. C. Blacklow
Department of Biological Chemistry and Molecular Pharmacology, Harvard Medical School, Boston, MA, USA

T. Borggrefe, B. D. Giaimo (eds.), *Molecular Mechanisms of Notch Signaling*, Advances in Experimental Medicine and Biology 1066, https://doi.org/10.1007/978-3-319-89512-3_3

1 Overview

Notch signaling facilitates communication between two cells to control cell fate in many contexts during development, adult homeostasis and disease pathogenesis (Bray 2006; Kovall et al. 2017). Normally, Notch receptors transmit signals by undergoing regulated intramembrane proteolysis (RIP) in response to engaging with transmembrane ligands presented on the surface of adjacent cells. Proteases of the ADAM (A Disintegrin and Metalloprotease) family first cleave Notch at a site called S2 located about 10 amino acids outside of the transmembrane region to shed the Notch ectodomain (Brou et al. 2000; Mumm and Kopan 2000), generating a truncated substrate for intramembrane cleavage by γ-secretase (Kopan and Goate 2000), which releases the Notch intracellular domain (NICD) from the membrane. Release of NICD from the membrane produces an active transcriptional effector that travels to the nucleus and participates in transcription of target genes (Kopan and Ilagan 2009).

This Notch proteolytic cascade is conformationally regulated by a proteolytic switch called the Negative Regulatory Region (NRR), a part of the protein that is sandwiched between the ligand binding EGF repeats and the transmembrane domain. X-ray structures (Gordon et al. 2007; Gordon et al. 2009a; Xiang et al. 2015) of the NRR have revealed that the ADAM protease S2 site is masked intramolecularly by interdomain interactions, demanding that events associated with ligand/receptor engagement induce a conformational change within the NRR to expose the cleavage site to its protease. Activating mutations of the Notch1 NRR that result in ligand-independent proteolysis are also found frequently in human leukemias (Weng et al. 2004), highlighting the importance of tight control of metalloprotease access to the S2 site. Indeed, that the NRR acts as the "proteolytic switch" for Notch signaling has led to substantial efforts to control the conformation of the NRR with therapeutic antibodies (Li et al. 2008; Aste-Amézaga et al. 2010; Gordon and Aster 2014). Moreover, the question of how conformational exposure of the protease site is achieved during physiologic activation, and thus how normal activation is bypassed in disease pathogenesis, has been the subject of intense study in the last several years, and is the subject of this chapter.

2 Models for Conformational Exposure: Allostery or Mechanical Force?

How ligand engagement relieves autoinhibition of Notch has been a longstanding question in the Notch signaling field. The most common mechanism for inducing conformational changes upon protein-protein binding is by allostery, where the effect of binding at one site is transmitted to another, often distal site, allowing for regulation of activity. However, because the binding site for Notch ligands is centered on EGF repeats 8–12, more than 20 EGF modules away (Rebay et al. 1991; Luca et al. 2015, 2017) and since genetic and biochemical studies have established a requirement for endocytosis of ligand into signal sending cells (Musse et al. 2012)–see also "Endocytic Trafficking of the Notch Receptor"– the favored model in the field first proposed more than 15 years ago is the "pulling", or mechanotransduction, model for Notch activation (Parks et al. 2000). Generally, mechanotransduction (Fig. 1) involves a stimulus that is sensed by a mechanosensor domain to convert the stimulus into a signal that allows the cell to respond (Vogel 2006). In the case of Notch, (Fig. 1) the putative stimulus is internalization of the ligand-Notch complex by receptor-mediated endocytosis into the signal-sending cell (discussed in "Endocytic Trafficking of the Notch Receptors"), which generates a pulling force to trigger Notch proteolysis and subsequent transcriptional activation in the signal-receiving cell. Other data consistent with a pulling model include the observation of "trans-endocytosis" in which the ligand-binding domain of Notch co-localizes with ligands in the signal-sending cell (Klueg and Muskavitch 1999; Parks et al. 2000; Nichols et al. 2007; Shaya et al. 2017) and the fact that soluble ligands generally do not activate Notch (Varnum-Finney et al. 2000).

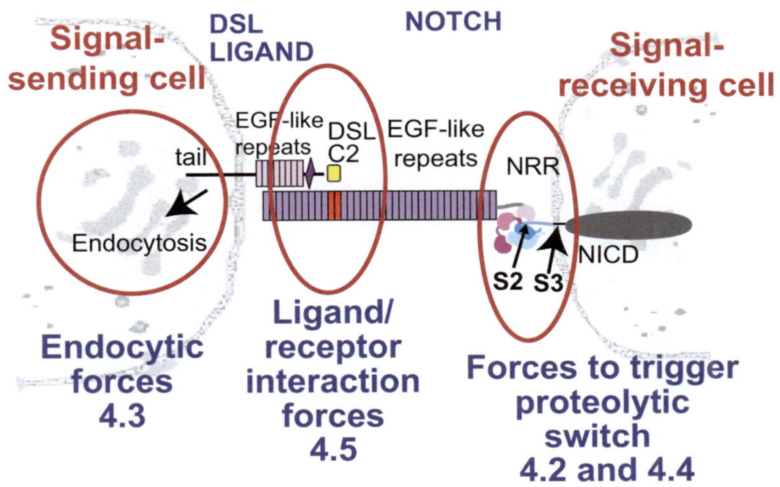

Fig. 1 Steps in the Notch signaling pathway where mechanical tensions have been measured in efforts to test the mechanotransduction mechanism for Notch activation. A ligand of the Delta-Serrate-Lag family (DSL) on the signal sending cell interacts with the Notch receptor on the signal receiving cell. Protein domains labeled in the figure: Epidermal Growth Factor (EGF) -like, Delta-Serrate-Lag (DSL), C2 domain (C2), Negative Regulatory Region (NRR), Notch Intracellular domain (NICD). S2 and S3 refer to two proteolytic cleavage sites involved in Notch activation

Over the last several years, researchers have attempted to test the pulling model, which makes a number of specific predictions:

1. if force is the stimulus for conformational exposure of the proteolytic site within the NRR, then the NRR must be the mechanosensor and application of force must induce its proteolysis,
2. the force required to induce cleavage sensitivity must be within the physiologic force regime,
3. ligand binding alone must not be sufficient to activate Notch,
4. forces generated by endocytosis must suffice to induce the conformational exposure of the protease site, and
5. the ligand-receptor bond must be able to withstand the force required to induce cleavage (*i.e.* not rupture during delivery of activation forces).

Here, we first summarize the structural features of the NRR domains of Notch receptors that establish the autoinhibited state and then review a number of recent studies aimed at testing the mechanotransduction model for Notch signaling using force spectroscopy and molecular tension sensors.

3 The NRR Mechanosensor Domain

The NRR controls activation of the Notch receptor, restraining the receptor in a protease-resistant conformation until activation by ligand. The NRR encompasses a series of three LIN12-Notch repeats (LNRs, defined as A, B and C) and a heterodimerization domain (HD). The HD is cleaved during normal receptor maturation by a furin-like protease at a site called S1 (Logeat et al. 1998) but the NRR is resistant to further proteolysis in the absence of ligand (Gordon et al. 2007). Crystal structures of the Notch1, Notch2, and Notch3 NRR have been solved (Gordon et al. 2007; Gordon et al. 2009a, b), as well as complexes of the Notch1 and Notch3 NRRs bound to inhibitory antibodies (Yan et al. 2010; Bernasconi-Elias et al. 2016). All of the NRR structures adopt a similar conformation (Fig. 2), with the three LNR domains wrapped tightly around the HD

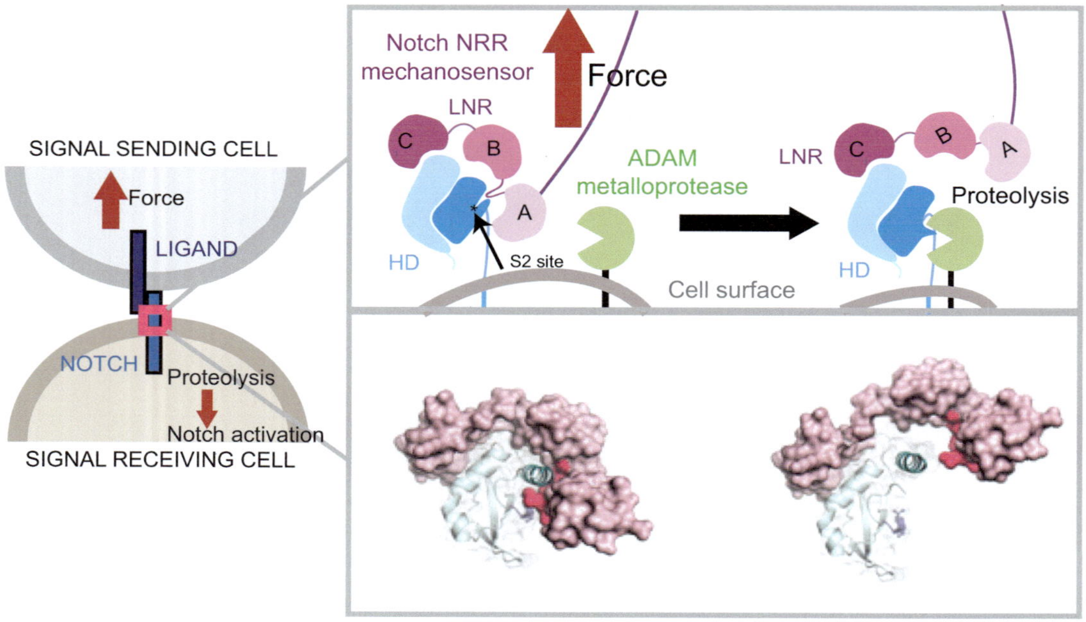

Fig. 2 The Negative Regulatory Region (NRR) of Notch is the putative mechanosensing domain in the mechanotransduction model for Notch signaling. Cartoon and structural models of the proteolytic resistant and sensitive states of NRR are shown. This figure is adapted from a Graphical Abstract (Gordon et al. 2015)

domain, protecting the S2 site from cleavage and preventing activation. The overall structure resembles a mushroom, with the LNR domain capping and protecting the HD "stem". The LNR domains each bind one calcium ion and are stabilized by three disulfide bonds. The HD domain is an alpha-beta sandwich with a substantial and highly conserved hydrophobic core. Known disease-related activating mutations are found inside the hydrophobic cores of the NRR of the Notch1 and Notch3 receptors. These mutations typically disturb the stability of the HD domain and lead to aberrant ligand-independent activation (Malecki et al. 2006; Gordon et al. 2007).

The extensive contacts between the LNR and HD domains bury a total surface area of ~3000 square angstroms and include highly complementary hydrophobic and electrostatic interactions. Hydrophobic residues derived from the linker connecting the first and second LNR repeats sterically occlude the S2 site, and hydrophobic interactions between the second LNR and the HD encircle the α-helix that sits above the strand containing the S2 site, clamping it in place. Though the overall architecture of the NRR is the same in all three structures, interdomain packing details diverge among the three NRRs and such variation may tune the sensitivity of a particular homolog to activation forces. For example, the Notch3 receptor exhibits increased basal activity *in vitro*, which may be explained by divergent interacting residues (Xiang et al. 2015). In all NRRs, the LNR-A/B interface is stabilized by three conserved tryptophan residues; in Notch1 and Notch2, this arrangement is further reinforced by an LNR-A histidine residue, which engages the third tryptophan of the cluster in an aromatic π-stacking interaction. In Notch3, a proline residue (P1408) substitutes for histidine and makes only limited van der Waals contact with W1434. The Notch3 structure is further altered by the replacement of a salt bridge between LNR-C and the first helix of the HD domain with a hydrogen bond, allowing LNR-C to pack closer to the HD domain.

X-ray structures, together with signaling assays mapping the minimum requirements for ligand-independent signaling, reveal that a substantial displacement of the LNRs must occur to expose the S2 site (Fig. 2). This conformational change must disrupt a large, buried surface between domains and likely requires substantial energy, a requirement more consistent with mechanical opening than allostery.

4 Force Spectroscopy Applied to Notch Signaling

4.1 Primer on Molecular Level Forces and Current Toolkit to Probe them

The X-ray structures have revealed the nature of the conformational change that must occur within the putative mechanosensing domain of Notch to expose the S2 site, leading to many recent studies aimed at measuring the mechanical forces generated and sensed by the Notch signaling system (Fig. 3). At a molecular level, proteins in cells are exposed to picoNewton (pN) forces. The thermal energy that must be overcome to do work on a protein in a cell is 4.1 pN-nm. The dimensions of proteins are on the order of nanometers, meaning that the relevant range of forces that induce conformational changes in proteins is sub-pN to tens of pN. Moreover the stall forces of molecular motors in the cell, which are involved in the internalization processes that likely provide the physiologic stimulus for Notch signaling, range from 1 pN to 10 pN (Mallik and Gross 2004). Application of forces to molecules *in vitro* or in cells can be achieved using several different techniques, including optical and magnetic tweezers and biomembrane force probes discussed in this chapter, each with pros and cons. These methods generally involve tethering proteins to a bead with properties that allow experimental manipulation of its position to "pull" on it. Techniques to measure tensions sensed by molecules have also been developed and the methods described in this chapter involve digital tension sensors comprised of duplex DNA or a protein/DNA complex designed to rupture at known forces.

4.2 Forces Associated with S2 Exposure *in vitro*

The measurements of forces involved in cell-based Notch activation (Fig. 3a) are associated with a myriad of complex and uncontrollable variables and involve events downstream of the ADAM proteolysis that triggers the cascade. Therefore, attempts have been made to use single molecule force spectroscopy to measure forces required to convert isolated, recombinant NRR molecules from a protease-resistant to protease-sensitive conformation. First, a study in which the Notch2 NRR was pulled using atomic force microscopy showed that multiple structural transitions occur in the 100 pN range of forces. The tethers formed between the AFM tip and surface-immobilized NRRs were disrupted when ADAM17 was added but the forces applied to the NRR likely drove complete unfolding of the NRR and thus how these findings relate to physiologic context is unclear (Stephenson and Avis 2012).

In an *in vitro* single molecule proteolysis assay based on magnetic tweezers (Fig. 3d), which can apply forces in the sub-pN to tens of pN range, the intrinsic force required to expose the S2 site within an isolated Notch1 NRR was measured directly for the first time. This study showed that the NRR becomes sensitive to proteolysis by ADAM17 in the range of 3.5–5.4 pN of force (Gordon et al. 2015). This force is in line with expected physiologic forces and is similar to forces required to induce protease sensitivity in *in vitro* studies optical trapping studies of the von Willebrand factor A2 domain, which requires 8 pN of force (Zhang et al. 2009). The metalloprotease inhibitor BB94 and conformation specific blocking antibodies abrogate force-induced proteolysis, indicative of site-specificity. This assay should be useful in future experiments comparing forces required for proteolytic sensitivity in other Notch homologs, Notch molecules containing the neighboring ligand binding domain and Notch receptors harboring disease mutations.

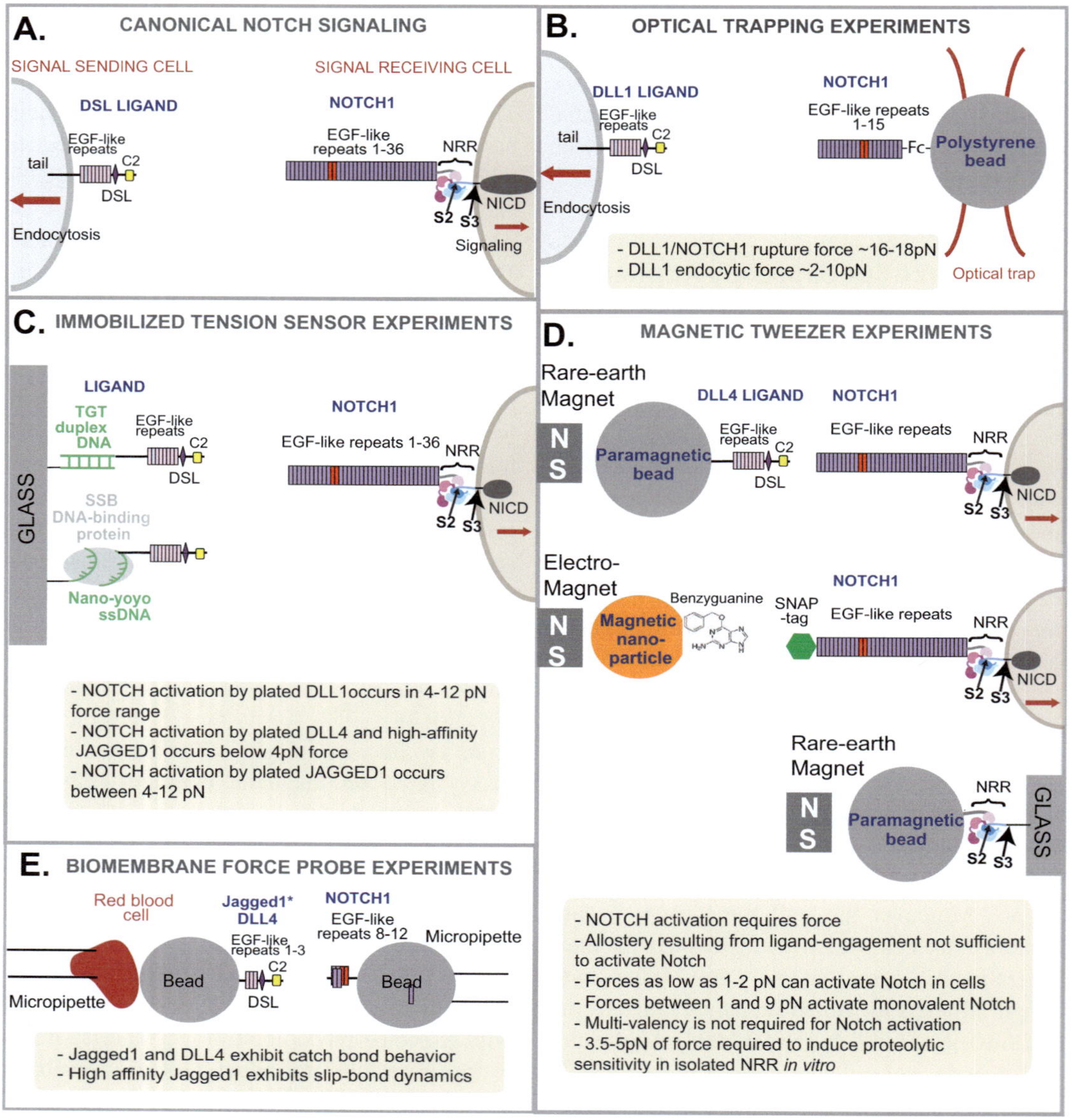

Fig. 3 Overview of force spectroscopy and molecular tension sensor experiments discussed in "The Molecular Mechanism of Notch Activation"

4.3 Forces Generated by Endocytosis

The first attempt to measure forces involved in DSL ligand endocytosis was performed using laser tweezers to optically trap polystyrene beads coated in Notch ectodomain added to cells expressing DLL1 (Meloty-Kapella et al. 2012). In this experiment (Fig. 3b), the presumption is that forces associated with endocytic processes act to pull the bead into the cell while an opposing force is applied to the beads by varying the intensity of the laser trap. This study found that the applied force required to balance the force of endocytic internalization of the ligand (the so-called stall force) varies across a broad range of 2–10 pN, with an average force of 2.8 pN. To address the issue of whether the measurements were detecting endocytosis of bound receptor by ligand-bearing cells, the authors showed that dominant negative dynamin mutants (Seugnet et al. 1997) which abrogate dynamin's ability to pinch off the endocytic vesicle thus blocking endocytosis, interfered with this process. Resistance to applied force also required ligand ubiquitination, which allows the endocytic adaptor protein Epsin to bind, (Wang and Struhl 2005), and active actin dynamics. In a completely separate study of endocytosis of the epidermal growth factor receptor (EGFR), which is also believed to be internalized by a similar receptor-mediated endocytosis pathway, Stabley and colleagues genetically-encoded a FRET-based molecular tension sensor into the EGFR and overexpressed it in cells. The tension sensed by the receptor was measured in its cellular context using FRET as a readout of extension of a previously calibrated flexible peptide "spring" sensor (Stabley et al. 2012). This study found that EGFR endocytosis generates approximately 4 pN of force.

4.4 Forces Involved in Notch Activation in Cells

If internalization by receptor-mediated endocytosis provides the physiologic mechanical force that induces exposure of the S2 protease site, then the forces involved in Notch activation must be of a similar magnitude to endocytic forces. The Ha laboratory first tried to measure forces associated with Notch activation using tension-gauge-tether (TGT) force sensors (Wang and Ha 2013). In this system (Fig. 3c), a DSL (Delta, Serrate, LAG-2) ligand is linked to a DNA strand which is annealed to a complementary strand fixed to a glass surface. By varying the DNA sequence, the "unzipping" force of the resulting DNA duplex can be modulated, providing a series of digital force sensors that rupture at different forces. When cells expressing Notch receptors are plated on these sensors, DNA unzipping as read out by a lack of Notch activation can only occur if the force generated between the plated ligand and the Notch expressing cell is greater than the rupture force of the DNA duplex. The Ha lab did not find that any of their DNA duplexes were unzipped in the Notch/plated DLL1 experiment. Since their lowest-magnitude force sensor ruptured at 12 pN, this put an upper limit on forces involved in Notch activation of 12 pN.

Gordon and colleagues (Gordon et al. 2015) designed an experiment where pN forces could be applied to Notch receptors on the surface of cells using magnetic tweezers. In this experiment (Fig. 3d), Notch expressing cells were plated in 96-well format on polymer "steps" of variable height, magnetic particles coated with DLL4 ligand were added to the cells and a plate of magnets was placed over the cells. This stepped setup allowed the bead-tethered Notch receptors to experience multiple different forces in the same experiment, depending on their distance from the magnet, and thus activation as a function of force could be measured using a standard luciferase transcriptional readout of Notch signaling. In wells where very low forces were applied, even though beads coated with soluble ligands were present, no Notch activation was observed. At forces on the order of 2 pN, Notch activation was instead observed. These data showed, for the first time, that soluble ligands are competent for activation of Notch when force is applied and that the force required for activation is in line with previously measured endocytic forces. One limitation of the magnetic tweezer experiments is that multiple receptors could potentially be engaged with a single magnetic bead, altering the force/activation profile.

This limitation was largely addressed in research by Seo and colleagues, who tagged Notch with magnetic plasmonic nanoparticles (MPN) specifically synthesized for monovalent interaction and control (Seo et al. 2016). Application of force *via* electromagnet to MPNs tethered to single Notch1 receptors on the cell surface, showed that application of ~9 pN of force causes the disappearance of the receptor from the surface and increased intracellular signaling (Fig. 3d); γ-secretase inhibitors completely abolished the effect, whereas application of 1 pN of force caused no loss of cell-surface Notch receptors or transcriptional activity. This effect was identical using MPNs conjugated with DLL1 targeting native Notch1 and benzylguanine-conjugated MPNs targeting a SNAP-tag Notch1 fusion. The researchers also created multivalent DLL1-conjugated MPNs to assess the effect of receptor oligomerization. They observed the same behavior as with monovalent ligand, indicating that oligomerization is not sufficient for activation.

Finally, in a variation of the Ha lab's TGT force sensor, called the 'Nano Yoyo', instead of rupturing DNA duplexes which require high magnitudes of force out of range for many physiologic processes, the DNA is instead wrapped around the *E. coli* single-strand binding protein (SSB) and 'unspooled' at known force (Chowdhury et al. 2016). This method (Fig. 3c) is sensitive to lower magnitudes of force than TGTs and measurements of Notch activation using the "unspooling" Nano-Yoyo deduced an activation force range between 4 pN and 12 pN for activation of the Notch1 receptor by DLL1, consistent with the other studies discussed here.

Interestingly, recent studies by Luca and Garcia in collaboration with the Ha lab using a combination of Nano-Yoyo and TGT tension sensors showed that the tension produced in the complex between immobilized wild type Jagged1 tension sensors and Notch1 expressing cells also were measured to be between 4 pN and 12 pN, similar to DLL1/Notch complexes (Luca et al. 2017). However, when tension sensors presenting the ligands DLL4 and a high affinity version of Jagged1 used to solve the co-crystal structure of Notch/Jagged1 were treated with Notch expressing cells, tension sensors ruptured at forces lower than 4 pN. This suggests that the nature of the ligand-receptor interaction plays a role in tuning Notch activation forces in that higher affinity interactions result in longer engagement times, permitting activation to occur at lower intercellular tensions.

4.5 Force Response of Ligand/Receptor Bonds

If the forces generated by endocytosis and forces required to activate Notch are in the 1–9 pN range as the studies above suggest, then the nature of the ligand receptor bond must be able to withstand these forces and thus have rupture forces of higher magnitudes than activation forces. One study used optical tweezers to measure forces required to rupture ligand-receptor complexes (Shergill et al. 2012), Using polystyrene beads coated with recombinant Notch1-Fc fusion molecules comprising the ligand-binding region, the authors probed the forces required to rupture the association of the beads with DLL1-expressing cells (Fig. 3b). Notwithstanding the caveat that the Notch1-Fc fusion is dimeric, the median force associated with tether rupture was in the range ~17–19 pN, consistent with the conclusion that the binding interaction can remain intact under the force required to expose the S2 site.

More recently, Luca and Garcia in collaboration with the Zhu lab used Biomembrane Force Probe (BFP) spectroscopy to probe the adhesion strength between a fragment of Notch1 and several different ligand constructs (Luca et al. 2017). The biomembrane force probe (Fig. 3e) consists of a ligand-coated red blood cell (RBC) aspirated by a micropipette, which acts as a spring with a known force constant. A second micropipette positions a receptor-coated bead such that receptor-ligand interactions can occur. When the receptor bead is pulled away, the receptor-ligand complex exerts a measurable force on the RBC; the bond lifetime is simply the length of time for which the force persists. Surprisingly, the researchers observed that the bond lifetime between Notch1 and both Jagged1 and DLL4 increased as the tension force was increased from 0 pN to 10 pN, which corresponds to so-called "catch-bond" behavior, which has been observed in selectins, T-cell receptor signaling and other cell-surface receptors (Chen et al. 2017). Protein interactions occurring under force can exhibit either catch bond behavior, in which the interaction lifetime increases in response to increasing force, or the more typical slip bond, in which force reduces bond lifetime. Catch bond behavior could help to explain how the low-affinity Notch-ligand interaction is able to lead to a significant response.

5 Insights from Synthetic Notch Signaling Systems

It is widely known that soluble ligands generally do not activate Notch, suggesting that the allostery from ligand binding is not sufficient to drive conformational exposure of the S2 site to activate Notch. Further evidence of the ligand binding event not being required for Notch activation comes from the advent of synthetic Notch signaling systems. In these systems, the regions of Notch and ligand that are responsible for binding to each other are removed and replaced with a pair of interacting proteins with high affinity for each other. In the first of these systems, engineered Notch1 and DLL4 receptors retained the NRR and intracellular signaling portions of Notch and the tail of the DLL4 ligand that recruits the endocytic machinery but could artificially be connected by FRB/FKBP domains that heterodimerize in the presence of the small molecule rapamycin. Robust Notch signaling occurs in this synthetic system, which relies on the normal proteolytic cascade. These signals are also suppressed by inhibitors of endocytosis or by removing the tail of the DLL4 ligand, as expected. A synthetic system using fly-Notch and the GFP/GFP-nanobody interaction was also presented in this work (Gordon et al. 2015). Other synthetic Notch systems have since been developed using a variety of protein-protein interactions to connect signal sending and receiving cells, such as antibody/antigen interactions. The lack of requirement for the native ligand binding interaction has been recently exploited to engineer T cell responses with a view toward CAR-T immunotherapy (Morsut et al. 2016; Roybal et al. 2016a; Roybal et al. et al. 2016b). In these systems, the ligand-binding domain of Notch is replaced with a single chain antibody against an antigen on a cancer cell and the synthetic receptor is expressed in T-cells. When the cancer cell encounters the engineered T-cell, Notch signaling is triggered, which induces expression of a chimeric T-cell receptor that recognizes a second antigen on the cancer cell, thus imparting specific engagement of antigen-bearing target cells. Interestingly, these synthetic systems react with completely unrelated ligands on the tumor cells, in which the endocytosis status and thus the origin of a mechanical stimulus is unknown.

6 Remaining Question/Future Directions

The preponderance of current evidence tends to support a mechanotransduction model for Notch signaling, in which internalization of the Notch-ligand complex via receptor mediated endocytosis induces a conformational change in the NRR of Notch to expose the S2 proteolytic site and thus drive Notch activation. However, there are dissenting studies that argue for endocytically-driven heterodimer dissociation prior to internalization of the transmembrane subunit (NTM) and subsequent ADAM proteolysis of NTM in an intracellular compartment (Chastagner et al. 2017). Moreover, confounding facts such as that worm Notch is activated by soluble ligand (Chen and Greenwald 2004) and that plated ligand ectodomains robustly activate Notch receptors have not been definitively explained in the context of the mechanotransduction model. Thus, many questions about activation still remain. Though forces generated by endocytosis, forces required to activate Notch and forces necessary to rupture ligand/receptor interactions have been measured, there have been no measurements of tensions sensed in the context of Notch signaling between two cells. A recent study from the Struhl lab showed that the NRR could be replaced by the von Willebrand factor (vwf) A2 domain, which reveals a cryptic proteolysis site in response to mechanical forces in the bloodstream. Interestingly, Notch endocytic forces were not sufficient to activate wildtype vwf domains, which require activation forces around 8 pN, but could induce proteolysis in disease mutants characterized by lower force threshholds (Langridge and Struhl 2017). Moreover, the extent to which the NRR unfolds when the minimal proteolytic force is applied has not yet been measured and may lead to important insights into targeting this domain with therapeutic antibodies. Though some studies suggest that the affinity of the ligand may tune its ability to activate Notch, it is not clear whether the intrinsic sensitivity to force varies among the various Notch NRRs. For example, Notch4 is missing a portion of LNR-A, which might radically alter its force-sensitivity profile.

Moreover, the increasing realization that glycosylation is a critical factor in controlling Notch-ligand receptor pairings/affinities suggests that glycosylation may also alter mechanical forces involved in Notch activation. Many of the assays described in this chapter could be used to probe effects of pathogenic mutations in Notch signaling components on ligand/receptor interaction strength and mechanosensitivity of the NRR. The fact that Notch is mechanosensitive also generates interest in understanding how altered mechanical microenvironments affect Notch signaling propensity, observed in many disease states such as breast cancer (Mouw et al. 2014). Could altered mechanical microenvironments in disease lead to an avenue for ligand-independent activation that does not require activating mutations? Finally, if other cell-surface receptors present in force-sensing structures in cells also undergo RIP, could they also transmit mechanical cues into the cell *via* a mechanotransduction mechanism?

References

Aste-Amézaga M, Zhang N, Lineberger JE et al (2010) Characterization of Notch1 antibodies that inhibit signaling of both normal and mutated Notch1 receptors. PLoS One 5:e9094. https://doi.org/10.1371/journal.pone.0009094

Bernasconi-Elias P, Hu T, Jenkins D, Firestone B, Gans S, Kurth E, Capodieci P, Deplazes-Lauber J, Petropoulos K, Thiel P, Ponsel D, Hee Choi S, LeMotte P, London A, Goetcshkes M, Nolin E, Jones MD, Slocum K, Kluk MJ, Weinstock DM, Christodoulou A, Weinberg O, Jaehrling J, Ettenberg SA, Buckler A, Blacklow SC, Aster JC, Fryer CJ (2016) Characterization of activating mutations of NOTCH3 in T-cell acute lymphoblastic leukemia and anti-leukemic activity of NOTCH3 inhibitory antibodies. Oncogene 35(47):6077–6086

Bray SJ (2006) Notch signalling: a simple pathway becomes complex. Nat Rev Mol Cell Biol 7(9):678–689

Brou C, Logeat F, Gupta N, Bessia C, LeBail O, Doedens JR, Cumano A, Roux P, Black RA, Israël A (2000) A novel proteolytic cleavage involved in Notch signaling. Mol Cell 5(2):207–216

Chastagner P, Rubinstein E, Brou C (2017) Ligand-activated Notch undergoes DTX4-mediated ubiquitylation and bilateral endocytosis before ADAM10 processing. Sci Signal 10(483):eaag2989

Chen N, Greenwald I (2004) The lateral signal for LIN-12/Notch in C. elegans vulval development comprises redundant secreted and transmembrane DSL proteins. Dev Cell 6(2):183–192

Chen Y, Ju L, Rushdi M et al (2017) Receptor-mediated cell mechanosensing. Mol Biol Cell 28:3134–3155. https://doi.org/10.1091/mbc.E17-04-0228

Chowdhury F, Li ITS, Ngo TTM, Leslie BJ, Kim BC, Sokoloski JE, Weiland E, Wang X, Chemla YR, Lohman TM, Ha T (2016) Defining single molecular forces required for Notch activation using Nano Yoyo. Nano Lett 16(6):3892–3897

Gordon WR, Aster JC (2014) Application and evaluation of anti-Notch antibodies to modulate Notch signaling. Methods Mol Biol 1187:323–333. https://doi.org/10.1007/978-1-4939-1139-4_24

Gordon WR, Vardar-Ulu D, Histen G et al (2007) Structural basis for autoinhibition of Notch. Nat Struct Mol Biol 14:295–300. https://doi.org/10.1038/nsmb1227

Gordon WR, Roy M, Vardar-Ulu D et al (2009a) Structure of the Notch1-negative regulatory region: implications for normal activation and pathogenic signaling in T-ALL. Blood 113:4381–4390. https://doi.org/10.1182/blood-2008-08-174748

Gordon WR, Vardar-Ulu D, L'Heureux S et al (2009b) Effects of S1 cleavage on the structure, surface export, and signaling activity of human Notch1 and Notch2. PLoS One 4:e6613. https://doi.org/10.1371/journal.pone.0006613

Gordon WR, Zimmerman B, He L et al (2015) Mechanical Allostery: evidence for a force requirement in the proteolytic activation of Notch. Dev Cell 33:729–736. https://doi.org/10.1016/j.devcel.2015.05.004

Klueg KM, Muskavitch MA (1999) Ligand-receptor interactions and trans-endocytosis of delta, serrate and Notch: members of the Notch signalling pathway in Drosophila. J Cell Sci 112(19): 3289–3297

Kopan R, Goate A (2000) A common enzyme connects Notch signaling and Alzheimer's disease. Genes Dev 14(22):2799–2806

Kopan R, Ilagan MXG (2009) The canonical Notch signaling pathway: unfolding the activation mechanism. Cell 137(2):216–233

Kovall RA, Gebelein B, Sprinzak D, Kopan R (2017) The canonical Notch signaling pathway: structural and biochemical insights into shape, sugar, and force. Dev Cell 41(3):228–241

Langridge PD, Struhl G (2017) Epsin-dependent ligand endocytosis activates Notch by force. Cell 171(6):1383–1396.e12

Li K, Li Y, Wu W et al (2008) Modulation of Notch signaling by antibodies specific for the extracellular negative regulatory region of NOTCH3. J Biol Chem 283:8046–8054. https://doi.org/10.1074/jbc.M800170200

Logeat F, Bessia C, Brou C, LeBail O, Jarriault S, Seidah NG, Israel A (1998) The Notch1 receptor is cleaved constitutively by a furin-like convertase. Proc Natl Acad Sci 95(14):8108–8112

Luca VC, Jude KM, Pierce NW, Nachury MV, Fischer S, Garcia KC (2015) Structural basis for Notch1 engagement of Delta-like 4. Science 347(6224):847–853

Luca VC, Kim BC, Ge C, Kakuda S, Di W, Roein-Peikar M, Haltiwanger RS, Zhu C, Ha T, Christopher Garcia K (2017) Notch-Jagged complex structure implicates a catch bond in tuning ligand sensitivity. Science 355(6331):1320–1324

Mallik R, Gross SP (2004) Molecular motors: strategies to get along. Curr Biol 14(22):R971–R982

Malecki MJ, Sanchez-Irizarry C, Mitchell JL et al (2006) Leukemia-associated mutations within the NOTCH1 heterodimerization domain fall into at least two distinct mechanistic classes. Mol Cell Biol 26:4642–4651. https://doi.org/10.1128/MCB.01655-05

Meloty-Kapella L, Shergill B, Kuon J, Botvinick E, Weinmaster G (2012) Notch ligand endocytosis generates mechanical pulling force dependent on dynamin, epsins, and actin. Dev Cell 22(6):1299–1312

Morsut L, Roybal KT, Xiong X, Gordley RM, Coyle SM, Thomson M, Lim WA (2016) Engineering customized cell sensing and response behaviors using synthetic Notch receptors. Cell 164(4):780–791

Mouw JK, Yui Y, Damiano L et al (2014) Tissue mechanics modulate microRNA-dependent PTEN expression to regulate malignant progression. Nat Med 20:360–367. https://doi.org/10.1038/nm.3497

Mumm JS, Kopan R (2000) Notch signaling: from the outside in. Dev Biol 228(2):151–165

Musse AA, Meloty-Kapella L, Weinmaster G (2012) Notch ligand endocytosis: mechanistic basis of signaling activity. Semin Cell Dev Biol 23(4):429–436

Nichols JT, Miyamoto A, Olsen SL, D'Souza B, Yao C, Weinmaster G (2007) DSL ligand endocytosis physically dissociates Notch1 heterodimers before activating proteolysis can occur. J Cell Biol 176(4):445–458

Parks AL et al (2000) Ligand endocytosis drives receptor dissociation and activation in the Notch pathway. Development 127(7):1373–1385

Rebay I, Fleming RJ, Fehon RG, Cherbas L, Cherbas P, Artavanis-Tsakonas S (1991) Specific EGF repeats of Notch mediate interactions with delta and serrate: implications for notch as a multifunctional receptor. Cell 67(4):687–699

Roybal KT, Williams JZ, Morsut L, Rupp LJ, Kolinko I, Choe JH, Walker WJ, McNally KA, Lim WA (2016a) Engineering T cells with customized therapeutic response programs using synthetic Notch receptors. Cell 167(2):419–432.e16

Roybal KT, Rupp LJ, Morsut L, Walker WJ, McNally KA, Park JS, Lim WA (2016b) Precision tumor Vrecognition by T cells with combinatorial antigen-sensing circuits. Cell 164(4):770–779

Seo D, Southard KM, Kim J-w, Lee HJ, Farlow J, Lee J-u, Litt DB, Haas T, Paul Alivisatos A, Cheon J, Gartner ZJ, Jun Y-w (2016) A mechanogenetic toolkit for interrogating cell signaling in space and time. Cell 165(6):1507–1518

Seugnet L, Simpson P, Haenlin M (1997) Requirement for dynamin during Notch signaling in Drosophila neurogenesis. Dev Biol 192:585–598. https://doi.org/10.1006/dbio.1997.8723

Shaya O, Binshtok U, Hersch M, Rivkin D, Weinreb S, Amir-Zilberstein L, Khamaisi B, Oppenheim O, Desai RA, Goodyear RJ, Richardson GP, Chen CS, Sprinzak D (2017) Cell-cell contact area affects Notch signaling and Notch-dependent patterning. Dev Cell 40(5):505–511.e6

Shergill B, Meloty-Kapella L, Musse AA, Weinmaster G, Botvinick E (2012) Optical tweezers studies on Notch: single-molecule interaction strength is independent of ligand endocytosis. Dev Cell 22(6):1313–1320

Stabley DR, Jurchenko C, Marshall SS, Salaita KS (2012) Visualizing mechanical tension across membrane receptors with a fluorescent sensor. Nat Methods 9:64–67. https://doi.org/10.1038/nmeth.1747

Stephenson NL, Avis JM (2012) Direct observation of proteolytic cleavage at the S2 site upon forced unfolding of the Notch negative regulatory region. Proc Natl Acad Sci 109(41):E2757–E2765

Varnum-Finney B et al (2000) Immobilization of Notch ligand, Delta-1, is required for induction of notch signaling. J Cell Sci 113(23):4313–4318

Vogel V (2006) Mechanotransduction involving multimodular proteins: converting force into biochemical signals. Annu Rev Biophys Biomol Struct 35(1):459–488

Wang X, Ha T (2013) Defining single molecular forces required to activate integrin and notch signaling. Science 340:991–994. https://doi.org/10.1126/science.1231041

Wang W, Struhl G (2005) Distinct roles for Mind bomb, Neuralized and Epsin in mediating DSL endocytosis and signaling in Drosophila. Development 132:2883–2894. https://doi.org/10.1242/dev.01860

Weng AP et al (2004) Activating mutations of NOTCH1 in human T cell acute lymphoblastic leukemia. Science 306(5694):269–271

Xiang X, Choi SH, Tiancen H, Tiyanont K, Habets R, Groot AJ, Vooijs M, Aster JC, Chopra R, Fryer C, Blacklow SC (2015) Insights into autoregulation of Notch3 from structural and functional studies of its negative regulatory region. Structure 23(7):1227–1235

Yan W, Cain-Hom C, Choy L, Hagenbeek TJ, de Leon GP, Chen Y, Finkle D, Venook R, Xiumin W, Ridgway J, Schahin-Reed D, Dow GJ, Shelton A, Stawicki S, Watts RJ, Zhang J, Choy R, Howard P, Kadyk L, Yan M, Zha J, Callahan CA, Hymowitz SG, Siebel CW (2010) Therapeutic antibody targeting of individual Notch receptors. Nature 464(7291):1052–1057

Zhang X, Halvorsen K, Zhang C-Z et al (2009) Mechanoenzymatic cleavage of the ultralarge vascular protein von Willebrand factor. Science 324:1330–1334. https://doi.org/10.1126/science.1170905

Regulation of Notch Function by *O*-Glycosylation

Beth M. Harvey and Robert S. Haltiwanger

Abstract

The Notch receptor initiates a unique intercellular signaling pathway that is evolutionarily conserved across all metazoans and contributes to the development and maintenance of numerous tissues. Consequently, many diseases result from aberrant Notch signaling. Emerging roles for Notch in disease are being uncovered as studies reveal new information regarding various components of this signaling pathway. Notch activity is regulated at several levels, but *O*-linked glycosylation of Epidermal Growth Factor (EGF) repeats in the Notch extracellular domain has emerged as a major regulator that, depending on context, can increase or decrease Notch activity. Three types of *O*-linked glycosylation occur at consensus sequences found within the EGF repeats of Notch: *O*-fucosylation, *O*-glucosylation, and *O*-GlcNAcylation. Recent studies have investigated the site occupancy of these types of glycosylation and also defined specific roles for these glycans on Notch structure and function. Nevertheless, there are many functional aspects to each type of *O*-glycosylation that remain unclear. Here, we will discuss molecular mechanisms of how *O*-glycosylation regulates Notch signaling and describe disorders associated with defects in Notch *O*-glycosylation.

Keywords

Notch signaling · EGF repeat · *O*-fucose · *O*-glucose · *O*-GlcNAc

Abbreviations

β4GalT-1	β4-Galactosyltransferase-1
GXYLT	Glucoside α3-Xylosyltransferase
XXYLT	Xyloside α3-Xylosyltransferase
A	Alanine
AOS	Adams-Oliver Syndrome
C	Cysteine
DDD	Dowling-Degos Disease
DLL	Delta-like ligand
DSL	Delta/Serrate/LAG2
ECD	Extracellular Domain
EGF	Epidermal Growth Factor-like
EOGT	EGF-domain-specific *O*-GlcNAc Transferase

B. M. Harvey
Department of Biochemistry and Cell Biology, Stony Brook University, Stony Brook, NY, USA

Present Address: Department of Cell and Developmental Biology, University of Pennsylvania, Philadelphia, PA, USA

R. S. Haltiwanger (✉)
Department of Biochemistry and Cell Biology, Stony Brook University, Stony Brook, NY, USA

Complex Carbohydrate Research Center, University of Georgia, Athens, GA, USA
e-mail: rhalti@uga.edu

T. Borggrefe, B. D. Giaimo (eds.), *Molecular Mechanisms of Notch Signaling*, Advances in Experimental Medicine and Biology 1066, https://doi.org/10.1007/978-3-319-89512-3_4

ER	Endoplasmic Reticulum
FNG	Fringe
Fuc	Fucose
G	Glycine
Gal	Galactose
Glc	Glucose
GlcNAc	*N*-Acetylglucosamine
JAG	Jagged
OGT	*O*-GlcNAc Transferase
P	Proline
POFUT1	Protein *O*-Fucosyltransferase 1
POGLUT1	Protein *O*-Glucosyltransferase 1
S	Ser, Serine; Sia, Sialic Acid
T	Thr, Threonine
Xyl	Xylose

1 Introduction

Glycosylation is arguably the most complex form of post-translational modification of proteins. The discovery that Fringe, a known modulator of Notch activity, is a glycosyltransferase that alters glycan structures in the Notch extracellular domain (ECD) provided definitive evidence that cell-specific glycosylation can regulate Notch activity (Bruckner et al. 2000; Moloney et al. 2000a). To this day, a number of different types of glycan structures have been identified on the Notch ECD, all of which affect Notch function. In this chapter, we will summarize what is currently known about the *O*-glycans that modify Notch, their biological effects, and the molecular mechanisms through which they function.

2 Types of *O*-Linked Glycosylation on NOTCH EGF Repeats

The Notch receptor is a large (~300 kDa) single-pass Type 1 transmembrane protein. There is one Notch receptor expressed in *Drosophila* and four mammalian homologs [NOTCH1–4, (Kopan and Ilagan 2009)]. To activate canonical Notch signaling, the Notch receptor binds to Delta, Serrate, LAG-2 (DSL) family ligands expressed on an adjacent cell designated as the signal-sending cell. In *Drosophila*, there is a single Delta ligand and Serrate ligand, while there are three DELTA-like ligand (DLL) homologs (DLL 1, 3, and 4) and two Serrate homologs called JAGGED (JAG) 1 and 2 in mammals (Fehon et al. 1990; Rebay et al. 1991). Various aspects of the Notch signaling pathway have been extensively reviewed previously (Bray 2006; Rana and Haltiwanger 2011; Chillakuri et al. 2012; Hori et al. 2013).

The Notch ECD contains up to 36 tandem Epidermal Growth Factor-like (EGF) repeats. EGF repeats are common motifs found in secreted and cell surface proteins, often involved in adhesion, receptor-ligand interactions and blood coagulation (Lin et al. 2001). They are approximately 40 amino acids in length and have six conserved cysteine residues that form three disulfide bonds. The secondary structure of an EGF repeat consists of two anti-parallel beta strands with corresponding loops that vary depending on primary sequences [(Hambleton et al. 2004), Fig. 1A, modified from (Luca et al. 2017)]. EGF repeats can also be classified as calcium-binding EGF repeats, based on consensus sequences within individual EGF repeats, and it is known that bound calcium imparts rigidity to the tertiary protein structure (Hambleton et al. 2004). However, one of the most consequential characteristics of EGF repeats is that they can be modified with *O*-linked glycans, which are sugar modifications of the hydroxy groups on serine or threonine residues (Harris and Spellman 1993). Three major types of *O*-glycosylation modify EGF repeats at consensus sequences between the conserved cysteine residues: *O*-fucosylation, *O*-glucosylation and *O*-GlcNAcylation. Fig. 1A shows the structure of NOTCH1 EGF11–12 modified with *O*-fucose, *O*-glucose and *O*-GlcNAc glycans.

2.1 *O*-Fucosylation and Fringe Elongation

O-Fucosylation was first reported as a novel post-translational modification in 1990 on the EGF repeat of the urokinase-type plasminogen activator protein (Kentzer et al. 1990). Ten years

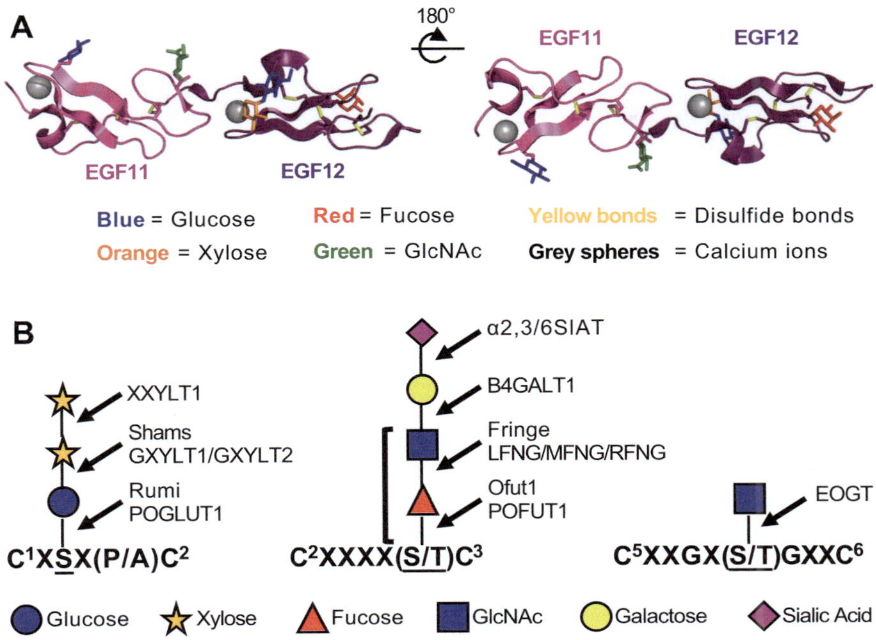

Fig. 1 Types of *O*-glycosylation modifying Notch EGF repeats. (**A**) Structure of EGF11–12 from NOTCH1, modified from PDB ID:5UK5 (Luca et al. 2017). The structure of an EGF repeat consists mainly of two anti-parallel β-sheets. EGF11 is shown with an *O*-GlcNAc modification (green) between conserved cysteines 5 and 6, in addition to the novel hexose modification between cysteines 3 and 4 (blue). EGF12 is modified with *O*-fucose (red) between cysteines 2 and 3 and with an *O*-glucose disaccharide (glucose, blue; xylose, orange) between cysteines 1 and 2. Disulfide bonds between conserved cysteine residues are shown in yellow. Calcium ions are depicted as grey spheres. (**B**) Current consensus sequences of each type of *O*-glycosylation are listed, in which the modified hydroxyl residues are underlined and conserved. Cysteine residues are numbered. X denotes any amino acid. *Drosophila* and mammalian glycosyltransferases responsible for adding each monosaccharide to the glycan are listed, respectively. *O*-Fucose has not been observed to be elongated past the disaccharide form in *Drosophila,* (Bruckner et al. 2000; Harvey et al. 2016), indicated by the bracket

later in 2000, NOTCH1 was shown to be *O*-fucosylated and to contain more putative sites of *O*-fucosylation than any other protein found in databases (Moloney et al. 2000b; Rampal et al. 2007). The enzyme responsible for adding fucose to EGF repeats is Protein *O*-Fucosyltransferase 1 [POFUT1 in mammals, Ofut1 in *Drosophila,* Fig. 1B, (Wang et al. 1996; Wang and Spellman 1998; Wang et al. 2001; Okajima and Irvine 2002)]. Unlike most glycosyltransferases, POFUT1 is an Endoplasmic Reticulum (ER) resident enzyme with a C-terminal ER retention sequence (Luo and Haltiwanger 2005). Ofut1/POFUT1 adds a fucose to a serine or threonine within a consensus sequence that lies between the second and third conserved cysteine residues of an EGF repeat: C^2xxxx(<u>S/T</u>)C^3 (Fig. 1A and B) (Shao et al. 2003; Muller et al. 2014); and it only modifies properly folded EGF repeats (Wang and Spellman 1998). Recently, the structure of POFUT1 with EGF repeat substrates was determined and provides insight into the interactions of POFUT1 with the *O*-fucosylation consensus sequence (Li et al. 2017). The *O*-fucose can be elongated to a GlcNAcβ1-3Fuc-*O*-Ser/Thr disaccharide after the addition of a β1-3*N*-acetylglucosamine by the Golgi-localized glycosyltransferase Fringe [Fig. 1B, (Bruckner et al. 2000; Moloney et al. 2000a)]. Three homologs of *Drosophila* Fringe exist in mammals: Lunatic, Manic and Radical Fringe [LFNG, MFNG and RFNG, (Johnston et al. 1997)]. In mammals, the disaccharide can be further elongated to a tetrasaccharide, Siaα2-6Galβ1-4GlcNAcβ1-3Fuc-*O*-

Ser/Thr, by the sequential modifications by a galactosyltransferase (β4GalT-1) and either an α2-3- or α2-6-sialyltransferases [Fig. 1B, (Moloney et al. 2000b)]. In flies, however, elongation of the *O*-fucose disaccharide on Notch to the tri- or tetrasaccharide has not been observed (Xu et al. 2007; Harvey et al. 2016).

The majority of the EGF repeats in the Notch ECD contain consensus sequences for *O*-fucosylation (22 of 36 for *Drosophila* Notch, 20 of 36 for mouse NOTCH1). To determine whether these predicted sites are actually modified, several studies have investigated their occupancy in *Drosophila* Notch and mouse NOTCH1, observing *O*-fucosylation and elongation by Fringe at a specific subset of predicted sites (Moloney et al. 2000a; Shao et al. 2003; Rampal et al. 2005; Xu et al. 2007). However, these data did not provide a complete analysis of all the predicted *O*-fucose sites nor the stoichiometry of *O*-fucosylation. The recent development of semi-quantitative mass spectral methods for site-specific analysis of Notch glycans has allowed mapping of all 22 *O*-fucose predicted sites in *Drosophila* Notch and the 20 predicted sites in mouse NOTCH1 (Harvey et al. 2016; Kakuda and Haltiwanger 2017). While a few EGF repeats retained unmodified *O*-fucose sites, the majority of sites were modified to high stoichiometries, indicating a high efficiency of modification by Ofut1/POFUT1. Interestingly, endogenous Notch isolated from *Drosophila* embryos was modified similarly to Notch produced in S2 cells (Harvey et al. 2016). Furthermore, upon co-expression of *Drosophila* Notch with Fringe, varying amounts of fucose elongation were detected (Harvey et al. 2016). Some EGF repeats were more heavily modified by Fringe, while others were not (Fig. 2B). These results indicate that Fringe selectively modifies *O*-fucose on certain EGF repeats more efficiently than others. Similarly, LFNG and MFNG modified only certain EGF repeats on NOTCH1, while RFNG modified an even smaller subset of those sites (Fig. 2A), suggesting distinct roles for LFNG, MFNG, and RFNG on NOTCH1 activity (Kakuda and Haltiwanger 2017). Interestingly, β-hydroxylation has been proposed to be a potential modulator of Notch activity, possibly by altering Fringe elongation (Lavaissiere et al. 1996; Dinchuk et al. 2002). While mapping *O*-fucosylation, β-hydroxylation modifications on EGF repeats 27 and 30 were also detected, though did not appear to affect the efficiency by which Fringe elongated those *O*-fucose sites (Kakuda and Haltiwanger 2017).

2.2 *O*-Glucosylation and Xylosylation

O-Glucose was first observed linked to serine residues on the EGF repeats of the blood coagulation factors VII and IX (Hase et al. 1988; Nishimura et al. 1989; Hase et al. 1990). *O*-Glucose was later detected on the NOTCH1 ECD in the same study that first described *O*-fucosylation on NOTCH1 (Moloney et al. 2000b). Similar to *O*-fucosylation, Notch proteins contain more predicted *O*-glucose sites than any other protein (Fernandez-Valdivia et al. 2011). The enzymes that mediate the addition of *O*-linked glucose to EGF repeats are *Drosophila* Rumi and its mammalian homolog Protein *O*-Glucosyltransferase 1 (POGLUT1) (Fig. 1B) (Acar et al. 2008). Like POFUT1, POGLUT1 is an ER resident enzyme with a C-terminal ER retention sequence (Acar et al. 2008) that also only modifies properly folded EGF repeats (Takeuchi et al. 2012). Rumi/POGLUT1 adds a glucose to a serine residue in a consensus sequence that lies between the first and second conserved cysteine residues of an EGF repeat: $C^1x\underline{S}x(P/A)C^2$ [Fig. 1A and B, (Rana et al. 2011)]. Recently, the structure of Rumi was solved in a complex with a folded EGF repeat, and revealed several interactions between Rumi and the *O*-glucose consensus sequence, as well as with a conserved hydrophobic region of the EGF repeat (Yu et al. 2016). The *O*-glucose monosaccharide can be elongated by the addition of a xylose residue to form a Xylα1-3Glcβ-*O*-Ser disaccharide by glucoside α3-xylosyltransferases [GXYLT1 and GXYLT2 in mammals, Shams in *Drosophila*, Fig. 1A and B]. The disaccharide can be further elongated by the addition of a second xylose by xyloside α3-xylosyltransferases (XXYLT1) to form a Xylα1-3Xylα1-3Glcβ-*O*-

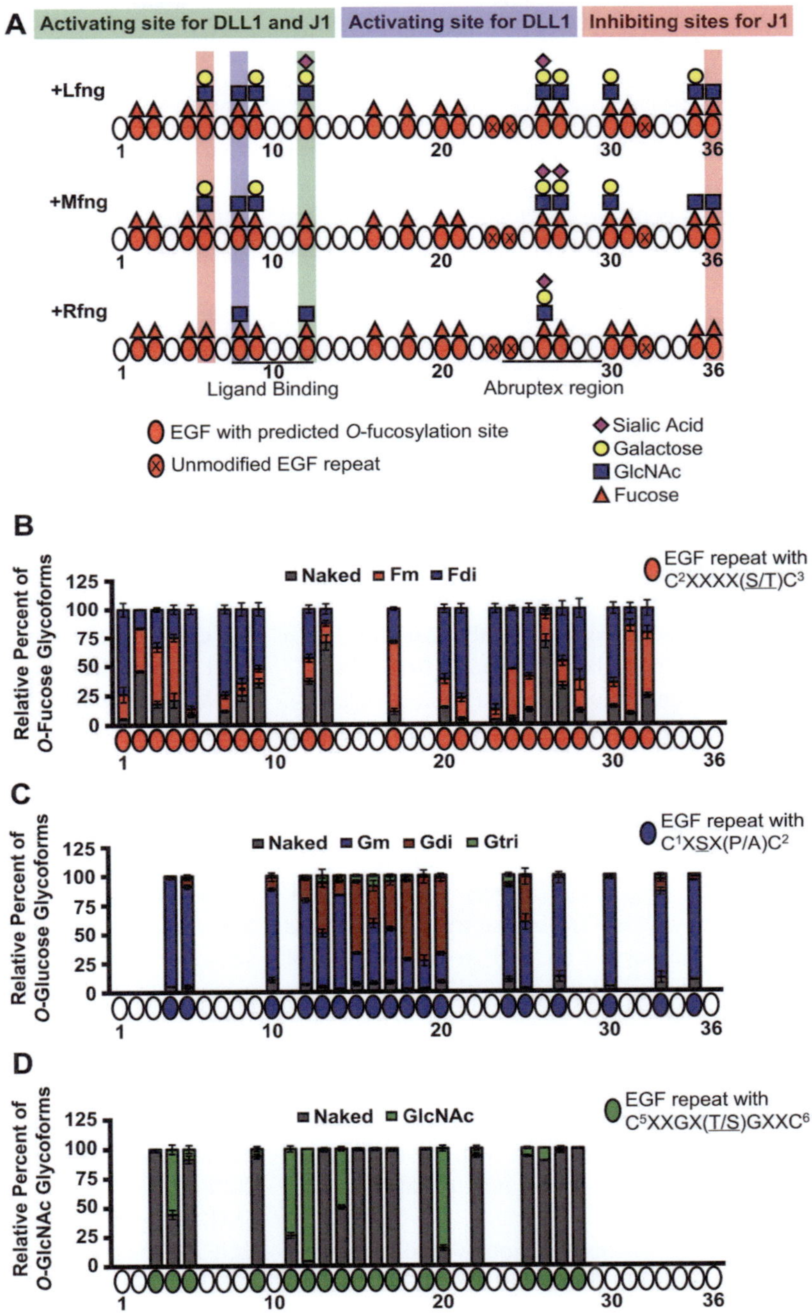

Fig. 2 Summary of *O*-glycosylation on mouse NOTCH1 and *Drosophila* Notch. (**A**) The Fringe-mediated Notch code, modified with permission from (Kakuda and Haltiwanger 2017). Fringe elongation of *O*-fucose at EGF12 enhances NOTCH1 binding and activation by both DLL1 and JAG1 (highlighted green), while Fringe elongation of at EGF8 enhances binding and activation to only DLL1 (highlighted blue). Fringe elongation at EGF6 and 36 inhibits NOTCH1 activation from JAG1. Note that RFNG does not modify *O*-fucose on EGF6 or 36, consistent with lack of inhibition of JAG1-mediated NOTCH1 activation by RFNG. The most abundant *O*-fucose glycoform detected at each predicted site is shown in A. Quantifications of each glycoform detected at predicted sites of *O*-fucosylation (**B**), *O*-glucosylation (**C**) and *O*-GlcNAcylation (**D**) on *Drosophila* Notch, modified with permission from (Harvey et al. 2016)

Ser trisaccharide [Fig. 1B, (Sethi et al. 2010; Sethi et al. 2012; Lee et al. 2013)]. Yu and colleagues solved the structure of XXYLT1 in complex with an Xylα1-3Glcβ-*O*-Ser modified EGF repeat and provided a detailed view of the mechanisms by which XXYLT1 retains the stereochemistry of the donor xylose upon transfer to the acceptor xylose on the EGF repeat (Yu et al. 2015). From the complexed structure, Yu and colleagues not only determined that XXYLT1 recognizes both the *O*-glucose disaccharide and the EGF repeat itself, but also that the EGF repeat undergoes an unexpected and substantial conformational change upon binding to the enzyme. Until these two structures were solved (Rumi/POGLUT1 and XXYLT1), the mechanisms of how folded EGF repeats were recognized and *O*-glycosylated by glycosyltransferases had been unclear. In contrast, other previously published structures of POFUT1 [*C. elegans* (Lira-Navarrete et al. 2011) and human (McMillan et al. 2017)] and MFNG (Jinek et al. 2006) were not in complex with EGF repeat acceptor substrates.

As with *O*-fucosylation, several mass spectral analyses have been published to determine which predicted *O*-glucosylation sites are modified, and also which are elongated by xylosylation (Acar et al. 2008; Rana et al. 2011; Lee et al. 2013; Harvey et al. 2016). Rana and colleagues observed that all *O*-glucose sites on mouse NOTCH1 were predominantly modified with *O*-glucose trisaccharide, apart from EGF27 which was underglucosylated (Rana et al. 2011). Conversely, Harvey and colleagues showed that, although Rumi modifies each *O*-glucose site on *Drosophila* Notch to high stoichiometries, only some sites were modified with *O*-glucose disaccharide, most notably in the region from EGF15 to EGF20 [Fig. 2C, (Harvey et al. 2016)]. Additionally, while *O*-glucose trisaccharide was previously observed on EGF16 and EGF18 of *Drosophila* Notch (Lee et al. 2013), semi-quantitative mass spectral analyses reveal the trisaccharide glycoform as only a minor modification at those sites [Fig. 2C, (Harvey et al. 2016)]. Based on these data, Rumi/POGLUT1 appears to modify sites of *O*-glucosylation with high efficiency but elongation with xylose may be site and species specific.

In addition to these previously studied modifications, a novel *O*-linked hexose modification at the serine residue between conserved cysteines 3 and 4 was recently identified on EGF11 from human NOTCH1 (Andrawes et al. 2013). The same modification on EGF11 was separately observed in the co-crystal structure of NOTCH1–DLL4, potentially mediating direct binding interactions and was tentatively identified as an *O*-glucose (Luca et al. 2015). This modification was also observed in the co-crystal structure of NOTCH1-JAG1 [Fig. 1A, (Luca et al. 2017)]. Interestingly, there are two other mammalian glycosyltransferases named KDELC1 and KDELC2 that share homology to Rumi/POGLUT1. Although KDELC1 and KDELC2 do not add glucose to an EGF repeat from human factor VII containing a classic Rumi/POGLUT1 site (Takeuchi et al. 2011), their ability to modify EGF11 of NOTCH1 has not yet been confirmed. Therefore, this novel modification could potentially prove to be another regulator of Notch activity.

2.3 *O*-GlcNAcylation

Many nuclear and cytoplasmic proteins are modified with an *O*-GlcNAc modification catalyzed by *O*-GlcNAc transferase (OGT) (Torres and Hart 1984; Kreppel et al. 1997). However, in 2008, *Drosophila* Notch was shown to be *O*-GlcNAcylated not by OGT, but by a distinct extracellular EGF-domain specific *O*-GlcNAc Transferase (EOGT) that localizes to the ER and functions independently of the intracellular OGT (Matsuura et al. 2008; Sakaidani et al. 2011; Sakaidani et al. 2012). EOGT adds a GlcNAc to a serine or threonine in a putative consensus sequence between the fifth and sixth conserved cysteine of an EGF repeat: $C^5xxGx(\underline{T/S})GxxC^6$ [Fig. 1A and B, (Alfaro et al. 2012; Harvey et al. 2016)]. The first site of *O*-GlcNAcylation established on an EGF repeat was on EGF20 of *Drosophila* Notch, a site that corroborates the current consensus sequence (Matsuura et al. 2008). However, of the eighteen predicted *O*-GlcNAc sites on *Drosophila* Notch ECD, EGF20 was one of only five sites found to be modified with *O*-GlcNAc in recent mass spectral analyses [Fig. 2D, (Harvey et al. 2016)]. Although sequence comparisons between modified and unmodified sites do not reveal significant differences, we speculate that the current consensus sequence for *O*-GlcNAcylation is not yet precisely defined. Additionally, further studies may reveal that increased EOGT expression leads to more extensive *O*-GlcNAcylation, although the five sites presently observed to be modified are potentially those most efficiently modified by EOGT.

3 Biological Significance of *O*-Glycosylation on Notch Activity

3.1 *O*-Fucosylation and *O*-Glucosylation Are Essential for Optimal Notch Activity

POFUT1 is ubiquitously expressed in adult tissue and deletion of *Ofut1/Pofut1* in flies or mice results in severe embryonic defects (Okajima and Irvine 2002; Shi and Stanley 2003). In the *Pofut1* knockout mice, or in flies with RNAi-mediated knock down of *Ofut1*, the phenotype is similar to those seen with inactive Notch signaling (Okajima and Irvine 2002; Shi and Stanley 2003). *Pofut1*$^{-/-}$ mice are severely defective in somitogenesis, vasculogenesis, cardiogenesis and neurogenesis (Shi and Stanley 2003). *Notch*-like phenotypes observed in *Ofut1* knockdown flies include loss of tissue from the wing margin, leg segment fusions, thickened wing veins and inappropriate bristle formation (Okajima and Irvine 2002). However, since Notch is not the sole substrate for Ofut1/POFUT1 (Vasudevan and Haltiwanger 2014), the lethal phenotypes associated with the *Ofut1/Pofut1* knockouts serve to emphasize the more global essential roles of *O*-fucosylation in embryonic development. Ofut1 has chaperone activity on Notch and is required for proper trafficking of Notch from the ER to the cell surface (Okajima et al. 2005; Matsumoto et al. 2016). POFUT1 also has effects on Notch trafficking (Okamura and Saga 2008; Yao et al. 2011), although only in certain contexts (Stahl et al. 2008). Therefore, it is not entirely clear if POFUT1 exhibits chaperone activity in mammals as it does in flies. Nevertheless, together these studies establish that *O*-fucosylation by POFUT1 is required for Notch activity.

O-Glucosylation is also essential for Notch activity (Acar et al. 2008). *Rumi* mutants in flies show *Notch*-like phenotypes affecting microchaete, eye and leg development in a temperature-sensitive manner (Acar et al. 2008). *Rumi* mutants exhibit these phenotypes when raised at 25 °C but not at 18 °C, although *rumi* mutants are sensitive to the loss of one copy of *Notch* at 18 °C (Acar et al. 2008). *Poglut1* knockout mice are embryonic lethal and also display *Notch*-like phenotypes, such as defects in somitogenesis and cardiogenesis (Fernandez-Valdivia et al. 2011). Interestingly, the defects contributing to the lethality observed in *Poglut1* mutants were more severe than those of *Notch* mutants, indicating that other targets also require *O*-glucosylation by POGLUT1 for development. For example, recent studies suggest that *O*-glucosylation of CRUMBS2 is essential for its function, and mutants in *Crumbs2* also result in early embryonic lethality (Ramkumar et al. 2015).

3.2 Extension of *O*-Fucose or *O*-Glucose Beyond the Monosaccharide Modulates NOTCH Function

Like *Notch,* the *fringe* gene was first characterized in *Drosophila* and is required for normal wing, eye and leg development (Irvine and Wieschaus 1994; Panin et al. 1997). The three mammalian *fringe* homolog genes, *Lfng*, *Mfng* and *Rfng*, are expressed in specific developmentally regulated patterns in mice (Johnston et al. 1997). To examine biological functions of the three mammalian Fringes, individual and combined knockout mice have been generated (Evrard et al. 1998; Zhang et al. 2002; Moran et al. 2009; Svensson et al. 2009; Song et al. 2016). Homozygous *Lfng* mutants have reduced viability at birth and before weaning; however, surviving mice have skeletal abnormalities and display truncated tails with shortened body axes due to severe somitogenesis defects (Evrard et al. 1998). Subsequent studies have demonstrated that *Lfng* plays an important role in the "somitogenesis clock", regulating the timing of Notch activation during somitogenesis (Wahi et al. 2016). Elimination of either *Mfng* or *Rfng* show no defects in embryonic development (Zhang et al. 2002; Moran et al. 2009) and no additional embryonic effects were observed in mice lacking all three *Fringe* genes not seen in *Lfng* null mice (Zhang et al. 2002; Moran et al. 2009; Svensson et al. 2009). However, mice lacking all three *Fringe* genes have furthered our understanding of Notch modulation by Fringe in various other aspects of development, most recently demonstrating Fringe function in T and B cell development (Song et al. 2016). LFNG has been implicated as a regulator of Notch activity during angiogenesis (Benedito et al. 2009) and kidney development (Liu et al. 2013). MFNG and RFNG have also been implicated in a variety of Notch-dependent processes, such as bile duct remodeling and heart development (Ryan et al. 2008; D'Amato et al. 2016). Lastly, although mice with mutations in β4GalT-1, the galactosyltransferase that generates the *O*-fucose trisaccharide, did not initially show any *Notch* phenotype (Asano et al. 1997; Lu et al. 1997), a later study specifically inspecting for embryonic deficiencies revealed subtle somitogenic and skeletal defects in mouse embryos lacking β4GalT-1 (Chen et al. 2006).

In contrast to *O*-fucose elongation, the roles of the di- and trisaccharide forms of the *O*-glucose observed on Notch are less understood (Rana et al. 2011; Lee et al. 2013). Interestingly, *shams* mutants show gain-of-function of certain *Notch*-like phenotypes (wing vein and bristle formation) and overexpression of human glucoside xylosyltransferase, GXYLT1, inhibited Notch signaling in flies (Lee et al. 2013). Lee and colleagues also found that while mutations of individual *O*-glucose sites did not affect Notch cell surface expression in the larval wing, they did increase Notch surface expression in the pupal wing and attributed the cause to a difference in expression of Shams between those stages of development (Lee et al. 2013). Another study found that Rumi/POGLUT1 negatively regulates JAG1-induced Notch signaling in the liver and proposed it as a unique setting for inhibition of Notch activity mediated by xylosylation in the development of a specific organ system (Thakurdas et al. 2016). Overall, xylosylation of the *O*-glucose appears to serve as another modulator of Notch activity, especially considering that *O*-glucose promotes Notch activity while subsequent xylosylation inhibits it (Lee et al. 2013).

3.3 Diseases Caused by Mutations in Glycosyltransferases that Modify Notch

Because the Notch signaling pathway controls several developmental processes and the maintenance of various tissues, there are many diseases that are caused by inappropriate Notch signaling. Alagille Syndrome and CADASIL are diseases attributed to mutations in the *NOTCH2* and *NOTCH3* genes, respectively (Joutel et al. 1996; McDaniell et al. 2006; Kamath et al. 2012). Additionally, the roles of the Notch signaling pathway in various cancers have been well studied, and were recently summarized in an in-depth review (Aster et al. 2016). However, there are other diseases that are caused by defects in the glycosyltransferases that modify Notch. Dowling-Degos disease (DDD) is a rare autosomal-dominant disease that presents with hyper- and hypopigmentation in flexure areas, as well as erythematous macules and papules on the neck, chest and abdomen. Originally, loss of function mutations in *KERATIN5* were found to cause DDD, but *KERATIN5* mutations were only observed in fewer than 50% of patients. Since 2013, mutations in one copy of *POFUT1* or *POGLUT1* (heterozygous) have been identified in DDD patients, and new mutations continue to be reported (Li et al. 2013; Basmanav et al. 2014; Chen et al. 2014; Li et al. 2016). *POFUT1* itself has also been shown to be upregulated in colorectal cancer (Loo et al. 2013), oral squamous cell carcinomas (Yokota et al. 2013), glioblastomas (Kroes et al. 2007), more aggressive hepatocellular carcinomas (Sawey et al. 2011; Ma et al. 2016) and gastric cancers (Dong et al. 2017). Mutations in *POGLUT1* and *XXYLT1* that specifically alter enzyme activity have been associated with various cancers (Yu et al. 2015; Yu et al. 2016). Additionally, it has been suggested that *POGLUT1* overexpression contributes to the pathogenesis of acute myelogenous leukemia and

T cell acute lymphoblastic leukemia (Wang et al. 2010). More recently, a family with an autosomal recessive limb-girdle muscular dystrophy was found to harbor a homozygous missense mutation in *POGLUT1* (Servian-Morilla et al. 2016). This mutation greatly reduces (but does not eliminate) the *O*-glucosyltransferase activity of the enzyme, reducing Notch activity and negatively affecting satellite cell renewal. Such studies reveal new biological roles for these glycosyltransferases and more are likely to be discovered.

LFNG, MFNG and RFNG have also been found to cause disease when their expression levels or activities are altered. Spondylocostal Dystostosis is characterized by congenital vertebral segmentation and rib defects. During embryogenesis in vertebrates, skeletal muscle, spinal vertebrae and ribs are formed from somites, produced during somitogenesis, a developmental process that is largely controlled by the Notch signaling pathway (Weinmaster and Kintner 2003). Although other proteins within the Notch signaling pathway are mutated in cases of the Spondylocostal Dysostosis, particularly severe cases are caused by a homozygous missense mutation in *LFNG* (Sparrow et al. 2006). A deficiency in *LFNG* and higher expression of *MFNG* have been shown to induce basal-like and claudin-low breast cancers (Xu et al. 2012; Zhang et al. 2015). *LFNG* also plays a tumor suppressive role in both prostate and pancreatic cancers (Zhang et al. 2014; Zhang et al. 2016).

The role of *O*-GlcNAcylation on Notch is not entirely clear, since *Eogt* mutants do not show any significant *Notch* phenotypes in flies, but do show phenotypes associated with the protein Dumpy, an EOGT target protein containing 306 EGF repeats (Sakaidani et al. 2011; Muller et al. 2013). However, wing-blistering phenotypes observed in *Eogt* mutants are suppressed upon removal or mutation of Notch signaling pathway members (Muller et al. 2013). Interestingly, mutations in *EOGT* in humans have been recently associated with Adams-Oliver syndrome (AOS), which is a rare congenital disorder typically characterized by aplasia cutis congenita of the scalp, as well as terminal limb defects (Shaheen et al. 2013; Cohen et al. 2014). *EOGT* mutations associated with AOS were tested in HEK293T cells, yielding decreased EOGT expression, altered cellular localization and impaired enzymatic activity (Ogawa et al. 2015). Further analyses of AOS patients have uncovered additional mutations in members of the Notch signaling pathway that may contribute to the disease (Stittrich et al. 2014; Meester et al. 2015). These studies implicate EOGT as a modulator of Notch signaling in the pathogenesis of AOS. Additionally, a very recent study confirms the importance of EOGT for vascular development in mice and also provides new insight to the role of *O*-GlcNAc and EOGT on Notch activation and ligand binding to DLL1 and DLL4 (Sawaguchi et al. 2017).

4 Molecular Mechanisms for the Effects of *O*-Glycosylation on Notch

4.1 *O*-Fucosylation of the Ligand-Binding Domain Is Essential for Optimal Notch-Ligand Binding

Elimination of *Ofut1* in flies or *Pofut1* in mice inhibits Notch activity in all contexts examined (Okajima and Irvine 2002; Shi and Stanley 2003), suggesting that *O*-fucosylation of Notch is essential for its function. In the fly system, Ofut1 has a chaperone activity not linked to its fucosyltransferase activity that is required for cell-surface expression of Notch (Okajima et al. 2005; Matsumoto et al. 2016). As mentioned above, the chaperone activity of POFUT1 is less clear in the mouse, where elimination of *Pofut1* in embryonic stem cells does not affect cell-surface expression of Notch, but is required for ligand binding (Stahl et al. 2008). This is consistent with studies that showed that elimination of *O*-fucose sites in EGF repeats reduced Notch-ligand interactions or ligand-dependent Notch activation (Rampal et al. 2005; Xu et al. 2005; Shi et al. 2007). In particular, elimination of the conserved *O*-fucose site in EGF12, which is part of the classically defined ligand-binding domain [EGF11–12, (Rebay et al.

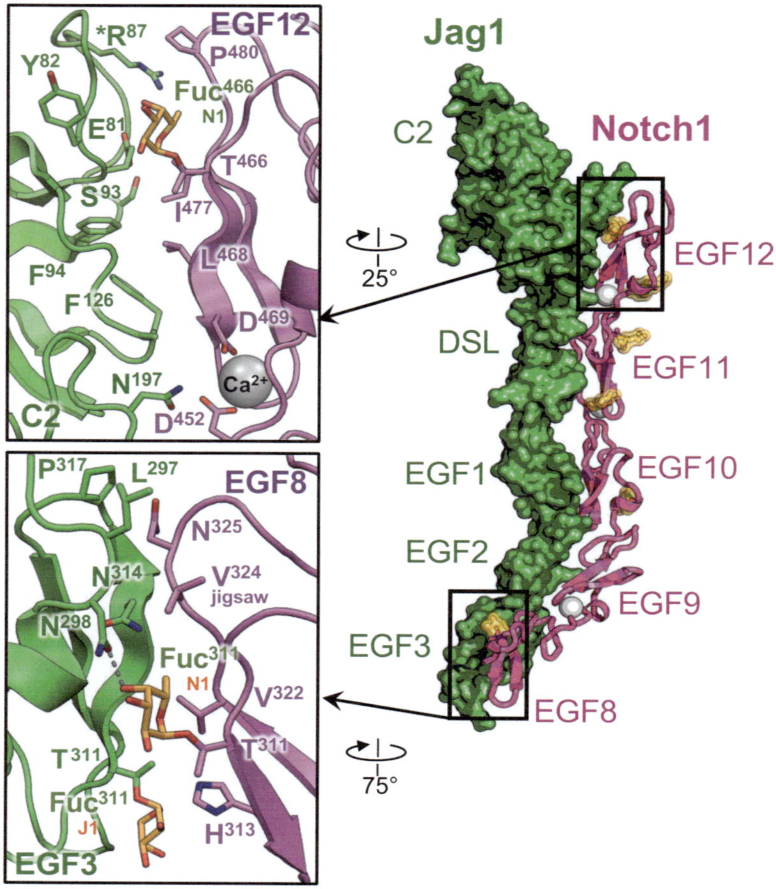

Fig. 3 *O*-Fucose glycans on NOTCH1 EGF8 and EGF12 directly interact with ligand. Structure of NOTCH1 (EGF8–12) and JAG1 (N-terminus to EGF3) are shown in magenta and green, respectively, in complex, modified with permission from (Luca et al. 2017). Selected interfaces between NOTCH1 EGF8 and EGF12 with JAG1 are shown in zoomed windows. The *O*-fucose modifying T466 on NOTCH1 EGF12 forms a hydrogen bond with the backbone carbonyl of Y82 in the JAG1 C2 domain. In addition, the *O*-fucose on T311 of NOTCH1 EGF8 forms a hydrogen bond with the side chain of JAG1 N298. Calcium ions are represented as gray spheres. *O*-Glycans are highlighted in orange

1991)] reduced ligand-binding and Notch activation (Rampal et al. 2005; Shi et al. 2007). Mutation of this site in EGF12 in endogenous *Drosophila* Notch or mouse NOTCH1 also had effects (Lei et al. 2003; Ge and Stanley 2008) but less severe than those caused by elimination of either *Ofut1* or *Pofut1*, suggesting that *O*-fucose modifications on other EGF repeats must also play important roles in Notch function. Several recent studies have extended the "ligand-binding domain" to EGF8–12 of Notch (Andrawes et al. 2013; Kakuda and Haltiwanger 2017; Luca et al. 2017). In particular, the analysis of other mutated *O*-fucose sites showed that *O*-fucose on EGF8 is important for NOTCH1 function (Kakuda and Haltiwanger 2017). Also, the *O*-fucose on EGF8 and 12 on NOTCH1 have been shown to interact with JAG1 in a recent co-crystal [Fig. 3, (Luca et al. 2017)]. Thus, *O*-fucosylation of the ligand-binding domain plays an important role in Notch-ligand interactions.

4.2 Fringe Modifications "Mark" Regions of the Notch ECD to Activate or Inhibit Notch Activity

Since the establishment of the modulatory roles of Fringe-mediated *O*-fucose elongation on Notch activity in *Drosophila* (Panin et al. 1997; Moloney et al. 2000a; Shao et al. 2003), several groups have sought to uncover the specific molecular mechanisms of how Fringe enhances Notch activation by Delta but inhibits activation by Serrate. Bruckner and colleagues were the first to show that Fringe modulates Notch activity by altering its ability to bind Delta in a cell based binding assay (Bruckner et al. 2000). Bruckner and colleagues were unable to detect any binding to Serrate in their assays, although Xu and colleagues would later show that Fringe indeed enhances binding of Notch to Delta while inhibiting Notch binding to Serrate in an *in vitro* binding assay (Xu et al. 2007). Thus, in flies, Fringe appears to modulate Notch activity by enhancing Notch-Delta binding and reducing Notch-Serrate binding. In an attempt to identify specific Fringe-elongated *O*-fucose sites that affect ligand interactions, the *O*-fucose consensus sequences of specific EGF repeats in *Drosophila* Notch were mutated and analyzed for Fringe effects on ligand binding (Xu et al. 2005). No single *O*-fucose site was found to be individually responsible for mediating the Fringe effect on ligand binding. The authors concluded that Fringe modification of multiple sites was responsible for the ligand binding effects that Fringe has on Notch.

Similar experiments have been performed with mammalian Notch proteins, although the mammalian system is more complicated with four Notch receptors and three Fringe proteins. While experiments consistently show that Notch modifications by any of the Fringes enhance binding to and activation by DLL ligands (Hicks et al. 2000; Yang et al. 2005; Hou et al. 2012; LeBon et al. 2014; Kakuda and Haltiwanger 2017), inconsistent effects of Fringe proteins on binding and activation from JAG ligands have been reported. One possible cause for these inconsistent results is that many of the cells used in these experiments express multiple Notch receptors [e.g. CHO cells express all four (Hou et al. 2012)]. Since each Fringe has distinct effects on different Notch receptors (Hicks et al. 2000; Yang et al. 2005), it could be difficult to determine which Notch receptor is being affected. Studies focusing on the effects of individual Fringes on a single Notch receptor (NOTCH1) yield more consistent results (Yang et al. 2005; LeBon et al. 2014; Kakuda and Haltiwanger 2017). In these reports, LFNG, MFNG and RFNG activate NOTCH1 from DLL1 by enhancing DLL1-NOTCH1 binding. The *O*-fucose modifications on EGF8 and 12 of NOTCH1 are required for this effect, indicating these are the most critical *O*-fucose sites responsible for Fringe-mediated enhancement of DLL1-NOTCH1 binding (Kakuda and Haltiwanger 2017). Interestingly, LFNG and MFNG inhibit NOTCH1 activation from JAG1, while RFNG enhances NOTCH1 activation from JAG1. In addition, contrary to what has been observed in *Drosophila* studies, all three Fringe proteins enhance JAG1-NOTCH1 binding (Yang et al. 2005; Taylor et al. 2014; Kakuda and Haltiwanger 2017). This suggests that inhibition of JAG1-NOTCH1 activation by LFNG and MFNG is not caused by reduced binding, but by some events downstream of ligand-binding. Elimination of the *O*-fucose residues on EGF6 and 36 abrogated the ability of either LFNG or MFNG to inhibit NOTCH1 activation by JAG1, suggesting these are the sites contributing most to this effect on NOTCH1-JAG1 activation (Kakuda and Haltiwanger 2017). Also, elimination of the *O*-fucose sites on EGF6 and 36 had no effect on NOTCH1-JAG1 binding, consistent with the fact that they are not part of the ligand-binding domain. Thus, EGF6 and 36 affect Notch activation through a mechanism other than ligand binding. The main conclusions from these studies are summarized in Fig. 2A. Ultimately, these results show that all three Fringe proteins "mark" *O*-fucose residues on EGF8 and 12 of NOTCH1 as those that enhance binding to and activation by DLL1. In addition, LFNG and MFNG, but not RFNG, "mark" the *O*-fucose residues on EGF6 and 36 to inhibit NOTCH1 activation from JAG1, although the mechanism by which this occurs is currently unknown.

In mammals, the *O*-fucose disaccharide can be further elongated to a tetrasaccharide (Moloney et al. 2000b) and subsequent studies have explored the function of the *O*-fucose glycan when elongated past the disaccharide. Chen and colleagues demonstrated that the galactose present in the *O*-fucose trisaccharide is required for LFNG or MFNG to inhibit Notch signaling from JAG1 (Chen et al. 2001). The *O*-fucose on EGF6 is elongated to the trisaccharide, indicating that the addition of galactose to this site is important for the effects of LFNG and MFNG to inhibit JAG1-NOTCH1 activation (Kakuda and

Haltiwanger 2017). Interestingly, Hou and colleagues later showed that although both LFNG and MFNG increase Notch signaling from DLL1, the *O*-fucose trisaccharide is required for the LFNG effect, but inhibits the MFNG effect (Hou et al. 2012). However, elongation of the *O*-fucose past the disaccharide glycoform on EGF12 of an EGF11–13 fragment had no additional effect on DLL1 binding (Taylor et al. 2014). Although the addition of the galactose does not appear to be important ligand binding on EGF12, elongation of other sites to the trisaccharide glycoform may mediate these differential effects of LFNG and MFNG.

4.3 *O*-Glucosylation Affects Notch Protein Stability and Trafficking

Although *O*-glucosylation is required for proper Notch activity (Acar et al. 2008; Fernandez-Valdivia et al. 2011), the molecular mechanisms by which *O*-glucosylation affects Notch function continue to be explored. Interestingly, *rumi* mutants in *Drosophila* are temperature sensitive for *Notch* phenotypes, regardless of whether the mutant is a simple point mutant or a complete loss of coding sequence (Acar et al. 2008). This suggests that the loss of *O*-glucose from Notch potentially destabilizes the receptor, causing a loss of function at higher temperatures. Loss of Notch-ligand binding was not observed in *rumi* mutants, though mutants did show localization defects at non-permissive temperatures (Acar et al. 2008). This result complements several other studies that have indicated a role for *O*-glucosylation in the proper folding and export of surface expressed proteins through the secretory pathway. Eyes shut, another Rumi/POGLUT1 target, showed intracellular accumulation in *rumi* mutants, affecting rhabdomere separation in the developing eye in flies (Haltom et al. 2014). Similarly, mouse CRUMBS2, another Rumi/POGLUT1 target protein, was shown to accumulate within the ER in *Poglut1* mutant mice, thus inhibiting normal gastrulation (Ramkumar et al. 2015). Although these data suggest that *O*-glucosylation by Rumi/POGLUT1 may affect the trafficking of Notch, it was found that the accumulation of Notch in *rumi* mutants was not due to ER entrapment and that the levels of Notch on the surface of the larval *rumi* mutant cells were still comparable to those of wild-type levels (Acar et al. 2008). Interestingly, a later study found that *O*-glucosylation plays a critical role in Notch trafficking, specifically in the export of Notch from the ER, but noted that loss of *O*-glucosylation could be compensated for by *O*-fucosylation on Notch (Matsumoto et al. 2016). Therefore, Matsumoto and colleagues concluded that while *O*-glucose is important for Notch trafficking and expression, its role may be masked by *O*-fucosylation redundancy. Another hint as to the role of *O*-glucose on Notch was shown by the expression of the Notch receptor with a deletion of the LNR repeats from the Notch ECD, causing ligand-independent S2 cleavage and activation (Leonardi et al. 2011). The gain-of-function phenotypes seen with this deletion were fully suppressed after the loss of *rumi*, suggesting that *O*-glucosylation of the Notch is a prerequisite for S2 cleavage at high temperatures. While this certainly suggests a unique mechanism for modulation of Notch activity, the precise role of *O*-glucose on Notch is still unclear.

Similar to studies on *O*-fucose, *O*-glucosylation sites have been mutated and evaluated for effects on Notch function in both flies and mice. Studies examining *O*-glucose site mutants in groups of different EGF repeats in *Drosophila* Notch showed that modification of multiple EGF repeats with *O*-glucose (in particular EGF10–20) serve to buffer against the temperature sensitive phenotypes observed in *rumi* mutant flies (Leonardi et al. 2011). As with the elimination of *rumi*, mutation of the *O*-glucose sites did not affect the ability of Notch to bind to ligands (Acar et al. 2008; Leonardi et al. 2011). The mutation of individual *O*-glucose sites on mouse NOTCH1 revealed that *O*-glucosylation of EGF28 is required for efficient activation by DLL1 but not by JAG1 (Rana et al. 2011). Together, these results indicate that while there may be some functional redundancy between

O-fucose and *O*-glucose for the folding of Notch, the function of *O*-glucosylation on Notch is distinct from that of *O*-fucosylation.

4.4 Structural Studies Provide Functional Significance to *O*-Glycosylation on Notch

Structural studies have the potential to offer significant insight into Notch receptor function. However, one of the greatest limitations impeding structural analyses of the Notch receptor is simply its large size. Therefore, small fragments or individual protein domains of the Notch receptor have been used for various structural studies. Since the role of glycosylation on Notch has remained poorly understood, several groups have investigated whether glycosylation affects the protein structure of Notch (Hiruma-Shimizu et al. 2010; Taylor et al. 2014; Luca et al. 2015; Hayakawa et al. 2016). Taylor and colleagues found that the backbone structure of EGF11–13 from human NOTCH1 remained unchanged between the unglycosylated, *O*-fucosylated or Fringe-elongated glycoforms, despite intramolecular contacts made between the sugars and amino acid residues of EGF12 (Taylor et al. 2014). In attempt to better understand the interactions between Notch receptors and their ligands, structures of the individual ligands have been solved (Cordle et al. 2008; Chillakuri et al. 2013; Kershaw et al. 2015). However, Luca and colleagues were the first to obtain the structure of NOTCH1 in complex with a ligand (Luca et al. 2015). Interestingly, Luca and colleagues accomplished this only after performing multiple rounds of ligand affinity maturation to optimize NOTCH1-DLL4 stability for their structural studies. The NOTCH1-DLL4 complex was collinear in an anti-parallel orientation, and they too observed that the calcium ions rigidified the binding platform, although calcium did not directly participate in the NOTCH1-DLL4 interface. They also found that the *O*-fucose on EGF12 was embedded directly in the NOTCH1-DLL4 interface, enhancing NOTCH1-DLL4 interactions. Although the *O*-glucose modifications mediated by POGLUT1 on NOTCH1 had no direct contacts with DLL4, they shielded hydrophobic residues on NOTCH1. More recently, Luca and colleagues have used the same methodologies to generate a co-crystal between NOTCH1 and JAG1 [Fig. 3, (Luca et al. 2017)]. This structure revealed multiple interactions between each of the NOTCH1 EGF repeats in an anti-parallel orientation with JAG1. Of particular note, the *O*-fucose residues on EGF8 and EGF12 were in direct contact with JAG1. As mentioned above, elimination of these sites reduces binding to and activation by both DLL1 and JAG1 (Kakuda and Haltiwanger 2017; Luca et al. 2017). These results provide explicit evidence for the roles of *O*-fucose modification of the newly defined ligand-binding domain (EGF8–12) on NOTCH1 in mediating ligand interactions.

5 Development of Small Molecules to Alter Notch *O*-Glycosylation and Function

Since Notch can function as either an oncogene or a tumor suppressor depending on context (Aster et al. 2016), the identification of small molecules that can increase or decrease Notch activity may be useful for therapeutics. Since *O*-glycosylation is not only essential for Notch function but also modulates Notch activity, molecules that target the glycosyltransferases that modify Notch have the potential to be very effective. Thus, inhibitors of POFUT1 or POGLUT1 could potentially be Notch inhibitors, while inhibitors of the xylosyltransferases (GXYLT1/2 or XXYLT1) could function as Notch activators. Similarly, Fringe inhibitors could increase or decrease Notch activity depending on the dominant ligand in a given system. The recent determination of structures for POGLUT1 (Yu et al. 2016), POFUT1 (Li et al. 2017), and XXYLT1 (Yu et al. 2015) will greatly aid the development of inhibitors for these enzymes. In addition, a recent report suggests that fucose analogs (chemically modified fucose molecules) can be utilized as substrates by POFUT1 and incorporated onto

Notch EGF repeats in cells to inhibit Notch binding of DLL ligands but not JAG ligands (Schneider et al. 2018). These sugar analogs provide a novel means of inhibiting Notch activity in a ligand-specific manner. Ultimately, these analogs can used as tools for better understanding how *O*-glycans affect Notch activity or themselves be developed into novel therapeutics for Notch-related diseases.

6 Conclusions

Despite significant advancements to the field of Notch signaling, complicated questions regarding ligand binding and Notch activation still remain unanswered. Due to the complexity of the signaling pathway (Hori et al. 2013), the size of the receptor and the influence of multiple types of glycosylation, the scope of individual studies is often limited to a small aspect of the actual mechanisms occurring *in vivo*. Nevertheless, such studies provide a necessary foundation for future advancements. Recent site mapping studies have revealed the site occupancy of *O*-glucosylation, *O*-GlcNAcylation, *O*-fucosylation and Fringe elongation, although additional studies are required to determine how and why certain sites are modified while others are not (Shao et al. 2003; Rampal et al. 2005; Takeuchi et al. 2012; Harvey et al. 2016; Kakuda and Haltiwanger 2017). In mapping the sites of *O*-fucosylation on Notch, several *O*-fucose sites were found to be efficiently elongated by Fringe but did not have great effects on Notch activity (Kakuda and Haltiwanger 2017). This raises questions regarding the effects of site-specificity in relation to other mechanisms of Notch activity. Harvey and colleagues also found sites of Fringe elongation adjacent to conserved non-calcium binding EGF repeats located outside the conventional ligand-binding domain, possibly serving to alter the structure of those EGF repeats (Harvey et al. 2016). Many publications have speculated on the global conformation of the Notch ECD, but additional studies are needed to fully clarify the effects of glycosylation on Notch structure and function. Lastly, there exist seemingly unknown events affected by Fringe elongation of *O*-fucose and *O*-glucosylation that couple ligand binding to Notch activation and are yet to be explained. Whatever the mechanisms, they appear to differ from those established in the *Drosophila* Notch signaling pathway and pose interesting evolutionary questions.

Despite the tremendous amount of progress since the *Notch* gene was first sequenced, certain aspects of the Notch signaling pathway, though extensively studied to date, are yet to be fully understood. *O*-Glycosylation is an essential modification for Notch activation and regulation. The effects of glycosylation on Notch exemplify the direct impacts of glycosylation on protein function and emphasize that glycosylation should not be overlooked. The ability to manipulate these glycans to modulate Notch activity may provide a novel class of research tools or potential therapeutics for Notch-related diseases.

Acknowledgments We would like to thank the members of the Haltiwanger laboratory for helpful comments and discussions. Original work was supported by NIH grant GM061126.

References

Acar M, Jafar-Nejad H, Takeuchi H, Rajan A, Ibrani D, Rana NA, Pan H, Haltiwanger RS, Bellen HJ (2008) Rumi is a CAP10 domain glycosyltransferase that modifies Notch and is required for Notch signaling. Cell 132(2):247–258. https://doi.org/10.1016/j.cell.2007.12.016

Alfaro JF, Gong CX, Monroe ME, Aldrich JT, Clauss TR, Purvine SO, Wang Z, Camp DG 2nd, Shabanowitz J, Stanley P, Hart GW, Hunt DF, Yang F, Smith RD (2012) Tandem mass spectrometry identifies many mouse brain O-GlcNAcylated proteins including EGF domain-specific O-GlcNAc transferase targets. Proc Natl Acad Sci U S A 109(19):7280–7285. https://doi.org/10.1073/pnas.1200425109

Andrawes MB, Xu X, Liu H, Ficarro SB, Marto JA, Aster JC, Blacklow SC (2013) Intrinsic selectivity of Notch 1 for Delta-like 4 over Delta-like 1. J Biol Chem 288(35):25477–25489. https://doi.org/10.1074/jbc.M113.454850

Asano M, Furukawa K, Kido M, Matsumoto S, Umesaki Y, Kochibe N, Iwakura Y (1997) Growth retardation and early death of beta-1,4-galactosyltransferase knockout mice with augmented proliferation and abnormal differentiation of epithelial cells. EMBO J 16(8):1850–1857. https://doi.org/10.1093/emboj/16.8.1850

Aster JC, Pear WS, Blacklow SC (2016) The varied roles of Notch in Cancer. Annu Rev Pathol. https://doi.org/10.1146/annurev-pathol-052016-100127

Basmanav FB, Oprisoreanu AM, Pasternack SM, Thiele H, Fritz G, Wenzel J, Grosser L, Wehner M, Wolf S, Fagerberg C, Bygum A, Altmuller J, Rutten A, Parmentier L, El Shabrawi-Caelen L, Hafner C, Nurnberg P, Kruse R, Schoch S, Hanneken S, Betz RC (2014) Mutations in POGLUT1, encoding protein O-glucosyltransferase 1, cause autosomal-dominant Dowling-Degos disease. Am J Hum Genet 94(1):135–143. https://doi.org/10.1016/j.ajhg.2013.12.003

Benedito R, Roca C, Sorensen I, Adams S, Gossler A, Fruttiger M, Adams RH (2009) The notch ligands Dll4 and Jagged1 have opposing effects on angiogenesis. Cell 137(6):1124–1135. https://doi.org/10.1016/j.cell.2009.03.025

Bray SJ (2006) Notch signalling: a simple pathway becomes complex. Nat Rev Mol Cell Biol 7(9):678–689. https://doi.org/10.1038/nrm2009

Bruckner K, Perez L, Clausen H, Cohen S (2000) Glycosyltransferase activity of Fringe modulates Notch-Delta interactions. Nature 406(6794):411–415. https://doi.org/10.1038/35019075

Chen J, Moloney DJ, Stanley P (2001) Fringe modulation of Jagged1-induced Notch signaling requires the action of beta 4galactosyltransferase-1. Proc Natl Acad Sci U S A 98(24):13716–13721. https://doi.org/10.1073/pnas.241398098

Chen J, Lu L, Shi S, Stanley P (2006) Expression of Notch signaling pathway genes in mouse embryos lacking beta4galactosyltransferase-1. Gene Expr Patterns 6(4):376–382. https://doi.org/10.1016/j.modgep.2005.09.009

Chen M, Li Y, Liu H, Fu X, Yu Y, Yu G, Wang C, Bao F, Liany H, Wang Z, Shi Z, Zhang D, Zhou G, Liu J, Zhang F (2014) Analysis of POFUT1 gene mutation in a Chinese family with Dowling-Degos disease. PLoS One 9(8):e104496. https://doi.org/10.1371/journal.pone.0104496

Chillakuri CR, Sheppard D, Lea SM, Handford PA (2012) Notch receptor-ligand binding and activation: insights from molecular studies. Semin Cell Dev Biol 23(4):421–428. https://doi.org/10.1016/j.semcdb.2012.01.009

Chillakuri CR, Sheppard D, Ilagan MX, Holt LR, Abbott F, Liang S, Kopan R, Handford PA, Lea SM (2013) Structural analysis uncovers lipid-binding properties of Notch ligands. Cell Rep 5(4):861–867. https://doi.org/10.1016/j.celrep.2013.10.029

Cohen I, Silberstein E, Perez Y, Landau D, Elbedour K, Langer Y, Kadir R, Volodarsky M, Sivan S, Narkis G, Birk OS (2014) Autosomal recessive Adams-Oliver syndrome caused by homozygous mutation in EOGT, encoding an EGF domain-specific O-GlcNAc transferase. Eur J Hum Genet 22(3):374–378. https://doi.org/10.1038/ejhg.2013.159

Cordle J, Johnson S, Tay JZ, Roversi P, Wilkin MB, de Madrid BH, Shimizu H, Jensen S, Whiteman P, Jin B, Redfield C, Baron M, Lea SM, Handford PA (2008) A conserved face of the Jagged/Serrate DSL domain is involved in Notch trans-activation and cis-inhibition. Nat Struct Mol Biol 15(8):849–857. https://doi.org/10.1038/nsmb.1457

D'Amato G, Luxan G, del Monte-Nieto G, Martinez-Poveda B, Torroja C, Walter W, Bochter MS, Benedito R, Cole S, Martinez F, Hadjantonakis AK, Uemura A, Jimenez-Borreguero LJ, de la Pompa JL (2016) Sequential Notch activation regulates ventricular chamber development. Nat Cell Biol 18(1):7–20. https://doi.org/10.1038/ncb3280

Dinchuk JE, Focht RJ, Kelley JA, Henderson NL, Zolotarjova NI, Wynn R, Neff NT, Link J, Huber RM, Burn TC, Rupar MJ, Cunningham MR, Selling BH, Ma J, Stern AA, Hollis GF, Stein RB, Friedman PA (2002) Absence of post-translational aspartyl beta-hydroxylation of epidermal growth factor domains in mice leads to developmental defects and an increased incidence of intestinal neoplasia. J Biol Chem 277(15):12970–12977. https://doi.org/10.1074/jbc.M110389200

Dong S, Wang Z, Huang B, Zhang J, Ge Y, Fan Q, Wang Z (2017) Bioinformatics insight into glycosyltransferase gene expression in gastric cancer: POFUT1 is a potential biomarker. Biochem Biophys Res Commun 483(1):171–177. https://doi.org/10.1016/j.bbrc.2016.12.172

Evrard YA, Lun Y, Aulehla A, Gan L, Johnson RL (1998) Lunatic fringe is an essential mediator of somite segmentation and patterning. Nature 394(6691):377–381. https://doi.org/10.1038/28632

Fehon RG, Kooh PJ, Rebay I, Regan CL, Xu T, Muskavitch MA, Artavanis-Tsakonas S (1990) Molecular interactions between the protein products of the neurogenic loci Notch and Delta, two EGF-homologous genes in Drosophila. Cell 61(3):523–534

Fernandez-Valdivia R, Takeuchi H, Samarghandi A, Lopez M, Leonardi J, Haltiwanger RS, Jafar-Nejad H (2011) Regulation of mammalian Notch signaling and embryonic development by the protein O-glucosyltransferase Rumi. Development 138(10):1925–1934. https://doi.org/10.1242/dev.060020

Ge C, Stanley P (2008) The O-fucose glycan in the ligand-binding domain of Notch1 regulates embryogenesis and T cell development. Proc Natl Acad Sci U S A 105(5):1539–1544. https://doi.org/10.1073/pnas.0702846105

Haltom AR, Lee TV, Harvey BM, Leonardi J, Chen YJ, Hong Y, Haltiwanger RS, Jafar-Nejad H (2014) The protein O-glucosyltransferase Rumi modifies eyes shut to promote rhabdomere separation in Drosophila. PLoS Genet 10(11):e1004795. https://doi.org/10.1371/journal.pgen.1004795

Hambleton S, Valeyev NV, Muranyi A, Knott V, Werner JM, McMichael AJ, Handford PA, Downing AK (2004) Structural and functional properties of the human notch-1 ligand binding region. Structure 12(12):2173–2183. https://doi.org/10.1016/j.str.2004.09.012

Harris RJ, Spellman MW (1993) O-linked fucose and other post-translational modifications unique to EGF modules. Glycobiology 3(3):219–224

Harvey BM, Rana NA, Moss H, Leonardi J, Jafar-Nejad H, Haltiwanger RS (2016) Mapping sites of O-glycosylation and fringe elongation on Drosophila Notch. J Biol Chem 291(31):16348–16360. https://doi.org/10.1074/jbc.M116.732537

Hase S, Kawabata S, Nishimura H, Takeya H, Sueyoshi T, Miyata T, Iwanaga S, Takao T, Shimonishi Y, Ikenaka T (1988) A new trisaccharide sugar chain linked to a serine residue in bovine blood coagulation factors VII and IX. J Biochem 104(6):867–868

Hase S, Nishimura H, Kawabata S, Iwanaga S, Ikenaka T (1990) The structure of (xylose)2glucose-O-serine 53 found in the first epidermal growth factor-like domain of bovine blood clotting factor IX. J Biol Chem 265(4):1858–1861

Hayakawa S, Koide R, Hinou H, Nishimura S (2016) Synthetic human NOTCH1 EGF modules unraveled molecular mechanisms for the structural and functional roles of calcium ions and O-Glycans in the ligand-binding region. Biochemistry 55(5):776–787. https://doi.org/10.1021/acs.biochem.5b01284

Hicks C, Johnston SH, diSibio G, Collazo A, Vogt TF, Weinmaster G (2000) Fringe differentially modulates Jagged1 and Delta1 signalling through Notch1 and Notch2. Nat Cell Biol 2(8):515–520. https://doi.org/10.1038/35019553

Hiruma-Shimizu K, Hosoguchi K, Liu Y, Fujitani N, Ohta T, Hinou H, Matsushita T, Shimizu H, Feizi T, Nishimura S (2010) Chemical synthesis, folding, and structural insights into O-fucosylated epidermal growth factor-like repeat 12 of mouse Notch-1 receptor. J Am Chem Soc 132(42):14857–14865. https://doi.org/10.1021/ja105216u

Hori K, Sen A, Artavanis-Tsakonas S (2013) Notch signaling at a glance. J Cell Sci 126(Pt 10):2135–2140. https://doi.org/10.1242/jcs.127308

Hou X, Tashima Y, Stanley P (2012) Galactose differentially modulates lunatic and manic fringe effects on Delta1-induced NOTCH signaling. J Biol Chem 287(1):474–483. https://doi.org/10.1074/jbc.M111.317578

Irvine KD, Wieschaus E (1994) Fringe, a boundary-specific signaling molecule, mediates interactions between dorsal and ventral cells during Drosophila wing development. Cell 79(4):595–606

Jinek M, Chen YW, Clausen H, Cohen SM, Conti E (2006) Structural insights into the Notch-modifying glycosyltransferase fringe. Nat Struct Mol Biol 13(10):945–946. https://doi.org/10.1038/nsmb1144

Johnston SH, Rauskolb C, Wilson R, Prabhakaran B, Irvine KD, Vogt TF (1997) A family of mammalian fringe genes implicated in boundary determination and the Notch pathway. Development 124(11):2245–2254

Joutel A, Corpechot C, Ducros A, Vahedi K, Chabriat H, Mouton P, Alamowitch S, Domenga V, Cecillion M, Marechal E, Maciazek J, Vayssiere C, Cruaud C, Cabanis EA, Ruchoux MM, Weissenbach J, Bach JF, Bousser MG, Tournier-Lasserve E (1996) Notch3 mutations in CADASIL, a hereditary adult-onset condition causing stroke and dementia. Nature 383(6602):707–710. https://doi.org/10.1038/383707a0

Kakuda S, Haltiwanger RS (2017) Deciphering the fringe-mediated Notch code: identification of activating and inhibiting sites allowing discrimination between ligands. Dev Cell 40(2):193–201. https://doi.org/10.1016/j.devcel.2016.12.013

Kamath BM, Bauer RC, Loomes KM, Chao G, Gerfen J, Hutchinson A, Hardikar W, Hirschfield G, Jara P, Krantz ID, Lapunzina P, Leonard L, Ling S, Ng VL, Hoang PL, Piccoli DA, Spinner NB (2012) NOTCH2 mutations in Alagille syndrome. J Med Genet 49(2):138–144. https://doi.org/10.1136/jmedgenet-2011-100544

Kentzer EJ, Buko A, Menon G, Sarin VK (1990) Carbohydrate composition and presence of a fucose-protein linkage in recombinant human pro-urokinase. Biochem Biophys Res Commun 171(1):401–406

Kershaw NJ, Church NL, Griffin MD, Luo CS, Adams TE, Burgess AW (2015) Notch ligand delta-like1: X-ray crystal structure and binding affinity. Biochem J 468(1):159–166. https://doi.org/10.1042/BJ20150010

Kopan R, Ilagan MX (2009) The canonical Notch signaling pathway: unfolding the activation mechanism. Cell 137(2):216–233. https://doi.org/10.1016/j.cell.2009.03.045

Kreppel LK, Blomberg MA, Hart GW (1997) Dynamic glycosylation of nuclear and cytosolic proteins Cloning and characterization of a unique O-GlcNAc transferase with multiple tetratricopeptide repeats. J Biol Chem 272(14):9308–9315

Kroes RA, Dawson G, Moskal JR (2007) Focused microarray analysis of glyco-gene expression in human glioblastomas. J Neurochem 103(Suppl 1):14–24. https://doi.org/10.1111/j.1471-4159.2007.04780.x

Lavaissiere L, Jia S, Nishiyama M, de la Monte S, Stern AM, Wands JR, Friedman PA (1996) Overexpression of human aspartyl(asparaginyl)beta-hydroxylase in hepatocellular carcinoma and cholangiocarcinoma. J Clin Invest 98(6):1313–1323. https://doi.org/10.1172/JCI118918

LeBon L, Lee TV, Sprinzak D, Jafar-Nejad H, Elowitz MB (2014) Fringe proteins modulate Notch-ligand cis and trans interactions to specify signaling states. elife 3:e02950. https://doi.org/10.7554/eLife.02950

Lee TV, Sethi MK, Leonardi J, Rana NA, Buettner FF, Haltiwanger RS, Bakker H, Jafar-Nejad H (2013) Negative regulation of notch signaling by xylose. PLoS Genet 9(6):e1003547. https://doi.org/10.1371/journal.pgen.1003547

Lei L, Xu A, Panin VM, Irvine KD (2003) An O-fucose site in the ligand binding domain inhibits Notch activation. Development 130(26):6411–6421. https://doi.org/10.1242/dev.00883

Leonardi J, Fernandez-Valdivia R, Li YD, Simcox AA, Jafar-Nejad H (2011) Multiple O-glucosylation sites on Notch function as a buffer against temperature-dependent loss of signaling. Development 138(16):3569–3578. https://doi.org/10.1242/dev.068361

Li M, Cheng R, Liang J, Yan H, Zhang H, Yang L, Li C, Jiao Q, Lu Z, He J, Ji J, Shen Z, Li C, Hao F, Yu H, Yao Z (2013) Mutations in POFUT1, encoding protein O-fucosyltransferase 1, cause generalized Dowling-Degos disease. Am J Hum Genet 92(6):895–903. https://doi.org/10.1016/j.ajhg.2013.04.022

Li CR, Brooks YS, Jia WX, Wang DG, Xiao XM, Li Q, Li M, Mao QX, Wang B (2016) Pathogenicity of POFUT1 mutations in two Chinese families with Dowling-Degos disease. J Eur Acad Dermatol Venereol 30(10):e79–e81. https://doi.org/10.1111/jdv.13323

Li Z, Han K, Pak JE, Satkunarajah M, Zhou D, Rini JM (2017) Recognition of EGF-like domains by the notch-modifying O-fucosyltransferase POFUT1. Nat Chem Biol 13(7):757–763. https://doi.org/10.1038/nchembio.2381

Lin HH, Stacey M, Saxby C, Knott V, Chaudhry Y, Evans D, Gordon S, McKnight AJ, Handford P, Lea S (2001) Molecular analysis of the epidermal growth factor-like short consensus repeat domain-mediated protein-protein interactions: dissection of the CD97-CD55 complex. J Biol Chem 276(26):24160–24169. https://doi.org/10.1074/jbc.M101770200

Lira-Navarrete E, Valero-Gonzalez J, Villanueva R, Martinez-Julvez M, Tejero T, Merino P, Panjikar S, Hurtado-Guerrero R (2011) Structural insights into the mechanism of protein O-fucosylation. PLoS One 6(9):e25365. https://doi.org/10.1371/journal.pone.0025365

Liu Z, Chen S, Boyle S, Zhu Y, Zhang A, Piwnica-Worms DR, Ilagan MX, Kopan R (2013) The extracellular domain of Notch2 increases its cell-surface abundance and ligand responsiveness during kidney development. Dev Cell 25(6):585–598. https://doi.org/10.1016/j.devcel.2013.05.022

Loo LW, Tiirikainen M, Cheng I, Lum-Jones A, Seifried A, Church JM, Gryfe R, Weisenberger DJ, Lindor NM, Gallinger S, Haile RW, Duggan DJ, Thibodeau SN, Casey G, Le Marchand L (2013) Integrated analysis of genome-wide copy number alterations and gene expression in microsatellite stable, CpG island methylator phenotype-negative colon cancer. Genes Chromosom Cancer 52(5):450–466. https://doi.org/10.1002/gcc.22043

Lu Q, Hasty P, Shur BD (1997) Targeted mutation in beta1,4-galactosyltransferase leads to pituitary insufficiency and neonatal lethality. Dev Biol 181(2):257–267. https://doi.org/10.1006/dbio.1996.8444

Luca VC, Jude KM, Pierce NW, Nachury MV, Fischer S, Garcia KC (2015) Structural biology. Structural basis for Notch1 engagement of Delta-like 4. Science 347(6224):847–853. https://doi.org/10.1126/science.1261093

Luca VC, Kim BC, Ge C, Kakuda S, Wu D, Roein-Peikar M, Haltiwanger RS, Zhu C, Ha T, Garcia KC (2017) Notch-Jagged complex structure implicates a catch bond in tuning ligand sensitivity. Science 355(6331):1320–1324. https://doi.org/10.1126/science.aaf9739

Luo Y, Haltiwanger RS (2005) O-fucosylation of notch occurs in the endoplasmic reticulum. J Biol Chem 280(12):11289–11294. https://doi.org/10.1074/jbc.M414574200

Ma L, Dong P, Liu L, Gao Q, Duan M, Zhang S, Chen S, Xue R, Wang X (2016) Overexpression of protein O-fucosyltransferase 1 accelerates hepatocellular carcinoma progression via the Notch signaling pathway. Biochem Biophys Res Commun 473(2):503–510. https://doi.org/10.1016/j.bbrc.2016.03.062

Matsumoto K, Ayukawa T, Ishio A, Sasamura T, Yamakawa T, Matsuno K (2016) Dual roles of O-glucose Glycans redundant with monosaccharide O-Fucose on Notch in Notch trafficking. J Biol Chem 291(26):13743–13752. https://doi.org/10.1074/jbc.M115.710483

Matsuura A, Ito M, Sakaidani Y, Kondo T, Murakami K, Furukawa K, Nadano D, Matsuda T, Okajima T (2008) O-linked N-acetylglucosamine is present on the extracellular domain of notch receptors. J Biol Chem 283(51):35486–35495. https://doi.org/10.1074/jbc.M806202200

McDaniell R, Warthen DM, Sanchez-Lara PA, Pai A, Krantz ID, Piccoli DA, Spinner NB (2006) NOTCH2 mutations cause Alagille syndrome, a heterogeneous disorder of the Notch signaling pathway. Am J Hum Genet 79(1):169–173. https://doi.org/10.1086/505332

McMillan BJ, Zimmerman B, Egan ED, Lofgren M, Xu X, Hesser A, Blacklow SC (2017) Structure of human POFUT1, its requirement in ligand-independent oncogenic Notch signaling, and functional effects of Dowling-Degos mutations. Glycobiology:1–10. https://doi.org/10.1093/glycob/cwx020

Meester JA, Southgate L, Stittrich AB, Venselaar H, Beekmans SJ, den Hollander N, Bijlsma EK, Helderman-van den Enden A, Verheij JB, Glusman G, Roach JC, Lehman A, Patel MS, de Vries BB, Ruivenkamp C, Itin P, Prescott K, Clarke S, Trembath R, Zenker M, Sukalo M, Van Laer L, Loeys B, Wuyts W (2015) Heterozygous loss-of-function mutations in DLL4 cause Adams-Oliver syndrome. Am J Hum Genet 97(3):475–482. https://doi.org/10.1016/j.ajhg.2015.07.015

Moloney DJ, Panin VM, Johnston SH, Chen J, Shao L, Wilson R, Wang Y, Stanley P, Irvine KD, Haltiwanger RS, Vogt TF (2000a) Fringe is a glycosyltransferase that modifies Notch. Nature 406(6794):369–375. https://doi.org/10.1038/35019000

Moloney DJ, Shair LH, Lu FM, Xia J, Locke R, Matta KL, Haltiwanger RS (2000b) Mammalian Notch1 is modified with two unusual forms of O-linked glycosylation found on epidermal growth factor-like modules. J Biol Chem 275(13):9604–9611

Moran JL, Shifley ET, Levorse JM, Mani S, Ostmann K, Perez-Balaguer A, Walker DM, Vogt TF, Cole SE

(2009) Manic fringe is not required for embryonic development, and fringe family members do not exhibit redundant functions in the axial skeleton, limb, or hindbrain. Dev Dyn 238(7):1803–1812. https://doi.org/10.1002/dvdy.21982

Muller R, Jenny A, Stanley P (2013) The EGF repeat-specific O-GlcNAc-transferase Eogt interacts with notch signaling and pyrimidine metabolism pathways in Drosophila. PLoS One 8(5):e62835. https://doi.org/10.1371/journal.pone.0062835

Muller J, Rana NA, Serth K, Kakuda S, Haltiwanger RS, Gossler A (2014) O-fucosylation of the notch ligand mDLL1 by POFUT1 is dispensable for ligand function. PLoS One 9(2):e88571. https://doi.org/10.1371/journal.pone.0088571

Nishimura H, Kawabata S, Kisiel W, Hase S, Ikenaka T, Takao T, Shimonishi Y, Iwanaga S (1989) Identification of a disaccharide (Xyl-Glc) and a trisaccharide (Xyl2-Glc) O-glycosidically linked to a serine residue in the first epidermal growth factor-like domain of human factors VII and IX and protein Z and bovine protein Z. J Biol Chem 264(34):20320–20325

Ogawa M, Sawaguchi S, Kawai T, Nadano D, Matsuda T, Yagi H, Kato K, Furukawa K, Okajima T (2015) Impaired O-linked N-acetylglucosaminylation in the endoplasmic reticulum by mutated epidermal growth factor (EGF) domain-specific O-linked N-acetylglucosamine transferase found in Adams-Oliver syndrome. J Biol Chem 290(4):2137–2149. https://doi.org/10.1074/jbc.M114.598821

Okajima T, Irvine KD (2002) Regulation of notch signaling by o-linked fucose. Cell 111(6):893–904

Okajima T, Xu A, Lei L, Irvine KD (2005) Chaperone activity of protein O-fucosyltransferase 1 promotes notch receptor folding. Science 307(5715):1599–1603. https://doi.org/10.1126/science.1108995

Okamura Y, Saga Y (2008) Pofut1 is required for the proper localization of the Notch receptor during mouse development. Mech Dev 125(8):663–673. https://doi.org/10.1016/j.mod.2008.04.007

Panin VM, Papayannopoulos V, Wilson R, Irvine KD (1997) Fringe modulates Notch-ligand interactions. Nature 387(6636):908–912. https://doi.org/10.1038/43191

Ramkumar N, Harvey BM, Lee JD, Alcorn HL, Silva-Gagliardi NF, McGlade CJ, Bestor TH, Wijnholds J, Haltiwanger RS, Anderson KV (2015) Protein O-Glucosyltransferase 1 (POGLUT1) promotes mouse gastrulation through modification of the apical polarity protein CRUMBS2. PLoS Genet 11(10):e1005551. https://doi.org/10.1371/journal.pgen.1005551

Rampal R, Arboleda-Velasquez JF, Nita-Lazar A, Kosik KS, Haltiwanger RS (2005) Highly conserved O-fucose sites have distinct effects on Notch1 function. J Biol Chem 280(37):32133–32140. https://doi.org/10.1074/jbc.M506104200

Rampal R, Luther KB, Haltiwanger RS (2007) Notch signaling in normal and disease states: possible therapies related to glycosylation. Curr Mol Med 7(4):427–445

Rana NA, Haltiwanger RS (2011) Fringe benefits: functional and structural impacts of O-glycosylation on the extracellular domain of Notch receptors. Curr Opin Struct Biol 21(5):583–589. https://doi.org/10.1016/j.sbi.2011.08.008

Rana NA, Nita-Lazar A, Takeuchi H, Kakuda S, Luther KB, Haltiwanger RS (2011) O-glucose trisaccharide is present at high but variable stoichiometry at multiple sites on mouse Notch1. J Biol Chem 286(36):31623–31637. https://doi.org/10.1074/jbc.M111.268243

Rebay I, Fleming RJ, Fehon RG, Cherbas L, Cherbas P, Artavanis-Tsakonas S (1991) Specific EGF repeats of Notch mediate interactions with Delta and Serrate: implications for Notch as a multifunctional receptor. Cell 67(4):687–699

Ryan MJ, Bales C, Nelson A, Gonzalez DM, Underkoffler L, Segalov M, Wilson-Rawls J, Cole SE, Moran JL, Russo P, Spinner NB, Kusumi K, Loomes KM (2008) Bile duct proliferation in Jag1/fringe heterozygous mice identifies candidate modifiers of the Alagille syndrome hepatic phenotype. Hepatology 48(6):1989–1997. https://doi.org/10.1002/hep.22538

Sakaidani Y, Nomura T, Matsuura A, Ito M, Suzuki E, Murakami K, Nadano D, Matsuda T, Furukawa K, Okajima T (2011) O-linked-N-acetylglucosamine on extracellular protein domains mediates epithelial cell-matrix interactions. Nat Commun 2:583. https://doi.org/10.1038/ncomms1591

Sakaidani Y, Ichiyanagi N, Saito C, Nomura T, Ito M, Nishio Y, Nadano D, Matsuda T, Furukawa K, Okajima T (2012) O-linked-N-acetylglucosamine modification of mammalian Notch receptors by an atypical O-GlcNAc transferase Eogt1. Biochem Biophys Res Commun 419(1):14–19. https://doi.org/10.1016/j.bbrc.2012.01.098

Sawaguchi S, Varshney S, Ogawa M, Sakaidani Y, Yagi H, Takeshita K, Murohara T, Kato K, Sundaram S, Stanley P, Okajima T (2017) O-GlcNAc on NOTCH1 EGF repeats regulates ligand-induced Notch signaling and vascular development in mammals. elife 6. https://doi.org/10.7554/eLife.24419

Sawey ET, Chanrion M, Cai C, Wu G, Zhang J, Zender L, Zhao A, Busuttil RW, Yee H, Stein L, French DM, Finn RS, Lowe SW, Powers S (2011) Identification of a therapeutic strategy targeting amplified FGF19 in liver cancer by Oncogenomic screening. Cancer Cell 19(3):347–358. https://doi.org/10.1016/j.ccr.2011.01.040

Schneider M, Kumar V, Nordstrom LU, Feng L, Takeuchi H, Stanley P, Wu P, Haltiwanger RS. Inhibition of Delta-induced Notch signaling using fucose analogs (In Review).

Servian-Morilla E, Takeuchi H, Lee TV, Clarimon J, Mavillard F, Area-Gomez E, Rivas E, Nieto-Gonzalez JL, Rivero MC, Cabrera-Serrano M, Gomez-Sanchez L, Martinez-Lopez JA, Estrada B, Marquez C, Morgado Y, Suarez-Calvet X, Pita G, Bigot A, Gallardo E, Fernandez-Chacon R, Hirano M, Haltiwanger RS, Jafar-Nejad H, Paradas C (2016) A POGLUT1 mutation causes a muscular dystrophy with reduced Notch signaling and satellite cell loss. EMBO Mol Med 8(11):1289–1309. https://doi.org/10.15252/emmm.201505815

Sethi MK, Buettner FF, Krylov VB, Takeuchi H, Nifantiev NE, Haltiwanger RS, Gerardy-Schahn R, Bakker H (2010) Identification of glycosyltransferase 8 family members as xylosyltransferases acting on O-glucosylated notch epidermal growth factor repeats. J Biol Chem 285(3):1582–1586. https://doi.org/10.1074/jbc.C109.065409

Sethi MK, Buettner FF, Ashikov A, Krylov VB, Takeuchi H, Nifantiev NE, Haltiwanger RS, Gerardy-Schahn R, Bakker H (2012) Molecular cloning of a xylosyltransferase that transfers the second xylose to O-glucosylated epidermal growth factor repeats of notch. J Biol Chem 287(4):2739–2748. https://doi.org/10.1074/jbc.M111.302406

Shaheen R, Aglan M, Keppler-Noreuil K, Faqeih E, Ansari S, Horton K, Ashour A, Zaki MS, Al-Zahrani F, Cueto-Gonzalez AM, Abdel-Salam G, Temtamy S, Alkuraya FS (2013) Mutations in EOGT confirm the genetic heterogeneity of autosomal-recessive Adams-Oliver syndrome. Am J Hum Genet 92(4):598–604. https://doi.org/10.1016/j.ajhg.2013.02.012

Shao L, Moloney DJ, Haltiwanger R (2003) Fringe modifies O-fucose on mouse Notch1 at epidermal growth factor-like repeats within the ligand-binding site and the Abruptex region. J Biol Chem 278(10):7775–7782. https://doi.org/10.1074/jbc.M212221200

Shi S, Stanley P (2003) Protein O-fucosyltransferase 1 is an essential component of Notch signaling pathways. Proc Natl Acad Sci U S A 100(9):5234–5239. https://doi.org/10.1073/pnas.0831126100

Shi S, Ge C, Luo Y, Hou X, Haltiwanger RS, Stanley P (2007) The threonine that carries fucose, but not fucose, is required for Cripto to facilitate Nodal signaling. J Biol Chem 282(28):20133–20141. https://doi.org/10.1074/jbc.M702593200

Song Y, Kumar V, Wei HX, Qiu J, Stanley P (2016) Lunatic, manic, and radical fringe each promote T and B cell development. J Immunol 196(1):232–243. https://doi.org/10.4049/jimmunol.1402421

Sparrow DB, Chapman G, Wouters MA, Whittock NV, Ellard S, Fatkin D, Turnpenny PD, Kusumi K, Sillence D, Dunwoodie SL (2006) Mutation of the LUNATIC FRINGE gene in humans causes spondylocostal dysostosis with a severe vertebral phenotype. Am J Hum Genet 78(1):28–37. https://doi.org/10.1086/498879

Stahl M, Uemura K, Ge C, Shi S, Tashima Y, Stanley P (2008) Roles of Pofut1 and O-fucose in mammalian notch signaling. J Biol Chem 283(20):13638–13651. https://doi.org/10.1074/jbc.M802027200

Stittrich AB, Lehman A, Bodian DL, Ashworth J, Zong Z, Li H, Lam P, Khromykh A, Iyer RK, Vockley JG, Baveja R, Silva ES, Dixon J, Leon EL, Solomon BD, Glusman G, Niederhuber JE, Roach JC, Patel MS (2014) Mutations in NOTCH1 cause Adams-Oliver syndrome. Am J Hum Genet 95(3):275–284. https://doi.org/10.1016/j.ajhg.2014.07.011

Svensson P, Bergqvist I, Norlin S, Edlund H (2009) MFng is dispensable for mouse pancreas development and function. Mol Cell Biol 29(8):2129–2138. https://doi.org/10.1128/MCB.01644-08

Takeuchi H, Fernandez-Valdivia RC, Caswell DS, Nita-Lazar A, Rana NA, Garner TP, Weldeghiorghis TK, Macnaughtan MA, Jafar-Nejad H, Haltiwanger RS (2011) Rumi functions as both a protein O-glucosyltransferase and a protein O-xylosyltransferase. Proc Natl Acad Sci U S A 108(40):16600–16605. https://doi.org/10.1073/pnas.1109696108

Takeuchi H, Kantharia J, Sethi MK, Bakker H, Haltiwanger RS (2012) Site-specific O-glucosylation of the epidermal growth factor-like (EGF) repeats of notch: efficiency of glycosylation is affected by proper folding and amino acid sequence of individual EGF repeats. J Biol Chem 287(41):33934–33944. https://doi.org/10.1074/jbc.M112.401315

Taylor P, Takeuchi H, Sheppard D, Chillakuri C, Lea SM, Haltiwanger RS, Handford PA (2014) Fringe-mediated extension of O-linked fucose in the ligand-binding region of Notch1 increases binding to mammalian Notch ligands. Proc Natl Acad Sci U S A 111(20):7290–7295. https://doi.org/10.1073/pnas.1319683111

Thakurdas SM, Lopez MF, Kakuda S, Fernandez-Valdivia R, Zarrin-Khameh N, Haltiwanger RS, Jafar-Nejad H (2016) Jagged1 heterozygosity in mice results in a congenital cholangiopathy which is reversed by concomitant deletion of one copy of Poglut1 (Rumi). Hepatology 63(2):550–565. https://doi.org/10.1002/hep.28024

Torres CR, Hart GW (1984) Topography and polypeptide distribution of terminal N-acetylglucosamine residues on the surfaces of intact lymphocytes. Evidence for O-linked GlcNAc. J Biol Chem 259(5):3308–3317

Vasudevan D, Haltiwanger RS (2014) Novel roles for O-linked glycans in protein folding. Glycoconj J 31(6–7):417–426. https://doi.org/10.1007/s10719-014-9556-4

Wahi K, Bochter MS, Cole SE (2016) The many roles of Notch signaling during vertebrate somitogenesis. Semin Cell Dev Biol 49:68–75. https://doi.org/10.1016/j.semcdb.2014.11.010

Wang Y, Spellman MW (1998) Purification and characterization of a GDP-fucose:polypeptide fucosyltransferase from Chinese hamster ovary cells. J Biol Chem 273(14):8112–8118

Wang Y, Lee GF, Kelley RF, Spellman MW (1996) Identification of a GDP-L-fucose:polypeptide fucosyltransferase and enzymatic addition of O-linked fucose to EGF domains. Glycobiology 6(8):837–842

Wang Y, Shao L, Shi S, Harris RJ, Spellman MW, Stanley P, Haltiwanger RS (2001) Modification of epidermal growth factor-like repeats with O-fucose Molecular cloning and expression of a novel GDP-fucose protein O-fucosyltransferase. J Biol Chem 276(43):40338–40345. https://doi.org/10.1074/jbc.M107849200

Wang Y, Chang N, Zhang T, Liu H, Ma W, Chu Q, Lai Q, Liu L, Wang W (2010) Overexpression of human CAP10-like protein 46 KD in T-acute lymphoblastic leukemia and acute myelogenous leukemia. Genet Test Mol Biomarkers 14(1):127–133. https://doi.org/10.1089/gtmb.2009.0145

Weinmaster G, Kintner C (2003) Modulation of notch signaling during somitogenesis. Annu Rev Cell Dev Biol 19:367–395. https://doi.org/10.1146/annurev.cellbio.19.111301.115434

Xu A, Lei L, Irvine KD (2005) Regions of Drosophila Notch that contribute to ligand binding and the modulatory influence of Fringe. J Biol Chem 280(34):30158–30165. https://doi.org/10.1074/jbc.M505569200

Xu A, Haines N, Dlugosz M, Rana NA, Takeuchi H, Haltiwanger RS, Irvine KD (2007) In vitro reconstitution of the modulation of Drosophila Notch-ligand binding by Fringe. J Biol Chem 282(48):35153–35162. https://doi.org/10.1074/jbc.M707040200

Xu K, Usary J, Kousis PC, Prat A, Wang DY, Adams JR, Wang W, Loch AJ, Deng T, Zhao W, Cardiff RD, Yoon K, Gaiano N, Ling V, Beyene J, Zacksenhaus E, Gridley T, Leong WL, Guidos CJ, Perou CM, Egan SE (2012) Lunatic fringe deficiency cooperates with the Met/Caveolin gene amplicon to induce basal-like breast cancer. Cancer Cell 21(5):626–641. https://doi.org/10.1016/j.ccr.2012.03.041

Yang LT, Nichols JT, Yao C, Manilay JO, Robey EA, Weinmaster G (2005) Fringe glycosyltransferases differentially modulate Notch1 proteolysis induced by Delta1 and Jagged1. Mol Biol Cell 16(2):927–942. https://doi.org/10.1091/mbc.E04-07-0614

Yao D, Huang Y, Huang X, Wang W, Yan Q, Wei L, Xin W, Gerson S, Stanley P, Lowe JB, Zhou L (2011) Protein O-fucosyltransferase 1 (Pofut1) regulates lymphoid and myeloid homeostasis through modulation of Notch receptor ligand interactions. Blood 117(21):5652–5662. https://doi.org/10.1182/blood-2010-12-326074

Yokota S, Ogawara K, Kimura R, Shimizu F, Baba T, Minakawa Y, Higo M, Kasamatsu A, Endo-Sakamoto Y, Shiiba M, Tanzawa H, Uzawa K (2013) Protein O-fucosyltransferase 1: a potential diagnostic marker and therapeutic target for human oral cancer. Int J Oncol 43(6):1864–1870. https://doi.org/10.3892/ijo.2013.2110

Yu H, Takeuchi M, LeBarron J, Kantharia J, London E, Bakker H, Haltiwanger RS, Li H, Takeuchi H (2015) Notch-modifying xylosyltransferase structures support an SNi-like retaining mechanism. Nat Chem Biol 11(11):847–854. https://doi.org/10.1038/nchembio.1927

Yu H, Takeuchi H, Takeuchi M, Liu Q, Kantharia J, Haltiwanger RS, Li H (2016) Structural analysis of Notch-regulating Rumi reveals basis for pathogenic mutations. Nat Chem Biol 12(9):735–740. https://doi.org/10.1038/nchembio.2135

Zhang N, Norton CR, Gridley T (2002) Segmentation defects of Notch pathway mutants and absence of a synergistic phenotype in lunatic fringe/radical fringe double mutant mice. Genesis 33(1):21–28. https://doi.org/10.1002/gene.10081

Zhang S, Chung WC, Wu G, Egan SE, Xu K (2014) Tumor-suppressive activity of Lunatic Fringe in prostate through differential modulation of Notch receptor activation. Neoplasia 16(2):158–167. https://doi.org/10.1593/neo.131870

Zhang S, Chung WC, Wu G, Egan SE, Miele L, Xu K (2015) Manic fringe promotes a claudin-low breast cancer phenotype through notch-mediated PIK3CG induction. Cancer Res 75(10):1936–1943. https://doi.org/10.1158/0008-5472.CAN-14-3303

Zhang S, Chung WC, Xu K (2016) Lunatic Fringe is a potent tumor suppressor in Kras-initiated pancreatic cancer. Oncogene 35(19):2485–2495. https://doi.org/10.1038/onc.2015.306

Modeling the Notch Response

Udi Binshtok and David Sprinzak

Abstract

NOTCH signaling regulates developmental processes in all tissues and all organisms across the animal kingdom. It is often involved in coordinating the differentiation of neighboring cells into different cell types. As our knowledge on the structural, molecular and cellular properties of the NOTCH pathway expands, there is a greater need for quantitative methodologies to get a better understanding of the processes controlled by NOTCH signaling. In recent years, theoretical and computational approaches to NOTCH signaling and NOTCH mediated patterning are gaining popularity. Mathematical models of NOTCH mediated patterning provide insight into complex and counterintuitive behaviors and can help generate predictions that can guide experiments. In this chapter, we review the recent advances in modeling NOTCH mediated patterning processes. We discuss new modeling approaches to lateral inhibition patterning that take into account cis-interactions between NOTCH receptors and ligands, signaling through long cellular protrusions, cell division processes, and coupling to external signals. We also describe models of somitogenesis, where NOTCH signaling is used for synchronizing cellular oscillations. We then discuss modeling approaches that consider the effect of cell morphology on NOTCH signaling and NOTCH mediated patterning. Finally, we consider models of boundary formation and how they are influenced by the combinatorial action of multiple ligands. Together, these topics cover the main advances in the field of modeling the NOTCH response.

Keywords

NOTCH signaling · Mathematical modeling · Pattern formation · Lateral inhibition · Boundary formation · Cis-inhibition · Cell morphology · Cell division · Filopodia

Abbreviations

SOP	Sensory organ precursors
NICD	NOTCH intra-cellular domain
Dll1	DELTA-LIKE-1
Dll4	DELTA-LIKE-4
EGF	Epidermal growth factor
VEGF	Vascular endothelial growth factor
VEGFr	Vascular endothelial growth factor receptor
ISCs	Intestinal stem cells
EBs	Enteroblasts
PSM	Presomitic Mesoderm

U. Binshtok · D. Sprinzak (✉)
Department of Biochemistry and Molecular Biology, George S. Wise Faculty of Life Science, Tel Aviv University, Tel Aviv, Israel
e-mail: davidsp@post.tau.ac.il

T. Borggrefe, B. D. Giaimo (eds.), *Molecular Mechanisms of Notch Signaling*, Advances in Experimental Medicine and Biology 1066, https://doi.org/10.1007/978-3-319-89512-3_5

1 Introduction

NOTCH signaling is the canonical signaling pathway used to coordinate the differentiation between neighboring cells during animal development (Artavanis-Tsakonas et al. 1999; Artavanis-Tsakonas and Muskavitch 2010; Kovall et al. 2017). Not only the molecular mechanism of NOTCH is highly conserved from worms, through flies, to humans, but also many of the developmental processes and circuits in which NOTCH is involved, are highly conserved. A classic example for such a conserved process is that of lateral inhibition patterning that describes the transition from an initially uniform field of cell to an alternating pattern of differentiation. Examples for lateral inhibition processes include the selection of sensory organ precursors (SOP, sensory bristles) in the *Drosophila* notum (Heitzler and Simpson 1991), hair cell patterning in the vertebrate inner ear (Daudet and Lewis 2005), the differentiation of intestinal precursors into absorptive and secretory cells (Sancho et al. 2015) and more. Other prototypical processes known to involve NOTCH signaling include asymmetric cell division (e.g. during neurogenesis), defining boundary cells (e.g. wing veins and wing margin), and coordinating synchronized oscillations (e.g. somitogenesis) (Artavanis-Tsakonas et al. 1999; Lewis 2003).

NOTCH mediated lateral inhibition has been first modeled by Julian Lewis and co-workers in 1996 (Collier et al. 1996). Since then, a large body of theoretical works have been developed to describe various aspects of NOTCH mediated patterning processes, including different types of lateral inhibition, boundary formation, wavefront propagation and synchronized oscillations (Shaya and Sprinzak 2011). Such models are used to formalize heuristic concepts into a quantitative picture that can help explaining unintuitive behaviors and generate testable predictions.

As our molecular and cellular understanding of NOTCH signaling progresses and more quantitative data is gathered, so do the modeling approaches become more refined and account for a larger variety of phenomena. In this chapter, we review the recent advances in modeling NOTCH mediated processes. Our goal is to provide a comprehensive picture of the current works in the field and represent the main approaches used to mathematically describe NOTCH mediated developmental processes. We focus here on the mathematical framework used in different approaches and provide the basic equations used to for each approach. For those who are interested in getting more practical information on performing the simulations, we refer to the practical tutorial by Formosa-Jordan and Sprinzak (Formosa-Jordan and Sprinzak 2014).

The chapter has four main sections corresponding to four topics. The first topic (Sect. 2) is lateral inhibition and extensions of the basic model to take into account cis-inhibition, cell divisions, filopodia, and external signals. The second topic (Sect. 3) is modeling synchronized oscillations during somitogenesis. The third topic (Sect. 4) is the role of cell geometry on NOTCH signaling and NOTCH mediated patterning. The fourth topic (Sect. 5) is NOTCH signaling during boundary formation and the role of multiple ligands.

2 Models of Lateral Inhibition

2.1 The Basic Lateral Inhibition Model

While the general concept of lateral inhibition has been first discussed by Wigglesworth in 1940 (Wigglesworth 1940), it was not until the 1990s that these concepts were formalized into a well-defined mathematical model (Collier et al. 1996). At its core, lateral inhibition patterning is a symmetry breaking process where a group of initially identical cells differentiate into alternating patterns of cell fates. This process involves a local competition between neighboring cells, where at a certain developmental time, all cells "strive" to differentiate into one cell type and at the same time prevent their neighbors from becoming that cell type. Within each small group of cells, one cell prevails and subsequently suppresses all its direct neighbors through NOTCH signaling.

The essential symmetry breaking process during lateral inhibition patterning is achieved by an intercellular feedback loop, in which NOTCH signaling from one cell downregulates DELTA ligand activity in the neighboring cell (Fig. 1a). This feedback can amplify small initial differences between cells, so that one cell ends up expressing high levels of DELTA ligand while its neighbors express low levels of DELTA. This type of mechanism can in principle generate the typical checkerboard like patterns associated with lateral inhibition.

The first work to mathematically model this process was the work by Collier and colleagues (Collier et al. 1996). The Collier model contains differential equations describing the time evolution of two variables - activated NOTCH and DELTA in each cell. Here, we will use a slightly expanded model developed by Sprinzak et al. (Sprinzak et al. 2010) that directly accounts for NOTCH receptors, DELTA ligands and signal levels as well as for the intracellular feedback represented by a repressor which is activated by NOTCH signaling and represses DELTA production (Fig. 1a).

The mechanism underlying the model (Fig. 1a) is described by the following set of reactions between NOTCH receptors and DELTA ligands on a lattice of cells.

1. NOTCH receptors from one cell, denoted by N_i, interact with DELTA ligands on a neighboring cell, denoted by D_j, to produce a signal (representing the cleaved intracellular domain of NOTCH). The interaction is described by a Michaelis-Menten reaction:

$$N_i + D_j \underset{\kappa^-}{\overset{\kappa^+}{\rightleftharpoons}} \left[N_i D_j\right] \xrightarrow{k^S} s_{j\to i} \tag{1}$$

Where the index i represents one cell in a lattice of cells, and the index j represent a neighboring cell j of cell i. $[N_iD_j]$ denotes the NOTCH-DELTA complex between cell i and cell j. κ^+ and κ^- are the association and dissociation rates of NOTCH and DELTA, respectively. κ^S is the rate associated with conversion of the NOTCH-DELTA complex into a signal [namely, the inverse time it takes for the NOTCH intracellular domain (NICD) to get cleaved once it interacts with DELTA]. $s_{j\to i}$ denotes the signal generated in cell i by interaction with cell j. The total signal generated in cell i is the summation over the signals generated from all of its neighboring cells:

$$S_i = \sum_j s_{j\to i} \tag{2}$$

2. The total signal in each cell activates a repressor, denoted by R_i, that downregulates the DELTA production in that cell:

$$S_i \to R_i \dashv D_i \tag{3}$$

The activation of the repressor by the signal and the repression of DELTA by the repressor are phenomenologically described in terms of an increasing and decreasing sigmoidal Hill functions, respectively.

3. NOTCH production rate is assumed to be constant.
4. All variables are assumed to have constant degradation rates.

These reactions are then converted into a set of ordinary differential equations for the levels of NOTCH, N_i, DELTA, D_i and the repressor, R_i, in each cell i (Sprinzak et al. 2010):

$$\frac{dN_i}{dt} = \beta_N - \gamma_N N_i - \kappa_t^{-1} N_i \langle D \rangle_i \tag{4}$$

$$\frac{dD_i}{dt} = \beta_D \frac{p_R^l}{p_R^l + R_i^l} - \gamma_D D_i - \kappa_t^{-1} D_i \langle N \rangle_i \tag{5}$$

$$\frac{dR_i}{dt} = \beta_R \frac{\left(\frac{\kappa_t^{-1}}{\gamma_S} N_i \langle D \rangle_i\right)^m}{p_S^m + \left(\frac{\kappa_t^{-1}}{\gamma_S} N_i \langle D \rangle_i\right)^m} - \gamma_R R_i \tag{6}$$

where β_N, β_D and β_R are the maximal production rates of NOTCH, DELTA and the repressor, respectively. γ_N, γ_D, γ_R, and γ_S are the degradation

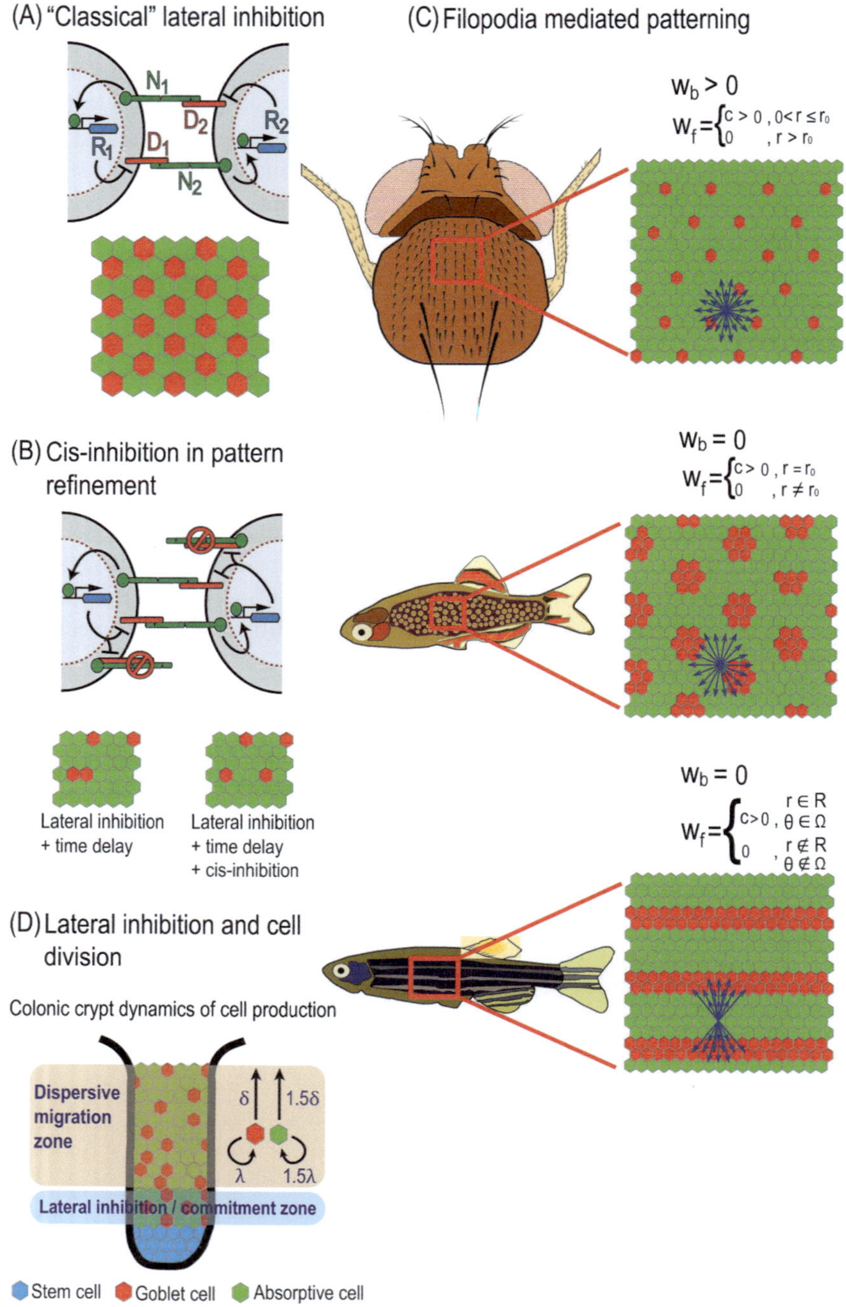

Fig. 1 Models of lateral inhibition. (**a**) "Classical" lateral inhibition. Top – Schematic representation of a lateral inhibition circuit in two cells. In this circuit NOTCH signaling in each cell is generated by the interactions between NOTCH receptors (N1 and N2) and NOTCH ligands (D1 and D2). NOTCH signaling in each cell activates a repressor (R1 and R2) that downregulates the expression or activity of DELTA in that cell. Bottom – A typical simulation result on a hexagonal lattice of cells. Simulation starting with uniform initial conditions (plus noise) results in a salt-and-pepper like pattern where each high DELTA cell (red) is surrounded by low DELTA cells (green). (**b**) Cis-inhibition in pattern refinement. Top – Schematic representation of a lateral inhibition model that includes cis-inhibition between receptors and ligands. In the cis-inhibition model, a ligand on one cell binds to a receptor

rates of NOTCH, DELTA, repressor and signal, respectively. $\kappa_t^{-1} \equiv \dfrac{\kappa^+ \kappa^s}{\kappa^- + \kappa^s}$ denotes the strength of trans-activation. The repression and activation reactions in Eqs. (5) and (6) are described in terms of Hill functions, where p_R and p_S describe the effective K_d for the repressor and the signal, respectively, and l and m describe the Hill coefficients for the two reactions. The terms $\langle D \rangle_i$ and $\langle N \rangle_i$ are the summation of DELTA and NOTCH, respectively, over the neighboring cells j that are in direct contact with cell i:

$$\langle D \rangle_i = \left(\Sigma_j D_j \right)_i$$

$$\langle N \rangle_i = \left(\Sigma_j N_j \right)_i \tag{7}$$

The total number of differential equations is three times the number of cells.

Equations (4)–(6), can then be solved using standard ordinary differential equation (ODE) solvers. These equations can be applied to study different geometries, e.g. two cells, a line of cells, or a cell lattice. Many simulations use hexagonal cell lattices as the one shown in Fig. 1a, to describe epithelial like tissues. The solution of these equations provides the level of NOTCH, DELTA and the repressor over time. It can be shown (Sprinzak et al. 2011) that depending on the parameters used, these equations can lead to two classes of steady states. One class is a homogeneous steady state, in which all cells have the same concentration of all variables. This situation corresponds to an unpatterned steady state. The second class is an alternating steady state solution such as the one shown in the bottom of Fig. 1a. In this solution, cells expressing high levels of DELTA and low levels of repressor ("high DELTA cells", red in Fig. 1a) are surrounded by cells expressing low levels of DELTA and high levels of repressor ("low DELTA cells", green in Fig. 1a). Several works have analyzed the types of possible patterns that can be generated from such a model and the dynamics leading to these patterns (Collier et al. 1996; Sprinzak et al. 2011; Formosa-Jordan and Sprinzak 2014). As can be seen in the examples below, the model can be expanded to include more complex situations such as taking into account additional interactions (e.g. cis-interactions), considering more complex cellular geometries (e.g. filopodia or cell shape) and including additional processes such as cell divisions.

Fig. 1 (continued) on the same cell leading to the formation of an inactive complex. These cell-autonomous interactions reduce the number of free ligands and receptors on the cell surface. Bottom – Cis-inhibition reduces the probability of defects (two neighboring red cells) for models of lateral inhibition that include time delays in the intracellular regulatory feedback. (**c**) Filopodia mediated patterning. Schematic of models that take into account long range NOTCH signaling mediated by filopodia. By controlling the relative weights of signaling through cell-cell body contacts, w_b, and filopodial contacts, w_f, a variety of patterns can be formed. Top – A model where signaling is mediated by both filopodia and cell body contacts. In this model filopodia are extended uniformly up to a radius r_0 from the center of the cell (blue arrows). This model corresponds to the large spacing pattern of bristles on the *Drosophila* notum. Middle – A model where NOTCH signaling is mediated only through filopodia and not through cell body contacts. Here, the filopodia extends only to cells that are at radius r_0 from the center of the cell. This model can produce a spotted pattern similar to the one observed in the skin of *pearl danio* fish. Bottom – A model where NOTCH signaling is limited to filopodia which are restricted to a radius range R and angular range Ω. Such a model can produce a striped pattern similar to the one observed in the *zebrafish* skin. (**d**) Lateral inhibition and cell division. A schematic model describing differentiation dynamics in the colonic crypt. Stem cells at the bottom of the crypt (blue) are differentiated into goblet cells (red, high DELTA) and absorptive cells (green, low DELTA), which migrate towards the top of the crypt (the lumen). Stem cells divide at the bottom of the crypt (stem cells layer), causing an upward movement of the entire lattice. Lateral inhibition occurs at the layer adjacent to the stem cells layer (lateral inhibition zone or commitment zone). At this layer, cell fates are determined by the lateral inhibition process as described in (**a**). Once differentiated, the cells migrate from the commitment zone to the dispersive migration zone. In this zone, goblet cells divide at a lower rate (λ) and migrate at a lower speed (δ) compared to absorptive cells (1.5λ and 1.5δ). The differences in division rates and migration speed result in a spaced salt-and-pepper like pattern in the lumen

2.2 Lateral Inhibition Patterning with Cis-Inhibition Between Receptors and Ligands

It's been known for quite some time that NOTCH receptors and ligands can interact not only when coming from neighboring cells (*in-trans*) but also within the same cell (*in-cis*) (de Celis and Bray 1997; Klein et al. 1997; Micchelli et al. 1997). Unlike trans-interactions which lead to activation of NOTCH receptors, cis-interactions lead to inhibition of NOTCH signaling. For example, overexpressing DELTA in wing margin cells leads to suppression of NOTCH signaling in those cells (Klein et al. 1997). Several studies in recent years addressed, using mathematical modeling, the question of how cis-inhibition affects lateral inhibition (Sprinzak et al. 2010; Sprinzak et al. 2011; Formosa-Jordan and Ibanes 2014). Introduction of cis-inhibition into the model is performed by adding the following reaction to the lateral inhibition model described in Eqs. (4)–(6):

$$N_i + D_i \underset{\kappa_c^-}{\overset{\kappa_c^+}{\rightleftharpoons}} [N_i D_i] \xrightarrow{k_c^S} \phi \tag{8}$$

Where $[N_iD_i]$ denotes the inactive cis-complex of NOTCH and DELTA in cell *i*. κ_c^+ and κ_c^- are the association and dissociation rates of NOTCH and DELTA, respectively. The cis-complex is typically assumed to be removed or endocytosed at a rate of k_c^S.

As in the basic model, these reactions are then converted into a set of ordinary differential equations for the levels of NOTCH, N_i, DELTA, D_i, and the repressor, R_i, in each cell *i*.

$$\frac{dN_i}{dt} = \beta_N - \gamma_N N_i - \kappa_t^{-1} N_i \langle D \rangle_i - \kappa_c^{-1} N_i D_i \tag{9}$$

$$\frac{dD_i}{dt} = \beta_D \frac{p_R^l}{p_R^l + R_i^l} - \gamma_D D_i - \kappa_t^{-1} D_i \langle N \rangle_i - \kappa_c^{-1} N_i D_i \tag{10}$$

$$\frac{dR_i}{dt} = \beta_R \frac{\left(\frac{\kappa_t^{-1}}{\gamma_S} N_i \langle D \rangle_i\right)^m}{p_S^m + \left(\frac{\kappa_t^{-1}}{\gamma_S} N_i \langle D \rangle_i\right)^m} - \gamma_R R_i \tag{11}$$

Where $\kappa_c^{-1} \equiv \frac{\kappa_c^+ \kappa_c^s}{\kappa_c^- + \kappa_c^s}$ denotes the strength of cis-inhibition.

Analysis of these equations shows that cis-inhibition can contribute to several aspects of pattern formation. First, cis-inhibition increases the ability to pattern, namely, there is a larger range of parameters that supports patterning (Sprinzak et al. 2011). Second, the dynamics towards a patterned state are faster with cis-inhibition (Barad et al. 2010; Sprinzak et al. 2011). These faster dynamics turn out to be important in suppressing errors during the selection of sensory organ precursors (SOP). Barad and colleagues showed that within the equipotential group of cells there are rare cases where two SOPs can be formed instead of one [Fig. 1b and (Barad et al. 2010)]. They argue that a model of lateral inhibition with time delays in the intracellular feedback mechanism is expected to produce higher error rates than observed. They then show that the faster dynamics associated with cis-inhibition, can suppress the errors associated with such time delays. As predicted from their model, heterozygous mutants of NOTCH, DELTA and SERRATE, exhibit higher frequency of errors. Interestingly, a recent theoretical paper by Glass and colleagues showed that under certain conditions, time delays may actually lead to less defects in lateral inhibition patterning on cell lattices (Glass et al. 2016). Hence, it remains to be elucidated whether time delays are helpful or unhelpful for generation of ordered patterns.

In a recent theoretical paper, Formosa-Jordan and colleagues generalized cis-interactions to include also weak cis-activation (Formosa-Jordan and Ibanes 2014). It was shown that under different parameter regimes, it is possible to get patterns that cannot be achieved using the standard lateral inhibition model. Although this result is currently theoretical, it would be interesting to check whether behaviors such as those predicted by this model may also appear in nature.

2.3 Filopodia in Lateral Inhibition Patterning

An interesting new concept, which has emerged in recent years, is that signaling between cells can be transduced through long cellular protrusions such as filopodia or cytonemes, enabling direct communication between distant cells (Kornberg and Roy 2014). Cellular protrusions such as filopodia can have diameters as small as 0.1 μm and extend dynamically over 100 μm length scale. Several recent papers provided experimental evidence for situations where developmental patterns involve NOTCH signaling through protrusions. These include bristle patterns on the *Drosophila* notum (Cohen et al. 2010; Hunter et al. 2016), spotted skin patterns on *pearl danio* fish and striped skin patterns on *zebrafish* (Hamada et al. 2014; Eom et al. 2015).

The possibility of long-distance signaling through protrusions provide means to expand the variety of possible lateral inhibition patterns. Two recent theoretical papers explored the potential patterns that can be formed when taking long-range filopodia into account (Hadjivasiliou et al. 2016; Vasilopoulos and Painter 2016). To consider signaling through filopodia in the lateral inhibition model, we need to add additional terms representing the receptors, ligands, and signals associated with filopodial signaling, which are different than signaling through cell body contacts. It is therefore useful to define the receptors and ligands contributing to the two distinct types of signaling in each cell: those from cell body contacts (*b*) and those from filopodial contacts (*f*). Following the paper form Hadjivasiliou and colleagues (Hadjivasiliou et al. 2016) we define:

$$\langle D\rangle_{i,b} = \left(\sum\nolimits_{j\in\{cell\ body\ contacts\}} D_j\right)_i$$

$$\langle D\rangle_{i,f} = \left(\sum\nolimits_{j\in\{filopodia\ contacts\}} D_j\right)_i$$

$$\langle N\rangle_{i,b} = \left(\sum\nolimits_{j\in\{cell\ body\ contacts\}} N_j\right)_i$$

$$\langle N\rangle_{i,f} = \left(\sum\nolimits_{j\in\{filopodia\ contacts\}} N_j\right)_i \tag{12}$$

Since the efficiency of signaling can be different between the two types of signals, relative weight factors need to be included. Thus, the dynamic Eqs. (4)–(6) are now modified in the following way:

$$\frac{dN_i}{dt} = \beta_N - \gamma_N N_i - \kappa_t^{-1} N_i \left[w_b \langle D\rangle_{i,b} + w_f \langle D\rangle_{i,f} \right] \tag{13}$$

$$\frac{dD_i}{dt} = \beta_D \frac{p_R^l}{p_R^l + R_i^l} - \gamma_D D_i - \kappa_t^{-1} D_i \left[w_b \langle N\rangle_{i,b} + w_f \langle N\rangle_{i,f} \right] \tag{14}$$

$$\frac{dR_i}{dt} = \beta_R \frac{\left[\frac{\kappa_t^{-1}}{\gamma_S} N_i \left(w_b \langle D\rangle_{i,b} + w_f \langle D\rangle_{i,f}\right)\right]^m}{p_S^m + \left[\frac{\kappa_t^{-1}}{\gamma_S} N_i \left(w_b \langle D\rangle_{i,b} + w_f \langle D\rangle_{i,f}\right)\right]^m} - \gamma_R R_i \tag{15}$$

where w_b and w_f are the weight factors for signals received from the cell body and filopodia contacts, respectively of cells *j* surrounding cell *i*. In the case of $w_b = 1$ and $w_f = 0$ we get the original dynamic Eqs. (4)–(6).

In general *w* can be a function of space and time, that is $w = w(r, \theta, t)$. Where *r* and θ are the polar coordinates in the two dimensions of the lattice of cells. For example, if the filopodia are a few cell diameters long and are dynamic [e.g. grow and shrink as in (Cohen et al. 2010)], the pattern generated is that of high DELTA cells with larger spacing (Fig. 1c, top panel). The dynamics of the filopodia in this case leads to an averaging effect which maintains equal distances between high DELTA cells. If body contact signaling is suppressed ($w_b = 0$), then spotted patterns can be formed (Fig. 1c, middle panel). Finally, if the filopodia have preferred directionality a striped solution can emerge (Fig. 1c, bottom panel). Recent evidence shows that these two latter examples correspond to situations which arise on fish skin patterns (Hamada et al. 2014; Eom et al. 2015).

It is worthwhile noting that skin patterns were usually described in term of reaction-diffusion Turing type models (Turing 1952). Turing patterning relies on feedback interactions between morphogens (long range diffusible ligands) and it has been shown that such models can generate a variety of patterns similar to the ones shown here (Meinhardt 1996; Meinhardt 2008; Kondo et al. 2009). In fact, it can be argued that lateral inhibition with long range filopodia is mathematically equivalent to reaction diffusion models. In particular, long range signaling through filopodia, can replace the role of morphogens in Turing patterning (Hamada et al. 2014). There are two main differences between reaction diffusion models and filopodia based lateral inhibition models. On the mathematical level, there is a difference between diffusion which is a linear process (i.e. flux is proportional to concentration gradient) and signaling which is often non-linear in nature (i.e. signaling is not necessarily proportional to the expression difference between cells). On the physical level, the typical diffusion rates are often too fast to account for the time scale of patterning. For example, striped patterns in *zebrafish* occur over days and weeks. Explaining such patterns with morphogens would require diffusion rates that are orders of magnitude smaller than known diffusion rates for biological molecules. Hence, filopodia based lateral inhibition provides a more realistic mechanism for skin patterning than classic reaction-diffusion processes.

2.4 Lateral Inhibition and Cell Division

In all the models described so far, it was assumed that the cellular morphology is fixed and does not change during the patterning processes. This is clearly a simplifying assumption that may be correct in some situations (e.g. when patterning is fast compared to other processes) but not always. In particular, cell division and cell growth can dynamically modify the connectivity among cells. Two recent papers discuss this issue in SOP patterning in the *Drosophila* notum (Hunter et al. 2016) and in patterning of secretory cells in the mammalian intestine (Toth et al. 2017).

The paper by Toth and colleagues (Toth et al. 2017) describes the interplay between lateral inhibition and cell divisions in the mammalian intestine. The epithelium of the intestine consists of crypts, which contain stem cells at the bottom of the crypts that continuously divide and eventually produce several types of intestinal cells including absorptive and secretory cells. Stem cell divisions yield stem cells which stay at the bottom of the crypt and progenitor cells which migrate upwards in the crypt and differentiate as they migrate. The differentiated cells keep dividing and migrating until they reach the intestine lumen. The final pattern in the lumen consists of a single secretory cells surrounded by a neighborhood of absorptive cells. Tóth et al. argued that due to the stochastic nature of cell divisions at the stem cells zone one would expect to get patched pattern in the lumen, where large groups of secretory cells and absorptive cells will be formed instead of the finely spaced salt and pepper like pattern observed. They suggest the following model for explaining this observation. In their model the authors assume that lateral inhibition takes place at a restricted zone, termed the commitment zone, right above the stem cell zone at the bottom of the crypt (Fig. 1d). As the progenitor cells differentiate at the commitment zone, the lateral inhibition process prevents the differentiation of two adjacent cells into secretory fates, as described in the dynamic Eqs. (4)–(6). In this way, the lateral inhibition process induces a pattern that maintains a ratio of 1:3 between the secretory cells (DELTA expressing cells) and the absorptive cells (NOTCH expressing cells). After passing the commitment zone, cells continue to migrate and divide on their way to the lumen without being subjected to lateral inhibition. To maintain a homogeneous distribution of secretory cells in the lumen (i.e. with secretory cells separated from each other), an additional mechanism was introduced. After leaving the commitment zone, the two types of differentiated cells further divide at different rates and migrate at different speeds towards the lumen. The dispersive behavior of the division and migration reduces the variability in the final pattern and leads to the observed equispaced pattern (see Fig. 1d).

Coupling between lateral inhibition and cell division was also introduced to explain robust bristle patterning in the fly notum (Hunter et al. 2016). Hunter and colleagues provided evidence that NOTCH signaling between DELTA expressing SOP and its neighbors affects cell division time and that the level of signaling is higher for direct neighbors (with cell body contacts) compared to secondary neighbors (with filopodial contacts). It was also shown that once a cell has divided, it is no longer inhibited by or inhibits other cells through NOTCH signaling. The authors then developed a lateral inhibition model that takes these observations into account. The model uses the basic dynamic equations shown in (13)–(15) and in addition includes a time-dependent probability term for cell division. This probability depends on the amount of signaling that a cell receives where a cell with higher signaling is more likely to divide. Analysis of the model showed that the coupling to cell division provides an internal clock for this process leading to a more ordered pattern.

We note that in both examples discussed here (e.g. the intestine and SOP patterning), lateral inhibition dynamics are still assumed to occur prior to cell division events. Namely, that fates which are determined by lateral inhibition on a fixed lattice are then used as an input for time dependent cell division processes. Although this makes modeling simpler, it is not clear whether this is indeed the case in all situations. Taking both processes into account at the same time may require alternative modeling approaches such as agent based modeling (as discussed in the next section).

2.5 Modulating Lateral Inhibition by External Signals

In many developmental systems, NOTCH mediated lateral inhibition is coupled to additional extracellular signals that can introduce a spatial bias, or pre-pattern, which modulate the process. An example for the effect of external pre-pattern on lateral inhibition is the organization of *Drosophila* bristles into organized rows on the *Drosophila* notum (Fig. 2a). A model of this process was recently described by Corson and colleagues (Corson et al. 2017). This work addresses the question of how a striped pattern of SOPs is formed in the presence of an initial pre-pattern. The initial pre-pattern in this system is generated by a spatially varying expression profile of proneural genes that control the expression of DELTA. The authors adopt a simplified modeling approach, where the cellular state is described by a single state variable, rather than the two variables in the lateral inhibition model described above. This state variable ranges from 0 to 1, where the value 0 corresponds to an epidermal fate and the value 1 corresponds to a SOP fate. The state variable in the model depends on signals from its neighbors (as in the lateral inhibition model above) and on an external signal which originates from the external pre-pattern of DELTA (blue lines in Fig. 2a). It is shown, that given the external pre-pattern, the same regulatory circuit can account for the resolution of multiple stripes and for the emergence of single SOPs in organized rows. We note that also in this model, the range of inhibition is taken to be larger than one cell diameter, potentially through the action of long-range filopodia.

Another notable example for lateral inhibition coupled to external signal is angiogenesis, the process by which new blood vessels are formed. During angiogenesis, vascular endothelial growth factor (VEGF) induces sprouting of new blood vessels by converting endothelial stalk cells into tip cells that spearhead branching points (Blanco and Gerhardt 2013). NOTCH signaling regulates the selection of tip cells through lateral inhibition, where high levels of VEGF induce DELTA-

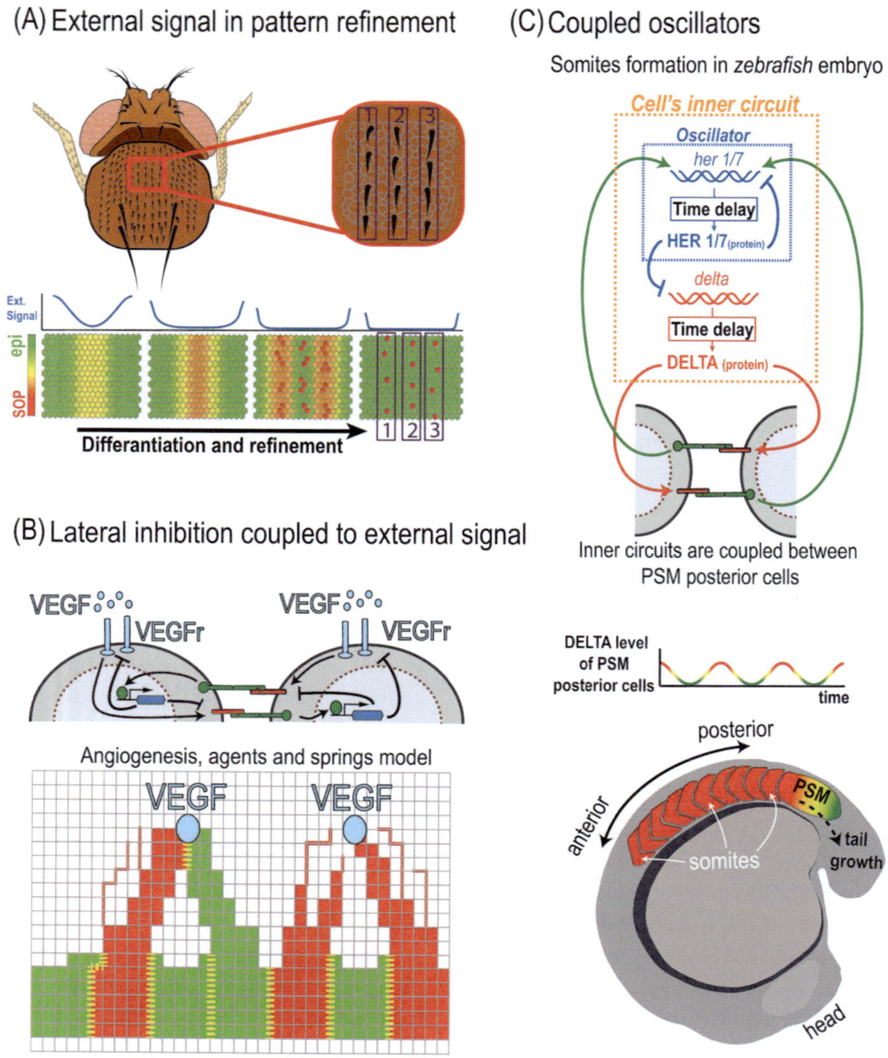

Fig. 2 Modulation of lateral inhibition by external signals and models for somitogenesis. (**a**) Lateral inhibition with external pre-pattern underlies refinement of *Drosophila* bristles into rows. Top – The small bristles on the *Drosophila* notum are arranged in distinct rows. Bottom – A model of the differentiation and refinement process of rows of bristles starting from a pre-pattern of pro-neural genes (external signal - blue lines). A continuous state variable ranges from an epidermal cell state (epi - green) and sensory organ precursor cell state (SOP - red). Pro-neural genes induce an external pre-pattern of DELTA expression which decreases at the center but remains high at the edges (blue lines). The combination of lateral inhibition and external signals underlies the formation of multiple rows as well as the emergence of distinct SOPs. (**b**) Lateral inhibition coupled to VEGF signals in angiogenesis. Top – Schematic representation of a sprouting angiogenesis model. In this model, a gradient of vascular endothelial growth factor (VEGF, light blue circles) induces a lateral inhibition circuit in endothelial cells. Here, activated VEGF receptors (VEGFr, light blue rods) induce DELTA-LIKE-4 (red) production to promote tip cell selection and repress tip cell fate in neighboring cells. The intracellular feedback (represented by the blue hexagon) also downregulates VEGFr production, making the stalk cells less sensitive to VEGF. Bottom – Schematic of an agent-based model of sprouting angiogenesis. An external VEGF source (light blue circles) creates a VEGF gradient which induces the sprouting of endothelial cells towards the source. Each cell in the model is represented by multiple finite element agents connected by springs (red and green intersections and squares in the grid). The model captures the elongation of tip cells (red) and stalk cells (green) towards the VEGF source. Lateral inhibition between neighboring cells determines tip versus stalk fates. The boundaries between cells are marked with

LIKE-4 (Dll4) at the tip cells. Dll4 activates NOTCH signaling that downregulates VEGF receptors (VEGFr) to suppress stalk cells from becoming tip cells themselves (Hellstrom et al. 2007). This regulatory feedback is coupled to the tip cell morphology, whereby tip cells extend filopodia towards the source of VEGF signal. As one cell stretches towards the source, it starts to gain a tip cell fate and pulls with it the neighboring cells that gain a stalk cell fate. One possible scenario observed is that two emerging tip cells will meet each other as they stretch towards the same source (Bentley et al. 2009). In this case, the two tip cells contact each other (anastomosis) and inhibit each other (through NOTCH signaling) so that one cell ends up changing back its fate to a stalk cell. This in turn can lead to emergence of new tip cells in other nearby positions.

This dynamic process has been modeled by Bentley and colleagues in several papers (Bentley et al. 2008; Bentley et al. 2009; Jakobsson et al. 2010). To take into account the morphological aspects of these processes, the authors used a finite element agent-based modeling approach (Fig. 1e). Unlike the lateral inhibition models described above in which the basic element is a cell, these models split each cell's membrane to small domains termed agents. These agents are located on a grid and are connected to each other by springs, representing the mechanical forces between them. During each time step in the simulation, the agents are free to move along the grid towards the source of the VEGF signal and retract according to a set of dynamical rules. As they move along the grid, new agents are generated at the space created between two adjacent agents, leading to extension of the membrane. This model considers dynamic extensions of the cells, where first filopodia are being extended from the membrane followed by the movement of the entire cell's membrane - if the conditions are right. In such an agent-based model, new connections between two emerging tip cells can be formed and NOTCH mediated lateral inhibition is initiated between them. Simulations of the model capture the typical branching dynamics during angiogenesis as well as the competition between attaching tip cells described above. Hence, this approach is particularly useful in describing situations where morphological dynamics are important.

Lateral inhibition is also coupled to diffusible ligands during the development of chick retina (Formosa-Jordan et al. 2012). In this system, lateral inhibition pattern controls neurogenesis. The area of active neurogenesis spreads through non-neurogenic regions in response to external morphogens (Sonic hedgehog), giving rise to a spreading wave front behavior. Formosa-Jordan and colleagues modeled the process by introducing a secreted morphogen from the already differentiated neurons. The morphogens spread and expand the region in which lateral inhibition is permitted, hence leading to the observed neurogenic wavefront. The authors also show that in order to get a robust propagating front, they have to assume that cells outside the neurogenic region express DELTA, which impose inhibiting boundary conditions and prevent random spreading of the propagating front. Hence, the combination of lateral inhibition and secreted morphogens gives rise to a robust pattern mediated by a propagating front.

Fig. 2 (continued) yellow dashed lines. Two tip cells can attach to one another as they migrate towards the source, initiating lateral inhibition at the new boundary. (**c**) A schematic of the *zebrafish* somitogenesis model. Top – An intracellular oscillator circuit (dashed blue rectangle) is coupled to an extracellular NOTCH-mediated lateral inhibition feedback (green and red arrows). Expression of *her1/7* and *delta* genes are down-regulated by the HER1/7 protein with a time delay. In parallel, DELTA activates NOTCH signaling in the neighboring cells. Finally, NOTCH signaling activates its target *her1/7* which closes the feedback loop. Bottom – DELTA levels in cells at the presomitic mesoderm (PSM) oscillate in synchrony and in phase. As the embryo elongates, somites (white arrows) are derived from the posterior end of the PSM, at the time where the cells are at high DELTA level phase

3 Modeling Somitogenesis

Although NOTCH signaling is typically involved in assigning distinct fates to neighboring cells, this is not always the case. A remarkable example for a situation where NOTCH is involved in synchronizing a population of cells occurs during somitogenesis in vertebrates. During somitogenesis, the future somites are sequentially formed from the anterior side of the presomitic mesoderm (PSM) in the embryo (Kimmel et al. 1995). The model which have been proposed to describe this process, is called the 'clock and wavefront' model (Cooke and Zeeman 1976). The basic idea of the model is that the cells in the PSM exhibit oscillations in gene expression, and at the same time respond to an FGF and retinoic acid gradients. During each oscillation period, cells which are at the right phase of the oscillation stop oscillating and become the new somite (Fig. 2c). This whole process repeats itself as the embryo elongates. Interestingly, this process crucially depends on NOTCH signaling, as both DELTA and other NOTCH pathway components exhibit oscillations in their expression levels (Jiang et al. 2000; Holley et al. 2002) .

The role of NOTCH signaling in this process was first elucidated in a seminal theoretical paper by Julien Lewis (Lewis 2003). Lewis proposed a model for *zebrafish* somitogenesis where each cell in the PSM contains an internal oscillator based on a delayed transactional feedback loop. He proposed that the NOTCH targets *her1* and *her7* (homologues of the Hes family), known to be transcriptional inhibitors, form the delayed negative feedback loop by transcriptionally repressing their own production (Fig. 2c). He then proposed that the role of NOTCH signaling is to synchronize the single cell oscillators leading to synchronized oscillations in the whole PSM.

The mathematical description of the Lewis model thus involves an internal delayed negative feedback, and coupling between cells through NOTCH signaling. The delay in this case is attributed to the time it takes to transcribe the *her1/7* mRNA and to translate the HER1/7 protein (we treat the Her1 and Her7 as a single entity in our model). The equations for the Her1/7 protein (H_i) and mRNA ($m_{H,i}$) in each cell i are therefore given by:

$$\frac{dH_i(t)}{dt} = \beta_H m_{H,i}(t-T_p) - \gamma_H H_i(t) \quad (16)$$

$$\frac{dm_{H,i}(t)}{dt} = \beta_m \left(\begin{array}{l} w_H \dfrac{p_H^l}{p_H^l + H_i(t-T_m)^l} \\ +w_D \dfrac{\langle D(t-T_m)\rangle_i^n}{p_D^n + \langle D(t-T_m)\rangle_i^n} \end{array} \right) - \gamma_m m_{H,i}(t) \quad (17)$$

where β_H and β_m are production rates of the protein and mRNA of Her1/7, respectively, and γ_H and γ_m are the degradation rates of the protein and mRNA of Her1/7, respectively. T_p and T_m are the delay times in translation and transcription, respectively. The production term in the equation of $m_{H,i}$ depends on a sum of two Hill functions representing the repression by H and the activation by DELTA (D) in neighboring cells, where w_H and w_D are the relative weights of these two contributions. We have used here a notation similar to the one in Eqs. (4)–(6) to describe these Hill functions. For simplicity, NOTCH is not taken explicitly into account here.

The equations for Delta mRNA, $m_{D,i}$, and protein, D_i, in each cell i are similarly given by:

$$\frac{dD_i(t)}{dt} = \beta_D m_{D,i}(t-T_p) - \gamma_D D_i(t) \quad (18)$$

$$\frac{dm_{D,i}(t)}{dt} = \beta_m \frac{p_H^l}{p_H^l + H_i(t-T_m)^l} - \gamma_m m_{D,i}(t) \quad (19)$$

where β_D and γ_D are the production and degradation rates of the DELTA protein, respectively.

Simulations of these equations show that under certain conditions on the parameters, a field of cells can maintain sustainable synchronized oscillations, where all the cells synchronize in phase. Out of phase oscillation are also possible for certain delay times. Interestingly, theses synchronized oscillation are quite robust to variability in the delay times and to perturbations of expression levels between neighboring cells. We note that without the delayed negative feedback these equations are reduced to the lateral inhibition model described above. More recent models of somitogenesis expanded the analysis to explain loss of global synchrony in the PSM (Riedel-Kruse et al. 2007) and the role of spatial gradients of expression within the PSM (Ay et al. 2014) on segmentation waves during somitogenesis.

4 NOTCH Signaling and Cell Morphology

4.1 Morphology Affects NOTCH Signaling and NOTCH Mediated Patterning

Despite the fact that changes in cellular and tissue morphology occur throughout development, their role during cell fate decision processes is usually neglected. In fact, changes in cell morphology are often considered as a downstream consequence of cell fate decisions rather than an integral part of these processes. In the context of NOTCH signaling, cell morphology can affect the magnitude of signaling between cells and thereby affecting cell fate decisions. Several recent papers address this issue and highlight the role of cell morphology on NOTCH mediated processes both experimentally and theoretically (Khait et al. 2015; Akanuma et al. 2016; Guisoni et al. 2017; Shaya et al. 2017).

One way in which cell morphology can influence NOTCH signaling is by affecting the contact area between cells. The question of how different contact geometries affect NOTCH signaling and NOTCH mediated patterning was recently addressed both theoretically and experimentally by Khait and colleagues (Khait et al. 2015) and Shaya and colleagues (Shaya et al. 2017). In the first work (Khait et al. 2015), the authors analyzed the interplay between membrane dynamics of NOTCH receptors and ligands and contact geometry. The authors developed a theoretical framework for analyzing how NOTCH signaling should depend on contact area, taking into account also membrane diffusion and endocytosis. The authors considered a simplified model of two cells, one expressing only NOTCH receptors and the other one expressing only DELTA ligands and analyzed how the membrane distribution of NOTCH and DELTA, as well as the magnitude of signaling, depend on the different parameters of the model. To take into account membrane diffusion and endocytosis, the authors used a reaction-diffusion model whose variables are the membrane concentrations of NOTCH (denoted by n), DELTA (denoted by d), NOTCH-DELTA complex (denoted by $[nd]$) and the total signal (denoted by S) (Fig. 3a). The equations for the two-cells case are given by:

$$\frac{dn}{dt} = D_n \nabla^2 n + k^n_{exo} n_0 - k^n_{endo}\, n + I_r(x) \left(\kappa^- [nd] - \kappa^+ nd\right) \tag{20}$$

$$\frac{dd}{dt} = D_d \nabla^2 d + k^d_{exo} d_0 - k^d_{endo} d + I_r(x) \left(\kappa^- [nd] - \kappa^+ nd\right) \tag{21}$$

$$\frac{d[nd]}{dt} = I_r(x) \begin{pmatrix} D_{[nd]} \nabla^2 [nd] + \kappa^+ nd \\ -\kappa^- [nd] - \kappa^s [nd] \end{pmatrix} \tag{22}$$

$$\frac{dS}{dt} = I_r(x) \kappa^s \int [nd] d^2 r - \gamma S \tag{23}$$

where D_n, D_d and D_{nd} are the diffusion rates for NOTCH, DELTA and the NOTCH-DELTA complex, respectively. n_0 and d_0, are the concentrations of the cytoplasmic pools of NOTCH and DELTA (assumed to be constant), respectively. The rates k^n_{exo}, k^n_{endo}, k^d_{exo} and k^d_{endo} denote endocytosis and exocytosis rates for NOTCH and DELTA, respectively. κ^+ and κ^- are the association and dissociation rates of NOTCH and DELTA, respectively. κ^S and γ are the rate associated with conversion of the NOTCH-DELTA complex into a signal and the signal degradation, respectively. $I_r(x)$ is a function that describes the spatial extent of the contact area where $I_r(x) = 1$ for $x < r_{contact}$ and zero elsewhere.

Analysis of these equations identified two possible scenarios. For relatively large contact areas and/or slow diffusion (e.g. in epithelial contacts), signaling strength is expected to be pro-

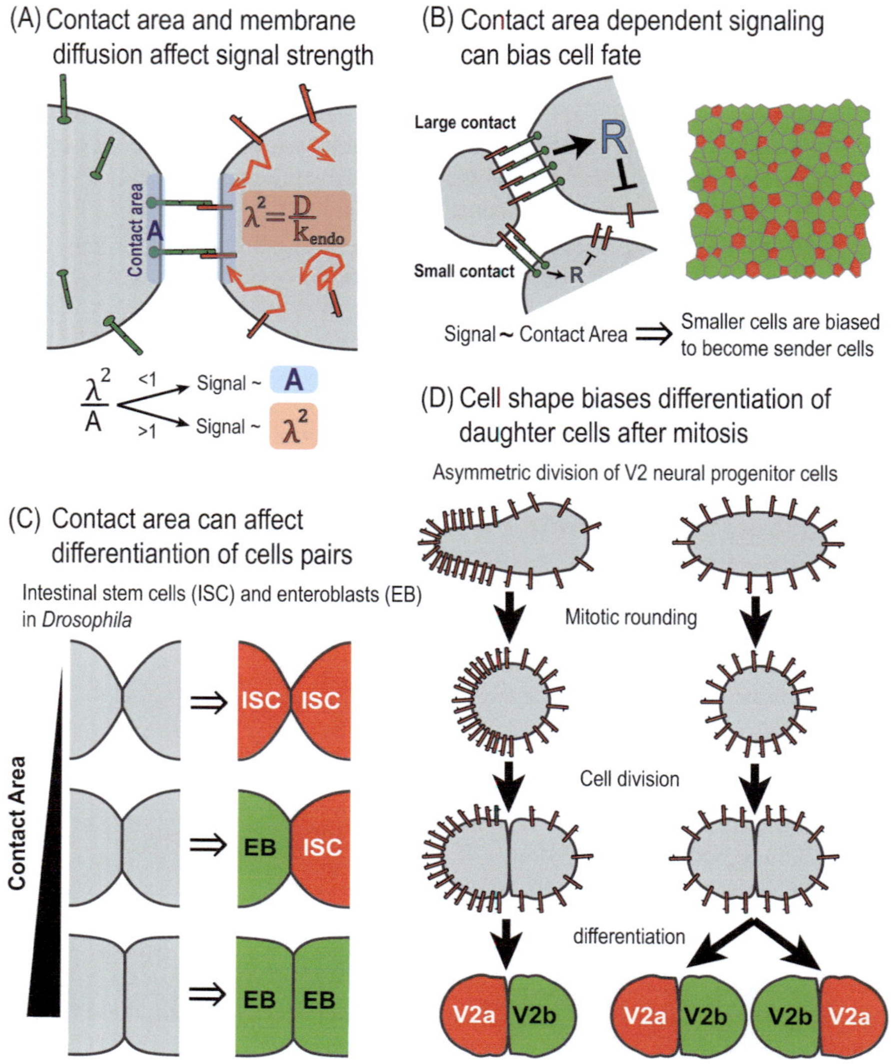

Fig. 3 NOTCH signaling and cell morphology. (**a**) Contact area and membrane diffusion affect signal strength. Schematic representation of a two-cell reaction-diffusion model that takes into account contact area and membrane dynamics. The model identified two distinct regimes depending on the ratio between the contact area, A, and the area defined by the diffusion length scale, λ^2. The diffusion length scale is defined by $\lambda = \sqrt{D / k_{endo}}$, where D and k_{endo} correspond to the diffusion coefficient and endocytosis rate of DELTA, respectively. While in one regime ($\lambda^2 < A$), signaling depends on the contact area, in the second regime ($\lambda^2 > A$), signaling is independent of the contact area. (**b**) Contact area dependent signaling can bias cell fate. Left - A schematic representation of a lateral inhibition model that takes into account the dependence of signaling on contact area. The model assumes that NOTCH signaling in each cell depends on the number of NOTCH-DELTA pairs formed on the boundaries with its neighbors, which is proportional to the length of the boundaries. Right – Simulations of the model over disordered cell lattices showed that smaller cells are more likely to become high DELTA cells (red). (**c**) Contact area can influence differentiation of cell pairs in the fly intestine. In the fly intestine, intestinal stem cells (ISC) divide and differentiate into either self-renewing ISC (red, high DELTA) or to enteroblasts (EB, green, low DELTA). The fates of the two daughter cells is determined by the lateral inhibition process and depends on the contact area between the two cells. The three possible final states are: ISC-ISC, in case of small contact area; EB-EB, in the case of large contact area; and ISC-EB, in the case of intermediate contact area. (**d**) Cell shape biases differentiation of daughter cells after mitosis in *zebrafish* neurogenesis. A schematic of a model for a lateral inhibition process which is biased by cell shape. Before mitosis, the concentration of DELTA is higher on the elongated side of the cell and lower on the round side of the cell. The asymmetry in DELTA concentration is maintained during mitosis and biases the lateral inhibition process so that the progenitor from the elongated side adopts the V2a fate (red) and the progenitor from the round side adopts the V2b fate (green)

portional to the contact area. By contrast, for relatively small contact areas and/or a fast diffusion regime (e.g. filopodia), signaling strength should be independent of contact area but dependent on the diffusion length scale of NOTCH receptors and ligands, defined by $\lambda = \sqrt{\frac{D}{k_{endo}}}$. Based on measurements of DELTA-LIKE-1 (Dll1) diffusion and endocytosis rates, the authors showed that the transition between the two regimes should occur for contact diameters on the order of 1-2 μm. An interesting possibility occurs in situations where signaling is proportional to the diffusion length scale (second scenario described above). In this case, the level of signaling can actually be regulated by controlling the effective diffusion properties of NOTCH ligands or receptors. Hence, both contact geometry and membrane dynamics of NOTCH receptors and ligands can play an important role in NOTCH-dependent processes.

But does signaling indeed correlates with contact area? This question was experimentally addressed by the work of Shaya and colleagues (Shaya et al. 2017) who used micropatterned devices to measure how NOTCH signaling depends on contact area. Consistent with the prediction of the Khait model, the authors found that NOTCH signaling indeed correlates with the contact width for contact diameters ranging from 1–40 μm.

Shaya and colleagues took this problem a step further and asked whether the dependence of NOTCH signaling on contact area can influence cell fate determination during lateral inhibition patterning. This was performed by expanding the lateral inhibition model to take into account the contact area and cell geometry (Fig. 3b). Instead of total NOTCH and DELTA levels at each cell, the expanded model followed the concentrations of NOTCH and DELTA on the cell boundaries (as in the Khait model above). For each boundary between cell i and cell j, we denote n_{ij} and d_{ij} as the concentrations of NOTCH and DELTA presented on the i-th cell, respectively. Similarly n_{ji} and d_{ji} denote the concentrations of NOTCH and DELTA presented on the j-th cell. The repressor level in each cell i, R_i, is then determined by the total signal received by cell i, from all its boundaries. Hence, the equations in this case are given by (Shaya et al. 2017):

$$\frac{dn_{ij}}{dt} = \frac{\beta_N}{L_i} - \gamma_N n_{ij} - \kappa_t^{-1} n_{ij} d_{ji} \quad (24)$$

$$\frac{dd_{ij}}{dt} = \frac{\beta_D}{L_i} \frac{p_R^l}{p_R^l + R_i^l} - \gamma_D d_{ij} - \kappa_t^{-1} n_{ji} d_{ij} \quad (25)$$

$$\frac{dR_i}{dt} = \beta_R \frac{\left(\sum_j \frac{\kappa_t^{-1}}{\gamma_S} n_{ij} d_{ji} l_{ij} \right)^m}{p_S^m + \left(\sum_j \frac{\kappa_t^{-1}}{\gamma_S} n_{ij} d_{ji} l_{ij} \right)^m} - \gamma_R R_i \quad (26)$$

Here, l_{ij} is the length of the i-j boundary and L_i is the perimeter of cell i. Cell size is also taken into account by normalizing the production rate of NOTCH and DELTA by the perimeter of the cell, L_i, meaning that the proteins distribute uniformly on the cell's membrane once produced. Overall, the number of Eqs. (24)–(26) is two times the number of boundaries plus one time the number of cells.

A major consequence derived from this model is how cell geometry affects cell fate. Simulating the model over a disordered lattice of cells revealed that smaller cells are more likely to become the high DELTA cells. This bias arises from an initial bias in inhibitory NOTCH signaling due to the differences in cell sizes across the lattice. Consistent with this prediction, Shaya and colleagues found that hair cell precursors in the developing chick inner ear are indeed smaller on average than non-hair cells (Shaya et al. 2017).

4.2 Contact Area Affects Differentiation of Cell Pairs

Another interesting example for the effect of contact area on cell fate was recently reported by Guisoni and colleagues (Guisoni et al. 2017). The authors described the differentiation dynamics of intestinal stem cells in the adult *Drosophila* using a two-cell lateral inhibition model. In this system, the intestinal stem cells (ISCs) divide and the daughter cells can adopt one of two different cell fates – either remain in ISC fate or differentiate into enteroblasts (EBs, precursors of enterocytes). The model taken in this work is the original Collier model (Collier et al. 1996). The dependence of the contact area on NOTCH signaling is introduced by assuming that the effective K_d in the Hill function, that describes activated NOTCH, is inversely proportional to the contact area [equivalent to assuming $p_S \sim \frac{1}{contact\ area}$ in Eq.(6)]. By defining a threshold activity level and analyzing the parameter space, the authors identified three distinct differentiation states: (i) both cells are high DELTA cells (ISCs); (ii) one cell is a high DELTA cell (ISC) and the other is a low DELTA cell (EB); and (iii) both cells are low DELTA cells (EBs). By setting the range of parameters to fit the experimental results from the fly's intestine development, the authors showed that the outcome of differentiation crucially depends on the contact area between the cells (Fig. 3c). As contact area increases, there is a transition from ISC:ISC pairs [state (i)] to EB:EB pairs [state (iii)]. Consistent with their prediction, the author showed that the observed differentiation states in *Drosophila* intestine indeed correlate with contact areas.

4.3 Cell Shape Biases Asymmetric Cell Division despite Mitotic Rounding

Another example for the effect of cell morphology on cell fate processes was recently reported in the context of asymmetric cell division in *zebrafish* neurogenesis (Akanuma et al. 2016). In this work, the authors showed that the asymmetric division of neural progenitor cells, termed V2 cells, in the developing *zebrafish* nervous system is affected by asymmetric cell elongation. They suggest that the DELTAC ligand is asymmetrically enriched to the more elongated side of the V2 cell, creating a bias in ligand concentration that is maintained during mitosis (Fig. 3d). This bias in local DELTAC concentration is translated to a bias in NOTCH signaling which is sufficient to define distinct cell fates for the two daughter cells. In order to model this process the authors coupled a lateral inhibition model to a cellular Potts model which simulates DELTAC ligands on a discrete grid representing the cell membrane [similar to the work described by Bentley and colleagues (Bentley et al. 2008)]. By running simulations in which cortical tension affects membrane dynamics of DELTAC, they show that sufficiently asymmetric cell shapes lead to asymmetry in DELTAC distributions that are sufficient to bias cell fates.

5 Boundary Formation and Multiple Ligands

5.1 Defining Sharp Boundaries with NOTCH Signaling

Another prototypical process known to be mediated by NOTCH signaling is that of boundary formation. Two examples where NOTCH signaling regulates boundaries are the wing margins (de Celis and Bray 1997; Klein et al. 1997; Micchelli et al. 1997) and wing veins in *Drosophila* (de Celis and Garcia-Bellido 1994; de Celis 1997). Although the role of NOTCH signaling in these two scenarios is to define boundary cells, the mechanism by which it operates is quite different. During vein boundary formation, NOTCH reads out a morphogen gradient to define the vein boundary. In the wing margin case on the other hand, NOTCH is used to specify boundary cells between two predefined compartments. Three recent works used mathematical models to understand the role of cis-inhibition in defining sharp boundaries and to elucidate the role of multiple NOTCH ligands in defining boundary cells (Sprinzak et al. 2010; Sprinzak et al. 2011; LeBon et al. 2014).

Wing vein boundaries are known to form in the wing imaginal disk through the interaction between epidermal growth factor (EGF) gradients and the NOTCH signaling. While the EGF gradients define the position of the veins, NOTCH

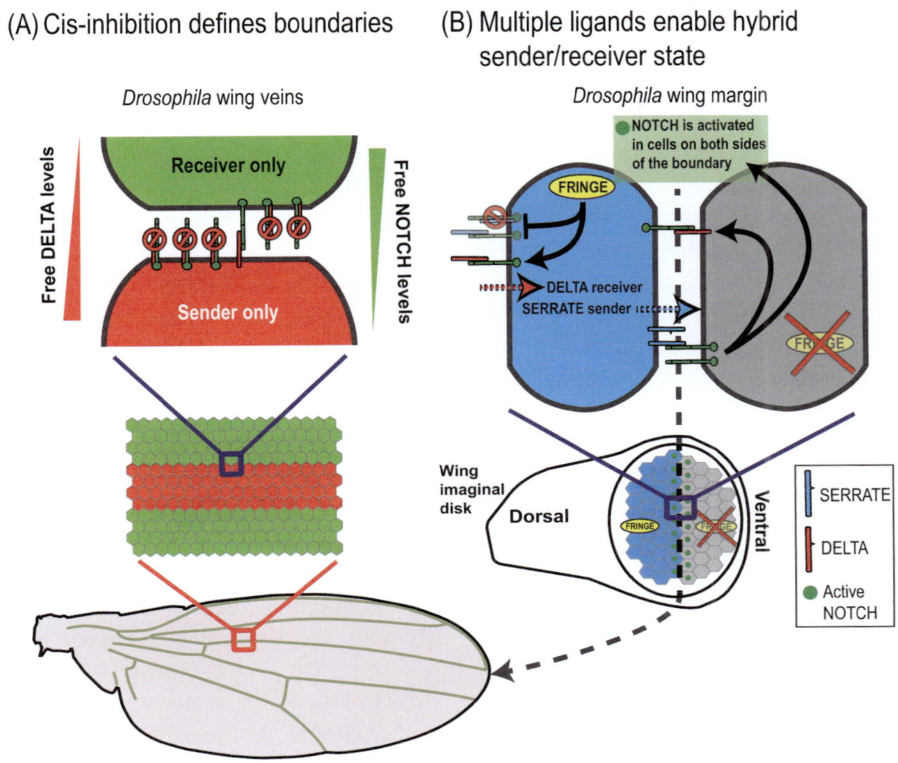

Fig. 4 NOTCH signaling in boundary formation. (**a**) Cis-inhibition defines boundaries. A schematic overview of the cis-inhibition model for the wing vein boundary in *Drosophila*. In this model, wing vein boundaries are defined by the interactions between "sender" cells and "receiver" cells. The cells in the vein region (red) express more DELTA than NOTCH while the cells in the inter-vein region (green) express more NOTCH than DELTA. Due to cis-interactions between NOTCH receptors and ligands, vein cells can send but not receive signals, while inter-vein cells can receive but not send signals. (**b**) Multiple ligands enable hybrid sender/receiver states. A schematic model for wing margin cells in *Drosophila* taking into account multiple ligands and modulation by FRINGE. The cells on the dorsal side (blue) express both DELTA, SERRATE and FRINGE. The FRINGE glycosyltransferase modulates the cis- and trans-interactions between NOTCH receptors and ligands. Expression of FRINGE promotes NOTCH-DELTA interactions and suppresses NOTCH-SERRATE interactions (both in cis and in trans). The model predicts that the dorsal boundary cells (blue) can simultaneously receive signals from DELTA expressing cells and send signals to the ventral boundary cells (gray) using the SERRATE ligands. At a later stage, the ventral boundary cells activate DELTA leading to NOTCH activation in the dorsal boundary cells. This situation leads to activation of cells only on the wing margin (green dots in bottom image)

signaling is involved in setting up the boundary between vein and inter-vein regions (de Celis 1997). One of the main questions in this process is how a graded concentration of a morphogen can be converted into a sharp all-or-none signal that defines the vein boundary. Sprinzak and colleagues used mathematical modeling to show that cis-inhibition between receptors and ligands plays a crucial role in generating the sharp response required for defining the boundary (Sprinzak et al. 2010; Sprinzak et al. 2011). The key insight obtained from the model was that cis-inhibition creates a situation where cells that express both receptors and ligands can either be in a "sender only" state or "receiver only" state, depending on the relative levels of NOTCH and DELTA that they express (Fig. 4a). The authors modeled the vein by assuming that DELTA production was graded (controlled by graded EGF signaling), while NOTCH production was constant (at least initially). The simulations showed that the resulting signaling profile exhibited two sharp stripes defining the vein boundaries that occurred at the positions in which the transition from sender to receiver states occurs (Fig. 4a). This pattern was consistent with the observed expression pattern of NOTCH transcriptional reporters [E(sp)] in the wing.

The model also provided an insight into a long-standing unintuitive observation of the system. It has been shown, that while heterozygous mutants of both NOTCH and DELTA exhibited mutant wing phenotypes (albeit different ones), the double heterozygous mutant restored the wildtype scenario (de Celis 2000). This observation is readily explained by the model, since in the double heterozygous mutant the relative levels of NOTCH and DELTA are maintained and so are the transition points between sender cells and receiver cells.

5.2 Combination of NOTCH Ligands Expand the Repertoire of Signaling States

More recently, LeBon and colleagues (LeBon et al. 2014) expanded the analysis to include multiple NOTCH ligands as well as the modulation of the receptor-ligand interactions by FRINGE glycosyltranferases. Experimental analysis in both mammalian cell culture and *Drosophila* showed that while glycosylation by FRINGE upregulates both cis- and trans-interactions between NOTCH1 and Dll1, it had an opposite effect on cis- and trans-interactions between NOTCH1 and JAG1. The combined effect of these interactions revealed that cells can be in different cellular states depending on the combination of ligands and FRINGE modulators they express. For example, cells that express both DELTA, SERRATE (JAG1 homolog in *Drosophila*), and FRINGE can receive signals from DELTA expressing cells (e.g. they are "DELTA receivers") while at the same time they can send out signals with their SERRATE ligands (e.g. "SERRATE senders") (Fig. 4b). This dual SERRATE sender/DELTA receiver cellular state can explain the bidirectional signaling observed in the wing margin cells (Fig. 4b). More generally, the model provided a framework for defining the possible sender/receiver states (based on the combination and levels of Notch receptors, Notch ligands, and FRINGES) as well as their ability to activate or get activated by other cellular states.

The effect of combination of multiple Notch ligands was also discussed in two other recent works by Petrovic and colleagues (Petrovic et al. 2014) and by Boreato and colleagues (Boareto et al. 2015). Petrovic and colleagues (Petrovic et al. 2014) showed that a circuit in which NOTCH signaling downregulates Dll1 but activates JAG1 can explain the transition from a lateral induction process (e.g. NOTCH signal induced higher ligand expression in the neighboring cell) that defines the chick inner ear sensory epithelium, to a lateral inhibition process that establish the alternating patterns of hair cells and supporting cells. A theoretical work by Boareto and colleagues (Boareto et al. 2015) showed that under certain conditions, when NOTCH signaling oppositely regulates DELTA and JAGGED, it is possible to obtain a three stable state solution corresponding to a full sender, a full receiver and a hybrid sender/receiver state. It remains to be seen whether such hybrid states are indeed observed experimentally. Overall, it is clear that the combinatorial action of multiple NOTCH receptors and ligands introduces another level of complexity which calls for additional theoretical and experimental works.

6 Future Perspectives

Despite the significant progress in modeling NOTCH mediated developmental processes as described here, it is clear that many questions still remain open. Some of the topics that still need be elucidated include the integration of morphological, regulatory and cell division processes, the role of multiple NOTCH receptors and ligands and the combined interaction between NOTCH and other signaling pathways. As more quantitative experimental data becomes available, it is expected that novel modeling approaches and deeper refinement of existing models will follow.

Acknowledgments This manuscript was supported by a grant from the European Research Council (grant no. 682161).

References

Akanuma T, Chen C, Sato T, Merks RM, Sato TN (2016) Memory of cell shape biases stochastic fate decision-making despite mitotic rounding. Nat Commun 7:11963. https://doi.org/10.1038/ncomms11963. ncomms11963 [pii]

Artavanis-Tsakonas S, Muskavitch MA (2010) Notch: the past, the present, and the future. Curr Top Dev Biol 92:1–29. doi: S0070-2153(10)92001-2 [pii]. https://doi.org/10.1016/S0070-2153(10)92001-2

Artavanis-Tsakonas S, Rand M, Lake R (1999) Notch signaling: cell fate control and signal integration in development. Science 284:770

Ay A et al (2014) Spatial gradients of protein-level time delays set the pace of the traveling segmentation clock waves. Development 141:4158–4167. https://doi.org/10.1242/dev.111930. 141/21/4158 [p ii]

Barad O, Rosin D, Hornstein E, Barkai N (2010) Error minimization in lateral inhibition circuits. Sci Signal 3:ra51. doi: 3/129/ra51 [pii]. https://doi.org/10.1126/scisignal.2000857

Bentley K, Gerhardt H, Bates PA (2008) Agent-based simulation of notch-mediated tip cell selection in angiogenic sprout initialisation. J Theor Biol 250:25–36. doi: S0022-5193(07)00443-2 [pii]. https://doi.org/10.1016/j.jtbi.2007.09.015

Bentley K, Mariggi G, Gerhardt H, Bates PA (2009) Tipping the balance: robustness of tip cell selection, migration and fusion in angiogenesis. PLoS Comput Biol 5:e1000549. https://doi.org/10.1371/journal.pcbi.1000549

Blanco R, Gerhardt H (2013) VEGF and Notch in tip and stalk cell selection. Cold Spring Harb Perspect Med 3:a006569. https://doi.org/10.1101/cshperspect.a006569. a006569 [pii], cshperspect.a006569 [pii]

Boareto M, Jolly MK, Lu M, Onuchic JN, Clementi C, Ben-Jacob E (2015) Jagged-Delta asymmetry in Notch signaling can give rise to a Sender/Receiver hybrid phenotype. Proc Natl Acad Sci U S A 112:E402–E409. https://doi.org/10.1073/pnas.1416287112. 1416287112 [pii]

Cohen M, Georgiou M, Stevenson NL, Miodownik M, Baum B (2010) Dynamic filopodia transmit intermittent Delta-Notch signaling to drive pattern refinement during lateral inhibition. Dev Cell 19:78–89. doi: S1534-5807(10)00296-0 [pii]. https://doi.org/10.1016/j.devcel.2010.06.006

Collier JR, Monk NA, Maini PK, Lewis JH (1996) Pattern formation by lateral inhibition with feedback: a mathematical model of delta-notch intercellular signalling. J Theor Biol 183:429–446. doi:S0022-5193(96)90233-7 [pii]. https://doi.org/10.1006/jtbi.1996.0233

Cooke J, Zeeman EC (1976) A clock and wavefront model for control of the number of repeated structures during animal morphogenesis. J Theor Biol 58:455–476

Corson F, Couturier L, Rouault H, Mazouni K, Schweisguth F (2017) Self-organized Notch dynamics generate stereotyped sensory organ patterns in Drosophila. Science 356. doi: eaai7407 [pii], science.aai7407 [pii]. https://doi.org/10.1126/science.aai7407

Daudet N, Lewis J (2005) Two contrasting roles for Notch activity in chick inner ear development: specification of prosensory patches and lateral inhibition of hair-cell differentiation. Development 132:541–551. doi: dev.01589 [pii]. https://doi.org/10.1242/dev.01589

de Celis J (1997) Notch signalling regulates veinlet expression and establishes boundaries between veins and interveins in the Drosophila wing. Development 124:1919–1928

de Celis J (2000) The Abruptex domain of Notch regulates negative interactions between Notch, its ligands and Fringe. Development 127:1291–1302

de Celis JF, Bray S (1997) Feed-back mechanisms affecting Notch activation at the dorsoventral boundary in the Drosophila wing. Development 124:3241–3251

de Celis J, Garcia-Bellido A (1994) Roles of the Notch gene in Drosophila wing morphogenesis. Mech Dev 46:109–122

Eom DS, Bain EJ, Patterson LB, Grout ME, Parichy DM (2015) Long-distance communication by specialized cellular projections during pigment pattern development and evolution. elife 4. https://doi.org/10.7554/eLife.12401. e12401 [pii]

Formosa-Jordan P, Ibanes M (2014) Competition in notch signaling with cis enriches cell fate decisions. PLoS One 9:e95744. https://doi.org/10.1371/journal.pone.0095744. PONE-D-13-50216 [pii]

Formosa-Jordan P, Sprinzak D (2014) Modeling Notch signaling: a practical tutorial. Methods Mol Biol 1187:285–310. https://doi.org/10.1007/978-1-4939-1139-4_22

Formosa-Jordan P, Ibanes M, Ares S, Frade JM (2012) Regulation of neuronal differentiation at the neurogenic wavefront. Development 139:2321–2329. https://doi.org/10.1242/dev.076406. 139/13/2321 [pii]

Glass DS, Jin X, Riedel-Kruse IH (2016) Signaling delays preclude defects in lateral inhibition patterning. Phys Rev Lett 116:128102. https://doi.org/10.1103/PhysRevLett.116.128102

Guisoni N, Martinez-Corral R, Garcia Ojalvo J, de Navascues J (2017) Diversity of fate outcomes in cell pairs under lateral inhibition. Development. https://doi.org/10.1242/dev.137950

Hadjivasiliou Z, Hunter GL, Baum B (2016) A new mechanism for spatial pattern formation via lateral and protrusion-mediated lateral signalling. J R Soc Interface 13. doi: 20160484 [pii], rsif.2016.0484 [pii]. https://doi.org/10.1098/rsif.2016.0484

Hamada H et al (2014) Involvement of Delta/Notch signaling in zebrafish adult pigment stripe patterning. Development 141:318–324. https://doi.org/10.1242/dev.099804. dev.099804 [pii]

Heitzler P, Simpson P (1991) The choice of cell fate in the epidermis of Drosophila. Cell 64:1083–1092

Hellstrom M et al (2007) Dll4 signalling through Notch1 regulates formation of tip cells during angiogenesis. Nature 445:776–780. doi: nature05571 [pii]. https://doi.org/10.1038/nature05571

Holley SA, Julich D, Rauch GJ, Geisler R, Nusslein-Volhard C (2002) her1 and the notch pathway function within the oscillator mechanism that regulates zebrafish somitogenesis. Development 129:1175–1183

Hunter GL et al (2016) Coordinated control of Notch/Delta signalling and cell cycle progression drives lateral inhibition-mediated tissue patterning. Development 143:2305–2310. https://doi.org/10.1242/dev.134213. dev.134213 [pii]

Jakobsson L et al (2010) Endothelial cells dynamically compete for the tip cell position during angiogenic sprouting. Nat Cell Biol 12:943–953. https://doi.org/10.1038/ncb2103. ncb2103 [pii]

Jiang YJ, Aerne BL, Smithers L, Haddon C, Ish-Horowicz D, Lewis J (2000) Notch signalling and the synchronization of the somite segmentation clock. Nature 408:475–479. https://doi.org/10.1038/35044091

Khait I et al. (2015) Quantitative analysis of Delta-like-1 membrane dynamics elucidates the role of contact geometry on Notch signaling. Cell Rep 14(2): 225–233. https://doi.org/10.1016/j.celrep.2015.12.040

Kimmel CB, Ballard WW, Kimmel SR, Ullmann B, Schilling TF (1995) Stages of embryonic development of the zebrafish. Dev Dyn 203:253–310. https://doi.org/10.1002/aja.1002030302

Klein T, Brennan K, Arias AM (1997) An intrinsic dominant negative activity of serrate that is modulated during wing development in Drosophila. Dev Biol 189:123–134. S0012-1606(97)98564-X [pii]. https://doi.org/10.1006/dbio.1997.8564

Kondo S, Iwashita M, Yamaguchi M (2009) How animals get their skin patterns: fish pigment pattern as a live Turing wave. Int J Dev Biol 53:851–856. https://doi.org/10.1387/ijdb.072502sk. 072502sk [pii]

Kornberg TB, Roy S (2014) Cytonemes as specialized signaling filopodia. Development 141:729–736. https://doi.org/10.1242/dev.086223. 141/4/729 [pii]

Kovall RA, Gebelein B, Sprinzak D, Kopan R (2017) The canonical notch signaling pathway: structural and biochemical insights into shape, sugar, and force. Dev Cell 41:228–241. S1534–5807(17)30294-0 [pii]. https://doi.org/10.1016/j.devcel.2017.04.001

LeBon L, Lee TV, Sprinzak D, Jafar-Nejad H, Elowitz MB (2014) Fringe proteins modulate Notch-ligand cis and trans interactions to specify signaling states. Elife 3:e02950. https://doi.org/10.7554/eLife.02950

Lewis J (2003) Autoinhibition with transcriptional delay: a simple mechanism for the zebrafish somitogenesis oscillator. Curr Biol 13:1398–1408. doi: S0960-9822(03)00534-7 [pii]

Meinhardt H (1996) Models of biological pattern formation: common mechanism in plant and animal development. Int J Dev Biol 40:123–134

Meinhardt H (2008) Models of biological pattern formation: from elementary steps to the organization of embryonic axes. Curr Top Dev Biol 81:1–63. S0070-2153(07)81001-5 [pii]. https://doi.org/10.1016/S0070-2153(07)81001-5

Micchelli CA, Rulifson EJ, Blair SS (1997) The function and regulation of cut expression on the wing margin of Drosophila: Notch, Wingless and a dominant negative role for Delta and Serrate. Development 124:1485–1495

Petrovic J et al (2014) Ligand-dependent Notch signaling strength orchestrates lateral induction and lateral inhibition in the developing inner ear. Development 141:2313–2324. https://doi.org/10.1242/dev.108100. dev.108100 [pii] [pii]

Riedel-Kruse IH, Muller C, Oates AC (2007) Synchrony dynamics during initiation, failure, and rescue of the segmentation clock. Science 317:1911–1915. 1142538 [pii]. https://doi.org/10.1126/science.1142538

Sancho R, Cremona CA, Behrens A (2015) Stem cell and progenitor fate in the mammalian intestine: Notch and lateral inhibition in homeostasis and disease. EMBO Rep 16:571–581. https://doi.org/10.15252/embr.201540188. embr.201540188 [pii]

Shaya O, Sprinzak D (2011) From Notch signaling to fine-grained patterning: Modeling meets experiments. Curr Opin Genet Dev 21:732–739. S0959-437X(11)00119-5 [pii]. https://doi.org/10.1016/j.gde.2011.07.007

Shaya O et al (2017) Cell-cell contact area affects Notch signaling and Notch dependent patterning. Dev Cell 40:1–7. https://doi.org/10.1016/j.devcel.2017.02.009

Sprinzak D et al (2010) Cis-interactions between Notch and Delta generate mutually exclusive signalling states. Nature 465:86–90. nature08959 [pii]. https://doi.org/10.1038/nature08959

Sprinzak D, Lakhanpal A, LeBon L, Garcia-Ojalvo J, Elowitz MB (2011) Mutual inactivation of Notch receptors and ligands facilitates developmental patterning. PLoS Comput Biol 7:e1002069. https://doi.org/10.1371/journal.pcbi.1002069. PCOMPBIOL-D-10-00231 [pii]

Toth B, Ben-Moshe S, Gavish A, Barkai N, Itzkovitz S (2017) Early commitment and robust differentiation in colonic crypts. Mol Syst Biol 13:902. https://doi.org/10.15252/msb.20167283

Turing AM (1952) The chemical basis of morphogenesis. Phil Trans R Soc London Ser B 237:37–72

Vasilopoulos G, Painter KJ (2016) Pattern formation in discrete cell tissues under long range filopodia-based direct cell to cell contact. Math Biosci 273:1–15. https://doi.org/10.1016/j.mbs.2015.12.008. S0025-5564(15)00270-9 [pii]

Wigglesworth VB (1940) The determination of characters at metamorphosis in Rhodnius prolixus. J Exp Biol 17:201–223

Endocytic Trafficking of the Notch Receptor

Björn Schnute, Tobias Troost, and Thomas Klein

Abstract

The endosomal pathway plays an important role in several aspects of Notch signalling ranging from ligand-dependent to independent activation and also degradation of the Notch receptor. Here, we will focus on its role during receptor degradation and describe the endosomal pathway with the components that are important for Notch degradation and the molecular machinery that orchestrates these events. Subsequently, we will describe the journey of Notch through the endosomal system and discuss the role of the genes involved. Mechanisms of the recently discovered ligand-independent activation of the Notch receptor in the endosomal pathway will be described and its contribution in physiologically Notch-dependent processes will be discussed. Last but not least, we will summarize the evidence for endosomal ligand-independent activation of the Notch pathway in vertebrates.

Keywords

Notch · Endocytosis · Endosomal trafficking · Endosome · Lysosome · ESCRT · Ubiquitination · Dx · Su(dx) · Lgd · Shrub · CHMP · Endosomal maturation · Degradation · Recycling

B. Schnute · T. Troost · T. Klein (✉)
Institute of Genetics, Heinrich-Heine-University, Duesseldorf, Germany
e-mail: thomas.klein@uni-duesseldorf.de

Abbreviations

AAA-ATPase	ATPase Associated with diverse cellular Activities
ADAM10	A Disintegrin and metalloprotease 10
Aki1	Akt Kinase-Interacting Protein 1
ANK	Ankyrin
AP	Adaptor Protein
CADASIL	Cerebral Autosomal Dominant Artheriopathy with Subcortical Infarcts and Leukoencephalopathy
Cbl	Casistas B-lineage lymphoma
CC2D1A	Coiled-Coil and C2 domain-containing protein 1A
CC2D1B	Coiled-Coil and C2 domain-containing protein 1B
CCZ1	Calcium-Caffeine-zinc sensitivity protein
CHMP	Charged multivesicular body protein
COMMD9	COMM (Copper metabolism) domain containing protein 9
CORVET	class C core vacuole/endosome tethering
Crb	Crumbs
CSL	CBF-1, Suppressor of Hairless, LAG-1
Dl	Delta
Dll3	Delta-like 3
DM14	Drosophila melanogaster 14

T. Borggrefe, B. D. Giaimo (eds.), *Molecular Mechanisms of Notch Signaling*, Advances in Experimental Medicine and Biology 1066, https://doi.org/10.1007/978-3-319-89512-3_6

dNedd4	Neural precursor cell expressed developmentally downregulated protein 4
DSL	(Delta/Serrate/LAG-2)
Dx	Deltex
EE	Early Endosome
EE2A	early endosomal antigene 2
EEVs	Early Endosomal Vesicles
EGF	Epidermal growth factor
ESCRT	Endosomal sorting complex required for transport
FGF	Fibroblast growth factor
Freud-1	FRE under Dual Repression-Binding-Protein 1
Freud-2	FRE under Dual Repression-Binding-Protein 2
FYVE	Fab1 YOTB VAC1 EEA1
GAP	GTPase activating protein
GDF	GTPase dissociation factor
GDI	GDP-dissociation inhibitor
GDP	Guanosine diphosphate
GEF	Guanine nucleotide exchange factor
GFP	Green fluorescent protein
GPI	Glycosylphosphatidylinositol
GTP	Guanosine triphosphate
HOPS	homotypic fusion and protein sorting
Hrs	Hepatocyte growth factor-regulated tyrosine kinase substrate
Hsc70	Heat shock cognate70
ICD	intracellular domain
ILV	intraluminal vesicle
Kuz	Kuzbanian
Lamp	Lysosome-associated membrane glycoprotein
Lgd	Lethal giant discs
LNR	Lin-12 / Notch repeat
ME	Maturing Endosome
Mon1	Monensin sensitivity protein 1
MVB	Multivesicular body
NECD	Notch extracellular domain
NEXT	Notch extracellular truncation
NICD	Notch intracellular domain
NRR	Negative Regulatory Region
PEST	Proline (P), Glutamic acid (E), Serine (S), Threonin(T)
PI(3,5)P2	Phosphatidylinositol 3,5-bisphoshphate
PI3P	Phosphatidylinositol 3-phosphate
Rab	ras-related in brain
RAM	RBPJ-associated molecule
RBPJ	recombination signal binding protein for immunoglobulin kappa J region
RE	Recycling Endosome
RME8	Receptor mediated Endocytosis 8
Ser	Serrate
Shrb	Shrub
SNARE	*s*oluble *N*-ethylmaleimide-sensitive-factor *a*ttachment *re*ceptor
SNX	Sorting nexin
Stam	Signal transducing adaptor molecule
Su(Dx)	Suppressor of Deltex
Su(H)	Suppressor of Hairless
TAPE	TBK1-associated Protein in Endolysosomes
TMPs	Transmembrane proteins
Tsg101	tumor susceptibility gene 101
Ub	Ubiquitin
UIM	ubiquitin interacting motif
Vps	Vacoular protein sorting

1 Introduction

The Notch signalling pathway is a fundamental pathway that mediates short-range communication between directly neighboured cells (Aster et al. 2017; Kovall et al. 2017). It is present in the genomes of all metazoans and involved in an uncountable number of developmental and homeostatic processes. Changes in the activity of the pathway during homeostasis results in various diseases ranging from inherited ones such as CADASIL (Cerebral Autosomal Dominant Artheriopathy with Subcortical Infarcts and Leukoencephalopathy) to various cancers in humans (Aster et al. 2017; Mašek and Andersson 2017). The pathway is activated by binding of a DSL (Delta/Serrate/LAG-2) ligand, in *Drosophila melanogaster* Delta (Dl) or Serrate (Ser), to the

Notch receptor. Dl and Ser are type I transmembrane proteins that can activate Notch only on adjacent cells. The result of this interaction is a sequence of two proteolytic cleavages that release the Notch intracellular domain (NICD) into the cytosol from which it translocates into the nucleus. In the nucleus, it associates with the CSL transcription factor Su(H) (Suppressor of Hairless; in mammals RBPJ (recombination signal binding protein for immunoglobulin kappa J region), also known as CSL (CBF-1, Suppressor of Hairless, LAG-1) and various co-factors to activate the target genes of the pathway.

The first ligand-dependent cleavage is performed by the metalloprotease ADAM10 (A Disintegrin and metalloprotease 10), encoded by *kuzbanian* (*kuz*) in *Drosophila*. This S2 cleavage removes the Notch extracellular domain (NECD) in a process of ecto-domain shedding. The freed NECD is endocytosed with the bound ligand into the signal-sending cell. The membrane-inserted truncated remnant, termed NEXT (Notch extracellular truncation), is cleaved by the γ-secretase complex in the transmembrane domain to release NICD (S3 cleavage). It appears that the γ-secretase is located in all intracellular membrane compartments, including endosomes and lysosomes (Schröder and Saftig 2016).

Whereas *Drosophila* has only one Notch receptor, mammals have four orthologs, termed Notch1–4 (Aster et al. 2017). All Notch receptors are type I transmembrane proteins that share the same general organisation with 29–36 EGF (Epidermal Growth Factor) repeats followed by 3 LNR (Lin-12 / Notch repeat) repeats and a hetero-dimerisation domain, together termed NRR (Negative Regulatory Region), in their extracellular domain. In Notch 1 and Notch 2 the EGF 11 and 12 are central for their interaction with the DSL domains of the ligands. In their ICDs, they share a RAM (RBPJ-associated molecule) domain followed by six Ankyrin (ANK) repeats and a PEST (enriched in proline (P), glutamic acid (E), serine (S), threonin(T)) domain.

Notch is a heterodimer consisting of the NECD and a membrane inserted intracellular domain (N^{TM}). Both parts are connected via non-covalent Ca^{2+} salt bridges located in the extracellular NRR close to the plasma membrane. The Notch heterodimer can be dissolved by depletion of Ca^{2+} from the culture medium of cell culture cells (Rand et al. 2000). The shedding of NECD results in a NEXT-like fragment that is subsequently cleaved by the γ-secretase-containing complex. Hence, depletion of Ca^{2+} can result in ligand-independent activation of the Notch pathway. In agreement with this finding it has been shown that γ-secretase cleaves many variants of Notch with a small extracellular domain in *Drosophila* (Struhl and Adachi 2000). Thus, also variants similar to NEXT that have been created by different mechanisms, e. g. by Ca^{2+} depletion, will be cleaved and will activate the pathway.

Work in the last decade revealed that the fundamental signalling pathways including the Notch pathway are tightly interwoven with the endosomal pathway (Dobrowolski and De Robertis 2012). It further revealed that the Notch pathway can be activated in a ligand-independent manner in some endosomal compartments.

In this chapter, we summarise recent findings of how Notch traffics through the endosomal pathway and how it is activated in particular endosomal compartments. We discuss whether and how this activation contributes to Notch signalling during development. We will focus on the analysis with the model *Drosophila* where most of the results are obtained. The endosomal pathway also plays an important but different role in generating a directed Notch signal during asymmetric cell division that is not discussed here. The reader is referred to some excellent recent reviews (Kandachar and Roegiers 2012; Schweisguth 2015) and "Notch and T Cell Function – A Complex Tale" within this book (Bigas and Porcheri).

2 The Endosomal Pathway

Transmembrane proteins (TMPs) like the Notch receptor are generally degraded in the lumen of the lysosome, which is located close to the nucleus in most cells (Fig. 1). To be degraded, a TMP has to be transported to and transferred into the lumen of the lysosome. These requirements

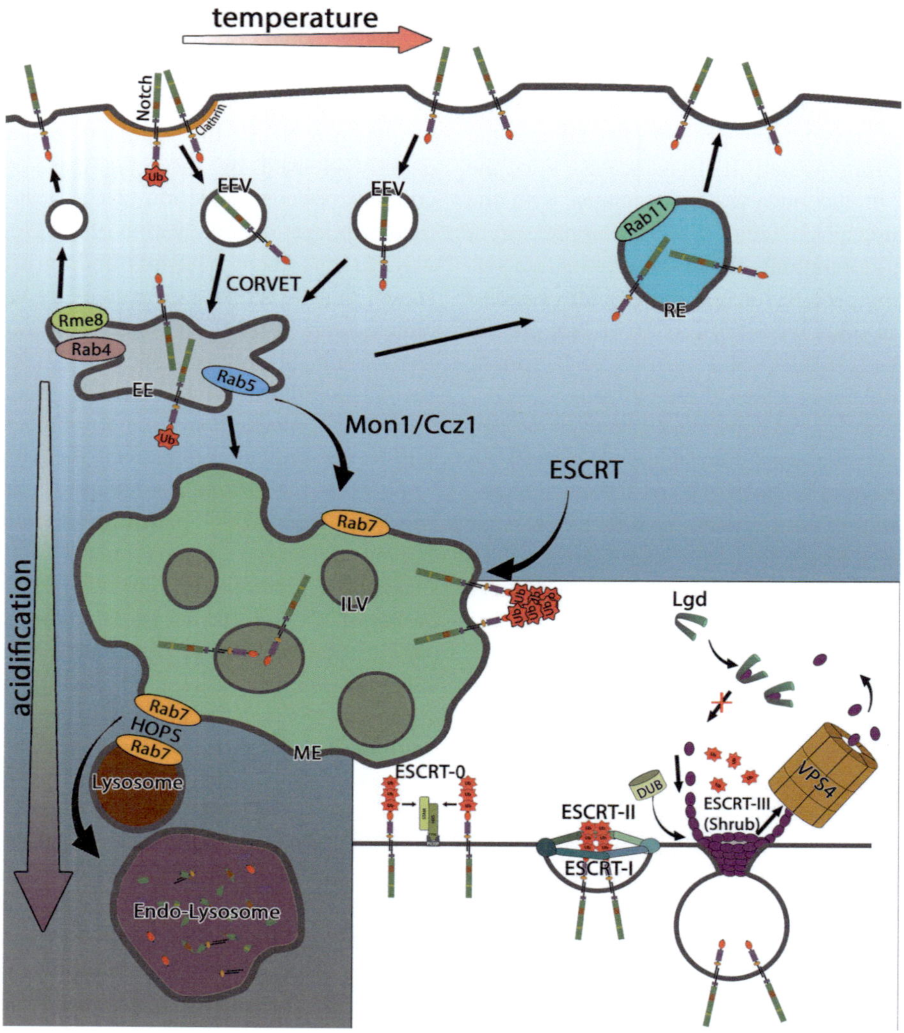

Fig. 1 Trafficking of Notch through the endosomal pathway to the lysosome. The journey is initiated by endocytosis which, depending on the temperature, can be clathrin-dependent or -independent. Endocytosis incorporates Notch into EEVs (early endocytic vesicles) that fuse together to form the EE (early endosome). EEVs can also fuse with already existing EEs. The fusion involves the CORVET tethering complex. The events in the early phase are organised by Rab5. Notch in the EE can be recycled back to the plasma membrane via a Rab4/Rme8-dependent pathway or via the slow Rab11-dependent pathway. The majority of Notch remains in the EE which matures and fuses with the lysosome where Notch and other cargo are degraded in the lumen of the endolysosome. Maturation includes the incorporation of Notch into ILVs, acidification by V-ATPase and replacement of Rab5 by Rab7 (Rab conversion). The Rab7 GEF Mon1/CCz1 is involved in Rab conversion. Rab7 organises the HOPS-mediated fusion with the lysosome. **Insert**: Formation of ILVs by the ESCRT machinery. ESCRT-O is recruited by the Rab5-induced production of PI(3)P on the cytosolic surface of the endosomal membrane. ESCRT-0 concentrates ubiquitylated cargo, including Notch. It also recruits ESCRT-I, which in turn includes ESCRT-II. The assembled super-complex induces a pit in the limiting membrane of the ME and also the polymerisation of the ESCRT-III core component Shrub/Chmp4/Snf7 into a filament. ESCRT-III, together with the Vps4 complex, pinches off the membrane in a so far not understood manner. In this way also the ICDs of Notch and other TMPs are translocated into the lumen of the endosome and separated from the cytosol. Vps4 also disassembles the ESCRT-III complex. In the cytosol, monomeric Shrub appears to bind to Lgd. This interaction might prevent uncontrolled polymerisation in the cytosol

are fulfilled upon trafficking of the TMPs through the endosomal pathway (Huotari and Helenius 2011). Entry into the pathway occurs through endocytosis, which is the pinching-off of a small part of the plasma membrane into the cytosol as early endosomal vesicles (EEVs, Fig.1). All TMPs that reside in the endocytosed membrane patch are cargo of the EEVs which subsequently fuse to form the early endosome (EE). Alternatively, EEVs fuse with already existing EEs. In these ways, cargo reaches the EE. Endocytosis can be dependent or independent of the formation of a clathrin coat that encases the initial membrane deformation and also the emerging indentation, termed clathrin coated pit. Clathrin-independent endocytosis occurs at lipid raft domains which are rich in cholesterol, GPI (glycosylphosphatidylinositol) -linked proteins and dependent on other coat proteins such as caveolin or flotillin (Huotari and Helenius 2011). The clathrin-dependent mechanism is the best-understood way of endocytosis and is involved in the endocytosis of a large spectrum of TMPs, including Notch (Windler and Bilder 2010). The basic unit of the clathrin coat is a triskelion made up of clathrin light and heavy chains (Fig. 2A). Many triskelions assemble a coat around the nascent clathrin-coated pit to stabilise it. The clathrin coat is anchored in the plasma membrane via adaptor proteins (APs). These APs not only bind to clathrin and the plasma membrane but also to different cargo proteins and thereby concentrate them at coated pits. Binding to cargo by APs occurs either via specific sorting sequences in the intracellular domain (ICD) of cargo proteins or via an ubiquitin label that is attached to lysine side chains of their ICDs. Examples for APs are AP-2 which directly binds to conserved sorting signals, or Epsin which binds to ubiquitin via two Ubiquitin interacting motifs (UIMs).

The EE is a central sorting station in the cell that sorts membrane and protein cargo to various destinations. Some of the TMPs, such as the

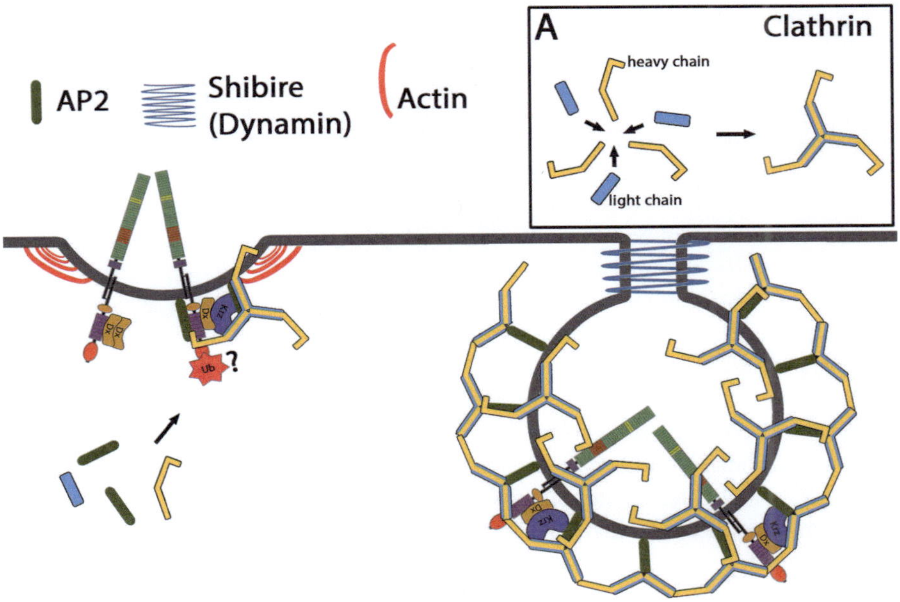

Fig. 2 A hypothetical model for Notch endocytosis by Dx/Krz. Dx binds to the Ankyrin repeats of NICD (depicted in purple) and also to Krz. Krz has binding sites for AP-2 and also Clathrin and might attach Notch to clathrin and AP-2 assembled in clathrin-coated pits. Su(dx) might have a similar role as an adaptor to a so far unidentified machinery that conduct the clathrin-independent endocytosis at higher temperature. The clathrin-coated pit elongates into the cytosol with involvement of an assembling actin network. The neck of the elongated pit is then severed through the activity of Shi, the Dynamin of *Drosophila*

mammalian transferrin receptor, are recycled back to the plasma membrane using either the slow or fast recycling pathway. In both ways, cargo is transported from the EE in vesicles that bud off from tubular extensions of the EE. In the fast way, cargo migrates from the EE directly to the plasma membrane whereas, in the slow way, it is first transported to the RE (recycling endosome) before reaching the plasma membrane. Another less prominent recycling pathway is mediated by the Retromer complex and takes the route via the Golgi apparatus to the plasma membrane. Recent work indicates that the TMP Crumbs (Crb) is recycled via this unusual route (Pocha et al. 2011; Zhou et al. 2011).

Cargo destined for degradation within the lysosomal lumen remains in the EE, which matures and eventually fuses with the lysosome where the luminal content is degraded by the hydrolases.

Maturation includes the acidification of the endosomal lumen and formation of intraluminal vesicles (ILVs) that transport TMPs from the limiting membrane into the lumen of the maturing endosome (ME) (Fig.1). During ILV formation, TMPs are concentrated at distinct spots of the limiting membrane that subsequently pinch off as vesicles into the endosomal lumen. The pinching off occurs in the opposite direction as during endocytosis and requires a different machinery. ILV formation is initiated already in EEs and continues during endosomal maturation. As a result, mature endosomes contain many ILVs and are recognisable at the electron microscope as multi-vesicular bodies (MVBs). ILV formation transports the ICD of TMPs into the endosomal lumen to enable their complete degradation upon fusion with the lysosome. The separation of the ICD from the cytosol also terminates signalling by activated signalling receptors and constitutes an important regulatory step during cell signalling (Wegner et al. 2011). In *Drosophila*, this step also prevents uncontrolled activation of the Notch pathway (see below).

The acidification of the endosomal lumen is a continuous process. The pH of the extracellular fluid is around 7 and drops to 6.8–6.1 in the lumen of the EE, to 6.4–4.8 of MEs and 4.5 of lysosomes (Huotari and Helenius 2011; Maxfield and Yamashiro 1987). In the lumen of the lysosome, hydrolases degrade the delivered cargo. These hydrolases are transported from the Golgi apparatus via carrier vesicles to the ME and from there, together with the cargo, to the lysosomal lumen. The dependency of their activity on low pH is a fail-safe mechanism that ensures that they become active only in mature endosomes and lysosomes which have a comparable pH. The acidification of the endosomal lumen is accompanied by a loss of luminal Ca^{2+} (Gerasimenko et al. 1998).

During maturation, the endosome also migrates from the periphery of the cell towards the perinuclear region where the lysosomes are located. This migration involves the recruitment of motor proteins and uses the microtubule cytoskeleton (Wang et al. 2011).

3 The Molecular Machinery of the Endosomal Pathway

Endocytosis requires the interplay of coat proteins, adaptors and cargo at specific sites of the plasma membrane (Huotari and Helenius 2011). Cargo binds either through sorting signals directly to AP-2, e.g. the di-leucine motif, or need to be ubiquitylated to be recognised by specific adaptors, such as Epsin (Nakatsu and Ohno 2003; Sen et al. 2012). Ubiquitin conjugation (or ubiquitylation) depends on an enzymatic cascade involving E1 Ubiquitin-activating enzymes, E2 Ubiquitin-conjugating enzymes and E3 ubiquitin ligases. The ubiquitin is attached by substrate-specific E3 ligases to the side chain of lysine located in the ICD of substrate TMPs (Moretti and Brou 2013).

The coated pit in which the TMPs are concentrated is subsequently pinched off by a polymer of the ATPase Dynamin that binds to the neck of the nascent EEV (Fig. 2). The abscission generates an EEV whose coat is immediately removed by the activities of the ATPase Hsc70 (Heat shock cognate 70) and auxillin (a Co-chaperone with ATPase activity). The removal of the coat allows EEVs to fuse with other EEVs or already existing EEs. The actual fusion is mediated by SNARE (*s*oluble *N*-ethylmaleimide-sensitive-factor *a*ttachment *re*ceptor) proteins. However, the specificity of the fusion is guaranteed by tethering factors that assure fusion of the correct membranes. Two of these tethering complexes are the CORVET (class C core vacuole/endosome tethering) and HOPS (homotypic fusion and protein sorting) complexes (Balderhaar and Ungermann 2013). CORVET mediates the fusion of EEVs with themselves or with existing EEs, while HOPS connects MEs with the lysosome (Fig. 1). Both complexes have the same core made up of the same four core components, Vps11 (Vacoular protein sorting), Vps16, Vps18 and Vps33. The CORVET complex additional includes VPS3 and VPS8, while the HOPS complex Vps39 and Vps41.

The events of the endosomal pathway are orchestrated by small GTPases of the Rab (Ras-related in brain) protein family (Zerial and McBride 2001). They are master regulators whose activity is determined by its cycle between the GDP (Guanosine diphosphate)- (inactive) and GTP (Guanosine triphosphate)-bound (active) states. Hence, they act as molecular switches. Rabs are poor enzymes that require additional factors to hydrolyse the γ-phosphate of GTP. These factors are termed GAPs (GTPase activating proteins). In addition, the exchange of the nucleotide requires factors of the GEF (guanine nucleotide exchange factor) family. Inactivated GDP bound Rabs are cytosolic and associated with proteins of the GDI (GDP-dissociation inhibitor) class. They are recruited to specific endosomal compartment in a not fully understood manner that is thought to require the activity of a GDF (GTPase dissociation factor) protein. Recent data show that also GEFs can take over this function (Cabrera and Ungermann 2013).

Each compartment has its characteristic set of Rabs which recruit effectors that ensures correct fusion of membrane compartments and maturation of the endosome. Most important for endosomal biogenesis and maturation are Rab5 and Rab7 which act in a sequence that is initiated at the plasma membrane by Rab5 and ends with the Rab7 mediated fusion of the ME with the lysosome (Fig. 1). Rab5 may be recruited to the plasma membrane at clathrin-coated pits through its GEF Rabex5 which possesses two ubiquitin-binding domains and can bind to ubiquitylated cargos (Raiborg et al. 2006). After EEV formation, activated Rab5 recruits tethering factors such as CORVET that assures correct SNARE mediated fusion. CORVET directly binds to Rab5 located at the donor and acceptor membrane thereby providing a bridge between both membranes. In the EE, Rab5 recruits the PI(3) kinase Vps34 to generate phosphatidylinositol 3-phosphate [PI(3)P], a characteristic and specific phospholipid of the EE. PI(3)P is a docking site for several Rab5 effectors such as the tethering factor EEA2 (early endosomal antigene 2). Moreover, it is the binding site for the ESCRT-0 (endosomal sorting complex required for transport) complex that initiates formation of the ILVs. The two Rab5 effectors bind to PI(3)P via a FYVE (Fab1 YOTB VAC1 EEA1) domain. PI(3)P is either hydrolysed or converted to PI(3,5)P2 (Phosphatidylinositol 3,5-bisphoshphate) by Fab1 during maturation of the endosome (Rusten et al. 2006). This may antagonise the activity of Rab5.

The EE also extends tubular structures from which recycling vesicles bud off to recycle to the plasma membrane via the Rab4 controlled fast pathway or, alternatively, to be transported to the recycling endosome and then to the plasma membrane in a Rab11-dependent manner (Fig. 1). After the recycling process is terminated, the EE prepares the fusion with the lysosome. During this maturation, Rab5 is gradually replaced by Rab7 in the process of Rab conversion (Rink et al. 2005). Rab conversion involves the recruitment of the GEF for Rab7 which is a dimer consisting of Mon1 and Ccz1 (Cabrera et al. 2014). The Mon1/Ccz1 GEF recruits Rab7 and might also interrupt a positive feedback loop that interrupts the activity of Rab5 (Kinchen and Ravichandran 2010; Nordmann et al. 2010; Poteryaev et al. 2010). The loss of activity of Mon1 or Ccz1 in *Drosophila* causes a phenotype that is very similar to the loss of Rab7 activity (Yousefian et al. 2013). It is caused by a failure of recruitment of Rab7 to the endosome. The defect in Rab7 recruitment causes the accumulation of enlarged MEs with a high number of ILVs. This phenotype is caused by the lack of Rab7-mediated fusion of the ME with the lysosome. Consequently, the lifetime of the endosomes is infinitely increased and the endosomes continuously enlarge by Rab5 mediated homotopic fusions with EEVs or other EEs. The high number of ILVs in the mutant MEs indicates that their formation is not affected by a failure of Rab conversion. Thus, Rab conversion is not coupled with ILV formation and both processes run in parallel during endosomal maturation. However, a coordination between the activation of Rab7 and ILV formation must occur in order to prevent uncontrolled signalling by the transported receptors (see below).

ILV formation continually occurs throughout maturation of the endosome and is mediated by four in sequence acting ESCRT (endosomal sorting complex required for transport) complexes: ESCRT-0, -I, -II, -III; and the Vps4-complex (Hurley 2015). All individual complexes cycle between the cytosol and the endosomal membrane either already in a complex (ESCRT-0, -I, -II) or as monomers (ESCRT-III). With the exception of ESCRT-0, each complex is recruited by the earlier acting one. ILV formation is initiated by the recruitment of ESCRT-0, consisting of Hrs (hepatocyte growth factor-regulated tyrosine kinase substrate) and STAM (Signal-transducing adaptor molecule), to a spot of the endosomal membrane in a Rab5/Vps34/PI(3)P-dependent manner. ESCRT-0 recruits ESCRT-I which in turn recruits ESCRT-II. The recruitment results in the assembly of a large super complex that concentrates cargo (Schmidt and Teis 2012). The label for incorporation of the cargo into this complex is represented by ubiquitin, often attached already at the plasma membrane by cargo-specific E3 ligases. The ESCRT-0, -I and -II contain 12 ubiquitin binding domains that concentrate ubiquitylated cargo at sites of ILV formation. *In vitro* data suggest that this super complex also induces inward membrane curvature (Wollert and Hurley 2010). However, the abscission occurs through ESCRT-III. It consists of four core factors which all belong to the CHMP (charged multi-vesicular body protein) protein family (Hurley 2015). Depending on the species they have different names (Table 1). In contrast to the previous acting ESCRT complexes, ESCRT-III is assembled only at the membrane. Assembly starts with the activation of Vps20 (Yeast and *Drosophila*)/CHMP6 (mammals) through binding to ESCRT-II. The binding induces a conformational change that causes the recruitment and polymerisation of Snf7/CHMP4 at the membrane. The polymerisation is terminated through capping by Vps24 followed by Vps2. *In vitro* and over-expression experiments suggest that the polymer forms a spiral around the cargo (Schöneberg et al. 2017). The spiral may act as a spring whose tension is released through membrane deformation. This deformation eventually results in the abscission of the ILVs into the endosomal lumen. ESCRT-III also recruits deubiquitinases that remove the ubiquitin label from the cargo before ILV incorporation (Fig. 1). The disassembly of the ESCRT-III occurs via the activity of the AAA-ATPase Vps4, which removes individual monomers in a sequential and ATP-consuming manner. In this way (manner), the ESCRT components are released into the cytosol for the next round of ILV formation and are not incorporated into the ILVs (Fig. 1, insert). Note that the loss of activity of each ESCRT complex results in a failure of ILV formation. Hence, the cargo is not transferred into the endosomal lumen but remains at the limiting membrane. In the case of the signalling receptors, their ICDs remain for a longer time in contact with the cytosol. This can result in uncontrolled cell signalling. The ESCRT machinery requires the ubiquitin label for incorporation of TMPs into ILVs (Ren and Hurley 2010).

While ubiquitylation is commonly accepted as a label for endocytosis, recent data cast some doubt that it is absolutely required. It has been shown that the lysine free ICD of mouse Dll3 (Delta-like 3) can induce endocytosis of a Dll1-Dll3 hybrid ligand (Heuss et al. 2008). Moreover, the EGF-receptor (EGF-R) variants that lack a binding site for the E3 ligase Cbl (Casitas B-lineage lymphoma) and therefore are not ubiquitylated are endocytosed but are not degraded (Huang et al. 2007). Instead, they are recycled to the plasma membrane. In case of the FGF

Table 1 Endocytic factors involved in trafficking of notch

Drosophila	Yeast & human/mammals	Protein complex / function
Hrs (Hepatocyte growth factor-regulated tyrosine kinase substrate)	Vps27p (Vacuolar protein sorting 27) (yeast) HGS (mammals)	Escrt 0
Stam (Signal- transducing adaptor molecule)	Hse1p (yeast) STAM1,2	Escrt 0
Tsg101 (tumor susceptibility gene 101)/Erupted	Vps23p (yeast) hVPS23/TSG101	Escrt I
Vps28 (Vacuolar protein sorting 28)	Vps28p (yeast) hVPS28	Escrt I
Vps37	Vps37p hVPS37A,B,C,D	Escrt I
Mvb12 (Multivesicular body sorting factor of 12 kilodaltons)	Mvb12p (yeast) MVB12a	Escrt I
Vps22	Vps22p (yeast) hVPS22 (EAP30)	Escrt II
Vps25	Vps25p (yeast) hVPS25 (EAP20)	Escrt II
Vps36	Vps36p (yeast) hVPS36 (EAP45)	Escrt II
Shrb (Shrub)	Vps32p/Snf7p (yeast) CHMP (chromatin modifying protein, later renamed charged multivesicular body protein)4A,B,C	Escrt III
Vps2	Vps2p (yeast) hVPS2A,B/CHMP2A,B	Escrt III
Vps20	Vps20p (yeast) hVPS20/CHMP6	Escrt III
Vps24	Vps24p (yeast) hVPS24/CHMP3	Escrt III
Vps4	Vps4p (yeast) hVPS4A,B (SKD1)	Vps4 Complex (AAA-ATPase)
Chmp5	Vps60p (yeast) hVPS60/CHMP5 (metazoan)	Accessory ESCRT
Lgd (Lethal giant discs)	No yeast homologues, CC2D1A,B/hLGD2,1	Regulator of Chmp4 activity
Dx (Deltex)	No yeast homologues DTX1,2,3,4,DTX3L	E3 Ubiquitin Ligase
Su(Dx) (Suppressor of Deltex)	Rsp5p (yeast) AIP4/Itch hNEDD4 hWWP1,2	E3 Ubiquitin Ligase
dNedd4 (Neural precursor cell expressed developmentally downregulated protein 4)	Rsp5 (yeast) AIP4/Itch hNEDD4 hWWP1,2	E3 Ubiquitin Ligase
Vps11(CG32350)	Vps11p hVPS11	HOPS/CORVET core
dVps16A (dVps16A)	Vps16p hVPS16	HOPS/CORVET core
Vps18/Deep orange (Dor)	Vps18p hVPS18	HOPS/CORVET core

(continued)

Table 1 (continued)

Drosophila	Yeast & human/mammals	Protein complex / function
Vps33 (dVps33A, Carnation (Car))	Vps33p hVPS33A	HOPS/CORVET core
Vps39 / CG7146	Vps39p hVPS39-1	HOPS – late endosomal fusion
Vps41/light(lt)	Vps41p hVPS41	HOPS – late endosomomal fusion
Vps3(missing in *Drosophila*?)	Vps3p hVPS39-2 (missing homolog?)	CORVET – early endosomal fusion
Vps8	Vps8p hVPS8	CORVET – early endosomal fusion
Dmon1/Dccz1	Mon1/ Ccz1 (yeast) Vacuolar fusion protein MON1 A / Vacuolar fusion protein CCZ1 homolog	Mon1/Ccz1 complex - Rab7 GEF
Lgl (Lethal giant larvae)	Sro7p (yeast) LLGL1,2,3,4	Scribble cell polarity complex component
Crb (Crumbs)	No yeast homologue CRB1,2,3	Cell polarity complex component
Rab4 (Ras related protein 4)	None in yeast RAB4A	Fast recycling
Rab5	Vps21 (yeast) RAB5A	Early endosomal fusion
Rab7	Ypt7 (yeast) RAB7A	Late endosomal fusion
Rab11	Ypt31p (yeast) RAB11A	Slow recycling
Shibire / Dynamin	Vps1p (yeast) DNM1,2,3	Membrane fission
Hsc70 (Heat shock cognate 71kDa protein)	Ssa1p (yeast) HSC70 / HSPA8	Clathrin coat disassembly
Auxilin	Swa2p (yeast) DNAJC6 (human)	Clathrin coat disassembly

(Fibroblast Growth Factor) -receptor 1, it has been shown that the recycling of the ubiquitylation deficient variant is caused by failure of incorporation into ILVs (Haugsten et al. 2008). These results suggest that the major function of ubiquitylation of TMPs at the plasma membrane may be their recognition by the ESCRT machinery to include them into ILVs.

The *Drosophila* ortholog of the central ESCRT-III component CHMP4 is Shrb (Shrub) (Sweeney et al. 2006). Recent work gave insight into the structural basis of its polymerisation at the endosomal membrane. Shrub has two complementary charged surfaces that allow the electrostatic interaction of each monomer with other monomers in a staggered arrangement via complementary charged surfaces (McMillan et al. 2016).

Experiments in *Drosophila* identified the tumor suppressor Lethal (2) giant discs (Lgd) as a vital positive regulator of the activity of Shrub (Troost et al. 2012). Lgd is a member of the Lgd family which is present in all metazoans but is absent from the genomes of unicellular organisms (Childress et al. 2006; Gallagher and Knoblich 2006; Jaekel and Klein 2006). Lgd and its mammalian orthologs CC2D1A and CC2D1B bind to Shrub/CHMP4 via their unique DM14 (Drosophila melanogaster 14) domain (McMillan et al. 2017; Troost et al. 2012). Members of the Lgd family usually possesses four repeats of the DM14 domain. Recent work shows that the odd-numbered DM14 domains mediate the interaction in a functionally redundant manner (McMillan et al. 2017; Troost et al. 2012). It also revealed the structural basis for the binding of the DM14 domain to Shrub. The DM14 domain is a helical hairpin with a positively charged lip that binds to the negatively charged surface of Shrub

also required for its homo-polymerisation. This suggests that the interaction of Lgd and Shrub and polymerisation of Shrub are mutually exclusive. Hence, it is likely that Lgd and Shrub interact in the cytosol where Shrub is in its monomeric form and that Lgd might be necessary to prevent uncontrolled and inappropriate polymerisation of Shrub (Fig. 1 insert).

The membrane of the EE is subdivided into several domains that organise recycling or degradation of cargo. These subdomains are kept separate through the activity of the DNAJ domain protein Rme8 (Receptor mediated endocytosis 8) and the sorting nexins (SNX) proteins (Norrisa et al. 2017).

During maturation, the lumen of the ME acidifies through the activity of the multi-protein complex V-ATPase (Cotter et al. 2015). This acidification is required for the activation of the hydrolases in late endosomes and lysosomes.

Another important process during maturation is the preparation of fusion of the mature endosome with the lysosome, which is initiated by the replacement of Rab5 through Rab7 on the ME. Activated Rab7 recruits the machinery required for fusion. Most important is the recruitment of the HOPS tethering complex (Balderhaar and Ungermann 2013). Rab7 is also present on the membrane of lysosomes and HOPS can bind to Rab7 on both membranes to connect the two organelles. The actual fusion is mediated by SNARE proteins and ends the lifetime of the endosome. The fusion also delivers the cargo to the lumen of the lysosome. The fusion must be coordinated with the ILV formation. It must be assured that the TMPs are incorporated before fusion of the ME in order to achieve their complete degradation.

4 The Journey of Notch Through the Endosomal Pathway

Past work in *Drosophila* indicates that full-length Notch is constitutively travelling through the pathway in a ligand-independent manner to be degraded in the lysosome (Jekely and Rorth 2003; Vaccari et al. 2008; Windler and Bilder 2010). This work is largely performed in imaginal discs that are epithelial organised tissues. Two kinds of experiments have been performed: the first type of experiment analysed the localisation of Notch in mutants that interrupt endocytosis at several stages which is based on the observation that Notch accumulates in the affected endosomal compartments. The second type of experiment was represented by uptake/pulse-chase assays with antibodies directed against epitopes of the ECD of Notch.

In imaginal discs cells, Notch localizes in EEs after 5 min chase (post endocytosis) and is completely degraded after 5 h (Vaccari et al. 2008; Windler and Bilder 2010). In mammalian cells, pulse-chase experiments revealed that Notch1 appears in not further specified endosomes after 30 min and is degraded already after 60–90 min (Chastagner et al. 2008).

4.1 Initiation of Endocytosis by E3 Ligases

Several E3 ligases have been identified that ubiquitylate NICD and trigger its endocytosis. Among them there are Deltex (Dx), Suppressor of Deltex [Su(dx), Itch in mouse and AIP4 (atrophin-1-interacting protein 4) in human] and neural precursor cell-expressed, developmentally down-regulated 4 (Nedd4) (Chastagner et al. 2008; Sakata et al. 2004; Wilkin et al. 2004; Yamada et al. 2011). Dx contains two WW domains that bind to the ANK repeat region of NICD and are crucial for its function. The ubiquitylation reaction is conducted by one of the C-terminal RING (Really interesting new gene) Fingers (RFs), which bind to the activated E2 conjugating enzyme. Su(dx) and Nedd4 belong to the Nedd4 family of HECT domain E3 ligases. They contain an N-terminal phospholipid binding C2 domain followed by a varying number of WW domains and specific target selection and a C-terminal HECT domain that transfers Ub. Members of the NEDD4 family have been shown to bind via the WW domains to a proline-rich motif [e. g. PP(X)Y] within their substrates

(Dodson et al. 2015). Such a motif is present in NICD and required for ubiquitylation by NEDD4 (Sakata et al. 2004).

The loss of *dx* function causes a strong reduction in the endocytosis of Notch and a corresponding increase in Notch levels at the plasma membrane (Yamada et al. 2011). As expected, the over-expression of Dx results in increased endocytosis of Notch and its removal from the plasma membrane (Hori et al. 2004). Further analysis and cell culture assays suggest that the activity of Dx induces clathrin-dependent endocytosis of Notch (Shimizu et al. 2014). Dx physically interacts with the non-visual arrestin Kurtz (Krz) in *Drosophila* (Mukherjee et al. 2005). Like in the case of *dx*, the loss of *krz* function results in an increase of Notch at the apical membrane, indicating that it acts in concert with Dx to induce Notch endocytosis. In this light, it has to be noted that the structure/function analysis of Dx indicates that the single RF is required for dimerization of Dx and can be replaced by other dimerisation domains, such as GST (Glutathion-S-Transferase) (Matsuno et al. 2002). This indicates that ubiquitylation is not required for Dx mediated endocytosis of Notch. Krz contains motifs at its C-terminus that bind to AP-2 and Clathrin (Mukherjee et al. 2005). It is recruited to the ICD of Notch via Dx. Thus, a likely scenario is that Krz and Dx form a ternary complex with the ICD of Notch and that the Dx/Krz complex acts as a specific adaptor for recruitment of Notch into clathrin-coated pits (Fig. 2).

Recent work shows that Su(dx) also induces endocytosis of Notch but in contrast to Dx, it takes place in a clathrin-independent manner (Shimizu et al. 2014). This endocytosis occurs from GPI-enriched membrane domains that are also enriched in cholesterol. Nothing is known about the machinery that performs this type of Notch endocytosis. Why Dx and Su(dx) direct Notch in different routes of endocytosis is puzzling at the first glimpse. However, recent findings indicate that, similar to Dx, the activity of the Ubiquitin transferring HECT domain of Su(dx) is not required for endocytosis (Shimizu et al. 2014). A likely possibility is therefore that also Su(dx) (and probably also Nedd4) acts as an adapter that specifically connects Notch to the in this case unknown endocytic machinery.

The emerging picture suggests that ubiquitylation of its ICD may not be required for initiation of Notch endocytosis. This is in agreement with recent studies of endocytosis of other signalling receptors, such as EGF-R and FGF-R1. In these cases the ubiquitin label appears to be required for later steps in the endosomal route to target the EGF-R for degradation (Haugsten et al. 2008; Huang et al. 2007). Su(dx) and Dx are also required for a later step of Notch trafficking and recent results indicate that this step requires ubiquitylation (Hori et al. 2011; Shimizu et al. 2014).

It is assumed that Notch is endocytosed at a constant rate without much regulation. However, a recent report indicates that Notch endocytosis can be modified in a tissue-specific manner. Crumbs (Crb) is a member of the polarity machinery required for the maintenance of polarity in epithelia (Nemetschke and Knust 2016). Crb and Notch co-localise at the apical adherens junctions and can physically interact. This interaction suppresses uncontrolled endocytosis of Notch. The absence of *crb* function results in a loss of a large proportion of Notch from the junctions. Thus, Crb regulates the level of Notch at the apical membrane in epithelia. Interestingly, the developmental consequences of this strong loss are very mild, suggesting that the consequences for Notch signalling are negligible. These observations raise the question at which site of the cell Notch is activated and how much Notch is required for normal signalling during imaginal disc development.

4.2 Endocytosis of Notch

A study by the Bilder lab using *Drosophila* imaginal disc cells revealed that Notch takes several entry routes into the endosomal pathway (Vaccari et al. 2009; Windler and Bilder 2010). These results fit nicely with the more recent analysis of the function of Dx and Su(dx) (Shimizu et al. 2014). Notch is endocytosed through a clathrin-dependent and -independent pathway. Most of the clathrin-dependent endocytosis is also dependent on AP2 (Adaptor related protein complex 2). However, a small fraction is independent of it. Interestingly, it is this AP-2-independent fraction that appears to be signalling competent in *Drosophila*.

It is not known how Notch is incorporated into clathrin-coated pits upon endocytosis. Two ways can be envisioned. Either an adapter protein other than AP-2 binds to ubiquitylated NICD or NICD contains a sorting motif. No classical adaptor protein emerged as a good candidate so far. Whether Epsin is required for Notch endocytosis has not been investigated and some catching up on this question should be done. Evidence exists that suggests that the Epsin-related Eps15 adapter is involved in endocytosis of NEXT during ligand-dependent activation of Notch in mammalian cells but it has not been investigated whether it is also involved in constitutive ligand-independent endocytosis (Gupta-Rossi et al. 2004). As suggested above, it is possible that Dx/Krz and Su(dx) might act as unusual adaptors. Notch contains a classical di-leucine motif (DIVRLL, consensus D/EXXXLL) in its ICD that is conserved among the ortholog receptors of most metazoans (Zhenga et al. 2013). It might be therefore possible that AP-2 binds directly to NICD through this motif to incorporate Notch into clathrin coated pits. A di-leucine signal has been shown to mediate EGF pathway induced endocytosis of the Notch ortholog Lin-12 in *Caenorhabditis elegans* during vulva development (Shaye and Greenwald 2005). However, this particular signal is not conserved in other Notch orthologs. Moreover, a recent study using mammalian cell culture suggests that the conserved di-leucine motif might be required for a later step in Notch trafficking (Zhenga et al. 2013). Thus, if this motif is required for endosomal trafficking of Notch also *in vivo* and at which step awaits further clarification. It is possible that the lack of knowledge is caused by a redundancy of mechanisms operating in parallel to incorporate Notch into pits and also the existence of clathrin-dependent and –independent ways of endocytosis. This complicates the analysis, since the loss of one mechanism might not cause detectable phenotypes. In agreement with this notion is that the known loss of function phenotypes of *dx*, *Su(dx)* and also *Nedd4* in *Drosophila* are relatively weak. Disregarding whether the endocytosis of Notch occurs in a clathrin-dependent or -independent manner, it requires the activity of Dynamin (encoded by *shibire* in *Drosophila*) for abscission of the forming vesicle (Vaccari et al. 2009; Windler and Bilder 2010).

4.3 Endosome Formation

After endocytosis, Notch travels in EEVs to the EE. This journey requires the activity of Rab5 and the SNARE Avalanche (Avl), a *Drosophila* Syntaxin family member (Lu and Bilder 2005; Vaccari et al. 2009). It has been recently reported that the tumour-suppressor Lethal giant larvae (Lgl) is also involved in endosomal trafficking of Notch in the *Drosophila* eye disc (Portela et al. 2015). Lgl is required for the organisation of the epithelial polarity of imaginal discs. Its loss of function results in loss of polarity and also in accumulation of Notch in an early endosomal compartment. The precise function of Lgl in trafficking of Notch is so far not clear and requires more investigations.

4.4 Recycling of Notch

In principle Notch can choose between two routes in the EE: it can enter the recycling pathway to travel back to the plasma membrane or it remains in the EE to be transported to the lysosome. Most of Notch appears to remain in the EE. However, a recent report shows that a small fraction of Notch is recycled in a manner that is dependent on the DNAJ protein Rme-8 in *Drosophila* via a Rab4 positive recycling pathway (Gomez-Lamarca et al. 2015). A recent report shows that Notch1 and Notch2 recycle in a COMMD9 (COMM domain-containing protein 9) and Retromer-dependent manner in mammalian cells (Li et al. 2015). However, the Retromer complex does not appear to be involved in recycling of Notch in *Drosophila* (Gomez-Lamarca et al. 2015).

4.5 Endosome Maturation

The majority of Notch is not recycled and remains in the limiting membrane of the EE (Early Endosome). Upon its way to the lysosome Notch has to be incorporated into ILVs (Intraluminal Vesicles). This step is absolutely required for the complete degradation of the receptor, since only then the ICD is located in the lumen of the ME (Maturing Endosome). It turns out that the Notch that remains at the limiting membrane is activated during endosomal maturation (see below). Thus, a second requirement of ILV incorporation is the prevention of uncontrolled activation of the pathway by Notch located in the limiting membrane.

Dx/Krz and Su(dx) are also located on MEs and interact to control the amount of Notch that is incorporated into ILVs (Hori et al. 2011; Shimizu et al. 2014). The function is temperature-dependent and imposes a buffering system that guarantees the correct level of Notch activation over the temperature span where *Drosophila* is viable [16–30 ° C, (Shimizu et al. 2014)]. This regulatory mechanism feeds into the ESCRT machinery via Shrub (Hori et al. 2011). In this process, the ubiquitylation function of Dx and Su(dx) is crucial: Dx mono-ubiquitylates the ICD of Notch, which appears to prevent Notch from incorporation into ILVs, probably because this label is not well-recognised by the ESCRT machinery (Ren and Hurley 2010). Krz binding appears to modify the activity of Dx and Notch is then poly-ubiquitylated. It has been shown that the poly-ubiquitin signal is preferred by the ESCRT machinery (Ren and Hurley 2010). Hence, the poly-ubiquitylation by Dx/Krz sends Notch into ILVs of MEs (Hori et al. 2011). Su(dx) antagonises the activity of Dx and also appears to induce ubiquitylation of Notch in a temperature-dependent manner but it is not known which type of ubiquitylation in involved in this process (Shimizu et al. 2014). Thus, all three factors appear to be part of a module that regulates the amount of Notch that is incorporated into ILVs or remains at the limiting membrane (Fig. 3).

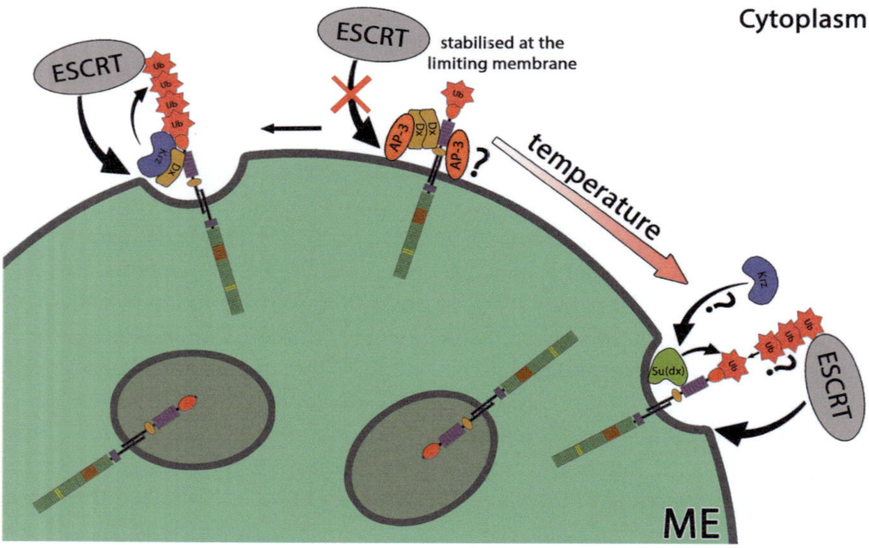

Fig. 3 Model for the second function of Dx and Su(dx) at the ME (maturing Endosome). The Dx/Krz/Su(dx) module regulates the amount of Notch in the limiting membrane of the ME that is incorporated into ILVs. Dx mono-ubiquitylates the ICD of Notch protruding from the membrane into the cytosol. This mono-ubiquitylation prevents the incorporation into ILV, probably because this signal is poorly recognised by the ESCRT machinery. AP-3 is required for this prevention in a so far unknown manner. It might bind to Dx and/or to Notch to do so. The binding of Krz to Dx and NICD changes the state of ubiquitylation of Notch to poly-ubiquitylation. This signal can be recognised by the ESCRT-machinery and Notch is incorporated into ILVs. Similarly, Su(dx) activity leads to poly-ubiquitylation of the NICD and subsequent incorporation of Notch into ILVs at higher temperature. It is not known whether Krz and Su(dx) act in the same or different pathways

Unfortunately, the relationship between Krz and Su(dx) has not been investigated so far. It is therefore not clear whether Krz and Su(dx) act in the same or in parallel pathways and the poly-ubiquitylation occurring through binding of Krz to Dx is conducted by Su(dx) (Fig. 3).

Su(dx) has been isolated in a genetic screen as a dominant suppressor of the phenotype of *dx* null mutants (Fostier et al. 1998). In over-expression experiments, it has been shown that the activity of Su(dx) is dominant over that of Dx and Notch is directed into ILVs of MEs (Shimizu et al. 2014). The different modes of ubiquitylation offer an explanation why Su(dx) overrides the activity of Dx if both were co-overexpressed. It is possible that Su(dx) adds further ubiquitin moieties to the Dx mono-ubiquitylated sites or to new sites. The poly-ubiquitin label can be recognised by the ESCRT machinery and extinguishes the Dx signal. According to this model, loss of Dx function results in too much incorporation of Notch into ILVs because of the activity of Su(dx). Interestingly, the loss of *dx* results in the translocation of a fraction of Notch in a so far unidentified endosomal compartment that lacks the classical endosomal markers (Yamada et al. 2011). Similar observations have been made of the endosomal compartment where Notch is located upon co-overexpression of Krz and Dx (Mukherjee et al. 2005). Thus, to fully understand the regulatory mechanism of the Dx/Krz/Su(dx) module, this compartment must be characterised. Moreover, older work suggests that Dx can perform its function without its HECT domain as long as it can oligomerise (Matsuno et al. 2002). Thus, it is possible that the detected mono-ubiquitylation of NICD is not important or not the only mechanism by which Dx prevents the incorporation of Notch into ILVs.

AP-3 is a poorly characterised AP required for the delivery of TMPs (transmembrane proteins) via sorting signals, including the di-leucine type also conserved in the ICD of most Notch receptors of metazoans (Dell'Angelica 2009). It has been shown that AP-3 is required for the activation of the Notch pathway by Dx over-expression and probably required to transport TMPs from EEs to the lysosome thereby avoiding the incorporation into ILVs (Wilkin et al. 2004). Unfortunately, it has not been tested whether AP-3 binds to the di-leucine signal of Notch1 so far. Moreover, many facets of the function of AP-3, such as the nature of its interaction partners, are still mysterious. It will be interesting to further explore the role of AP-3 in the trafficking of Notch and its relationship to Dx, Krz and Su(dx).

The incorporation of Notch into ILVs requires the activity of the ESCRT complexes and Lgd. The loss of the function of ESCRT machinery results in a dramatic loss or complete failure of ILV formation (Hurley 2010; Vaccari et al. 2008). Consequently, all TMPs remain in the limiting membrane of the ME and only their ECD, which protrudes into the endosomal lumen, will be degraded upon fusion with the lysosome. In addition, degradation of Notch and other endosomal cargo is strongly delayed. In the absence of *lgd* function the activity of Shrub is reduced (Troost et al. 2012). The reduced activity still allows the formation of ILVs but an elegant approach by the Schweisguth group, using a GFP (Green fluorescent protein) /Cherry dual tagged Notch receptor, showed that Notch is not completely incorporated into ILVs (Couturier et al. 2014).

4.6 Delivery of Notch to the Lysosome

Notch is delivered to the lysosome through fusion of the ME with this compartment. This fusion is mediated by Rab7 which recruits and activates the HOPS tethering complex. Either loss of function of Rab7 or its GEF that is also required for its recruitment results in very similar phenotypes, which are represented by dramatically enlarged endosomes that contain many ILVs (Yousefian et al. 2013). This indicates that ILV formation is not affected by the loss of the activity of *rab7*. Moreover, the analysis in *Drosophila* indicates that the endosome matures normal but fails to fuse with the lysosome (Yousefian et al. 2013). This suggests that the main function of Rab7 is to orchestrate the fusion of the ME with the lysosome.

5 Ligand-Independent Activation of the Notch Pathway in the Endosomal Pathway

Mutations in several genes involved in trafficking of the Notch pathway also show classical Notch loss or gain of function phenotypes, indicating that they contribute to the regulation of the activity of the pathway. The factors encoded by these genes appear to be involved in the regulation of the entry of Notch into ILVs. In all cases tested, the activation is independent of the ligands and requires only the components of the signal-receiving side of the pathway (Hori et al. 2012; Jaekel and Klein 2006; Moberg et al. 2005; Thompson et al. 2005). Mutations in genes that encode factors involved in earlier steps, e.g. *rab5*, or later steps, e.g. *mon1,* that is required for fusion of the ME with the lysosome, do not result in ligand-independent activation (Lu and Bilder 2005; Yousefian et al. 2013).

Perhaps the strongest ectopic activation of the Notch pathway is observed upon loss of function of ESCRT-I, -II and -III components. While ESCRT-III is involved in many membrane abscission events of the cell, ESCRT-I and –II are devoted to ILV formation, indicating that loss of the ILV formation causes the activation of the pathway. A similar Notch activation was observed in *lgd* mutant tissues (Childress et al. 2006; Gallagher and Knoblich 2006; Jaekel and Klein 2006; Klein 2003). In contrast to loss of ESCRT-function, loss of *lgd* function does not cause cell lethality. Therefore, more details about the activation of Notch upon loss of this ESCRT regulator should be investigated. The activation of Notch in *lgd* mutants requires the additional activity of Su(H) and the γ-secretase complex but is independent of Kuz/ADAM10 (Jaekel and Klein 2006; Schneider et al. 2012). Moreover, it requires the fusion of the ME with the lysosome, indicating that the activation occurs in the lysosome (Schneider et al. 2012). In addition, ILVs are still generated in *lgd* mutant cells but Notch is not completely incorporated (Couturier et al. 2014; Troost et al. 2012).

A working model has been proposed that suggests that in *lgd* mutant cells a fraction of Notch is not incorporated into ILVs but remains in the limiting membrane (Troost et al. 2012). The requirement for the γ-secretase indicates that ecto-domain shedding must occur for activation in a ligand-independent manner. Hence, the conditions in the lysosomal lumen favour ecto-domain shedding of Notch in an alternative manner. The alternative shedding of the ECD of Notch, which after fusion protrudes into the lumen of the lysosome, may occur either through degradation of ECD by the activated hydrolases. Other non-exclusive possibilities are the dissociation of the ECD from N^{TM} due to depletion of Ca^{2+} or low pH. It has been shown that Notch expressed in cell culture can be activated through depletion of Ca^{2+} from the culture medium (Rand et al. 2000). The acidification of the endosomal/lysosomal lumen, the activation of the hydrolases and also depletion of Ca^{2+} depend on the activity of vATPase. It has been shown that the function of vATPase is required for the ligand-independent activation of Notch in *lgd* mutant cells (Schneider et al. 2012).

Although a plausible scenario to explain activation of Notch, there are some findings that are not easily explained by the model and some parts have not been shown. Firstly, formally it has not been shown that Notch is really activated in a ligand-independent manner, if it remains in the limiting membrane of the ME. Secondly, the requirements for activation of Notch in *shrub*- and *lgd*-mutant cells differ. The activation of Notch in *shrub* mutant cells does not require the fusion of the ME with the lysosomes (Schneider et al. 2012). This difference in requirement can be explained by a difference in the amount of Notch that remains in the limiting membrane in the two kinds of mutants. In *shrub* mutants probably all Notch of the ME remains at the membrane, since ILV formation is abolished. In contrast, a smaller fraction of Notch probably remains in the limiting membrane of *lgd* mutant endosomes, since ILV formation still occurs. Thus, the activatable fraction of Notch in ESCRT mutant MEs is much higher than in *lgd* mutant ones and a sufficient fraction of Notch might be activated already in late endosomes where the luminal conditions in terms of pH and activity of the hydrolases are close to that of the lysosome. In contrast, in *lgd* mutants, some activation of Notch might also occur already in the late endosome but below the threshold of detection. Only if the complete membrane fraction is cleaved in the lysosome, Notch is activated at detectable levels. Thirdly, the loss of function of ESCRT-0 and also Vps4 does not result in ectopic activation of the Notch pathway and also suppresses the activation of the pathway in *lgd* mutant cells (Gallagher and Knoblich 2006; Jaekel and Klein 2006; Legent et al. 2015; Tognon et al. 2014). The requirement of the function of ESCRT-0 in activation of Notch in *lgd* cells indicates that ESCRT-0 has an additional role in protein trafficking besides the initiation of ILV formation. One important additional function of ESCRT-0 is the clustering of the cargo at the point of membrane ILV-invagination. It has been suggested that this function is required for activation of Notch in cells mutant for the other ESCRTs or *lgd* (Tognon et al. 2014). To protect the limiting membrane of the lysosome from attacks of the activated luminal hydrolases, the luminal side of the membrane is covered with highly glycosylated proteins, such as LAMP1 (Lysosome-associated membrane glycoprotein) and LAMP2 (Schwake et al. 2013). It is possible that only the clustered Notch interrupts this cover and allows the access of the hydrolases to its extracellular domain. Loss of *vps4* function results in a failure to remove the assembled ESCRT-III complex from the membrane. As a consequence, Shrub and the other ESCRT-III components accumulate at the limiting membrane of the ME. It is possible that the association of Notch with the arrested ESCRT-III complex prevents its activation upon fusion with the lysosome. Alternatively, the ME might not fuse with the endosome due to a regulatory process that prevents fusion of the ME with the lysosome before the ESCRT machinery has incorporated TMPs into ILVs. The danger of Notch activation indicates that such a mechanism must exist.

A qualitatively similar model for Notch activation has been suggested to explain the ectopic activation in *Su(dx)* mutant cells or over-expression of Dx (Hori et al. 2011; Wilkin et al. 2008, Fig. 4). As described above, Su(dx) appears to promote the incorporation of Notch into ILVs. According to this model, loss of function of *Su(dx)* results in an increased fraction of Notch remaining in the limiting membrane and its activation after fusion with the lysosome occurs in a similar manner as described for *ESCRT* and *lgd* mutants (Fig. 4). Over-expression of Dx results in an increased amount of mono-ubiquitylated Notch, which remains in the limiting membrane and gets activated (Fig. 4). The activation depends on the activity of the HOPS tethering complex (Hori et al. 2011; Wilkin et al. 2008). Hence, it is likely that, as in the case of *lgd* mutants, it occurs in the lysosome upon fusion of the ME. Unfortunately, the relationship between Su(dx), Dx and Lgd has not been explored so far. However, the available data suggest that Lgd is devoted to Shrub and is part of the general trafficking machinery, while Su(dx) and Dx are specific to Notch. Hence, they are probably involved in different pathways that are required for incorporation of Notch into ILVs. However, the loss of their function has similar consequences: the remaining of a larger fraction of Notch in the limiting membrane of the ME.

The recent description of the *dx* null phenotype revealed that its loss of function results in weak and tissue-specific Notch loss of function phenotypes, indicating that it is a positive modulator of Notch activity (Fuwa et al. 2006). The most prominent and penetrant phenotype is the broadening of the wing veins in adults, similar to Dl heterozygous flies. Wing notches and weak defects in eye and ocelli development are observed with incomplete penetrance. The affected developmental processes are controlled by ligand-dependent Notch signalling, suggesting that Dx might also be involved in ligand-dependent signalling. This notion is supported by a recent study in mammalian cells that shows that DTX4, one of the five mammalian Dx orthologs, enhances Dll1-dependent signalling of Notch1 upstream of

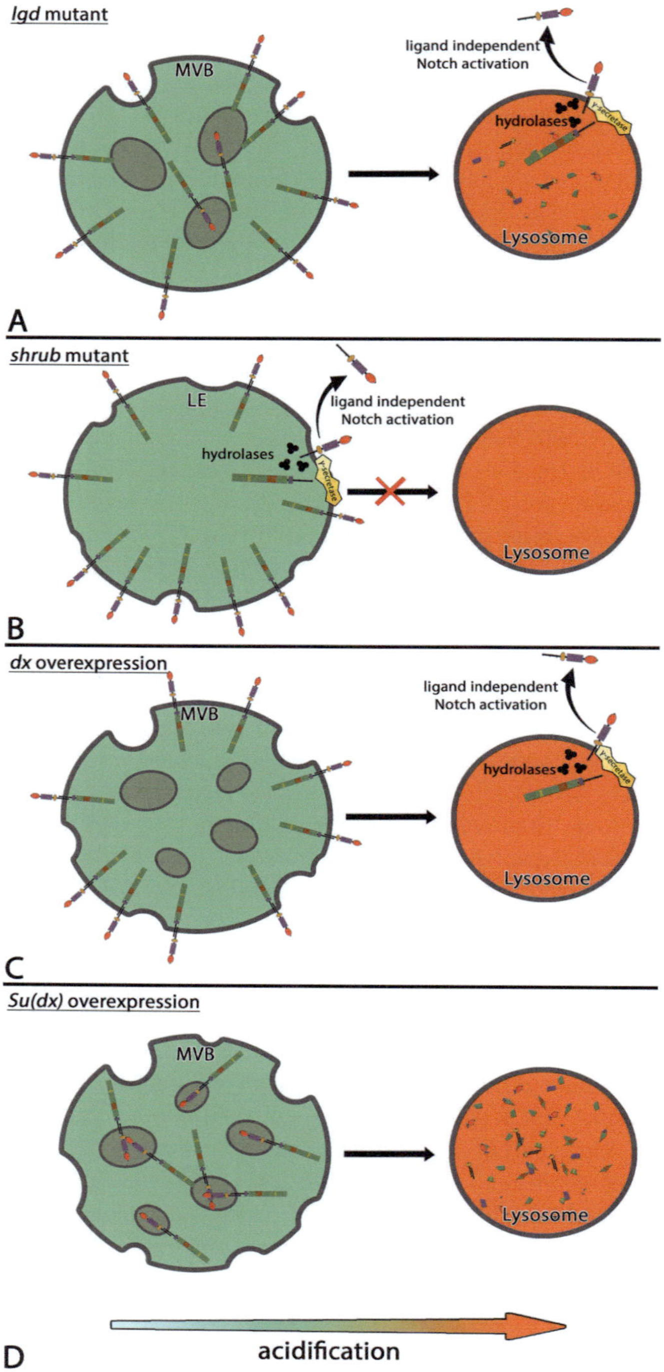

Fig. 4 Notch activity in different genetic backgrounds. (**a**) In *lgd* mutants, a fraction of Notch is not incorporated in ILVs and remains in the limiting membrane of the ME. After fusion, the ECD of Notch of this fraction is removed and the arising NEXT-like fragment is cleaved by the γ-secretase complex. This releases NICD into the cytosol from which it translocates into the nucleus and activates the transcription of the target genes. (**b**) In *shrub* mutants, ILV formation is abolished and all Notch remains in the limiting membrane of the ME. At the ends of maturation, the conditions in the lumen of the ME are similar to that of the lysosome and a large fraction of Notch

ADAM10 (Chastagner et al. 2017). However, it is possible that Dx produces ligand-independent Notch signalling as a basis on which ligand-dependent signalling is added. The sum of both is required in certain tissues for sufficient activation of the pathway, e. g. during wing development.

6 Contribution of Ligand-Independent Activation of Notch in the Endosome during Development in *Drosophila*

The observation that Notch can be activated in the endosome raises the question whether this ligand-independent mode of activation is used during development. In the most cases described so far, the mode of Notch activation is caused by the disruption of genes that control endosomal sorting and ubiquitylation, resulting in pathological pathway activity. However, the Dx/Krz/Su(dx) module appears to constitute a specific sorting system that has been evolved to regulate the amount of Notch at the limiting membrane of the ME. This system regulates the amount of Notch activity generated in the endosome in a temperature-dependent manner. According to the proposed model, the ligand-independent activity of Notch generated in the endosomal pathway is required in addition to ligand dependent activation in some developmental processes (Shimizu et al. 2014). While this might be the function of the Dx/Krz/Su(dx) module in poikilothermic animals such as *Drosophila*, the function of their orthologs in warm-blooded animals must be different. A caveat of the proposed model is that it is based on results largely obtained with cell culture and over-expression experiments.

Nevertheless, one developmental process has been identified so far that uses ligand-independent endosomal activation of Notch in *Drosophila*. The crystal cells, which are part of the immune system, require the activity of Notch for their survival. This activity is created in a ligand-independent manner that is dependent on the trafficking of Notch through the endosomal system (Mukherjee et al. 2011). Another recent work suggests that ligand-independent activation of the Notch pathway might be more common in *Drosophila* than anticipated (Palmer et al. 2014). Cis-inhibition is a regulatory mode where the ligands inhibit the activation of the Notch receptor located in the same cell, in a concentration-dependent manner. It has been shown that this mode of regulation is used in several developmental processes for the regulation of the activity of the Notch pathway (Yaron and Sprinzal 2011). The work of Palmer and colleagues (Palmer et al. 2014) suggests that cis-inhibition is a more common regulatory mode than anticipated. Moreover, it indicates that cis-inhibition also suppresses ligand-independent activation of Notch in the endosomal pathway. Hence, ligand-independent activation of the Notch pathway in the endosomal pathway might be present in many cells during development. It is probably regulated by cis-inhibition and the Dx/Krz/Su(dx) module adds significantly to the net pathway activity in several processes.

6.1 Notch Activation in the Endosomal Pathway of Mammals

While there are several examples for mutations in genes encoding regulators of ILV formation that cause activation of the Notch pathway in *Drosophila*, the evidence for this mode of activation in mammalian cells is scarce and allows no definitive verdict. Two orthologs exist in mam-

Fig. 4 (continued) is cleaved. The amount of the generated NICD leads to a detectable activity of the pathway before the ME fuses with the lysosome. The Dx/Krz/Su(dx) module regulates the amount of Notch that remains in the limiting membrane of the ME. (**c**) Over-expression of Dx increases the fraction of Notch that remains in the limiting membrane. After fusion of the ME with the lysosome, more Notch is cleaved by the described mechanism and results in an increase of the activity of the pathway. (**d**) Conversely the over-expression of Su(dx) results in the incorporation of more Notch into ILVs of the ME and less activation of the pathway after its fusion with the lysosome

mals for Lgd, termed CC2D1A (Coiled-Coil and C2 domain-containing protein 1A)/Lgd2/Aki1 (Akt Kinase-Interacting Protein 1) /TAPE (TBK1-associated Protein in Endolysosomes) and Freud-1 (FRE under Dual Repression-Binding-Protein 1) and CC2D1B/Lgd1/Freud-2 (Childress et al. 2006; Gallagher and Knoblich 2006; Jaekel and Klein 2006). The individual loss of function of each gene results in weak endosomal defects but not in activation of the Notch pathway (Drusenheimer et al. 2015; Manzini et al. 2014). However, experiments in *Drosophila* show that both LGD1 and LGD2 can substitute for Lgd (Drusenheimer et al. 2015; Jaekel and Klein 2006). Since both genes are also co-expressed in many tissues, they probably act in a functional redundant manner and must be concomitantly removed to reveal the function of Lgd in mammals. Therefore the final conclusion whether loss of Lgd function causes ectopic Notch activation in mammals awaits the analysis of the double mutant.

So far it has not been reported that loss of function of a core component of the ESCRT machinery causes activation of Notch in mammals. CHMP5 is an auxiliary ESCRT protein that is required for the full activity of ESCRT-III and its loss of function results in the activation of the Notch pathway in the follicle epithelium of *Drosophila* (Berns et al. 2014). However, the knockout in mouse does not result in a significant activation of the pathway (Shim et al. 2006). A hypomorphic mutation in the gene encoding the ESCRT-II component Vps25 also does not result in a significant activation of the Notch pathway in mice but results in distortion of FGF and Hedgehog signalling (Handschuh et al. 2014). These results do not support a role of the ESCRT pathway in prevention of ectopic activation of Notch in mammals. However, some evidence exists that argues for ligand-independent endosomal activation of the Notch pathway. The depletion of the ESCRT-I component Tsg101 (Tumor susceptibility gene 101) results in activation of Notch in ciliated HEK293 and hTERT-RPE1 cells, indicating that loss of ESCRT function can result in activation of the Notch pathway in the endosomal pathway at least in cell culture (Leitch et al. 2014).

The information about the function of Dx in mammals is incomplete and also contradicting. In mammals, five Dx orthologs exist, termed DTX1 to DTX4 and DTX3L (Deltex3-like) (Kishi et al. 2001; Matsuno et al. 1998). DTX3 and DTX3L lack of the WW domains necessary for interactions with Notch and cannot bind to it. The function of the orthologs appears to differ in a cell type-specific manner. However, the WW domain containing DTX1, 2 and 4 can bind to the intracellular domain of Notch1 (Chastagner et al. 2017; Kishi et al. 2001; Matsuno et al. 1998). While Dx is a positive regulator of Notch activity in *Drosophila*, the situation is more complex in mammals: DTX1, 2 and 4 appear to be negative regulators of ligand-dependent Notch signalling in 1010 thymoma cells and forced over-expression of DTX1 during T-cell development suppresses Notch signalling (Lehar and Bevan 2006; Maillard et al. 2004). Moreover, Notch signalling was more potent in DTX1 DTX2 double mutant cells that are also depleted of DTX4 (Lehar and Bevan 2006). These results suggest that DTX orthologs are negative regulators of ligand-dependent Notch signalling during T-cell development. On the other hand, cell culture experiments with mouse DTX orthologs suggest that they can also act as Notch activators (Kishi et al. 2001; Matsuno et al. 1998). Recent work using a cell culture system suggests that DTX4 is a potent enhancer of ligand-dependent Notch1 signalling upstream of ADAM10 (Chastagner et al. 2017). In the used system, DTX4 ubiquitylates the ICD of Notch1 to elicit its endocytosis. This endocytosis enhances the S2 cleavage by ADAM10. While the cell culture experiments suggest an involvement in the regulation of Notch signalling, no abnormalities have been found in the so far investigated developmental processes *in vivo*. For example, no defects in Notch dependent T-cell development were observed in DTX1 DTX2 double mutant mice (Lehar and Bevan 2006).

The antagonism between Su(dx) and Dx is conserved in mammals. The Su(dx) ortholog AIP4/Itch ubiquitylates Dx, which sends it for lysosomal degradation (Chastagner et al. 2006). Hence, AIP4/Itch antagonises Dx through initiation of its degradation. AIP4/Itch (AIP4 was introduced, Itch is not an abbreviation) has also been shown to ubiquitylate the ICD of Notch1 (Chastagner et al. 2008). This unusual poly-ubiquitylation on lysine (K) 29 of the Ubiquitin results in the degradation of the receptor in the lysosome. In contrast to *Drosophila*, the binding of AIP4 to Notch is indirect. In agreement with this, the PPSY binding motif for binding of WW domains, which is present in *Drosophila* Notch, is lacking in Notch1. It appears that one adapter that connects AIP4 to Notch is DTX (Chastagner et al. 2008). In summary, the evidence so far for the control of a ligand-independent endosomal activation mechanism of Notch by DTX and AIP4 in mammals is scarce. It appears that the two E3 ligases are involved in ligand-dependent signalling and Notch degradation.

While the investigations of the orthologs of *Drosophila* genes involved in endosomal activation of Notch did not result in a clear answer *in vivo*, work in zebrafish suggests that activation of Notch in the endosomal pathway occurs upon loss of function of genes that are required for cilial function (Leitch et al. 2014). The depletion of components of the cilial BBS (Bardet Biedl Syndrome) complex results in ligand-independent activation of the Notch pathway in zebrafish. This activation is accompanied by accumulation of Notch in MEs and is enhanced by concomitant depletion of TSG101. These results argue for an involvement of the ESCRT complex in the regulation of endosomal activity of the Notch pathway *in vivo* in vertebrates and indicate that additional components are important in vertebrates. Moreover, the activation of Notch in cells mutant for genes that encode factors important for the regulation of ILV incorporation of Notch might be tissue specific. The loss of function of the ESCRT-I component TSG-101 in mice results in early embryonic lethality and massive cell death. It is therefore possible that the activation of Notch in these mutants has been overlooked.

Acknowledgments Work in the Klein lab is supported by the Deutsche Forschungsgemeinschaft (DFG) through SFB 1208 "Identity and Dynamics of Membrane Systems-from Molecules to Cellular Functions "and Sachbeihilfe KL-1028/9-1 and KL-1028/8-1. We apologise to every researchers whose publications are not cited due to restrictions in space.

References

Aster JC, Pear WS, Blacklow SC (2017) The varied roles of Notch in cancer. Annu Rev Pathol Mech Dis 12:245–275

Balderhaar HJ, Ungermann C (2013) CORVET and HOPS tethering complexes–coordinators of endosome and lysosome fusion. J Cell Sci 126:1307–1316

Berns N, Woichansky I, Friedrichsen S, Kraft N, Riechmann V (2014) A genome-scale in vivo RNAi analysis of epithelial development in Drosophila identifies new proliferation domains outside of the stem cell niche. JCS 127:2736–2748

Cabrera M, Ungermann C (2013) Guanine nucleotide exchange factors (GEFs) have a critical but not exclusive role in organelle localization of Rab GTPases. JBC 288:28704–28712

Cabrera M, Engelbrecht-Vandré S, Ungermann C (2014) Function of the Mon1-Ccz1 complex on endosomes. Small GTPases 3:1–3

Chastagner P, Israel A, Brou C (2006) Itch/AIP4 mediates Deltex degradation through the formation of K29-linked polyubiquitin chains. EMBO Rep 7:1147–1153

Chastagner P, Israel A, Brou C (2008) AIP4/Itch regulates Notch receptor degradation in the absence of ligand. PLoS One 3:e2735

Chastagner P, Rubinstein E, Brou C (2017) Ligand-activated Notch undergoes DTX4-mediated ubiquitylation and bilateral endocytosis before ADAM10 processing. Sci Signal 10:1–13

Childress JL, Acar M, Tao C, Halder G (2006) Lethal Giant discs, a novel C2-domain protein, restricts notch activation during endocytosis. Curr Biol 16:1–6

Cotter K, Stransky L, McGuire C, Forgac M (2015) Recent insights into the structure, regulation, and function of the V-ATPases. Trends Biochem Sci 10:611–622

Couturier L, Trylinski M, Mazouni K, Darnet L, Schweisguth F (2014) A fluorescent tagging approach in Drosophila reveals late endosomal trafficking of Notch and Sanpodo. J Cell Biol 207:351–363

Dell'Angelica EC (2009) AP-3-dependent trafficking and disease: the first decade. Curr Opin Cell Biol 21:552–559

Dobrowolski R, De Robertis E (2012) Endocytic control of growth factor signalling: multivesicular bodies as signalling organelles. Nat Rev Mol Biol 13:53–60

Dodson EJ, Fishbain-Yoskovitz V, Rotem-Bamberger S, Schueler-Furman O (2015) Versatile communication strategies among tandem WW domain repeats. Exp Biol Med (Maywood) 3:351–360

Drusenheimer N, Migdal B, Jäckel S, Tveriakhina L, Scheider K, Schulz K, Gröper J, Köhrer K, Klein T (2015) The mammalian orthologs of Drosophila Lgd, CC2D1A and CC2D1B, function in the endocytic pathway, but their individual loss of function does not affect notch signalling. PLoS Genet 11(12):e1005749

Fostier M, Evans DA, Artavanis-Tsakonas S, Baron M (1998) Genetic characterization of the Drosophila melanogaster suppressor of deltex gene: a regulator of notch signaling. Genetics 150:1477–1485

Fuwa TJ, Hori K, Sasamura T, Higgs J, Baron M, Matsuno K (2006) The first deltex null mutant indicates tissue-specific deltex-dependent Notch signaling in Drosophila. Mol Gen Genomics 275:251–263

Gallagher CM, Knoblich J (2006) The conserved c2 domain protein lethal (2) giant discs regulates protein trafficking in Drosophila. Dev Cell 11:641–653

Gerasimenko JV, Tepikin AV, Petersen OH, Gerasimenko OV (1998) Calcium uptake via endocytosis with rapid release from acidifying endosomes. Curr Biol 8:1335–1338

Gomez-Lamarca MJ, Snowdon LA, Seib E, Klein T, Bray SJ (2015) Rme-8 depletion perturbs Notch recycling and predisposes to pathogenic signaling. JBC 210:303–318

Gupta-Rossi N, Six E, LeBail O, Logeat F, Chastagner P, Olry A, Israel A, Brou C (2004) Monoubiquitination and endocytosis direct gamma-secretase cleavage of activated Notch receptor. J Cell Biol 166:73–83

Handschuh K, Feenstra J, Koss M, Ferretti E, Risolino M, Zewdu R, Sahai MA, Bénazet JD, Peng XP, Depew MJ et al (2014) ESCRT-II/Vps25 constrains digit number by endosome-mediated selective modulation of FGF-SHH signaling. Cell Rep 9:674–687

Haugsten EM, Malecki J, Bjørklund SM, Olsnes S, Wesche J (2008) Ubiquitination of fibroblast growth factor receptor 1 is required for its intracellular sorting but not for its endocytosis. Mol Biol Cell 8:3390–3403

Heuss SF, Ndiaye-Lobry D, Six E, Israel A, Logeat F (2008) The intracellular region of Notch ligands Dll1 and Dll3 regulates their trafficking and signalling activity. PNAS 105:11212–11217

Hori K, Fostier M, Ito M, Fuwa TJ, Go MJ, Okano H, Baron M, Matsuno K (2004) Drosophila deltex mediates suppressor of Hairless-independent and late-endosomal activation of notch signaling. Development 131:5527–5537

Hori K, Sen A, Kirchhausen T, Artavanis-Tsakonas S (2011) Synergy between the ESCRT-III complex and Deltex defines a ligand-independent Notch signal. J Cell Biol 195:1005–1015

Hori K, Kirchhausen T, Artavanis-Tsakonas S (2012) Regulation of ligand-independent Notch signal through intracellular trafficking. Comm Integr Biol 5:374–376

Huang F, Goh LK, Sorkin A (2007) EGF receptor ubiquitination is not necessary for its internalisation. PNAS 104:16904–16909

Huotari J, Helenius A (2011) Endosome maturation. EMBO J 30:3481–3500

Hurley JH (2010) The ESCRT complexes. Crit Rev Biochem Mol Biol 45:463–487

Hurley JH (2015) ESCRTs are everywhere. EMBO J 34:2398–2407

Jaekel R, Klein T (2006) The Drosophila Notch inhibitor and tumor suppressor gene lethal (2) giant discs encodes a conserved regulator of endosomal trafficking. Dev Cell 11:655–669

Jekely G, Rorth P (2003) Hrs mediates downregulation of multiple signalling receptors in *Drosophila*. EMBO Rep 4:1164–1168

Kandachar V, Roegiers F (2012) Endocytosis and control of Notch signaling. Curr Opin Cell Biol 4:534–540

Kinchen JM, Ravichandran KS (2010) Identification of two evolutionarily conserved genes regulating processing of engulfed apoptotic cells. Nature 464:778–782

Kishi N, Tang Z, Maeda Y, Hirai A, Mo R, Ito M, Suzuki S, Nakao K, Kinoshita T, Kadesch T et al (2001) Murine homologs of deltex define a novel gene family involved in vertebrate Notch signaling and neurogenesis. Int J Dev Neurosci 19:21–35

Klein T (2003) The tumour suppressor gene l(2)giant discs is required to restrict the activity of Notch to the dorsoventral boundary during Drosophila wing development. Dev Biol 255:313–333

Kovall RA, Gebelein B, Sprinzak D, Kopan R (2017) The canonical Notch signalling pathway: structural and biochemical insights into shape, sugar, and force. Dev Cell 41:228–241

Legent K, Liu HH, Treisman JE (2015) Drosophila Vps4 promotes epidermal growth factor receptor signaling independently of its role in receptor degradation. Development 142:1480–1491

Lehar SM, Bevan MJ (2006) T cells develop normally in the absence of both Deltex1 and Deltex2. Mol Cell Biol 26:7358–7371

Leitch CC, Lodh S, Prieto-Echagüe V, Badano JL, Zaghloul NA (2014) Basal body proteins regulate Notch signaling through endosomal trafficking. J Cell Sci 127:2407–2019

Li H, Koo Y, Mao X, Sifuentes-Dominguez L, Morris LL, Jia D, Miyata N, Faulkner RA, van Deursen JM, Vooijs M et al (2015) Endosomal sorting of Notch receptors through COM MD9-dependent pathways modulates Notch signaling. JBC 211:605–617

Lu H, Bilder D (2005) Endocytic control of epithelial polarity and proliferation in *Drosophila*. Nat Cell Biol 7:1132–1139

Maillard I, Weng AP, Carpenter AC, Rodriguez CG, Sai H, Xu L, Allman D, Aster JC, Pear WS (2004) Mastermind critically regulates Notch-mediated lymphoid cell fate decisions. Blood 104:1696–1702

Manzini MC, Xiong L, Shaheen R, Tambunan DE, Di Costanzo S, Mitisalis V, Tischfield DJ, Cinquino A, Ghaziuddin M, Christian M et al (2014) CC2D1A regulates human intellectual and social function as well as NF-κB signaling homeostasis. Cell Rep 8:647–655

Mašek J, Andersson ER (2017) The developmental biology of genetic Notch disorders. Development 144:1743–1763

Matsuno K, Eastman D, Mitsiades T, Quinn AM, Carcanciu ML, Ordentlich P, Kadesch T, Artavanis-Tsakonas S (1998) Human deltex is a conserved regulator of Notch signalling. Nat Genet 19:74–78

Matsuno K, Ito M, Hori K, Miyashita F, Suzuki S, KIshi N, Artavanis-Tsakonas S, Okano H (2002) Involvement of a proline rich motif and RING-H2 finger of Deltex in the regulation of Notch-signaling. Development 129:1049–1159

Maxfield FR, Yamashiro DJ (1987) Endosome acidification and the pathways of receptor-mediated endocytosis. Adv Exp Med Biol 225:189–198

McMillan BJ, Tibbe C, Jeon H, Drabek AA, Klein T, Blacklow SC (2016) Electrostatic interactions between elongated monomers drive filamentation of Drosophila shrub, a Metazoan ESCRT-III Protein. Cell Rep 16:1211–1217

McMillan BJ, Tibbe C, Drabek AA, Seegar TCM, Blacklow SC, Klein T (2017) Structural basis for regulation of ESCRT-III complexes by Lgd. Cell Rep 19:1–8

Moberg KH, Schelble S, Burdick SK, Hariharan IK (2005) Mutations in erupted, the Drosophila ortholog of mammalian tumor susceptibility gene 101, elicit non-cell-autonomous overgrowth. Dev Cell 9:699–710

Moretti J, Brou C (2013) Ubiquitinations in the Notch signalling pathway. Int J Mol Sci 14:6359–6381

Mukherjee A, Veraksa A, Bauer A, Rosse C, Camonis J, Artavanis-Tsakonas S (2005) Regulation of Notch signalling by non-visual ß-arrestin. Nat Cell Biol 7:1191–1201

Mukherjee T, Kim WS, Mandal L, Banerjee U (2011) Interaction between Notch and Hif-alpha in development and survival of Drosophila blood cells. Science 332:1210–1213

Nakatsu F, Ohno H (2003) The epsin protein family: coordinators of endocytosis and signaling. Cell Struct Funct 28:419–429

Nemetschke L, Knust E (2016) Drosophila Crumbs prevents ectopic Notch activation in developing wings by inhibiting ligand-independent endocytosis. Development 143:4543–4553

Nordmann M, Cabrera M, Perz A, Bröcker C, Ostrowicz C, Engelbrecht-Vandré S, Ungermann C (2010) The Mon1-Ccz1 complex is the GEF of the late endosomal Rab7 homolog Ypt7. Curr Biol 20:1654–1659

Norrisa A, Tamminenib P, Wanga S, Gerdesa J, Murra A, Kwanc KY, Caib I, Granta BD (2017) SNX-1 and RME-8 oppose the assembly of HGRS-1/ESCRT-0 degradative microdomains on endosomes. PNAS 3:E307–E316

Palmer WH, Jia D, Deng W (2014) Cis-interactions between Notch and its ligands block ligand-independent Notch activity. elife 3:e04415

Pocha SM, Wassmer T, Niehage C, Hoflack B, Knust E (2011) Retromer controls epithelial cell polarity by trafficking the apical determinant Crumbs. Curr Biol 21(13):1111–1117

Portela M, Parsons LM, Grzeschik NA, Richardson HE (2015) Regulation of Notch signaling and endocytosis by the Lgl neoplastic tumor suppressor. Cell Cycle 14:1496–1506

Poteryaev D, Datta S, Ackema K, Zerial M, Spang A (2010) Identification of the switch in early-to-late endosome transition. Cell 14:497–508

Raiborg C, Slagsvold T, Stenmark H (2006) A new side to ubiquitin. Trends Biochem Sci 10:541–544

Rand MD, Grimm LM, Artavanis-Tsakonas S, Patriub V, Blacklow SC, Sklar J, Aster JC (2000) Calcium depletion dissociates and activates heterodimeric notch receptors. Mol Cell Biol 5:1825–1835

Ren X, Hurley JH (2010) VHS domains of ESCRT-0 cooperate in high-avidity binding to polyubiquitinated cargo. EMBO J 29:1045–1054

Rink J, Ghigo E, Kalaidzidis Y, Zerial M (2005) Rab conversion as a mechanism of progression from early to late endosomes. Cell 122:735–749

Rusten TE, Rodahl LM, Pattni K, Englund C, Samakovlis C, Dove S, Brech A, Stenmark H (2006) Fab1 phosphatidylinositol 3-phosphate 5-kinase controls trafficking but not silencing of endocytosed receptors. Mol Biol Cell 9:3989–4001

Sakata T, Sakaguchi H, Tsuda L, Higashitani A, Aigaki T, Matsuno K, Hayashi S (2004) Drosophila Nedd4 regulates endocytosis of notch and suppresses its ligand-independent activation. Curr Biol 14:2228–2236

Schmidt O, Teis D (2012) The ESCRT machinery. Curr Biol 22:R116–R120

Schneider M, Troost T, Grawe F, Martinez-Arias A, Klein T (2012) Activation of Notch in lgd mutant cells requires the fusion of late endosomes with the lysosome. J Cell Sci 126:645–656

Schöneberg J, Lee IH, Iwasa JH, Hurley JH (2017) Reverse-topology membrane scission by the ESCRT proteins. Nat Rev Mol Cell Biol 18:5–17

Schröder B, Saftig P (2016) Intramembrane proteolysis within lysosomes. Ageing Res Rev 32:51–64

Schwake M, Schröder B, Saftig P (2013) Lysosomal membrane proteins and their central role in physiology. Traffic 14:739–748

Schweisguth F (2015) Asymmetric cell division in the Drosophila bristle lineage: from the polarization of sensory organ precursor cells to Notch-mediated binary fate decision. Wiley Interdiscip Rev Dev Biol. https://doi.org/10.1002/wdev.175

Sen A, Madhivanan K, Mukherjee D, Aguilar RC (2012) The epsin protein family: coordinators of endocytosis and signaling. Biomol Concepts 2:117–126

Shaye DD, Greenwald I (2005) LIN-12/Notch trafficking and regulation of DSL ligand activity during vulval induction in Caenorhabditis elegans. Development 132:5081–5092

Shim JH, Xiao C, Hayden MS, Lee KY, Trombetta ES, Pypaert M, Nara A, Yoshimori T, Wilm B, Erdjument-Bromage H et al (2006) CHMP5 is essential for late endosome function and down-regulation of receptor signaling during mouse embryogensis. JCB 172:1045–1056

Shimizu H, Woodcock SA, Wilkin MB, Trubenová B, Monk NA, Baron M (2014) Compensatory flux changes within an endocytic trafficking network maintain thermal robustness of Notch signaling. Cell 5:1160–1174

Struhl G, Adachi A (2000) Requirements for Presenillin-dependent cleavage of Notch and other transmembrane proteins. Mol Cell 6:625–636

Sweeney NT, Brennan JE, Jan YN, Gao FB (2006) The coiled-coil protein shrub controls neuronal morphogenesis in Drosophila. Curr Biol 16:1006–1011

Thompson BJ, Mathieu J, Sung HH, Loeser E, Rorth P, Cohen SM (2005) Tumor suppressor properties of the

ESCRT-II complex component Vps25 in Drosophila. Dev Cell 9:711–720

Tognon E, Wollscheid N, Cortese K, Tacchetti C, Vaccari T (2014) ESCRT-0 is not required for ectopic Notch activation and tumor suppression in Drosophila. PLoS One 9:e93987

Troost T, Jaeckel S, Ohlenhard N, Klein T (2012) The tumour suppressor Lethal (2) giant discs is required for the function of the ESCRT-III component Shrub/CHMP4. JCS 125:763–776

Vaccari T, Lu H, Kanwar R, Fortini M, Bilder D (2008) Endosomal entry regulates Notch receptor activation in Drosophila melanogaster. J Cell Biol 180:755–762

Vaccari T, Rusten TE, Menut L, Nezis IP, Brech A, Stenmark H, Bilder D (2009) Comparative analysis of ESCRT-I, ESCRT-II and ESCRT-III function in Drosophila by efficient isolation of ESCRT mutants. J Cell Sci 122:2413–2423

Wang T, Ming Z, Xiaochun W, Hong W (2011) Rab7: role of its protein interaction cascades in endo-lysosomal traffic. Cell Signal 3:516–521

Wegner CS, Rodahl LM, Stenmark H (2011) ESCRT proteins and cell signalling. Traffic 10:1291–1997

Wilkin MB, Carbery AM, Fostier M, Aslam H, Mazaleyrat SL, Higgs J, Myat A, Evans DA, Cornell M, Baron M (2004) Regulation of notch endosomal sorting and signaling by Drosophila Nedd4 family proteins. Curr Biol 14:2237–2244

Wilkin M, Tongngok P, Gensch N, Clemence S, Motoki M, Yamada K, Hori K, Taniguchi-Kanai M, Franklin E, Matsuno K et al (2008) Drosophila HOPS and AP-3 complex genes are required for a Deltex-regulated activation of notch in the endosomal trafficking pathway. Dev Cell 15:762–772

Windler SL, Bilder D (2010) Endocytic internalization routes required for Delta/Notch signaling. Curr Biol 20:538–543

Wollert T, Hurley JH (2010) Molecular mechanism of multivesicular body biogenesis by ESCRT complexes. Nature 484:864–869

Yamada K, Fuwa TJ, Ayukawa T, Tanaka T, Nakamura A, Wilkin M, Baron M, Matsuno K (2011) Roles of Drosophila Deltex in Notch receptor endocytic trafficking and activation. Genes Cells 16:261–272

Yaron A, Sprinzak D (2011) The cis side of juxtacrine signaling: a new role in the development of the nervous system. Trends Neurosci 35:230–239

Yousefian J, Troost T, Grawe F, Sasamura T, Fortini M, Klein T (2013) Dmon1 controls recruitment of Rab7 to maturing endosomes in Drosophila. JCS 126:1583–1594

Zerial M, McBride H (2001) Rab proteins as membrane organizers. Nat Rev Cell Biol 2:107–119

Zhenga L, Saundersa CA, Sorensenb EB, Waxmonskya NC, Conner SD (2013) Notch signaling from the endosome requires a conserved dileucine motif. MBoC 24:297–307

Zhou B, Wu Y, Lin X (2011) Retromer regulates apical-basal polarity through recycling Crumbs. Dev Biol 360:87–95

Part III

Intracellular Signaling Mechanisms

The Notch Interactome: Complexity in Signaling Circuitry

Diana M. Ho, K. G. Guruharsha, and Spyros Artavanis-Tsakonas

Abstract

The Notch pathway controls a very broad spectrum of cell fates in metazoans during development, influencing proliferation, differentiation and cell death. Given its central role in normal development and homeostasis, misregulation of Notch signals can lead to various disorders including cancer. How the Notch pathway mediates such pleiotropic and differential effects is of fundamental importance. It is becoming increasingly clear through a number of large-scale genetic and proteomic studies that Notch interacts with a staggeringly large number of other genes and pathways in a context-dependent, complex, and highly regulated network, which determines the ultimate biological outcome. How best to interpret and analyze the continuously increasing wealth of data on Notch interactors remains a challenge. Here we review the current state of genetic and proteomic data related to the Notch interactome.

Keywords

Genome-wide screens · Genetic modifiers · Notch signal integration · Protein-protein interactions · Signaling interactome

Abbreviations

BioPlex	biophysical interactions of ORFeome-based complexes
CADASIL	cerebral autosomal dominant arteriopathy with subcortical infarcts and leukoencephalopathy
DPiM	*Drosophila* protein interaction map
NICD	Notch intracellular domain
PPI	protein-protein interaction
T-ALL	T-cell acute lymphoblastic leukemia
Y2H	yeast two-hybrid

1 The Developmental Logic of the Core Notch Signaling Pathway

At first glance, the core Notch pathway may seem deceptively simple but a quick glance into the literature reveals that it is in fact staggeringly complicated, with multiple levels of regulation and

D. M. Ho
Department of Cell Biology, Harvard Medical School, Boston, MA, USA

K. G. Guruharsha
Neurology Research, Biogen Inc., Cambridge, MA, USA

S. Artavanis-Tsakonas (✉)
Neurology Research, Biogen Inc., Cambridge, MA, USA

Department of Cell Biology, Harvard Medical School, Boston, MA, USA
e-mail: artavanis@hms.harvard.edu

T. Borggrefe, B. D. Giaimo (eds.), *Molecular Mechanisms of Notch Signaling*, Advances in Experimental Medicine and Biology 1066, https://doi.org/10.1007/978-3-319-89512-3_7

complex context-dependent crosstalk with other pathways in the cell. It is essential to keep in mind that the biological outcome of a Notch signal depends on the cellular and developmental context (Bray 2016; Artavanis-Tsakonas and Muskavitch 2010). For example, Notch can induce proliferation or differentiation (Ho and Artavanis-Tsakonas 2016), drive apoptosis or cell survival (Arya and White 2015) and act as either a tumor suppressor or as an oncogene depending on specific cellular contexts (Ranganathan et al. 2011; Lobry et al. 2014). It is also highly dosage-dependent, such that both too much and too little Notch both cause dramatic effects (Guruharsha et al. 2012).

The core components and the canonical mechanism of the Notch pathway have been well studied and as there are several excellent reviews [for example, see (Artavanis-Tsakonas and Muskavitch 2010; Kopan and Ilagan 2009; Andersson et al. 2011; Kovall et al. 2017) summarizing the core mechanisms, we will not go into the details here. Suffice it to say that Notch serves a rather unique developmental role as it links the fate of one cell to that of its neighbor. The single pass transmembrane Notch receptor expressed on the surface of one cell interacts physically with a membrane bound ligand on the adjacent cell, triggering a cascade of proteolytic cleavages that results in the release of the intracellular domain of Notch followed by its translocation into the nucleus where it drives Notch-dependent transcription (Kovall and Blacklow 2010; Kopan and Ilagan 2009). The Notch signal is devoid of a catalytic step as the signal depends on stoichiometric relationships, affording Notch its biochemical distinctiveness (Bray 2016; Kopan and Ilagan 2009). Moreover, given that usually both the receptor and the ligand are expressed on both neighboring cells, various factors including inhibitory cis interactions between the receptor and ligand will define which is the signal receiving vs the sending cell (Bray 2016; del Álamo et al. 2011).

The cell controls Notch signals at multiple levels: upstream regulation of ligand and receptor expression (Bray 2016), extracellular modifications that affect ligand-receptor interactions (Rana and Haltiwanger 2011), membrane trafficking factors and post-translational modifications that influence the number of signal-competent receptors and ligands on the surface of the cell (Musse et al. 2012; Conner 2016) and finally nuclear factors that influence the transcriptional activity of the Notch intracellular domain (Bray and Bernard 2010; Bray 2016; Giaimo et al. 2017); all can and have been shown to modulate the quality and the strength of the Notch signal. Considering all these elements one can plausibly predict that there will be many genes whose activity may influence Notch signaling. In the past ten years, a number of large-scale studies have been conducted to identify modifiers of Notch, using both genetic and proteomic approaches, revealing the extraordinary complexity of the genetic circuitry capable of modulating Notch signals.

2 Genetic Modifiers of Notch

A number of genetic screens have been conducted to identify genes that interact with Notch to affect different aspects of Notch biology. These screens primarily utilized *Drosophila melanogaster* as the experimental system, as it is amenable to large-scale genetic screening, contains well-characterized Notch components and mutations and has clear molecular and phenotypic readouts for Notch activation. It is also noteworthy that, to date, screens in other model organisms have revealed relatively few Notch modifiers compared to *Drosophila*. Nevertheless, important studies involving *Caenorhabditis elegans* have revealed the link between Notch and the *presenilin* gene, which encodes a crucial element of the γ-secretase complex that cleaves the intracellular domain of Notch (Levitan and Greenwald 1995; Kovall et al. 2017). Genetic screens using the worm Notch genes, *lin-12* and *glp-1*, as well as the presenilin gene *sel-12* also revealed a handful of interactions with novel modifiers [reviewed in (Greenwald and Kovall 2013)].

Additionally, and perhaps surprisingly, no large-scale studies in mammalian cells have been published to date. However, with the advent of new RNAi and CRISPR based technologies this may soon change. The current state of the field is such that any network analysis of the Notch interactome must out of necessity rely heavily on the data from model systems. While these data are highly valuable in identifying and dissecting genes and pathways capable of crosstalking with Notch signals, they cannot always be simply interpreted in the context of human disease and development. However, these functional screens do define working hypotheses for human biology, as more often than not a functional relationship between Notch and another gene in *Drosophila* is also valid in mammals.

In principle, an unbiased genetic screen uncovers modifiers at all levels of the Notch pathway (e.g. upstream regulators, genes that interact with the core pathway itself, downstream targets, and other functional modulators). Additionally, they reveal both positive and negative regulators and, of course, both direct and indirect interactors. The genetic screens that have been conducted to date in *Drosophila* comprise two main types – those that utilized the Exelixis collection of insertional mutants and those that utilized an RNAi-based loss of function approach.

2.1 Exelixis Collection Screens

The Exelixis collection of transposon-mediated mutations, which consists of over 15,000 individual fly lines that are together estimated to disrupt approximately 53% of the protein-coding genes in *Drosophila* (Parks et al. 2004; Thibault et al. 2004), was utilized in several screens for genetic modifiers of the Notch pathway. The Exelixis collection was the first genome-wide *Drosophila* mutant collection in which the location of each insertion was sequenced, thus enabling the rapid identification of affected genes and facilitating large-scale genetic screens (Parks et al. 2004).

Kankel and colleagues (Kankel et al. 2007) performed the first screen to systematically interrogate the Exelixis collection for Notch modifiers. They tested modifiers in combination with overexpression of a dominant negative form of the nuclear Notch cofactor Mastermind in the fly wing, which produces a notched wing margin phenotype due to loss of Notch signaling and, remarkably, they identified a very large number (408!) of potential modifier genes. These genes included several known Notch interactors, but the majority had not previously been linked to Notch. Gene ontology analysis identified novel links between Notch and RNA processing genes, cell cycle genes and also genes involved in other signaling pathways. 175 of the 408 were further shown to interact with at least three additional *Notch* or *deltex* mutants, making them strong candidates as *bona fide* Notch interactors.

Shalaby and colleagues (Shalaby et al. 2009) reported a screen of the Exelixis collection in a *GMR-Gal4;UAS-delta (GMR > delta)* background, which causes Notch-dependent cell fate changes in the fly retina. The authors identified 274 candidate genes that modify the *GMR > delta* eye phenotype. Interestingly, of these modifiers, a majority was unable to modify the wing notching phenotype caused by Delta overexpression in the wing. This finding suggests that many non-core Notch interactors may modulate Notch only in specific contexts (e.g. specific tissues, cell types, or biological processes). We note also that they only confirmed a subset (284/798) of mutant lines identified in their primary screen, so the true number of modifiers is likely to be significantly higher.

A third screen was performed using the Exelixis collection to interrogate Notch-induced proliferation phenotypes (Ho et al. 2015; Pallavi et al. 2012). Expression of the constitutively active Notch intracellular domain (NICD) causes a large eye phenotype that can be enhanced or suppressed by genes that modulate the proliferative activity of Notch. Note that in contrast to the Shalaby and colleagues screen (Shalaby et al. 2009) where the Notch pathway was activated in the post-mitotic cells of the eye, this screen focused on the undifferentiated, actively dividing cells. This study identified 360 potential genetic modifiers of Notch, 206 of which have clear human orthologs. This diverse list of modifiers is enriched for developmental/morphogenetic factors, cell cycle/cell division genes and genes involved in transcription, but also contains a large number (84) of genes with no GO or INTERPRO annotation.

Finally, we note that other insertional mutant collections can be interrogated in much the same way as the Exelixis collection for Notch pathway modulation. For example, the laboratory of Maria Dominguez has utilized the GeneSearch system (Toba et al. 1999) to screen for genes involved in Delta-dependent tumorigenesis in the eye (Da Ros et al. 2013; Ferres-Marco et al. 2006); however, the full results of this screen have not yet been published.

2.2 RNAi Based Genetic Screens for Notch Modifiers

The advent of RNAi-based technology provided another means by which large numbers of genes could be systematically disrupted and interrogated in Notch signaling backgrounds.

Saj and colleagues (Saj et al. 2010) performed a combined *in vitro/in vivo* RNAi screen using transcriptional reporters. They first screened an RNAi library in an S2 *Drosophila* cell line containing a constitutively active Notch-VP16 construct along with an NRE-GFP luciferase reporter for Notch activity and identified 900 *in vitro* modifiers. These are enriched for muscle, eye and nervous system genes, as well as for transcriptional regulation and vesicular transport processes. Further testing of 501 of these genes *in vivo* revealed that 283 of these genes affected at least one Notch-dependent phenotype in the fly wing and 167 affected Notch activation in the fly retina; interestingly, 102 genes scored positively in both the wing and retina assays. Overall, these assays identified a total of 333 *in vivo* Notch interactors, representing a large diversity of biological processes. The authors also show that the network generated by their interactors (both *in vitro* and *in vivo*) has predictive value for identifying additional Notch interactors.

Mummery-Widmer and colleagues (Mummery-Widmer et al. 2009) performed a morphology-based *in vivo* screen with >20,000 transgenic RNAi fly lines predicted to target 82.2% of the genes in the genome, by assessing changes in Notch-dependent external sensory organ morphology in the notum. The resultant phenotypes were divided into those that represent effects on asymmetric cell division (226 genes) and lateral inhibition (233 genes). Nuclear import pathways and the COP9 signallosome were among the novel gene classes that were enriched in this list of modifiers. A small-scale, targeted RNAi-based screen was later performed in the same tissue by Le Bras and colleagues, who tested 418 candidate genes (consisting only of known Notch pathway components and endolysosomal pathway genes) and recovered 113 potential Notch interactors (Le Bras et al. 2012). Interestingly, three-quarters of these genes were not identified in the earlier, broader screen (Mummery-Widmer et al. 2009), possibly due to the more thorough approach taken with a smaller, targeted screen.

Mourikis and colleagues (Mourikis et al. 2010) performed a cell-based RNAi screen in KC167 cells using a constitutively active, membrane bound form of Notch and a transcriptional activation readout, and found 399 modifiers. Chromatin-associated factors, transcription factors, mRNA processing factors, and ribosomal proteins were among the classes significantly enriched. Interestingly, proteomic analysis showed an unexpected link between the mRNA processing factors and chromatin components identified in this study. The authors note that only one gene identified by Mummery-Widmer and colleagues (Mummery-Widmer et al. 2009) was also found in their study, which they suggest may reflect the different readouts used (*in vitro*, transcriptional readout vs *in vivo*, morphological readout).

Most recently, the Deng lab reported the results of an RNAi screen for genes that affect differentiation and cell cycle in *Drosophila* follicle cells. Out of 2205 RNAi lines, they recovered 33 modifiers (Jia et al. 2015). Again, most of these genes (20/33) had not been identified in the earlier RNAi screens described above.

Additionally, a broad genome-wide RNAi-based screen for modifiers of signaling cascades in *C. elegans* found 15 genes that interacted with *lin-12* and approximately 50 additional genes that interacted with *sel-12* (Lehner et al. 2006).

In summary, hundreds of genes have been functionally linked to the Notch pathway through these studies, revealing a very complex gene circuitry that can modulate the signal. An emerging theme when considering these genetic screens is that the modifier lists from each of them have surprisingly little overlap, even though several of the studies screened the same mutant collection (Guruharsha et al. 2012). Of the genes that do overlap across multiple screens, a relatively high proportion are core complex members. This suggests that there is a great deal of spatio-temporal/process-specific regulation and that the Notch circuitry is highly dynamic and context-responsive. It also suggests that the peripheral Notch interactome (as distinct from the core complex) in any given cell is an important determinant of pleiotropic function. The specific readout and experimental design for each screen undoubtedly affects the identification of modifier genes (e.g. a transcriptional reporter readout and a downstream phenotypic readout may enrich for different types of genes). Finally, no mutant collection (Exelixis or RNAi collections) contains 100% coverage of the protein-coding genes in the *Drosophila* genome. Thus, any single screen does not provide a full view of the entire Notch network. However, a meta-analysis of the data from multiple screens, especially when combined with other large data sets (such as proteomic data) produces a far more comprehensive picture of the entire Notch network.

3 Proteomic Interactions in the Notch Pathway

While genetic interactions reveal per definition functional associations, the underlying molecular relationships can be either direct or indirect. A direct way to examine molecular kinships is to examine protein-protein interactions; for example, co-immunoprecipitation and mass spectrometry experiments can be employed. A significant challenge in using proteomic approaches to study Notch, or in fact any other signaling pathway, is that many interactions may be highly transient and thus undetectable under the given experimental conditions. Nonetheless, clear detection of protein interactions may provide important mechanistic insights into the role a given protein plays in the cell. Given the complexity of the genetic circuitry revealed by the genetic interaction screens, probing protein interactions provides a powerful tool to identify direct molecular interactions.

The first effort to identify protein interactors of Notch pathway members through mass spectrometry analysis was performed using a handful of Notch core members from *Drosophila* (full length Notch, Notch ICD, Mastermind and Deltex) as bait (Veraksa et al. 2005). However, a systematic and comprehensive study of all core pathway members, such as that described for insulin signaling (Vinayagam et al. 2016), has not been reported for the Notch pathway. We note however that a detailed protocol has been described to achieve this goal using affinity purification followed by mass spectrometry analysis (Guruharsha et al. 2014).

Given the complexity of the Notch circuitry, what constitutes a Notch pathway element is certainly ill defined. Keeping this in mind, more than one hundred genes/proteins are listed as Notch pathway members in pathway databases such as KEGG (http://www.genome.jp/kegg/) (Kanehisa et al. 2017) and Reactome (http://www.reactome.org/) [(Milacic et al. 2012; Fabregat et al. 2016), Table 1]. Notably, the list of genes identified as pathway members by KEGG and Reactome only partially overlap, highlighting the fact that individual databases often contain incomplete information. Furthermore, these proteins have more than two thousand reported interactions (of which nearly one-third are physical) in publicly available databases such as GeneMania (Warde-Farley et al. 2010), STRING (Szklarczyk et al. 2015), IREF (Turinsky et al. 2011), IntAct (Orchard et al. 2014) and others that curate and consolidate data on protein interactions across the scientific literature. Not surprisingly, the majority of these interactions were discovered through small-scale and targeted studies; large-

Table 1 List of human Notch pathway genes from KEGG (Kanehisa et al. 2017) and Reactome (Fabregat et al. 2016)

Gene	KEGG	Reactome	Gene	KEGG	Reactome	Gene	KEGG	Reactome
ADAM17	Yes	Yes	DVL1	Yes	No	HEY1	No	Yes
APH1A	Yes	Yes	DVL2	Yes	No	HEY2	No	Yes
APH1B	Yes	Yes	DVL3	Yes	No	HEYL	No	Yes
CREBBP	Yes	Yes	NUMBL	Yes	No	HIF1A	No	Yes
DLL1	Yes	Yes	PSEN1	Yes	No	ITCH	No	Yes
DLL4	Yes	Yes	PTCRA	Yes	No	JUN	No	Yes
DTX1	Yes	Yes	RBPJL	Yes	No	MAMLD1	No	Yes
DTX2	Yes	Yes	ADAM10	No	Yes	MDK	No	Yes
DTX4	Yes	Yes	AGO1	No	Yes	MIB1	No	Yes
EP300	Yes	Yes	AGO2	No	Yes	MIB2	No	Yes
HDAC1	Yes	Yes	AGO3	No	Yes	MOV10	No	Yes
HDAC2	Yes	Yes	AGO4	No	Yes	MYC	No	Yes
HES1	Yes	Yes	ARRB1	No	Yes	NCOR1	No	Yes
HES5	Yes	Yes	ARRB2	No	Yes	NEURL1	No	Yes
JAG1	Yes	Yes	ATP2A1	No	Yes	NEURL1B	No	Yes
JAG2	Yes	Yes	ATP2A2	No	Yes	POFUT1	No	Yes
KAT2A	Yes	Yes	ATP2A3	No	Yes	POGLUT1	No	Yes
KAT2B	Yes	Yes	B4GALT1	No	Yes	RAB6A	No	Yes
LFNG	Yes	Yes	CCNC	No	Yes	RBX1	No	Yes
MAML1	Yes	Yes	CCND1	No	Yes	RPS27A	No	Yes
MAML2	Yes	Yes	CDK8	No	Yes	SEL1L	No	Yes
MAML3	Yes	Yes	CNTN1	No	Yes	SKP1	No	Yes
MFNG	Yes	Yes	CREB1	No	Yes	ST3GAL3	No	Yes
NCOR2	Yes	Yes	CUL1	No	Yes	ST3GAL4	No	Yes
NCSTN	Yes	Yes	DLK1	No	Yes	ST3GAL6	No	Yes
NOTCH1	Yes	Yes	DNER	No	Yes	TBL1X	No	Yes
NOTCH2	Yes	Yes	E2F1	No	Yes	TBL1XR1	No	Yes
NOTCH3	Yes	Yes	E2F3	No	Yes	TFDP1	No	Yes
NOTCH4	Yes	Yes	FBXW7	No	Yes	TFDP2	No	Yes
NUMB	Yes	Yes	FCER2	No	Yes	TLE1	No	Yes
PSEN2	Yes	Yes	FURIN	No	Yes	TLE2	No	Yes
PSENEN	Yes	Yes	GZMB	No	Yes	TLE3	No	Yes
RBPJ	Yes	Yes	HDAC10	No	Yes	TLE4	No	Yes
RFNG	Yes	Yes	HDAC11	No	Yes	TMED2	No	Yes
SNW1	Yes	Yes	HDAC3	No	Yes	TNRC6A	No	Yes
CIR1	Yes	No	HDAC4	No	Yes	TNRC6B	No	Yes
CTBP1	Yes	No	HDAC5	No	Yes	TNRC6C	No	Yes
CTBP2	Yes	No	HDAC6	No	Yes	TP53	No	Yes
DLL3	Yes	No	HDAC7	No	Yes	UBA52	No	Yes
DTX3	Yes	No	HDAC8	No	Yes	UBB	No	Yes
DTX3L	Yes	No	HDAC9	No	Yes	UBC	No	Yes

scale studies only represent a relatively small subset of the data. Duplication of the same information from different sources is also an issue when analyzing information from these databases. Fig. 1 visually summarizes the current status of Notch pathway connections from existing databases (Fig. 1A), encompassing all of the various interaction types. Navigating this labyrinth of connections from different types of studies with varying degrees of rigor, redundancy and confidence is undoubtedly daunting. Even breaking this network down to distinct interaction types (Fig. 1B-E) only marginally simplifies this complexity.

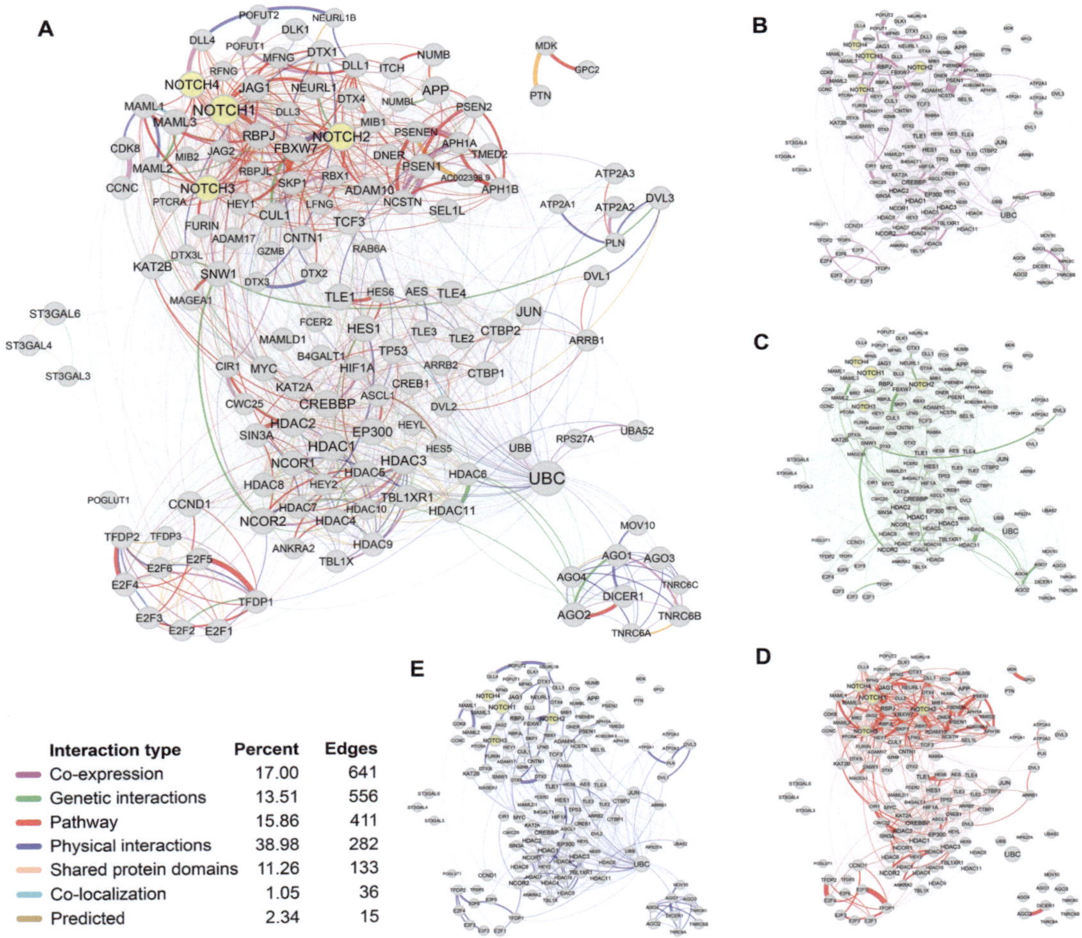

Fig. 1 Summary of Notch pathway connections from the literature. (A) Network from Genemania (Warde-Farley et al. 2010) using the Notch genes from Table 1, visualized using Gephi (Bastian et al. 2009). Human Notch genes are highlighted in light yellow. Different types of interactions (edges) are shown with specific colors and thickness is relative to the confidence score. The "score" is defined by Genemania differently for each type of interaction as well as each experimental approach. The size of the circles depicting each node is proportional to the number of edges i.e. interactions with other genes. The table in the figure lists the types of experimental methodology that defined connections. The contribution of each individual approach to the entire network in A is indicated as well as the number of edges. "Edges" represent unique interaction between any two proteins derived from similar data type and may include more than one study. For example, all PPIs between two proteins reported by multiple studies are combined as one edge. Additional panels (B, C, D and E) show snapshots of four richest interaction types (co-expression, genetic interactions, pathway and physical interactions) isolated from the full network to simplify the depiction and highlight the differences in coverage

Focusing on proteomic interactions with the Notch receptor itself, two groups have recently attempted to tease out the complexity of Notch signaling in the nucleus by defining interacting partners of NOTCH1 and NOTCH3. Yatim and colleagues (Yatim et al. 2012) hoped to identify transcriptional regulatory complexes interacting with the NOTCH1 intracellular domain (N1ICD) in the nucleus beyond the well-characterized core activation complex. To accomplish this, they used a human NOTCH1-dependent T-cell acute lymphoblastic leukemia (T-ALL) cell line engineered to express tagged human N1ICD and performed immunoaffinity purification experiments from nuclear extracts. They identified 127 N1ICD-interacting proteins (including 27 core Notch pathway members), which were involved in a wide range of molecular functions including DNA repair/replication, chromatin regulation, transcriptional coactivation, RNA processing, protein modification, trafficking and signaling crosstalk. They also further validated several hits and found that the core transcriptional complex physically associates with the transcription coactivator AF4p12 (ALL1-Fused Gene From Chromosome 4p12 Protein), the PBAF (polybromo-associated BRG1- or HBRM-associated factors) nucleosome remodeling complex, and the histone demethylases LSD1 (lysine (K)-specific demethylase 1A) and PHF8 (PHD Finger Protein 8).

Jung and colleagues (Jung et al. 2014) screened for NOTCH3-intracellular domain (N3ICD)-interacting proteins using a human proteome microarray. The 27 interactors they identified are enriched in gene expression, cell cycle, cell signaling and cellular developmental networks with a specific focus on ubiquitination pathways. Based on this pathway analysis, they followed up and validated WWP2 (WW Domain Containing E3 Ubiquitin Protein Ligase 2), an E3 ubiquitin-protein ligase, as a novel protein binding partner of N3ICD. Interestingly, WWP2 did not interact with the intracellular domains of the other three Notch receptors, indicating, not surprisingly, that the interactomes of the different human Notch paralogues are not identical.

It is important to note that as transcriptomes and proteomes differ considerably among individual cell types (Cherbas et al. 2011; Guruharsha et al. 2011), the outcome of the same signaling event might be quite different. For example, analyzing the interactome in different cell types or under different signaling conditions (e.g. + or – Notch) may provide insights into understanding the differential outcomes and pleiotropic nature of Notch signaling. The proteomic chip approach (Jung et al. 2014) is in a sense an "unbiased" approach in that it is essentially a cell-free system that should therefore be free of any context-dependent effects; on the other hand, such a truly *in vitro* approach may overlook potentially important context-dependent, dynamic or transient effects. Indeed, the dynamic and highly tunable nature of Notch signaling is crucial to its biological function. As with the genetic data, it is clear that the more individual studies that we can integrate into a single map, the more complete our picture of the Notch interactome will be.

Unbiased, proteome-scale studies have been very informative in providing molecular context to unstudied proteins, including the hundreds of genetic modifiers of Notch signaling as well as the proteins involved in the complex crosstalk with other signaling pathways (Guruharsha et al. 2012). In the following sections, we provide an overview of such studies and discuss how they can augment Notch interactome studies.

4 Proteome-Scale Interaction Maps

Large-scale protein interaction maps generated using the affinity purification and mass spectrometry (AP-MS) method published for *Drosophila* (Guruharsha et al. 2011) and human (Huttlin et al. 2015, 2017) provide an unbiased blueprint of protein complexes in these cells. The *Drosophila* protein interaction map (DPiM) was generated from analyzing ~5000 baits and the latest human map (termed BioPlex 2.0 (biophysical interactions of ORFeome-based complexes version 2.0)) from ~5900 baits. A smaller, complementary proteomic study was performed using transcription factors (which are underrepresented in the DPiM due to experimental design) as bait to interrogate nuclear proteins from the same cell type (S2R+) that was used for the DPiM; approximately 50% of the annotated transcription factors in the fly genome are represented in this map (Rhee et al. 2014). These maps are far from complete as they only represent slightly more than half of the proteome in each of those organisms. Nevertheless, each has already revealed thousands of protein-protein interactions that were not previously reported in the literature. Importantly, preliminary comparisons indicate that the fly and human maps are consistent to a large degree but not surprisingly, do not completely overlap. Apart from genuine differences between the fly and human proteomes, it is worth pointing out that the proteome coverage of clone sets used to generate the fly (Yu et al. 2011) and human interactomes (The Orfeome Collaboration 2016) are different. Additionally, the differences in the overall cellular proteome of the cell lines used in those studies (S2R+ for flies, HEK-293T for human) also affect the possible interactions that can be observed. Lastly, the assignation of orthologues between fly and human genes of necessity introduces some uncertainty. Orthology mapping between flies and humans has improved vastly in recent years and tools such as the DRSC Integrative Ortholog Prediction Tool [http://www.flyrnai.org/cgi-bin/DRSC_orthologs.pl, (Hu et al. 2011)] are extremely useful. Still, there will always be genes on both sides for which no clear ortholog exists, judging just from sequence comparisons. Moreover, as a rule, one protein in fly may have multiple orthologs in human (e.g. flies have a single Notch gene while humans have four). One final caveat when analyzing the existing DPiM proteomic map in the context of Notch is that S2R+ cells do not express the Notch receptor itself but can still respond to exogenous Notch, indicating that the downstream components of the pathway seem to be intact (Hori et al. 2011). We note however that because there is essentially no endogenous Notch background, S2R+ cells would be an excellent system in which to assess how the proteome changes in the presence of Notch.

Using the complementary yeast two-hybrid (Y2H) approach, proteomic-scale maps have been produced for several model organisms (Rual et al. 2005; Li et al. 2004; Yu et al. 2008; Giot et al. 2003) as well as for human (Stelzl et al. 2005; Rolland et al. 2014). The human Y2H interactome contains 14,000 high-quality binary human protein-protein interactions. The Y2H analysis takes advantage of uniform coverage of the complete interactome space surpassing any current map by other approaches. While the Y2H approach benefits from assessment of pairwise interactions independent of background proteome and purification conditions, it does not faithfully recapitulate multi-subunit complexes that require more than two proteins to form stable structures. Given the distinctions of individual methods, the expected overlap between AP-MS and Y2H is also limited but both provide important information.

Recently, another group (Sahni et al. 2015) described a large-scale comparison of 1140 human wild type and 2890 mutant proteins associated with disease. The mutations consisted of single nucleotide changes compared to wild type proteins, largely assumed not to impair protein folding or stability. Y2H-based analyses indicated that these missense mutations led to variations in the protein-protein interaction profiles over 60% of the time. For transcription factors, protein-DNA interactions were disrupted in ~80% of the mutant cases. Such analyses could provide detailed insights into the etiology of diseases such as CADASIL (cerebral autosomal dominant arteriopathy with subcortical infarcts and leukoencephalopathy), which is largely caused by dominant missense mutations in the Notch3 extracellular domain (Louvi and Artavanis-Tsakonas 2012), although we note that experimental modifications would have to be made because the analysis of extracellular domains is difficult with traditional Y2H approaches.

5 Integrating the Genetic and Proteomic Data

We had previously shown the utility of proteomic maps to serve as a scaffold on which we can place genetic interactors in order to reveal larger mechanistic insights (Guruharsha et al. 2012). In addition to the large scale genetic screens and protein-protein interaction (PPI) studies described above, some additional sources of genetic and proteomic interaction data can be overlaid onto the Notch interaction map. There is a wealth of genetic information in *Drosophila* derived from a range of individual functional studies done on the Notch pathway over many decades, which has been largely curated by FlyBase [http://flybase.org/, (Gramates et al. 2017)]. Perrimon and colleagues have also developed an analysis tool that uses protein complex information from yeast, fly and human to annotate, analyze and visualize results from functional RNAi screens, overexpression screens, genome-wide association studies or exome sequencing projects (Vinayagam et al. 2013). Finally, attempts have been made to computationally discern the "directionality" (e.g. the direction of information flow) of PPI data using structural insights (Liu et al. 2009) or genetic relationships (Vinayagam et al. 2014), which can provide additional clues as to function.

In an attempt to refine the Notch network (Fig. 1A), we chose to integrate the *Drosophila* and human data and filter them by taking into account only large scale unbiased studies [DPiM, Bioplex, Signed Network, curated literature and fly genetic interactions, (Guruharsha et al. 2011; Huttlin et al. 2015; Vinayagam et al. 2014; Rolland et al. 2014; Gramates et al. 2017)] and the results are shown in Fig. 2. It is evident that *Drosophila* Notch occupies a central place in the network, given that *Drosophila* studies provide the most evidence to connect the rest of the pathway members. Notably, the combination of information from both *Drosophila* and human provides additional data points as well as corroborative evidence of interactions between Notch genetic modifiers.

Combining modifier data from multiple genetic screens with proteomic data reveals many enriched complexes that are linked to Notch (Guruharsha et al. 2012). Given that many of the clusters in proteome maps are enriched for functional annotation (e.g. gene ontology) terms, this analysis also suggests potential functional roles for many uncharacterized Notch interacting proteins. We posit that additional members of enriched complexes would also interact genetically with Notch pathway alleles. Indeed, Rhee and colleagues tested 88 physical interactors from their nuclear proteome (that were not themselves known to be Notch genetic interactors) of proteins identified in the *mastermind* Exelixis genetic screen (Kankel et al. 2007) and found that 35% of these physical interactors could themselves modify the *mastermind* wing phenotype. This represented a 7-fold enrichment over the predicted hit rate for random screening (Rhee et al. 2014). These findings validate the predictive power of the Notch interactome map and are being further extended and corroborated by further studies involving the entire DPiM map that will be reported elsewhere.

Although we have focused on genetic and proteomic data in this review, we note that several other classes of data can also be mapped onto the Notch interactome. Many groups have examined transcriptional regulation and transcriptional target genes downstream of Notch/CSL [reviewed in (Bray and Bernard 2010)]. Chemical biology screens utilizing small molecule inhibitors may also provide another source of relevant data; for example, a screen of chemical kinase inhibitors revealed that ERK/MAPK signaling regulates Notch in breast cancer cells (Izrailit et al. 2013). Integrating these data in the Notch interactome provides an additional element of functionality that may increase the predictive power of the interaction map.

The integration of PPI networks with relevant datasets (e.g. human genetics and gene expression data) has been applied to understanding disease mechanisms. Mutations in Notch itself are associated with several diseases, including T-ALL, CADASIL, and Alagille syndrome (Louvi and Artavanis-Tsakonas 2012) and perturbations in Notch signaling have also been correlated with additional solid cancers. This latter class of Notch-related cancers is of particular interest with respect to the Notch interactome, as direct mutations in Notch itself tend to be rare in such cases (Louvi and Artavanis-Tsakonas 2012; Koch and Radtke 2010); rather, changes in the regulatory network surrounding Notch seem often to be responsible. Furthermore, experimental evidence indicates that while Notch per se may not be oncogenic, the synergy of Notch signals with many other individual gene activities can induce tumorigenesis (Kiaris et al. 2004; Klinakis et al. 2006; Fre et al. 2009; Brumby and Richardson 2003; Pallavi et al. 2012; Ho et al. 2015). Barabási and colleagues (Goh et al. 2007) found that genes associated with similar human diseases/disorders tend to cluster into neighborhoods in the network, suggesting the existence of

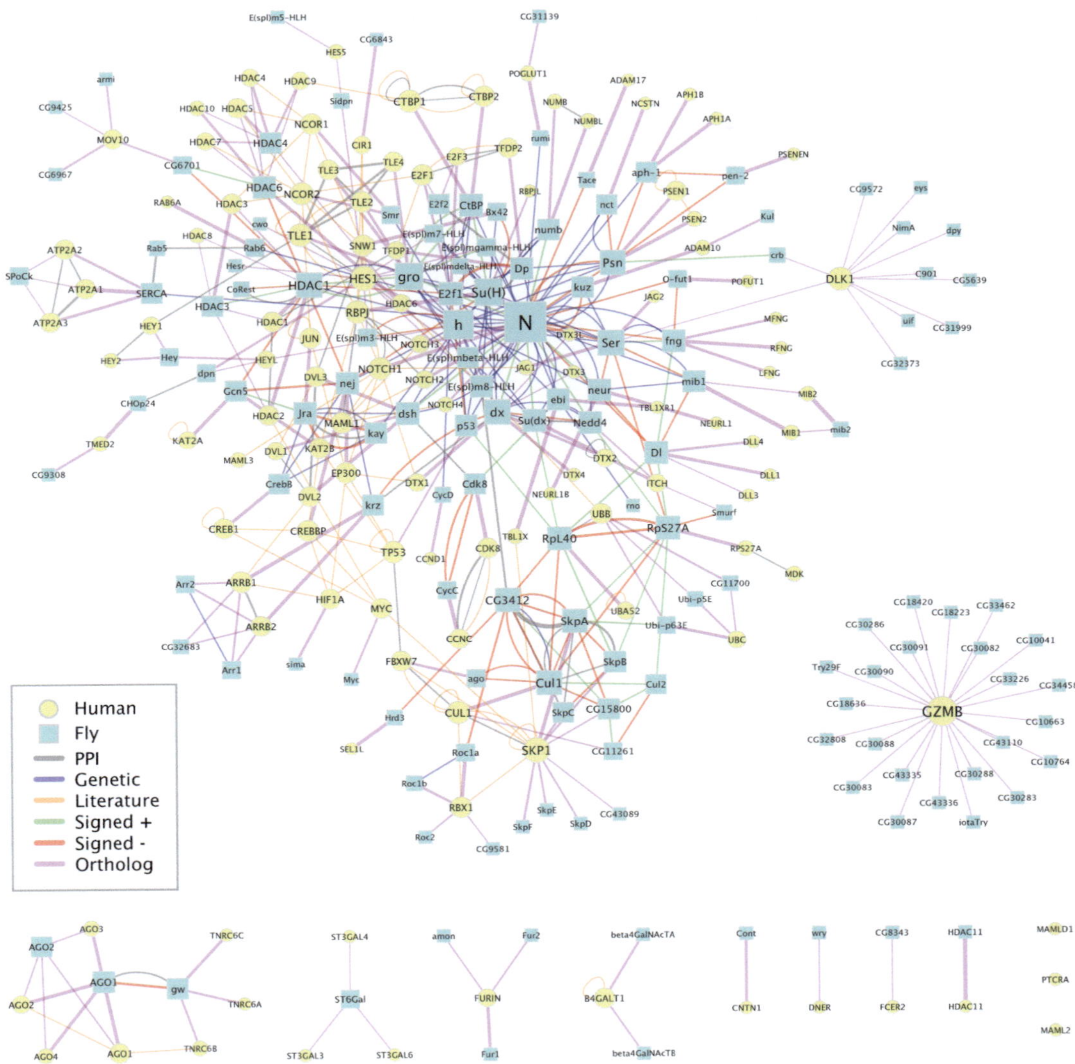

Fig. 2 Fly and Human integrated network. The integrated Notch interaction network (human and fly), limited to large scale unbiased studies [DPiM, Bioplex, Signed Network, curated literature and fly genetic interactions (Guruharsha et al. 2011; Huttlin et al. 2015; Vinayagam et al. 2014; Rolland et al. 2014; Gramates et al. 2017)], shows that *Drosophila* Notch is at the core of the network. Human genes/nodes are shown as yellow circles (from Table 1 and Fig. 1) and fly genes are shown as cyan squares. The size of the node is proportional to the number of connections. Colored lines connecting nodes indicate different types of interactions (explanatory box) and thickness of the lines is proportional to the strength or score as described by the original authors/sources. "PPI" is from DPiM and BioPlex 2.0, while "Genetic" refers to all genetic interactions from independent studies curated by FlyBase. "Literature" interactions are from 11,045 high-quality protein pairs that have multiple sources of evidence in the literature as curated by Rolland and colleagues (Rolland et al. 2014) in 2014. All "Signed" interactions are from Vinayagam and colleagues (Vinayagam et al. 2014) and the "+" or "-" corresponds to interaction with positive or negative correlation, respectively in 42 genome-wide RNAi screens. "Ortholog" connects all the fly orthologs of human Notch genes from Table 1 that was mapped using DIOPT tool (Hu et al. 2011)

distinct disease-specific functional modules which can themselves be integrated into an overall human disease network, or "Diseasome" (Ghiassian et al. 2015; Menche et al. 2015). Comparison of the Notch interactome between normal and disease states, or perhaps even simply comparison of the Notch interactome with existing Diseasome networks, will undoubtedly be informative.

Data from genome-wide association studies (GWAS) and exome sequencing have been integrated with PPIs to narrow down and refine putative disease causal genes (Lundby et al. 2014; Arking et al. 2014) and gain insights into molecular processes underlying disease (Lage 2014). We envision combining the Notch interactome with GWAS, exome sequencing and gene expression data in order to identify regulatory nodes within the circuitry that may contribute to Notch-dependent disorders, thus identifying better therapeutic targets.

6 Concluding Remarks

The many Notch-related studies over the years have revealed the staggering complexity of the Notch signaling circuitry. Consequently, specific binary relationships between Notch pathway elements and other genes should always be considered in the light of this complexity. The remarkable improvements in technology have afforded us unprecedented tools to identify relationships between Notch signals and other cellular elements and to delve deeply into the Notch regulatory network, providing us also with a paradigm of cellular complexity. It is likely that a similar degree of complexity underlies the other fundamental cellular signaling pathways such as Wnt, Hedgehog, Receptor Tyrosine Kinase, Transforming Growth Factor-beta, etc. Indeed, the context dependent integration of Notch with these pathways determines cell fates and drives development and morphogenesis (Borggrefe et al. 2016). Thus, examining Notch within an interaction map that takes into account proteomic, genetic and other relevant information across multiple data sets and different experimental approaches is essential if we are to understand Notch function. Only then will we be able to provide a reliable roadmap of the molecular interactions in a cell that can be used to predict downstream biological responses.

Acknowledgements The authors thank the members of the Artavanis-Tsakonas lab for helpful discussions and comments. This work was supported by NIH grants 5R01CA098402 and 5RO1HG003616 to S.A.-T.

References

Andersson ER, Sandberg R, Lendahl U (2011) Notch signaling: simplicity in design, versatility in function. Development 138:3593–3612

Arking DE, Pulit SL, Crotti L, van der Harst P, Munroe PB, Koopmann TT, Sotoodehnia N, Rossin EJ, Morley M, Wang X, Johnson AD, Lundby A, Gudbjartsson DF, Noseworthy PA, Eijgelsheim M, Bradford Y, Tarasov KV, Dorr M, Muller-Nurasyid M, Lahtinen AM, Nolte IM, Smith AV, Bis JC, Isaacs A, Newhouse SJ, Evans DS, Post WS, Waggott D, Lyytikainen LP, Hicks AA, Eisele L, Ellinghaus D, Hayward C, Navarro P, Ulivi S, Tanaka T, Tester DJ, Chatel S, Gustafsson S, Kumari M, Morris RW, Naluai AT, Padmanabhan S, Kluttig A, Strohmer B, Panayiotou AG, Torres M, Knoflach M, Hubacek JA, Slowikowski K, Raychaudhuri S, Kumar RD, Harris TB, Launer LJ, Shuldiner AR, Alonso A, Bader JS, Ehret G, Huang H, Kao WH, Strait JB, Macfarlane PW, Brown M, Caulfield MJ, Samani NJ, Kronenberg F, Willeit J, Consortium CA, Consortium C, Smith JG, Greiser KH, Meyer Zu Schwabedissen H, Werdan K, Carella M, Zelante L, Heckbert SR, Psaty BM, Rotter JI, Kolcic I, Polasek O, Wright AF, Griffin M, Daly MJ, Dcct/Edic, Arnar DO, Holm H, Thorsteinsdottir U, MC E, Denny JC, Roden DM, Zuvich RL, Emilsson V, Plump AS, Larson MG, O'Donnell CJ, Yin X, Bobbo M, D' Adamo AP, Iorio A, Sinagra G, Carracedo A, Cummings SR, Nalls MA, Jula A, Kontula KK, Marjamaa A, Oikarinen L, Perola M, Porthan K, Erbel R, Hoffmann P, Jockel KH, Kalsch H, Nothen MM, Consortium H, den Hoed M, Loos RJ, Thelle DS, Gieger C, Meitinger T, Perz S, Peters A, Prucha H, Sinner MF, Waldenberger M, de Boer RA, Franke L, van der Vleuten PA, Beckmann BM, Martens E, Bardai A, Hofman N, Wilde AA, Behr ER, Dalageorgou C, Giudicessi JR, Medeiros-Domingo A, Barc J, Kyndt F, Probst V, Ghidoni A, Insolia R, Hamilton RM, Scherer SW, Brandimarto J, Margulies K, Moravec CE, del Greco MF, Fuchsberger C, O'Connell JR, Lee WK, Watt GC, Campbell H, Wild SH, El Mokhtari NE, Frey N, Asselbergs FW, Mateo Leach I, Navis G, van den Berg MP, van Veldhuisen DJ, Kellis M, Krijthe BP, Franco OH, Hofman A, Kors JA, Uitterlinden AG, Witteman JC, Kedenko L, Lamina C, Oostra BA, Abecasis GR, Lakatta EG, Mulas A, Orru M, Schlessinger D, Uda M, Markus MR, Volker U, Snieder H, Spector TD, Arnlov J, Lind L, Sundstrom J, Syvanen AC, Kivimaki M, Kahonen M, Mononen N, Raitakari OT, Viikari JS, Adamkova V, Kiechl S, Brion M, Nicolaides AN, Paulweber B, Haerting J, Dominiczak AF, Nyberg F, Whincup PH, Hingorani AD, Schott JJ, Bezzina CR, Ingelsson E,

Ferrucci L, Gasparini P, Wilson JF, Rudan I, Franke A, Muhleisen TW, Pramstaller PP, Lehtimaki TJ, Paterson AD, Parsa A, Liu Y, van Duijn CM, Siscovick DS, Gudnason V, Jamshidi Y, Salomaa V, Felix SB, Sanna S, Ritchie MD, Stricker BH, Stefansson K, Boyer LA, Cappola TP, Olsen JV, Lage K, Schwartz PJ, Kaab S, Chakravarti A, Ackerman MJ, Pfeufer A, de Bakker PI, Newton-Cheh C (2014) Genetic association study of QT interval highlights role for calcium signaling pathways in myocardial repolarization. Nat Genet 46(8):826–836. https://doi.org/10.1038/ng.3014

Artavanis-Tsakonas S, Muskavitch MAT (2010) Chapter One: Notch: the past, the present, and the future. Curr Top Dev Biol 92:1–29

Arya R, White K (2015) Cell death in development: signaling pathways and core mechanisms. Semin Cell Dev Biol 39:12–19

Bastian M, Heymann S, Jacomy M (2009) Gephi: an open source software for exploring and manipulating networks

Borggrefe T, Lauth M, Zwijsen A, Huylebroeck D, Oswald F, Giaimo BD (2016) The Notch intracellular domain integrates signals from Wnt, Hedgehog, TGFbeta/BMP and hypoxia pathways. Biochim Biophys Acta 1863(2):303–313. https://doi.org/10.1016/j.bbamcr.2015.11.020. S0167-4889(15)00404-8 [pii]

Bray S, Bernard F (2010) Chapter Eight: Notch targets and their regulation. Curr Top Dev Biol 92:253–275

Bray SJ (2016) Notch signalling in context. Nat Rev Mol Cell Biol 17:722–735

Brumby AM, Richardson HE (2003) Scribble mutants cooperate with oncogenic Ras or Notch to cause neoplastic overgrowth in Drosophila. EMBO J 22:5769–5779

Cherbas L, Willingham A, Zhang D, Yang L, Zou Y, Eads BD, Carlson JW, Landolin JM, Kapranov P, Dumais J, Samsonova A, Choi J-H, Roberts J, Davis CA, Tang H, Van Baren MJ, Ghosh S, Dobin A, Bell K, Lin W, Langton L, Duff MO, Tenney AE, Zaleski C, Brent MR, Hoskins RA, Kaufman TC, Andrews J, Graveley BR, Perrimon N, Celniker SE, Gingeras TR, Cherbas P (2011) The transcriptional diversity of 25 Drosophila cell lines. Genome Res 21:301–314

Conner SD (2016) Regulation of notch signaling through intracellular transport. Int Rev Cell Mol Biol 323:107–127

Da Ros VG, Gutierrez-Perez I, Ferres-Marco D, Dominguez M (2013) Dampening the signals transduced through hedgehog via microRNA miR-7 facilitates notch-induced tumourigenesis. PLoS Biol 11(5):e1001554. https://doi.org/10.1371/journal.pbio.1001554. PBIOLOGY-D-12-03081 [pii]

del Álamo D, Rouault H, Schweisguth F (2011) Mechanism and significance of cis-inhibition in Notch signalling. Curr Biol 21:R40–R47

Fabregat A, Sidiropoulos K, Garapati P, Gillespie M, Hausmann K, Haw R, Jassal B, Jupe S, Korninger F, McKay S, Matthews L, May B, Milacic M, Rothfels K, Shamovsky V, Webber M, Weiser J, Williams M, Wu G, Stein L, Hermjakob H, D'Eustachio P (2016) The Reactome pathway knowledgebase. Nucleic Acids Res 44(D1):D481–D487. https://doi.org/10.1093/nar/gkv1351

Ferres-Marco D, Gutierrez-Garcia I, Vallejo DM, Bolivar J, Gutierrez-Avino FJ, Dominguez M (2006) Epigenetic silencers and Notch collaborate to promote malignant tumours by Rb silencing. Nature 439(7075):430–436. doi:nature04376 [pii]. https://doi.org/10.1038/nature04376

Fre S, Pallavi SK, Huyghe M, Laé M, Janssen K-P, Robine S, Artavanis-Tsakonas S, Louvard D (2009) Notch and Wnt signals cooperatively control cell proliferation and tumorigenesis in the intestine. Proc Natl Acad Sci U S A 106:6309–6314

Ghiassian SD, Menche J, Barabasi AL (2015) A DIseAse MOdule Detection (DIAMOnD) algorithm derived from a systematic analysis of connectivity patterns of disease proteins in the human interactome. PLoS Comput Biol 11(4):e1004120. https://doi.org/10.1371/journal.pcbi.1004120

Giaimo BD, Oswald F, Borggrefe T (2017) Dynamic chromatin regulation at Notch target genes. Transcription 8(1):61–66. https://doi.org/10.1080/21541264.2016.1265702

Giot L, Bader JS, Brouwer C, Chaudhuri A, Kuang B, Li Y, Hao YL, Ooi CE, Godwin B, Vitols E, Vijayadamodar G, Pochart P, Machineni H, Welsh M, Kong Y, Zerhusen B, Malcolm R, Varrone Z, Collis A, Minto M, Burgess S, McDaniel L, Stimpson E, Spriggs F, Williams J, Neurath K, Ioime N, Agee M, Voss E, Furtak K, Renzulli R, Aanensen N, Carrolla S, Bickelhaupt E, Lazovatsky Y, DaSilva A, Zhong J, Stanyon CA, Finley RL Jr, White KP, Braverman M, Jarvie T, Gold S, Leach M, Knight J, Shimkets RA, McKenna MP, Chant J, Rothberg JM (2003) A protein interaction map of Drosophila melanogaster. Science 302(5651):1727–1736. https://doi.org/10.1126/science.1090289. 1090289 [pii]

Goh KI, Cusick ME, Valle D, Childs B, Vidal M, Barabasi AL (2007) The human disease network. Proc Natl Acad Sci U S A 104(21):8685–8690. https://doi.org/10.1073/pnas.0701361104

Gramates LS, Marygold SJ, Santos GD, Urbano JM, Antonazzo G, Matthews BB, Rey AJ, Tabone CJ, Crosby MA, Emmert DB, Falls K, Goodman JL, Hu Y, Ponting L, Schroeder AJ, Strelets VB, Thurmond J, Zhou P, The FlyBase C (2017) FlyBase at 25: looking to the future. Nucleic Acids Res 45(D1):D663–D671. https://doi.org/10.1093/nar/gkw1016

Greenwald I, Kovall R (2013) Notch signaling: genetics and structure. WormBook

Guruharsha KG, Hori K, Obar RA, Artavanis-Tsakonas S (2014) Proteomic analysis of the Notch interactome. Methods Mol Biol 1187:181–192. https://doi.org/10.1007/978-1-4939-1139-4_14

Guruharsha KG, Kankel MW, Artavanis-Tsakonas S (2012) The Notch signalling system: recent insights into the complexity of a conserved pathway. Nat Rev Genet 13:654–666

Guruharsha KG, Rual J-F, Zhai B, Mintseris J, Vaidya P, Vaidya N, Beekman C, Wong C, Rhee DY, Cenaj O, Mckillip E, Shah S, Stapleton M, Wan KH, Yu C, Parsa B, Carlson JW, Chen X, Kapadia B, Vijayraghavan K, Gygi SP, Celniker SE, Obar RA, Artavanis-Tsakonas S (2011) A protein complex network of Drosophila melanogaster. Cell 147:690–703

Ho DM, Artavanis-Tsakonas S (2016) The notch-mediated proliferation circuitry. Curr Top Dev Biol 116:17–33

Ho DM, Pallavi SK, Artavanis-Tsakonas S (2015) The Notch-mediated hyperplasia circuitry in Drosophila reveals a Src-JNK signaling axis. elife 4:e05996

Hori K, Sen A, Kirchhausen T, Artavanis-Tsakonas S (2011) Synergy between the ESCRT-III complex and Deltex defines a ligand-independent Notch signal. J Cell Biol 195:1005–1015

Hu Y, Flockhart I, Vinayagam A, Bergwitz C, Berger B, Perrimon N, Mohr SE (2011) An integrative approach to ortholog prediction for disease-focused and other functional studies. BMC Bioinf 12:357. https://doi.org/10.1186/1471-2105-12-357

Huttlin EL, Ting L, Bruckner RJ, Gebreab F, Gygi MP, Szpyt J, Tam S, Zarraga G, Colby G, Baltier K, Dong R, Guarani V, Vaites LP, Ordureau A, Rad R, Erickson BK, Wühr M, Chick J, Zhai B, Kolippakkam D, Mintseris J, Obar RA, Harris T, Artavanis-Tsakonas S, Sowa ME, De Camilli P, Paulo JA, Wade Harper J, Gygi SP (2015) The BioPlex network: a systematic exploration of the human interactome. Cell 162(2):425–440

Huttlin EL, Bruckner RJ, Paulo JA, Cannon JR, Ting L, Baltier K, Colby G, Gebreab F, Gygi MP, Parzen H, Szpyt J, Tam S, Zarraga G, Pontano-Vaites L, Swarup S, White AE, Schweppe DK, Rad R, Erickson BK, Obar RA, Guruharsha KG, Li K, Artavanis-Tsakonas S, Gygi SP, Harper JW (2017) Architecture of the human interactome defines protein communities and disease networks. Nature 545(7655):505–509. https://doi.org/10.1038/nature22366. nature22366 [pii]

Izrailit J, Berman HK, Datti A, Wrana JL, Reedijk M (2013) High throughput kinase inhibitor screens reveal TRB3 and MAPK-ERK/TGFβ pathways as fundamental Notch regulators in breast cancer. Proc Natl Acad Sci U S A 110:1714–1719

Jia D, Soylemez M, Calvin G, Bornmann R, Bryant J, Hanna C, Huang Y-C, Deng W-M (2015) A large-scale in vivo RNAi screen to identify genes involved in Notch-mediated follicle cell differentiation and cell cycle switches. Sci Rep 5. https://doi.org/10.1038/srep12328

Jung JG, Stoeck A, Guan B, Wu RC, Zhu H, Blackshaw S, Shih Ie M, Wang TL (2014) Notch3 interactome analysis identified WWP2 as a negative regulator of Notch3 signaling in ovarian cancer. PLoS Genet 10(10):e1004751. https://doi.org/10.1371/journal.pgen.1004751

Kankel MW, Hurlbut GD, Upadhyay G, Yajnik V, Yedvobnick B, Artavanis-Tsakonas S (2007) Investigating the genetic circuitry of mastermind in Drosophila, a notch signal effector. Genetics 177:2493–2505

Kanehisa M, Furumichi M, Mao T, Sato Y, Morishima K (2017) KEGG: new perspectives on genomes, pathways, diseases and drugs. Nucleic Acids Res 45(D1):D353–D361

Kiaris H, Politi K, Grimm LM, Szabolcs M, Fisher P, Efstratiadis A, Artavanis-Tsakonas S (2004) Modulation of notch signaling elicits signature tumors and inhibits hras1-induced oncogenesis in the mouse mammary epithelium. Am J Pathol 165:695–705

Klinakis A, Szabolcs M, Politi K, Kiaris H, Artavanis-Tsakonas S, Efstratiadis A (2006) Myc is a Notch1 transcriptional target and a requisite for Notch1-induced mammary tumorigenesis in mice. Proc Natl Acad Sci U S A 103:9262–9267

Koch U, Radtke F (2010) Notch signaling in solid tumors. Curr Top Dev Biol 92:411–455

Kopan R, Ilagan MXG (2009) The canonical Notch signaling pathway: unfolding the activation mechanism. Cell 137:216–233

Kovall RA, Blacklow SC (2010) Mechanistic insights into Notch receptor signaling from structural and biochemical studies. Curr Top Dev Biol 92:31–71

Kovall RA, Gebelein B, Sprinzak D, Kopan R (2017) The Canonical Notch Signaling Pathway: Structural and Biochemical Insights into Shape, Sugar, and Force. Dev Cell 41(3):228–241. doi:S1534-5807(17)30294-0 [pii]. https://doi.org/10.1016/j.devcel.2017.04.001

Lage K (2014) Protein-protein interactions and genetic diseases: the interactome. Biochim Biophys Acta 1842(10):1971–1980. https://doi.org/10.1016/j.bbadis.2014.05.028

Le Bras S, Rondanino C, Kriegel-Taki G, Dussert A, Le Borgne R (2012) Genetic identification of intracellular trafficking regulators involved in Notch-dependent binary cell fate acquisition following asymmetric cell division. J Cell Sci 125:4886–4901

Lehner B, Crombie C, Tischler J, Fortunato A, Fraser AG (2006) Systematic mapping of genetic interactions in Caenorhabditis elegans identifies common modifiers of diverse signaling pathways. Nat Genet 38:896–903

Levitan D, Greenwald I (1995) Facilitation of lin-12-mediated signalling by sel-12, a Caenorhabditis elegans S182 Alzheimer's disease gene. Nature 377:351–354

Li S, Armstrong CM, Bertin N, Ge H, Milstein S, Boxem M, Vidalain PO, Han JD, Chesneau A, Hao T, Goldberg DS, Li N, Martinez M, Rual JF, Lamesch P, Xu L, Tewari M, Wong SL, Zhang LV, Berriz GF, Jacotot L, Vaglio P, Reboul J, Hirozane-Kishikawa T, Li Q, Gabel HW, Elewa A, Baumgartner B, Rose DJ, Yu H, Bosak S, Sequerra R, Fraser A, Mango SE, Saxton WM, Strome S, Van Den Heuvel S, Piano F, Vandenhaute J, Sardet C, Gerstein M, Doucette-Stamm L, Gunsalus KC, Harper JW, Cusick ME, Roth FP, Hill DE, Vidal M (2004) A map of the interactome network of the metazoan C. elegans. Science 303(5657):540–543. https://doi.org/10.1126/science.1091403. 1091403 [pii]

Liu W, Li D, Wang J, Xie H, Zhu Y, He F (2009) Proteome-wide prediction of signal flow direction in protein interaction networks based on interacting domains. Mol Cell Proteomics 8(9):2063–2070. https://doi.org/10.1074/mcp.M800354-MCP200

Lobry C, Oh P, Mansour MR, Look AT, Aifantis I (2014) Notch signaling: switching an oncogene to a tumor suppressor. Blood 123:2451–2459

Louvi A, Artavanis-Tsakonas S (2012) Notch and disease: a growing field. Semin Cell Dev Biol 23:473–480

Lundby A, Rossin EJ, Steffensen AB, Acha MR, Newton-Cheh C, Pfeufer A, Lynch SN, Consortium QTIIG, Olesen SP, Brunak S, Ellinor PT, Jukema JW, Trompet S, Ford I, Macfarlane PW, Krijthe BP, Hofman A, Uitterlinden AG, Stricker BH, Nathoe HM, Spiering W, Daly MJ, Asselbergs FW, van der Harst P, Milan DJ, de Bakker PI, Lage K, Olsen JV (2014) Annotation of loci from genome-wide association studies using tissue-specific quantitative interaction proteomics. Nat Methods 11(8):868–874. https://doi.org/10.1038/nmeth.2997

Menche J, Sharma A, Kitsak M, Ghiassian SD, Vidal M, Loscalzo J, Barabasi AL (2015) Disease networks. Uncovering disease-disease relationships through the incomplete interactome. Sci 347(6224):1257601. https://doi.org/10.1126/science.1257601

Milacic M, Haw R, Rothfels K, Wu G, Croft D, Hermjakob H, D'Eustachio P, Stein L (2012) Annotating cancer variants and anti-cancer therapeutics in reactome. Cancers (Basel) 4(4):1180–1211. https://doi.org/10.3390/cancers4041180

Mourikis P, Lake RJ, Firnhaber CB, DeDecker BS (2010) Modifiers of notch transcriptional activity identified by genome-wide RNAi. BMC Dev Biol 10:107

Mummery-Widmer JL, Yamazaki M, Stoeger T, Novatchkova M, Bhalerao S, Chen D, Dietzl G, Dickson BJ, Knoblich JA (2009) Genome-wide analysis of Notch signalling in Drosophila by transgenic RNAi. Nature 458:987–992

Musse AA, Meloty-Kapella L, Weinmaster G (2012) Notch ligand endocytosis: mechanistic basis of signaling activity. Semin Cell Dev Biol 23:429–436

Orchard S, Ammari M, Aranda B, Breuza L, Briganti L, Broackes-Carter F, Campbell NH, Chavali G, Chen C, del-Toro N, Duesbury M, Dumousseau M, Galeota E, Hinz U, Iannuccelli M, Jagannathan S, Jimenez R, Khadake J, Lagreid A, Licata L, Lovering RC, Meldal B, Melidoni AN, Milagros M, Peluso D, Perfetto L, Porras P, Raghunath A, Ricard-Blum S, Roechert B, Stutz A, Tognolli M, van Roey K, Cesareni G, Hermjakob H (2014) The MIntAct project--IntAct as a common curation platform for 11 molecular interaction databases. Nucleic Acids Res 42(Database issue):D358–D363. https://doi.org/10.1093/nar/gkt1115

Pallavi SK, Ho DM, Hicks C, Miele L, Artavanis-Tsakonas S (2012) Notch and Mef2 synergize to promote proliferation and metastasis through JNK signal activation in Drosophila. EMBO J 31:2895–2907

Parks AL, Cook KR, Belvin M, Dompe NA, Fawcett R, Huppert K, Tan LR, Winter CG, Bogart KP, Deal JE, Deal-Herr ME, Grant D, Marcinko M, Miyazaki WY, Robertson S, Shaw KJ, Tabios M, Vysotskaia V, Zhao L, Andrade RS, Edgar KA, Howie E, Killpack K, Milash B, Norton A, Thao D, Whittaker K, Winner MA, Friedman L, Margolis J, Singer MA, Kopczynski C, Curtis D, Kaufman TC, Plowman GD, Duyk G, Francis-Lang HL (2004) Systematic generation of high-resolution deletion coverage of the Drosophila melanogaster genome. Nat Genet 36:288–292

Rana NA, Haltiwanger RS (2011) Fringe benefits: functional and structural impacts of O-glycosylation on the extracellular domain of Notch receptors. Curr Opin Struct Biol 21:583–589

Ranganathan P, Weaver KL, Capobianco AJ (2011) Notch signalling in solid tumours: a little bit of everything but not all the time. Nat Rev Cancer 11:338–351

Rhee DY, Cho D-Y, Zhai B, Slattery M, Ma L, Mintseris J, Wong CY, White KP, Celniker SE, Przytycka TM, Gygi SP, Obar RA, Artavanis-Tsakonas S (2014) Transcription factor networks in Drosophila melanogaster. Cell Rep 8:2031–2043

Rolland T, Taşan M, Charloteaux B, Pevzner SJ, Zhong Q, Sahni N, Yi S, Lemmens I, Fontanillo C, Mosca R, Kamburov A, Ghiassian SD, Yang X, Ghamsari L, Balcha D, Begg BE, Braun P, Brehme M, Broly MP, Carvunis A-R, Convery-Zupan D, Corominas R, Coulombe-Huntington J, Dann E, Dreze M, Dricot A, Fan C, Franzosa E, Gebreab F, Gutierrez BJ, Hardy MF, Jin M, Kang S, Kiros R, Lin GN, Luck K, MacWilliams A, Menche J, Murray RR, Palagi A, Poulin MM, Rambout X, Rasla J, Reichert P, Romero V, Ruyssinck E, Sahalie JM, Scholz A, Shah AA, Sharma A, Shen Y, Spirohn K, Tam S, Tejeda AO, Trigg SA, Twizere J-C, Vega K, Walsh J, Cusick ME, Yu X, Barabási A-L, Iakoucheva LM, Aloy P, Rivas JDL, Tavernier J, Calderwood MA, Hill DE, Hao T, Roth FP, Vidal M (2014) A proteome-scale map of the human interactome network. Cell 159(5):1212–1226

Rual JF, Venkatesan K, Hao T, Hirozane-Kishikawa T, Dricot A, Li N, Berriz GF, Gibbons FD, Dreze M, Ayivi-Guedehoussou N, Klitgord N, Simon C, Boxem M, Milstein S, Rosenberg J, Goldberg DS, Zhang LV, Wong SL, Franklin G, Li S, Albala JS, Lim J, Fraughton C, Llamosas E, Cevik S, Bex C, Lamesch P, Sikorski RS, Vandenhaute J, Zoghbi HY, Smolyar A, Bosak S, Sequerra R, Doucette-Stamm L, Cusick ME, Hill DE, Roth FP, Vidal M (2005) Towards a proteome-scale map of the human protein-protein interaction network. Nature 437(7062):1173–1178. doi:nature04209 [pii]. https://doi.org/10.1038/nature04209

Sahni N, Yi S, Taipale M, Fuxman Bass JI, Coulombe-Huntington J, Yang F, Peng J, Weile J, Karras GI, Wang Y, Kovacs IA, Kamburov A, Krykbaeva I, Lam MH, Tucker G, Khurana V, Sharma A, Liu YY, Yachie N, Zhong Q, Shen Y, Palagi A, San-Miguel A, Fan C, Balcha D, Dricot A, Jordan DM, Walsh JM, Shah AA, Yang X, Stoyanova AK, Leighton A, Calderwood MA, Jacob Y, Cusick ME, Salehi-Ashtiani K, Whitesell LJ,

Sunyaev S, Berger B, Barabasi AL, Charloteaux B, Hill DE, Hao T, Roth FP, Xia Y, Walhout AJ, Lindquist S, Vidal M (2015) Widespread macromolecular interaction perturbations in human genetic disorders. Cell 161(3):647–660. https://doi.org/10.1016/j.cell.2015.04.013

Saj A, Arziman Z, Stempfle D, van Belle W, Sauder U, Horn T, Dürrenberger M, Paro R, Boutros M, Merdes G (2010) A combined ex vivo and in vivo RNAi screen for notch regulators in Drosophila reveals an extensive notch interaction network. Dev Cell 18:862–876

Shalaby NA, Parks AL, Morreale EJ, Osswalt MC, Pfau KM, Pierce EL, Muskavitch MAT (2009) A screen for modifiers of notch signaling uncovers Amun, a protein with a critical role in sensory organ development. Genetics 182:1061–1076

Stelzl U, Worm U, Lalowski M, Haenig C, Brembeck FH, Goehler H, Stroedicke M, Zenkner M, Schoenherr A, Koeppen S, Timm J, Mintzlaff S, Abraham C, Bock N, Kietzmann S, Goedde A, Toksoz E, Droege A, Krobitsch S, Korn B, Birchmeier W, Lehrach H, Wanker EE (2005) A human protein-protein interaction network: a resource for annotating the proteome. Cell 122(6):957–968. doi:S0092-8674(05)00866-4 [pii]. https://doi.org/10.1016/j.cell.2005.08.029

Szklarczyk D, Franceschini A, Wyder S, Forslund K, Heller D, Huerta-Cepas J, Simonovic M, Roth A, Santos A, Tsafou KP, Kuhn M, Bork P, Jensen LJ, von Mering C (2015) STRING v10: protein-protein interaction networks, integrated over the tree of life. Nucleic Acids Res 43 (Database issue):D447–452. doi:10.1093/nar/gku1003

The ORFeome Collaboration (2016) A genome-scale human ORF-clone resource. Nature Methods 13(3):191–192

Thibault ST, Singer MA, Miyazaki WY, Milash B, Dompe NA, Singh CM, Buchholz R, Demsky M, Fawcett R, Francis-Lang HL, Ryner L, Cheung LM, Chong A, Erickson C, Fisher WW, Greer K, Hartouni SR, Howie E, Jakkula L, Joo D, Killpack K, Laufer A, Mazzotta J, Smith RD, Stevens LM, Stuber C, Tan LR, Ventura R, Woo A, Zakrajsek I, Zhao L, Chen F, Swimmer C, Kopczynski C, Duyk G, Winberg ML, Margolis J (2004) A complementary transposon tool kit for Drosophila melanogaster using P and piggyBac. Nat Genet 36:283–287

Toba G, Ohsako T, Miyata N, Ohtsuka T, Seong KH, Aigaki T (1999) The gene search system. A method for efficient detection and rapid molecular identification of genes in Drosophila melanogaster. Genetics 151(2):725–737

Turinsky AL, Razick S, Turner B, Donaldson IM, Wodak SJ (2011) Interaction databases on the same page. Nat Biotechnol 29(5):391–393. https://doi.org/10.1038/nbt.1867

Veraksa A, Bauer A, Artavanis-Tsakonas S (2005) Analyzing protein complexes in Drosophila with tandem affinity purification-mass spectrometry. Dev Dyn 232(3):827–834. https://doi.org/10.1002/dvdy.20272

Vinayagam A, Hu Y, Kulkarni M, Roesel C, Sopko R, Mohr SE, Perrimon N (2013) Protein complex-based analysis framework for high-throughput data sets. Sci Signal 6(264):rs5. https://doi.org/10.1126/scisignal.2003629

Vinayagam A, Kulkarni MM, Sopko R, Sun X, Hu Y, Nand A, Villalta C, Moghimi A, Yang X, Mohr SE, Hong P, Asara JM, Perrimon N (2016) An integrative analysis of the InR/PI3K/Akt network identifies the dynamic response to insulin signaling. Cell Rep 16(11):3062–3074. https://doi.org/10.1016/j.celrep.2016.08.029

Vinayagam A, Zirin J, Roesel C, Hu Y, Yilmazel B, Samsonova AA, Neumuller RA, Mohr SE, Perrimon N (2014) Integrating protein-protein interaction networks with phenotypes reveals signs of interactions. Nat Methods 11(1):94–99. https://doi.org/10.1038/nmeth.2733

Warde-Farley D, Donaldson SL, Comes O, Zuberi K, Badrawi R, Chao P, Franz M, Grouios C, Kazi F, Lopes CT, Maitland A, Mostafavi S, Montojo J, Shao Q, Wright G, Bader GD, Morris Q (2010) The GeneMANIA prediction server: biological network integration for gene prioritization and predicting gene function. Nucleic Acids Res 38(suppl_2):W214–W220

Yatim A, Benne C, Sobhian B, Laurent-Chabalier S, Deas O, Judde JG, Lelievre JD, Levy Y, Benkirane M (2012) NOTCH1 nuclear interactome reveals key regulators of its transcriptional activity and oncogenic function. Mol Cell 48(3):445–458. https://doi.org/10.1016/j.molcel.2012.08.022

Yu H, Braun P, Yildirim MA, Lemmens I, Venkatesan K, Sahalie J, Hirozane-Kishikawa T, Gebreab F, Li N, Simonis N, Hao T, Rual J-F, Dricot A, Vazquez A, Murray RR, Simon C, Tardivo L, Tam S, Svrzikapa N, Fan C, de Smet A-S, Motyl A, Hudson ME, Park J, Xin X, Cusick ME, Moore T, Boone C, Snyder M, Roth FP, Barabasi A-L, Tavernier J, Hill DE, Vidal M (2008) High-quality binary protein interaction map of the yeast interactome network. Science 322(5898):104–110

Yu C, Wan KH, Hammonds AS, Stapleton M, Carlson JW, Celniker SE (2011) Development of expression-ready constructs for generation of proteomic libraries. Methods Mol Biol 723:257–272. https://doi.org/10.1007/978-1-61779-043-0_17

Integration of *Drosophila* and Human Genetics to Understand Notch Signaling Related Diseases

Jose L. Salazar and Shinya Yamamoto

Abstract

Notch signaling research dates back to more than one hundred years, beginning with the identification of the *Notch* mutant in the fruit fly *Drosophila melanogaster*. Since then, research on *Notch* and related genes in flies has laid the foundation of what we now know as the Notch signaling pathway. In the 1990s, basic biological and biochemical studies of Notch signaling components in mammalian systems, as well as identification of rare mutations in Notch signaling pathway genes in human patients with rare Mendelian diseases or cancer, increased the significance of this pathway in human biology and medicine. In the 21st century, *Drosophila* and other genetic model organisms continue to play a leading role in understanding basic Notch biology. Furthermore, these model organisms can be used in a translational manner to study underlying mechanisms of Notch-related human diseases and to investigate the function of novel disease associated genes and variants. In this chapter, we first briefly review the major contributions of *Drosophila* to Notch signaling research, discussing the similarities and differences between the fly and human pathways. Next, we introduce several biological contexts in *Drosophila* in which Notch signaling has been extensively characterized. Finally, we discuss a number of genetic diseases caused by mutations in genes in the Notch signaling pathway in humans and we expand on how *Drosophila* can be used to study rare genetic variants associated with these and novel disorders. By combining modern genomics and state-of-the art technologies, *Drosophila* research is continuing to reveal exciting biology that sheds light onto mechanisms of disease.

Keywords

Notch signaling · *Drosophila* · Mendelian diseases · Functional genomics · Translational research

J. L. Salazar
Department of Molecular and Human Genetics, Baylor College of Medicine (BCM), Houston, TX, USA

S. Yamamoto (✉)
Department of Molecular and Human Genetics, Baylor College of Medicine (BCM), Houston, TX, USA

Program in Developmental Biology, BCM, Houston, TX, USA

Department of Neuroscience, BCM, Houston, TX, USA

Jan and Dan Duncan Neurological Research Institute, Texas Children's Hospital, Houston, TX, USA
e-mail: yamamoto@bcm.edu

T. Borggrefe, B. D. Giaimo (eds.), *Molecular Mechanisms of Notch Signaling*, Advances in Experimental Medicine and Biology 1066, https://doi.org/10.1007/978-3-319-89512-3_8

Abbreviations

AD	Alzheimer's disease
ADAM10	*A Disintegrin and Metalloprotease 10*
AES	Amino-terminal Enhancer of Split
ago	*archipelago*
amx	*almondex*
AOS	Adams-Oliver Syndrome
AP-3	Adaptor Protein-3
Aph	Anterior pharynx defective
APOE	*APOlipoprotein E*
APP	Amyloid Precursor Protein
aPKC	atypical Protein Kinase C
ARHGAP31	*Rho GTPase-activating protein 31*
Arp2/3	Actin-related protein 2/3
AS-C	Achaete-Scute Complex
BAC	Bacterial Artificial Chromosome
bib	*big brain*
BDSC	Bloomington *Drosophila* Stock Center
bHLH	basic Helix-Loop-Helix
C. elegans	*Caenorhabditis elegans*
CADASIL	Cerebral Autosomal Dominant Arteriopathy with Subcortical Infarcts and Leukoencephalopathy
CBP/CREBBP	C-Adenosine Mono Phosphate Responsive Element (cAMP-RE)-Binding protein (CREB)-Binding Protein
CDK8	*Cyclin-Dependent Kinase 8*
CREB	cAMP response element binding protein
cDNA	complementary Deoxyribo Nucleic Acid
CHARGE	Cohorts for Heart and Aging Research in Genomic Epidemiology
CRISPR	Clustered Regularly Interspaced Short Palindromic Repeats
CtBP	C-terminal Binding Protein
DECHIPHER	DatabasE of genomiC varIation and Phenotype in Humans using Ensembl Resources
DFS	Dominant Female Sterile
DGGR	*Drosophila* Genomics and Genetic Resources
DGRC	*Drosophila* Genomics Resource Center
DIOPT	*Drosophila* RNAi Screening Center Integrative Ortholog Prediction Tool
Dl	*Delta*
DLL	DeLta-Like
DOCK6	*Dedicator Of Cytokinesis 6*
DSHB	Developmental Studies Hybridoma Bank
dx	*deltex*
E(spl)-C	*Enhancer of split-Complex*
EGF	Epidermal Growth Factor
EHBP-1	*EH (Eps15 Homology) domain Binding Protein-1*
elav	*embryonic lethal abnormal vision*
EMS	Ethyl MethaneSulfonate
Eogt	*EGF-domain O-GlcNAc transferase*
EP300	*E1A binding protein P300*
ER	Endoplasmic Reticulum
ESCRT	Endosomal Sorting Complex Required for Transport
FAD	Familial Alzheimer's Disease
FBXW7	*F-BoX and WD repeat domain containing 7*
FHL1	*Four and a Half LIM domains 1*
FLP	FLiPpase
Fng	*Fringe*
FRT	Flippase Recognition Target
GAP	GTPase-Activating Protein
GEF	Guanine nucleotide Exchange Factor
Geno2MP	Genotype to Mendelian Phenotype Browser
GFI1	Growth Factor Independent 1 transcriptional repressor
glcNAc	N-Acetylglucosamine
glp-1	*abnormal germ line proliferation-1*
GOM	Granular Osmophilic Material
Gro	*Groucho*
GWAS	Genome-Wide Association Studies
H	*Hairless*
hAPF	hours After Puparium Formation

HCOP	Human genome organization gene nomenclature committee Comparison of Orthology Predictions search
HDR	Homology Directed Repair
HES	Hairy and Enhancer of Split
HEY	Hairy/Enhancer-of-split related with YRPW motif
HOPS	HOmotypic fusion and Protein Sorting
IMF	Infantile MyoFibromatosis
JAG	JAGged
kuz	*kuzbanian*
l(2)gd1	*lethal (2) giant discs 1*
LFNG	*Lunatic FriNGe*
LIM	*Lin11, Isl-1 and Mec-3*
lin-12	*cell lineage defective-12*
LMNS	Lateral MeNingocele Syndrome
LNR	Lin-12/Notch Repeat
LOAD	Late-Onset Alzheimer's Disease
LVNC	Left Ventricular Non Compaction
mam	*mastermind*
MAML	MAsterMind-Like
MARRVEL	Model organism Aggregated Resources for Rare Variant ExpLoration
MESP2	*Mesoderm posterior bHLH transcription factor 2*
MFNG	*Manic FriNGe*
MGI	Mouse Genome Informatics
mib	mindbomb
MiMIC	*Minos*-Mediated Integration Cassette
mRNA	messenger RiboNucleic Acid
Nct/NCSTN	*Nicastrin*
nej	*nejire*
neur/NEURL	*neuralized*
NEXT	Notch EXtracellular Truncation
NICD	Notch IntraCellular Domain
NRR	Negative Regulatory Region
O-fut1	*O-fucosyltransferase-1*
OMIM	Online Mendelian Inheritance in Man
PDGFRB	*Platelet Derived Growth Factor Receptor Beta*
pen/PSENEN	*presenilin enhancer*
PEST	proline (P), glutamic acid (E), serine (S) and threonine (T)-rich
POFUT1	*Protein O-fucosyltransferase 1*
POGLUT1	*Protein O-glucosyltransferase 1*
PTM	Post-Translational Modification
Psn/PSEN	*Presenilin*
RBPJ	*Recombination signal Binding Protein for immunoglobulin kappa J region*
RFNG	*Radical FriNGe*
RIPPLY2	*RIPPLY transcriptional repressor 2*
RMCE	Recombinase Mediated Cassette Exchange
RNAi	RNA interference
SA	Splice Acceptor
SCDO	SpondyloCostal DysOstosis
Sec15	*Secretory 15*
Ser	*Serrate*
SGD	Saccharomyces Genome Database
SHARP	SMRT/HDAC1 Associated Repressor Protein
shi	*shibire*
SOP	Sensory Organ Precursor
spdo	*sanpodo*
SPEN	*SPlit ENds family transcriptional repressor*
spl	*split*
Su(dx)	*Suppressor of deltex*
Su(H)	*Suppressor of Hairless*
TBX6	*T-BoX 6*
Temp	*Tempura*
TLE	Transducin Like Enhancer protein
TM2D3	*TM2 Domain containing 3*
UAS	Upstream Activation Sequence
V-ATPase	Vacuolar-Adenosine TriPhosphatase
VDRC	Vienna *Drosophila* Resource Center
VUS	Variant of Unknown Significance
WASp	Wiskott-Aldrich Syndrome protein
wg	*wingless*
ZFIN	ZebraFish Information Network

1 Discovery and Expansion of the Notch Signaling Pathway in *Drosophila*

The first fly *Notch* (gene symbol: *N*) mutant was discovered in the laboratory of Thomas Hunt Morgan in 1913, and is named so based on its dominant wing notching phenotype (Artavanis-Tsakonas and Muskavitch 2010; Dexter 1914). In addition to notched wings, *Notch* null mutant flies exhibit additional dominant wing vein and mechanosensory bristle density abnormalities, as well as recessive embryonic lethality (Lindsley and Zimm 1992). This lethality is primarily caused by the conversion of epidermal cells into neurons due to defects in lateral inhibition during neurogenesis, a developmental defect known as the "neurogenic" phenotype (Lehmann et al. 1981; Lehmann et al. 1983; Poulson 1936; Poulson 1937). The *Notch* gene was cloned and sequenced in the mid-1980s and was found to encode a large transmembrane receptor-like protein (Wharton et al. 1985; Kidd et al. 1986). Several genes that exhibit similar phenotypes when mutated such as *Delta (Dl)* (Kopczynski et al. 1988), *Serrate (Ser)* (Thomas et al. 1991; Fleming et al. 1990), *neuralized (neur)* (Boulianne et al. 1991), *mastermind (mam)* (Smoller et al. 1990), *Hairless (H)* (Maier et al. 1992) and *deltex (dx)* (Xu and Artavanis-Tsakonas 1990; Morgan et al. 1922) were cloned and characterized around the same time. Additional core genes of the pathway, such as *Suppressor of Hairless* [*Su(H)*)] (Fortini and Artavanis-Tsakonas 1994; Plunkett 1926), and genes in the *Enhancer of split-Complex* [*E(spl)-C*, *spl* is a hypomorphic allele of *Notch*] (Welshons 1956), were identified through genetic interaction screens with other genes in the pathway and were also cloned in the 1990s (Busseau et al. 1994; Go and Artavanis-Tsakonas 1998; Furukawa et al. 1991; Schweisguth and Posakony 1992; Delidakis and Artavanis-Tsakonas 1992; Schrons et al. 1992). Interestingly, epistatic analyses laid the basic outline of the pathway prior to the molecular cloning of many of these genes (Artavanis-Tsakonas et al. 1995; Artavanis-Tsakonas and Muskavitch 2010), demonstrating the power of pure genetic studies.

Technological advances allowed investigators to look for additional regulators of the pathway that were missed by previous genetic screens. One drawback of classic mutagenesis screens using X-rays and chemical mutagens such as EMS (Ethyl methanesulfonate) is that it is often challenging and labor intensive to map and clone the affected gene and to identify the molecular lesions. Development of transposon (e.g. *P*-elements, *PiggyBac*)-based techniques provided a new tool to perform random mutagenesis screens to quickly identify new mutants that exhibit Notch signaling related phenotypes or to molecularly clone previously identified mutants that were left unmapped (Goto et al. 2001; Irvine and Wieschaus 1994; Periz and Fortini 1999; Tian et al. 2004).

In *Drosophila*, embryonic developmental defects can be masked if maternal mRNAs and/or proteins that are deposited into the oocyte by the mother during oogenesis are sufficient for the animals to progress through embryogenesis. Hence, for genes that are abundantly expressed in oocytes, null mutants do not exhibit classical embryomic neurogenic phenotypes but typically die at a later developmental stage and were therefore missed in classic embryonic screens. Such maternal effect genes can be uncovered by generating oocytes that are homozygous for the mutation by combining a FLP/FRT (FLiPpase/Flippase Recognition Target) system-based site directed mitotic recombination technique (Xu and Rubin 1993) with a germline-specific dominant female sterile (DFS) mutation (Chou and Perrimon 1992; Perrimon et al. 1996). Using this FLP-DFS technique, several novel maternal effect genes were identified (Sasamura et al. 2003; Selva et al. 2001; Goode et al. 1996). In addition, development of reverse genetic strategies based on knowledge of the molecular map of the fly genome allowed to generate mutations in genes that have been implicated in Notch signaling in other systems but have not been studied in *Drosophila* such as *Presenilin (Pns)* (Struhl and Greenwald 1999; Ye et al. 1999; Guo et al. 1999a). Finally, additional modifier screens (Mahoney et al. 2006; Eun et al. 2007;

Hing et al. 1999; Rottgen et al. 1998; Royet et al. 1998; Schreiber et al. 2002; Shalaby et al. 2009; van de Hoef et al. 2009; Verheyen et al. 1996; Yedvobnick et al. 2001; Mishra et al. 2015; Bray et al. 2005; Hori et al. 2011), somatic mutagenesis screens (Acar et al. 2008; Jafar-Nejad et al. 2005; Rajan et al. 2009; Tien et al. 2008; Charng et al. 2014; Giagtzoglou et al. 2013; Giagtzoglou et al. 2012; Vaccari and Bilder 2005; Vaccari et al. 2009; Domanitskaya and Schupbach 2012; Sun et al. 2011; Yan et al. 2009; Berdnik et al. 2002; Hutterer and Knoblich 2005; Gallagher and Knoblich 2006; Haberman et al. 2010), genome-wide or targeted transgenic RNAi (RNA interference) based screens (Dornier et al. 2012; Jia et al. 2015; Berns et al. 2014; Zhang et al. 2012; Saj et al. 2010; Mummery-Widmer et al. 2009; Gomez-Lamarca et al. 2015) and UAS (Upstream Activation Sequence)/GAL4 system (Brand and Perrimon 1993)-mediated over-expression screens (Vallejo et al. 2011; Hall et al. 2004; Da Ros et al. 2013; Pi et al. 2011; Adam and Montell 2004) have increased our knowledge of genes that regulate Notch signaling *in vivo*. These genetic screens, along with cell culture based assays (Francis et al. 2002; Mourikis et al. 2010; Goodfellow et al. 2007; Li et al. 2014) and systems biology driven approaches including transcriptomics (Bernard et al. 2010; Slaninova et al. 2016; Pezeron et al. 2014; Babaoglan et al. 2013; Housden et al. 2013; Djiane et al. 2013; Krejci et al. 2009; Pines et al. 2010; Terriente-Felix et al. 2013; Zacharioudaki et al. 2016) and proteomics (Mukherjee et al. 2005; Guruharsha et al. 2011; Guruharsha et al. 2014; Moshkin et al. 2009) have allowed fly researchers to continue to discover new genes that regulate Notch signaling in diverse contexts. Because diagrams that illustrate Notch signaling now look more like a complicated intertwined web (Ilagan and Kopan 2007) rather than a simple linear pathway (Artavanis-Tsakonas et al. 1995), the pathway is now occasionally referred to as the "Notch Signaling Network (Hurlbut et al. 2007; Artavanis-Tsakonas and Muskavitch 2010)" or the "Notch Signaling System (Guruharsha et al. 2012)" to emphasize the complexity and dynamic nature of the pathway.

2 The *Drosophila* Notch Signaling Pathway and Its Relationship to the Mammalian Pathway.

Studies of the Notch signaling pathway in *Drosophila* have provided the framework for subsequent studies in other model organisms, including human (Artavanis-Tsakonas and Muskavitch 2010). One key advantage of studying Notch signaling in fruit flies is the genetic simplicity of the pathway compared to other organisms. Most core Notch pathway components are encoded by single genes in the fly genome while the structure and function of these factors remain largely conserved between flies and mammals. For example, the *Drosophila* genome contains one gene (*Notch*) that encodes for the Notch receptor, whereas the human genome contains four (*NOTCH1-4*; Table 1) (Kopan and Ilagan 2009). Even the simple *C. elegans* genome encodes two Notch receptors [*lin-12 (cell LINeage defective-12)* and *glp-1 (abnormal Germ Line Proliferation-1)*] (Yochem et al. 1988; Yochem and Greenwald 1989), giving *Drosophila* an advantage when trying to determine whether certain biological phenomena depend on Notch signaling or when performing structure-function studies of Notch in an *in vivo* setting (Leonardi et al. 2011; Yamamoto et al. 2012). In this section, we provide an outline of the Notch signaling pathway as currently understood in *Drosophila melanogaster* (Fig. 1), while pointing out some key differences found between the fly and mammalian pathways.

2.1 Biosynthesis and Trafficking of the Notch Receptor

For Notch signaling to be activated in a canonical fashion, two cells, one signal receiving and one signal sending, need to be juxtaposed (juxtacrine signaling). The Notch receptor is synthesized in the signal receiving cell and undergoes a number of post-translational modifications (PTMs) in both the endoplasmic reticulum (ER) and the Golgi apparatus (Fortini 2009). In the ER, the

Table 1 List of *Drosophila* genes discussed in this chapter along with their human homologs, disease association and functions. Human genes in **bold** have been linked diseases based on OMIM (Online Mendelian Inheritance in Man, 2017). The fly genes that are homologous to these disease genes are also shown in **bold**. Fly genes with "?" have mammalian homologs that have been shown to be involved in the Notch pathway but their role in Notch signaling in *Drosophila* have not been studied or are not clear. See (FlyBase-Arp2/3 2017; FlyBase-ESCRT, 2017; FlyBase-AP3 2017; FlyBase-HOPS 2017; HGNC-V-ATPase 2017; HGNC-Arp2/3 2017; HGNC-ESCRT 2017; HGNC-V-ATPase 2017; Dell'Angelica 2009; Solinger and Spang 2013) to find the full list of genes involved in Arp2/3-WASp, ESCRT, AP-3, HOPS and V-ATPase complexes in *Drosophila* and human

Drosophila gene (symbol)	*Human homolog(s)* (OMIM disease #)	Protein Functions
Notch (N)	***NOTCH1* (#109730, #616028), *NOTCH2* (#102500, #610205), *NOTCH3* (#125310, #130720, #615293),** *NOTCH4*	**Receptor**
Delta (Dl)	*DLL1*, ***DLL3* (#277300), *DLL4* (#616589)**	**Ligand**
Serrate (Ser)	***JAG1* (#118450, #187500),** *JAG2*	
rumi	***POGLUT1* (#615696, 617232)**	**Receptor glycosylation**
O-fucosyltransferase 1 (O-fut1)	***POFUT1* (#615327)**	
shams	*GXYLT1, GXYLT2*	
fringe	***LFNG* (#609813),** *MFNG, RFNG*	
EGF-domain O-GlcNAc transferase (Eogt)	***EOGT* (#615297)**	
neuralized (neur)	*NEURL1, NEURL1B*	**E3 ligase for ligand**
mind bomb 1 (mib1)	***MIB1* (#615092)**	
Furin1 (Fur1)?, Furin2 (Fur2)?	*FUR*	**S1 cleavage**
kuzbanian (kuz)	***ADAM10* (#615537)**	**S2 cleavage**
Presenilin (Psn)	***PSEN1* (#172700, #600274, #607822, #613694, #613737), *PSEN2* (#606889, #613697)**	**S3 cleavage**
aph-1	*APH1A, APH1B*	
nicastrin (nct)	***NCSTN* (#142690)**	
pen-2	***PSENEN* (#613736)**	
shibire (shi)	***DNM1* (#616346), *DNM2* (#160150, #606482, #615368),** *DNM3*	**Receptor and ligand endocytosis**
Sec15	*EXOC6, EXOC6B*	**Ligand trafficking**
Rab11	*RAB11A, RAB11B*	
Arp2/3 Complex: 8 genes (e.g. *Arp2, Arp3*) (FlyBase-Arp2/3 2017)	Arp2/3 Complex: 9 genes (HGNC-Arp2/3 2017)	
WASp	***WAS* (#301000), *WASL***	
Ehbp1	***EHBP1* (#611868)**, *EHBP1L*	
temp	*PTAR1*	
Numb	*NUMB, NUMBL*	**Receptor trafficking**
Sanpodo (Spdo)	-	
deltex (dx)	*DTX1, DTX2, DTX3, DTX3L, DTX4*	
supressor of deltex (su(dx))	***ITCH* (#613385)**, *WWP1, WWP2*	
lethal(2)giant discs 1 (l(2)gd1)	***CC2D1A* (#608443),** *CC2D1B*	
ESCRT complex: 20 genes (e.g. *shrub, Vps25*) (FlyBase-ESCRT 2017)	**ESCRT complex: 30 genes (#114480, #600795, #605387, #614898, #614696)** (HGNC-ESCRT 2017)	
AP-3 complex: 4 genes (e.g. *carmine (cm), ruby (rb)*) (Flybase-AP3 2017)	**AP-3 complex: 7 genes (#608233, #617050, #617276))** (Dell'Angelica 2009)	
HOPS complex: 7 genes (e.g. *carnation (car), deep orange (dor)*) (Flybase-HOPS 2017)	**HOPS complex: 8 genes (#208085, #616683, #617303)** (Solinger and Spang 2013)	
V-ATPase complex: 33 genes (e.g. *VhaAC39-1, Vha68-2*) (Flybase-VATPase 2017)	**V-ATPase complex: 23 genes (#124480, #219200, #259700, #267300, #278250, #259700, #616455, #617402, #617403)** (HGNC-VATPase 2017)	**Vesicle acidification**

(continued)

Table 1 (continued)

Drosophila gene (symbol)	*Human homolog(s)* (OMIM disease #)	Protein Functions
Supressor of Hairless (Su(H))	***RBPJ*** **(#614814)**	**Transcription factor**
Hairless (H)	-	**Corepressor**
-	*SPEN (SHARP/Mint)*	
-	***FHL1 (KyoT2)*** **(#300695, #300696, #300717, #300718)**	
groucho (gro)	*TLE1, TLE2, TLE3, TLE4, TLE5,* ***TLE6*** **(#616814)**	
C-terminal Binding Protein (CtBP)	*CTBP1, CTBP2*	
mastermind (mam)	*MAML1, MAML2, MAML3*	**Coactivator**
nejire (nej)	***EP300*** **(#114500, #613684),** ***CREBBP*** **(#180849)**	
Enhancer of split **complex [E(spl)-C]: 7 bHLH repressor genes: e.g.** ***E(spl)-m8)***	*HES1, HES2, HES3, HES4, HES5, HES6,* ***HES7*** **(#613686)**	**Target Genes**
sage?	***MESP2*** **(#608681)**	
Doc?1, Doc2?, Doc3?	***TBX6*** **(#122600)**	
Cdk8?	*CDK8*	**NICD degradation**
archipelago (ago)?	*FBXW7*	

extracellular domain of Notch becomes heavily O-glucosylated by Rumi (protein O-glucosyltransferase) (Acar et al. 2008) and O-fucosylated by O-fut1 (protein O-fucosyltransferase) (Okajima and Irvine 2002; Sasamura et al. 2003). The monosaccharides added by these enzymes to selective EGF (epidermal growth factor)-like repeats of Notch can further be elongated by Shams (glucoside xylosyltransferase) (Sethi et al. 2010; Lee et al. 2013) and Fringe (Fng, O-Fucosylpeptide β3-N-acetylglucosaminyltransferase) (Bruckner et al. 2000; Panin et al. 1997; Moloney et al. 2000; Irvine and Wieschaus 1994) in the ER or Golgi apparatus. Experiments using enzymatically inactive mutants and transgenic over-expression strains have revealed that glycosylation of Notch is critical for ligand selectivity as well as for proper signal activation upon ligand-receptor interaction (Jafar-Nejad et al. 2010; Rana and Haltiwanger 2011). Both *Drosophila* and mammalian Notch receptors have also been shown to undergo additional glycosylation by Eogt [EGF-domain O-GlcNAc (N-Acetylglucosamine) transferase] in the ER (Muller et al. 2013; Sakaidani et al. 2012). *Eogt* mutant flies do not exhibit obvious Notch signaling defects but genetically interact with other members of the pathway, indicating that this gene plays a modulatory role.

In addition, Notch undergoes the first (S1) proteolytic cleavage mediated by an unknown (potentially a furin-like) protease, in the Golgi. It has been reported that in *Drosophila* cells the majority of the Notch receptor found at the cell membrane consists of the ~300 kDa full-length protein (Johansen et al. 1989; Kidd et al. 1989) while in mammals, most Notch at the cell surface has undergone S1 cleavage (Blaumueller et al. 1997; Logeat et al. 1998). Although there has been some controversy in the *Drosophila* literature (Kidd and Lieber 2002), S1 cleavage is not absolutely required for signal activation but rather it seems to facilitate the transport of the receptor to the cell surface contributing to signal strength (Lake et al. 2009), similar to the effect of S1 cleavage on mammalian Notch receptors (Gordon et al. 2009a).

2.2 Biosynthesis and Trafficking of the Ligands and Ligand-Receptor Interaction

Notch receptors that have undergone proper PTMs in the ER and Golgi are exported to the plasma membrane where they can physically interact with ligands presented by the neighboring signal-sending cells. While the Notch recep-

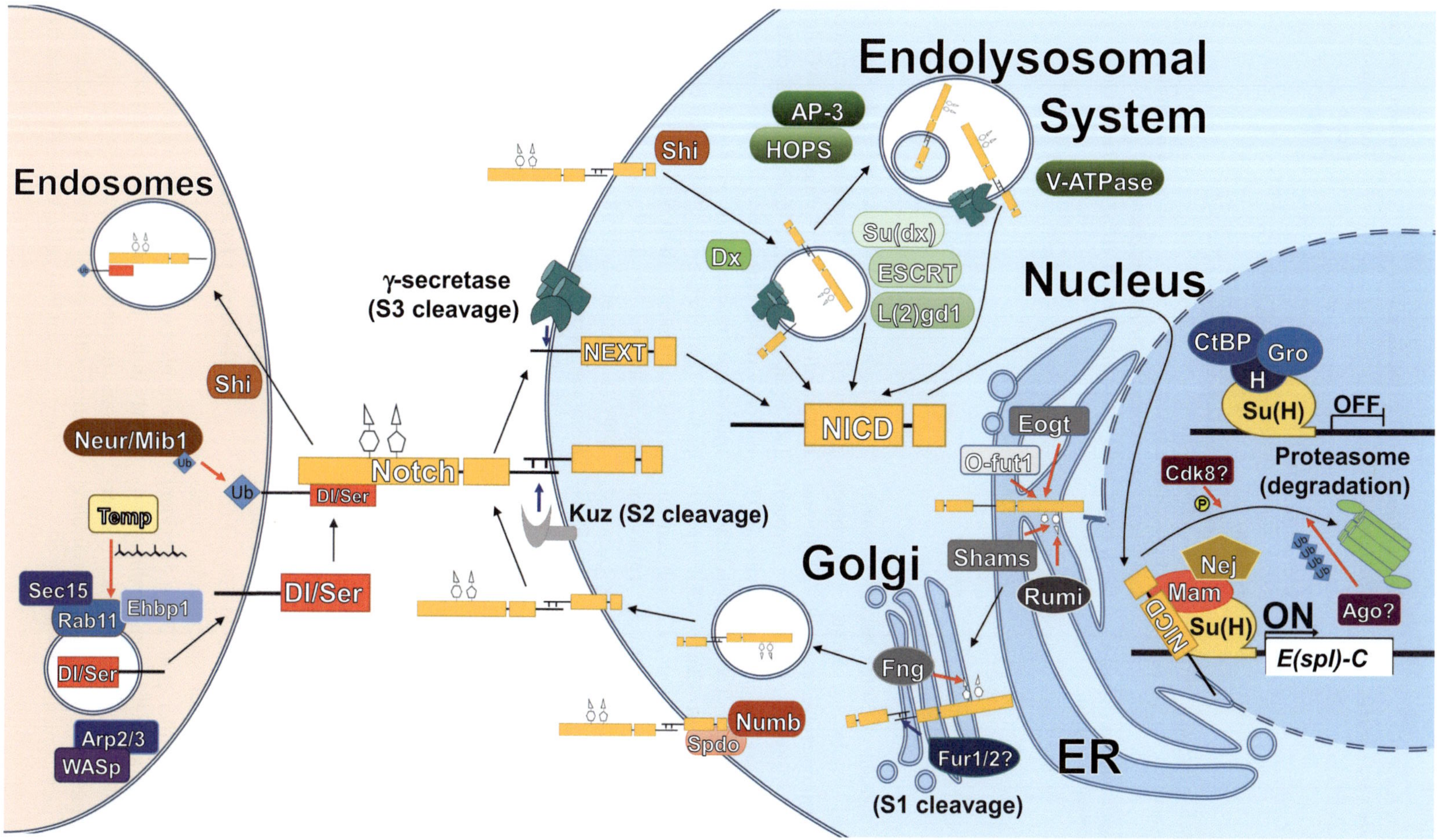

Fig. 1 Schematic diagram of the *Drosophila* Notch signaling pathway. Canonical Notch signaling takes place between two juxtaposed cells (left: signal sending cell, right: signal receiving cell). Different steps of signal activation and functions of molecules depicted here are described in detail in the main text. Note that some players depicted here only regulate Notch signaling in specific contexts. Abbreviations for proteins shown are based on FlyBase gene symbol nomenclature (also see Table 1)

tor is expressed relatively broadly (Fehon et al. 1991; Hartley et al. 1987), the ligands Delta (Vassin et al. 1987; Bender et al. 1993; Kopczynski and Muskavitch 1989; Parks et al. 1995; Kooh et al. 1993) and Serrate (Thomas et al. 1991; Fleming et al. 1990; Bachmann and Knust 1998a) exhibit unique and dynamic patterns of expressions during development. Together with the selective expression of Fng (Irvine and Wieschaus 1994), which facilitates Notch-Delta interactions while suppressing Notch-Serrate interactions (Xu et al. 2007; Panin et al. 1997), the spatial and temporal pattern of ligand expression plays a critical role in determining where Notch signaling becomes activated. In mammals, three orthologs of *Delta* [*Delta-like* (*DLL*)*1,3,4*] and two orthologs of *Serrate* [*Jagged* (*JAG*) *1,2*] are present. The existence of multiple *DLL* and *JAG* ligands, together with the presence of three *fng* orthologs (*Lunatic (LFNG)*, *Manic (MFNG)* and *Radical (RFNG) Fringe*), increases the complexity of Notch signaling in mammals compared to *Drosophila* (Kopan and Ilagan 2009). Multiple ligands can be expressed in the same tissue and can bind/activate the four Notch receptors with varying affinities. Furthermore, *DLL3*, the most divergent of the *DLL* paralogs, functions as a decoy ligand due to the lack of monoubiquitination sites in the cytoplasmic domain required for receptor activation (Heuss et al. 2008). Hence, this protein has been proposed to inhibit rather than activate Notch signaling in a cell autonomous manner by binding to the Notch receptors *in cis* (cis-inhibition) and preventing them from binding to ligands presented *in trans* (Ladi et al. 2005). Interestingly, three mammalian orthologs of *fng* have recently been shown to modify the same Notch receptor in different manners; some modifications can inhibit certain ligand-receptor interactions, others can potentiate them (Kakuda and Haltiwanger 2017). Hence, four receptors x five (four primarily activating and one inhibiting) ligands x 3 Fng enzymes generates a much more complicated scenario in mammals, compared to the one receptor x two (primarily activating) ligands x one Fng enzyme system in *Drosophila*.

Ligand-receptor interaction is necessary but not sufficient for canonical Notch signaling activation. After ligand and receptor bind to each other, the signal sending cell endocytoses the ligand-receptor complex, generating a physical force that unravels a second (S2) cleavage site embedded in the negative regulatory region (NRR) of the Notch receptor. Without the pulling force, three LNR (Lin-12/Notch Repeat) domains within the NRR limit the access of ADAM (A Disintegrin and Metalloprotease) proteases and prevent them from cleaving the S2 site (Gordon et al. 2007; Gordon et al. 2009b). Upon the conformational change mediated by ligand-endocytosis and force generation, Kuzbanian (Kuz, ADAM10 in human) cleaves the S2 site, shedding the majority of the extracellular domain and leaving behind a membrane-tethered portion of the Notch receptor referred to as the NEXT (Notch extracellular truncation) (Kovall and Blacklow 2010). In order to endocytose the ligands and generate the pulling force, cytoplasmic domains of Delta or Serrate must be mono-ubiquitinated by E3 ubiquitin ligases Neuralized (Neur) or Mindbomb 1 (Mib1) (Le Bras et al. 2011; Weinmaster and Fischer 2011). *neur* and *mib1* are differentially expressed and function in different Notch dependent biological processes during *Drosophila* development (Wang and Struhl 2005; Lai et al. 2005; Le Borgne et al. 2005; Pitsouli and Delidakis 2005; Lai et al. 2001; Pavlopoulos et al. 2001). Although *mind bomb 2* (*mib2*, *MIB2* in human) is present in the fly genome (Koo et al. 2005), its *in vivo* role in Notch signaling is not clear (Nguyen et al. 2007). The human genome contains two *neur* orthologs (*NEURL1*, *NEURL1B*) and one *mib1* ortholog (*MIB1*). Although studies based on cultured cells indicate that these genes can all regulate Notch activity (Rullinkov et al. 2009; Koutelou et al. 2008; Teider et al. 2010), only *Mib1* has been reported to exhibit a strong Notch signaling defect *in vivo* when mutated in mice (Koo et al. 2007). Hence, the dependence of the Notch pathway on ligand mono-ubiquitination by Neur and Mib family proteins seems to have diverged and/or acquired a high degree of redundancy during evolution.

2.3 Proteolytic Cleavages of the Notch Receptor and Transcriptional Regulation

After S2 cleavage of Notch, NEXT is further processed by the γ-secretase complex, an intramembrane protease composed of Presenilin (Psn), Nicastrin (Nct), Anterior pharynx defective 1 (Aph-1), and Presenilin enhancer-2 (Pen-2). Two *Psn* orthologs (*PSEN1* and *PSEN2*) and two *aph-1* orthologs (*APH1A* and *APH1B*) together with single orthologs for *Nct* (*NCSTN*) and *pen-2* (*PSENEN*) exist in the human genome. γ-secretase performs the S3 cleavage of NEXT to release the Notch intracellular domain (NICD) from the membrane (Tagami et al. 2008). It remains unclear whether γ-secretase primarily processes NEXT at the cell membrane, within endocytic vesicles or both. The requirement of the genes that primarily function in Clathrin-dependent endocytosis [e.g. Dynamin encoded by the *shi (shibire)* gene] for Notch activation in signal-receiving cells in certain contexts supports that S3 cleavage takes place in endocytic vesicles (Vaccari et al. 2008; Seugnet et al. 1997). However, other studies argue that S3 cleavage primarily occurs at the plasma membrane and that endocytosis is not required (Struhl and Adachi 2000), suggesting that this may be a context-specific issue. Indeed, proteins and molecular machineries that regulate endocytic trafficking and degradation of Notch receptors such as Dx (E3 ligase) (Yamada et al. 2011; Fuwa et al. 2006; Hori et al. 2011), Suppressor of dx [Su(dx), E3 ligase] (Mazaleyrat et al. 2003; Cornell et al. 1999), ESCRT (Endosomal Sorting Complex Required for Transport) complex (multivesicular body formation) (Vaccari et al. 2009; Herz et al. 2009), Lethal (2) giant discs 1 [L(2)gd1, adaptor protein] (Schneider et al. 2013; Troost et al. 2012), AP-3 (Adaptor Protein-3) complex (late endosomal trafficking) and HOPS (HOmotypic fusion and Protein Sorting, endosome-lysosome fusion) complexes (Wilkin et al. 2008; Takats et al. 2014) and Vacuolar-ATPase (V-ATPase, vesicle acidification) complex (Yan et al. 2009; Vaccari et al. 2010) can fine-tune Notch activity, likely by regulating the efficiency of Notch cleavage in different subcellular compartments and/or modulating the balance between ligand-dependent and -independent signaling activities (Baron 2012; Shimizu et al. 2014).

After being released from the membrane, NICD trafficks to the nucleus and forms a transcriptional activation complex with Su(H) [RBPJ (Recombination signal Binding Protein for immunoglobulin kappa J region) in human] (Furukawa et al. 1992; Fortini and Artavanis-Tsakonas 1994) and Mam [MAML (Mastermind-like)1-3 in human] (Wu et al. 2002; Kitagawa et al. 2001). In the absence of NICD, Su(H) is bound to the co-repressor Hairless (H) (Bang and Posakony 1992; Maier et al. 1992; Bailey and Posakony 1995) which in turn recruits additional co-repressors such as Groucho [TLE (Transducin-like enhancer protein) 1-6 in human, TLE5 is also referred to as AES (Amino-terminal Enhancer Of Split)] and CtBP (C-terminal Binding Protein, CTBP1-2 in human) to silence target genes (Barolo et al. 2002; Nagel et al. 2005; Paroush et al. 1994; Stifani et al. 1992; Morel et al. 2001). Once NICD enters the nucleus and binds to Su(H), H is no longer able to bind to Su(H) (Yuan et al. 2016). The active NICD-Su(H)-Mam complex further recruits transcriptional co-activators such as the histone acetyltransferase CBP [CREB (cAMP responsive element binding protein)-binding protein]/p300 [*nejire (nej)* in *Drosophila, EP300* (*E1A binding protein P300*) and *CREBBP (CREB-binding protein)* in human] to initiate transcription of downstream target genes (Jia et al. 2015; Skalska et al. 2015). While most genes that are involved in the transcriptional activation complex are conserved between flies and mammals, no direct homolog of *Hairless* exists in mammalian genomes. Instead, two structurally unrelated co-repressors, KyoT2 (encoded by the *FHL1* gene in human) (Taniguchi et al. 1998) and SHARP [SMRT/HDAC1 Associated Repressor Protein, encoded by the *SPEN* (*SPlit ENds family transcriptional repressor*) gene in human, also called Mint] (Oswald et al. 2002; Kuroda et al. 2003) play the same function, binding to RBPJ and further recruiting additional corepressors to silence transcription (Borggrefe and Oswald 2014). Interestingly, Hairless and KyoT2/SHARP bind to RBPJ through different molecular mechanisms (Yuan et al. 2016; Collins et al. 2014; Borggrefe and Oswald 2016), suggesting that these genes were integrated into the Notch pathway independently through convergent evolution.

A number of Notch target genes have been identified to date (Bray and Bernard 2010), but the best characterized target genes are found in the *E(spl)-C* (Bailey and Posakony 1995; Jennings et al. 1994; Lecourtois and Schweisguth 1995; Celis et al. 1996a). *E(spl)-C* encodes seven basic helix-loop-helix (bHLH) proteins that function as transcriptional repressors (E(spl)-m3, m5, m7, m8, mβ, mγ, mδ) (Knust et al. 1992; Delidakis et al. 1991; Klambt et al. 1989; Delidakis et al. 2014). In addition, the gene that encodes Gro as well as four Bearded family proteins (E(spl)-m2, m4, m6, mα), a group of small proteins that inhibit Neur function, are also found at this locus (Hartley et al. 1988; Lai et al. 2000; Ziemer et al. 1988; Wurmbach et al. 1999). bHLH E(spl) proteins antagonize the activity of proneural bHLH proteins such as Achaete and Scute during neurogenesis and this relationship is generally conserved in mammals (Oellers et al. 1994; Heitzler et al. 1996; Gigliani et al. 1996; Nakao and Campos-Ortega 1996). Homologs of bHLH *E(spl)* genes are known as *HES* (*Hairy and Enhancer of Split*) genes in mammals (*HES1-7* human) (Akazawa et al. 1992; Sasai et al. 1992). Together with the structurally and evolutionarily related *HEY* (*Hairy/Enhancer-of-split related with YRPW motif*) genes (*HEY1, HEY2, HEYL* in human) which are also under the control of Notch signaling in many contexts (Kokubo et al. 1999; Leimeister et al. 1999), these factors play critical roles in developmental events that involve proneural transcription factors as well as in a number of other Notch-dependent contexts (Weber et al. 2014; Kobayashi and Kageyama 2014).

Finally, signal termination of Notch is mediated by ubiquitin-proteasomal degradation of the NICD. Based on experiments in mammalian systems, the PEST [proline (P), glutamic acid (E), serine (S) and threonine (T)-rich] domain near the C-terminus of NICD is required for phosphorylation by CDK8 (Cyclin-dependent kinase 8) and subsequent poly-ubiquitination by the ubiquitin E3 ligase FBXW7 (F-BoX and WD repeat domain containing) 7. However, whether *Cdk8* and *archipelago* (*ago*, *FBXW7* homolog) also play similar roles in the *Drosophila* Notch signaling pathway *in vivo* waits further confirmation.

2.4 Non-canonical Activation of the Pathway and Species Specificity

In addition to the canonical signaling pathway described above, a number of studies have revealed non-canonical ways by which Notch signal can be activated (e.g. non-canonical ligands, Su(H)-independent signaling, signal crosstalk). Due to space limitations, we will not discuss these alternative pathways here and refer the readers to the following review articles (Yamamoto et al. 2010; Palmer and Deng 2015; Heitzler 2010; D'Souza et al. 2010; Johnson 2011).

As we have seen, there are a number of similarities between the *Drosophila* and human (mammalian) Notch signaling pathways but there are a number of differences we referred to that one should keep in mind. As we have already discussed, duplication (*JAG1/2*), triplication (*DLL1/3/4* and *LFNG/MFNG/RFNG*) and quadruplication (*NOTCH1-4*) of core genes in the pathway during mammalian evolution have increased the complexity of the pathway compared to *Drosophila*. Some genes maintained redundancy while others acquired novel functions or became subfunctionalized to fine-tune the pathway in mammals. In addition, there have been new genes that have been incorporated into the pathway, some of which do not have an obvious ortholog in *Drosophila* (e.g. KyoT2, SHARP). In addition, one should also keep in mind that there are other key biological differences [e.g. minimal role of CpG DNA methylation in gene regulation in *Drosophila* (Schwanbeck 2015; Takayama et al. 2014), lack of primary cilia in most somatic cells (Jana et al. 2016; Ezratty et al. 2011)] that are known to exist between insects and mammals and that may impact the translation of some findings from *Drosophila* to human.

3 Notch Signaling in *Drosophila* Development

Since the identification of *Notch* and other members of the canonical signaling pathway as fundamental genes involved in the embryonic and post-embryonic development of *Drosophila*, numerous studies have focused on elucidating the role of Notch signaling during the development of diverse organs in the fly (Bray 1998; Bray 2006). Over the years, several tissues that exhibit characteristic morphological defects when Notch signaling activity is altered have been used as models to understand how the pathway works and to further identify novel pathway members. These include the embryonic central nervous system (brain and ventral nerve cord), the adult peripheral nervous system (mechanosensory and chemosensory bristles, chordotonal organs and eyes), adult appendages (wings and legs), hematopoietic organs (lymph gland) and reproductive organs (ovary and testis). In addition, studies on post-developmental functions of Notch signaling, such as its role in synaptic plasiticity (Kidd et al. 2015; Lieber et al. 2011) and stem cell maintenance/differentiation (Koch et al. 2013; Udolph 2012; Xie et al. 2008), are also being explored. Here, we will focus on the role of Notch signaling during the development of the adult mechanosensory organs (bristles) and the wing margin during *Drosophila* development. These two tissues are well established model systems to study three conceptually distinct modes of Notch signaling that are reiteratively used during morphogenesis and organogenesis across evolution: lateral inhibition, lineage decisions and inductive signaling (Bray 2006; Hartenstein and Posakony 1989; Hartenstein and Posakony 1990; Lai and Orgogozo 2004).

3.1 Notch Signal-Mediated Lateral Inhibition During Early Development of Mechanosensory Bristles

Mechanosensory bristles are part of the peripheral nervous system that allow the fly to sense mechanical forces and provide proprioception for coordinated movement and behavior (Fig. 2a, b) (Garcia-Bellido and Santamaria 1978; Campuzano and Modolell 1992). The bristles in the notum (dorsal thorax) are formed in a highly reproducible and stereotypical fashion (Hartenstein and Posakony 1989) and their development can be easily traced using fixed or live imaging strategies (Couturier and Schweisguth 2014). Bristle precursor cells, which are called sensory organ precursor (SOP) cells, are selected out from a group of cells referred to as the proneural cluster (Fig. 2c). Proneural clusters are specific groups of ectodermal cells that begin to express proneural bHLH transcription factors of the Achaete-Scute Complex (AS-C, *ASCL1*-5 in human (Huang et al. 2014)). A single SOP is selected from a proneural cluster through Notch-mediated lateral inhibition (Hartenstein and Posakony 1990). Lateral inhibition is achieved through a genetic circuitry that works through a feedback loop that involves inductive and repressive transcriptional relationships between Notch signaling components and several transcription factors (Fig. 2d) (Fortini 2009; Furman and Bukharina 2008; Heitzler et al. 1996). In a proneural cluster, all cells initially express both Notch and Delta and have equal potential to either become an SOP or an epithelial cell. Delta activates Notch in neighboring cells, which leads to expression of downstream target genes in the *E(spl)-C*. E(spl) proteins function as transcriptional repressors and down-regulate the expression of AS-C, which are positive regulators of *Delta* transcription. Thus, decrease in AS-C expression due to upregulation of *E(spl)* leads to the reduction of Delta expression in the signal-receiving cell.

In addition to inducing the expression of Delta, AS-C bHLH transcription factors positively regulate the expression of a zinc finger nuclear protein called Senseless (Sens) [GFI1 (Growth Factor Independent 1 transcriptional repressor) in human] (Nolo et al. 2000; Acar et al. 2006; Jafar-Nejad et al. 2003). Sens participates in this genetic circuitry by promoting the transcription of AS-C target genes by working as a transcriptional coactivator through physical interactions with AS-C bHLH proteins. In addition, at low expression levels Sens functions as a

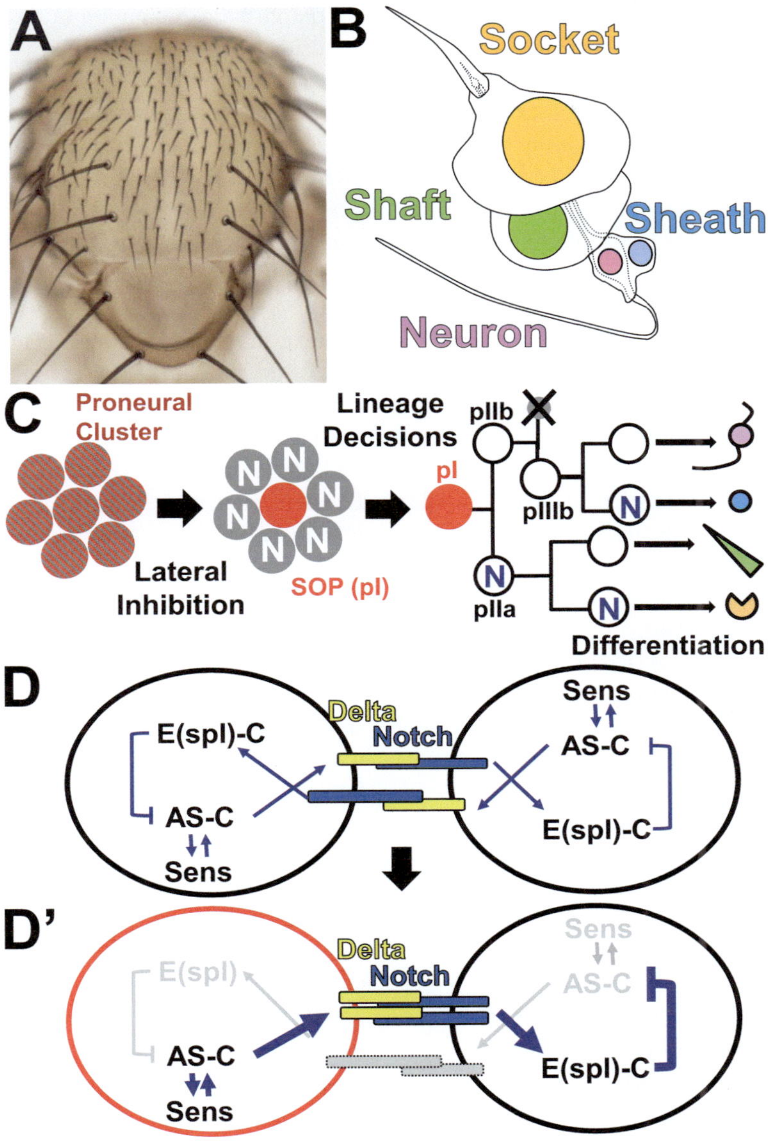

Fig. 2 Notch signaling is required for lateral inhibition and lineage decisions during mechanosensory organ development. (**a**) Photograph of the fly notum. Large (macrochaetae) and small (microchaetae) bristles are organized in a stereotypical fashion. (**b**) Schematic diagram of a single mechanosensory organ (bristle). (**c**) Schematic diagram representing the development of a single bristle. "N" indicates cells that activate Notch signaling. (**d-d'**) Schematic diagrams of lateral inhibition during the selection of a sensory organ precursor (SOP) cell. In the beginning both cells have the potential to become an SOP. As development progresses, two cells acquire distinct fates through amplification of small differences through transcriptional feedback loops built into the stem. Cells that become the net signal sending cell becomes the SOP (labeled in red), and the net signal receiving cell(s) takes the epithelial cell fate. Panels B and C were adapted and mofidied from (Yamamoto et al. 2012)

transcriptional repressor through direct binding to DNA, thus acting as a binary switch to further amplify the feedback loop that is established by AS-C, Delta and E(spl) (Acar et al. 2006). Within the young proneural cluster, the expression level of AS-C, Sens and Notch signaling components are similar among the cells. However, at some time point during development, the equilibrium of Notch-Delta signaling becomes disrupted which is thought to be through a stochastic event (Barad et al. 2011; Corson et al. 2017; Troost et al. 2015). When one cell receives less Notch signal, expression of E(spl) within this cell is reduced and Delta expression becomes derepressed. Thus, cells receiving less Notch signal begin to express higher levels of Delta, which in turn can send stronger Delta mediated Notch signals to neighboring cells. Through this feed-forward loop, one cell that continues to send the signal to neighboring cell eventually becomes selected out as the SOP (low Notch activity), while the other cells remain in the epithelial cell fate [high Notch activity, (Fig. 2c, d)]. This mechanism allows the bristles on the fly notum to be formed in an evenly spaced manner. Loss of Notch signaling during this process, which occurs between 0 to 14 hours after puparium formation (hAPF), leads to generation of more SOPs at the expense of epithelial cells (Fig. 3a, b) (Hartenstein and Posakony 1990). During lateral inhibition of the SOPs, only Notch-Delta signaling is essential and Notch-Serrate signaling does not seem to be required (Yamamoto et al. 2012; Zeng et al. 1998). Thus, the loss of Serrate does not show any defect in bristle spacing, whereas the loss of Delta in mutant clones exhibits bristle tufting in the adult notum (Parks and Muskavitch 1993).

3.2 Notch Signaling-Mediated Lineage/Cell Fate Decisions Upon Asymmetric Cell Division of Sensory Organ Precursor Cells

Each bristle is composed of four cells: a socket, a shaft, a sheath and a mechanosensory neuron. These four cells are generated by a series of asymmetric cell divisions of the SOP and subsequent lineage specification through Notch signaling (Fig. 2b, c) (Jan and Jan 2001; Schweisguth 2015). Socket and shaft cells are located externally and provide the mechanical apparatus for mechanosensation. The neuron and the sheath cell, which is thought to function as a glial cell, are located internally and cannot be observed by simple visual examination of the fly notum. The dendrite of the mechanosensory neuron is located at the base of the bristle and is thought to contain mechanosensitive ion channels that open to depolarize the neuron upon deflection of the shaft cell (Barolo et al. 2000). The axon of the neuron targets the central nervous system to transmit the signal to higher nervous system centers (Chen et al. 2006).

The SOP, also referred to as the pI cell in this context, first divides along the anterior-posterior axis of the body to give rise to the posterior pIIa cell, the precursor cell of the external cells, and the anterior pIIb cell which gives rise to the internal cells (Fig. 2c). When the SOP divides, the cell fate determinants Neur and Numb are segregated into the pIIb cell but not into the pIIa cell (Le Borgne and Schweisguth 2003; Rhyu et al. 1994). This unequal inheritance of cell fate determinants, mediated by the Par3 (encoded by the *bazooka* gene in *Drosophila*)-Par6-aPKC (atypical Protein Kinase C) complex, determines the subsequent direction of Notch signaling between the pIIa and pIIb cells in order to specify distinct fates. Both pIIa and pIIb express comparable levels of Notch, Delta and Serrate but Neur, essential for ligand activity by promoting their mono-ubiquitination and endocytosis, is apportioned to the pIIb cell. Hence, ligands in the pIIb cell have the ability to signal, whereas ligands in the pIIa cell do not (Le Borgne and Schweisguth 2003). In addition, Numb (NUMB and NUMBL in human), an endocytic adaptor protein, acts in the pIIb cell to block signal reception by promoting the endocytosis of Notch and Sanpodo (Spdo) (Roegiers et al. 2005; Hutterer and Knoblich 2005; Langevin et al. 2005). Spdo encodes a transmembrane protein required for cell fate specification at the cell surface by further modulating the trafficking of the Notch receptor (O'Connor-Giles and Skeath 2003). Unlike Numb, Spdo has no obvious human homolog. Furthermore, proteins that regulate the proper trafficking of the ligands to the apical signaling interface, such as Sec15 (Secretory 15, component of the Exocyst complex) (Jafar-Nejad et al. 2005), Rab11 (small GTPase involved in vesicle recycling and exocytosis) (Emery et al. 2005),

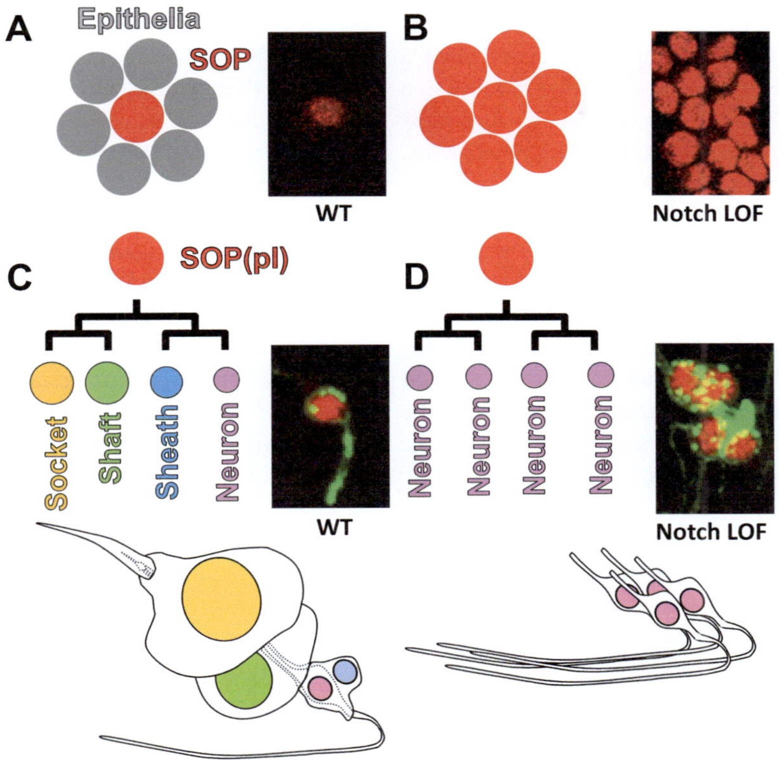

Fig. 3 Phenotypic consequences of Notch signaling loss during mechanosensory organ development. (**a**) Notch signaling mediates the lateral inhibition to specify an SOP from a proneural cluster. Cells that receive high Notch signaling becomes epidermal cells. (**b**) Upon loss of Notch signaling during lateral inhibition, all cells takes the SOP cell fate. Photographs show SOPs marked by Senseless expression (Red). (**c**) Reiterative Notch signaling is required to specify the four cell fates of the mechanosensory organ. The cells that receives the highest amount of Notch signaling becomes the Socket cells while the cells that receive the least becomes the neuron. (**d**) Upon loss of Notch signaling during lineage decisions all cells take on the neuronal fate. Photographs show neuronal nuclei and membrane, labeled by antibodies against Elav (Red) and Hrp (Green). Panels A and B were adapted and mofidied from (Yamamoto et al. 2012)

EHBP-1 [EH (Eps15 Homology) domain Binding Protein-1, adaptor protein that binds to Sec15 and Rab11) (Giagtzoglou et al. 2012), Tempura (geranylgeranyltransferase for certain Rabs including Rab11) (Charng et al. 2014) and the Arp (Actin-related protein) 2/3-WASp (Wiskott-Aldrich Syndrome protein) complex (regulator of cytoskeleton and vesicle trafficking through Actin polymerization) (Rajan et al. 2009; Ben-Yaacov et al. 2001) are also critical for proper communication between the two cells. Together, these mechanisms create a bias so that the pIIb cell becomes the signal-sending cell while the pIIa cell becomes the signal-receiving cell.

The pIIa and pIIb cells further undergo several rounds of asymmetric cell divisions and signaling to specify the four distinct cell types. An additional glial cell is formed through asymmetric division of the pIIb cell but this cell undergoes apoptosis and does not contribute to the mechanosensory organ in the adult notum (Reddy and Rodrigues 1999; Gho et al. 1999). A complete loss of Notch signaling during this process, which occurs between 16 to 24 hAPF, leads to a neurogenic phenotype (Fig. 3c, d) (Hartenstein and Posakony 1990). As a result, external socket and shaft cells, as well as the internal sheath cells, are lost leading to a balding phenotype on the notum.

In contrast, gain of Notch signaling during this process leads to generation of more external cells at the expense of internal cells, thus exhibiting a multiple socket cell phenotype in the most extreme case (Guo et al. 1996). The two ligands, Delta and Serrate, act redundantly during linage decisions to form the bristle. The loss of function of either ligand alone does not show lineage specification defects but cells mutant for both ligands exhibit a strong balding defect similar to loss of Notch (Zeng et al. 1998). In summary, Notch is used for both lateral inhibition and cell fate specification during the development of the mechanosensory organ, and is regulated by a number of distinct factors.

3.3 Notch Mediated Inductive Signaling During the Formation of the Wing Margin

The wing of a fly is a bilayered structure composed of dorsal and ventral wing blades that are bound together via integrin mediated attachment (Fig. 4a, b) (Blair 2007). The two surfaces of the wing blades meet at the wing margin to form the rim of the wing. Mechanosensory and chemosensory bristles are located along the anterior wing margin, whereas non-innervated bristles align the posterior wing margin. The wing margin is specified during the larval stage within the wing imaginal disc, which gives rise to the future wing and notum tissue of the adult fly. The dorsal domain of the wing imaginal disc expresses the selector gene *apterous*, which encodes a homeodomain transcription factor with two LIM (Lin11, Isl-1 and Mec-3) domains (Bourgouin et al. 1992). Apterous turns on the expression of *Serrate* and *Fng* specifically in the dorsal domain, whereas *Notch* and *Delta* are expressed in both compartments (Klein et al. 1998; Kim et al. 1995; Bachmann and Knust 1998b). Serrate can signal to the ventral compartment but cannot signal within the dorsal compartment due to differences in Fng modification of Notch in the two compartments (Fleming et al. 1997; Rauskolb et al. 1999; Klein and Arias 1998; Celis et al. 1996b; Panin et al. 1997). Conversely, Delta can signal to the dorsal cells but cannot signal to the ventral compartment (Panin et al. 1997; Klein and Arias 1998; Doherty et al. 1996). This bidirectional signaling through Delta and Serrate along the dorsal-ventral boundary leads to the activation of Notch, which in turn activates genes specific for the wing margin such as *wingless* (*wg*) and *cut* (Fig. 3c) (Jack and DeLotto 1992; Micchelli et al. 1997; Rulifson and Blair 1995; Neumann and Cohen 1997; Kim et al. 1995). Wg (WNT1 in human), a Wnt signaling ligand, acts as a morphogen to pattern the wing along the dorsal-ventral axis (Neumann and Cohen 1997; Zecca et al. 1996; Alexandre et al. 2014) whereas Cut is a homeodomain transcription factor that is involved in maintaining the expression of Wg as well as repressing the expression of Delta and Serrate within the future wing margin tissue (Blochlinger et al. 1988; Celis and Bray 1997; Micchelli et al. 1997). At later stages in wing margin development, the cells that co-express Wg and Cut down-regulate the expression of Notch ligands, whereas cells flanking the wing margin cells express high levels of Delta and Serrate via high Wnt signaling activation. Thus, Delta and Serrate from the flanking cells continue to signal to the wing margin cells to maintain the expression of Wg and Cut, reinforcing the establishment of a solid compartmental boundary through a positive feedback loop (Michel et al. 2016; Celis and Bray 1997). Loss of Notch signaling during this induction leads to the loss of wing margin tissue (Fig. 4d, e). Unlike decisions in the bristle lineage where Delta and Serrate act redundantly, both ligands are necessary for wing margin specification. Hence, loss of either Delta or Serrate alone leads to a reduction in Wg and Cut expression, resulting in the notching of the wing (Celis et al. 1996b). Mild loss of wing margin tissue at the distal tip of the wing can even be seen in flies that are heterozygous for a null mutation of *Notch* (Lindsley and Zimm 1992). This haploinsufficiency phenotype of *Notch*, which originally gave the name "Notch" to the gene and the pathway, emphasizes the strict dosage sensitivity of inductive signaling during wing margin formation.

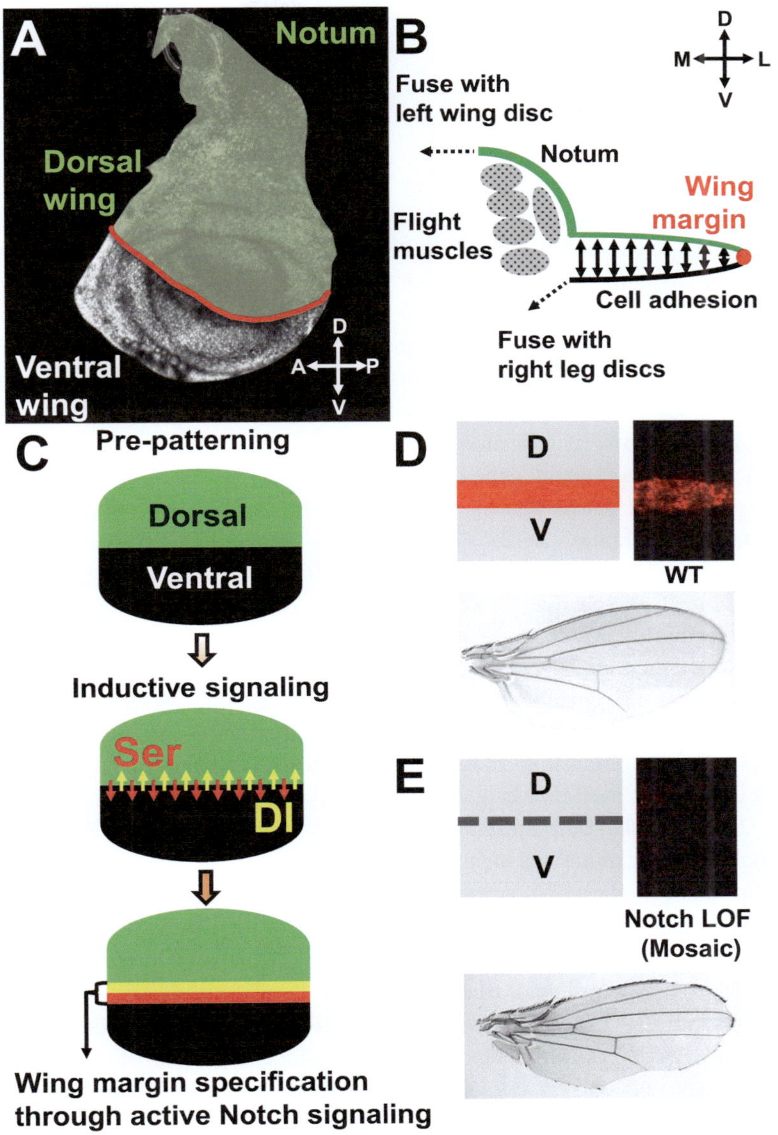

Fig. 4 Notch signaling is required for inductive signaling during wing margin development. (**a**) Photograph of a DAPI-stained wing imaginal disc that is pseudocolored for the dorsal domain (green) and the future wing margin (red). (**b**) Schematic diagram of a transverse section of a part of the fly thorax. The dorsal wing imaginal disc (green) gives rise to the notum (dorsal thorax) and the dorsal wing blade. The ventral wing imaginal disc forms the ventral wing blade. The boundary between the dorsal and ventral compartment becomes the wing margin (red). (**c**) Notch mediated inductive signaling specifies the future wing margin during imaginal disc development. Serrate (Ser) signals from the dorsal to the ventral compartment (red arrows) whereas Delta (Dl) signals from the ventral to the dorsal compartment (yellow). (**d**) Upon Notch signaling activation at the dorsal-ventral boundary, genes such as Wingless (red) and Cut (not shown) are expressed, specifying the wing margin. (**e**) Upon loss of Notch signaling during inductive signaling, these genes fail to be expressed and the wings exhibit a "notching" phenotype. Abbreviation of axes in panels A-B: D (Dorsal), V (Ventral), A (Anterior), P(Posterior), M (Medial), L (Lateral). Panel C was adapted and mofidied from (Yamamoto et al. 2012).

4 Human Diseases Caused by Rare Mutations in Notch Pathway Genes

In parallel to efforts to reveal the genes and mechanisms that coordinate the Notch signaling pathway using model organisms and cultured cell lines, medical research has uncovered a strong link between Notch and many human diseases (Stanley and Okajima 2010; Gridley 2010; MacGrogan et al. 2010; Liu et al. 2010; Gridley 2003; Gridley 1997). To date, inherited or *de novo* mutations in human genes that encode core components of the pathway such as the receptors, ligands, transcription factors and downstream target genes have been shown to cause diverse Mendelian disorders (Louvi and Artavanis-Tsakonas 2012; Penton et al. 2012). By studying these rare diseases and patients from a clinical perspective, physicians and scientists made discoveries that had major impacts on basic Notch research. In addition, there is growing evidence that misregulation of Notch signaling may participate in more common disorders, ranging from neuropsychiatric to metabolic disorders (Ables et al. 2011; Pierfelice et al. 2011; Bi and Kuang 2015; Geisler and Strazzabosco 2015). Furthermore, somatic mutations in genes in the pathway and/or misregulation of Notch signaling activity has also been linked to oncogenesis and tumor progression in different cancer types (Koch and Radtke 2010; Kovall and Blacklow 2010; Ranganathan et al. 2011). Here, we will provide an overview of Mendelian disorders caused by mutations in genes that encode core Notch signaling components in human, most of which are catalogued in OMIM (Online Mendelian Inheritance in Man 2017), an online database of human genotypes and phenotypes. The role of Notch signaling in cancer will be further discussed in other chapters of this book (e.g. Chaps. "Notch and Senescence" and "Notch in Leukemia").

4.1 Adams-Oliver Syndrome

Adams-Oliver syndrome (AOS) is a developmental disorder characterized by aplasia cutis congenital (a congenital skin defect, typically of the scalp) and transverse limb defects (typically digital amputations) (Hassed et al. 2017; Adams and Oliver 1945). In addition, some AOS patients exhibit nervous system and cardiac/vascular abnormalities. Dominant mutations in *NOTCH1* (OMIM #616028), *DLL4* (OMIM #616589), *RBPJ* (OMIM #614814) and recessive mutations in *EOGT* (OMIM #615297) are known to cause this condition. Additional mutations in *DOCK6* [*Dedicator Of CytoKinesis 6*, guanine nucleotide exchange factor (GEF) for Rho-GTPases, OMIM #614219] and *ARHGAP31* [*RHo GTPase Activating Protein 31*, GTPase-activating protein (GAP) for Rho-GTPases, OMIM #100300] have also been linked to AOS but the relationship between these genes and Notch signaling is currently unknown. A number of missense, nonsense and frameshift mutations in *NOTCH1* (Stittrich et al. 2014; Southgate et al. 2015) and *DLL4* (Meester et al. 2015) have been found in patients with this condition, suggesting that haploinsufficiency is the underlying mechanism of the dominant inheritance for these genes. AOS-linked mutations identified in *RBPJ* have been shown to impair the DNA binding capacity of the encoded protein (Hassed et al. 2012).

4.2 Alagille Syndrome and Hajdu-Cheney Syndrome

Alagille Syndrome is a developmental disorder that affects a number of organ systems including the liver (paucity of intrahepatic bile ducts), cardiovascular system (stenosis of the pulmonic valve), kidney (renal dysplasia), skeleton (abnormal "butterfly" vertebrae), eye (posterior embryotoxon, a characteristic defect in the layers of the eye called the ring of Schwalbe) and dysmorphic facial features (Saleh et al. 2016). The main manifestation of disease is seen in the liver where bile duct formation is defective, resulting in chronic cholestasis (Alagille et al. 1987). Dominant nonsense, frameshift and missense mutations in *JAG1* (~90% of cases, OMIM #118450) (Oda et al. 1997; Li et al. 1997) or *NOTCH2* (a few % of cases, OMIM #610205) (McDaniell et al. 2006) cause this condition, suggesting that haploinsufficiency of these genes is the underlying genetic mechanism. Dominant mutations in *NOTCH2* are also associated with a different congenital disease called Hajdu-Cheney syndrome that primarily manifests as a skeletal disease (OMIM #102500) (Majewski et al. 2011; Simpson et al. 2011; Isidor et al. 2011). Variants associated with this disease are late truncating mutations that remove the C-terminal PEST domain of *NOTCH2*, resulting in production of a more stable protein. Hence, both loss- and gain-of-function mutations in *NOTCH2* cause rare genetic disorders that are phenotypically and mechanistically distinct from each other.

4.3 Aortic Valve Diseases and Tetralogy of Fallot

Notch signaling plays a key role during the development of the cardiovascular system (MacGrogan et al. 2010). Notch is used reiteratively in cardiac development: during cardiomyocyte specification and differentiation, atrioventricular canal development, cardiac valve development, ventricular trabeculation, and outflow tract development. Cardiac defects are often seen in patients with AOS, Alagille Syndrome and Hajdu-Cheney syndrome with a number of different cardiac lesions. Dominant mutations in *NOTCH1* and *JAG1* have also been linked to primary congenital heart diseases such as Aortic Valve Disease 1 (OMIM #109730) (McBride et al. 2008; Garg et al. 2005) and Tetralogy of Fallot (OMIM #187500) (Krantz et al. 1999), respectively. *NOTCH1* mutations linked to Aortic Valve Disease are nonsense and frameshift mutations, suggesting a haploinsufficient mechanism (Garg 2006). Why some patients with loss-of-function mutations in *NOTCH1* exhibit Adams-Oliver syndrome while others only present cardiac symptoms is unclear. To date, all mutations in *JAG1* that are linked to Tetralogy of Fallot are missense alleles (Krantz et al. 1999; Kola et al. 2011; Digilio et al. 2013; Guida et al. 2011), which may have different consequences from the Alagille syndrome-linked mutations in this gene. In addition, dominant mutations (one nonsense and one missense, respectively) in *MIB1* (OMIM #615092) (Luxan et al. 2013) and *MIB2* (OMIM #N/A) (Piccolo et al. 2017) have been linked to Left Ventricular Noncompaction (LVNC), a form of cardiomyopathy. Patients with a mutation in *MIB2* also exhibit gastrointestinal phenotypes and reminiscent of Ménétrier disease (Piccolo et al. 2017). In sum, cardiac defects are often associated with mutations affecting Notch signaling, which is likely due to the fact that Notch signaling plays a number of critical roles during cardiovascular development in vertebrates (D'Amato et al. 2016). These phenotypes can be presented together with defects in other organ systems, reflecting the highly pleiotropic nature of this pathway.

4.4 Spondylocostal Dysostosis

Notch signaling also affects skeletal development, and mutations in several core signaling components and downstream target genes have been associated with rare skeletal disorders (Tao et al. 2010; Zanotti and Canalis 2010). Spondylocostal Dysostosis (SCDO) is primarily an autosomal recessive disorder, presenting with abnormal vertebra formation and patterning defects. Five of the six types of SCDO identified to date are caused by recessive mutations in core Notch signaling pathway components and downstream target genes: SCDO1; *DLL3* (OMIM #277300) (Bulman et al. 2000), SCDO2; *MESP2* (*Mesoderm posterior bHLH transcription factor 2*, Notch target gene, OMIM #608681) (Whittock et al. 2004), SCDO3; *LFNG* (OMIM #609813) (Sparrow et al. 2006), SCDO4; *HES7* (Notch target gene, OMIM #613686) (Sparrow et al. 2008), and SCDO5; *TBX6* "[*T-box 6*, Notch target gene, OMIM #122600 (White et al. 2005; Yasuhiko et al. 2006), autosomal dominant forms have also been reported (Wu et al. 2015; Sparrow et al. 2013)]. The sixth SCDO gene, *RIPPLY2* (*Ripply transcriptional repressor 2*, OMIM #616566), lies downstream of the pathway and regulates the expression and/or function of *MESP2* and *TBX6* (Karaca et al. 2015). In mice, many of these genes have been shown to play a critical role in somitogenesis (Sparrow et al. 2011; Wahi et al. 2016), indicating that SCDO is caused by misregulation of an evolutionarily conserved transcriptional pathway that regulates somite (precursor of vertebra and other tissues) formation (Pourquie and Kusumi 2001; Kageyama et al. 2010).

4.5 CADASIL and *NOTCH3*-related Disorders

Cerebral Autosomal Dominant Arteriopathy with Subcortical Infarcts and Leukoencephalopathy (CADASIL) is the most common heritable cause of stroke and vascular dementia, characterized by five main symptoms: migraine with aura, subcortical ischemia, mood disorders, apathy, and cognitive decline generally found in families with an autosomal dominant pattern of inheritance (Chabriat et al. 2009). Accumulation of granular osmophilic material (GOM), which accompanies vascular smooth muscle degeneration and arteriopathy in postmortem CADASIL patient brain tissue, is a characteristic pathological feature of the disease (Bergmann et al. 1996). Over 90% of CADASIL patients carry a dominant mutation in *NOTCH3* (OMIM #125310) and over 170 mutations have been identified to date (Joutel et al. 1996; Monet-Lepretre et al. 2009). Interestingly, the majority of the mutations involve loss or gain of cysteine residues in one of the 34 EGF repeats in the extracellular domain of NOTCH3 (Joutel et al. 1997). The odd numbers of cysteines (5 or 7) per EGF repeat caused by CADASIL *NOTCH3* mutations are thought to disrupt the formation of proper intra-molecular disulfide bonds. Although no logical explanation has been proposed, it is interesting to note that the majority of the mutations are clustered between EGFr-1-5 and the distribution of CADASIL associated missense mutations along the extracellular domain of NOTCH3 is uneven (Monet-Lepretre et al. 2009).

Whether CADASIL is caused by loss or gain of function of *NOTCH3* has been under extensive debate (Rutten et al. 2014). Some CADASIL mutations behave as loss-of-function alleles of *NOTCH3* based on ligand-receptor binding and signaling assays performed in cultured cells and in mouse models (Peters et al. 2004; Joutel 2011; Ayata 2010; Haritunians et al. 2005; Joutel et al. 2004). However, considering that heterozygous deletions of the *NOTCH3* locus have not been associated with CADASIL in human patients, and that *Notch3* knockout mice do not exhibit pathological phenotypes that are characteristic for the disease (Domenga et al. 2004), the degree to which defects in Notch signaling contributes to the pathogenesis of this disorder remains unclear. Others propose that the pathogenesis of CADASIL is due to a toxic-gain-of-function (neomorphic effect) of *NOTCH3* and that non-physiological intermolecular disulfide bonds formed between the free cysteine residues of NOTCH3 and other transmembrane and/or secreted proteins is the primary cause of disease (Dichgans et al. 2000; Donahue and Kosik 2004; Opherk et al. 2009). The extracellular portion of NOTCH3 has been found to be associated with or included in the GOM (Joutel et al. 2000; Joutel et al. 2001; Ishiko et al. 2006), which also consists of numerous proteins including Clusterin and Collagen18α1/Endostatin (Arboleda-Velasquez et al. 2011). However, it remains to be determined whether extracellular accumulation/aggregation of secreted and cell surface proteins in the GOM is due to direct interaction of these factors with mutant NOTCH3 protein. In addition, whether there is a causal connection between GOM formation and clinical symptoms found in CADASIL patients still needs to be investigated and clarified. Furthermore, since most studies have been performed only on a small subset of pathogenic mutations in *NOTCH3*, further studies on a spectrum of mutations are needed to reveal the full molecular pathology of the disease.

Mutations in *NOTCH3* have also been found in patients with lateral meningocele syndrome (LMNS, OMIM #130720) (Gripp et al. 2015) where *de novo NOTCH3* variants are identified, and in a single family with an autosomal dominantly inherited infantile myofibromatosis 2

(IMF2, OMIM #615293) (Martignetti et al. 2013). The former disease is characterized by distinct facial characteristics, hypotonia, hyperextensibility and meningocele-related neurologic phenotypes such as bladder dysfunction, while the latter disorder is characterized by formation of benign tumors in connective tissues that arise due to excessive mescenchymal cell proliferation. Other reported cases of infantile myofibromatosis have been linked to the *PDGFRB* (*Platelet Derived Growth Factor Receptor Beta*) gene (OMIM #228550), and the role Notch signaling in the pathogenesis of this disease is unknown. Both disorders have been proposed to be caused through gain-of-function mechanisms [late truncating mutations that delete the PEST domain for LMNS (Gripp et al. 2015); missense mutation in the NRR domain for IMF2 (Martignetti et al. 2013)], but further functional studies and additional patient identification are necessary to reveal a clear genotype-phenotype relationship.

4.6 Other Mendelian Diseases Caused by Mutations in Notch Signaling Pathway Genes: γ-Secretase Complex Related Disorders as an Example

In addition to the disorders described above, there are a number of Mendelian diseases that are caused by mutations in homologs of *Drosophila* genes that are known to be critical for Notch signaling. However, since many genes are pleiotropic and have functions outside of the Notch signaling pathway, it is not clear which aspect, if any, of the patient's phenotypes can be explained by defects in Notch signaling.

For example, dominant missense mutations in *PSEN1* (OMIM #607822, 600274, 172700) and *PSEN2* (OMIM #606889), that encode catalytic subunits of the γ-secretase complex, cause rare early onset familial forms of Alzheimer's disease (AD) and other forms of dementia. Although several studies have implicated the role of Notch signaling in AD pathogenesis (Woo et al. 2009), the primary mechanism by which mutations in *PSENs* cause AD seems to be through altered proteolytic processing of Amyloid Precursor Protein (APP), another well characterized substrate of the γ-secretase complex (Selkoe and Hardy 2016). Additional dominant missense mutations in *PSEN1* (OMIM #613694) and *PSEN2* (OMIM #613697) have also been found in patients with dilated cardiomyopathy (Li et al. 2006; Gianni et al. 2010; Gianni et al. 2011). The functional consequences of these missense mutations are unclear and whether defects in Notch signaling may be contributing to this condition has not been investigated. Furthermore, loss-of-function mutations in *PSEN1* and other components of the γ-secretase complex cause another type of disease known as familial acne inversa. This condition, also known as hidradenitis suppurativa, is a chronic relapsing skin inflammatory disease that has been linked to haploinsufficiency of *PSEN1* (OMIM #613737), *NCSTN* (*Nicastrin*, OMIM #142690) and *PSENEN* (*Presenilin enhancer gamma-secretase subunit*, OMIM #613736). Since Notch signaling plays multiple key roles in the development and maintenance of the skin (Nowell and Radtke 2013) and immune system (Yuan et al. 2011), it has been proposed that defects in Notch signaling contributes to the pathogenesis (Pink et al. 2013), but additional experimental evidence is needed to strengthen this model.

Similarly, dominant mutations in *POFUT1* (*Protein O-fucosyltransferase 1*, OMIM #615327) (Li et al. 2013), *POGLUT1* [*Protein O-glucosyltransferase 1*, OMIM #615696, this gene is also linked to muscular dystrophy (OMIM #617232)] Basmanav et al. 2014) and *ADAM10* (OMIM #615537) (Kono et al. 2013) cause skin disorders that results in pigmentation defects (Dowling-Degos disease or reticulate acropigmentation of Kitamura). Considering that Notch regulates multiple aspects of melanocyte development (Liu et al. 2015), it is likely that defects in Notch signaling contribute to the pathogenesis of these diseases (Yu et al. 2016; McMillan et al. 2017). However, direct experimental evidence is necessary to test this hypothesis. Likewise, mutations in a number of genes encoding general cellular machineries that affect Notch receptor trafficking and activation (e.g. Clathrin-Dynamin mediated endocytosis, ESCRT, AP-3, HOPS, V-ATPase complexes) are also linked to diverse diseases but additional work is required to determine the degree by which Notch signaling defects contribute to the pathology of these disorders.

In summary, genes that have been well established to function in Notch signaling are linked to a number of Mendelian diseases. The fact that the Notch pathway is pleiotropic likely contributes to the broad range of human phenotypes affecting a wide range of organ systems. In addition, the strict dosage dependence of the pathway may explain the involvement of both gain- and loss-of-function mechanisms and both recessive and dominant modes of inheritance leading to disease. Further investigations that focus on the functional impacts of each pathogenic mutation will likely provide better mechanistic understandings of how specific phenotypes associated with these disorders may be explained by defects in Notch signaling.

5 Using *Drosophila* to Study Rare Functional Variants in Genes Linked to Notch Signaling Pathway and Beyond

Advances in sequencing technologies are accelerating the pace of disease gene discovery (Gonzaga-Jauregui et al. 2012). As of February 2017, of 149 genes that have been linked to Notch signaling in *Drosophila melanogaster*, 130 are conserved in human (87%) and 48 (37%) have human homologs that are linked to Mendelian diseases based on FlyBase (2017), a manually curated database for *Drosophila* genetics and biology, and OMIM (2017). Identification of some of these disease genes was made possible through whole-exome sequencing (Majewski et al. 2011; Simpson et al. 2011; Isidor et al. 2011; Piccolo et al. 2017; Basmanav et al. 2014; Kono et al. 2013). As more and more exomes and genomes are sequenced in research and clinical diagnostic laboratories using high-throughput sequencing technologies (Ramoni et al. 2017; Chong et al. 2015; Green et al. 2016), new human diseases that are caused by mutations in genes that have been previously linked to Notch signaling in flies are likely to be identified. In addition, the list of novel rare variants of unknown significance (VUS) in known Notch-related disease genes will also likely to continue to expand. Proper interpretation of the functional consequences of these VUS will become critical for researchers to identify the underlying causes of disease and for clinicians to make medical decisions for patients in the era of personalized medicine.

For Notch-related disorders, a number of *in vitro* and *in vivo* assays in mammalian systems have been established to assess the function of disease-associated variants. For example, several Alagille syndrome linked mutations in *JAG1* (p.L37S, p.P163L and p.R184H) were shown to affect subcellular localization, glycosylation, and signaling capability of JAG1 using skeletal muscle derived cell lines, leading to the proposal of a haploinsufficient (loss-of-function) model of disease pathogenesis (Morrissette et al. 2001; Tada et al. 2012). In another study, however, two Alagille syndrome linked mutations in *JAG1* (p.C284F and p.E1003X) were reported to exhibit a dominant-negative effect on Notch signaling when overexpressed in NIH3T3 cells (Boyer-Di Ponio et al. 2007).

Conflicting results obtained through *in vitro* experiments are typically resolved using *in vivo* model systems. To date, most *in vivo* studies that attempt to understand the functional consequences of disease-associated variants in Notch related diseases have been performed in the mouse (*Mus musculus*). One key advantage of mouse models is that one can screen for phenotypic similarities between the mutant mice and disease patients. For example, heterozygous inactivation of *Rbpj* in mice causes cardiac phenotypes that are often seen in human diseases (Nus et al. 2011). Similarly, cardiac phenotypes seen in LVNC patients that carry mutations in *MIB1* were successfully phenocopied in heart specific *Mib1* knockout mice (Luxan et al. 2013). Importantly, reduced Notch1 signaling in the developing heart was observed in these animals suggesting that loss-of-function of *MIB1* and subsequent reduction in Notch activation is likely to be associated with LVNC. Some studies in mice have used gene knock-in strategies to introduce analogous mutations into the mouse ortholog of the human gene to understand the function of specific disease-linked mutations. For example, one study modeled two CADASIL-linked mutations (p.C455R and p.R1031C) in mice and showed that these mutations behave as hypomorphic alleles (Arboleda-Velasquez et al. 2011). Furthermore, Clusterin and Collagen18α1/Endostatin, materials found in GOM in CADASIL patient brain vessels, accumulated in brain blood vessels of the mice, proving a phenotypic link between the human patients and the mouse models. Although important insights into the role of disease associated *NOTCH3* variants in vascular biology can be obtained by these types of studies, a potential confounding factor of these mouse mutants is that they do not exhibit key features of CADASIL such as development of spontaneous stroke (Arboleda-Velasquez et al. 2011; Lundkvist et al. 2005). Similar arguments have been made for mouse knock-in models for AD-linked mutations in *PSEN1* (Xia et al. 2015; Xia et al. 2016; Veugelen et al. 2016; Guo et al. 1999b; Siman et al. 2000; Flood et al. 2002). Nevertheless, these models provide valuable information about the role of the genes and variants in a physiological setting, a complex systemic environment that cannot be easily mimicked in cell or tissue culture based studies.

One large drawback for gene modification based experiments in mice is the cost and time that is required to generate reagents and to complete the analysis of a given variant. When hundreds of novel VUS are identified from large sequencing efforts, it is unrealistic and uneconomical to use the mice to study all of these variants *in vivo*. *In vitro* experiments can be used as a first line of screening prior to the generation of mouse models, but slight defects that may be amplified through intercellular feedback loops *in vivo* (e.g. during lateral inhibition and inductive signaling) may be missed through simple cell based assays. Furthermore, if the disease-linked variants affect animo acids that are not conserved between human and mouse, a knock-in strategy cannot be applied. Based on the deep biological knowledge of Notch signaling and rich genetic toolkits the community has generated to characterize this pathway (Bloomington *Drosophila* Stock Center 2017; *Drosophila* Genomics Resource Center 2017; Developmental Studies Hybridoma Bank 2017), *Drosophila* can be a powerful tool to bridge this gap. Here, we will discuss several strategies to functionally characterize disease-associated variants using *Drosophila*, starting with the identification of the potential fly ortholog of a gene of interest. We will close this section by providing examples of such *Drosophila* studies that have been performed to study disease-associated genes involved in Notch signaling.

5.1 Using Bioinformatics to Aggregate Existing Knowledge and Resources

The first step in disease-linked variant functional studies using *Drosophila* is to perform bioinformatics analyses to identify the strongest *Drosophila* ortholog candidate for the human gene of interest. There are a number of orthology prediction programs that use different algorithms and criteria to predict the most likely ortholog candidate (Altenhoff and Dessimoz 2012). User-friendly online tools such as DIOPT (*Drosophila* RNAi Screening Center Integrative Ortholog Prediction Tool 2017) and HCOP (Human genome organization gene nomenclature committee Comparison of Orthology Prediction search 2017) integrates a number of these programs to provide the users with an arbitrary score. The higher the DIOPT or HCOP scores are for a given gene combination, more likely the two genes are to be true orthologs. Due to the two rounds of whole genome duplication evens that likely to have occurred in ancestral vertebrates (Dehal and Boore 2005) [although there is still some debate (Kasahara 2007)], there are many cases in which multiple human genes are orthologous to a single fly gene as seen for *Notch* (*NOTCH1-4*), *Delta* (*DLL1,3,4*) and *Serrate*

(*JAG1,2*) (Table 1). Once a fly gene of interest is identified, one can determine whether the gene has been linked to Notch signaling in *Drosophila* by using PubMed (2017) or FlyBase (2017). Information such as known gene function, expression patterns, physical interactors, available reagents and publication records can be obtained through these websites. A newly developed integrative online resource called MARRVEL (Model organism Aggregated Resources for Rare Variant ExpLoration) (Wang et al. 2017; MARRVEL 2011) integrates DIOPT (*Drosophila* RNAi Screening Center Integrative Ortholog Prediction Tool 2017), Flybase (2017) as well as additional human genomics OMIM (Online Mendelian Inheritance in Man 2017); DGV (Database of Genomic Variants 2017); ExAC (Exome Aggregation Consortium) Browser 2017; Geno2MP (Genotype to Mendelian Phenotype Browser 2017); DECHIPHER (DatabasE of genomiC varIation and Phenotype in Humans using Ensembl Resources 2017); ClinVar 2017) and model organisms databases SGD (Saccharomyces Genome Database 2017); PomBase 2017; WormBase 2017; ZFIN (ZebraFish Information Network 2017); MGI (Mouse Genome Informatics 2017); RGD (Rat Genome Database 2017) to help the users to perform a wide survey of the gene and variant of interest. These searches are important to confirm that the gene/variant of interest is worth investigating in-depth prior to initiating any experiments in model organisms.

For genes that have been linked to Notch signaling in *Drosophila*, it is important to determine the context in which this link has been made and to find out the tools and experimental strategies that were used to make the conclusion. One gene may have been functionally studied using a clean null allele in the context of embryonic central nervous system development, while another gene may have been studied using tissue-specific RNAi expression in the developing wing margin without proper control experiments. By obtaining information about the biological context and experimental strategy that was used in previous studies, one can determine how to design a set of experiments to test the function of the variant of interest. It is also important to determine whether the reagents used in the previous studies are available through stock centers or individual laboratories. If the mutant or transgenic stocks are available from public stock centers such as BDSC (Bloomington *Drosophila* Stock Center 2017), DGGR (*Drosophila* Genomics and Genetic Resources 2017), VDRC (Vienna *Drosophila* Resource Center 2017) or from individual labs upon request, this will save time and resources. Additional genetic tools such as Notch signaling reporters (Housden et al. 2014) and classical alleles of core Notch signaling pathway genes that can be useful for signaling and genetic interaction studies are also available from some of these stock centers. Many monoclonal antibodies (e.g. anti-Notch, anti-Delta) and constructs/plasmids that are useful for Notch signaling studies in *Drosophila* [e.g. transgenic vectors, cDNA clones and BACs (Bacterial Artificial Chromosomes)] are also available from DSHB (Developmental Studies Hybridoma Bank 2017) and DGRC (*Drosophila* Genomics Resource Center 2017), respectively. In summary, by performing a thorough search of the existing knowledge and resources using online tools and databases, one can obtain sufficient information to design a set of experiments to test the functional significance of a variant of interest in *Drosophila*.

5.2 Selecting the Best Strategy to Study the Variant of Interest in Flies

One important consideration when studying a human missense variant in *Drosophila* is whether the mutated/altered amino acid is conserved or not. While there are some exceptional cases (see the *TM2D3* case discussed below in Sect. 5.3.2), conserved amino acids tends to be functionally more important (Katsonis et al. 2014). In addition, the conservation of the residue allows one to test the function of the variant in the context of the fly gene/protein. By introducing the analogous variant in a fly cDNA or genomic rescue construct and expressing them in the mutant background in *Drosophila*, one can test if the variant behaves differently from the wild-type/reference fly gene. Also, if the variant of interest is conserved, site-directed mutagenesis using CRISPR (Clustered Regularly Interspaced Short Palindromic Repeat)-Cas9 System can be employed to edit the endogenous fly gene to create a clean knock-in allele via homology directed repair (HDR) (Gratz et al. 2015).

If the amino acid is not conserved, one needs to somehow "humanize" the fly gene to be able to test the impact of the variant. There are a number of strategies to achieve this, and one powerful strategy that our laboratory has been using recently is based on the T2A-GAL4 system (Fig. 5a) (Diao and White 2012). This method allows one to generate a convenient "2-in-1" strain that can dramatically facilitate gene humanization experiments in *Drosophila* (Bellen and Yamamoto 2015). The first step is to introduce an artificial exon consisting of a splice acceptor (SA), ribosomal skipping T2A sequence, GAL4 [Yeast transcription factor that binds to and activates UAS (Upstream Activating Sequence)] coding sequence and a transcriptional termination (polyA) signal into a coding intron (introns flanked by two coding exons) of a gene of interest. Introduction of this cassette can be performed via CRISPR/HDR or through recombinase mediated cassette exchange (RMCE) of MiMIC (*Minos*-Mediated Integration Cassette) insertions (Diao et al. 2015; Gnerer et al. 2015). MiMIC is an engineered versatile transposable element that has been extensively mobilized in the fly genome and can be used as an entry point to manipulate genes in many sophisticated ways (Nagarkar-Jaiswal et al. 2015a, b; Venken et al. 2011). By flanking the T2A-GAL4 cassette with inverted *attP* sites, one can further convert this line into a GFP-tagged protein trap line via RMCE to enable a number of sophisticated biochemical, cell biological and genetic experiments (Bellen and Yamamoto 2015). If the gene lacks an intron, the GAL4-polyA cassette can be knocked into the first coding exon via the CRISPR/HDR (Fig. 5b). The T2A-GAL4 and GAL4 knock-in strains function as loss-of-function alleles due to the polyA signal. At the same time, these cassettes produce a GAL4 protein that is expressed in the same temporal and spatial manner as the host gene. The T2A ribosomal skipping peptide ensures that the protein is physically separated from the host protein so that GAL4 can enter the nucleus (Diao and White 2012; Daniels et al. 2014). Upon nuclear entry, GAL4 will drive the expression of a cDNA of interest under the control of UAS elements (Brand and Perrimon 1993). Hence, by combining the T2A-GAL4/GAL4 knock-in lines with a UAS-human cDNA transgenic line, one can replace the fly protein with a human protein to determine whether the two are interchangeable. Easily scorable phenotypes such as lethality or sterility can be used as crude assays to determine whether the human protein can function in the fly body. If one observes a complete or partial rescue with a reference (wild-type) human cDNA, one can use this as a reference point to compare how well the variant cDNA functions (Chao et al. 2017; Yoon et al. 2017). Further rescue experiments of Notch related phenotypes (e.g. neurogenic, wing notching) or signaling defects (e.g. activation of Notch reporters or target genes) will provide information on whether the variant impacts Notch signaling *in vivo*.

In addition to this T2A-GAL4 strategy, one can also make use of mutant fly strains that have been previously characterized and try to rescue the mutant phenotypes using UAS-human cDNA transgenic lines and ubiquitous- or tissue-specific GAL4 drivers. To date, we have rescued a number of fly mutants by ubiquitous expression of human cDNAs (Charng et al. 2014; Sandoval et al. 2014; Yamamoto et al. 2014; Wang et al. 2015; David-Morrison et al. 2016; Zhang et al. 2013; Xiong et al. 2012), suggesting that many human genes have shared molecular functions and can replace the fly orthologs *in vivo*. A more rudimentary humanization experiment can be performed by co-expressing an RNAi against a fly gene together with a human cDNA. Furthermore, in addition to rescue/replacement based functional studies, one can perform over-expression based experiments in a wild-type background using the GAL4/UAS system to determine whether there are any differences observed when reference and variant forms

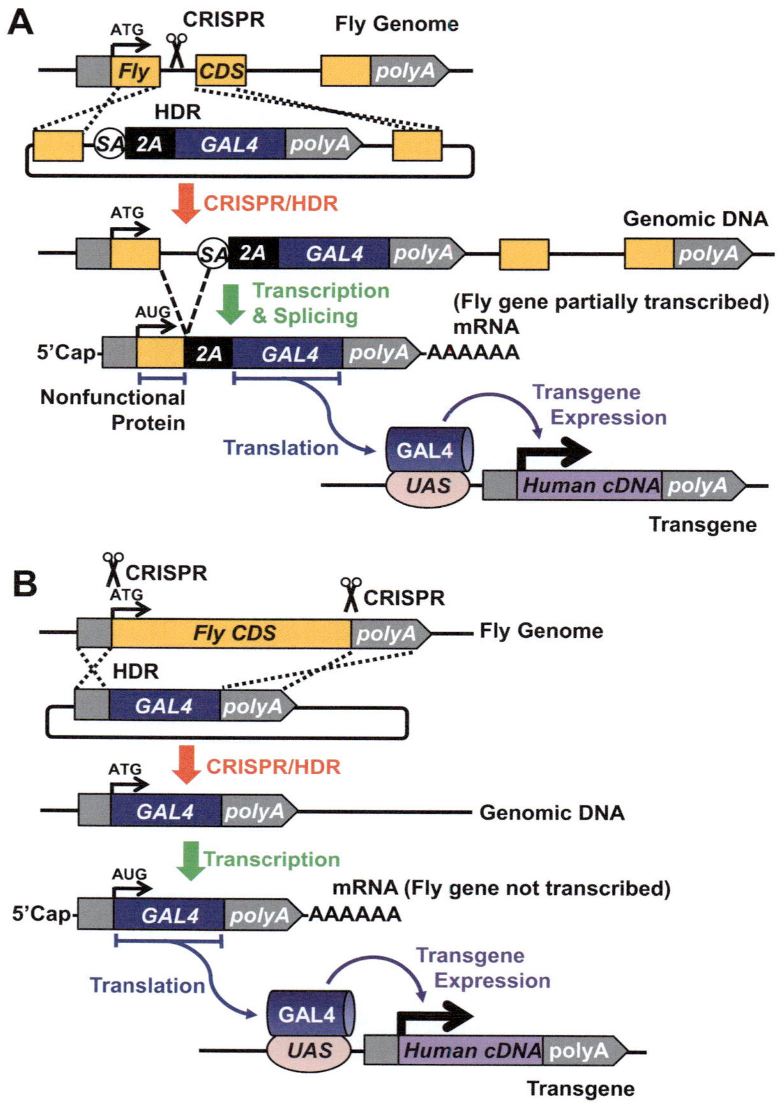

Fig. 5 Strategies to "humanize" *Drosophila* genes *in vivo*. (**a**) For genes that have coding introns (introns flanking two coding exons), one can insert a T2A-GAL4 cassette via CRISPR and HDR (homology directed repair). When the gene of interest is transcribed, the splice acceptor (SA) forces the splicing machinery to include the T2A-GAL4 cassette. The transcriptional termination site (polyA) stops the transcription, preventing the rest of the gene to be transcribed. When the transcript (mRNA) is translated, N-terminal of the fly protein is made but is prematurely truncated due to the T2A (2A) ribosomal skipping sequence, leading to generation of nonfunctional proteins in most cases. T2A sequence further allows the GAL4 protein to be translated, which in turn translocates to the nucleus to activate the expression of human cDNAs (wild-type/reference or mutant/variant) under the control of UAS elements. (**b**) For genes that do not have a coding intron, one can knock-in a GAL4 in the fly gene of interest. GAL4 will be transcribed and translated in the same temporal and special manner as the fly gene, allowing one to express the human cDNA in a mutant background. Grey boxes: 5' and 3' untranslated regions. Orange box: Fly coding sequence (CDS).

are compared. This could be especially useful for cases in which hypermorphic (gain-of-function), antimorphic (dominant negative), or neomorphic scenarios are suspected. However, ectopic overexpression based phenotypes observed through these studies need to be assessed with caution since they may not reflect the physiological function of the variant. Similarly, negative data may simply be due to the lack of sensitivity of the phenotype or assay that is being performed. Hence, positive data are strongly indicative that the variant has a functional impact in human, but one cannot rule out a candidate gene/variant due to negative data obtained from *Drosophila* studies.

5.3 Functional Studies of Disease Associated Variants in Notch Signaling Genes in *Drosophila*

Although functional studies of human disease associated variants in genes linked to Notch signaling have not been performed extensively in *Drosophila*, the following two examples related to Alzheimer's disease illustrate the value of assessing Notch signaling related phenotypes in flies to elucidate the possible impact of disease-associated variants.

5.3.1 Early Onset Familial Alzheimer's Disease and *PSEN1*

Autosomal dominant mutations in *PSEN1* are found in a number of families that exhibit rare early-onset forms of familial Alzheimer's disease (FAD) (Sherrington et al. 1995). There are >240 missense variants that have been identified to date (Alzforum 2017) but potential impacts on PSEN1 function have not been experimentally defined for most of them. The age of onset of FAD varies from 24-65 years, suggesting that some alleles maybe more detrimental than others. By introducing 14 representative *PSEN1* mutations found in conserved amino acids into the fly *Psn* homolog cDNA driven by an endogenous promoter, Seidner and colleagues performed a series of rescue experiments to determine whether there is any genotype-phenotype correlation that they can identify that parallel the severity in human patients (Seidner et al. 2006). By assessing rescue of lethality, examining morphological defects in the wing margin, bristle and eye, and performing *in vivo* reporter assays in a *Psn* null mutant background, they were able to group the FAD-linked variants into three distinct classes, which correlated well with the severity of disease presentation in human patients. It is interesting to note that the authors also attempted rescue of the fly *Psn* null mutant with human *PSEN1* or *PSEN2* transgenes but they failed, suggesting that human *PSEN1* and *PSEN2* cannot function in the context of the fly γ-secretase complex. Humanization of the entire γ-secretase complex (*Psn*, *nct*, *pen-2*, *aph-1*) may circumvent the problem but this needs to be experimentally tested.

5.3.2 Late Onset Sporadic Alzheimer's Disease and *TM2D3*

Compared to FAD, the genetic causes of late-onset Alzheimer's disease (LOAD) remain to be defined. Since LOAD is much more common than FAD and found sporadically, a number of Genome-Wide Association Studies (GWAS) primarily focusing on common variants have been performed and a number of loci that increase the risk of LOAD have been identified (Cuyvers and Sleegers 2016; Wangler et al. 2017). Other than the well-established coding variants (p.C112R and p.R158C) in *APOE* (*APOlipoprotein E*, OMIM #104310) (Poirier et al. 1993; Mahoney-Sanchez et al. 2016; Corder et al. 1993), however, the significance of most of these variants awaits to be experimentally verified. Through a recent exome-wide association study, the CHARGE Consortium (Cohorts for Heart and Aging Research in Genomic Epidemiology Consortium 2017; Psaty et al. 2009) identified a rare coding variant in a previously uncharacterized gene that has a strong effect size on LOAD (Jakobsdottir et al. 2016). A p.P155L variant in *TM2D3* (*TM2 domain containing 3*) was associated with a ~7.5 fold chance of developing LOAD and a 10-year earlier age of onset in an Icelandic population. However, the study was not able to replicate this finding in other populations since this allele was 10 times less frequent in non-Icelanders. *TM2D3* encodes a transmembrane protein whose function has never been studied in vertebrate species. Furthermore, the p.P155L variant was predicted to be benign using multiple mutation prediction programs, leading to skepticism that this variant has functional consequences. The fly *TM2D3* homolog *almondex (amx)* was one of the earliest genes together with *Notch*, *Delta*, *neur*, *mam*, *E(spl)* and *big brain (bib,* endocing an Aquaporin-family protein*)*, to be linked to Notch signaling based on its strong embryonic neurogenic phenotype when maternally mutated (Lehmann et al. 1981; Lehmann et al. 1983). *amx* mutants undergo normal development, likely due to a large maternal contribution, but all embryos that are laid by homozygous or hemizygous (mutant over a deficiency of the locus) *amx* mutant females exhibit a strong neurogenic phenotype and die as embryos (Shannon 1972; Lehmann et al. 1983). Although the molecular function of Amx is still unknown, genetic epistasis experiments have suggested that Amx likely functions at the S3 cleavage step of Notch activation (Michellod and Randsholt 2008). Considering that *PSEN1* and *PSEN2*, two genes that cause FAD also act at the same step in the Notch pathway, and that maternal-zygotic *Psn* null mutant embryos phenocopy the maternal *amx* mutant phenotype in *Drosophila* (Struhl and Greenwald 1999; Ye et al. 1999), the p.P155L in *TM2D3* was an excellent candidate variant that may increase the risk of LOAD through regulation of the γ-secretase complex. Since the variant amino acid (p.P155) is not conserved between human and *Drosophila*, we humanized the fly *amx* gene by generating a genomic rescue construct in which the fly *amx* coding region has been replaced by the human sequence. Interestingly, the reference *TM2D3* was able to partially rescue the neurogenic phenotype and lethality of the maternal *amx* mutant embryos, whereas *TM2D3* with the p.P155L variant was not able to do so (Jakobsdottir et al. 2016). Hence, p.P155L associated with LOAD was shown to be a functional variant based on Notch-signaling related phenotypic assay performed *in vivo*, and further functional studies are ongoing to determine the precise molecular function of TM2D3/Amx *in vivo*.

In summary, genetic tools and phenotypic assays in *Drosophila* provide valuable information to assess the functional consequences of disease-linked variants *in vivo*. Even for conditions such as AD for which the pathogenic involvement of Notch signaling is still obscure, Notch signaling related phenotypes in *Drosophila* tissues such as the wing, bristle, and embryonic nervous system can be used as robust and reproducible phenotypic readouts to determine the functionality of disease-associated human variants of interest. Similar strategies can be employed to determine the functionalities of many VUS that are identified through massive sequencing efforts.

6 Conclusions

Notch signaling is a unique pathway that regulates diverse biological processes through a relatively simple and straightforward signal transduction mechanism. Studies in *Drosophila* have played a pioneering role in assembling the Notch signaling pathway, elucidating numerous factors that fine-tune it in diverse contexts. By combining a number of genetic tools and resources, fly researchers have uncovered a number of biological functions of this pathway in a variety of developmental and post-developmental settings.

Although Notch signaling in *Drosophila* has been extensively studied over the last century, exciting new discoveries continue to be made in the fly field that impact the larger biomedical community. Large-scale screens using newer technologies will likely continue to expand the list of genes that regulate Notch signaling *in vivo* in *Drosophila*, which could subsequently be used as a starting point when studying the function of orthologous genes in vertebrates. Through efforts of clinicians and human geneticists, a number of human diseases that are caused by mutations in genes linked to Notch signaling have been discovered, increasing the significance of the pathway in human health. We foresee that such gene/disease lists will continue to expand through efforts in the human genomics field.

Understanding the functional consequences of VUS is critical for these studies, and experiments in *Drosophila* can accelerate such efforts.

Moving forward, close communications and collaborations among *Drosophila* researchers, human geneticists and clinicians will provide a synergistic effect to quickly identify novel human disease linked genes, study the function of variants of interest, and begin to understand the biological function of these genes *in vivo* (Wangler et al. 2015). The rich knowledge regarding various biological contexts that are regulated by Notch signaling and the extensive genetic tools that are available to *Drosophila* researchers provide a unique advantage when studying novel human disease genes in the context of Notch signaling. Through further information exchange and collaborations with vertebrate biologists, biochemists, molecular biologists, structural biologists and bioinformaticians, such knowledge can further be translated to develop potential therapies for patients. *Drosophila* will continue to be at the forefront of the Notch signaling field, and discoveries made using the fly will provide valuable information to understand human pathology and possibly tame this pathway when necessary.

Acknowledgements We apologize to our colleagues whose works we were not able to include. We thank Drs. Andrew K. Groves, Hamed Jafar-Nejad, Hillary K. Graves and Michael F. Wangler for constructive comments and helpful suggestions. S.Y. is supported by the Jan and Dan Duncan Neurological Research Institute at Texas Children's Hospital (NRI Fellowship), the Naman Family Fund for Basic Research and the Caroline Wiess Law Fund for Research in Molecular Medicine (BCM Junior Faculty Seed Funding Program), Alzheimer's Association (NIRH-15-364099), Simons Foundation Autism Research Initiative (Award#368479), and the National Institutes of Health (NIH, U54 NS093793). J.L.S. received support from the NIH (GMR2556929).

References

Ables JL, Breunig JJ, Eisch AJ, Rakic P (2011) Not(ch) just development: Notch signalling in the adult brain. Nat Rev Neurosci 12(5):269–283

Acar M, Jafar-Nejad H, Giagtzoglou N, Yallampalli S, David G, He Y, Delidakis C, Bellen HJ (2006) Senseless physically interacts with proneural proteins and functions as a transcriptional co-activator. Development 133(10):1979–1989

Acar M, Jafar-Nejad H, Takeuchi H, Rajan A, Ibrani D, Rana NA, Pan H, Haltiwanger RS, Bellen HJ (2008) Rumi is a CAP10 domain glycosyltransferase that modifies Notch and is required for Notch signaling. Cell 132(2):247–258

Adam JC, Montell DJ (2004) A role for extra macrochaetae downstream of Notch in follicle cell differentiation. Development 131(23):5971–5980

Adams FH, Oliver CP (1945) Hereditary deformities in man due to arrested development. J Hered 36:3–7

Akazawa C, Sasai Y, Nakanishi S, Kageyama R (1992) Molecular characterization of a rat negative regulator with a basic helix-loop-helix structure predominantly expressed in the developing nervous system. J Biol Chem 267(30):21879–21885

Alagille D, Estrada A, Hadchouel M, Gautier M, Odievre M, Dommergues JP (1987) Syndromic paucity of interlobular bile ducts (Alagille syndrome or arteriohepatic dysplasia): review of 80 cases. J Pediatr 110(2):195–200

Alexandre C, Baena-Lopez A, Vincent JP (2014) Patterning and growth control by membrane-tethered wingless. Nature 505(7482):180–185

Altenhoff AM, Dessimoz C (2012) Inferring orthology and paralogy. Methods Mol Biol 855:259–279

Alzforum (2017) http://www.alzforum.org/mutations

Arboleda-Velasquez JF, Manent J, Lee JH, Tikka S, Ospina C, Vanderburg CR, Frosch MP, Rodriguez-Falcon M, Villen J, Gygi S, Lopera F, Kalimo H, Moskowitz MA, Ayata C, Louvi A, Artavanis-Tsakonas S (2011) Hypomorphic Notch 3 alleles link Notch signaling to ischemic cerebral small-vessel disease. Proc Natl Acad Sci U S A 108(21):E128–E135

Artavanis-Tsakonas S, Muskavitch MA (2010) Notch: the past, the present, and the future. Curr Top Dev Biol 92:1–29

Artavanis-Tsakonas S, Matsuno K, Fortini ME (1995) Notch signaling. Science 268(5208):225–232

Ayata C (2010) CADASIL: experimental insights from animal models. Stroke 41(10 Suppl):S129–S134

Babaoglan AB, Housden BE, Furriols M, Bray SJ (2013) Deadpan contributes to the robustness of the notch response. PLoS One 8(9):e75632

Bachmann A, Knust E (1998a) Dissection of cis-regulatory elements of the Drosophila gene Serrate. Dev Genes Evol 208(6):346–351

Bachmann A, Knust E (1998b) Positive and negative control of Serrate expression during early development of the Drosophila wing. Mech Dev 76(1–2):67–78

Bailey AM, Posakony JW (1995) Suppressor of hairless directly activates transcription of enhancer of split complex genes in response to Notch receptor activity. Genes Dev 9(21):2609–2622

Bang AG, Posakony JW (1992) The Drosophila gene Hairless encodes a novel basic protein that controls alternative cell fates in adult sensory organ development. Genes Dev 6(9):1752–1769

Barad O, Hornstein E, Barkai N (2011) Robust selection of sensory organ precursors by the Notch-Delta pathway. Curr Opin Cell Biol 23(6):663–667

Barolo S, Walker RG, Polyanovsky AD, Freschi G, Keil T, Posakony JW (2000) A notch-independent activity of suppressor of hairless is required for normal mechanoreceptor physiology. Cell 103(6):957–969

Barolo S, Stone T, Bang AG, Posakony JW (2002) Default repression and Notch signaling: Hairless acts as an adaptor to recruit the corepressors Groucho and dCtBP to Suppressor of Hairless. Genes Dev 16(15):1964–1976

Baron M (2012) Endocytic routes to Notch activation. Semin Cell Dev Biol 23(4):437–442

Basmanav FB, Oprisoreanu AM, Pasternack SM, Thiele H, Fritz G, Wenzel J, Grosser L, Wehner M, Wolf S, Fagerberg C, Bygum A, Altmuller J, Rutten A, Parmentier L, El Shabrawi-Caelen L, Hafner C, Nurnberg P, Kruse R, Schoch S, Hanneken S, Betz RC (2014) Mutations in POGLUT1, encoding protein O-glucosyltransferase 1, cause autosomal-dominant Dowling-Degos disease. Am J Hum Genet 94(1):135–143

Bellen HJ, Yamamoto S (2015) Morgan's legacy: fruit flies and the functional annotation of conserved genes. Cell 163(1):12–14

Bender LB, Kooh PJ, Muskavitch MA (1993) Complex function and expression of Delta during Drosophila oogenesis. Genetics 133(4):967–978

Ben-Yaacov S, Le Borgne R, Abramson I, Schweisguth F, Schejter ED (2001) Wasp, the Drosophila Wiskott-Aldrich syndrome gene homologue, is required for cell fate decisions mediated by Notch signaling. J Cell Biol 152(1):1–13

Berdnik D, Torok T, Gonzalez-Gaitan M, Knoblich JA (2002) The endocytic protein alpha-Adaptin is required for numb-mediated asymmetric cell division in Drosophila. Dev Cell 3(2):221–231

Bergmann M, Ebke M, Yuan Y, Bruck W, Mugler M, Schwendemann G (1996) Cerebral autosomal dominant arteriopathy with subcortical infarcts and leukoencephalopathy (CADASIL): a morphological study of a German family. Acta Neuropathol 92(4):341–350

Bernard F, Krejci A, Housden B, Adryan B, Bray SJ (2010) Specificity of Notch pathway activation: twist controls the transcriptional output in adult muscle progenitors. Development 137(16):2633–2642

Berns N, Woichansky I, Friedrichsen S, Kraft N, Riechmann V (2014) A genome-scale in vivo RNAi analysis of epithelial development in Drosophila identifies new proliferation domains outside of the stem cell niche. J Cell Sci 127(Pt 12):2736–2748

Bi P, Kuang S (2015) Notch signaling as a novel regulator of metabolism. Trends Endocrinol Metab 26(5):248–255

Blair SS (2007) Wing vein patterning in Drosophila and the analysis of intercellular signaling. Annu Rev Cell Dev Biol 23:293–319

Blaumueller CM, Qi H, Zagouras P, Artavanis-Tsakonas S (1997) Intracellular cleavage of Notch leads to a heterodimeric receptor on the plasma membrane. Cell 90(2):281–291

Blochlinger K, Bodmer R, Jack J, Jan LY, Jan YN (1988) Primary structure and expression of a product from cut, a locus involved in specifying sensory organ identity in Drosophila. Nature 333(6174):629–635

Bloomington Drosophila Stock Center (2017) http://flystocks.bio.indiana.edu/

Borggrefe T, Oswald F (2014) Keeping notch target genes off: a CSL corepressor caught in the act. Structure 22(1):3–5

Borggrefe T, Oswald F (2016) Setting the stage for notch: the Drosophila Su(H)-Hairless repressor complex. PLoS Biol 14(7):e1002524

Boulianne GL, de la Concha A, Campos-Ortega JA, Jan LY, Jan YN (1991) The Drosophila neurogenic gene neuralized encodes a novel protein and is expressed in precursors of larval and adult neurons. EMBO J 10(10):2975–2983

Bourgouin C, Lundgren SE, Thomas JB (1992) Apterous is a Drosophila LIM domain gene required for the development of a subset of embryonic muscles. Neuron 9(3):549–561

Boyer-Di Ponio J, Wright-Crosnier C, Groyer-Picard MT, Driancourt C, Beau I, Hadchouel M, Meunier-Rotival M (2007) Biological function of mutant forms of JAGGED1 proteins in Alagille syndrome: inhibitory effect on Notch signaling. Hum Mol Genet 16(22):2683–2692

Brand AH, Perrimon N (1993) Targeted gene expression as a means of altering cell fates and generating dominant phenotypes. Development 118(2):401–415

Bray S (1998) Notch signalling in Drosophila: three ways to use a pathway. Semin Cell Dev Biol 9(6):591–597

Bray SJ (2006) Notch signalling: a simple pathway becomes complex. Nat Rev Mol Cell Biol 7(9):678–689

Bray S, Bernard F (2010) Notch targets and their regulation. Curr Top Dev Biol 92:253–275

Bray S, Musisi H, Bienz M (2005) Bre1 is required for Notch signaling and histone modification. Dev Cell 8(2):279–286

Bruckner K, Perez L, Clausen H, Cohen S (2000) Glycosyltransferase activity of Fringe modulates Notch-Delta interactions. Nature 406(6794): 411–415

Bulman MP, Kusumi K, Frayling TM, McKeown C, Garrett C, Lander ES, Krumlauf R, Hattersley AT, Ellard S, Turnpenny PD (2000) Mutations in the human delta homologue, DLL3, cause axial skeletal defects in spondylocostal dysostosis. Nat Genet 24(4):438–441

Busseau I, Diederich RJ, Xu T, Artavanis-Tsakonas S (1994) A member of the Notch group of interacting loci, deltex encodes a cytoplasmic basic protein. Genetics 136(2):585–596

Campuzano S, Modolell J (1992) Patterning of the Drosophila nervous system: the achaete-scute gene complex. Trends Genet 8(6):202–208 de Celis JF, Bray S (1997) Feed-back mechanisms affecting Notch activation at the dorsoventral boundary in the Drosophila wing. Development 124(17):3241–3251

de Celis JF, de Celis J, Ligoxygakis P, Preiss A, Delidakis C, Bray S (1996a) Functional relationships between Notch, Su(H) and the bHLH genes of the E(spl) complex: the E(spl) genes mediate only a subset of Notch activities during imaginal development. Development 122(9):2719–2728

de Celis JF, Garcia-Bellido A, Bray SJ (1996b) Activation and function of Notch at the dorsal-ventral boundary of the wing imaginal disc. Development 122(1):359–369

Chabriat H, Joutel A, Dichgans M, Tournier-Lasserve E, Bousser MG (2009) Cadasil. Lancet Neurol 8(7):643–653

Chao HT, Davids M, Burke E, Pappas JG, Rosenfeld JA, McCarty AJ, Davis T, Wolfe L, Toro C, Tifft C, Xia F, Stong N, Johnson TK, Warr CG, Yamamoto S, Adams DR, Markello TC, Gahl WA, Bellen HJ, Wangler MF, Malicdan MC (2017) A syndromic neurodevelopmental disorder caused by de novo variants in EBF3. Am J Hum Genet 100(1):128–137

Charng WL, Yamamoto S, Jaiswal M, Bayat V, Xiong B, Zhang K, Sandoval H, David G, Gibbs S, Lu HC, Chen K, Giagtzoglou N, Bellen HJ (2014) Drosophila Tempura, a novel protein prenyltransferase alpha subunit, regulates notch signaling via Rab1 and Rab11. PLoS Biol 12(1):e1001777

Chen BE, Kondo M, Garnier A, Watson FL, Puettmann-Holgado R, Lamar DR, Schmucker D (2006) The molecular diversity of Dscam is functionally required for neuronal wiring specificity in Drosophila. Cell 125(3):607–620

Chong JX, Buckingham KJ, Jhangiani SN, Boehm C, Sobreira N, Smith JD, Harrell TM, McMillin MJ, Wiszniewski W, Gambin T, Coban Akdemir ZH, Doheny K, Scott AF, Avramopoulos D, Chakravarti A, Hoover-Fong J, Mathews D, Witmer PD, Ling H, Hetrick K, Watkins L, Patterson KE, Reinier F, Blue E, Muzny D, Kircher M, Bilguvar K, Lopez-Giraldez F, Sutton VR, Tabor HK, Leal SM, Gunel M, Mane S, Gibbs RA, Boerwinkle E, Hamosh A, Shendure J, Lupski JR, Lifton RP, Valle D, Nickerson DA, Bamshad MJ (2015) The genetic basis of Mendelian phenotypes: discoveries, challenges, and opportunities. Am J Hum Genet 97(2):199–215

Chou TB, Perrimon N (1992) Use of a yeast site-specific recombinase to produce female germline chimeras in Drosophila. Genetics 131(3):643–653

ClinVar (2017) https://www.ncbi.nlm.nih.gov/clinvar/

Cohorts for Heart and Aging Research in Genomic Epidemiology Consortium (2017) http://www.charge-consortium.com/

Collins KJ, Yuan Z, Kovall RA (2014) Structure and function of the CSL-KyoT2 corepressor complex: a negative regulator of Notch signaling. Structure 22(1):70–81

Corder EH, Saunders AM, Strittmatter WJ, Schmechel DE, Gaskell PC, Small GW, Roses AD, Haines JL, Pericak-Vance MA (1993) Gene dose of apolipoprotein E type 4 allele and the risk of Alzheimer's disease in late onset families. Science 261(5123):921–923

Cornell M, Evans DA, Mann R, Fostier M, Flasza M, Monthatong M, Artavanis-Tsakonas S, Baron M (1999) The Drosophila melanogaster Suppressor of deltex gene, a regulator of the Notch receptor signaling pathway, is an E3 class ubiquitin ligase. Genetics 152(2):567–576

Corson F, Couturier L, Rouault H, Mazouni K, Schweisguth F (2017) Self-organized Notch dynamics generate stereotyped sensory organ patterns in Drosophila. Science 356(6337):501

Couturier L, Schweisguth F (2014) Antibody uptake assay and in vivo imaging to study intracellular trafficking of Notch and Delta in Drosophila. Methods Mol Biol 1187:79–86

Cuyvers E, Sleegers K (2016) Genetic variations underlying Alzheimer's disease: evidence from genome-wide association studies and beyond. Lancet Neurol 15(8):857–868

Da Ros VG, Gutierrez-Perez I, Ferres-Marco D, Dominguez M (2013) Dampening the signals transduced through hedgehog via microRNA miR-7 facilitates notch-induced tumourigenesis. PLoS Biol 11(5):e1001554

D'Amato G, Luxan G, de la Pompa JL (2016) Notch signalling in ventricular chamber development and cardiomyopathy. FEBS J 283(23):4223–4237

Daniels RW, Rossano AJ, Macleod GT, Ganetzky B (2014) Expression of multiple transgenes from a single construct using viral 2A peptides in Drosophila. PLoS One 9(6):e100637

Database of Genomic Variants (2017) http://dgv.tcag.ca/dgv/app/home/

DatabasE of genomiC varIation and Phenotype in Humans using Ensembl Resources (2017) https://decipher.sanger.ac.uk/

David-Morrison G, Xu Z, Rui YN, Charng WL, Jaiswal M, Yamamoto S, Xiong B, Zhang K, Sandoval H, Duraine L, Zuo Z, Zhang S, Bellen HJ (2016) WAC regulates mTOR activity by acting as an adaptor for the TTT and Pontin/Reptin complexes. Dev Cell 36(2):139–151

Dehal P, Boore JL (2005) Two rounds of whole genome duplication in the ancestral vertebrate. PLoS Biol 3(10):e314

Delidakis C, Artavanis-Tsakonas S (1992) The Enhancer of split [E(spl)] locus of Drosophila encodes seven independent helix-loop-helix proteins. Proc Natl Acad Sci U S A 89(18):8731–8735

Delidakis C, Preiss A, Hartley DA, Artavanis-Tsakonas S (1991) Two genetically and molecularly distinct functions involved in early neurogenesis reside within the Enhancer of split locus of Drosophila melanogaster. Genetics 129(3):803–823

Delidakis C, Monastirioti M, Magadi SS (2014) E(spl): genetic, developmental, and evolutionary aspects of a group of invertebrate Hes proteins with close ties to Notch signaling. Curr Top Dev Biol 110:217–262

Dell'Angelica EC (2009) AP-3-dependent trafficking and disease: the first decade. Curr Opin Cell Biol 21(4):552–559

Developmental Studies Hybridoma Bank (2017) http://dshb.biology.uiowa.edu/

Dexter JS (1914) The analysis of a case of continuous variation in Drosophila by a study of its linkage relations. Am Nat 48:712–758

Diao F, White BH (2012) A novel approach for directing transgene expression in Drosophila: T2A-Gal4 in-frame fusion. Genetics 190(3):1139–1144

Diao F, Ironfield H, Luan H, Diao F, Shropshire WC, Ewer J, Marr E, Potter CJ, Landgraf M, White BH (2015) Plug-and-play genetic access to drosophila cell types using exchangeable exon cassettes. Cell Rep 10(8):1410–1421

Dichgans M, Ludwig H, Muller-Hocker J, Messerschmidt A, Gasser T (2000) Small in-frame deletions and missense mutations in CADASIL: 3D models predict misfolding of Notch3 EGF-like repeat domains. Eur J Hum Genet 8(4):280–285

Digilio MC, Luca AD, Lepri F, Guida V, Ferese R, Dentici ML, Angioni A, Marino B, Dallapiccola B (2013) JAG1 mutation in a patient with deletion 22q11.2 syndrome and tetralogy of Fallot. Am J Med Genet A 161A(12):3133–3136

Djiane A, Krejci A, Bernard F, Fexova S, Millen K, Bray SJ (2013) Dissecting the mechanisms of Notch induced hyperplasia. EMBO J 32(1):60–71

Doherty D, Feger G, Younger-Shepherd S, Jan LY, Jan YN (1996) Delta is a ventral to dorsal signal complementary to Serrate, another Notch ligand, in Drosophila wing formation. Genes Dev 10(4):421–434

Domanitskaya E, Schupbach T (2012) CoREST acts as a positive regulator of Notch signaling in the follicle cells of Drosophila melanogaster. J Cell Sci 125(Pt 2):399–410

Domenga V, Fardoux P, Lacombe P, Monet M, Maciazek J, Krebs LT, Klonjkowski B, Berrou E, Mericskay M, Li Z, Tournier-Lasserve E, Gridley T, Joutel A (2004) Notch3 is required for arterial identity and maturation of vascular smooth muscle cells. Genes Dev 18(22):2730–2735

Donahue CP, Kosik KS (2004) Distribution pattern of Notch3 mutations suggests a gain-of-function mechanism for CADASIL. Genomics 83(1):59–65

Dornier E, Coumailleau F, Ottavi JF, Moretti J, Boucheix C, Mauduit P, Schweisguth F, Rubinstein E (2012) TspanC8 tetraspanins regulate ADAM10/Kuzbanian trafficking and promote Notch activation in flies and mammals. J Cell Biol 199(3):481–496

Drosophila Genomics and Genetic Resources (2017) http://www.dgrc.kit.ac.jp/

Drosophila Genomics Resource Center (2017) https://dgrc.bio.indiana.edu/

Drosophila RNAi Screening Center Integrative Ortholog Prediction Tool (2017) http://www.flyrnai.org/cgi-bin/DRSC_orthologs.pl

D'Souza B, Meloty-Kapella L, Weinmaster G (2010) Canonical and non-canonical Notch ligands. Curr Top Dev Biol 92:73–129

Emery G, Hutterer A, Berdnik D, Mayer B, Wirtz-Peitz F, Gaitan MG, Knoblich JA (2005) Asymmetric Rab 11 endosomes regulate delta recycling and specify cell fate in the Drosophila nervous system. Cell 122(5):763–773

Eun SH, Lea K, Overstreet E, Stevens S, Lee JH, Fischer JA (2007) Identification of genes that interact with Drosophila liquid facets. Genetics 175(3):1163–1174

Exome Aggregation Consortium Browser (2017) http://exac.broadinstitute.org/

Ezratty EJ, Stokes N, Chai S, Shah AS, Williams SE, Fuchs E (2011) A role for the primary cilium in Notch signaling and epidermal differentiation during skin development. Cell 145(7):1129–1141

Fehon RG, Johansen K, Rebay I, Artavanis-Tsakonas S (1991) Complex cellular and subcellular regulation of notch expression during embryonic and imaginal development of Drosophila: implications for notch function. J Cell Biol 113(3):657–669

Fleming RJ, Scottgale TN, Diederich RJ, Artavanis-Tsakonas S (1990) The gene Serrate encodes a putative EGF-like transmembrane protein essential for proper ectodermal development in Drosophila melanogaster. Genes Dev 4(12A):2188–2201

Fleming RJ, Gu Y, Hukriede NA (1997) Serrate-mediated activation of Notch is specifically blocked by the product of the gene fringe in the dorsal compartment of the Drosophila wing imaginal disc. Development 124(15):2973–2981

Flood DG, Reaume AG, Dorfman KS, Lin YG, Lang DM, Trusko SP, Savage MJ, Annaert WG, De Strooper B, Siman R, Scott RW (2002) FAD mutant PS-1 gene-targeted mice: increased A beta 42 and A beta deposition without APP overproduction. Neurobiol Aging 23(3):335–348

FlyBase (2017) http://flybase.org/

FlyBase-AP3 (2017) http://flybase.org/reports/FBgg0000158.html

Flybase-AP3 (2017) http://flybase.org/reports/FBgg0000136.html

FlyBase-ESCRT (2017) http://flybase.org/reports/FBgg0000215.html

Flybase-HOPS (2017) http://flybase.org/reports/FBgg0000106.html

Flybase-VATPase (2017) http://flybase.org/reports/FBgg0000111.html

Fortini ME (2009) Notch signaling: the core pathway and its posttranslational regulation. Dev Cell 16(5):633–647

Fortini ME, Artavanis-Tsakonas S (1994) The suppressor of hairless protein participates in notch receptor signaling. Cell 79(2):273–282

Francis R, McGrath G, Zhang J, Ruddy DA, Sym M, Apfeld J, Nicoll M, Maxwell M, Hai B, Ellis MC, Parks AL, Xu W, Li J, Gurney M, Myers RL, Himes CS, Hiebsch R, Ruble C, Nye JS, Curtis D (2002) aph-1 and pen-2 are required for Notch pathway signaling, gamma-secretase cleavage of betaAPP, and presenilin protein accumulation. Dev Cell 3(1):85–97

Furman DP, Bukharina TA (2008) How Drosophila melanogaster forms its mechanoreceptors. Curr Genomics 9(5):312–323

Furukawa T, Kawaichi M, Matsunami N, Ryo H, Nishida Y, Honjo T (1991) The Drosophila RBP-J kappa gene encodes the binding protein for the immunoglobulin J kappa recombination signal sequence. J Biol Chem 266(34):23334–23340

Furukawa T, Maruyama S, Kawaichi M, Honjo T (1992) The Drosophila homolog of the immunoglobulin recombination signal-binding protein regulates peripheral nervous system development. Cell 69(7):1191–1197

Fuwa TJ, Hori K, Sasamura T, Higgs J, Baron M, Matsuno K (2006) The first deltex null mutant indicates tissue-specific deltex-dependent Notch signaling in Drosophila. Mol Gen Genomics 275(3):251–263

Gallagher CM, Knoblich JA (2006) The conserved c2 domain protein lethal (2) giant discs regulates protein trafficking in Drosophila. Dev Cell 11(5):641–653

Garcia-Bellido A, Santamaria P (1978) Developmental analysis of the achaete-scute system of Drosophila melanogaster. Genetics 88(3):469–486

Garg V (2006) Molecular genetics of aortic valve disease. Curr Opin Cardiol 21(3):180–184

Garg V, Muth AN, Ransom JF, Schluterman MK, Barnes R, King IN, Grossfeld PD, Srivastava D (2005) Mutations in NOTCH1 cause aortic valve disease. Nature 437(7056):270–274

Geisler F, Strazzabosco M (2015) Emerging roles of Notch signaling in liver disease. Hepatology 61(1):382–392

Geno2MP (2017) Genotype to Mendelian phenotype browser. http://geno2mp.gs.washington.edu/Geno2MP/

Gho M, Bellaiche Y, Schweisguth F (1999) Revisiting the Drosophila microchaete lineage: a novel intrinsically asymmetric cell division generates a glial cell. Development 126(16):3573–3584

Giagtzoglou N, Yamamoto S, Zitserman D, Graves HK, Schulze KL, Wang H, Klein H, Roegiers F, Bellen HJ (2012) dEHBP1 controls exocytosis and recycling of Delta during asymmetric divisions. J Cell Biol 196(1):65–83

Giagtzoglou N, Li T, Yamamoto S, Bellen HJ (2013) Drosophila EHBP1 regulates Scabrous secretion during Notch-mediated lateral inhibition. J Cell Sci 126(Pt 16):3686–3696

Gianni D, Li A, Tesco G, McKay KM, Moore J, Raygor K, Rota M, Gwathmey JK, Dec GW, Aretz T, Leri A, Semigran MJ, Anversa P, Macgillivray TE, Tanzi RE, del Monte F (2010) Protein aggregates and novel presenilin gene variants in idiopathic dilated cardiomyopathy. Circulation 121(10):1216–1226

Gianni D, Li A, Tesco G, McKay KM, Moore J, Raygor K, Rota M, Gwathmey JK, Dec GW, Aretz T, Leri A, Semigran MJ, Anversa P, Macgillivray TE, Tanzi RE, del Monte F (2011) Protein aggregates and novel presenilin gene variants in idiopathic dilated cardiomyopathy. Circulation 121(10):1216–1226

Gigliani F, Longo F, Gaddini L, Battaglia PA (1996) Interactions among the bHLH domains of the proteins encoded by the Enhancer of split and achaete-scute gene complexes of Drosophila. Mol Gen Genet 251(6):628–634

Gnerer JP, Venken KJ, Dierick HA (2015) Gene-specific cell labeling using MiMIC transposons. Nucleic Acids Res 43(8):e56

Go MJ, Artavanis-Tsakonas S (1998) A genetic screen for novel components of the notch signaling pathway during Drosophila bristle development. Genetics 150(1):211–220

Gomez-Lamarca MJ, Snowdon LA, Seib E, Klein T, Bray SJ (2015) Rme-8 depletion perturbs Notch recycling and predisposes to pathogenic signaling. J Cell Biol 210(2):303–318

Gonzaga-Jauregui C, Lupski JR, Gibbs RA (2012) Human genome sequencing in health and disease. Annu Rev Med 63:35–61

Goode S, Melnick M, Chou TB, Perrimon N (1996) The neurogenic genes egghead and brainiac define a novel signaling pathway essential for epithelial morphogenesis during Drosophila oogenesis. Development 122(12):3863–3879

Goodfellow H, Krejci A, Moshkin Y, Verrijzer CP, Karch F, Bray SJ (2007) Gene-specific targeting of the histone chaperone asf1 to mediate silencing. Dev Cell 13(4):593–600

Gordon WR, Vardar-Ulu D, Histen G, Sanchez-Irizarry C, Aster JC, Blacklow SC (2007) Structural basis for autoinhibition of Notch. Nat Struct Mol Biol 14(4):295–300

Gordon WR, Vardar-Ulu D, L'Heureux S, Ashworth T, Malecki MJ, Sanchez-Irizarry C, McArthur DG, Histen G, Mitchell JL, Aster JC, Blacklow SC (2009a) Effects of S1 cleavage on the structure, surface export, and signaling activity of human Notch1 and Notch2. PLoS One 4(8):e6613

Gordon WR, Roy M, Vardar-Ulu D, Garfinkel M, Mansour MR, Aster JC, Blacklow SC (2009b) Structure of the Notch1-negative regulatory region: implications for normal activation and pathogenic signaling in T-ALL. Blood 113(18):4381–4390

Goto S, Taniguchi M, Muraoka M, Toyoda H, Sado Y, Kawakita M, Hayashi S (2001) UDP-sugar transporter implicated in glycosylation and processing of Notch. Nat Cell Biol 3(9):816–822

Gratz SJ, Rubinstein CD, Harrison MM, Wildonger J, O'Connor-Giles KM (2015) CRISPR-Cas9 Genome editing in Drosophila. Curr Protoc Mol Biol 111:31 32 31–20

Green RC, Goddard KA, Jarvik GP, Amendola LM, Appelbaum PS, Berg JS, Bernhardt BA, Biesecker LG, Biswas S, Blout CL, Bowling KM, Brothers KB, Burke W, Caga-Anan CF, Chinnaiyan AM, Chung WK, Clayton EW, Cooper GM, East K, Evans JP, Fullerton SM, Garraway LA, Garrett JR, Gray SW, Henderson

GE, Hindorff LA, Holm IA, Lewis MH, Hutter CM, Janne PA, Joffe S, Kaufman D, Knoppers BM, Koenig BA, Krantz ID, Manolio TA, McCullough L, McEwen J, McGuire A, Muzny D, Myers RM, Nickerson DA, Ou J, Parsons DW, Petersen GM, Plon SE, Rehm HL, Roberts JS, Robinson D, Salama JS, Scollon S, Sharp RR, Shirts B, Spinner NB, Tabor HK, Tarczy-Hornoch P, Veenstra DL, Wagle N, Weck K, Wilfond BS, Wilhelmsen K, Wolf SM, Wynn J, Yu JH (2016) Clinical sequencing exploratory research consortium: accelerating evidence-based practice of genomic medicine. Am J Hum Genet 98(6):1051–1066

Gridley T (1997) Notch signaling in vertebrate development and disease. Mol Cell Neurosci 9(2):103–108

Gridley T (2003) Notch signaling and inherited disease syndromes. Hum Mol Genet 12(1):R9–R13

Gridley T (2010) Notch signaling in the vasculature. Curr Top Dev Biol 92:277–309

Gripp KW, Robbins KM, Sobreira NL, Witmer PD, Bird LM, Avela K, Makitie O, Alves D, Hogue JS, Zackai EH, Doheny KF, Stabley DL, Sol-Church K (2015) Truncating mutations in the last exon of NOTCH3 cause lateral meningocele syndrome. Am J Med Genet A 167A(2):271–281

Guida V, Chiappe F, Ferese R, Usala G, Maestrale G, Iannascoli C, Bellacchio E, Mingarelli R, Digilio MC, Marino B, Uda M, De Luca A, Dallapiccola B (2011) Novel and recurrent JAG1 mutations in patients with tetralogy of Fallot. Clin Genet 80(6):591–594

Guo M, Jan LY, Jan YN (1996) Control of daughter cell fates during asymmetric division: interaction of Numb and Notch. Neuron 17(1):27–41

Guo Y, Livne-Bar I, Zhou L, Boulianne GL (1999a) Drosophila presenilin is required for neuronal differentiation and affects notch subcellular localization and signaling. J Neurosci 19(19):8435–8442

Guo Q, Fu W, Sopher BL, Miller MW, Ware CB, Martin GM, Mattson MP (1999b) Increased vulnerability of hippocampal neurons to excitotoxic necrosis in presenilin-1 mutant knock-in mice. Nat Med 5(1):101–106

Guruharsha KG, Rual JF, Zhai B, Mintseris J, Vaidya P, Vaidya N, Beekman C, Wong C, Rhee DY, Cenaj O, McKillip E, Shah S, Stapleton M, Wan KH, Yu C, Parsa B, Carlson JW, Chen X, Kapadia B, Vijayraghavan K, Gygi SP, Celniker SE, Obar RA, Artavanis-Tsakonas S (2011) A protein complex network of Drosophila melanogaster. Cell 147(3):690–703

Guruharsha KG, Kankel MW, Artavanis-Tsakonas S (2012) The Notch signalling system: recent insights into the complexity of a conserved pathway. Nat Rev Genet 13(9):654–666

Guruharsha KG, Hori K, Obar RA, Artavanis-Tsakonas S (2014) Proteomic analysis of the Notch interactome. Methods Mol Biol 1187:181–192

Haberman AS, Akbar MA, Ray S, Kramer H (2010) Drosophila acinus encodes a novel regulator of endocytic and autophagic trafficking. Development 137(13):2157–2166

Hall LE, Alexander SJ, Chang M, Woodling NS, Yedvobnick B (2004) An EP overexpression screen for genetic modifiers of Notch pathway function in Drosophila melanogaster. Genet Res 83(2):71–82

Haritunians T, Chow T, De Lange RP, Nichols JT, Ghavimi D, Dorrani N, St Clair DM, Weinmaster G, Schanen C (2005) Functional analysis of a recurrent missense mutation in Notch3 in CADASIL. J Neurol Neurosurg Psychiatry 76(9):1242–1248

Hartenstein V, Posakony JW (1989) Development of adult sensilla on the wing and notum of Drosophila melanogaster. Development 107(2):389–405

Hartenstein V, Posakony JW (1990) A dual function of the Notch gene in Drosophila sensillum development. Dev Biol 142(1):13–30

Hartley DA, Xu TA, Artavanis-Tsakonas S (1987) The embryonic expression of the Notch locus of Drosophila melanogaster and the implications of point mutations in the extracellular EGF-like domain of the predicted protein. EMBO J 6(11):3407–3417

Hartley DA, Preiss A, Artavanis-Tsakonas S (1988) A deduced gene product from the Drosophila neurogenic locus, enhancer of split, shows homology to mammalian G-protein beta subunit. Cell 55(5):785–795

Hassed SJ, Wiley GB, Wang S, Lee JY, Li S, Xu W, Zhao ZJ, Mulvihill JJ, Robertson J, Warner J, Gaffney PM (2012) RBPJ mutations identified in two families affected by Adams-Oliver syndrome. Am J Hum Genet 91(2):391–395

Hassed S, Li S, Mulvihill J, Aston C, Palmer S (2017) Adams-Oliver syndrome review of the literature: Refining the diagnostic phenotype. Am J Med Genet A 173(3):790–800

HCOP (Human Genome Organization Gene Nomenclature Committee Comparison of Orthology Predictions Search) (2017) http://www.genenames.org/cgi-bin/hcop

Heitzler P (2010) Biodiversity and noncanonical Notch signaling. Curr Top Dev Biol 92:457–481

Heitzler P, Bourouis M, Ruel L, Carteret C, Simpson P (1996) Genes of the Enhancer of split and achaete-scute complexes are required for a regulatory loop between Notch and Delta during lateral signalling in Drosophila. Development 122(1):161–171

Herz HM, Woodfield SE, Chen Z, Bolduc C, Bergmann A (2009) Common and distinct genetic properties of ESCRT-II components in Drosophila. PLoS One 4(1):e4165

Heuss SF, Ndiaye-Lobry D, Six EM, Israel A, Logeat F (2008) The intracellular region of Notch ligands Dll1 and Dll3 regulates their trafficking and signaling activity. Proc Natl Acad Sci U S A 105(32):11212–11217

HGNC-Arp2/3 (2017) http://www.genenames.org/cgi-bin/genefamilies/set/39

HGNC-ESCRT (2017) http://www.genenames.org/cgi-bin/genefamilies/set/1111

HGNC-VATPase (2017) http://www.genenames.org/cgi-bin/genefamilies/set/415

Hing HK, Bangalore L, Sun X, Artavanis-Tsakonas S (1999) Mutations in the heatshock cognate 70 protein (hsc4) modulate Notch signaling. Eur J Cell Biol 78(10):690–697

Hori K, Sen A, Kirchhausen T, Artavanis-Tsakonas S (2011) Synergy between the ESCRT-III complex and Deltex defines a ligand-independent Notch signal. J Cell Biol 195(6):1005–1015

Housden BE, Fu AQ, Krejci A, Bernard F, Fischer B, Tavare S, Russell S, Bray SJ (2013) Transcriptional dynamics elicited by a short pulse of notch activation involves feed-forward regulation by E(spl)/Hes genes. PLoS Genet 9(1):e1003162

Housden BE, Li J, Bray SJ (2014) Visualizing Notch signaling in vivo in Drosophila tissues. Methods Mol Biol 1187:101–113

Huang C, Chan JA, Schuurmans C (2014) Proneural bHLH genes in development and disease. Curr Top Dev Biol 110:75–127

Hurlbut GD, Kankel MW, Lake RJ, Artavanis-Tsakonas S (2007) Crossing paths with Notch in the hyper-network. Curr Opin Cell Biol 19(2):166–175

Hutterer A, Knoblich JA (2005) Numb and alpha-Adaptin regulate Sanpodo endocytosis to specify cell fate in Drosophila external sensory organs. EMBO Rep 6(9):836–842

Ilagan MX, Kopan R (2007) SnapShot: notch signaling pathway. Cell 128(6):1246

Irvine KD, Wieschaus E (1994) fringe, a Boundary-specific signaling molecule, mediates interactions between dorsal and ventral cells during Drosophila wing development. Cell 79(4):595–606

Ishiko A, Shimizu A, Nagata E, Takahashi K, Tabira T, Suzuki N (2006) Notch3 ectodomain is a major component of granular osmiophilic material (GOM) in CADASIL. Acta Neuropathol 112(3): 333–339

Isidor B, Lindenbaum P, Pichon O, Bezieau S, Dina C, Jacquemont S, Martin-Coignard D, Thauvin-Robinet C, Le Merrer M, Mandel JL, David A, Faivre L, Cormier-Daire V, Redon R, Le Caignec C (2011) Truncating mutations in the last exon of NOTCH2 cause a rare skeletal disorder with osteoporosis. Nat Genet 43(4):306–308

Jack J, DeLotto Y (1992) Effect of wing scalloping mutations on cut expression and sense organ differentiation in the Drosophila wing margin. Genetics 131(2):353–363

Jafar-Nejad H, Acar M, Nolo R, Lacin H, Pan H, Parkhurst SM, Bellen HJ (2003) Senseless acts as a binary switch during sensory organ precursor selection. Genes Dev 17(23):2966–2978

Jafar-Nejad H, Andrews HK, Acar M, Bayat V, Wirtz-Peitz F, Mehta SQ, Knoblich JA, Bellen HJ (2005) Sec15, a component of the exocyst, promotes notch signaling during the asymmetric division of Drosophila sensory organ precursors. Dev Cell 9(3):351–363

Jafar-Nejad H, Leonardi J, Fernandez-Valdivia R (2010) Role of glycans and glycosyltransferases in the regulation of Notch signaling. Glycobiology 20(8):931–949

Jakobsdottir J, van der Lee SJ, Bis JC, Chouraki V, Li-Kroeger D, Yamamoto S, Grove ML, Naj A, Vronskaya M, Salazar JL, DeStefano AL, Brody JA, Smith AV, Amin N, Sims R, Ibrahim-Verbaas CA, Choi SH, Satizabal CL, Lopez OL, Beiser A, Ikram MA, Garcia ME, Hayward C, Varga TV, Ripatti S, Franks PW, Hallmans G, Rolandsson O, Jansson JH, Porteous DJ, Salomaa V, Eiriksdottir G, Rice KM, Bellen HJ, Levy D, Uitterlinden AG, Emilsson V, Rotter JI, Aspelund T, O'Donnell CJ, Fitzpatrick AL, Launer LJ, Hofman A, Wang LS, Williams J, Schellenberg GD, Boerwinkle E, Psaty BM, Seshadri S, Shulman JM, Gudnason V, van Duijn CM (2016) Rare Functional Variant in TM2D3 is Associated with Late-Onset Alzheimer's Disease. PLoS Genet 12(10):e1006327

Jan YN, Jan LY (2001) Asymmetric cell division in the Drosophila nervous system. Nat Rev Neurosci 2(11):772–779

Jana SC, Bettencourt-Dias M, Durand B, Megraw TL (2016) Drosophila melanogaster as a model for basal body research. Cilia 5:22

Jennings B, Preiss A, Delidakis C, Bray S (1994) The Notch signalling pathway is required for Enhancer of split bHLH protein expression during neurogenesis in the Drosophila embryo. Development 120(12):3537–3548

Jia D, Soylemez M, Calvin G, Bornmann R, Bryant J, Hanna C, Huang YC, Deng WM (2015) A large-scale in vivo RNAi screen to identify genes involved in Notch-mediated follicle cell differentiation and cell cycle switches. Sci Rep 5:12328

Johansen KM, Fehon RG, Artavanis-Tsakonas S (1989) The notch gene product is a glycoprotein expressed on the cell surface of both epidermal and neuronal precursor cells during Drosophila development. J Cell Biol 109(5):2427–2440

Johnson EA (2011) HIF takes it up a notch. Sci Signal 4(181):pe33

Joutel A (2011) Pathogenesis of CADASIL: transgenic and knock-out mice to probe function and dysfunction of the mutated gene, Notch3, in the cerebrovasculature. Bioessays 33(1):73–80

Joutel A, Corpechot C, Ducros A, Vahedi K, Chabriat H, Mouton P, Alamowitch S, Domenga V, Cecillion M, Marechal E, Maciazek J, Vayssiere C, Cruaud C, Cabanis EA, Ruchoux MM, Weissenbach J, Bach JF, Bousser MG, Tournier-Lasserve E (1996) Notch3 mutations in CADASIL, a hereditary adult-onset condition causing stroke and dementia. Nature 383(6602):707–710

Joutel A, Vahedi K, Corpechot C, Troesch A, Chabriat H, Vayssiere C, Cruaud C, Maciazek J, Weissenbach J, Bousser MG, Bach JF, Tournier-Lasserve E (1997) Strong clustering and stereotyped nature of Notch3 mutations in CADASIL patients. Lancet 350(9090):1511–1515

Joutel A, Andreux F, Gaulis S, Domenga V, Cecillon M, Battail N, Piga N, Chapon F, Godfrain C, Tournier-Lasserve E (2000) The ectodomain of the Notch3 receptor accumulates within the cerebrovasculature of CADASIL patients. J Clin Invest 105(5):597–605

Joutel A, Favrole P, Labauge P, Chabriat H, Lescoat C, Andreux F, Domenga V, Cecillon M, Vahedi K, Ducros A, Cave-Riant F, Bousser MG, Tournier-Lasserve E (2001) Skin biopsy immunostaining with a Notch3 monoclonal antibody for CADASIL diagnosis. Lancet 358(9298):2049–2051

Joutel A, Monet M, Domenga V, Riant F, Tournier-Lasserve E (2004) Pathogenic mutations associated with cerebral autosomal dominant arteriopathy with subcortical infarcts and leukoencephalopathy differently affect Jagged1 binding and Notch3 activity via the RBP/JK signaling Pathway. Am J Hum Genet 74(2):338–347

Kageyama R, Niwa Y, Shimojo H, Kobayashi T, Ohtsuka T (2010) Ultradian oscillations in Notch signaling regulate dynamic biological events. Curr Top Dev Biol 92:311–331

Kakuda S, Haltiwanger RS (2017) Deciphering the fringe-mediated Notch code: identification of activating and inhibiting sites allowing discrimination between ligands. Dev Cell 40(2):193–201

Karaca E, Yuregir OO, Bozdogan ST, Aslan H, Pehlivan D, Jhangiani SN, Akdemir ZC, Gambin T, Bayram Y, Atik MM, Erdin S, Muzny D, Gibbs RA, Lupski JR (2015) Rare variants in the notch signaling pathway describe a novel type of autosomal recessive Klippel-Feil syndrome. Am J Med Genet A 167A(11):2795–2799

Kasahara M (2007) The 2R hypothesis: an update. Curr Opin Immunol 19(5):547–552

Katsonis P, Koire A, Wilson SJ, Hsu TK, Lua RC, Wilkins AD, Lichtarge O (2014) Single nucleotide variations: biological impact and theoretical interpretation. Protein Sci 23(12):1650–1666

Kidd S, Lieber T (2002) Furin cleavage is not a requirement for Drosophila Notch function. Mech Dev 115(1–2):41–51

Kidd S, Kelley MR, Young MW (1986) Sequence of the notch locus of Drosophila melanogaster: relationship of the encoded protein to mammalian clotting and growth factors. Mol Cell Biol 6(9):3094–3108

Kidd S, Baylies MK, Gasic GP, Young MW (1989) Structure and distribution of the Notch protein in developing Drosophila. Genes Dev 3(8):1113–1129

Kidd S, Struhl G, Lieber T (2015) Notch is required in adult Drosophila sensory neurons for morphological and functional plasticity of the olfactory circuit. PLoS Genet 11(5):e1005244

Kim J, Irvine KD, Carroll SB (1995) Cell recognition, signal induction, and symmetrical gene activation at the dorsal-ventral boundary of the developing Drosophila wing. Cell 82(5):795–802

Kitagawa M, Oyama T, Kawashima T, Yedvobnick B, Kumar A, Matsuno K, Harigaya K (2001) A human protein with sequence similarity to Drosophila mastermind coordinates the nuclear form of notch and a CSL protein to build a transcriptional activator complex on target promoters. Mol Cell Biol 21(13):4337–4346

Klambt C, Knust E, Tietze K, Campos-Ortega JA (1989) Closely related transcripts encoded by the neurogenic gene complex enhancer of split of Drosophila melanogaster. EMBO J 8(1):203–210

Klein T, Arias AM (1998) Interactions among Delta, Serrate and Fringe modulate Notch activity during Drosophila wing development. Development 125(15):2951–2962

Klein T, Couso JP, Martinez Arias A (1998) Wing development and specification of dorsal cell fates in the absence of apterous in Drosophila. Curr Biol 8(7):417–420

Knust E, Schrons H, Grawe F, Campos-Ortega JA (1992) Seven genes of the Enhancer of split complex of Drosophila melanogaster encode helix-loop-helix proteins. Genetics 132(2):505–518

Kobayashi T, Kageyama R (2014) Expression dynamics and functions of Hes factors in development and diseases. Curr Top Dev Biol 110:263–283

Koch U, Radtke F (2010) Notch signaling in solid tumors. Curr Top Dev Biol 92:411–455

Koch U, Lehal R, Radtke F (2013) Stem cells living with a Notch. Development 140(4):689–704

Kokubo H, Lun Y, Johnson RL (1999) Identification and expression of a novel family of bHLH cDNAs related to Drosophila hairy and enhancer of split. Biochem Biophys Res Commun 260(2):459–465

Kola S, Koneti NR, Golla JP, Akka J, Gundimeda SD, Mundluru HP (2011) Mutational analysis of JAG1 gene in non-syndromic tetralogy of Fallot children. Clin Chim Acta 412(23-24):2232–2236

Kono M, Sugiura K, Suganuma M, Hayashi M, Takama H, Suzuki T, Matsunaga K, Tomita Y, Akiyama M (2013) Whole-exome sequencing identifies ADAM10 mutations as a cause of reticulate acropigmentation of Kitamura, a clinical entity distinct from Dowling-Degos disease. Hum Mol Genet 22(17):3524–3533

Koo BK, Yoon KJ, Yoo KW, Lim HS, Song R, So JH, Kim CH, Kong YY (2005) Mind bomb-2 is an E3 ligase for Notch ligand. J Biol Chem 280(23):22335–22342

Koo BK, Yoon MJ, Yoon KJ, Im SK, Kim YY, Kim CH, Suh PG, Jan YN, Kong YY (2007) An obligatory role of mind bomb-1 in notch signaling of mammalian development. PLoS One 2(11):e1221

Kooh PJ, Fehon RG, Muskavitch MA (1993) Implications of dynamic patterns of Delta and Notch expression for cellular interactions during Drosophila development. Development 117(2):493–507

Kopan R, Ilagan MX (2009) The canonical Notch signaling pathway: unfolding the activation mechanism. Cell 137(2):216–233

Kopczynski CC, Muskavitch MA (1989) Complex spatio-temporal accumulation of alternative transcripts from the neurogenic gene Delta during Drosophila embryogenesis. Development 107(3):623–636

Kopczynski CC, Alton AK, Fechtel K, Kooh PJ, Muskavitch MA (1988) Delta, a Drosophila neurogenic gene, is transcriptionally complex and encodes a protein related to blood coagulation factors and epidermal growth factor of vertebrates. Genes Dev 2(12b):1723–1735

Koutelou E, Sato S, Tomomori-Sato C, Florens L, Swanson SK, Washburn MP, Kokkinaki M, Conaway RC, Conaway JW, Moschonas NK (2008) Neuralized-like 1 (Neurl1) targeted to the plasma membrane by N-myristoylation regulates the Notch ligand Jagged1. J Biol Chem 283(7):3846–3853

Kovall RA, Blacklow SC (2010) Mechanistic insights into Notch receptor signaling from structural and biochemical studies. Curr Top Dev Biol 92:31–71

Krantz ID, Smith R, Colliton RP, Tinkel H, Zackai EH, Piccoli DA, Goldmuntz E, Spinner NB (1999) Jagged1 mutations in patients ascertained with isolated congenital heart defects. Am J Med Genet 84(1):56–60

Krejci A, Bernard F, Housden BE, Collins S, Bray SJ (2009) Direct response to Notch activation: signaling crosstalk and incoherent logic. Sci Signal 2(55):ra1

Kuroda K, Han H, Tani S, Tanigaki K, Tun T, Furukawa T, Taniguchi Y, Kurooka H, Hamada Y, Toyokuni S, Honjo T (2003) Regulation of marginal zone B cell development by MINT, a suppressor of Notch/RBP-J signaling pathway. Immunity 18(2):301–312

Ladi E, Nichols JT, Ge W, Miyamoto A, Yao C, Yang LT, Boulter J, Sun YE, Kintner C, Weinmaster G (2005) The divergent DSL ligand Dll3 does not activate Notch signaling but cell autonomously attenuates signaling induced by other DSL ligands. J Cell Biol 170(6):983–992

Lai EC, Orgogozo V (2004) A hidden program in Drosophila peripheral neurogenesis revealed: fundamental principles underlying sensory organ diversity. Dev Biol 269(1):1–17

Lai EC, Bodner R, Posakony JW (2000) The enhancer of split complex of Drosophila includes four Notch-regulated members of the bearded gene family. Development 127(16):3441–3455

Lai EC, Deblandre GA, Kintner C, Rubin GM (2001) Drosophila neuralized is a ubiquitin ligase that promotes the internalization and degradation of delta. Dev Cell 1(6):783–794

Lai EC, Roegiers F, Qin X, Jan YN, Rubin GM (2005) The ubiquitin ligase Drosophila Mind bomb promotes Notch signaling by regulating the localization and activity of Serrate and Delta. Development 132(10):2319–2332

Lake RJ, Grimm LM, Veraksa A, Banos A, Artavanis-Tsakonas S (2009) In vivo analysis of the Notch receptor S1 cleavage. PLoS One 4(8):e6728

Langevin J, Le Borgne R, Rosenfeld F, Gho M, Schweisguth F, Bellaiche Y (2005) Lethal giant larvae controls the localization of notch-signaling regulators numb, neuralized, and Sanpodo in Drosophila sensory-organ precursor cells. Curr Biol 15(10):955–962

Le Borgne R, Schweisguth F (2003) Unequal segregation of Neuralized biases Notch activation during asymmetric cell division. Dev Cell 5(1):139–148

Le Borgne R, Remaud S, Hamel S, Schweisguth F (2005) Two distinct E3 ubiquitin ligases have complementary functions in the regulation of delta and serrate signaling in Drosophila. PLoS Biol 3(4):e96

Le Bras S, Loyer N, Le Borgne R (2011) The multiple facets of ubiquitination in the regulation of notch signaling pathway. Traffic 12(2):149–161

Lecourtois M, Schweisguth F (1995) The neurogenic suppressor of hairless DNA-binding protein mediates the transcriptional activation of the enhancer of split complex genes triggered by Notch signaling. Genes Dev 9(21):2598–2608

Lee TV, Sethi MK, Leonardi J, Rana NA, Buettner FF, Haltiwanger RS, Bakker H, Jafar-Nejad H (2013) Negative regulation of notch signaling by xylose. PLoS Genet 9(6):e1003547

Lehmann R, Dietrich U, Jiménez F, Campos-Ortega JA (1981) Mutations of early neurogenesis in Drosophila. Wilehm Roux Arch Dev Biol (Dev Genes Evo) 190(4):226–229

Lehmann R, Jimenez F, Dietrich U, Campos-Ortega JA (1983) On the phenotype and development of mutants of early neurogenesis in Drosophila melanogaster. Wilehm Roux Arch Dev Biol (Dev Genes Evo) 192(2):62–74

Leimeister C, Externbrink A, Klamt B, Gessler M (1999) Hey genes: a novel subfamily of hairy- and Enhancer of split related genes specifically expressed during mouse embryogenesis. Mech Dev 85(1-2):173–177

Leonardi J, Fernandez-Valdivia R, Li YD, Simcox AA, Jafar-Nejad H (2011) Multiple O-glucosylation sites on Notch function as a buffer against temperature-dependent loss of signaling. Development 138(16):3569–3578

Li L, Krantz ID, Deng Y, Genin A, Banta AB, Collins CC, Qi M, Trask BJ, Kuo WL, Cochran J, Costa T, Pierpont ME, Rand EB, Piccoli DA, Hood L, Spinner NB (1997) Alagille syndrome is caused by mutations in human Jagged1, which encodes a ligand for Notch1. Nat Genet 16(3):243–251

Li D, Parks SB, Kushner JD, Nauman D, Burgess D, Ludwigsen S, Partain J, Nixon RR, Allen CN, Irwin RP, Jakobs PM, Litt M, Hershberger RE (2006) Mutations of presenilin genes in dilated cardiomyopathy and heart failure. Am J Hum Genet 79(6):1030–1039

Li M, Cheng R, Liang J, Yan H, Zhang H, Yang L, Li C, Jiao Q, Lu Z, He J, Ji J, Shen Z, Li C, Hao F, Yu H, Yao Z (2013) Mutations in POFUT1, encoding protein O-fucosyltransferase 1, cause generalized Dowling-Degos disease. Am J Hum Genet 92(6):895–903

Li J, Housden BE, Bray SJ (2014) Notch signaling assays in Drosophila cultured cell lines. Methods Mol Biol 1187:131–141

Lieber T, Kidd S, Struhl G (2011) DSL-Notch signaling in the Drosophila brain in response to olfactory stimulation. Neuron 69(3):468–481

Lindsley DL, Zimm GG (1992) The Genome of Drosophila melanogaster. Academic Press, San Diego

Liu J, Sato C, Cerletti M, Wagers A (2010) Notch signaling in the regulation of stem cell self-renewal and differentiation. Curr Top Dev Biol 92:367–409

Liu J, Fukunaga-Kalabis M, Li L, Herlyn M (2015) Developmental pathways activated in melanocytes and melanoma. Arch Biochem Biophys 563:13–21

Logeat F, Bessia C, Brou C, LeBail O, Jarriault S, Seidah NG, Israel A (1998) The Notch1 receptor is cleaved constitutively by a furin-like convertase. Proc Natl Acad Sci U S A 95(14):8108–8112

Louvi A, Artavanis-Tsakonas S (2012) Notch and disease: a growing field. Semin Cell Dev Biol 23(4):473–480

Lundkvist J, Zhu S, Hansson EM, Schweinhardt P, Miao Q, Beatus P, Dannaeus K, Karlstrom H, Johansson CB, Viitanen M, Rozell B, Spenger C, Mohammed A, Kalimo H, Lendahl U (2005) Mice carrying a R142C Notch 3 knock-in mutation do not develop a CADASIL-like phenotype. Genesis 41(1):13–22

Luxan G, Casanova JC, Martinez-Poveda B, Prados B, D'Amato G, MacGrogan D, Gonzalez-Rajal A, Dobarro D, Torroja C, Martinez F, Izquierdo-Garcia JL, Fernandez-Friera L, Sabater-Molina M, Kong YY, Pizarro G, Ibanez B, Medrano C, Garcia-Pavia P, Gimeno JR, Monserrat L, Jimenez-Borreguero LJ, de la Pompa JL (2013) Mutations in the NOTCH pathway regulator MIB1 cause left ventricular noncompaction cardiomyopathy. Nat Med 19(2):193–201

MacGrogan D, Nus M, de la Pompa JL (2010) Notch signaling in cardiac development and disease. Curr Top Dev Biol 92:333–365

Mahoney MB, Parks AL, Ruddy DA, Tiong SY, Esengil H, Phan AC, Philandrinos P, Winter CG, Chatterjee R, Huppert K, Fisher WW, L'Archeveque L, Mapa FA, Woo W, Ellis MC, Curtis D (2006) Presenilin-based genetic screens in Drosophila melanogaster identify novel notch pathway modifiers. Genetics 172(4):2309–2324

Mahoney-Sanchez L, Belaidi AA, Bush AI, Ayton S (2016) The complex role of apolipoprotein E in Alzheimer's disease: an overview and update. J Mol Neurosci 60(3):325–335

Maier D, Stumm G, Kuhn K, Preiss A (1992) Hairless, a Drosophila gene involved in neural development, encodes a novel, serine rich protein. Mech Dev 38(2):143–156

Majewski J, Schwartzentruber JA, Caqueret A, Patry L, Marcadier J, Fryns JP, Boycott KM, Ste-Marie LG, McKiernan FE, Marik I, Van Esch H, Michaud JL, Samuels ME (2011) Mutations in NOTCH2 in families with Hajdu-Cheney syndrome. Hum Mutat 32(10):1114–1117

Martignetti JA, Tian L, Li D, Ramirez MC, Camacho-Vanegas O, Camacho SC, Guo Y, Zand DJ, Bernstein AM, Masur SK, Kim CE, Otieno FG, Hou C, Abdel-Magid N, Tweddale B, Metry D, Fournet JC, Papp E, McPherson EW, Zabel C, Vaksmann G, Morisot C, Keating B, Sleiman PM, Cleveland JA, Everman DB, Zackai E, Hakonarson H (2013) Mutations in PDGFRB cause autosomal-dominant infantile myofibromatosis. Am J Hum Genet 92(6):1001–1007

MARRVEL (2017) http://marrvel.org/

Mazaleyrat SL, Fostier M, Wilkin MB, Aslam H, Evans DA, Cornell M, Baron M (2003) Down-regulation of Notch target gene expression by Suppressor of deltex. Dev Biol 255(2):363–372

McBride KL, Riley MF, Zender GA, Fitzgerald-Butt SM, Towbin JA, Belmont JW, Cole SE (2008) NOTCH1 mutations in individuals with left ventricular outflow tract malformations reduce ligand-induced signaling. Hum Mol Genet 17(18):2886–2893

McDaniell R, Warthen DM, Sanchez-Lara PA, Pai A, Krantz ID, Piccoli DA, Spinner NB (2006) NOTCH2 mutations cause Alagille syndrome, a heterogeneous disorder of the notch signaling pathway. Am J Hum Genet 79(1):169–173

McMillan BJ, Zimmerman B, Egan ED, Lofgren M, Xu X, Hesser A, Blacklow SC (2017) Structure of human POFUT1, its requirement in ligand-independent oncogenic Notch signaling, and functional effects of Dowling-Degos mutations. Glycobiology 27(8):777–786

Meester JA, Southgate L, Stittrich AB, Venselaar H, Beekmans SJ, den Hollander N, Bijlsma EK, Helderman-van den Enden A, Verheij JB, Glusman G, Roach JC, Lehman A, Patel MS, de Vries BB, Ruivenkamp C, Itin P, Prescott K, Clarke S, Trembath R, Zenker M, Sukalo M, Van Laer L, Loeys B, Wuyts W (2015) Heterozygous loss-of-function mutations in DLL4 cause Adams-Oliver syndrome. Am J Hum Genet 97(3):475–482

Micchelli CA, Rulifson EJ, Blair SS (1997) The function and regulation of cut expression on the wing margin of Drosophila: Notch, Wingless and a dominant negative role for Delta and Serrate. Development 124(8):1485–1495

Michel M, Aliee M, Rudolf K, Bialas L, Julicher F, Dahmann C (2016) The selector gene apterous and Notch are required to locally increase mechanical cell bond tension at the Drosophila dorsoventral compartment boundary. PLoS One 11(8):e0161668

Michellod MA, Randsholt NB (2008) Implication of the Drosophila beta-amyloid peptide binding-like protein AMX in Notch signaling during early neurogenesis. Brain Res Bull 75(2-4):305–309

Mishra AK, Sachan N, Mutsuddi M, Mukherjee A (2015) Kinase active Misshapen regulates Notch signaling in Drosophila melanogaster. Exp Cell Res 339(1):51–60

Moloney DJ, Panin VM, Johnston SH, Chen J, Shao L, Wilson R, Wang Y, Stanley P, Irvine KD, Haltiwanger RS, Vogt TF (2000) Fringe is a glycosyltransferase that modifies Notch. Nature 406(6794):369–375

Monet-Lepretre M, Bardot B, Lemaire B, Domenga V, Godin O, Dichgans M, Tournier-Lasserve E, Cohen-Tannoudji M, Chabriat H, Joutel A (2009) Distinct phenotypic and functional features of CADASIL mutations in the Notch3 ligand binding domain. Brain 132(Pt 6):1601–1612

Morel V, Lecourtois M, Massiani O, Maier D, Preiss A, Schweisguth F (2001) Transcriptional repression by suppressor of hairless involves the binding of a hairless-dCtBP complex in Drosophila. Curr Biol 11(10):789–792

Morgan TH, Sturtevant AH, Bridges CB (1922) Year B Carnegie Inst Wash 22:283–287

Morrissette JD, Colliton RP, Spinner NB (2001) Defective intracellular transport and processing of JAG1 missense mutations in Alagille syndrome. Hum Mol Genet 10(4):405–413

Moshkin YM, Kan TW, Goodfellow H, Bezstarosti K, Maeda RK, Pilyugin M, Karch F, Bray SJ, Demmers JA, Verrijzer CP (2009) Histone chaperones ASF1 and NAP1 differentially modulate removal of active histone marks by LID-RPD3 complexes during NOTCH silencing. Mol Cell 35(6):782–793

Mourikis P, Lake RJ, Firnhaber CB, DeDecker BS (2010) Modifiers of notch transcriptional activity identified by genome-wide RNAi. BMC Dev Biol 10:107

Mouse Genome Informatics (2017) http://www.informatics.jax.org/

Mukherjee A, Veraksa A, Bauer A, Rosse C, Camonis J, Artavanis-Tsakonas S (2005) Regulation of Notch signalling by non-visual beta-arrestin. Nat Cell Biol 7(12):1191–1201

Muller R, Jenny A, Stanley P (2013) The EGF repeat-specific O-GlcNAc-transferase Eogt interacts with notch signaling and pyrimidine metabolism pathways in Drosophila. PLoS One 8(5):e62835

Mummery-Widmer JL, Yamazaki M, Stoeger T, Novatchkova M, Bhalerao S, Chen D, Dietzl G, Dickson BJ, Knoblich JA (2009) Genome-wide analysis of Notch signalling in Drosophila by transgenic RNAi. Nature 458(7241):987–992

Nagarkar-Jaiswal S, DeLuca SZ, Lee PT, Lin WW, Pan H, Zuo Z, Lv J, Spradling AC, Bellen HJ (2015a) A genetic toolkit for tagging intronic MiMIC containing genes. elife 4:e08469

Nagarkar-Jaiswal S, Lee PT, Campbell ME, Chen K, Anguiano-Zarate S, Gutierrez MC, Busby T, Lin WW, He Y, Schulze KL, Booth BW, Evans-Holm M, Venken KJ, Levis RW, Spradling AC, Hoskins RA, Bellen HJ (2015b) A library of MiMICs allows tagging of genes and reversible, spatial and temporal knockdown of proteins in Drosophila. elife 4:e05338

Nagel AC, Krejci A, Tenin G, Bravo-Patino A, Bray S, Maier D, Preiss A (2005) Hairless-mediated repression of notch target genes requires the combined activity of Groucho and CtBP corepressors. Mol Cell Biol 25(23):10433–10441

Nakao K, Campos-Ortega JA (1996) Persistent expression of genes of the enhancer of split complex suppresses neural development in Drosophila. Neuron 16(2):275–286

Neumann CJ, Cohen SM (1997) Long-range action of Wingless organizes the dorsal-ventral axis of the Drosophila wing. Development 124(4):871–880

Nguyen HT, Voza F, Ezzeddine N, Frasch M (2007) Drosophila mind bomb2 is required for maintaining muscle integrity and survival. J Cell Biol 179(2):219–227

Nolo R, Abbott LA, Bellen HJ (2000) Senseless, a Zn finger transcription factor, is necessary and sufficient for sensory organ development in Drosophila. Cell 102(3):349–362

Nowell C, Radtke F (2013) Cutaneous Notch signaling in health and disease. Cold Spring Harb Perspect Med 3(12):a017772

Nus M, MacGrogan D, Martinez-Poveda B, Benito Y, Casanova JC, Fernandez-Aviles F, Bermejo J, de la Pompa JL (2011) Diet-induced aortic valve disease in mice haploinsufficient for the Notch pathway effector RBPJK/CSL. Arterioscler Thromb Vasc Biol 31(7):1580–1588

O'Connor-Giles KM, Skeath JB (2003) Numb inhibits membrane localization of Sanpodo, a four-pass transmembrane protein, to promote asymmetric divisions in Drosophila. Dev Cell 5(2):231–243

Oda T, Elkahloun AG, Pike BL, Okajima K, Krantz ID, Genin A, Piccoli DA, Meltzer PS, Spinner NB, Collins FS, Chandrasekharappa SC (1997) Mutations in the human Jagged1 gene are responsible for Alagille syndrome. Nat Genet 16(3):235–242

Oellers N, Dehio M, Knust E (1994) bHLH proteins encoded by the Enhancer of split complex of Drosophila negatively interfere with transcriptional activation mediated by proneural genes. Mol Gen Genet 244(5):465–473

Okajima T, Irvine KD (2002) Regulation of notch signaling by o-linked fucose. Cell 111(6):893–904

Online Mendelian Inheritance in Man (2017) https://www.omim.org/

Opherk C, Duering M, Peters N, Karpinska A, Rosner S, Schneider E, Bader B, Giese A, Dichgans M (2009) CADASIL mutations enhance spontaneous multimerization of NOTCH3. Hum Mol Genet 18(15):2761–2767

Oswald F, Kostezka U, Astrahantseff K, Bourteele S, Dillinger K, Zechner U, Ludwig L, Wilda M, Hameister H, Knochel W, Liptay S, Schmid RM (2002) SHARP is a novel component of the Notch/RBP-Jkappa signalling pathway. EMBO J 21(20):5417–5426

Palmer WH, Deng WM (2015) Ligand-independent mechanisms of notch activity. Trends Cell Biol 25(11):697–707

Panin VM, Papayannopoulos V, Wilson R, Irvine KD (1997) Fringe modulates Notch-ligand interactions. Nature 387(6636):908–912

Parks AL, Muskavitch MA (1993) Delta function is required for bristle organ determination and morphogenesis in Drosophila. Dev Biol 157(2):484–496

Parks AL, Turner FR, Muskavitch MA (1995) Relationships between complex Delta expression and the specification of retinal cell fates during Drosophila eye development. Mech Dev 50(2–3):201–216

Paroush Z, Finley RL Jr, Kidd T, Wainwright SM, Ingham PW, Brent R, Ish-Horowicz D (1994) Groucho is required for Drosophila neurogenesis, segmentation, and sex determination and interacts directly with hairy-related bHLH proteins. Cell 79(5):805–815

Pavlopoulos E, Pitsouli C, Klueg KM, Muskavitch MA, Moschonas NK, Delidakis C (2001) neuralized Encodes a peripheral membrane protein involved in delta signaling and endocytosis. Dev Cell 1(6):807–816

Penton AL, Leonard LD, Spinner NB (2012) Notch signaling in human development and disease. Semin Cell Dev Biol 23(4):450–457
Periz G, Fortini ME (1999) Ca(2+)-ATPase function is required for intracellular trafficking of the Notch receptor in Drosophila. EMBO J 18(21): 5983–5993
Perrimon N, Lanjuin A, Arnold C, Noll E (1996) Zygotic lethal mutations with maternal effect phenotypes in Drosophila melanogaster. II Loci on the second and third chromosomes identified by P-element-induced mutations. Genetics 144(4):1681–1692
Peters N, Opherk C, Zacherle S, Capell A, Gempel P, Dichgans M (2004) CADASIL-associated Notch3 mutations have differential effects both on ligand binding and ligand-induced Notch3 receptor signaling through RBP-Jk. Exp Cell Res 299(2):454–464
Pezeron G, Millen K, Boukhatmi H, Bray S (2014) Notch directly regulates the cell morphogenesis genes Reck, talin and trio in adult muscle progenitors. J Cell Sci 127(Pt 21):4634–4644
Pi H, Huang YC, Chen IC, Lin CD, Yeh HF, Pai LM (2011) Identification of 11-amino acid peptides that disrupt Notch-mediated processes in Drosophila. J Biomed Sci 18:42
Piccolo P, Attanasio S, Secco I, Sangermano R, Strisciuglio C, Limongelli G, Miele E, Mutarelli M, Banfi S, Nigro V, Pons T, Valencia A, Zentilin L, Campione S, Nardone G, Lynnes TC, Celestino-Soper PB, Spoonamore KG, D'Armiento FP, Giacca M, Staiano A, Vatta M, Collesi C, Brunetti-Pierri N (2017) MIB2 variants altering NOTCH signalling result in left ventricle hypertrabeculation/non-compaction and are associated with Ménétrierlike gastropathy. Hum Mol Genet 26(1):33–43
Pierfelice T, Alberi L, Gaiano N (2011) Notch in the vertebrate nervous system: an old dog with new tricks. Neuron 69(5):840–855
Pines MK, Housden BE, Bernard F, Bray SJ, Roper K (2010) The cytolinker Pigs is a direct target and a negative regulator of Notch signalling. Development 137(6):913–922
Pink AE, Simpson MA, Desai N, Trembath RC, Barker JN (2013) gamma-Secretase mutations in hidradenitis suppurativa: new insights into disease pathogenesis. J Invest Dermatol 133(3):601–607
Pitsouli C, Delidakis C (2005) The interplay between DSL proteins and ubiquitin ligases in Notch signaling. Development 132(18):4041–4050
Plunkett CR (1926) The interaction of genetic and environmental factors in development. J Exp Zool 46:181–244
Poirier J, Davignon J, Bouthillier D, Kogan S, Bertrand P, Gauthier S (1993) Apolipoprotein E polymorphism and Alzheimer's disease. Lancet 342(8873):697–699
PomBase (2017) https://www.pombase.org/
Poulson DF (1936) Chromosome deficiencies and embryonic development. Ph.D. thesis, California Institute of Technology, CA, USA
Poulson DF (1937) Chromosomal deficiencies and the embryonic development of Drosophila melanogaster. Proc Natl Acad Sci U S A 23(3):133–137
Pourquie O, Kusumi K (2001) When body segmentation goes wrong. Clin Genet 60(6):409–416
Psaty BM, O'Donnell CJ, Gudnason V, Lunetta KL, Folsom AR, Rotter JI, Uitterlinden AG, Harris TB, Witteman JC, Boerwinkle E (2009) Cohorts for Heart and Aging Research in Genomic Epidemiology (CHARGE) consortium: design of prospective meta-analyses of genome-wide association studies from 5 cohorts. Circ Cardiovasc Genet 2(1):73–80
PubMed (2017) https://www.ncbi.nlm.nih.gov/pubmed/
Rajan A, Tien AC, Haueter CM, Schulze KL, Bellen HJ (2009) The Arp2/3 complex and WASp are required for apical trafficking of Delta into microvilli during cell fate specification of sensory organ precursors. Nat Cell Biol 11(7):815–824
Ramoni RB, Mulvihill JJ, Adams DR, Allard P, Ashley EA, Bernstein JA, Gahl WA, Hamid R, Loscalzo J, McCray AT, Shashi V, Tifft CJ, Wise AL (2017) The undiagnosed diseases network: accelerating discovery about health and disease. Am J Hum Genet 100(2):185–192
Rana NA, Haltiwanger RS (2011) Fringe benefits: functional and structural impacts of O-glycosylation on the extracellular domain of Notch receptors. Curr Opin Struct Biol 21(5):583–589
Ranganathan P, Weaver KL, Capobianco AJ (2011) Notch signalling in solid tumours: a little bit of everything but not all the time. Nat Rev Cancer 11(5):338–351
Rat Genome Database (2017) http://rgd.mcw.edu/
Rauskolb C, Correia T, Irvine KD (1999) Fringe-dependent separation of dorsal and ventral cells in the Drosophila wing. Nature 401(6752):476–480
Reddy GV, Rodrigues V (1999) A glial cell arises from an additional division within the mechanosensory lineage during development of the microchaete on the Drosophila notum. Development 126(20):4617–4622
Rhyu MS, Jan LY, Jan YN (1994) Asymmetric distribution of numb protein during division of the sensory organ precursor cell confers distinct fates to daughter cells. Cell 76(3):477–491
Roegiers F, Jan LY, Jan YN (2005) Regulation of membrane localization of Sanpodo by lethal giant larvae and neuralized in asymmetrically dividing cells of Drosophila sensory organs. Mol Biol Cell 16(8):3480–3487
Rottgen G, Wagner T, Hinz U (1998) A genetic screen for elements of the network that regulates neurogenesis in Drosophila. Mol Gen Genet 257(4):442–451
Royet J, Bouwmeester T, Cohen SM (1998) Notchless encodes a novel WD40-repeat-containing protein that modulates Notch signaling activity. EMBO J 17(24):7351–7360
Rulifson EJ, Blair SS (1995) Notch regulates wingless expression and is not required for reception of the paracrine wingless signal during wing margin neurogenesis in Drosophila. Development 121(9):2813–2824

Rullinkov G, Tamme R, Sarapuu A, Lauren J, Sepp M, Palm K, Timmusk T (2009) Neuralized-2: expression in human and rodents and interaction with Delta-like ligands. Biochem Biophys Res Commun 389(3):420–425

Rutten JW, Haan J, Terwindt GM, van Duinen SG, Boon EM, Lesnik Oberstein SA (2014) Interpretation of NOTCH3 mutations in the diagnosis of CADASIL. Expert Rev Mol Diagn 14(5):593–603

Saccharomyces Genome Database (2017) http://www.yeastgenome.org/

Saj A, Arziman Z, Stempfle D, van Belle W, Sauder U, Horn T, Durrenberger M, Paro R, Boutros M, Merdes G (2010) A combined ex vivo and in vivo RNAi screen for notch regulators in Drosophila reveals an extensive notch interaction network. Dev Cell 18(5):862–876

Sakaidani Y, Ichiyanagi N, Saito C, Nomura T, Ito M, Nishio Y, Nadano D, Matsuda T, Furukawa K, Okajima T (2012) O-linked-N-acetylglucosamine modification of mammalian Notch receptors by an atypical O-GlcNAc transferase Eogt1. Biochem Biophys Res Commun 419(1):14–19

Saleh M, Kamath BM, Chitayat D (2016) Alagille syndrome: clinical perspectives. Appl Clin Genet 9:75–82

Sandoval H, Yao CK, Chen K, Jaiswal M, Donti T, Lin YQ, Bayat V, Xiong B, Zhang K, David G, Charng WL, Yamamoto S, Duraine L, Graham BH, Bellen HJ (2014) Mitochondrial fusion but not fission regulates larval growth and synaptic development through steroid hormone production. Elife 3: e03558.

Sasai Y, Kageyama R, Tagawa Y, Shigemoto R, Nakanishi S (1992) Two mammalian helix-loop-helix factors structurally related to Drosophila hairy and Enhancer of split. Genes Dev 6(12B):2620–2634

Sasamura T, Sasaki N, Miyashita F, Nakao S, Ishikawa HO, Ito M, Kitagawa M, Harigaya K, Spana E, Bilder D, Perrimon N, Matsuno K (2003) neurotic, a novel maternal neurogenic gene, encodes an O-fucosyltransferase that is essential for Notch-Delta interactions. Development 130(20):4785–4795

Schneider M, Troost T, Grawe F, Martinez-Arias A, Klein T (2013) Activation of Notch in lgd mutant cells requires the fusion of late endosomes with the lysosome. J Cell Sci 126(Pt 2):645–656

Schreiber SL, Preiss A, Nagel AC, Wech I, Maier D (2002) Genetic screen for modifiers of the rough eye phenotype resulting from overexpression of the Notch antagonist hairless in Drosophila. Genesis 33(3):141–152

Schrons H, Knust E, Campos-Ortega JA (1992) The Enhancer of split complex and adjacent genes in the 96F region of Drosophila melanogaster are required for segregation of neural and epidermal progenitor cells. Genetics 132(2):481–503

Schwanbeck R (2015) The role of epigenetic mechanisms in Notch signaling during development. J Cell Physiol 230(5):969–981

Schweisguth F (2015) Asymmetric cell division in the Drosophila bristle lineage: from the polarization of sensory organ precursor cells to Notch-mediated binary fate decision. Wiley Interdiscip Rev Dev Biol 4(3):299–309

Schweisguth F, Posakony JW (1992) Suppressor of Hairless, the Drosophila homolog of the mouse recombination signal-binding protein gene, controls sensory organ cell fates. Cell 69(7):1199–1212

Seidner GA, Ye Y, Faraday MM, Alvord WG, Fortini ME (2006) Modeling clinically heterogeneous presenilin mutations with transgenic Drosophila. Curr Biol 16(10):1026–1033

Selkoe DJ, Hardy J (2016) The amyloid hypothesis of Alzheimer's disease at 25 years. EMBO Mol Med 8(6):595–608

Selva EM, Hong K, Baeg GH, Beverley SM, Turco SJ, Perrimon N, Hacker U (2001) Dual role of the fringe connection gene in both heparan sulphate and fringe-dependent signalling events. Nat Cell Biol 3(9):809–815

Sethi MK, Buettner FF, Krylov VB, Takeuchi H, Nifantiev NE, Haltiwanger RS, Gerardy-Schahn R, Bakker H (2010) Identification of glycosyltransferase 8 family members as xylosyltransferases acting on O-glucosylated notch epidermal growth factor repeats. J Biol Chem 285(3):1582–1586

Seugnet L, Simpson P, Haenlin M (1997) Requirement for dynamin during Notch signaling in Drosophila neurogenesis. Dev Biol 192(2):585–598

Shalaby NA, Parks AL, Morreale EJ, Osswalt MC, Pfau KM, Pierce EL, Muskavitch MA (2009) A screen for modifiers of notch signaling uncovers Amun, a protein with a critical role in sensory organ development. Genetics 182(4):1061–1076

Shannon MP (1972) Characterization of the female-sterile mutant almondex of Drosophila melanogaster. Genetica 43(2):244–256

Sherrington R, Rogaev EI, Liang Y, Rogaeva EA, Levesque G, Ikeda M, Chi H, Lin C, Li G, Holman K et al (1995) Cloning of a gene bearing missense mutations in early-onset familial Alzheimer's disease. Nature 375(6534):754–760

Shimizu H, Woodcock SA, Wilkin MB, Trubenova B, Monk NA, Baron M (2014) Compensatory flux changes within an endocytic trafficking network maintain thermal robustness of Notch signaling. Cell 157(5):1160–1174

Siman R, Reaume AG, Savage MJ, Trusko S, Lin YG, Scott RW, Flood DG (2000) Presenilin-1 P264L knock-in mutation: differential effects on abeta production, amyloid deposition, and neuronal vulnerability. J Neurosci 20(23):8717–8726

Simpson MA, Irving MD, Asilmaz E, Gray MJ, Dafou D, Elmslie FV, Mansour S, Holder SE, Brain CE, Burton BK, Kim KH, Pauli RM, Aftimos S, Stewart H, Kim CA, Holder-Espinasse M, Robertson SP, Drake WM, Trembath RC (2011) Mutations in NOTCH2 cause Hajdu-Cheney syndrome, a disorder of severe and progressive bone loss. Nat Genet 43(4):303–305

Skalska L, Stojnic R, Li J, Fischer B, Cerda-Moya G, Sakai H, Tajbakhsh S, Russell S, Adryan B, Bray SJ (2015) Chromatin signatures at Notch-regulated enhancers reveal large-scale changes in H3K56ac upon activation. EMBO J 34(14):1889–1904

Slaninova V, Krafcikova M, Perez-Gomez R, Steffal P, Trantirek L, Bray SJ, Krejci A (2016) Notch stimulates growth by direct regulation of genes involved in the control of glycolysis and the tricarboxylic acid cycle. Open Biol 6(2):150155

Smoller D, Friedel C, Schmid A, Bettler D, Lam L, Yedvobnick B (1990) The Drosophila neurogenic locus mastermind encodes a nuclear protein unusually rich in amino acid homopolymers. Genes Dev 4(10):1688–1700

Solinger JA, Spang A (2013) Tethering complexes in the endocytic pathway: CORVET and HOPS. FEBS J 280(12):2743–2757

Southgate L, Sukalo M, Karountzos AS, Taylor EJ, Collinson CS, Ruddy D, Snape KM, Dallapiccola B, Tolmie JL, Joss S, Brancati F, Digilio MC, Graul-Neumann LM, Salviati L, Coerdt W, Jacquemin E, Wuyts W, Zenker M, Machado RD, Trembath RC (2015) Haploinsufficiency of the NOTCH1 receptor as a cause of Adams-Oliver syndrome with variable cardiac anomalies. Circ Cardiovasc Genet 8(4):572–581

Sparrow DB, Chapman G, Wouters MA, Whittock NV, Ellard S, Fatkin D, Turnpenny PD, Kusumi K, Sillence D, Dunwoodie SL (2006) Mutation of the LUNATIC FRINGE gene in humans causes spondylocostal dysostosis with a severe vertebral phenotype. Am J Hum Genet 78(1):28–37

Sparrow DB, Guillen-Navarro E, Fatkin D, Dunwoodie SL (2008) Mutation of Hairy-and-Enhancer-of-Split-7 in humans causes spondylocostal dysostosis. Hum Mol Genet 17(23):3761–3766

Sparrow DB, Chapman G, Dunwoodie SL (2011) The mouse notches up another success: understanding the causes of human vertebral malformation. Mamm Genome 22(7-8):362–376

Sparrow DB, McInerney-Leo A, Gucev ZS, Gardiner B, Marshall M, Leo PJ, Chapman DL, Tasic V, Shishko A, Brown MA, Duncan EL, Dunwoodie SL (2013) Autosomal dominant spondylocostal dysostosis is caused by mutation in TBX6. Hum Mol Genet 22(8):1625–1631

Stanley P, Okajima T (2010) Roles of glycosylation in Notch signaling. Curr Top Dev Biol 92:131–164

Stifani S, Blaumueller CM, Redhead NJ, Hill RE, Artavanis-Tsakonas S (1992) Human homologs of a Drosophila Enhancer of split gene product define a novel family of nuclear proteins. Nat Genet 2(2):119–127

Stittrich AB, Lehman A, Bodian DL, Ashworth J, Zong Z, Li H, Lam P, Khromykh A, Iyer RK, Vockley JG, Baveja R, Silva ES, Dixon J, Leon EL, Solomon BD, Glusman G, Niederhuber JE, Roach JC, Patel MS (2014) Mutations in NOTCH1 cause Adams-Oliver syndrome. Am J Hum Genet 95(3):275–284

Struhl G, Adachi A (2000) Requirements for presenilin-dependent cleavage of notch and other transmembrane proteins. Mol Cell 6(3):625–636

Struhl G, Greenwald I (1999) Presenilin is required for activity and nuclear access of Notch in Drosophila. Nature 398(6727):522–525

Sun Y, Yan Y, Denef N, Schupbach T (2011) Regulation of somatic myosin activity by protein phosphatase 1beta controls Drosophila oocyte polarization. Development 138(10):1991–2001

Tada M, Itoh S, Ishii-Watabe A, Suzuki T, Kawasaki N (2012) Functional analysis of the Notch ligand Jagged1 missense mutant proteins underlying Alagille syndrome. FEBS J 279(12):2096–2107

Tagami S, Okochi M, Yanagida K, Ikuta A, Fukumori A, Matsumoto N, Ishizuka-Katsura Y, Nakayama T, Itoh N, Jiang J, Nishitomi K, Kamino K, Morihara T, Hashimoto R, Tanaka T, Kudo T, Chiba S, Takeda M (2008) Regulation of Notch signaling by dynamic changes in the precision of S3 cleavage of Notch-1. Mol Cell Biol 28(1):165–176

Takats S, Pircs K, Nagy P, Varga A, Karpati M, Hegedus K, Kramer H, Kovacs AL, Sass M, Juhasz G (2014) Interaction of the HOPS complex with Syntaxin 17 mediates autophagosome clearance in Drosophila. Mol Biol Cell 25(8):1338–1354

Takayama S, Dhahbi J, Roberts A, Mao G, Heo SJ, Pachter L, Martin DI, Boffelli D (2014) Genome methylation in D. melanogaster is found at specific short motifs and is independent of DNMT2 activity. Genome Res 24(5):821–830

Taniguchi Y, Furukawa T, Tun T, Han H, Honjo T (1998) LIM protein KyoT2 negatively regulates transcription by association with the RBP-J DNA-binding protein. Mol Cell Biol 18(1):644–654

Tao J, Chen S, Lee B (2010) Alteration of Notch signaling in skeletal development and disease. Ann N Y Acad Sci 1192:257–268

Teider N, Scott DK, Neiss A, Weeraratne SD, Amani VM, Wang Y, Marquez VE, Cho YJ, Pomeroy SL (2010) Neuralized1 causes apoptosis and downregulates Notch target genes in medulloblastoma. Neuro-Oncology 12(12):1244–1256

Terriente-Felix A, Li J, Collins S, Mulligan A, Reekie I, Bernard F, Krejci A, Bray S (2013) Notch cooperates with Lozenge/Runx to lock haemocytes into a differentiation programme. Development 140(4):926–937

Thomas U, Speicher SA, Knust E (1991) The Drosophila gene Serrate encodes an EGF-like transmembrane protein with a complex expression pattern in embryos and wing discs. Development 111(3):749–761

Tian X, Hansen D, Schedl T, Skeath JB (2004) Epsin potentiates Notch pathway activity in Drosophila and C. elegans. Development 131(23):5807–5815

Tien AC, Rajan A, Schulze KL, Ryoo HD, Acar M, Steller H, Bellen HJ (2008) Ero1L, a thiol oxidase, is required for Notch signaling through cysteine bridge formation of the Lin12-Notch repeats in Drosophila melanogaster. J Cell Biol 182(6):1113–1125

Troost T, Jaeckel S, Ohlenhard N, Klein T (2012) The tumour suppressor Lethal (2) giant discs is required for the function of the ESCRT-III component Shrub/CHMP4. J Cell Sci 125(Pt 3):763–776

Troost T, Schneider M, Klein T (2015) A re-examination of the selection of the sensory organ precursor of the bristle sensilla of Drosophila melanogaster. PLoS Genet 11(1):e1004911

Udolph G (2012) Notch signaling and the generation of cell diversity in Drosophila neuroblast lineages. Adv Exp Med Biol 727:47–60

Vaccari T, Bilder D (2005) The Drosophila tumor suppressor vps25 prevents nonautonomous overproliferation by regulating notch trafficking. Dev Cell 9(5):687–698

Vaccari T, Lu H, Kanwar R, Fortini ME, Bilder D (2008) Endosomal entry regulates Notch receptor activation in Drosophila melanogaster. J Cell Biol 180(4):755–762

Vaccari T, Rusten TE, Menut L, Nezis IP, Brech A, Stenmark H, Bilder D (2009) Comparative analysis of ESCRT-I, ESCRT-II and ESCRT-III function in Drosophila by efficient isolation of ESCRT mutants. J Cell Sci 122(Pt 14):2413–2423

Vaccari T, Duchi S, Cortese K, Tacchetti C, Bilder D (2010) The vacuolar ATPase is required for physiological as well as pathological activation of the Notch receptor. Development 137(11):1825–1832

Vallejo DM, Caparros E, Dominguez M (2011) Targeting Notch signalling by the conserved miR-8/200 microRNA family in development and cancer cells. EMBO J 30(4):756–769

van de Hoef DL, Hughes J, Livne-Bar I, Garza D, Konsolaki M, Boulianne GL (2009) Identifying genes that interact with Drosophila presenilin and amyloid precursor protein. Genesis 47(4):246–260

Vassin H, Bremer KA, Knust E, Campos-Ortega JA (1987) The neurogenic gene Delta of Drosophila melanogaster is expressed in neurogenic territories and encodes a putative transmembrane protein with EGF-like repeats. EMBO J 6(11):3431–3440

Venken KJ, Schulze KL, Haelterman NA, Pan H, He Y, Evans-Holm M, Carlson JW, Levis RW, Spradling AC, Hoskins RA, Bellen HJ (2011) MiMIC: a highly versatile transposon insertion resource for engineering Drosophila melanogaster genes. Nat Methods 8(9):737–743

Verheyen EM, Purcell KJ, Fortini ME, Artavanis-Tsakonas S (1996) Analysis of dominant enhancers and suppressors of activated Notch in Drosophila. Genetics 144(3):1127–1141

Veugelen S, Saito T, Saido TC, Chavez-Gutierrez L, De Strooper B (2016) Familial Alzheimer's disease mutations in presenilin generate amyloidogenic abeta peptide seeds. Neuron 90(2):410–416

Vienna Drosophila Resource Center (2017) http://stockcenter.vdrc.at/

Wahi K, Bochter MS, Cole SE (2016) The many roles of Notch signaling during vertebrate somitogenesis. Semin Cell Dev Biol 49:68–75

Wang W, Struhl G (2005) Distinct roles for Mind bomb, Neuralized and Epsin in mediating DSL endocytosis and signaling in Drosophila. Development 132(12):2883–2894

Wang S, Tan KL, Agosto MA, Xiong B, Yamamoto S, Sandoval H, Jaiswal M, Bayat V, Zhang K, Charng WL, David G, Duraine L, Venkatachalam K, Wensel TG, Bellen HJ (2015) The retromer complex is required for rhodopsin recycling and its loss leads to photoreceptor degeneration. PLoS Biol 12(4):e1001847

Wang J, Al-Ouran R, Hu Y, Kim SY, Wan YW, Wangler MF, Yamamoto S, Chao HT, Comjean A, Mohr SE, Udn PN, Liu Z, Bellen HJ (2017) MARRVEL: integration of human and model organism genetic resources to facilitate functional annotation of the human genome. Am J Hum Genet 100(6):843–853

Wangler MF, Yamamoto S, Bellen HJ (2015) Fruit flies in biomedical research. Genetics 199(3):639–653

Wangler MF, Hu Y, Shulman JM (2017) Drosophila and genome-wide association studies: a review and resource for the functional dissection of human complex traits. Dis Model Mech 10(2):77–88

Weber D, Wiese C, Gessler M (2014) Hey bHLH transcription factors. Curr Top Dev Biol 110:285–315

Weinmaster G, Fischer JA (2011) Notch ligand ubiquitylation: what is it good for? Dev Cell 21(1):134–144

Welshons WJ (1956) Dosage experiments with split mutants in the presence of an enhancer of split. Drosophila Inf Serv 30:157–158

Wharton KA, Johansen KM, Xu T, Artavanis-Tsakonas S (1985) Nucleotide sequence from the neurogenic locus notch implies a gene product that shares homology with proteins containing EGF-like repeats. Cell 43(3 Pt 2):567–581

White PH, Farkas DR, Chapman DL (2005) Regulation of Tbx6 expression by Notch signaling. Genesis 42(2):61–70

Whittock NV, Sparrow DB, Wouters MA, Sillence D, Ellard S, Dunwoodie SL, Turnpenny PD (2004) Mutated MESP2 causes spondylocostal dysostosis in humans. Am J Hum Genet 74(6):1249–1254

Wilkin M, Tongngok P, Gensch N, Clemence S, Motoki M, Yamada K, Hori K, Taniguchi-Kanai M, Franklin E, Matsuno K, Baron M (2008) Drosophila HOPS and AP-3 complex genes are required for a Deltex-regulated activation of notch in the endosomal trafficking pathway. Dev Cell 15(5):762–772

Woo HN, Park JS, Gwon AR, Arumugam TV, Jo DG (2009) Alzheimer's disease and Notch signaling. Biochem Biophys Res Commun 390(4):1093–1097

WormBase (2017) http://www.wormbase.org/

Wu L, Sun T, Kobayashi K, Gao P, Griffin JD (2002) Identification of a family of mastermind-like transcriptional coactivators for mammalian notch receptors. Mol Cell Biol 22(21):7688–7700

Wu N, Ming X, Xiao J, Wu Z, Chen X, Shinawi M, Shen Y, Yu G, Liu J, Xie H, Gucev ZS, Liu S, Yang N, Al-Kateb H, Chen J, Zhang J, Hauser N, Zhang T, Tasic V, Liu P, Su X, Pan X, Liu C, Wang L, Shen J, Shen J, Chen Y, Zhang T, Zhang J, Choy KW, Wang J, Wang Q, Li S, Zhou W, Guo J, Wang Y, Zhang C, Zhao H, An Y, Zhao Y, Wang J, Liu Z, Zuo Y, Tian Y, Weng X, Sutton VR, Wang H, Ming Y, Kulkarni S, Zhong TP, Giampietro PF, Dunwoodie SL, Cheung SW, Zhang X, Jin L, Lupski JR, Qiu G, Zhang F (2015) TBX6 null variants and a common hypomorphic allele in congenital scoliosis. N Engl J Med 372(4):341–350

Wurmbach E, Wech I, Preiss A (1999) The Enhancer of split complex of Drosophila melanogaster harbors three classes of Notch responsive genes. Mech Dev 80(2):171–180

Xia D, Watanabe H, Wu B, Lee SH, Li Y, Tsvetkov E, Bolshakov VY, Shen J, Kelleher RJ 3rd (2015) Presenilin-1 knockin mice reveal loss-of-function mechanism for familial Alzheimer's disease. Neuron 85(5):967–981

Xia D, Kelleher RJ 3rd, Shen J (2016) Loss of Abeta43 production caused by Presenilin-1 mutations in the Knockin mouse brain. Neuron 90(2):417–422

Xie T, Song X, Jin Z, Pan L, Weng C, Chen S, Zhang N (2008) Interactions between stem cells and their niche in the Drosophila ovary. Cold Spring Harb Symp Quant Biol 73:39–47

Xiong B, Bayat V, Jaiswal M, Zhang K, Sandoval H, Charng WL, Li T, David G, Duraine L, Lin YQ, Neely GG, Yamamoto S, Bellen HJ (2012) Crag is a GEF for Rab11 required for rhodopsin trafficking and maintenance of adult photoreceptor cells. PLoS Biol 10(12):e1001438

Xu T, Artavanis-Tsakonas S (1990) deltex, a locus interacting with the neurogenic genes, Notch, Delta and mastermind in Drosophila melanogaster. Genetics 126(3):665–677

Xu T, Rubin GM (1993) Analysis of genetic mosaics in developing and adult Drosophila tissues. Development 117(4):1223–1237

Xu A, Haines N, Dlugosz M, Rana NA, Takeuchi H, Haltiwanger RS, Irvine KD (2007) In vitro reconstitution of the modulation of Drosophila Notch-ligand binding by Fringe. J Biol Chem 282(48):35153–35162

Yamada K, Fuwa TJ, Ayukawa T, Tanaka T, Nakamura A, Wilkin MB, Baron M, Matsuno K (2011) Roles of Drosophila deltex in Notch receptor endocytic trafficking and activation. Genes Cells 16(3):261–272

Yamamoto S, Charng WL, Bellen HJ (2010) Endocytosis and intracellular trafficking of Notch and its ligands. Curr Top Dev Biol 92:165–200

Yamamoto S, Charng WL, Rana NA, Kakuda S, Jaiswal M, Bayat V, Xiong B, Zhang K, Sandoval H, David G, Wang H, Haltiwanger RS, Bellen HJ (2012) A mutation in EGF repeat-8 of Notch discriminates between Serrate/Jagged and Delta family ligands. Science 338(6111):1229–1232

Yamamoto S, Jaiswal M, Charng WL, Gambin T, Karaca E, Mirzaa G, Wiszniewski W, Sandoval H, Haelterman NA, Xiong B, Zhang K, Bayat V, David G, Li T, Chen K, Gala U, Harel T, Pehlivan D, Penney S, Vissers LE, de Ligt J, Jhangiani SN, Xie Y, Tsang SH, Parman Y, Sivaci M, Battaloglu E, Muzny D, Wan YW, Liu Z, Lin-Moore AT, Clark RD, Curry CJ, Link N, Schulze KL, Boerwinkle E, Dobyns WB, Allikmets R, Gibbs RA, Chen R, Lupski JR, Wangler MF, Bellen HJ (2014) A drosophila genetic resource of mutants to study mechanisms underlying human genetic diseases. Cell 159(1):200–214

Yan Y, Denef N, Schupbach T (2009) The vacuolar proton pump, V-ATPase, is required for notch signaling and endosomal trafficking in Drosophila. Dev Cell 17(3):387–402

Yasuhiko Y, Haraguchi S, Kitajima S, Takahashi Y, Kanno J, Saga Y (2006) Tbx6-mediated Notch signaling controls somite-specific Mesp2 expression. Proc Natl Acad Sci U S A 103(10):3651–3656

Ye Y, Lukinova N, Fortini ME (1999) Neurogenic phenotypes and altered Notch processing in Drosophila Presenilin mutants. Nature 398(6727): 525–529

Yedvobnick B, Helms W, Barrett B (2001) Identification of chromosomal deficiencies that modify Drosophila mastermind mutant phenotypes. Genesis 30(4):250–258

Yochem J, Greenwald I (1989) glp-1 and lin-12, genes implicated in distinct cell-cell interactions in C. elegans, encode similar transmembrane proteins. Cell 58(3):553–563

Yochem J, Weston K, Greenwald I (1988) The Caenorhabditis elegans lin-12 gene encodes a transmembrane protein with overall similarity to Drosophila Notch. Nature 335(6190):547–550

Yoon WH, Sandoval H, Nagarkar-Jaiswal S, Jaiswal M, Yamamoto S, Haelterman NA, Putluri N, Putluri V, Sreekumar A, Tos T, Aksoy A, Donti T, Graham BH, Ohno M, Nishi E, Hunter J, Muzny DM, Carmichael J, Shen J, Arboleda VA, Nelson SF, Wangler MF, Karaca E, Lupski JR, Bellen HJ (2017) Loss of Nardilysin, a Mitochondrial Co-chaperone for alpha-Ketoglutarate Dehydrogenase, Promotes mTORC1 Activation and Neurodegeneration. Neuron 93(1): 115–131

Yu H, Takeuchi H, Takeuchi M, Liu Q, Kantharia J, Haltiwanger RS, Li H (2016) Structural analysis of Notch-regulating Rumi reveals basis for pathogenic mutations. Nat Chem Biol 12(9):735–740

Yuan JS, Kousis PC, Suliman S, Visan I, Guidos CJ (2011) Functions of notch signaling in the immune system: consensus and controversies. Annu Rev Immunol 28:343–365

Yuan Z, Praxenthaler H, Tabaja N, Torella R, Preiss A, Maier D, Kovall RA (2016) Structure and function of the Su(H)-Hairless repressor complex, the major antagonist of notch signaling in Drosophila melanogaster. PLoS Biol 14(7):e1002509

Zacharioudaki E, Housden BE, Garinis G, Stojnic R, Delidakis C, Bray SJ (2016) Genes implicated in stem cell identity and temporal programme are directly targeted by Notch in neuroblast tumours. Development 143(2):219–231

Zanotti S, Canalis E (2010) Notch and the skeleton. Mol Cell Biol 30(4):886–896

Zebrafish Information Network (2017) https://zfin.org/

Zecca M, Basler K, Struhl G (1996) Direct and long-range action of a wingless morphogen gradient. Cell 87(5):833–844

Zeng C, Younger-Shepherd S, Jan LY, Jan YN (1998) Delta and Serrate are redundant Notch ligands required for asymmetric cell divisions within the Drosophila sensory organ lineage. Genes Dev 12(8):1086–1091

Zhang J, Liu M, Su Y, Du J, Zhu AJ (2012) A targeted in vivo RNAi screen reveals deubiquitinases as new regulators of Notch signaling. G3 (Bethesda) 2(12):1563–1575

Zhang K, Li Z, Jaiswal M, Bayat V, Xiong B, Sandoval H, Charng WL, David G, Haueter C, Yamamoto S, Graham BH, Bellen HJ (2013) The C8ORF38 homologue Sicily is a cytosolic chaperone for a mitochondrial complex I subunit. J Cell Biol 200(6):807–820

Ziemer A, Tietze K, Knust E, Campos-Ortega JA (1988) Genetic analysis of enhancer of split, a locus involved in neurogenesis in Drosophila melanogaster. Genetics 119(1):63–74

Mechanisms of Non-canonical Signaling in Health and Disease: Diversity to Take Therapy up a Notch?

Victor Alfred and Thomas Vaccari

Abstract

Non-canonical Notch signaling encompasses a wide range of cellular processes, diverging considerably from the established paradigm. It can dispense of ligand, proteolytic or nuclear activity. Non-canonical Notch signaling events have been studied mostly in the fruit fly *Drosophila melanogaster,* the organism in which Notch was identified first and a powerful model for understanding signaling outcomes. However, non-canonical events are ill-defined and their involvement in human physiology is not clear, hampering our understanding of diseases arising from Notch signaling alterations. At a time in which therapies based on specific targeting of Notch signaling are still an unfulfilled promise, detailed understanding of non-canonical Notch events might be key to devising more specific and less toxic pharmacologic options. Based on the blueprint of non-canonical signaling in *Drosophila,* here, we review and rationalize current evidence about non-canonical Notch signaling. Our effort might inform Notch biologists developing new research avenues and clinicians seeking future treatment of Notch-dependent diseases.

Keywords

Drosophila melanogaster · Ligand-independent signaling · mTOR/Akt signaling · Non-canonical activation · NF-κB signaling · Notch signaling · Wnt/β-catenin signaling

Abbreviations

ATM	Ataxia Telangiectasia Mutated
crb	Crumbs
CSL	CBF1/Su(H)/Lag-1
DDR	DNA Damage Response
Dl	Delta
DSL	Delta/Serrate/Lag2
dx	Deltex
ESCRT	Endosomal Sorting Complex Required for Transport
Hes	Hairy and enhancer of split
krz	Kurtz
LE	Late endosome
lgd	Lethal (2) giant discs
Mash1	Mammalian achaete scute homolog-1
Nedd4	Neural precursor cell expressed developmentally down-regulated protein 4
NF-κB	Nuclear Factor-κB
NICD	Notch intracellular domain

V. Alfred
IFOM, Istituto FIRC di Oncologia Molecolare at IFOM-IEO Campus, Milan, Italy

T. Vaccari (✉)
IFOM, Istituto FIRC di Oncologia Molecolare at IFOM-IEO Campus, Milan, Italy

Dipartimento di Bioscienze, Università degli Studi di Milano, Milan, Italy
e-mail: thomas.vaccari@unimi.it

T. Borggrefe, B. D. Giaimo (eds.), *Molecular Mechanisms of Notch Signaling*, Advances in Experimental Medicine and Biology 1066, https://doi.org/10.1007/978-3-319-89512-3_9

PI3K	Phosphatidylinositol 4,5-Bisphosphate 3-kinase
Psn	Presenilin
Su(dx)	Suppressor of dx
Su(H)	Suppressor of Hairless
TSC1/2	Tuberous Sclerosis 1/2
Wg	Wingless

1 Introduction

More than three decades after the identification of Notch in *Drosophila melanogaster* (Artavanis-Tsakonas et al. 1983), study of the Notch pathway continues to reveal a complexity that extends far beyond that predicted from the identification of relatively few core components [see for review (Guruharsha et al. 2012)]. This complexity is reflected by a wide and multifaceted role of Notch in multiple human ailments, including congenital disease and cancer (Louvi and Artavanis-Tsakonas 2012). Core components of the Notch pathway are well known to take part in so-called canonical signaling events initiated by interaction of Delta/Serrate/Lag2 (DSL) ligands - expressed by signal-sending cells - with the Notch receptor at the plasma membrane of receiving cells. Canonical signaling requires emergence to the plasma membrane of Notch as a heavily glycosylated heterodimer. Such modifications and cleavage operated by the serine protease Furin (also called S1 cleavage) occur in the Golgi apparatus. Binding of DSL ligands in trans is thought to displace the extracellular domain of Notch, which is held in place by Ca^{2+} interactions (Fig. 1a). Shedding of extracellular Notch requires endocytosis of the ligand in the signal-sending cell. It has been proposed that the Notch receptor is deformed by the pulling forces of the endocytosis in such a way to reveal a site for cleavage by metalloproteases (called S2 cleavage). Metalloproteases cleavage turns Notch into a substrate for a final cleavage (S3 cleavage) by the γ-secretase complex on the cytoplasmic side of the plasma membrane (Fig. 1a). As Notch molecules are constitutively targeted to lysosomes for ubiquitin-dependent degradation, cleavage events during canonical signaling are likely to occur also on the endosomal membrane. In all cases of canonical signaling, once the intracellular domain of Notch (NICD) is liberated from membranes, it accesses the nucleus where it regulates transcription of target genes by de-repressing the CBF1/Su(H)/Lag-1 (CSL) [RBP-Jκ in mammals, Su(H) in *Drosophila*] transcription complex (Fig. 1b). Signal termination is ensured by ubiquitin-dependent proteasomal degradation of NICD [see for review (Bray 2016)].

Understanding the fine regulation of a transmembrane receptor turning into a transcription factor, in the context of canonical signaling events, as well as of an expanding set of seemingly non-canonical events, has proven challenging. Part of the challenge derives from the fact that Notch behaves as a recursive cell fate switch in countless developmental and homeostatic processes in metazoans. In addition, the existence of four Notch paralogs (Notch1–4) and five DSL ligands paralogs in mammalian cells further complicates the matter (Andersson et al. 2011). Although we currently do not have all the evidence, the nature of Notch signaling as a highly context-dependent cell fate switch suggests that non-canonical signaling events might be a norm rather than an exception. Thus, detailed understanding of non-canonical signaling holds the promise of unlocking our ability to develop innovative strategies to counteract diseases associated with altered Notch activity. Indeed, efficient inhibition of Notch signaling has proved too toxic for clinical use, mostly due to unwanted on-target effects (van Es et al. 2005; Purow 2012; Andersson and Lendahl 2014). Despite recent development of Notch paralog-specific inhibitors that will hopefully reduce toxicity (Wu et al. 2010), expanding our knowledge of Notch regulators beyond those required for canonical signaling might increase therapeutic options. In this chapter, we discuss studies that propose mechanisms of non-canonical signaling. Because most

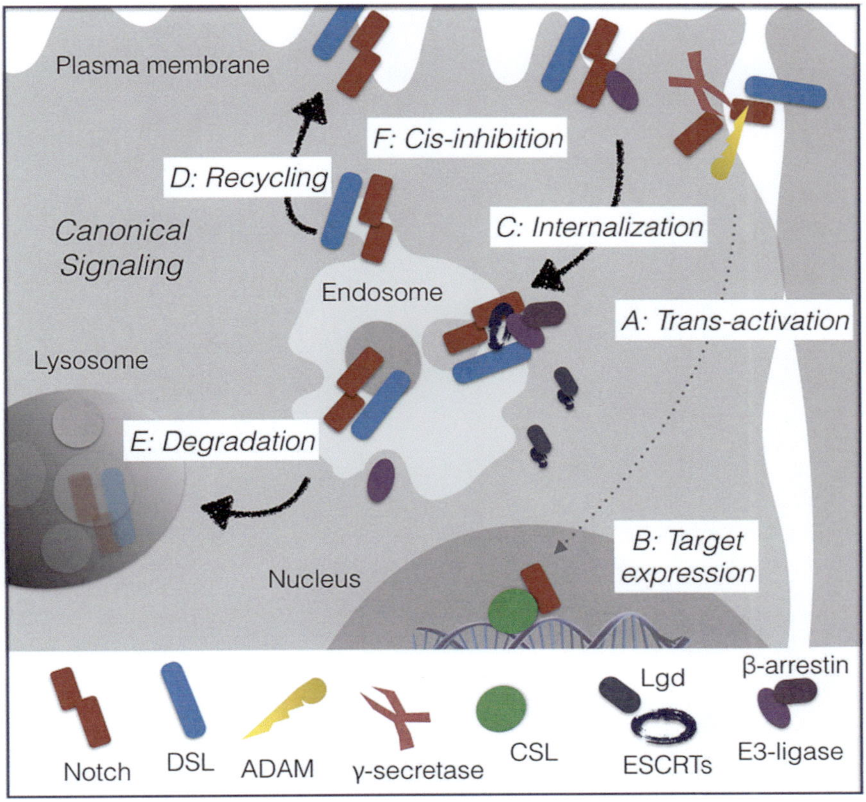

Fig. 1 Canonical Notch signaling. (**a**) Notch is cleaved to NICD by γ-secretase at the plasma membrane, or en-route to endosomes, upon trans-activation by a DSL ligand expressed in signal-sending cells. Extracellular cleavage by ADAM metallo-proteases is a prerequisite for γ-secretase processing. (**b**) Target gene expression depends on CSL-mediated transcription derepressed by NICD. (**c**) After ubiquitination by a number of E3-ligases, such pool is internalized to endosomes. (**d**) Once in endosomes, Notch can be recycled back to the plasma membrane. (**e**) Alternatively, it is sorted to internal vescicles of the endosomes by ESCRTs and Lgd and eventually degraded in the lysosome. (**f**) A large pool of Notch is kept inactive by cis-inhibition by DSL ligands.

of the initial and current evidence of non-canonical signal emerged from work in *Drosophila*, a prime model system for genetics and developmental studies, where possible we will describe first the existing evidence in *Drosophila* and then discuss how it relates to the findings in mammalian systems, which are currently less abundant. Based on such survey of the literature, we have attempted in the final part of the chapter to rationalize the logic of non-canonical signaling in the context of disease, an effort that we hope will facilitate future understanding of Notch signaling in pathology.

2 Notch Signaling That Is Independent of Ligand Interactions in Trans

2.1 dx-Dependent Activation

One of the first modifiers of Notch that appeared to act independently of ligands is the product of the *Drosophila deltex* (*dx*) gene. *dx* encodes a cytoplasmic RING domain-containing protein, called dx, that binds to the Ankyrin repeats of the Notch intracellular domain (Xu and Artavanis-Tsakonas 1990; Busseau et al. 1994; Diederich

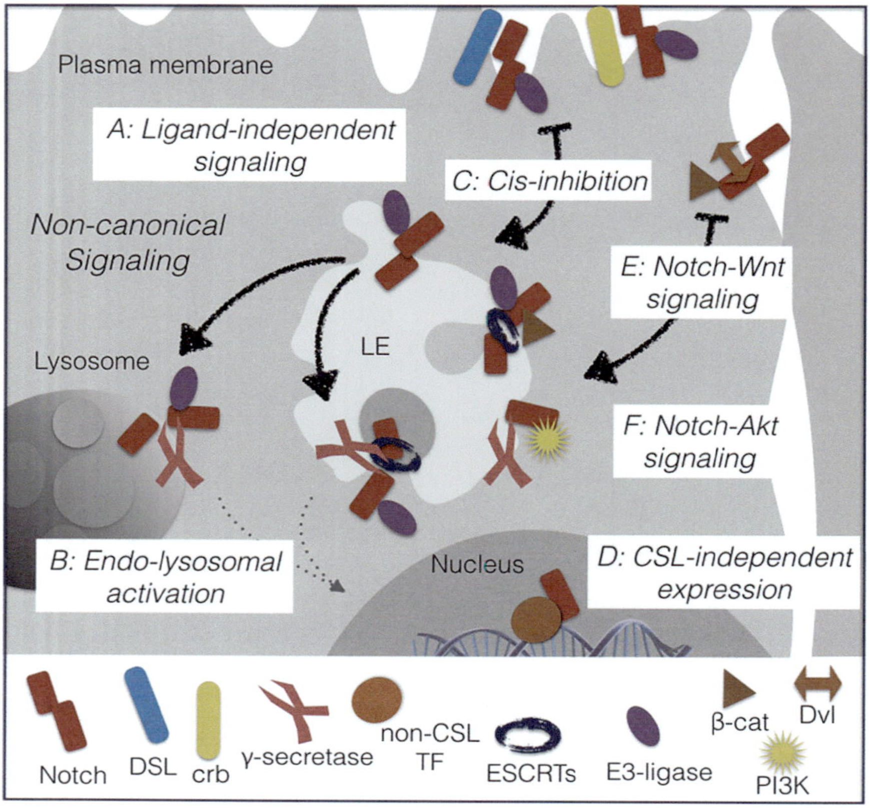

Fig. 2 Non-canonical Notch signaling. (**a**) A pool of Notch can be activated in a ligand-independent fashion. (**b**) Such pool of internalized Notch can be saved from endosomal sorting towards degradation by the activity of the E3 –ligase dx, and cleaved at the endosome, or upon fusion with the lysosome. β-arrestin, ESCRTs and other factors participate in the process. (**c**) Cis-inhibition prevents inappropriate ligand-independent activation. (**d**) In some cases this pathway can result in target gene expression that is CSL-independent. Inhibition of signaling can be achieved by direct binding to the apical determinant crb, or to Dvl, a Wnt signaling component (**e**). The level of a second Wnt component, β-catenin (β-cat), which also binds to Notch, is downregulated by endosomal sorting, thus preventing excess Wnt signaling. (**f**) Notch activation is inhibited by Akt signaling by directly binding the component PI3K, possibly in the endosomal system

et al. 1994; Matsuno et al. 1995). Indeed, ectopic activation of Notch signaling could be achieved by dx overexpression in *Drosophila* wing margin cells that lacked of both Delta (Dl) and Serrate ligands. dx generally functions as a positive regulator of Notch signaling because *dx* mutants suppress Notch gain-of-function wing phenotypes (Hori et al. 2004). Molecularly, dx was found to facilitate Notch mono-ubiquitination and relocalization from the cell surface towards the late endosome (LE) (Fig. 1c). It also retains Notch on the LE limiting membrane, possibly favoring ectodomain shedding in the endosomal environment (Fig. 2a–b). By depleting Notch from the cell surface, dx reduces the pool of receptors accessible to ligands; however, by retaining Notch on the LE surface, dx prevents some Notch receptors from degradation in lysosomes, making them available for signaling [Fig. 1b, (Hori et al. 2004; Wilkin et al. 2008; Yamada et al. 2011; Hori et al. 2011)]. In addition, when complexed with kurtz (krz), a non-visual β-arrestin identified as the first dx physical interactor, dx attenuates Notch signaling. The binding of dx-krz complex to Notch promotes the polyubiquitination of endocytosed Notch receptors leading to their degradation (Fig. 1c). This requires the presence of shrub (Charged Multivesicular Body Protein 4

[CHMP4] in mammals), a subunit of the Endosomal Sorting Complex Required for Transport (ESCRT)-III that directs cargoes towards the intraluminal vesicles of LEs ([Fig. 1c], (Mukherjee et al. 2005; Hori et al. 2011)]. Blocking the trafficking of Notch receptors towards LEs by mutations of Rab5, Rab7, AP-3 and HOPS complex genes inhibited dx-mediated Notch activation suggesting that ligand-independent Notch signaling requires Notch receptors to be localized to the LE limiting membrane (Vaccari et al. 2008; Wilkin et al. 2008; Zheng et al. 2013). Consistent with the fact that dx mostly affects ligand-independent signaling, ectopic Notch signaling in *Drosophila* ESCRT mutant tissues is not abolished by mutations that block ligand activity (Vaccari et al. 2008).

dx is conserved and also regulates Notch signaling in mammals. One of the human dx homologues, DTX1, interacts with the ankyrin repeat regions of NOTCH1 and NOTCH2, similarly to *Drosophila* dx. DTX1 over-expression has the same effect as over-expressing NICD on inhibiting the activity of E47 and Mammalian achaete scute homolog-1 (MASH1), which are the product of two different Notch target genes, suggesting that DTX1 also functions as a positive regulator of Notch signaling (Matsuno et al. 1998; Yamamoto et al. 2001). Misexpression of murine *dx* homologs, mouse *Deltex1* (*Dtx1*), *mouse Deltex2* (*Dtx2*) and *mouse Deltex3* (*Dtx3*) also inhibits E47 activity, similar to the effect of activated Notch (Kishi et al. 2001). Thus, the ability of Deltex to activate Notch signaling is conserved from *Drosophila* to mammals; unlike *Drosophila*, however, how much of this activity is dependent on Notch ligands is currently unclear. Mammalian *dx* homologs may also inhibit Notch signaling in specific developmental contexts (Sestan et al. 1999; Izon et al. 2002; Kiaris et al. 2004). In addition, DTX3L regulates the endosomal sorting of receptors other than Notch and may regulate general endosomal sorting via ubiquitination of ESCRT subunits (Holleman and Marchese 2014).

Drosophila Suppressor of Deltex [*Su(dx)*] mutants dominantly suppress the phenotypes of *dx* mutants, hence the name. *Su(dx)*, which encodes a Homologous to the E6-AP Carboxyl Terminus (HECT)-type E3 ubiquitin ligase of the Neural precursor cell expressed developmentally down-regulated protein 4 (Nedd4) family, has been described as a negative regulator of the Notch pathway by antagonizing Deltex and Notch (Busseau et al. 1994; Fostier et al. 1998; Cornell et al. 1999; Mazaleyrat et al. 2003). Su(dx) transiently interacts with Notch at the cell surface and subsequently sorts constitutively-internalized full-length Notch receptors away from the Rab11-positive recycling endosome and into an ESCRT/ubiquitin-positive compartment for degradation to downregulate signaling [Fig. 1d–e, (Wilkin et al. 2004; Djiane et al. 2011)]. Su(dx) also counteracts ligand-independent Notch signaling by directly ubiquitinating Notch receptors. *Drosophila* Nedd4, a second HECT-domain E3 ligase, also reduces ligand-independent Notch signaling using similar mechanisms proposed for Su(dx) [Fig. 1d, (Sakata et al. 2004)]. In addition, *Drosophila* Nedd4 family interacting protein (Ndfip) promotes ligand-independent Notch signaling using the same mechanism as dx (Dalton et al. 2011). The mouse homolog of Su(dx), called Itchy E3 Ubiquitin Protein Ligase (Itch), also physically interacts with Notch, promotes its ubiquitination and degradation, and subsequently downregulates Notch signaling (Qiu et al. 2000). Su(dx) may also antagonize Notch signaling by promoting the degradation of dx. Indeed, Atrophin-1 Interacting Protein 4 (AIP4), the human Su(dx) homolog, directly binds dx to promote its polyubiquitination and subsequent degradation (Chastagner et al. 2006).

Ligand-independent signaling in *Drosophila* is also prevented by *lethal (2) giant discs* (*lgd*) which was originally classified as a tumor suppressor gene because its deletion caused overproliferation of larval epithelial imaginal discs, eventually found to be due to ectopic activation of Notch signaling [Fig. 1c, (Bryant and Schubiger 1971; Watson et al. 1994)]. lgd encodes a C2-domain containing protein that binds phospholipids and interacts with the ESCRT-III subunit shrub. Analysis of *lgd* mutant tissues suggests that lgd functions in endosomal sorting towards degradation and that defects in lgd ultimately reduce shrub function, causing Notch receptors to accumulate on the limiting membrane of LEs and ectopically signal, even in absence of ligands (Fig. 2b). Importantly, such activation depends on the cleavage of Notch by the γ-secretase complex and requires fusion of LEs with lysosomes (Klein 2003; Childress et al. 2006; Gallagher and Knoblich 2006; Jaekel and Klein 2006; Troost et al. 2012; Schneider et al. 2013). This suggests that Ca^{2+} release associated with fusion events, or protein degradation in the lysosomal lumen, could substitute for ligands in shedding the extracellular part of the Notch heterodimer. lgd regulation of ESCRT activity might extend beyond Notch. Indeed, *Drosophila* lgd mutants display ectopic activation of the BMP/Dpp signaling receptor Thickveins (Morawa et al. 2015). Human lgd homologs Coiled-Coil And C2 Domain Containing 1A (CC2D1)A/B also play important roles in endosomal sorting by interacting with and regulating the ESCRT-III subunit CHMP4. However, CC2D1A and CC2D1B mutants do not display marked differences in Notch signaling (Usami et al. 2012; Martinelli et al. 2012; Drusenheimer et al. 2015). This, together with the evidence that human lgd paralogs also control Nuclear Factor-κB (NF-κB) and Epidermal Growth Factor Receptor (EGFR) signaling (Zhao et al. 2010; Deshar et al. 2016), indicates that aspects of endocytic ligand-independent Notch activation in human cells might differ from the *Drosophila* paradigm. A more general discussion of the endocytic control of Notch signaling, not limited to non-canonical signaling is covered in a dedicated chapter.

2.2 Cis-Inhibition of Signaling

In *Drosophila*, DSL ligands in signal-sending cells not only activate the Notch receptor in trans in nearby signal-receiving cells, but also are thought to repress Notch in cis at the plasma membrane [Fig. 1f, (Miller et al. 2009; Becam et al. 2010; Sprinzak et al. 2011)]. Recently, it has been proposed that cis-interactions function to prevent ligand-independent Notch activation in the endocytic compartment, thereby favoring ligand-directed signaling at the plasma membrane (Fig. 2c). Indeed, when all ligand are genetically removed from fly ovary cell, Notch signaling becomes activated in the nearby follicular epithelium (Palmer et al. 2014). Thus while cis-inhibition has been regarded as part of the canonical mechanism of activation, such new evidence indicates that it also controls the non-canonical ligand-independent regulation of signaling.

Interestingly, DSL ligands are not the only molecules that can repress ligand-independent signaling *in cis*. Indeed, in *Drosophila* the apico-basal polarity determinant crumbs (crb) appears to selectively prevent full-length Notch from being internalized from the cell surface by physically associating in its extracellular domain with the extracellular domain of Notch, resulting in repression of dx-dependent Notch activation [Fig. 2c, (Nemetschke and Knust 2016)]. The role of crb in Notch signaling might not be limited to regulation of ligand-independent Notch activation, as others have reported actions of crb on Notch signaling that requires ligands (Herranz et al. 2006; Richardson and Pichaud 2010). crb can be added to the growing list of potential unconventional Notch ligands. Some of these, including the extracellular domain of crb that interacts with Notch, contain tandem EGF-like repeats similar to those present in the Notch receptors and ligands. However, as is the case for the mammalian Dl and Notch-like epidermal growth factor-related receptor (DNER) or the *Drosophila* Notch-activating protein weary (wry), their ability to directly regulate Notch activity has not been fully tested and remains controversial (Eiraku et al. 2005; Kim et al. 2010; Hsieh et al. 2013; Greene et al. 2016).

2.3 Ligand-Independent Signaling in Physiology

What is the role of ligand-independent Notch signaling in normal physiology? In a study of *Drosophila* hemocyte development, Banerjee and colleagues (2015) observed that ligand-independent activation of the Notch receptor provides most of the Notch activation in crystal cells, a type of *Drosophila* blood cells. Because crystal cells are circulating and lack proximity to ligand-bearing cells, they circumvent this limitation by activating Notch non-canonically (Mukherjee et al. 2011). A second context that is likely to rely on ligand independent signaling is represented by imaginal disc proliferation. Indeed, lack of *lgd* cause over proliferation but do not affect inductive signaling at the wing margin, which is well known to rely on DSL ligands (Troost et al. 2012; Schneider et al. 2013). However, the effect *of dx, Su(dx)* and *crb* mutations are apparent also in wing margin and vein formation, suggesting that ligand-independent Notch signaling might also help to refine ligand-directed canonical signaling (Busseau et al. 1994; Fostier et al. 1998; Cornell et al. 1999; Mazaleyrat et al. 2003; Nemetschke and Knust 2016). Consistent with this, mutations of endo-lysosomal components and regulators affect Notch-dependent tissue proliferation as well as differentiation, as reported recently for sensory organ precursors (Vaccari et al. 2010; Couturier et al. 2013; Tognon et al. 2016). One compelling reason for ligand-independent refinement of signaling in *Drosophila* has been proposed recently. Baron and colleagues showed that Su(dx) activity on Notch signaling is temperature-dependent and suggested that the dx-Su(dx) system acts as a buffer against temperature fluctuations which are a critical feature of ectothermic organisms (Shimizu et al. 2014).

3 Protease-Independent Notch Signaling

Notch is constitutively processed by Furin at the S1 site to produce the heterodimer at the cell surface (Blaumueller et al. 1997; Logeat et al. 1998). Activation of CSL by Delta1 and Jagged is blocked by Furin inhibition, suggesting that only the Notch heterodimer can activate CSL in response to ligands (Jarriault et al. 1998; Bush et al. 2001). However, the myogenesis of C2C12 myoblast cells, which is usually inhibited by activated Notch signaling, occurs when Notch receptor is unprocessed and is CSL-independent because it proceeds in the absence of a functional CSL (Bush et al. 2001).

In *Drosophila*, after ligand binding, the S2 site of Notch is cleaved by Kuzbanian, a metalloprotease of the A Disintegrin And Metalloproteinase protein (ADAM) family (Lieber et al. 2002). While similar cleavages have been reported in mammalian cells by ADAM10 and the Tumor Necrosis Factor (TNF)-alpha-converting enzyme (TACE), interestingly in melanoma cells Notch1 can also be processed non-canonically by the Membrane-Tethered -Matrix Metallo-Protease 1 (MT1-MMP) (Ma et al. 2014).

Presenilin (Psn)-mediated cleavage of Notch, to release the NICD, is a hallmark of canonical Notch signaling. One study shows the existence of a non-canonical signaling that is activated in the absence of Psn. In mouse blastocysts simultaneously mutated for *Presenilin 1/2* (PS1/2), partial *Hairy and enhancer of split* 1 (*Hes1)* activation is still retained in the presence of Dl ligand (Berechid et al. 2002). A Psn- and CSL-independent regulation of cytokine production in response to ligand activation in human T-cells has also been observed (Stallwood et al. 2006). One recent report suggests that in rat neurons, expression of presynaptic proteins depends on DSL ligands but might not depend on Psn-mediated cleavage (Hayashi et al. 2016).

Taken together, these evidences are limited but they reveal that in particular instances, unprocessed or non-canonically-processed Notch can mediate downstream signaling events.

4 Non-canonical Notch Signaling in the Nucleus

4.1 CSL-Independent Notch Signaling

Several reports now indicate that NICD can regulate transcription factors without activating CSL (Fig. 2d). Initially, by analyzing a class of Notch gain-of-function mutants in *Drosophila* called *microchaetae defective* (*Notch*^{MCD}) characterized by the loss of the microchaetae (small bristles),

Heitzler and colleagues showed that the *Notch*^{MCD} phenotype is independent of the CSL homolog Su(H), of Dl and of the E(Spl) class of Notch targets; this indicates that the Notch-dependent process of lateral inhibition that drives microchaetae specification is in part non-canonical (Ramain et al. 2001). In one study, ectopic Notch signaling induced by dx overexpression in *Drosophila* tissues has also been proposed to be mostly in part Su(H)-independent because dx still activated Notch target genes in the absence of *Su(H)* (Hori et al. 2004).

Some forms of Notch signaling appear to occur independently of CSL also in vertebrates. During myogenesis, active Notch inhibits the activity of the basic helix-loop-helix (bHLH) transcription factor MyoD to prevent muscle differentiation via CSL activation (Kopan et al. 1994; Kuroda et al. 1999) and in murine C2C12 myoblast cells, expression of activated Notch or co-culture with Jagged1-expressing cells inhibits differentiation in myotubes. Other studies have however shown that the myogenic inhibition induced by Notch can also be CSL-independent. In fact, expression of a truncated form of Notch lacking of the CSL-interacting sequences, that is unable to activate CSL, still inhibited the differentiation of C2C12 into myotubes in presence of inhibitory Notch signals (Shawber et al. 1996; Nofziger et al. 1999). However, Honjo and colleagues have argued that these Notch constructs still bound CSL weakly and that the Notch-mediated myogenic block of C2C12 cells requires Hes1 transactivation (Kato et al. 1997). Despite this, other instances of CSL-independent Notch signaling have been reported: (1) activated Notch4, but not constitutively active CSL fused to the activator VP16 inhibits apoptosis of epithelial cells (MacKenzie et al. 2004); (2) a dominant negative CSL mutant, unable to bind DNA but still able to interact with NICD, blocks NICD-mediated *HES1* expression but not the anti-apoptotic activity of NICD on medium-deprived HeLa cells (Perumalsamy et al. 2009); (3) Notch-dependent proliferation, activation and differentiation of mouse peripheral CD4$^+$ T cells isolated from mice that had been conditionally deleted of CSL occurs normally, while these processes were blocked in Notch-deleted T cells (Dongre et al. 2014).

Interestingly, some of the downstream effects of Dtx overexpression on Notch target genes in vertebrates seem to be independent of CSL activity. In experiments performed on the rat neuroepithelial cell line MNS-70, expression of either activated Notch1 or human DTX1 prevents MNS-70 cells from differentiating into glia and neurons via inhibiting the downregulation of Nestin, which is controlled by the transcriptional factor MASH1. Notch1 interacts physically with DTX1, which in turns interact with p300 to inhibit MASH1. The DTX1-mediated inhibition of MASH1 is CSL-independent because dominant-negative CSL blocks Notch2-mediated inhibition of rat MASH1, but has no effect on DTX1-mediated MASH1 inhibition. In addition, DTX1 overexpression, unlike active Notch1, does not induce activity of a CSL-dependent promoter and *Hes1* expression (Yamamoto et al. 2001). Another case of Dtx-mediated CSL-independent activity is in the regulation of B-lymphocyte development. Either Dtx overexpression or activated Notch1/2 inhibit E47, a bHLH transcription factor essential for B-lymphocyte specification and this is independent of CSL based on two observations: (1) while activated Notch1/2 induce CSL activity, Dtx has no effect; (2) a Notch construct lacking the RAM domain required for CSL interaction still retains the ability to inhibit E47 (Ordentlich et al. 1998). Additionally, Dtx regulates Notch signaling during neural crest formation independently of CSL factors (Endo et al. 2002; Endo et al. 2003). Finally, induction of Phosphatidylinositol 4,5-Bisphosphate 3-kinase (PI3K) by Jagged1 in a cell model of Human Papilloma Virus (HPV) cervical tumors is mediated by Dtx but not by CSL (Veeraraghavalu et al. 2005). Such abundant evidence is often correlative and based mostly on overexpression of truncated Notch forms. While it seems safe to suggest that aspects of Notch-dependent myogenesis, apoptosis and T-cell development rely on CSL-independent signaling, a better understanding of which Notch signaling events require CSL-independent and/or ligand-independent signaling await more physiologic and in vivo set-ups studies (Fig. 1d).

4.2 Non-canonical Activation of Notch Target Genes

Several studies indicate that canonical Notch target genes of the Hes/Hey (Hairy/Enhancer-of-split related with YRPW motif protein) family of transcription factors (Iso et al. 2003) can be activated independently of Notch. Examples are the regulation of *Hes1* expression by Sonic Hedgehog, c-Jun N-terminal kinase (JNK), RAS/Mitogen-Activated Protein Kinase (MAPK), Fibroblast Growth Factor (FGF) and Wnt signaling pathways. In these contexts, removal of Notch, inhibition of Notch cleavage or ablation of CSL do not appear to interfere with *Hes1*, *Hes6*, *Hey2* and *Hairy-1* regulation and some transcriptional components of these pathways directly interact with the promoter of Notch targets to drive their expression (Stockhausen et al. 2005; Ingram et al. 2008; Doetzlhofer et al. 2009; Kubo and Nakagawa 2009; Wall et al. 2009; Sanalkumar et al. 2010; Carvalho et al. 2015).

4.3 Transcription-Independent Activity of Nuclear Notch

A function of Notch that relies on nuclear localization of the NICD but is independent of its transcriptional activity is regulation of DNA damage response (DDR). DDR coordinates cellular response to double strand breaks (DSBs) by recruiting to and activating the kinase Ataxia Telangiectasia Mutated (ATM) at sites of DNA damage [reviewed in (Maréchal and Zou 2013)]. Notch negatively regulates DDR by physically interacting with ATM and by inhibiting the binding of ATM to specific proteins required for ATM autophosphorylation and subsequent phosphorylation of its substrates. Also, this is not mediated by transcriptional activity of the intracellular domain of Notch because loss of its transactivation domain (TAD) does not impair its ability to inhibit ATM signaling. Finally, activated Notch and phosphorylated ATM show an inverse correlation in breast cancer tissues, suggesting that this non-canonical Notch-mediated regulation of DDR may be clinically relevant (Vermezovic et al. 2015; Adamowicz et al 2016).

5 Non-nuclear Functions of Notch

5.1 Notch and Wnt/β-Catenin Signaling

During the development of muscle progenitors in the mesoderm, signaling via *Drosophila* Wnt (called Wingless or Wg) is required for the expression of the S59 transcription factor, while Notch signaling restricts the number of S59-positive muscle progenitors via canonical lateral inhibition. An activity of Notch distinct from lateral inhibition, that is impaired upon *Notch* deletion but not upon deletion of *Su(H)* or *E(Spl)*, restores S59 expression in the absence of *Wg* (Brennan et al. 1999). In presence of Wg signaling, induced by a constitutively-active form of Armadillo (the *Drosophila* homolog of β-catenin), expression of either full length Notch or a transmembrane form of Notch that cannot be cleaved into NICD, suppresses the phenotypic effect of activated β-catenin. This effect is proposed to require neither the function of Notch ligands nor its transcriptional activity but rather to depend on the physical interaction between Notch and β-catenin [Fig. 2e, (Hayward et al. 2005; Sanders et al. 2009)].

Non-canonical Notch/Wnt regulation is conserved in vertebrates where the levels of Notch frequently correlate inversely with those of β-catenin. Indeed, in *Xenopus leavis* embryos, the interaction between Notch and β-catenin has been shown to regulate the size and anterior-posterior patterning of the brain. Reduction of Notch in these embryos re-localizes β-catenin from cell-cell junctions to the nucleus to activate Wnt signaling ectopically. On the other hand, expression of the NICD reduced the levels of β-catenin independently of CSL activity (Acosta et al. 2011). In mammalian cells, Notch1 physically associates with and prevents the accumulation of active (unphosphorylated) β-catenin, even in absence of CSL activity. Even though β-catenin levels are reduced by NICD in *Xenopus* embryos, the cleavage of Notch is not required because a membrane-tethered form of Notch that cannot be cleaved is able to bind and reduce the levels of active β-catenin, derepressing its transcriptional activity (Kwon et al. 2011).

In the canonical Wnt pathway, cytoplasmic β-catenin levels are controlled by the activity of a 'destruction complex' in absence of Wnt ligands. The destruction complex — composed of Ser/Thr kinases casein kinase 1 (CK1) and glycogen synthase kinase 3β (GSK3β), Adenomatous polyposis coli (APC), and the scaffolding protein Axin — phosphorylates β-catenin to induce its ubiquitination and eventual its degradation by the proteasome (Stamos and Weis 2013). Interestingly, GSK3β is not required for Notch-mediated reduction of active β-catenin because: (1) reduction of activated β-catenin by Notch occurs in the *Drosophila* tissues lacking *GSK3β* (*Shaggy* in flies); (2) a form of β-catenin that lacks the GSK3β target residues is still post-translationally downregulated by Notch (Acosta et al. 2011); (3) pharmacological inhibition of the destruction complex using a specific inhibitor of GSK3β does not influence the ability of Notch to titrate the levels of active β-catenin (Kwon et al. 2011).

How then does Notch regulate the levels of active β-catenin? The first hint came from observations that the genetic interaction between *Notch*^{MCD} alleles and *Wg* alleles in *Drosophila* required *dx*, suggesting that endo-lysosomal trafficking may play a role in this process (Ramain et al. 2001). Also, full length Notch and β-catenin were found to partially co-localize in endocytic vesicles (Sanders et al. 2009). This has been confirmed in mammalian cells in which knockdown of the endocytic adaptor proteins Numb and Numb-like (Numbl) or treatment with bafilomycin A1, a specific vacuolar H$^+$-ATPase (V-ATPase) inhibitor abrogate the ability of Notch to titrate levels of active β-catenin (Kwon et al. 2011). Taken together, Notch is thought to lower the protein levels of β-catenin by mediating its endo-lysosomal trafficking and eventual degradation in the lysosomes [Fig. 2e, (Kwon et al. 2011)]. It is worth mentioning here that the Wnt pathway may also modulate Notch signaling. GSK3β protects the intracellular domain of Notch from proteasomal degradation by direct binding and phosphorylation, thus positively regulating Notch signaling. Another Wnt/Wingless component, Disheveled (Dvl), binds the intracellular domain of Notch to inhibit signaling (Fig. 2e). However, these Wnt effects are generally observed on the canonical Notch signaling pathway that requires ligands and CSL activity (Axelrod et al. 1996; Foltz et al. 2002; Collu et al. 2012).

5.2 Notch and mTOR/Akt Signaling

Non-canonical Notch signaling is involved in regulating the Mechanistic Target Of Rapamycin/(mTOR)/AKT pathway in human cells (Fig. 2f). In T-cells, starved HeLa cells and cervical cancer cell lines, ectopic expression of Notch inhibits apoptosis by inducing the expression of anti-apoptotic proteins and this is blocked by overexpressing the Notch antagonist Numb. This anti-apoptotic function of Notch is mediated by activation of the mTOR/AKT pathway based on the following evidences: (1) the anti-apoptotic function of Notch is blocked when PI3K activity is chemically inhibited; (2) Notch1 physically associates with PI3K; (3) Notch expression leads to an enhancement of Akt phosphorylation (AktS473) and; (4) depletion of Akt or mTOR inhibition by rapamycin abrogates Notch anti-apoptotic activity (Rangarajan et al. 2001; Nair et al. 2003; Sade et al. 2004; Mungamuri et al. 2006; Perumalsamy et al. 2009). The Notch signaling that functions here requires the presence of ligands but appears to be independent of CSL activity. Interestingly, the crosstalk between Notch and mTOR/Akt for its anti-apoptotic activity does not require nuclear localisation but is mediated by a cytoplasmic pool of NICD. Moreover, a form of NICD that is tethered to the membrane recapitulated NICD anti-apoptotic effect while prolonged nuclear retention by additional nuclear localisation sequences (NLS) curtailed this effect (Perumalsamy et al. 2009). In addition, a non-canonical PI3K/AKT-dependent Notch activity that does not require γ-secretase activity mediates cytokine responses of dendritic cells in response to pathogenic stimuli [(Gentle et al. 2012), Fig. 1f].

5.3 Notch and NF-κB Signaling

Cytoplasmic Notch is important for the NF-κB activation in stimulated T cells by regulating the formation of the CBM (CARMA1, BCL10 and MALT1) complex. NICD, acting as a scaffold protein, physically interacts with components of the CBM signalosome in the cytosol, promotes nuclear accumulation of NF-κB, and stimulates NF-κB signaling (Shin et al. 2014). In breast cancer cells, non-canonical Notch signaling upregulates the expression of the proinflammatory cytokine interleukin-6 (IL-6), independently of CSL activity, by associating with cytoplasmic components of the NF-κB pathway (Jin et al. 2013).

5.4 Notch and APP

In *Drosophila*, zebrafish and human cells, a well-established cross-talk exists between the Amyloid Precursor Protein (APP), involved in Alzheimer disease, and Notch both of which are cleaved by γ-secretase (Merdes et al. 2004; Fischer et al. 2005; Banote et al. 2016). Intriguingly, several studies indicate that Notch proteins physically interact with APP and its cofactors (Roncarati et al. 2002; Fischer et al. 2005; Fassa et al. 2005; Chen et al. 2006; Kim et al. 2011; Banote et al. 2016). While these effects could also be explained by competition of Notch and APP for γ-secretase cleavage (Berezovska et al. 2001; Lleó et al. 2003), the recent realization that APP might act as an important endocytic regulator (Kim et al. 2016) suggests that APP might directly modulate activation of Notch from endosomes.

6 Conclusions and Importance of Non-canonical Notch Signaling in Pathology

Despite several years of study in multiple metazoan systems, the extent and relevance of non-canonical Notch signaling events are still unclear and require further study. Here, we have discussed the existing literature to differentiate types of non-canonical Notch signaling events. Because Notch signaling has been most extensively studied in *Drosophila*, whenever possible we have used evidence in flies as a blueprint to rationalize our current understanding of non-canonical signaling. Notwithstanding the clear mechanistic similarity with vertebrate and mammalian systems, for example in the extensive use of the endocytic system for Notch degradation and ligand-independent activation, the *Drosophila* paradigm might eventually represent an extreme adaptation of an ectothermic organism to cope with environmental changes. On the other hand, clear conservation of most of the non-canonical signaling machinery and the emergence of roles in mammalian physiology indicates that this form of signaling is likely to assist and integrate canonical signaling in a wide range of developmental and homeostatic contexts in several organs. Considering this, it comes as no surprise that in Notch-linked pathologies, non-canonical signaling is likely to play a role, as initial evidence seems to suggest. In fact, human cell assays designed to test the activity of Jagged1 missense mutations that cause Alagille syndrome, revealed that they do not work efficiently in cis-inhibition (Tada et al. 2012). In spondylocostal dysostosis, a congenital abnormal vertebral segmentation syndrome caused by mutation in Delta Like (Dll) 3, a defect in cis-inhibition ultimately leading to increased degradation of Notch1 has also been proposed using mouse as model system (Chapman et al. 2011). Endocytosis of Dll1 has

Table 1 Types of non-canonical signaling, their component requirements, localization in the cell and relevance for disease. Tentative requirements or relationships that awaits experimental validation are indicated by question marks. X: Not required; √: Required; ? Not tested; β-cat: β-catenin

Type of non-canonical signaling	Requirements					Cellular localisation	Relevance for disease
	DSL	dx	Furin	γ-secretase	CSL		
Ligand-independent	X	√	√?	√	√	Endosome and lysosome	Breast cancer, lung cancer?
Cis-regulation	X/√	√	√?	√	X	PM and endosomes	Alagille syndrome, Spondylocostal dysostosis, Familiar Parkinson disease
Protease-independent	√	?	–	–	X		Melanoma
Wnt -associated	X (β-cat)/√(Dvl)	√	X	X	X	PM and endosomes	Colon and cervical cancer?
Akt -associated	X	X?	√?	X	X	Endosome and lysosome?	T-ALL leukemia, Tuberous sclerosis
Nf-κb -associated	?	?	?	X?	X	Cytoplasm?	Breast cancer, Inflammation?
APP -associated	X?	√?	√?	√?	√?	PM and endosomes?	Alzheimer disease? Down syndrome

been proposed to affect cis-inhibition in a study of autosomal-dominant familial Parkinson disease with mutations in trafficking regulator Leucine Rich Repeat Kinase 2 (LRRK2) (Imai et al. 2015). Given the cross-talk of Notch with APP and the emerging involvement in endo-lysosomal degradation process, it is also predictable that non-canonical Notch regulation might prove relevant to Alzheimer pathogenesis and to management of Down syndrome, that entails ectopic APP activity (Doran et al. 2017).

Aspects of non-canonical signaling might be associated to Notch also in tumorigenesis. Indeed, considering that conserved endocytic regulators of Notch trafficking such as Numb have been implicated in non-canonical signaling (Kwon et al. 2011) and are mutated in Notch dependent cancers (Pece et al. 2004; Pece et al. 2011), we predict that non-canonical Notch signaling might often be disrupted alongside juxtacrine ligand-directed signaling operating canonically. Experiments in colon and cervical cancer models and in models of cancer chemoresistance indicate that non-canonical Notch signaling involving Wnt and Akt might be affected and also point to a prominent role of endocytosis in these processes (Maliekal et al. 2008; Bertrand et al. 2012). In tuberous sclerosis, a dominantly inherited multisystem disease characterized by the growth of benign tumors and caused by mutations in *Tuberous Sclerosis 1/2 (TSC1/2)*, Notch is upregulated. TSC1/2 are lysosome-associated regulators of the kinase mTOR but the Notch activation in *Drosophila* mutants in *Tsc1 and gigas* (fly TSC2) appear to be independent of other mTOR regulators, suggesting that TSC1/2 might be directly regulating Notch activity (Karbowniczek et al. 2010; Ma et al. 2010). Some aspects of Notch activation appear sensitive to endocytosis also in assays that use breast and T-acute lymphoblastic leukemia (T-ALL) cancer models (Kobia et al. 2014; Faronato et al. 2015). Examples of non-canonical regulation that do not involve endocytosis include non-canonical processing of Notch1 by alternative MMPs (Ma et al. 2014).

In closing, we submit to life scientists operating in the Notch field a simple table that catalogs instances of non-canonical Notch signaling and their relevance to pathologies (Table 1). This might be helpful to design future experiments to disprove occurrence of non-canonical signaling in their experimental systems. We hope our survey and rationalization of the existing evidence of non-canonical signaling might ultimately be helpful to devising treatments for Notch diseases with increased efficacy and minimal toxicity. Ideally, these should target specific non-canonical regulators in appropriate contexts, alone or in conjunction with existing inhibitors of core signaling components.

Acknowledgments The authors wish to thank E. Morelli for critically reading the manuscript.Work in T.V. laboratory is supported by Worldwide Cancer Research grant #18-0399 and by AIRC (Associazione Italiana Ricerca sul Cancro) Investigator grant #20661.

References

Acosta H, López SL, Revinski DR, Carrasco AE (2011) Notch destabilises maternal beta-catenin and restricts dorsal-anterior development in Xenopus. Development 138:2567–2579. https://doi.org/10.1242/dev.061143

Adamowicz M, Vermezovic J, d'Adda di Fagagna F (2016) NOTCH1 Inhibits Activation of ATM by Impairing the Formation of an ATM-FOXO3a-KAT5/Tip60 Complex. Cell Rep 16:2068–2076. https://doi.org/10.1016/j.celrep.2016.07.038

Andersson ER, Lendahl U (2014) Therapeutic modulation of Notch signalling — are we there yet? Nat Rev Drug Discov 13:357–378. https://doi.org/10.1038/nrd4252

Andersson ER, Sandberg R, Lendahl U (2011) Notch signaling: simplicity in design, versatility in function. Development 138:3593–3612. https://doi.org/10.1242/dev.063610

Artavanis-Tsakonas S, Muskavitch MA, Yedvobnick B (1983) Molecular cloning of Notch, a locus affecting neurogenesis in Drosophila melanogaster. Proc Natl Acad Sci U S A 80:1977–1981

Axelrod JD, Matsuno K, Artavanis-Tsakonas S, Perrimon N (1996) Interaction between wingless and Notch signaling pathways mediated by dishevelled. Science 271:1826–1832

Banote RK, Edling M, Eliassen F et al (2016) β-Amyloid precursor protein-b is essential for Mauthner cell development in the zebrafish in a Notch-dependent manner. Dev Biol 413:26–38. https://doi.org/10.1016/j.ydbio.2016.03.012

Becam I, Fiuza UM, Arias AM, Milán M (2010) A role of receptor Notch in ligand cis-inhibition in Drosophila. Curr Biol 20:554–560. https://doi.org/10.1016/j.cub.2010.01.058

Berechid BE, Kitzmann M, Foltz DR et al (2002) Identification and characterization of presenilin-independent Notch signaling. J Biol Chem 277:8154–8165. https://doi.org/10.1074/jbc.M108238200

Berezovska O, Jack C, Deng A et al (2001) Notch1 and amyloid precursor protein are competitive substrates for presenilin1-dependent gamma-secretase cleavage. J Biol Chem 276:30018–30023. https://doi.org/10.1074/jbc.M008268200

Bertrand FE, Angus CW, Partis WJ, Sigounas G (2012) Developmental pathways in colon cancer. Cell Cycle 11:4344–4351. https://doi.org/10.4161/cc.22134

Blaumueller CM, Qi H, Zagouras P (1997) Intracellular cleavage of notch leads to a heterodimeric receptor on the plasma membrane. Cell 90:281–291

Bray SJ (2016) Notch signalling in context. Nat Rev Mol Cell Biol 17:722–735. https://doi.org/10.1038/nrm.2016.94

Brennan K, Baylies M, Arias AM (1999) Repression by Notch is required before Wingless signalling during muscle progenitor cell development in Drosophila. Curr Biol 9:707–710. https://doi.org/10.1016/S0960-9822(99)80313-3

Bryant PJ, Schubiger G (1971) Giant and duplicated imaginal discs in a new lethal mutant of Drosophila melanogaster. Dev Biol 24:233–263

Bush G, Disibio G, Miyamoto A et al (2001) Ligand-induced signaling in the absence of furin processing of Notch1. Dev Biol 229:494–502. https://doi.org/10.1006/dbio.2000.9992

Busseau I, Diederich RJ, Xu T, Artavanis-Tsakonas S (1994) A member of the Notch group of interacting loci, deltex encodes a cytoplasmic basic protein. Genetics 136:585–596

Carvalho FLF, Marchionni L, Gupta A et al (2015) HES6 promotes prostate cancer aggressiveness independently of Notch signalling. J Cell Mol Med 19:1624–1636. https://doi.org/10.1111/jcmm.12537

Chapman G, Sparrow DB, Kremmer E, Dunwoodie SL (2011) Notch inhibition by the ligand Delta-Like 3 defines the mechanism of abnormal vertebral segmentation in spondylocostal dysostosis. Hum Mol Genet 20:905–916. https://doi.org/10.1093/hmg/ddq529

Chastagner P, Israël A, Brou C (2006) Itch/AIP4 mediates Deltex degradation through the formation of K29-linked polyubiquitin chains. EMBO Rep 7:1147–1153. https://doi.org/10.1038/sj.embor.7400822

Chen C-D, Oh S-Y, Hinman JD, Abraham CR (2006) Visualization of APP dimerization and APP-Notch2 heterodimerization in living cells using bimolecular fluorescence complementation. J Neurochem 97:30–43. https://doi.org/10.1111/j.1471-4159.2006.03705.x

Childress JL, Acar M, Tao C, Halder G (2006) Lethal Giant discs, a novel C2-domain protein, restricts Notch activation during endocytosis. Curr Biol 16:2228–2233. https://doi.org/10.1016/j.cub.2006.09.031

Collu GM, Hidalgo-Sastre A, Acar A et al (2012) Dishevelled limits Notch signalling through inhibition of CSL. Development 139:4405–4415. https://doi.org/10.1242/dev.081885

Cornell M, Evans DA, Mann R et al (1999) The Drosophila melanogaster suppressor of deltex gene, a regulator of the Notch receptor signaling pathway, is an E3 class ubiquitin ligase. Genetics 152:567–576

Couturier L, Mazouni K, Schweisguth F (2013) Numb localizes at endosomes and controls the endosomal sorting of Notch after asymmetric division in Drosophila. Curr Biol 23:588–593. https://doi.org/10.1016/j.cub.2013.03.002

Dalton HE, Denton D, Foot NJ et al (2011) Drosophila Ndfip is a novel regulator of Notch signaling. Cell Death Differ 18:1150–1160. https://doi.org/10.1038/cdd.2010.130

Deshar R, Cho E-B, Yoon SK, Yoon J-B (2016) CC2D1A and CC2D1B regulate degradation and signaling of EGFR and TLR4. Biochem Biophys Res Commun 480:280–287. https://doi.org/10.1016/j.bbrc.2016.10.053

Diederich RJ, Matsuno K, Hing H, Artavanis-Tsakonas S (1994) Cytosolic interaction between deltex and Notch ankyrin repeats implicates deltex in the Notch signaling pathway. Development 120:473–481

Djiane A, Shimizu H, Wilkin M et al (2011) Su(dx) E3 ubiquitin ligase-dependent and -independent functions of polychaetoid, the Drosophila ZO-1 homologue. J Cell Biol 192:189–200. https://doi.org/10.1083/jcb.201007023

Doetzlhofer A, Basch ML, Ohyama T et al (2009) Hey2 regulation by FGF provides a Notch-independent mechanism for maintaining pillar cell fate in the organ of Corti. Dev Cell 16:58–69. https://doi.org/10.1016/j.devcel.2008.11.008

Dongre A, Surampudi L, Lawlor RG et al (2014) Non-canonical Notch signaling drives activation and differentiation of peripheral CD4+ T cells. Front Immunol 5:1–14. https://doi.org/10.3389/fimmu.2014.00054

Doran E, Keator D, Head E et al (2017) Down syndrome, partial trisomy 21, and absence of Alzheimer's disease: the role of APP. J Alzheimers Dis 56:459–470. https://doi.org/10.3233/JAD-160836

Drusenheimer N, Migdal B, Jäckel S et al (2015) The mammalian Orthologs of Drosophila lgd, CC2D1A and CC2D1B, function in the endocytic pathway, but their individual loss of function does not affect Notch signalling. PLoS Genet 11:e1005749. https://doi.org/10.1371/journal.pgen.1005749

Eiraku M, Tohgo A, Ono K et al (2005) DNER acts as a neuron-specific Notch ligand during Bergmann glial development. Nat Neurosci 8:873–880. https://doi.org/10.1038/nn1492

Endo Y, Osumi N, Wakamatsu Y (2002) Bimodal functions of Notch-mediated signaling are involved in neural crest formation during avian ectoderm development. Development 129:863–873

Endo Y, Osumi N, Wakamatsu Y (2003) Deltex/Dtx mediates NOTCH signaling in regulation of Bmp4 expression in cranial neural crest formation during avian development. Develop Growth Differ 45:241–248. https://doi.org/10.1046/j.1524-4725.2003.693.x

Faronato M, Nguyen VTM, Patten DK et al (2015) DMXL2 drives epithelial to mesenchymal transition in hormonal therapy resistant breast cancer through notch hyper-activation. Oncotarget 6:22467–22479. https://doi.org/10.18632/oncotarget.4164

Fassa A, Mehta P, Efthimiopoulos S (2005) Notch 1 interacts with the amyloid precursor protein in a Numb-independent manner. J Neurosci Res 82:214–224. https://doi.org/10.1002/jnr.20642

Fischer DF, van Dijk R, Sluijs JA et al (2005) Activation of the Notch pathway in down syndrome: cross-talk of Notch and APP. FASEB J 19:1451–1458. https://doi.org/10.1096/fj.04-3395.com

Foltz DR, Santiago MC, Berechid BE, Nye JS (2002) Glycogen synthase kinase-3beta modulates notch signaling and stability. Curr Biol 12:1006–1011. doi: S0960982202008886 [pii]

Fostier M, Evans DA, Artavanis-Tsakonas S, Baron M (1998) Genetic characterization of the Drosophila melanogaster suppressor of deltex gene: a regulator of notch signaling. Genetics 150:1477–1485

Gallagher CM, Knoblich JA (2006) The conserved C2 domain protein lethal (2) Giant discs regulates protein trafficking in Drosophila. Dev Cell 11:641–653. https://doi.org/10.1016/j.devcel.2006.09.014

Gentle ME, Rose A, Bugeon L, Dallman MJ (2012) Noncanonical Notch signaling modulates cytokine responses of dendritic cells to inflammatory stimuli. J Immunol 189:1274–1284. https://doi.org/10.4049/jimmunol.1103102

Greene M, Lai Y, Pajcini K et al (2016) Delta/Notch-like EGF-related receptor (DNER) is not a Notch ligand. PLoS One 11:e0161157. https://doi.org/10.1371/journal.pone.0161157

Guruharsha KG, Kankel MW, Artavanis-Tsakonas S (2012) The Notch signalling system: recent insights into the complexity of a conserved pathway. Nat Rev Genet 13:654–666. https://doi.org/10.1038/nrg3272

Hayashi Y, Nishimune H, Hozumi K et al (2016) A novel non-canonical Notch signaling regulates expression of synaptic vesicle proteins in excitatory neurons. Sci Rep 6:23969. https://doi.org/10.1038/srep23969

Hayward P, Brennan K, Sanders P et al (2005) Notch modulates Wnt signalling by associating with Armadillo/beta-catenin and regulating its transcriptional activity. Development 132:1819–1830. https://doi.org/10.1242/dev.01724

Herranz H, Stamataki E, Feiguin F et al (2006) Self-refinement of Notch activity through the transmembrane protein crumbs: modulation of gamma-secretase activity. EMBO Rep 7:297–302. https://doi.org/10.1038/sj.embor.7400617

Holleman J, Marchese A (2014) The ubiquitin ligase deltex-3l regulates endosomal sorting of the G protein-coupled receptor CXCR4. Mol Biol Cell 25:1892–1904. https://doi.org/10.1091/mbc.E13-10-0612

Hori K, Fostier M, Ito M et al (2004) Drosophila deltex mediates suppressor of hairless-independent and late-endosomal activation of Notch signaling. Development 131:5527–5537. https://doi.org/10.1242/dev.01448

Hori K, Sen A, Kirchhausen T, Artavanis-Tsakonas S (2011) Synergy between the ESCRT-III complex and Deltex defines a ligand-independent Notch signal. J Cell Biol 195:1005–1015. https://doi.org/10.1083/jcb.201104146

Hsieh F-Y, Ma T-L, Shih H-Y et al (2013) Dner inhibits neural progenitor proliferation and induces neuronal and glial differentiation in zebrafish. Dev Biol 375:1–12. https://doi.org/10.1016/j.ydbio.2013.01.007

Imai Y, Kobayashi Y, Inoshita T et al (2015) The Parkinson's disease-associated protein kinase LRRK2 modulates Notch signaling through the endosomal pathway. PLoS Genet 11:e1005503. https://doi.org/10.1371/journal.pgen.1005503

Ingram WJ, McCue KI, Tran TH et al (2008) Sonic Hedgehog regulates Hes1 through a novel mechanism that is independent of canonical Notch pathway signalling. Oncogene 27:1489–1500. https://doi.org/10.1038/sj.onc.1210767

Iso T, Kedes L, Hamamori Y (2003) HES and HERP families: multiple effectors of the notch signaling pathway. J Cell Physiol 194:237–255. https://doi.org/10.1002/jcp.10208

Izon DJ, Aster JC, He Y et al (2002) Deltex1 redirects lymphoid progenitors to the B cell lineage by antagonizing Notch1. Immunity 16:231–243. https://doi.org/10.1016/S1074-7613(02)00271-6

Jaekel R, Klein T (2006) The Drosophila Notch inhibitor and tumor suppressor gene lethal (2) giant discs encodes a conserved regulator of endosomal trafficking. Dev Cell 11:655–669. https://doi.org/10.1016/j.devcel.2006.09.019

Jarriault S, Le Bail O, Hirsinger E et al (1998) Delta-1 activation of notch-1 signaling results in HES-1 transactivation. Mol Cell Biol 18:7423–7431

Jin S, Mutvei AP, Chivukula IV et al (2013) Non-canonical Notch signaling activates IL-6/JAK/STAT signaling in breast tumor cells and is controlled by p53 and IKKα/IKKβ. Oncogene 32:4892–4902. https://doi.org/10.1038/onc.2012.517

Karbowniczek M, Zitserman D, Khabibullin D et al (2010) The evolutionarily conserved TSC/Rheb pathway activates Notch in tuberous sclerosis complex and Drosophila external sensory organ development. J Clin Invest 120:93–102. https://doi.org/10.1172/JCI40221

Kato H, Taniguchi Y, Kurooka H et al (1997) Involvement of RBP-J in biological functions of mouse Notch1 and its derivatives. Development 124:4133–4141. https://doi.org/10.1126/science.7716513

Kiaris H, Politi K, Grimm LM et al (2004) Modulation of Notch signaling elicits signature tumors and inhibits Hras1-induced oncogenesis in the mouse mammary epithelium. Am J Pathol 165:695–705. https://doi.org/10.1016/S0002-9440(10)63333-0

Kim IM, Wolf MJ, Rockman HA (2010) Gene deletion screen for cardiomyopathy in adult Drosophila identifies a new Notch ligand. Circ Res 106:1233–1243. https://doi.org/10.1161/CIRCRESAHA.109.213785

Kim M-Y, Mo J-S, Ann E-J et al (2011) Regulation of Notch1 signaling by the APP intracellular domain facilitates degradation of the Notch1 intracellular domain and RBP-Jk. J Cell Sci 124:1831–1843. https://doi.org/10.1242/jcs.076117

Kim S, Sato Y, Mohan PS et al (2016) Evidence that the rab5 effector APPL1 mediates APP-βCTF-induced dysfunction of endosomes in down syndrome and Alzheimer's disease. Mol Psychiatry 21:707–716. https://doi.org/10.1038/mp.2015.97

Kishi N, Tang Z, Maeda Y et al (2001) Murine homologs of deltex define a novel gene family involved in vertebrate Notch signaling and neurogenesis. Int J Dev Neurosci 19:21–35. https://doi.org/10.1016/S0736-5748(00)00071-X

Klein T (2003) The tumour suppressor gene l(2)giant discs is required to restrictthe activity of Notch to the dorsoventral boundaryduring Drosophila wing development. Dev Biol 255:313–333. https://doi.org/10.1016/S0012-1606(02)00052-0

Kobia F, Duchi S, Deflorian G, Vaccari T (2014) Pharmacologic inhibition of vacuolar H+ ATPase reduces physiologic and oncogenic Notch signaling. Mol Oncol 8:207–220. https://doi.org/10.1016/j.molonc.2013.11.002

Kopan R, Nye JS, Weintraub H (1994) The intracellular domain of mouse Notch: a constitutively activated repressor of myogenesis directed at the basic helix-loop-helix region of MyoD. Development 120:2385–2396. https://doi.org/10.1016/0168-9525(91)90220-k

Kubo F, Nakagawa S (2009) Hairy1 acts as a node downstream of Wnt signaling to maintain retinal stem cell-like progenitor cells in the chick ciliary marginal zone. Development 136:1823–1833. https://doi.org/10.1242/dev.029272

Kuroda K, Tani S, Minoguchi S et al (1999) Delta-induced Notch signaling mediated by RBP-J inhibits MyoD expression and myogenesis delta-induced Notch signaling mediated by RBP-J inhibits MyoD expression and myogenesis. J Biol Chem 274:7238–7244

Kwon C, Cheng P, King IN et al (2011) Notch post-translationally regulates β-catenin protein in stem and progenitor cells. Nat Cell Biol 13:1244–1251. https://doi.org/10.1038/ncb2313

Lieber T, Kidd S, Young MW (2002) Kuzbanian-mediated cleavage of Drosophila Notch. Genes Dev 16:209–221. https://doi.org/10.1101/gad.942302

Lleó A, Berezovska O, Ramdya P et al (2003) Notch1 competes with the amyloid precursor protein for γ-secretase and down-regulates Presenilin-1 gene expression. J Biol Chem 278:47370–47375. https://doi.org/10.1074/jbc.M308480200

Logeat F, Bessia C, Brou C et al (1998) The Notch1 receptor is cleaved constitutively by a furin-like convertase. Proc Natl Acad Sci U S A 95:8108–8112

Louvi A, Artavanis-Tsakonas S (2012) Notch and disease: a growing field. Semin Cell Dev Biol 23:473–480. https://doi.org/10.1016/j.semcdb.2012.02.005

Ma J, Meng Y, Kwiatkowski DJ et al (2010) Mammalian target of rapamycin regulates murine and human cell differentiation through STAT3/p63/Jagged/Notch cascade. J Clin Invest 120:103–114. https://doi.org/10.1172/JCI37964

Ma J, Tang X, Wong P et al (2014) Noncanonical activation of Notch1 protein by membrane type 1 matrix metalloproteinase (MT1-MMP) controls melanoma cell proliferation. J Biol Chem 289:8442–8449. https://doi.org/10.1074/jbc.M113.516039

MacKenzie F, Duriez P, Wong F et al (2004) Notch4 inhibits endothelial apoptosis via RBP-Jkappa-dependent and -independent pathways. J Biol Chem 279:11657–11663. https://doi.org/10.1074/jbc.M312102200

Maliekal TT, Bajaj J, Giri V et al (2008) The role of Notch signaling in human cervical cancer: implications for solid tumors. Oncogene 27:5110–5114. https://doi.org/10.1038/onc.2008.224

Maréchal A, Zou L (2013) DNA damage sensing by the ATM and ATR kinases. Cold Spring Harb Perspect Biol. https://doi.org/10.1101/cshperspect.a012716

Martinelli N, Hartlieb B, Usami Y et al (2012) CC2D1A is a regulator of ESCRT-III CHMP4B. J Mol Biol 419:75–88. https://doi.org/10.1016/j.jmb.2012.02.044

Matsuno K, Diederich RJ, Go MJ et al (1995) Deltex acts as a positive regulator of Notch signaling through interactions with the Notch ankyrin repeats. Development 121:2633–2644

Matsuno K, Eastman D, Mitsiades T et al (1998) Human deltex is a conserved regulator of Notch signalling. Nat Genet 19:74–78. https://doi.org/10.1038/ng0598-74

Mazaleyrat SL, Fostier M, Wilkin MB et al (2003) Down-regulation of Notch target gene expression by suppressor of deltex. Dev Biol 255:363–372

Merdes G, Soba P, Loewer A et al (2004) Interference of human and Drosophila APP and APP-like proteins with PNS development in Drosophila. EMBO J 23:4082–4095. https://doi.org/10.1038/sj.emboj.7600413

Miller AC, Lyons EL, Herman TG (2009) Cis-inhibition of Notch by Endogenous delta biases the outcome of lateral inhibition. Curr Biol 19:1378–1383. https://doi.org/10.1016/j.cub.2009.06.042

Morawa KS, Schneider M, Klein T (2015) lgd regulates the activity of the BMP/Dpp signalling pathway during Drosophila oogenesis. Development 142:1325–1335. https://doi.org/10.1242/dev.112961

Mukherjee A, Veraksa A, Bauer A et al (2005) Regulation of Notch signalling by non-visual beta-arrestin. Nat Cell Biol 7:1191–1201. https://doi.org/10.1038/ncb1327

Mukherjee T, Kim WS, Mandal L et al (2011) Interaction between Notch and Hif-alpha in development and survival of Drosophila blood cells. Science 332:1210–1213. https://doi.org/10.1126/science.1199643

Mungamuri SK1, Yang X, Thor AD, Somasundaram K (2006) Survival signaling by Notch1: mammalian target of rapamycin (mTOR)-dependent inhibition of p53. Cancer Res 66:4715–4724

Nair P, Somasundaram K, Krishna S (2003) Activated Notch1 inhibits p53-induced apoptosis and sustains transformation by human papillomavirus type 16 E6 and E7 oncogenes through a PI3K-PKB/Akt-dependent pathway. J Virol 77(12):7106. https://doi.org/10.1128/jvi.77.12.7106-7112.2003

Nemetschke L, Knust E (2016) Drosophila crumbs prevents ectopic Notch activation in developing wings by inhibiting ligand-independent endocytosis. Development 143:4543–4553

Nofziger D, Miyamoto A, Lyons KM, Weinmaster G (1999) Notch signaling imposes two distinct blocks in the differentiation of C2C12 myoblasts. Development 126:1689–1702

Ordentlich P, Lin A, Shen C et al (1998) Notch inhibition of E47 supports the existence of a novel signaling pathway. Mol Cell Biol 18:2230–2239. https://doi.org/10.1128/MCB.18.4.2230

Palmer WH, Jia D, Deng W-M (2014) Cis-interactions between Notch and its ligands block ligand-independent Notch activity. Elife 3. https://doi.org/10.7554/eLife.04415

Pece S, Serresi M, Santolini E et al (2004) Loss of negative regulation by Numb over Notch is relevant to human breast carcinogenesis. J Cell Biol 167:215–221. https://doi.org/10.1083/jcb.200406140

Pece S, Confalonieri S, Romano P R, Di Fiore PP (2011) NUMB-ing down cancer by more than just a NOTCH. Biochim Biophys Acta–Rev Cancer 1815:26–43. https://doi.org/10.1016/j.bbcan.2010.10.001

Perumalsamy LR, Nagala M, Banerjee P, Sarin A (2009) A hierarchical cascade activated by non-canonical Notch signaling and the mTOR-Rictor complex regulates neglect-induced death in mammalian cells. Cell Death Differ 16:879–889. https://doi.org/10.1038/cdd.2009.20

Purow B (2012) Notch inhibition as a promising new approach to cancer therapy. Adv Exp Med Biol 727:305–319. https://doi.org/10.1007/978-1-4614-0899-4_23

Qiu L, Joazeiro C, Fang N et al (2000) Recognition and ubiquitination of Notch by Itch, a hect-type E3 ubiquitin ligase. J Biol Chem 275:35734–35737

Ramain P, Khechumian K, Seugnet L et al (2001) Novel Notch alleles reveal a Deltex-dependent pathway repressing neural fate. Curr Biol 11:1729–1738. https://doi.org/10.1016/S0960-9822(01)00562-0

Rangarajan A, Syal R, Selvarajah S et al (2001) Activated Notch1 signaling cooperates with papillomavirus oncogenes in transformation and generates resistance to apoptosis on matrix withdrawal through PKB/Akt. Virology 286:23–30. https://doi.org/10.1006/viro.2001.0867

Richardson ECN, Pichaud F (2010) Crumbs is required to achieve proper organ size control during Drosophila head development. Development 137:641–650. https://doi.org/10.1242/dev.041913

Roncarati R, Sestan N, Scheinfeld MH et al (2002) The gamma-secretase-generated intracellular domain of beta-amyloid precursor protein binds Numb and inhibits Notch signaling. Proc Natl Acad Sci U S A 99:7102–7107. https://doi.org/10.1073/pnas.102192599

Sade H, Krishna S, Sarin A (2004) The anti-apoptotic effect of Notch-1 requires p56lck-dependent, Akt/PKB-mediated signaling in T cells. J Biol

Chem 279:2937–2944. https://doi.org/10.1074/jbc.M309924200

Sakata T, Sakaguchi H, Tsuda L et al (2004) Drosophila Nedd4 regulates endocytosis of Notch and suppresses its ligand-independent activation. Curr Biol 14:2228–2236. https://doi.org/10.1016/j.cub.2004.12.028

Sanalkumar R, Indulekha CL, Divya TS et al (2010) ATF2 maintains a subset of neural progenitors through CBF1/Notch independent Hes-1 expression and synergistically activates the expression of Hes-1 in Notch-dependent neural progenitors. J Neurochem 113:807–818. https://doi.org/10.1111/j.1471-4159.2010.06574.x

Sanders PGT, Muñoz-Descalzo S, Balayo T et al (2009) Ligand-independent traffic of notch buffers activated armadillo in Drosophila. PLoS Biol 7:e1000169. https://doi.org/10.1371/journal.pbio.1000169

Schneider M, Troost T, Grawe F et al (2013) Activation of Notch in lgd mutant cells requires the fusion of late endosomes with the lysosome. J Cell Sci 126:645–656. https://doi.org/10.1242/jcs.116590

Sestan N, Artavanis-Tsakonas S, Rakic P (1999) Contact-dependent inhibition of cortical neurite growth mediated by notch signaling. Science 286:741–746

Shawber C, Nofziger D, Hsieh JJ et al (1996) Notch signaling inhibits muscle cell differentiation through a CBF1-independent pathway. Development 122:3765–3773

Shimizu H, Woodcock SA, Wilkin MB et al (2014) Compensatory flux changes within an l'endocytic trafficking network maintain thermal robustness of notch signaling. Cell 157:1160–1174. https://doi.org/10.1016/j.cell.2014.03.050

Shin HM, Tilahun ME, Cho OH et al (2014) NOTCH1 can initiate NF-kB activation via cytosolic interactions with components of the T cell signalosome. Front Immunol 5:1–15. https://doi.org/10.3389/fimmu.2014.00249

Sprinzak D, Lakhanpal A, LeBon L et al (2011) Mutual inactivation of Notch receptors and ligands facilitates developmental patterning. PLoS Comput Biol 7:e1002069. https://doi.org/10.1371/journal.pcbi.1002069

Stallwood Y, Briend E, Ray KM et al (2006) Small interfering RNA-mediated knockdown of notch ligands in primary CD4+ T cells and dendritic cells enhances cytokine production. J Immunol 177:885–895

Stamos JL, Weis WI (2013) The β-catenin destruction complex. Cold Spring Harb Perspect Biol 5:a007898. https://doi.org/10.1101/cshperspect.a007898

Stockhausen MT, Sjölund J, Axelson H (2005) Regulation of the Notch target gene Hes-1 by TGFα induced Ras/MAPK signaling in human neuroblastoma cells. Exp Cell Res 310:218–228. https://doi.org/10.1016/j.yexcr.2005.07.011

Tada M, Itoh S, Ishii-Watabe A et al (2012) Functional analysis of the Notch ligand Jagged1 missense mutant proteins underlying Alagille syndrome. FEBS J 279:2096–2107. https://doi.org/10.1111/j.1742-4658.2012.08595.x

Tognon E, Kobia F, Busi I et al (2016) Control of lysosomal biogenesis and Notch-dependent tissue patterning by components of the TFEB-V-ATPase axis in Drosophila melanogaster. Autophagy 12:499–514

Troost T, Jaeckel S, Ohlenhard N, Klein T (2012) The tumour suppressor Lethal (2) giant discs is required for the function of the ESCRT-III component Shrub/CHMP4. J Cell Sci 125:763–776. https://doi.org/10.1242/jcs.097261

Usami Y, Popov S, Weiss ER et al (2012) Regulation of CHMP4/ESCRT-III function in human immunodeficiency virus type 1 budding by CC2D1A. J Virol 86:3746–3756. https://doi.org/10.1128/JVI.06539-11

Vaccari T, Lu H, Kanwar R et al (2008) Endosomal entry regulates Notch receptor activation in Drosophila melanogaster. J Cell Biol 180:755–762. https://doi.org/10.1083/jcb.200708127

Vaccari T, Duchi S, Cortese K et al (2010) The vacuolar ATPase is required for physiological as well as pathological activation of the Notch receptor. Development 137:1825–1832. https://doi.org/10.1242/dev.045484

van Es JH, van Gijn ME, Riccio O et al (2005) Notch/γ--secretase inhibition turns proliferative cells in intestinal crypts and adenomas into goblet cells. Nature 435:959–963. https://doi.org/10.1038/nature03659

Veeraraghavalu K, Subbaiah VK, Srivastava S et al (2005) Complementation of human papillomavirus type 16 E6 and E7 by Jagged1-specific Notch1-phosphatidylinositol 3-kinase signaling involves pleiotropic oncogenic functions independent of CBF1;Su(H);Lag-1 activation. J Virol 79:7889–7898. https://doi.org/10.1128/JVI.79.12.7889-7898.2005

Vermezovic J, Adamowicz M, Santarpia L et al (2015) Notch is a direct negative regulator of the DNA-damage response. Nat Struct Mol Biol 22:417–424. https://doi.org/10.1038/nsmb.3013

Wall DS, Mears AJ, McNeill B et al (2009) Progenitor cell proliferation in the retina is dependent on Notch-independent Sonic hedgehog/Hes1 activity. J Cell Biol 184:101–112. https://doi.org/10.1083/jcb.200805155

Watson KL, Justice RW, Bryant PJ (1994) Drosophila in cancer research: the first fifty tumor suppressor genes. J Cell Sci Suppl 18:19–33

Wilkin MB, Carbery A-M, Fostier M et al (2004) Regulation of Notch endosomal sorting and signaling by Drosophila Nedd4 family proteins. Curr Biol 14:2237–2244. https://doi.org/10.1016/j.cub.2004.11.030

Wilkin M, Tongngok P, Gensch N et al (2008) Drosophila HOPS and AP-3 complex genes are required for a Deltex-regulated activation of Notch in the endosomal trafficking pathway. Dev Cell 15:762–772. https://doi.org/10.1016/j.devcel.2008.09.002

Wu Y, Cain-Hom C, Choy L et al (2010) Therapeutic antibody targeting of individual Notch receptors. Nature 464:1052–1057. https://doi.org/10.1038/nature08878

Xu T, Artavanis-Tsakonas S (1990) Deltex, a locus interacting with the neurogenic genes, Notch, Delta and mastermind in Drosophila melanogaster. Genetics 126:665–677

Yamada K, Fuwa TJ, Ayukawa T et al (2011) Roles of Drosophila Deltex in Notch receptor endocytic trafficking and activation. Genes Cells 16:261–272. https://doi.org/10.1111/j.1365-2443.2011.01488.x

Yamamoto N, Yamamoto SI, Inagaki F et al (2001) Role of Deltex-1 as a transcriptional regulator downstream of the Notch receptor. J Biol Chem 276:45031–45040. https://doi.org/10.1074/jbc.M105245200

Zhao M, Li X-D, Chen Z (2010) CC2D1A, a DM14 and C2 domain protein, activates NF-kappaB through the canonical pathway. J Biol Chem 285:24372–24380. https://doi.org/10.1074/jbc.M109.100057

Zheng L, Saunders CA, Sorensen EB et al (2013) Notch signaling from the endosome requires a conserved dileucine motif. Mol Biol Cell 24:297–307. https://doi.org/10.1091/mbc.E12-02-0081

The Notch3 Receptor and Its Intracellular Signaling-Dependent Oncogenic Mechanisms

Diana Bellavia, Saula Checquolo, Rocco Palermo, and Isabella Screpanti

Abstract

During evolution, gene duplication of the Notch receptor suggests a progressive functional diversification. The Notch3 receptor displays a number of structural differences with respect to Notch1 and Notch2, most of which have been reported in the transmembrane and in the intracellular regions, mainly localized in the negative regulatory region (NRR) and trans-activation domain (TAD). Targeted deletion of Notch3 does not result in embryonic lethality, which is in line with its highly restricted tissue expression pattern. Importantly, deregulated Notch3 expression and/or activation, often results in disrupted cell differentiation and/or pathological development, most notably in oncogenesis in different cell contexts. Mechanistically this is due to Notch3-related genetic alterations or epigenetic or posttranslational control mechanisms. In this chapter we discuss the possible relationships between the structural differences and the pathological role of Notch3 in the control of mouse and human cancers. In future, targeting the unique features of Notch3-oncogenic mechanisms could be exploited to develop anticancer therapeutics.

Keywords

Notch3 · T cell leukemogenesis · Epigenetic regulation · Post-translational modifications

Abbreviations

ADAM	A Disintegrin And Metalloprotease
DSL	Delta, Serrate
ECD	Extracellular domain
HD	Heterodimerization domain
NICD	Notch intracellular domain
NRR	Negative regulatory region
TAD	Trans-activation domain
LAG-2	Notch Ligands or DSL ligands
T-ALL	T-cell acute lymphoblastic leukemia
Ptcra	Invariant preTα chain of the pre-T cell receptor
Tregs	T regulatory cells
IKK	Inhibitor of KAPPA-B kinase complex

D. Bellavia · I. Screpanti (✉)
Department of Molecular Medicine, Sapienza University of Rome, Rome, Italy
e-mail: isabella.screpanti@uniroma1.it; diana.bellavia@uniroma1.it

S. Checquolo
Department of Medico-Surgical Sciences and Biotechnologies, Sapienza University of Rome, Rome, Italy
e-mail: saula.checquolo@uniroma1.it

R. Palermo
Center for Life Nano Science@Sapienza, Istituto Italiano di Tecnologia, Rome, Italy
e-mail: rocco.palermo@iit.it

T. Borggrefe, B. D. Giaimo (eds.), *Molecular Mechanisms of Notch Signaling*, Advances in Experimental Medicine and Biology 1066, https://doi.org/10.1007/978-3-319-89512-3_10

PTM	Post-translational modification
IK	Ikaros
IK-DN	Ikaros dominant negative isoforms
HUD	RNA-binding protein D of the ELAV/Hu family
BORIS/CTCFL	Brother Of Regulator of Imprinted Sites/CTCF-like protein
TCR	T-cell receptor
NF-κB	Nuclear factor-κB
N1ICD	Notch1 intracellular domain
N2ICD	Notch2 intracellular domain
N3ICD	Notch3 intracellular domain.

1 Is Notch-3 Only a Structurally Different Notch?

The evolutionarily conserved Notch signalling pathway functions as a major mediator of cell-fate determination during development and regulates a diverse set of biological functions, including cell differentiation, proliferation and survival. The Notch family consists of one member in *Drosophila*, two receptors LIN-12 and GLP-1 in *Caenorhabditis elegans* and four Notch homologs (Notch1–4) in mammals (Kopan and Weintraub 1993; Lardelli et al. 1994; Weinmaster et al. 1991, 1992), which share a similar modular organization. The increasing number of Notch homologs during the evolution process suggests a progressive functional diversification of different Notch proteins, which may be related to their structural differences. Therefore, the four mammalian Notch proteins show both redundant and distinct function and tissue distribution (Wu and Bresnick 2007). Notch1 and Notch2 are expressed in a wide variety of tissue and are essential for mammalian development (McCright et al. 2001; Swiatek et al. 1994). On the contrary, Notch3 and Notch4 expression is largely restricted. Notch3 is expressed predominantly in vasculature smooth muscle (Joutel et al. 2000), in the central nervous system (Lardelli et al. 1994) and in subsets of thymocytes and T lymphoid cells (Anastasi et al. 2003; Felli et al. 1999), while Notch4 is expressed preferentially in vascular endothelial cells (Uyttendaele et al. 1996). This different tissue distribution reveals a different impact of Notch paralogs in sustaining cell and tissue development. Indeed, targeted deletion of Notch1 and − 2 results in embryonic lethality, owing essential functions for mammalian development (Conlon et al. 1995; McCright et al. 2001; Swiatek et al. 1994), while targeted deletions of Notch3 (Domenga et al. 2004) and − 4 do not yield embryonic lethality (Krebs et al. 2000; Krebs et al. 2003).

The *Notch3* gene was identified as the third mammalian Notch and was initially described as being expressed in proliferating neuroepithelium (Lardelli et al. 1994).

Despite Notch1, 2 and 3 share a similar basic structure, Notch3 displays a number of structural differences with respect to Notch1 and Notch2. Significantly, several differences have been reported in the shorter intracellular region of Notch3 (N3ICD), that includes the trans-activation domain (TAD), which could in part justify the weak transactivation activity of the N3ICD, when compared to the Notch1 intracellular domain (N1ICD) and the Notch2 intracellular domain (N2ICD) on specific target genes (Beatus et al. 2001). Moreover, slight differences with respect to Notch1 and Notch2 have been reported both in transmembrane (TM) and extracellular domain of Notch3 (Notch3-ECD). In this regard, Notch3 contains only 34 EGF-like repeats, whereas the Notch1-ECD consists of 36 EGF-like repeats. Notch3 lacks both the equivalent of EGF repeat 21 and a region comprising parts of EGF-repeats 2 and 3 (Lardelli et al. 1994). Notably, mutations of the extracellular domain of Notch3 have been specifically associated with cerebral autosomal-dominant arteriopathy with subcortical infarcts and leukoencephalopathy (CADASIL), a human inherited small vessel disease causing stroke and dementia (Joutel et al. 2000). Moreover, the generation of mouse models bearing archetypal CADASIL-Notch3 mutations genetically demonstrated the pathogenicity of such mutations (Lacombe et al. 2005; Monet et al. 2007).

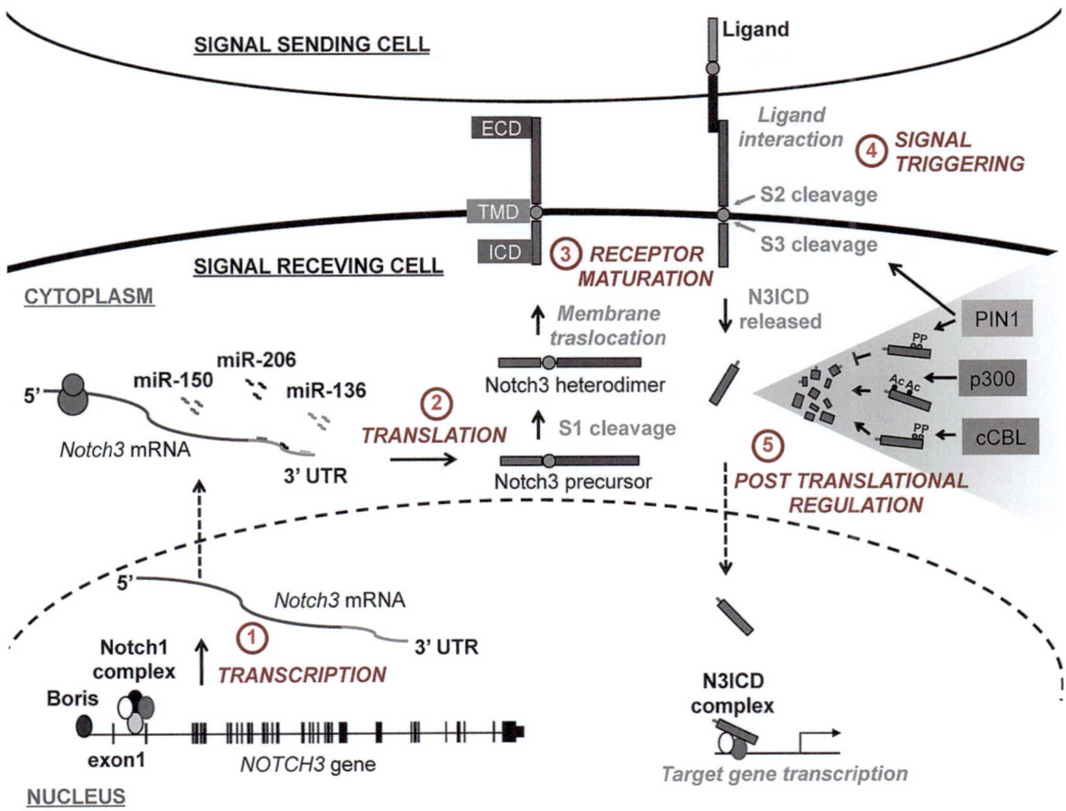

Fig. 1 Multilevel regulation of Notch3 gene and protein expression and function. (1) *Notch3* transcription and epigenetic regulation; (2) Notch3 translation; (3) Notch3 receptor maturation and membrane translocation; (4) Notch3 signaling triggering and (5) multiprotein post-translational regulation and N3ICD function

The canonical Notch signalling pathway, shared by all Notch receptors, appears as a paradigm of simplicity. After its synthesis in the ER, the Notch receptor is transported through the secretory pathway to the *trans*-Golgi network, where it is constitutively cleaved by a furin-like convertase. Following this proteolytic processing event, referred as S1 cleavage, the Notch receptor proceeds to the cell surface and forms a bipartite receptor, in which the heterodimers are held together by non covalent interactions within the heterodimerization domain (HD) (Fig. 1, step (3)). This domain is closed to the membrane on the extracellular side of the cell and is flanked by site 2 (S2) towards the C-terminus and a negative regulatory region (NRR) towards the N-terminus. In the unstimulated state, the NRR prevents access to S2 site and the subsequent cleavage (Fig. 1, step (4)). In response to ligand binding, a conformational change of the NRR induces to successive additional cleavage catalysed by A Disintegrin And Metalloprotease (ADAM)-type metalloproteases and γ-secretase, respectively. Notably, in a recent work S. Blacklow and coworkers (Xu et al. 2015) demonstrated that the autoinhibited conformation of the Notch3 NRR is less stable with respect to Notch1 and 2 NRRs and suggested an increased basal activity of Notch3.

The presence of four mammalian Notch receptors and many ligands suggests that each Notch protein is able to target a distinct set of downstream genes. Notch paralogs show, in fact, a functional diversity and a specificity for transcription of the target genes (Amsen et al. 2004; Shimizu et al. 2002) and they may even play contrasting roles in the same tissue (Fan et al. 2004). N1ICD is considered a potent transcriptional activator of the *HES* promoter, while N3ICD is a much weaker activator, when compared to N1ICD, and can repress N1ICD-mediated HES1 activation in certain contexts (Beatus et al. 2001). Interestingly, Urban Lendahl's group (Beatus et al. 2001) demonstrated that the different transcriptional activation capacity is based on structural differences that involve the ankyrin repeat regions in the N1- and N3-ICD. Moreover, the authors identified a novel important region in the N3ICD, named RE/AC (for repression/activation), located immediately C-terminal to the ankyrin repeat region and required for N1ICD-dependent activation and N3ICD-dependent repression of *HES-1* and *− 5* gene promoters. In addition to these observations, it has been suggested that the different Notch proteins read binding site orientation and distribution on the promoter differently: N1ICD activates paired CSL-binding sites (in both head-to-head and tail-to-tail orientation) very efficiently, whereas N3ICD works better in activating promoters with a single CSL, but requires additional cis-elements. Moreover, a specialized TAD domain has been identified in the C-terminal region of Notch3, which preferentially activates promoters with zinc finger binding sites near a CSL binding site, as in the case of the *HES5* promoter (Ong et al. 2006). Altogether these observations suggest that each Notch intracellular domain (NICD) protein differently reads binding site orientation and distribution on the promoters of responsive genes explaining, at least in part, the variations in transcriptional activity of Notch paralogs.

2 Preferential Notch3 and Jagged1 Receptor/Ligand Pairing Results in Synergistic Activity

Despite the diversification along the phylogenesis, the main structural features of Notch receptors and their specific ligands, belonging to the Delta, Serrate/Jagged and LAG-2 ligand family (DSL proteins), as well as the activation mechanisms of Notch signaling are conserved. Notch ligands contain a DSL motif involved in Notch receptor binding, a variable number of EGF repeats and a short intracellular tail required for endocytosis. Notably, diversity in binding affinity and/or processing triggering of specific receptor or ligands may exist.

Like Notch receptors, DSL ligands, after binding, become a substrate for proteolysis by a member of the ADAM-metalloproteases. The ADAM17 activity allows the shedding of the ectodomain fragment and generates a membrane-tethered transmembrane-intracellular domain (TM-ICD) (Ikeuchi and Sisodia 2003; Qi et al. 1999), which then undergoes an intramembrane cleavage mediated by Presenilin/γ-secretase complex activity that releases a soluble intracellular fragment (Ikeuchi and Sisodia 2003; LaVoie and Selkoe 2003; Nehring et al. 2005), which moves into the nucleus. The shedding of DSL ligands has been described as an important event for the activation of Notch signaling in neighboring cells, even if the biological function of the ligand soluble form is controversial, as it has been shown to be able to act as both agonist and antagonist of the Notch receptor activity (Li and Baker 2004; Sakamoto et al. 2002). Interestingly, it was demonstrated that the soluble form of the extracellular domain of Jagged1 may be shed in the culture medium of cells overexpressing it and that this soluble form of Jagged1 may interact with Notch1, Notch2 and Notch3 in binding assays, with a higher affinity for Notch3 (LaVoie and Selkoe 2003; Shimizu et al. 1999; Shimizu et al. 2000). Moreover, a bidirectional function has been suggested for Notch ligands, in particular for Jagged1, which has been shown to be able not only to trigger Notch signaling in neighbouring cells, but also to signal intrinsically through the soluble cytoplasmic domain (Jag1-ICD), released by γ-secretase-dependent cleavage, which is able to translocate to the nucleus and to activate gene expression (Jeffries and Capobianco 2000; LaVoie and Selkoe 2003). Intriguingly, the intracellular domain of Jagged1 has been shown to be able to increase the expression of both Jagged1 itself and Notch3 mRNAs and to include a C-terminal PDZ (PSD-95/DLG/ZO-1) domain which has been demonstrated to be essential for neoplastic transformation, although through an unknown molecular mechanism (Ascano et al. 2003). Consistently with this original observation, it was reported that the overexpression of Jagged1 protein was associated with a poor prognosis in several human tumors, such as prostate cancer (Santagata et al. 2004), tongue squamous cell carcinoma (Zhang et al. 2013) and renal cell carcinoma (Wu et al. 2011). Moreover, it has been suggested that Notch3/Jagged1 co-expression could be important for the malignant transformation of several human tumors, including ovarian tumors (Choi et al. 2008) and lung cancer (Konishi et al. 2007). Starting from these observations, we recently demonstrated the existence of a direct relationship between Notch3 and Jagged1 in T-cell acute lymphoblastic leukemia (T-ALL) context (Pelullo et al. 2014). Our data suggest that Jagged1 is a novel Notch3 target gene and the enforced expression of N3ICD is able to determine a cell membrane lipid raft–associated constitutive processing of Jagged1. Moreover, we observed a Notch3/Jagged1 cis-interaction within the same cell, which results in the autocrine reinforcement of Notch3 signaling, since once Jag1-ICD moves to the nucleus it can behave as a co-activator of the N3ICD-driven transcriptional complex, empowering the transcriptional activation of Notch target genes (e.g. the invariant preTα chain of the pre-T cell receptor, *Ptcra)* involved in the onset and progression of T-ALL. Finally, we demonstrated that the shedding of Jagged1 extracellular region, triggered by ADAM17 ends in the paracrine amplification of Notch signaling in adjacent cells. Collectively, our observations suggest a new molecular mechanism, in which a Notch3-dependent dysregulated expression and processing of Jagged1 takes part in a multistep oncogenic process, playing a role in controlling cell growth, apoptosis and migration, finally favouring tumor aggressiveness and progression.

3 Notch3, Pre-T-cell Receptor, NF-κB and Ikaros Relationships: The Functional Peculiarity of the Third Notch in T Cell Development and Leukemogenesis

Among the four members of the family, Notch1 and 3 receptors have been mainly involved in both T cell development and leukemogenesis. To this regard, Notch3 is included among Notch1-target genes in T-ALL (Wang et al. 2014; Yatim et al. 2012). A significant role for Notch1 has been suggested in the initial T cell lineage commitment of bone marrow-derived common lymphoid precursors (Pui et al. 1999; Radtke et al. 1999) and in intrathymic T cell lineage choices, by favoring the CD8 versus CD4 and αβ versus γδ T cell lineage decision (Fowlkes and Robey 2002; Robey et al. 1996; Washburn et al. 1997). Conversely, a specific role of Notch3 at the pre-T cell receptor (pre-TCR) checkpoint has been suggested. Indeed, Notch3 expression has been demonstrated to be preferentially upregulated by thymic stromal cell-derived signals in DN immature thymocytes prior to their transition to more mature DP cells. Subsequently, it is downregulated across the DN to DP transition and stays at very low to undetectable levels in mature T lymphocytes (Felli et al. 1999), while remaining significantly expressed in T regulatory cells (Tregs) (Anastasi et al. 2003), in which we previously showed that N3ICD and nuclear factor κB (NF-κB) cooperate to positively regulate the transcription of mouse *Foxp3*, the master control gene of Tregs, resulting in its increased protein expression (Barbarulo et al. 2011). The transition across DN to DP is a critical step of thymocyte differentiation, mediated by pre-TCR signalling pathway and it is characterised by activated NF-κB and intense proliferation (Aifantis et al. 2001; Voll et al. 2000; von Boehmer et al. 1998). Notably, *Lck* promoter-driven Notch3-IC transgenic (*N3ICD tg*) mice display a peculiar phenotype of dysregulated early T cell development

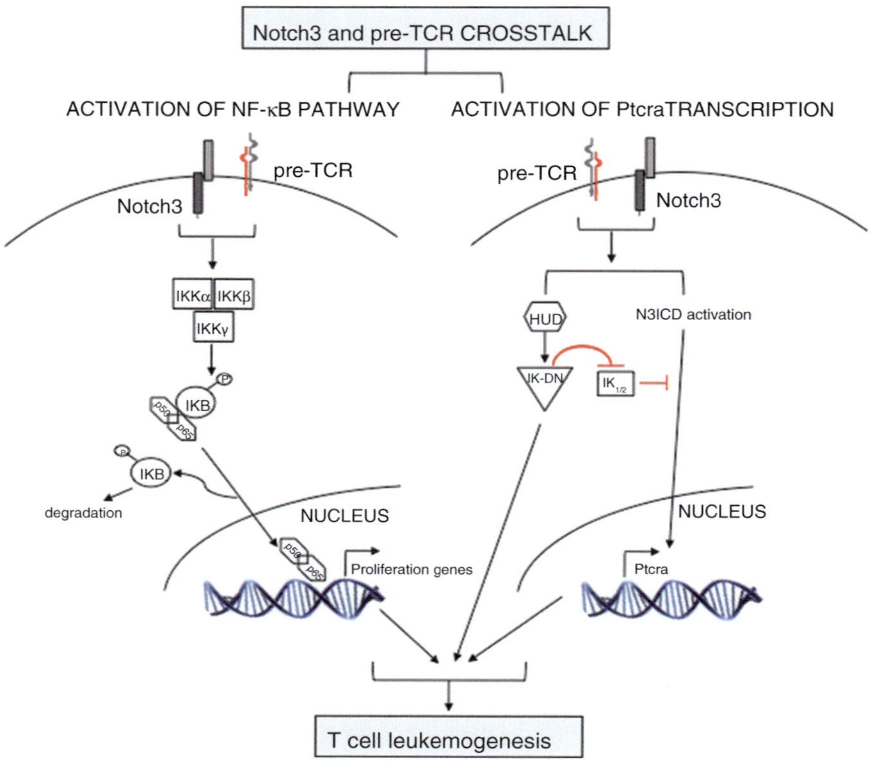

Fig. 2 The strict interdependency of Notch3 and pre-TCR signaling pathways plays a crucial role in T cell leukemogenesis. The cartoon illustrates the central role played by the relationship between Notch3 and pre-TCR in sustaining the activation of NF-κB canonical signaling pathway, mediated by the activation of p50/p65 heterodimer, in turn triggered by the IKKαβγ complex. Moreover, it is indicated the regulation of *Ikaros* alternative splicing, by HUD, involved in the development of T cell leukemogenesis. IK 1/2, Ikaros 1 and 2 isoforms DNA binding; IK-DN Ikaros dominant negative isoforms

characterised by a significant expansion of CD25$^+$ DN thymocytes, sustained expression of surface CD25 protein and *Ptcra* gene and protein and constitutively activated NF-κB in all thymocyte subsets and in peripheral T cells (Bellavia et al. 2000). Increased expression of Ptcra, leading to the constitutive activation of pre-TCR signalling, represents a critical event in T-cell leukemogenesis. Indeed, *Ptcra* gene disruption in *N3ICD tg* mice, obtained by the generation of double mutant mice (*N3ICD tg/Ptcra knockout* ($^{-/-}$)), prevents the development of T-cell leukemia (Bellavia et al. 2002).

Consistently, we demonstrated that in *N3ICD tg* mice the N3ICD and pre-TCR crosstalk promotes the expression of distinct oncogenic gene clusters, by sustaining NF-κB canonical pathway activity (active p65/p50 complexes), through the regulation of assembling of the Inhibitor of KAPPA-B Kinase (IKK) complex, by favoring the generation of IKKα/IKKβ/IKKγ trimeric complex (IKKα/β/γ) (Fig. 2) (Vacca et al. 2006). Conversely, the abrogation of *Ptcra* gene, thus of the pre-TCR function in *N3ICD tg/Ptcra*$^{-/-}$ double mutant mice, leads to the expression of pro-survival and pro-differentiative genes, by favoring the generation of IKKα/IKKα homodimeric complex, and the activation of the NF-κB alternative pathway (active NF-κB2/RelB complexes) (Vacca et al. 2006).

Similarly to what happens in *N3ICD tg* mice, mice heterozygous for gene mutations of *Ikaros*, a master regulator of lymphopoiesis and leukemogenesis, show an apparently normal lymphoid cell distribution at birth, but early in life they develop a very aggressive lymphoblastic leukemia with a concomitant loss of heterozygosity, resulting in predominant synthesis of short Ikaros isoforms (IK4 to IK9) (Winandy et al. 1995; Yoshida et al. 2006). Such shorter isoforms include fewer than three N-terminal zinc-fingers, are unable to bind DNA and behave as dominant negative (IK-DN) isoforms upon heterodimerization with longer Ikaros isoforms (IK1 to IK3) that have an intact DNA-binding domain (Sun et al. 1996). Intriguingly, leukemia development in Ikaros mutant mice is strictly dependent on pre-TCR/TCR signaling which sustains the expansion of immature thymocyte populations (Winandy et al. 1999).

Together, these observations support the hypothesis that pre-TCR signalling may represent a functional link justifying a possible preferential relationship between Notch3 and Ikaros pathways in the onset of T cell leukemia.

Interestingly, it was previously observed that Ikaros and the Notch effector protein CSL/RBP-Jκ recognize the same DNA-binding sequence *in vitro*, the core motif TGGGAA, leading the authors to hypothesize a model in which Ikaros may antagonize NICD/CSL-induced transcriptional activation by competing with the similar consensus DNA-binding motif (Beverly and Capobianco 2003). Intriguingly, thymocytes and lymphoma cells from *N3ICD tg* mice display an inappropriate expression of non-DNA binding short Ikaros isoforms, whereas the deletion of *Ptcra* prevents the altered expression pattern of Ikaros isoforms both at mRNA and protein levels (Bellavia et al. 2007). Therefore, pre-TCR-triggered signals and non DNA-binding IK-DN isoforms could depend on each other and cooperate with Notch signalling in T cell transformation. Indeed, by utilizing thymocytes from *N3ICD tg*, *Ptcra*$^{-/-}$ and *N3ICD tg/Ptcra*$^{-/-}$ mice, we observed that Ikaros short isoforms are virtually absent in thymocytes from mouse strains of *Ptcra*$^{-/-}$ background. We thus hypothesized that the presence of a functional pre-TCR is necessary to sustain an inappropriate expression of non DNA-binding Ikaros isoforms, even in the presence of an overexpressed N3ICD.

Moreover, only in cells with a functional pre-TCR, N3ICD but not N1ICD was able to unequivocally alter the Ikaros isoform expression pattern, generating an increased expression of shorter IK-DN isoforms (Fig. 2) (Bellavia et al. 2007), clearly highlighting a specific role for N3ICD that is non redundant with respect to N1ICD in affecting a pathway specifically involved in lymphopoiesis and leukemia. Notwithstanding such Notch3 specific role in regulating the generation of pro-oncogenic IK-DN isoforms, it has been subsequently shown that while the deletion of *Notch3* had little effect on the development of T cell leukemia arising in mice bearing a knock-down mutation in the *Ikaros* gene, the deletion of floxed *Notch1* promoter/exon 1 sequences, which results in the accumulation of constitutively active *Notch1* transcripts, significantly accelerates leukemogenesis (Jeannet et al. 2010). This observation on the one hand may suggest that the absence or gene mutations of *Ikaros*, may behave as a master oncogenic driver in T-ALL, being able to sustain the generation of activating mutations of *Notch1* but, on the other hand, that in the presence of a functional *Ikaros* gene, Notch3 activation, by increasing IK-DN isoforms, may trigger a cascade of events, starting with the functional inactivation of Ikaros and ending with the enforced activation of Notch signaling, irrespectively of either Notch1 or Notch3. It is important to this regard the observation that deletion of *Ikaros,* specifically targeted to thymocytes, leads to the persistent expression of Notch target genes that are essential for the rapid development of T cell leukemias in mice (Geimer Le Lay et al. 2014). If we consider that Ikaros is strongly upregulated only after DN3 stage of intrathymic T cell development, when instead Notch3 should start decreasing, the persistence of Notch3 activation after this differentiation step may sustain Ikaros functional inactivation and the subsequent leukemogenesis. Notably, we have to consider that while *Ikaros* gene mutations are rare in human T-ALL, the increased expression of shorter *Ikaros* mRNA transcripts has been reported by several authors (reviewed in (Kastner and Chan 2011).

Together the observations above may also suggest that N3ICD may exert its effect on *Ptcra* transcription through two different ways: a direct mechanism in which N3ICD is able to drive a strong transcriptional activity of the *Ptcra* promoter (Talora et al. 2003), and an indirect one, in which N3ICD is able to remove the Ikaros inhibitory effect by positively regulating the generation of IK-DN isoforms. The latter effect appears to be mediated by the upregulation of the RNA binding protein, HuD, which is able to regulate both RNA-alternative splicing and RNA stability. In addition, the mechanism through which N3ICD determines a HuD overexpression seems to be pre-TCR dependent, since in *N3ICDtg/Ptcra*$^{-/-}$ double mutant mice, Notch3 activation was not able to induce HuD overexpression (Bellavia et al. 2007), further highlighting the importance of the strict interdependence of Notch3 and pre-TCR in sustaining the activation or regulation of different regulatory pathways (i.e. NF-κB or Ikaros), possibly involved in specific stages of T cell differentiation and leukemogenesis. The cartoon depicted in Fig. 2 illustrates the central role played by the relationship between Notch3 and pre-TCR in T cell leukemogenesis, through the activation of canonical NF-κB pathway (p50/p65 heterodimer) and the triggering of *Ikaros* alternative splicing.

4 Multilevel Regulation of Notch3 Expression and Function

Notch receptors are susceptible of multiple modifications. A major challenge in the field is how to correlate Notch gene expression and its multiple interconnected modifications, at both transcriptional and post-translational and gene and protein levels, with different biological behaviors in cells.

Transcription of new mRNAs, subjected to genetic and epigenetic control mechanisms, alternative RNA splicing and their translation into a protein that can be further modified by different posttraslational modifications create a continuous finely tuned regulatory network.

As reported above, N3ICD expression and activity deregulation drives oncogenic signals in cancer cells of various origins, including breast, lung, ovary and colon, as well as T lymphoid cells (Chen et al. 2012; Konishi et al. 2007; Pierfelice et al. 2011; Serafin et al. 2011; Yamaguchi et al. 2008; Bellavia et al. 2000). Since gene mutation and/or amplification have been rarely reported, how NOTCH3 gene and protein expression and/or function is upregulated in these cancers and the mechanisms used for the possible Notch3-mediated and/or -sustained carcinogenesis represent a challenging issue. Indeed, while several data have been reported in literature demonstrating that NOTCH1 can directly activate NOTCH3 expression in T-ALL cells, in which *NOTCH1* often displays activating mutations, the same has not been clarified for solid tumors (i.e. lung and breast cancers), where *NOTCH1* activating mutations are rarely observed, despite the overexpression and/or the constitutive activation of NOTCH3 (Haruki et al. 2005; Choy et al. 2017).

4.1 Notch3 Epigenetic Regulation

Epigenetic mechanisms involving methylation process at the *Notch3* locus have been reported to contribute to the aberrant expression/activation of Notch3 signaling in the development and progression of some cancers.

The analysis of the DNA methylation status of Notch pathway genes in different leukemia cell lines and patient-derived primary samples has shown that the *NOTCH3* locus is preferentially hyper-methylated in lymphoblastic B cell lines and in primary B cell acute lymphoblastic leukemia (B-ALL), while being very low methylated in T-ALL cell lines or in T-ALL-bearing patient-derived primary samples. Consequently, high *NOTCH3* expression was found in several primary or established T-ALL cells, while any or very low expression levels were detected in B-cell lineage leukemias. These observations suggested that the epigenetic regulation of *NOTCH3* expression as well as of other Notch pathway genes could correlate with their specific expression and function in human leukemias (Kuang et al. 2013). Consistently, we recently demonstrated that the cancer testis antigen BORIS/CTCFL (Brother Of Regulator of Imprinted Sites/CTCF-like protein) plays a key role in the epigenetic processes that cause *NOTCH3* overexpression in cancer cells (Zampieri et al. 2014) (Fig. 1, step (1)). In this work we demonstrated that the binding of BORIS/CTCFL to the *NOTCH3* promoter region sustains high *NOTCH3* gene expression in T-ALL, by maintaining the permissive chromatin configuration through the tri-methylation of the histone H3 lysine 4 (H3K4me3) at the *NOTCH3* gene regulatory region. Conversely, BORIS inhibition reverts H3K4me3 mark at the putative promoter region, leading to transcriptionally un-permissive chromatin status. We thus unveiled a direct correlation between BORIS/CTCFL and *NOTCH3* expression/activity in both T-ALL cell lines and patient-derived primary samples (Zampieri et al. 2014).

Besides the above described promoter region, the intron1 region of *NOTCH3* gene has been indicated as an enhancer regulatory region involved in the Notch1-dependent mechanisms driving Notch3 expression in T-ALL cells (Wang et al. 2014). Moreover, an independent study demonstrated that active N1ICD sustains Notch3 expression in T-ALL cells and in hematopoietic stem cells derived from human umbilical cord blood co-cultured with OP9 bone marrow stromal cells ectopically expressing the Notch ligand DLL-1. The authors demonstrated that Notch1 drives a multifunctional chromatin-remodelling complex, including the Lysine-specific histone demethylase 1A (LSD1) and the histone lysine demethylase PHF8 to the Notch3 enhancer region. The authors concluded that the recruitment of the Notch1-driven complex to the Notch3 intron1 is required to ensure Notch3 expression by promoting efficient levels of H3K9me2 demethylation by LSD1 and of H3K27me2 by PHF8 at the regulatory *Notch3* gene regions (Yatim et al. 2012).

As well as in oncogenesis, the regulation of *Notch3* locus by DNA methylation has been identified as a key mechanism controlling *Notch3* expression/function during hepatic stellate cells activation. *Notch1* and *Notch3* gene expression has been found inversely regulated in quiescent versus activated hepatic stellate cells and it has been demonstrated that the switch of expression of the two Notch receptors is finely regulated during this process by chromatin regulation (Reister et al. 2011).

In addition to DNA methylation, increasing evidence highlights the cross-talk between miRNA machinery and Notch3 signaling as a key factor in the regulation of physiological embryogenesis and of homeostasis in different tissues and deregulation of this balance has been linked to carcinogenesis, tumor metastatization and development of drug-resistance in some cancers.

Among miRNAs affecting Notch3 signaling, miR-206 is one of the most studied. It has been demonstrated to promote apoptosis and to inhibit cell migration and foci formation in HeLa cells by targeting Notch3 expression (Song et al. 2009). Subsequently, miR-206/Notch3 crosstalk has been intensively studied in other contexts: it has been demonstrated that miR-1 and miR-206 allow myogenic differentiation in skeletal muscle by directly down-regulating Notch3 expression and therefore sustaining the pro-differentiative activity of the myogenic transcription factor Mef2c (Gagan et al. 2012). Consistently, higher Notch3 levels of expression have been detected in skeletal muscle satellite cells deleted for miR-206 when compared to the wild type counterparts (Liu et al. 2012). Together these studies suggest that miR-206 promotes skeletal muscle differentiation by selectively repressing key anti-differentiation pathways in muscle, including Notch3 signaling (Gagan et al. 2012; Liu et al. 2012). miR-206/Notch3 relationship has been also identified as critical in cancer contexts in which it has been suggested that miR-206 acts as a tumor suppressor by inhibiting Notch3 expression. Finally, an inverse correlation between miR-206 and Notch3 expression levels has been revealed among 49 primary samples of colorectal cancer patients and decreased Notch3 expression induced by overexpressing mimic miR-206 in colon cancer cells has been shown to be associated with impaired cancer cell proliferation and migration and also activation of apoptosis (Wang et al. 2015). Notch3 has been revealed as a direct miR-206 target gene also in human hepatocellular carcinoma HepG2 cells and in the same report the authors proposed a model in which, by targeting Notch3, miR-206 overexpression suppresses *in vitro* tumor growth and metastasis (Liu et al. 2014).

Apart from miR-206/Notch3 axis, it has been suggested that lymphoid specific miR-150 regulates Notch3 expression during T cells maturation and enforced expression of miR-150 interferes with DN3 to DN4 transition of T-cell progenitors. Accordingly, miR-150 overexpression exerts pro-apoptotic and anti-proliferative effects in T-ALL cell lines by directly targeting Notch3 (Ghisi et al. 2011).

A very recent study unveiled the tumor suppressor function of miR-136, exerted by repressing the Notch3-dependent oncogenic signaling in ovarian cancer, and consistently it has been shown an inverse correlation between miR-136 and Notch3 expression in ovarian serous carcinoma samples (Jeong et al. 2017).

The data above describe Notch3 as a miRNA target (Fig. 1). Conversely, we recently explored whether activated N3ICD may directly regulate miRNA network in leukemias. In that study, by Notch3 gain-of-function and loss-of-function approaches in human and mouse T-cell lines, we identified miR-223 as a direct N3ICD target gene. We were able to reveal the oncogenic role of the miR-223/Notch signaling axis that, via the repression of the onco-suppressor FBXW7, sustains the Notch-dependent tumor promoting program in T-ALL (Kumar et al. 2014). Interestingly, we demonstrated that miR-223 is involved in the mechanisms behind the T-ALL cell resistance to γ-secretase inhibitor treatments, thus suggesting its inhibition in novel target therapy protocols (Kumar et al. 2014). The Fig. 1 schematically represents the mechanisms regulating *NOTCH3* gene transcription (step (1)) and protein translation (step (2)), as well as receptor maturation (step (3)), signaling triggering (step (4)) and N3ICD post-translational modifications (step (5)).

4.2 N3ICD Post-translational Modifications

Post-translational modifications (PTMs), such as phosphorylation, acetylation or ubiquitination are integral parts of gene regulation through which specific gene products are turned on or off.

The characterization of post-translational alterations to which Notch3 receptor is subjected in different cellular contexts is essential for understanding how PTMs, in turn regulated by other signaling pathways, selectively modulate Notch3 receptor activity in a context-specific manner.

Protein ubiquitination is a PTM that regulates the protein trafficking specific fate: mono-ubiquitination of membrane proteins triggers their endocytosis, by targeting them to endosomes/lysosomes, whereas K48-linked poly-ubiquitination is a signal for targeting cytosolic proteins to the proteasome for degradation (Deribe et al. 2010). The ubiquitination is critical for maintaining appropriate Notch protein levels and plays a critical role in different step of the Notch signaling maintenance. Altered ubiquitination and degradation of Notch3 protein contribute to sustain Notch3 expression and Notch3-dependent leukemia in *N3ICD tg* mice (Checquolo et al. 2010). Notably genetic ablation of *Ptcra* expression in *N3ICD tg* mice abrogates tumor development, by retaining the E3 ubiquitin ligase c-Cbl in the cytoplasm, where it is able to target Notch3 protein to the proteasomal degradative pathway (Fig 1, step (5)). These data suggest that the E3 ligase c-Cbl may represent an additional regulator of Notch3 and pre-TCR relationship with respect to T-cell leukemogenesis (Checquolo et al. 2010).

In an effort to identify the molecular modulators of the Notch3 signaling pathway, by using a human proteome microarray screening, it has been demonstrated that another E3 ubiquitin ligase named WWP2 negatively regulates NOTCH3 in ovarian cancer (Jung et al. 2014). WWP2 promotes a strong mono-ubiquitination pattern of the membrane tethered fragment of NOTCH3, N3-NEXT, which occurs to a lesser extent in the "resting form" N3-TM or the cytosolic soluble form, N3ICD, finally counteracting the Notch3-promoted cancer stem cell-like cell population (CSCs) survival/self renewal and platinum resistance. These data provided new insights into how Notch3 and its oncogenic signaling could be suppressed: reagents upregulating the expression of WWP2 or gene therapy approaches, by re-introducing WWP2 into cancer cells, represent promising strategies for targeting Notch3-dependent carcinogenesis. In non-small cell lung cancers (NSCLC), the serine threonine receptor-associated protein (STRAP) is able to bind N3ICD through the ankyrin repeat region, and this binding is enhanced in a proteasomal inhibition-dependent manner. Indeed, in *vitro* ubiquitination studies indicate that STRAP reduces the ubiquitination of N3ICD, thus suggesting its important role in N3ICD stabilization (Kashikar et al. 2011). This is further supported by the highly significant correlation (59%) between STRAP and N3ICD levels observed in patients-derived lung tissue microarray (Kashikar et al. 2011).

More recently, it has been demonstrated that N-acetylcysteine (NAC) decreases Notch3 levels through a lysosome-dependent degradation pathway, thereby negatively regulating Notch3 oncogenic signaling in cancer cells (Zhang et al. 2016). NAC modifies the non-covalent binding region of Notch3 receptor, thus decreasing protein stability and leading to protein degradation. Interestingly, NAC does not affect Notch1 expression: this is possibly due to the differences between Notch1 and Notch3 proteins along the heterodimerization region, which only share 41% homology.

Notch3 activation has been recently reported to promote invasive glioma formation (Pierfelice et al. 2011) and such feature has been suggested as a prognostic marker for high-grade glioma therapy follow up (Alqudah et al. 2013). Notably, Temozolomide (TMZ), an alkylating agent used as a first-line chemotherapy drug for gliomas, upregulates its known target gene CHAC1 (Cation transport regulator-like protein 1), which in turn binds to the Notch3 protein, resulting in the inhibition of N3ICD formation and activation, thus finally influencing TMZ-mediated cytotoxicity (Chen et al. 2016).

While the ubiquitination appears to be critical for maintaining appropriate Notch3 protein expression and function, increasing evidence indicates that acetylation of Notch receptors is an equally important PTM for either stabilizing or activating the protein (Palermo et al. 2014).

In this regard, we recently reported that Notch3 is acetylated and deacetylated at lysines 1692 and 1731 by p300 and HDAC1, respectively, and this balance is impaired by HDAC inhibitors (HDACi) that promote hyperacetylation (Palermo et al. 2012). Notch3 acetylation primes ubiquitination and proteasomal-mediated degradation of the protein (Fig. 1, step (5)), thus preventing the development of T-cell leukemia/lymphoma in *N3ICD tg* mice (Palermo et al. 2012), indicating that the balance of Notch receptor acetylation/deacetylation represents a key regulatory switch in Notch3-dependent T-ALL.

Together the observations above suggest that Notch3 can be modified by more than one type of PTM, which control Notch3 protein expression and function. A crucial aspect of signaling through PTMs is the presence of modular protein domains that recognize different types of PTMs located on specific residues. This coupling of PTMs with PTM-recognition domains creates an attractive 'decoding' mechanism for monitoring and responding to alterations in the cellular microenvironment (Deribe et al. 2010). Addition of phosphate group to the serine, threonine, tyrosine residues is an ubiquitous regulatory mechanism and was one of the first PTMs to be described. The phospho Ser/Thr/Pro residues represent the target sites of the prolyl isomerase Pin1, a peptidyl-prolyl isomerase that can alter the conformation of phosphoproteins and affect protein function and/or stability (Yeh and Means 2007). Recent studies suggested a pivotal role of Pin1 in increasing the oncogenic activity of Notch1 protein in breast cancer development and progression (Rustighi et al. 2009; Rustighi et al. 2014). We next have shown that Notch3 represents an additional target protein of Pin1 isomerase. Mechanistically, the Notch3-Pin1 binding correlates with the regulation of Notch3 protein expression and signaling, through a dual mechanism that impinges on its S3 cleavage at the cell membrane and on the stability of its cleaved product, leading to increased expression of the N3ICD. We demonstrated that Notch3 protein is subjected to phosphorylation in Ser/Thr/Pro motifs and this PTM has an effect on its interaction with other protein, such as Pin1, which results in Notch3 stabilization (Fig. 1, step (5)), thus enhancing Notch3-dependent invasiveness properties. Consistently, Pin1 deletion in *N3ICD tg* mice, by reducing N3ICD stability and signaling activity, impairs the expansion/invasiveness of $CD4^+CD8^+$ DP T cells in peripheral lymphoid and non-lymphoid organs and in circulating blood, finally preventing the progression of T-ALL in this mouse model (Franciosa et al. 2016).

Notably, more than 50% of human T-ALL patient samples show activating *NOTCH1* mutations (Mansour et al. 2006; Weng et al. 2004), whereas overexpression of *NOTCH3*, which is mutated only in a subset of T-ALL (Bernasconi-Elias et al. 2016), irrespectively of gross abnormalities in the *Notch3* locus, is a common finding in human T-ALL (Bellavia et al. 2002), raising the possibility that in this context high Pin1 expression might contribute to sustain increased stability and function of N3ICD protein, similarly to what happens for Notch1 in breast cancer where it is rarely mutated (Rustighi et al. 2014; Santarpia et al. 2012). Interestingly, we observed a statistically significant direct correlation between *NOTCH3* and *PIN1* gene expression in a cohort of T-ALL patients (Franciosa et al. 2016).

In addition to its Pin1-dependent regulation involving specific phosphorylated Ser/Thr/Pro motifs, it has been recently demonstrated that NOTCH3 is tyrosine phosphorylated in an EGFR kinase-dependent fashion. Indeed, the treatment of *EGFR*-mutated lung cancer cell lines with the tyrosin kinase inhibitor (TKI) Erlotinib increases their clonogenicity, resulting in an enriched stem cell-like population dependent on NOTCH3 but not on NOTCH1 (Arasada et al. 2014). The kinase-dependent physical association between the Notch3 and EGFR receptors and subsequent EGFR-dependent activation of NOTCH3 after Erlotinib treatment suggests that selective NOTCH3 inhibition combined with TKI therapy should be preferentially explored in treating patients with lung tumors (Arasada et al. 2014).

5 Conclusions and Future Perspectives

After our first observation that constitutive activation of Notch3, sustained by the enforced expression of *N3ICD*, was able to induce an aggressive T-ALL in mice (Bellavia et al. 2000), a number of reports suggested a role of NOTCH3 in human T-ALL (Agnusdei et al. 2014; Bellavia et al. 2002; Kumar et al. 2014; Lu et al. 2013; Masiero et al. 2011). More importantly, activating mutations of *NOTCH3* have been recently identified in primary T-ALL patient-derived xenografts even in the absence of NOTCH1 activation, as it is also the case in the established TALL-1 cell line, suggesting the possible existence of a human T-ALL cluster characterized by *NOTCH3* mutations (Bernasconi-Elias et al. 2016). Moreover, the authors report the presence of *NOTCH3* mutations in a number of cell lines derived from different human solid cancers. In addition, Notch3 expression/deregulation has been identified as a key factor in sustaining uncontrolled CSCs self renewal and/or expansion and in characterizing poor prognosis and chemoresistance in several solid tumors.

In ovarian cancer, up-regulated Notch3 signaling has been associated with poor outcome (Hu et al. 2014; Jung et al. 2010; Liu et al. 2016; Rahman et al. 2012) and its inhibition has been shown to enhance tumor sensitivity to paclitaxel, platinum and carboplatin in experimental models of NOTCH3-positive ovarian cancer tumors (Hu et al. 2014; Kang et al. 2016; McAuliffe et al. 2012; Park et al. 2010). NOTCH3 up-regulation has been indicated also as a feature of aggressive colorectal cancer and *NOTCH3* silencing as well as treatments with the antagonist Notch2/3 antibody decreased tumor burden in mice bearing colorectal cancer xenografts (Ozawa et al. 2014; Pasto et al. 2014; Serafin et al. 2011) .

Comparable studies linked Notch3 activity to the maintenance of CSCs in liver cancer and

NOTCH3 expression was strongly associated with more aggressive traits and poor survival in patients bearing hepato- and cholangiocarcinoma (Guest et al. 2016; Hu et al. 2013; Zhang et al. 2015). Additional work demonstrated that Notch3 inhibition reverts sorafenib resistance in mice xenograft models of hepatocellular carcinoma (Giovannini et al. 2013) and enhances the cell death in response to doxorubicin treatments (Giovannini et al. 2009).

In line with these observations Notch signaling blockade, obtained with anti-Notch2/3 antibody either as a single agent or in combination with chemotherapeutic agents, reduced tumor-initiating cell frequency, overcome treatment resistance and resulted in cancer regression also in patient-derived xenografts from pancreatic, breast, ovarian and lung cancers (Yen et al. 2015).

Together these observations suggest that deregulated expression and/or activating mutations of Notch3 may behave as oncogenic drivers and suggest the targeting of Notch3 in such cancers as a compelling additional anticancer therapeutic possibility to overcome chemoresistance and to reinforce conventional therapies.

Acknowledgments We thank members of Screpanti's laboratory, whose comments and work have contributed to the realization of this chapter. Work in the author's laboratories is supported by Grants from the Associazione Italiana per la Ricerca sul Cancro (AIRC), Institut Pasteur - Fondazione Cenci-Bolognetti and Sapienza University of Rome. This chapter is dedicated to the memory of Professor Alberto Gulino.

References

Agnusdei V, Minuzzo S, Frasson C, Grassi A, Axelrod F, Satyal S, Gurney A, Hoey T, Seganfreddo E, Basso G, Valtorta S, Moresco RM, Amadori A, Indraccolo S (2014) Therapeutic antibody targeting of Notch1 in T-acute lymphoblastic leukemia xenografts. Leukemia 28(2):278–288. https://doi.org/10.1038/leu.2013.183

Aifantis I, Gounari F, Scorrano L, Borowski C, von Boehmer H (2001) Constitutive pre-TCR signaling promotes differentiation through Ca2+ mobilization and activation of NF-kappaB and NFAT. Nat Immunol 2(5):403–409. https://doi.org/10.1038/87704

Alqudah MA, Agarwal S, Al-Keilani MS, Sibenaller ZA, Ryken TC, Assem M (2013) NOTCH3 is a prognostic factor that promotes glioma cell proliferation, migration and invasion via activation of CCND1 and EGFR. PLoS One 8(10):e77299. https://doi.org/10.1371/journal.pone.0077299

Amsen D, Blander JM, Lee GR, Tanigaki K, Honjo T, Flavell RA (2004) Instruction of distinct CD4 T helper cell fates by different notch ligands on antigen-presenting cells. Cell 117(4):515–526

Anastasi E, Campese AF, Bellavia D, Bulotta A, Balestri A, Pascucci M, Checquolo S, Gradini R, Lendahl U, Frati L, Gulino A, Di Mario U, Screpanti I (2003) Expression of activated Notch3 in transgenic mice enhances generation of T regulatory cells and protects against experimental autoimmune diabetes. J Immunol 171(9):4504–4511

Arasada RR, Amann JM, Rahman MA, Huppert SS, Carbone DP (2014) EGFR blockade enriches for lung cancer stem-like cells through Notch3-dependent signaling. Cancer Res 74(19):5572–5584. https://doi.org/10.1158/0008-5472.CAN-13-3724

Ascano JM, Beverly LJ, Capobianco AJ (2003) The C-terminal PDZ-ligand of JAGGED1 is essential for cellular transformation. J Biol Chem 278(10):8771–8779. https://doi.org/10.1074/jbc.M211427200

Barbarulo A, Grazioli P, Campese AF, Bellavia D, Di Mario G, Pelullo M, Ciuffetta A, Colantoni S, Vacca A, Frati L, Gulino A, Felli MP, Screpanti I (2011) Notch3 and canonical NF-kappaB signaling pathways cooperatively regulate Foxp3 transcription. J Immunol 186(11):6199–6206. https://doi.org/10.4049/jimmunol.1002136

Beatus P, Lundkvist J, Oberg C, Pedersen K, Lendahl U (2001) The origin of the ankyrin repeat region in Notch intracellular domains is critical for regulation of HES promoter activity. Mech Dev 104(1–2):3–20

Bellavia D, Campese AF, Alesse E, Vacca A, Felli MP, Balestri A, Stoppacciaro A, Tiveron C, Tatangelo L, Giovarelli M, Gaetano C, Ruco L, Hoffman ES, Hayday AC, Lendahl U, Frati L, Gulino A, Screpanti I (2000) Constitutive activation of NF-kappaB and T-cell leukemia/lymphoma in Notch3 transgenic mice. EMBO J 19(13):3337–3348. https://doi.org/10.1093/emboj/19.13.3337

Bellavia D, Campese AF, Checquolo S, Balestri A, Biondi A, Cazzaniga G, Lendahl U, Fehling HJ, Hayday AC, Frati L, von Boehmer H, Gulino A, Screpanti I (2002) Combined expression of pTalpha and Notch3 in T cell leukemia identifies the requirement of preTCR for leukemogenesis. Proc Natl Acad Sci U S A 99(6):3788–3793. https://doi.org/10.1073/pnas.062050599

Bellavia D, Mecarozzi M, Campese AF, Grazioli P, Talora C, Frati L, Gulino A, Screpanti I (2007) Notch3 and the Notch3-upregulated RNA-binding protein HuD regulate Ikaros alternative splicing. EMBO J 26(6):1670–1680. https://doi.org/10.1038/sj.emboj.7601626

Bernasconi-Elias P, Hu T, Jenkins D, Firestone B, Gans S, Kurth E, Capodieci P, Deplazes-Lauber J, Petropoulos K, Thiel P, Ponsel D, Hee Choi S, LeMotte P, London A, Goetcshkes M, Nolin E, Jones MD, Slocum K, Kluk MJ, Weinstock DM, Christodoulou A, Weinberg O, Jaehrling J, Ettenberg SA, Buckler A, Blacklow SC,

Aster JC, Fryer CJ (2016) Characterization of activating mutations of NOTCH3 in T-cell acute lymphoblastic leukemia and anti-leukemic activity of NOTCH3 inhibitory antibodies. Oncogene 35(47):6077–6086. https://doi.org/10.1038/onc.2016.133

Beverly LJ, Capobianco AJ (2003) Perturbation of Ikaros isoform selection by MLV integration is a cooperative event in Notch(IC)-induced T cell leukemogenesis. Cancer Cell 3(6):551–564

Checquolo S, Palermo R, Cialfi S, Ferrara G, Oliviero C, Talora C, Bellavia D, Giovenco A, Grazioli P, Frati L, Gulino A, Screpanti I (2010) Differential subcellular localization regulates c-Cbl E3 ligase activity upon Notch3 protein in T-cell leukemia. Oncogene 29(10):1463–1474. https://doi.org/10.1038/onc.2009.446

Chen X, Thiaville MM, Chen L, Stoeck A, Xuan J, Gao M, Shih Ie M, Wang TL (2012) Defining NOTCH3 target genes in ovarian cancer. Cancer Res 72(9):2294–2303. https://doi.org/10.1158/0008-5472.CAN-11-2181

Chen PH, Shen WL, Shih CM, Ho KH, Cheng CH, Lin CW, Lee CC, Liu AJ, Chen KC (2016) The CHAC1-inhibited Notch3 pathway is involved in temozolomide-induced glioma cytotoxicity. Neuropharmacology 116:300–314. https://doi.org/10.1016/j.neuropharm.2016.12.011

Choi JH, Park JT, Davidson B, Morin PJ, Shih Ie M, Wang TL (2008) Jagged-1 and Notch3 juxtacrine loop regulates ovarian tumor growth and adhesion. Cancer Res 68(14):5716–5723. https://doi.org/10.1158/0008-5472.CAN-08-0001

Choy L, Hagenbeek TJ, Solon M, French D, Finkle D, Shelton A, Venook R, Brauer MJ, Siebel CW (2017) Constitutive NOTCH3 signaling promotes the growth of basal breast cancers. Cancer Res 77(6):1439–1452. https://doi.org/10.1158/0008-5472.CAN-16-1022

Conlon RA, Reaume AG, Rossant J (1995) Notch1 is required for the coordinate segmentation of somites. Development 121(5):1533–1545

Deribe YL, Pawson T, Dikic I (2010) Post-translational modifications in signal integration. Nat Struct Mol Biol 17(6):666–672. https://doi.org/10.1038/nsmb.1842

Domenga V, Fardoux P, Lacombe P, Monet M, Maciazek J, Krebs LT, Klonjkowski B, Berrou E, Mericskay M, Li Z, Tournier-Lasserve E, Gridley T, Joutel A (2004) Notch3 is required for arterial identity and maturation of vascular smooth muscle cells. Genes Dev 18(22):2730–2735. https://doi.org/10.1101/gad.308904

Fan X, Mikolaenko I, Elhassan I, Ni X, Wang Y, Ball D, Brat DJ, Perry A, Eberhart CG (2004) Notch1 and notch2 have opposite effects on embryonal brain tumor growth. Cancer Res 64(21):7787–7793. https://doi.org/10.1158/0008-5472.CAN-04-1446

Felli MP, Maroder M, Mitsiadis TA, Campese AF, Bellavia D, Vacca A, Mann RS, Frati L, Lendahl U, Gulino A, Screpanti I (1999) Expression pattern of notch1, 2 and 3 and Jagged1 and 2 in lymphoid and stromal thymus components: distinct ligand-receptor interactions in intrathymic T cell development. Int Immunol 11(7):1017–1025

Fowlkes BJ, Robey EA (2002) A reassessment of the effect of activated Notch1 on CD4 and CD8 T cell development. J Immunol 169(4):1817–1821

Franciosa G, Diluvio G, Gaudio FD, Giuli MV, Palermo R, Grazioli P, Campese AF, Talora C, Bellavia D, D'Amati G, Besharat ZM, Nicoletti C, Siebel CW, Choy L, Rustighi A, Sal GD, Screpanti I, Checquolo S (2016) Prolyl-isomerase Pin1 controls Notch3 protein expression and regulates T-ALL progression. Oncogene 35(36):4741–4751. https://doi.org/10.1038/onc.2016.5

Gagan J, Dey BK, Layer R, Yan Z, Dutta A (2012) Notch3 and Mef2c proteins are mutually antagonistic via Mkp1 protein and miR-1/206 microRNAs in differentiating myoblasts. J Biol Chem 287(48):40360–40370. https://doi.org/10.1074/jbc.M112.378414

Geimer Le Lay AS, Oravecz A, Mastio J, Jung C, Marchal P, Ebel C, Dembele D, Jost B, Le Gras S, Thibault C, Borggrefe T, Kastner P, Chan S (2014) The tumor suppressor Ikaros shapes the repertoire of notch target genes in T cells. Sci Signal 7(317):ra28. https://doi.org/10.1126/scisignal.2004545

Ghisi M, Corradin A, Basso K, Frasson C, Serafin V, Mukherjee S, Mussolin L, Ruggero K, Bonanno L, Guffanti A, De Bellis G, Gerosa G, Stellin G, D'Agostino DM, Basso G, Bronte V, Indraccolo S, Amadori A, Zanovello P (2011) Modulation of microRNA expression in human T-cell development: targeting of NOTCH3 by miR-150. Blood 117(26):7053–7062. https://doi.org/10.1182/blood-2010-12-326629

Giovannini C, Gramantieri L, Chieco P, Minguzzi M, Lago F, Pianetti S, Ramazzotti E, Marcu KB, Bolondi L (2009) Selective ablation of Notch3 in HCC enhances doxorubicin's death promoting effect by a p53 dependent mechanism. J Hepatol 50(5):969–979. https://doi.org/10.1016/j.jhep.2008.12.032

Giovannini C, Baglioni M, Baron Toaldo M, Ventrucci C, D'Adamo S, Cipone M, Chieco P, Gramantieri L, Bolondi L (2013) Notch3 inhibition enhances sorafenib cytotoxic efficacy by promoting GSK3b phosphorylation and p21 down-regulation in hepatocellular carcinoma. Oncotarget 4(10):1618–1631

Guest RV, Boulter L, Dwyer BJ, Kendall TJ, Man TY, Minnis-Lyons SE, Lu WY, Robson AJ, Gonzalez SF, Raven A, Wojtacha D, Morton JP, Komuta M, Roskams T, Wigmore SJ, Sansom OJ, Forbes SJ (2016) Notch3 drives development and progression of cholangiocarcinoma. Proc Natl Acad Sci U S A 113(43):12250–12255. https://doi.org/10.1073/pnas.1600067113

Haruki N, Kawaguchi KS, Eichenberger S, Massion PP, Olson S, Gonzalez A, Carbone DP, Dang TP (2005) Dominant-negative Notch3 receptor inhibits mitogen-activated protein kinase pathway and the growth of human lung cancers. Cancer Res 65(9):3555–3561. https://doi.org/10.1158/0008-5472.CAN-04-3132

Hu L, Xue F, Shao M, Deng A, Wei G (2013) Aberrant expression of Notch3 predicts poor survival for hepatocellular carcinomas. Biosci Trends 7(3):152–156

Hu W, Liu T, Ivan C, Sun Y, Huang J, Mangala LS, Miyake T, Dalton HJ, Pradeep S, Rupaimoole R, Previs RA, Han HD, Bottsford-Miller J, Zand B, Kang Y, Pecot CV, Nick AM, Wu SY, Lee JS, Sehgal V, Ram P, Liu J, Tucker SL, Lopez-Berestein G, Baggerly KA, Coleman RL, Sood AK (2014) Notch3 pathway alterations in ovarian cancer. Cancer Res 74(12):3282–3293. https://doi.org/10.1158/0008-5472.CAN-13-2066

Ikeuchi T, Sisodia SS (2003) The notch ligands, Delta1 and Jagged2, are substrates for presenilin-dependent "gamma-secretase" cleavage. J Biol Chem 278(10):7751–7754. https://doi.org/10.1074/jbc.C200711200

Jeannet R, Mastio J, Macias-Garcia A, Oravecz A, Ashworth T, Geimer Le Lay AS, Jost B, Le Gras S, Ghysdael J, Gridley T, Honjo T, Radtke F, Aster JC, Chan S, Kastner P (2010) Oncogenic activation of the Notch1 gene by deletion of its promoter in Ikaros-deficient T-ALL. Blood 116(25):5443–5454. https://doi.org/10.1182/blood-2010-05-286658

Jeffries S, Capobianco AJ (2000) Neoplastic transformation by Notch requires nuclear localization. Mol Cell Biol 20(11):3928–3941

Jeong JY, Kang H, Kim TH, Kim G, Heo JH, Kwon AY, Kim S, Jung SG, An HJ (2017) MicroRNA-136 inhibits cancer stem cell activity and enhances the anti-tumor effect of paclitaxel against chemoresistant ovarian cancer cells by targeting Notch3. Cancer Lett 386:168–178. https://doi.org/10.1016/j.canlet.2016.11.017

Joutel A, Andreux F, Gaulis S, Domenga V, Cecillon M, Battail N, Piga N, Chapon F, Godfrain C, Tournier-Lasserve E (2000) The ectodomain of the Notch3 receptor accumulates within the cerebrovasculature of CADASIL patients. J Clin Invest 105(5):597–605. https://doi.org/10.1172/JCI8047

Jung SG, Kwon YD, Song JA, Back MJ, Lee SY, Lee C, Hwang YY, An HJ (2010) Prognostic significance of Notch 3 gene expression in ovarian serous carcinoma. Cancer Sci 101(9):1977–1983. https://doi.org/10.1111/j.1349-7006.2010.01641.x

Jung JG, Stoeck A, Guan B, Wu RC, Zhu H, Blackshaw S, Shih Ie M, Wang TL (2014) Notch3 interactome analysis identified WWP2 as a negative regulator of Notch3 signaling in ovarian cancer. PLoS Genet 10(10):e1004751. https://doi.org/10.1371/journal.pgen.1004751

Kang H, Jeong JY, Song JY, Kim TH, Kim G, Huh JH, Kwon AY, Jung SG, An HJ (2016) Notch3-specific inhibition using siRNA knockdown or GSI sensitizes paclitaxel-resistant ovarian cancer cells. Mol Carcinog 55(7):1196–1209. https://doi.org/10.1002/mc.22363

Kashikar ND, Zhang W, Massion PP, Gonzalez AL, Datta PK (2011) Role of STRAP in regulating GSK3beta function and Notch3 stabilization. Cell Cycle 10(10):1639–1654. https://doi.org/10.4161/cc.10.10.15630

Kastner P, Chan S (2011) Role of Ikaros in T-cell acute lymphoblastic leukemia. World J Biol Chem 2(6):108–114. https://doi.org/10.4331/wjbc.v2.i6.108

Konishi J, Kawaguchi KS, Vo H, Haruki N, Gonzalez A, Carbone DP, Dang TP (2007) Gamma-secretase inhibitor prevents Notch3 activation and reduces proliferation in human lung cancers. Cancer Res 67(17):8051–8057. https://doi.org/10.1158/0008-5472.CAN-07-1022

Kopan R, Weintraub H (1993) Mouse notch: expression in hair follicles correlates with cell fate determination. J Cell Biol 121(3):631–641

Krebs LT, Xue Y, Norton CR, Shutter JR, Maguire M, Sundberg JP, Gallahan D, Closson V, Kitajewski J, Callahan R, Smith GH, Stark KL, Gridley T (2000) Notch signaling is essential for vascular morphogenesis in mice. Genes Dev 14(11):1343–1352

Krebs LT, Xue Y, Norton CR, Sundberg JP, Beatus P, Lendahl U, Joutel A, Gridley T (2003) Characterization of Notch3-deficient mice: normal embryonic development and absence of genetic interactions with a Notch1 mutation. Genesis 37(3):139–143. https://doi.org/10.1002/gene.10241

Kuang SQ, Fang Z, Zweidler-McKay PA, Yang H, Wei Y, Gonzalez-Cervantes EA, Boumber Y, Garcia-Manero G (2013) Epigenetic inactivation of Notch-Hes pathway in human B-cell acute lymphoblastic leukemia. PLoS One 8(4):e61807. https://doi.org/10.1371/journal.pone.0061807

Kumar V, Palermo R, Talora C, Campese AF, Checquolo S, Bellavia D, Tottone L, Testa G, Miele E, Indraccolo S, Amadori A, Ferretti E, Gulino A, Vacca A, Screpanti I (2014) Notch and NF-kB signaling pathways regulate miR-223/FBXW7 axis in T-cell acute lymphoblastic leukemia. Leukemia 28(12):2324–2335. https://doi.org/10.1038/leu.2014.133

Lacombe P, Oligo C, Domenga V, Tournier-Lasserve E, Joutel A (2005) Impaired cerebral vasoreactivity in a transgenic mouse model of cerebral autosomal dominant arteriopathy with subcortical infarcts and leukoencephalopathy arteriopathy. Stroke 36(5):1053–1058. https://doi.org/10.1161/01.STR.0000163080.82766.eb

Lardelli M, Dahlstrand J, Lendahl U (1994) The novel Notch homologue mouse Notch 3 lacks specific epidermal growth factor-repeats and is expressed in proliferating neuroepithelium. Mech Dev 46(2):123–136

LaVoie MJ, Selkoe DJ (2003) The Notch ligands, Jagged and Delta, are sequentially processed by alpha-secretase and presenilin/gamma-secretase and release signaling fragments. J Biol Chem 278(36):34427–34437. https://doi.org/10.1074/jbc.M302659200

Li Y, Baker NE (2004) The roles of cis-inactivation by Notch ligands and of neuralized during eye and bristle patterning in Drosophila. BMC Dev Biol 4:5. https://doi.org/10.1186/1471-213X-4-5

Liu N, Williams AH, Maxeiner JM, Bezprozvannaya S, Shelton JM, Richardson JA, Bassel-Duby R, Olson EN (2012) microRNA-206 promotes skeletal muscle regeneration and delays progression of

Duchenne muscular dystrophy in mice. J Clin Invest 122(6):2054–2065. https://doi.org/10.1172/JCI62656

Liu W, Xu C, Wan H, Liu C, Wen C, Lu H, Wan F (2014) MicroRNA-206 overexpression promotes apoptosis, induces cell cycle arrest and inhibits the migration of human hepatocellular carcinoma HepG2 cells. Int J Mol Med 34(2):420–428. https://doi.org/10.3892/ijmm.2014.1800

Liu Z, Yun R, Yu X, Hu H, Huang G, Tan B, Chen T (2016) Overexpression of Notch3 and pS6 is associated with poor prognosis in human ovarian epithelial cancer. Mediat Inflamm 2016:5953498. https://doi.org/10.1155/2016/5953498

Lu C, Li JY, Ge Z, Zhang L, Zhou GP (2013) Par-4/THAP1 complex and Notch3 competitively regulated pre-mRNA splicing of CCAR1 and affected inversely the survival of T-cell acute lymphoblastic leukemia cells. Oncogene 32(50):5602–5613. https://doi.org/10.1038/onc.2013.349

Mansour MR, Linch DC, Foroni L, Goldstone AH, Gale RE (2006) High incidence of Notch-1 mutations in adult patients with T-cell acute lymphoblastic leukemia. Leukemia 20(3):537–539. https://doi.org/10.1038/sj.leu.2404101

Masiero M, Minuzzo S, Pusceddu I, Moserle L, Persano L, Agnusdei V, Tosello V, Basso G, Amadori A, Indraccolo S (2011) Notch3-mediated regulation of MKP-1 levels promotes survival of T acute lymphoblastic leukemia cells. Leukemia 25(4):588–598. https://doi.org/10.1038/leu.2010.323

McAuliffe SM, Morgan SL, Wyant GA, Tran LT, Muto KW, Chen YS, Chin KT, Partridge JC, Poole BB, Cheng KH, Daggett J Jr, Cullen K, Kantoff E, Hasselbatt K, Berkowitz J, Muto MG, Berkowitz RS, Aster JC, Matulonis UA, Dinulescu DM (2012) Targeting Notch, a key pathway for ovarian cancer stem cells, sensitizes tumors to platinum therapy. Proc Natl Acad Sci U S A 109(43):E2939–E2948. https://doi.org/10.1073/pnas.1206400109

McCright B, Gao X, Shen L, Lozier J, Lan Y, Maguire M, Herzlinger D, Weinmaster G, Jiang R, Gridley T (2001) Defects in development of the kidney, heart and eye vasculature in mice homozygous for a hypomorphic Notch2 mutation. Development 128(4):491–502

Monet M, Domenga V, Lemaire B, Souilhol C, Langa F, Babinet C, Gridley T, Tournier-Lasserve E, Cohen-Tannoudji M, Joutel A (2007) The archetypal R90C CADASIL-NOTCH3 mutation retains NOTCH3 function in vivo. Hum Mol Genet 16(8):982–992. https://doi.org/10.1093/hmg/ddm042

Nehring LC, Miyamoto A, Hein PW, Weinmaster G, Shipley JM (2005) The extracellular matrix protein MAGP-2 interacts with Jagged1 and induces its shedding from the cell surface. J Biol Chem 280(21):20349–20355. https://doi.org/10.1074/jbc.M500273200

Ong CT, Cheng HT, Chang LW, Ohtsuka T, Kageyama R, Stormo GD, Kopan R (2006) Target selectivity of vertebrate notch proteins. Collaboration between discrete domains and CSL-binding site architecture determines activation probability. J Biol Chem 281(8):5106–5119. https://doi.org/10.1074/jbc.M506108200

Ozawa T, Kazama S, Akiyoshi T, Murono K, Yoneyama S, Tanaka T, Tanaka J, Kiyomatsu T, Kawai K, Nozawa H, Kanazawa T, Yamaguchi H, Ishihara S, Sunami E, Kitayama J, Morikawa T, Fukayama M, Watanabe T (2014) Nuclear Notch3 expression is associated with tumor recurrence in patients with stage II and III colorectal cancer. Ann Surg Oncol 21(8):2650–2658. https://doi.org/10.1245/s10434-014-3659-9

Palermo R, Checquolo S, Giovenco A, Grazioli P, Kumar V, Campese AF, Giorgi A, Napolitano M, Canettieri G, Ferrara G, Schinina ME, Maroder M, Frati L, Gulino A, Vacca A, Screpanti I (2012) Acetylation controls Notch3 stability and function in T-cell leukemia. Oncogene 31(33):3807–3817. https://doi.org/10.1038/onc.2011.533

Palermo R, Checquolo S, Bellavia D, Talora C, Screpanti I (2014) The molecular basis of notch signaling regulation: a complex simplicity. Curr Mol Med 14(1):34–44

Park JT, Chen X, Trope CG, Davidson B, Shih Ie M, Wang TL (2010) Notch3 overexpression is related to the recurrence of ovarian cancer and confers resistance to carboplatin. Am J Pathol 177(3):1087–1094. https://doi.org/10.2353/ajpath.2010.100316

Pasto A, Serafin V, Pilotto G, Lago C, Bellio C, Trusolino L, Bertotti A, Hoey T, Plateroti M, Esposito G, Pinazza M, Agostini M, Nitti D, Amadori A, Indraccolo S (2014) NOTCH3 signaling regulates MUSASHI-1 expression in metastatic colorectal cancer cells. Cancer Res 74(7):2106–2118. https://doi.org/10.1158/0008-5472.CAN-13-2022

Pelullo M, Quaranta R, Talora C, Checquolo S, Cialfi S, Felli MP, te Kronnie G, Borga C, Besharat ZM, Palermo R, Di Marcotullio L, Capobianco AJ, Gulino A, Screpanti I, Bellavia D (2014) Notch3/Jagged1 circuitry reinforces notch signaling and sustains T-ALL. Neoplasia 16(12):1007–1017. https://doi.org/10.1016/j.neo.2014.10.004

Pierfelice TJ, Schreck KC, Dang L, Asnaghi L, Gaiano N, Eberhart CG (2011) Notch3 activation promotes invasive glioma formation in a tissue site-specific manner. Cancer Res 71(3):1115–1125. https://doi.org/10.1158/0008-5472.CAN-10-0690

Pui JC, Allman D, Xu L, DeRocco S, Karnell FG, Bakkour S, Lee JY, Kadesch T, Hardy RR, Aster JC, Pear WS (1999) Notch1 expression in early lymphopoiesis influences B versus T lineage determination. Immunity 11(3):299–308

Qi H, Rand MD, Wu X, Sestan N, Wang W, Rakic P, Xu T, Artavanis-Tsakonas S (1999) Processing of the notch ligand delta by the metalloprotease Kuzbanian. Science 283(5398):91–94

Radtke F, Wilson A, Stark G, Bauer M, van Meerwijk J, MacDonald HR, Aguet M (1999) Deficient T cell fate specification in mice with an induced inactivation of Notch1. Immunity 10(5):547–558

Rahman MT, Nakayama K, Rahman M, Katagiri H, Katagiri A, Ishibashi T, Ishikawa M, Iida K, Nakayama S, Otsuki Y, Miyazaki K (2012) Notch3 overexpression as potential therapeutic target in advanced stage chemoresistant ovarian cancer. Am J Clin Pathol 138(4):535–544. https://doi.org/10.1309/AJCPKDLRQ8F3EWNS

Reister S, Kordes C, Sawitza I, Haussinger D (2011) The epigenetic regulation of stem cell factors in hepatic stellate cells. Stem Cells Dev 20(10):1687–1699. https://doi.org/10.1089/scd.2010.0418

Robey E, Chang D, Itano A, Cado D, Alexander H, Lans D, Weinmaster G, Salmon P (1996) An activated form of Notch influences the choice between CD4 and CD8 T cell lineages. Cell 87(3):483–492

Rustighi A, Tiberi L, Soldano A, Napoli M, Nuciforo P, Rosato A, Kaplan F, Capobianco A, Pece S, Di Fiore PP, Del Sal G (2009) The prolyl-isomerase Pin1 is a Notch1 target that enhances Notch1 activation in cancer. Nat Cell Biol 11(2):133–142. https://doi.org/10.1038/ncb1822

Rustighi A, Zannini A, Tiberi L, Sommaggio R, Piazza S, Sorrentino G, Nuzzo S, Tuscano A, Eterno V, Benvenuti F, Santarpia L, Aifantis I, Rosato A, Bicciato S, Zambelli A, Del Sal G (2014) Prolyl-isomerase Pin1 controls normal and cancer stem cells of the breast. EMBO Mol Med 6(1):99–119. https://doi.org/10.1002/emmm.201302909

Sakamoto K, Ohara O, Takagi M, Takeda S, Katsube K (2002) Intracellular cell-autonomous association of Notch and its ligands: a novel mechanism of Notch signal modification. Dev Biol 241(2):313–326. https://doi.org/10.1006/dbio.2001.0517

Santagata S, Demichelis F, Riva A, Varambally S, Hofer MD, Kutok JL, Kim R, Tang J, Montie JE, Chinnaiyan AM, Rubin MA, Aster JC (2004) JAGGED1 expression is associated with prostate cancer metastasis and recurrence. Cancer Res 64(19):6854–6857. https://doi.org/10.1158/0008-5472.CAN-04-2500

Santarpia L, Qi Y, Stemke-Hale K, Wang B, Young EJ, Booser DJ, Holmes FA, O'Shaughnessy J, Hellerstedt B, Pippen J, Vidaurre T, Gomez H, Valero V, Hortobagyi GN, Symmans WF, Bottai G, Di Leo A, Gonzalez-Angulo AM, Pusztai L (2012) Mutation profiling identifies numerous rare drug targets and distinct mutation patterns in different clinical subtypes of breast cancers. Breast Cancer Res Treat 134(1):333–343. https://doi.org/10.1007/s10549-012-2035-3

Serafin V, Persano L, Moserle L, Esposito G, Ghisi M, Curtarello M, Bonanno L, Masiero M, Ribatti D, Sturzl M, Naschberger E, Croner RS, Jubb AM, Harris AL, Koeppen H, Amadori A, Indraccolo S (2011) Notch3 signalling promotes tumour growth in colorectal cancer. J Pathol 224(4):448–460. https://doi.org/10.1002/path.2895

Shimizu K, Chiba S, Kumano K, Hosoya N, Takahashi T, Kanda Y, Hamada Y, Yazaki Y, Hirai H (1999) Mouse jagged1 physically interacts with notch2 and other notch receptors. Assessment by quantitative methods. J Biol Chem 274(46):32961–32969

Shimizu K, Chiba S, Saito T, Kumano K, Hirai H (2000) Physical interaction of Delta1, Jagged1, and Jagged2 with Notch1 and Notch3 receptors. Biochem Biophys Res Commun 276(1):385–389. https://doi.org/10.1006/bbrc.2000.3469

Shimizu K, Chiba S, Saito T, Kumano K, Hamada Y, Hirai H (2002) Functional diversity among Notch1, Notch2, and Notch3 receptors. Biochem Biophys Res Commun 291(4):775–779. https://doi.org/10.1006/bbrc.2002.6528

Song G, Zhang Y, Wang L (2009) MicroRNA-206 targets notch3, activates apoptosis, and inhibits tumor cell migration and focus formation. J Biol Chem 284(46):31921–31927. https://doi.org/10.1074/jbc.M109.046862

Sun L, Liu A, Georgopoulos K (1996) Zinc finger-mediated protein interactions modulate Ikaros activity, a molecular control of lymphocyte development. EMBO J 15(19):5358–5369

Swiatek PJ, Lindsell CE, del Amo FF, Weinmaster G, Gridley T (1994) Notch1 is essential for postimplantation development in mice. Genes Dev 8(6):707–719

Talora C, Campese AF, Bellavia D, Pascucci M, Checquolo S, Groppioni M, Frati L, von Boehmer H, Gulino A, Screpanti I (2003) Pre-TCR-triggered ERK signalling-dependent downregulation of E2A activity in Notch3-induced T-cell lymphoma. EMBO Rep 4(11):1067–1072. https://doi.org/10.1038/sj.embor.embor7400013

Uyttendaele H, Marazzi G, Wu G, Yan Q, Sassoon D, Kitajewski J (1996) Notch4/int-3, a mammary proto-oncogene, is an endothelial cell-specific mammalian Notch gene. Development 122(7):2251–2259

Vacca A, Felli MP, Palermo R, Di Mario G, Calce A, Di Giovine M, Frati L, Gulino A, Screpanti I (2006) Notch3 and pre-TCR interaction unveils distinct NF-kappaB pathways in T-cell development and leukemia. EMBO J 25(5):1000–1008. https://doi.org/10.1038/sj.emboj.7600996

Voll RE, Jimi E, Phillips RJ, Barber DF, Rincon M, Hayday AC, Flavell RA, Ghosh S (2000) NF-kappa B activation by the pre-T cell receptor serves as a selective survival signal in T lymphocyte development. Immunity 13(5):677–689

von Boehmer H, Aifantis I, Azogui O, Feinberg J, Saint-Ruf C, Zober C, Garcia C, Buer J (1998) Crucial function of the pre-T-cell receptor (TCR) in TCR beta selection, TCR beta allelic exclusion and alpha beta versus gamma delta lineage commitment. Immunol Rev 165:111–119

Wang H, Zang C, Taing L, Arnett KL, Wong YJ, Pear WS, Blacklow SC, Liu XS, Aster JC (2014) NOTCH1-RBPJ complexes drive target gene expression through dynamic interactions with superenhancers. Proc Natl Acad Sci U S A 111(2):705–710. https://doi.org/10.1073/pnas.1315023111

Wang XW, Xi XQ, Wu J, Wan YY, Hui HX, Cao XF (2015) MicroRNA-206 attenuates tumor proliferation and migration involving the downregulation of NOTCH3 in colorectal cancer. Oncol Rep 33(3):1402–1410. https://doi.org/10.3892/or.2015.3731

Washburn T, Schweighoffer E, Gridley T, Chang D, Fowlkes BJ, Cado D, Robey E (1997) Notch activity influences the alphabeta versus gammadelta T cell lineage decision. Cell 88(6):833–843

Weinmaster G, Roberts VJ, Lemke G (1991) A homolog of Drosophila Notch expressed during mammalian development. Development 113(1):199–205

Weinmaster G, Roberts VJ, Lemke G (1992) Notch2: a second mammalian Notch gene. Development 116(4):931–941

Weng AP, Ferrando AA, Lee W, JPt M, Silverman LB, Sanchez-Irizarry C, Blacklow SC, Look AT, Aster JC (2004) Activating mutations of NOTCH1 in human T cell acute lymphoblastic leukemia. Science 306(5694):269–271. https://doi.org/10.1126/science.1102160

Winandy S, Wu P, Georgopoulos K (1995) A dominant mutation in the Ikaros gene leads to rapid development of leukemia and lymphoma. Cell 83(2):289–299

Winandy S, Wu L, Wang JH, Georgopoulos K (1999) Pre-T cell receptor (TCR) and TCR-controlled checkpoints in T cell differentiation are set by Ikaros. J Exp Med 190(8):1039–1048

Wu J, Bresnick EH (2007) Bare rudiments of notch signaling: how receptor levels are regulated. Trends Biochem Sci 32(10):477–485. https://doi.org/10.1016/j.tibs.2007.09.002

Wu K, Xu L, Zhang L, Lin Z, Hou J (2011) High Jagged1 expression predicts poor outcome in clear cell renal cell carcinoma. Jpn J Clin Oncol 41(3):411–416. https://doi.org/10.1093/jjco/hyq205

Xu X, Choi SH, Hu T, Tiyanont K, Habets R, Groot AJ, Vooijs M, Aster JC, Chopra R, Fryer C, Blacklow SC (2015) Insights into autoregulation of Notch3 from structural and functional studies of its negative regulatory region. Structure 23(7):1227–1235. https://doi.org/10.1016/j.str.2015.05.001

Yamaguchi N, Oyama T, Ito E, Satoh H, Azuma S, Hayashi M, Shimizu K, Honma R, Yanagisawa Y, Nishikawa A, Kawamura M, Imai J, Ohwada S, Tatsuta K, Inoue J, Semba K, Watanabe S (2008) NOTCH3 signaling pathway plays crucial roles in the proliferation of ErbB2-negative human breast cancer cells. Cancer Res 68(6):1881–1888. https://doi.org/10.1158/0008-5472.CAN-07-1597

Yatim A, Benne C, Sobhian B, Laurent-Chabalier S, Deas O, Judde JG, Lelievre JD, Levy Y, Benkirane M (2012) NOTCH1 nuclear interactome reveals key regulators of its transcriptional activity and oncogenic function. Mol Cell 48(3):445–458. https://doi.org/10.1016/j.molcel.2012.08.022

Yeh ES, Means AR (2007) PIN1, the cell cycle and cancer. Nat Rev Cancer 7(5):381–388. https://doi.org/10.1038/nrc2107

Yen WC, Fischer MM, Axelrod F, Bond C, Cain J, Cancilla B, Henner WR, Meisner R, Sato A, Shah J, Tang T, Wallace B, Wang M, Zhang C, Kapoun AM, Lewicki J, Gurney A, Hoey T (2015) Targeting notch signaling with a notch2/notch3 antagonist (tarextumab) inhibits tumor growth and decreases tumor-initiating cell frequency. Clin Cancer Res Off J Am Assoc Cancer Res 21(9):2084–2095. https://doi.org/10.1158/1078-0432.CCR-14-2808

Yoshida T, Ng SY, Zuniga-Pflucker JC, Georgopoulos K (2006) Early hematopoietic lineage restrictions directed by Ikaros. Nat Immunol 7(4):382–391. https://doi.org/10.1038/ni1314

Zampieri M, Ciccarone F, Palermo R, Cialfi S, Passananti C, Chiaretti S, Nocchia D, Talora C, Screpanti I, Caiafa P (2014) The epigenetic factor BORIS/CTCFL regulates the NOTCH3 gene expression in cancer cells. Biochim Biophys Acta 1839(9):813–825. https://doi.org/10.1016/j.bbagrm.2014.06.017

Zhang TH, Liu HC, Liang YJ, Liang LZ, Zheng GS, Huang HZ, Wu JN, Liao GQ (2013) Suppression of tongue squamous cell carcinoma growth by inhibition of Jagged1 in vitro and in vivo. J Oral Pathol Med Off Public Int Assoc Oral Pathol Am Acad Oral Pathol 42(4):322–331. https://doi.org/10.1111/jop.12013

Zhang Q, Lu C, Fang T, Wang Y, Hu W, Qiao J, Liu B, Liu J, Chen N, Li M, Zhu R (2015) Notch3 functions as a regulator of cell self-renewal by interacting with the beta-catenin pathway in hepatocellular carcinoma. Oncotarget 6(6):3669–3679

Zhang X, Wang YN, Zhu JJ, Liu XX, You H, Gong MY, Zou M, Cheng WH, Zhu JH (2016) N-acetylcysteine negatively regulates Notch3 and its malignant signaling. Oncotarget 7(21):30855–30866. https://doi.org/10.18632/oncotarget.8806

Notch and Neurogenesis

Anna Engler, Runrui Zhang, and Verdon Taylor

Abstract

Neurogenesis is the process of forming neurons and is essential during vertebrate development to produce most of the neurons of the adult brain. However, neurogenesis continues throughout life at distinct locations in the vertebrate brain. Neural stem cells (NSCs) are the origin of both embryonic and adult neurogenesis, but their activity and fate are tightly regulated by their local milieu or niche. In this chapter, we will discuss the role of Notch signaling in the control of neurogenesis and regeneration in the embryo and adult. Notch-dependence is a common feature among NSC populations, we will discuss how differences in Notch signaling might contribute to heterogeneity among adult NSCs. Understanding the fate of multiple NSC populations with distinct functions could be important for effective brain regeneration.

Keywords

Neurogenesis · Development · Notch · Rbpj · Hes · Central nervous system · Maintenance · Neural stem cells · Subventricular zone · Dentate gyrus

Abbreviations

BMP	Bone Morphogenic Protein
CNS	Central Nervous System
DG	Dentate Gyrus
Dll	Delta-like
E	Embryonic day
FGF	Fibroblast Growth Factor
Hes	Hairy and enhancer of split
IP	Intermediate progenitor
IPC	Intermediate progenitor cell
LV	Lateral ventricle
NEPs	Neuroepithelial Cells
NICD	Notch intracellular domain
NSCs	Neural Stem Cells
OB	Olfactory Bulb
P	Postnatal day
RA	Retinoic Acid
RGC	radial glia cell
RMS	Rostral Migratory Stream
SGZ	Subgranular Zone
Shh	sonic hedgehog
SVZ	Subventricular Zone
TAP	transient amplifying progenitor
VZ	Ventricular Zone
Wnt	Wingless

Author contributed equally with all other contributors. Anna Engler and Runrui Zhang

A. Engler · R. Zhang · V. Taylor (✉)
Embryology and Stem Cell Biology, Department of Biomedicine, University of Basel, Basel, Switzerland
e-mail: verdon.taylor@unibas.ch

T. Borggrefe, B. D. Giaimo (eds.), *Molecular Mechanisms of Notch Signaling*, Advances in Experimental Medicine and Biology 1066, https://doi.org/10.1007/978-3-319-89512-3_11

1 Neural Stem Cells at the Base of Neurogenesis

The development of the central nervous system (CNS) is an intricate process precisely regulated in time and space. The majority of the cells in the adult brain are produced during embryonic development. However, a small pool of stem cells remains in the adult brain. Adult neural stem cells (NSCs) are found in two distinct niches and give rise to new neurons throughout life. The mouse has been instrumental as a model system to study the mechanism controlling developmental and adult neurogenesis. This is in part due to the powerful genetics and also the wealth of knowledge about the anatomy and function of distinct neuronal populations in the developing and adult mouse brain. It has become clear, and is widely accepted that NSCs are the origins of the neurogenic lineage. In the developing brain, radial glia in the ventricular zone (VZ) are the stem cells (Greig et al. 2013). They divide rapidly, giving rise to intermediate progenitors that amplify the progenitor pool and neuronal progeny (Noctor et al. 2004). These committed neural progenitors migrate out of the VZ and colonize the subventricular zone where they divide and generate neuroblasts, thus amplifying the progenitor pool and consequently the number of neuronal progeny. The neuroblasts migrate radially through the cortex forming layers in an inside-out fashion according to their birthdate. While neuroblasts of projection neurons migrate along the radial glial fibers of the radial glia to the superficial layers in the pallium, the interneurons are generated in the ganglionic eminences of the subpallium and undergo a long-range tangential migration to reach the cortex (Marin 2013). During the neurogenic phase of brain development (embryonic neurogenesis) NSCs have to divide to generate differentiated progeny but also maintain the stem cell pool. Hence, following cell division one of the daughter cells must remain as a NSC. Some of these retained NSCs are maintained from the embryo even into the adult brain where they later can function as a source for adult born neurons (Fuentealba et al. 2015; Furutachi et al. 2015). Adult NSCs are found in two distinct niches, the lateral wall (LW) of the subventricular zone (SVZ) and the subgranular zone (SGZ) of the dentate gyrus (DG). Adult NSCs generate differentiated neurons through intermediate progenitors/transient amplifying progenitors that rapidly divide and consecutively give rise to neuroblasts and neurons. Neuroblasts of the LW migrate along the rostral migratory stream (RMS) into the olfactory bulb (OB) where they differentiate and integrate into local circuits, while the adult born neurons in the SGZ integrate locally.

Neurogenesis has been studied extensively in *Drosophila melanogaster*, zebrafish and the mouse as model systems. Here we will mainly focus on mammals and refer the reader to excellent reviews covering neurogenesis in *Drosophila* and zebrafish (Alunni and Bally-Cuif 2016; Homem and Knoblich 2012). The focus of this chapter will be on the role of Notch signaling in NSC maintenance and differentiation in embryonic and adult neurogenesis.

2 Role of Notch Components in Embryonic Neurogenesis

2.1 NSCs of the Embryonic Central Nervous System

NSCs generate all neurons of the developing and adult brain (Kazanis et al. 2008; Fuentealba et al. 2015). In the developing mouse, neurulation, the process of forming the neural plate, starts around embryonic day 8 (E8) and is initiated through a combination of growth factors and inhibitory signals secreted by the notochord, the dorsal ectoderm and the Spemann organizer (Tam and Loebel 2007). Once the neural plate is formed on the dorsal side of the embryo, the neural ectodermal cells that form the neural plate start to become regionalized. These neural ectodermal cells, also called neuroepithelial cells (NEPs), are radial progenitors spanning the thickness of the neural plate. The NEPs at the lateral edges of the plate become the multipotent neural crest stem cells, the precursors of the peripheral nervous system, melanocytes and, in some regions of the embryo, specific muscles and bone (Bhatt et al.

2013; Sauka-Spengler and Bronner 2010). The neural plate invaginates into the embryo at the midline and the lateral edges of the neural plate fold dorsally. The two lateral edges of the neural plate meet and fuse at the dorsal midline, zippering closed in the anterior and posterior directions starting at the future hindbrain region. The neural plate now forms the neural tube.

NEPs in the neuroepithelium of the neural tube are the first stem cells of the central nervous system. The interplay between morphogen gradients and signaling pathways, including sonic hedgehog (Shh), retinoic acid (RA), fibroblast growth factor (FGF), wingless (Wnt) and bone morphogenic protein (BMP) regionalize the neural tube (Greig et al. 2013; Franco and Muller 2013; Lupo et al. 2006). Due to this patterning, the NEPs of the neural tube become more specified and defined structural domains start to appear.

The four most important segments of the regionalized tube are the forebrain, the midbrain, the hindbrain and the spinal cord. The forebrain contains two cortical structures – the neocortex and the hippocampus. Both of these structures are derived embryonically and in the early postnatal period. The neocortex starts to be formed by E11.5 and, at least in terms of neuronal architecture, is finished by birth whereas the hippocampus starts to be formed by E17.5 and is anatomically complete around postnatal day 14 (P14) in the mouse (Nicola et al. 2015; Rolando and Taylor 2014). These two regions also harbor the NSC niches and contribute stem cells to the known neurogenic regions of the adult brain.

At E9 in the mouse, the neuroepithelium is a pseudostratified single layer of NEPs. Initially, the NEPs proliferate and increase in number without discernable differentiation. As development progresses and the production of neurons from the neuroepithelium starts, the NEPs transform into radial glia cells (RGCs), that form the VZ and function as NSCs during development (Noctor et al. 2004). The precursor and progenitor populations have distinct features of the embryonic brain. RGCs span the thickness of the cortex from the apical to the basal surface with their radial process and their soma remains in the VZ (Gotz and Huttner 2005). RGCs have a polarized structure spanning the thickness of the neural tube with an apical process anchored at the lumen of the tube and a long basolateral process to the forming surface of the brain. The progeny of RGCs are intermediate progenitor cells (IPCs). These cells are not connected to either surface of the neural tube and reside in the SVZ. RGCs are the NSCs of the mammalian brain whereas IPCs are short-lived intermediate cells. Formation of the neocortex is divided into an initial expansion period where NSCs undergo symmetric cell divisions followed by a neurogenic period during mid-late embryogenesis (E9-E18) and then by a gliogenic period. During the neurogenic period of development, NSCs primarily divide asymmetrically with one daughter cell remaining as a stem cell and the other becoming an IPC and committing to differentiation (Noctor et al. 2007; Noctor et al. 2004). Before E15.5, NSCs can also undergo direct neurogenesis producing neurons without an obvious IPC (Telley et al. 2016). The excitatory neurons of the neocortex are produced in a sequential manner by the NSCs to form the six individual layers of the isocortex in an inside out fashion (Franco and Muller 2013; Guo et al. 2013). Towards the end of the neurogenic period the NSCs switch in their fate and start to generate glial cells. Some of the NSCs are put aside during embryonic development of the neocortex to contribute to the adult NSC pools. These NSCs exit cell cycle during the embryogenesis and remain relatively inactive until postnatal and adult periods (Fuentealba et al. 2015; Liu et al. 2011; Greig et al. 2013; Furutachi et al. 2015). NSCs in the nervous systems are conserved throughout phylogeny from *Drosophila* to human. Interestingly, the maintenance mechanisms of NSCs in all of these organisms are highly conserved (Artavanis-Tsakonas et al. 1999).

2.2 Dependency of Embryonic Neural Stem Cells on Notch Signaling

One of the pivotal control mechanisms of NSCs maintenance is Notch signaling. NSCs in the embryo express Notch receptors and active signaling is evident based on the expression of the canonical Notch target *Hes5* (Hatakeyama et al. 2004; Basak and Taylor 2007) (see "s"). IPCs and neuroblasts also express Notch receptors, however, they do not express *Hes5*. The Notch receptors in these cells seem to signal through a mainly non-canonical pathway (Alberi et al. 2011; Franklin et al. 1999; Louvi and Artavanis-Tsakonas 2006; Stump et al. 2002). The cells of the neuronal lineage not only express Notch receptors and ligands, but genetic evidence underlines the importance of Notch signaling in NSCs maintenance, differentiation and fate choice (Louvi and Artavanis-Tsakonas 2006; Kageyama et al. 2007).

In all animals, Notch signaling is associated with the maintenance of NSCs and the control of their fate (Lutolf et al. 2002; Gaiano et al. 2000; Mason et al. 2006; Alunni and Bally-Cuif 2016; Artavanis-Tsakonas et al. 1999; Homem and Knoblich 2012). In mice, expression of Notch receptors and downstream components of the pathway starts in the neural tube around E8–9 and continues into the VZ and SVZ during neural development (Weinmaster et al. 1991). The essential role of Notch signaling in embryonic development is evident from the phenotypes of Notch mutant mice. Deletion of *Rbpj*, the transcription factor and central component of the canonical Notch signal pathway, results in a developmental block and embryonic lethality at E9.5 (de la Pompa et al. 1997). *Rbpj*-null embryos also show a delay in nervous system development (de la Pompa et al. 1997). Similarly, *Notch1*-deficient mice die around E9.5 showing similar morphological defects as the *Rbpj* mutants (Conlon et al. 1995). This made the functional analysis of Notch function in the nervous system difficult as differentiation in the neural tube only starts after this point. Therefore, elegant gain-of-function experiments were instrumental to reveal the potential functions of Notch signaling in the embryonic brain (Gaiano et al. 2000). Expression of active Notch signaling in the NSCs of the embryonic telencephalon blocked neurogenesis and promoted glial fate indicating an important role for Notch in cell fate decision making (Gaiano et al. 2000). Similarly, conditional loss of function experiments revealed the role of Notch1 in NSCs of the embryonic central nervous system (Lutolf et al. 2002). *Notch1*-deficient NSCs precociously exited cell cycle and started neurogenesis at the expense of gliogenesis, indicating an essential role for Notch1 in regulating the onset of neuronal differentiation by embryonic NSCs (Lutolf et al. 2002; Mason et al. 2006). Similarly, Notch signaling regulates cell viability in the developing CNS. Conditional gene ablation of *Notch1* or simultaneous *Notch1* and *Notch3* deletion from embryonic NSCs in the neural tube increased cell death in both the neural progenitor and differentiating neuronal populations (Mason et al. 2006). In contrast, genetic ablation experiments revealed that although Notch signaling is required for the maintenance and regulation of NSCs differentiation during embryonic development, Notch signaling is not required for the formation of definitive NSCs (Hitoshi et al. 2002). These studies have clearly illustrated that Notch signaling components are essential in NSC maintenance and survival in the embryonic brain (Table 1).

Notch signaling is activated by cell-cell interaction via canonical transmembrane Notch ligands of the Delta-like (Dll) and Jagged, DSL family, that are widely expressed, with overlapping distribution in the embryonic brain (Stump et al. 2002). DLL1, DLL3 and JAGGED1 are all expressed by NEPs, RGCs and/or differentiating cells in the neural tube (Lindsell et al. 1996; Stump et al. 2002; Nyfeler et al. 2005). Loss of *Jagged1* results in an embryonic lethal phenotype and mice die shortly after E10.5 (Xue et al. 1999). Conditional deletion of *Jagged1* from NSCs of the developing hindbrain showed its requirement during granule cell formation (Weller et al. 2006). *Jagged1* loss results in accumulation of granule cell precursors and a delayed granule neuron differentiation as well as their aberrant migration (Weller et al. 2006) (Table 1).

The classical downstream effectors of Notch signaling in the embryonic mammalian brain are hairy and enhancer of split (Hes) transcription factors (Ohtsuka et al. 2001). Upon ligand-receptor interaction and activation of the Notch signaling cascade, the Notch intracellular domain

Table 1 Notch in Embryonic Neurogenesis; Effects of Notch signaling manipulation during brain development

Development		
Notch1	KO: Embryonic lethal (E9.5)	Conlon et al. (1995)
	GOF: Glial, instead of neuronal fate	
	cKO: Precocious cell cycle exit, neurogenesis increased	Gaiano et al. (2000)
		Lutolf et al. (2002)
		Mason et al. (2006)
Notch2	KO: Embryonic lethal (E9.5)	Hamada et al. (1999)
Rbpj	KO: Embryonic lethal (E10.5), delayed CNS development	de la Pompa et al. (1997)
Hes1	KO: Redundancy by Hes5	Hatakeyama et al. (2004)
	Compound KO Hes1 and Hes5: loss of NSCs	
Hes5	KO: Redundancy by Hes1	
Jagged	KO: Embryonic lethal (E10.5)	Xue et al. (1999)
	cKO: Defects in migration, differentiation, survival	Weller et al. (2006)
Dll1	Dll1 expressed till E13.5	Stump et al. (2002)
Dll4	Weak expression in embryo	Stump et al. (2002)

(NICD)-Rbpj–Maml complex induces *Hes1* and *Hes5* expression in the brain (Honjo 1996; Artavanis-Tsakonas et al. 1999; Ohtsuka et al. 1999; Kageyama et al. 2007). The basic-helix-loop-helix factors HES1 and HES5 repress the expression of the proneural genes (*Ascl1*, *Atoh1*, *Neurog1* and *Neurog2*) and thereby inhibit NSCs differentiation and neuron production (Lutolf et al. 2002; Hatakeyama et al. 2004; Hatakeyama et al. 2001) (Table 1).

It has now become clear that, as in the formation of the somites, Notch signaling and *Hes1* expression oscillates in neural stem/progenitor cells of the developing embryo (Masamizu et al. 2006; Shimojo et al. 2008). The cyclical production and degradation of Hes1 mRNA and protein play a critical role in these oscillations. In addition, HES1 binds its own promoter and thereby negatively regulates its own expression, counteracting Notch activation. In neural stem/progenitor cells the period of *Hes1* oscillatory expression is around 2–3 h, and this periodicity of oscillation projects onto the expression of the proneural genes and *Dll1*, which is a repressed target of Notch via the Hes-mediated repression (Hirata et al. 2002). However, in differentiating neurons *Hes* expression is repressed leading to sustained expression of *Ngn2* and *Dll1* (Shimojo et al. 2008). Together, the Hes-mediated Notch signaling feedback mechanism ensures neural fate separation into stem/progenitor and neurons (Shimojo et al. 2011) (see "Oscillatory Control of Notch Signaling in Development"). During embryonic development, deletion of either *Hes1* or *Hes5* has no apparent effects on neural development and NSCs activity. However, the double deletion of *Hes1* and *Hes5* causes severe phenotypes leading to disorganization of the neural tube, premature neuronal differentiation and loss of radial glia in the embryo (Hatakeyama et al. 2004) (Table 1). In addition, *Hes1* is a target of other signaling pathways including BMP, where it may act as a signal integrator (Kageyama et al. 2007). Hence, the studies discussed above and many others illustrate the essential role of Notch signaling in NSCs maintenance during the development of the brain.

3 Notch Signaling in Adult Neurogenesis

In order to guarantee a life-long reservoir of progenitors for neurogenesis, NSCs are set-aside during late stages of development (E15 onwards in the mouse) (Fuentealba et al. 2015; Furutachi et al. 2015). At this point, NSCs can be found in two distinct states – quiescent and active NSCs (Doetsch et al. 1999; Codega et al. 2014; Giachino et al. 2014b). In the adult brain, NSCs are mainly quiescent and divide infrequently. However, quiescent NSCs likely enter cell cycle and become active NSCs before exiting cell cycle again to reenter the quiescent state (Urban et al. 2016; Bonaguidi et al. 2011). Activated NSCs may go through multiple cell cycles generating dividing daughter progenitors by asymmetric cell division

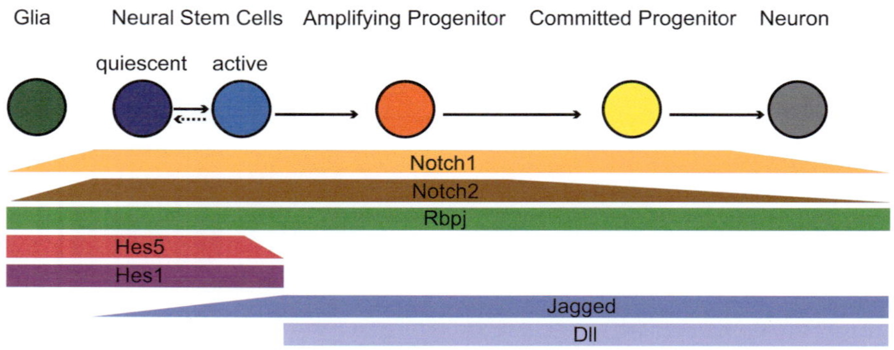

Fig. 1 Notch Signaling along the Neurogenic Lineage; NSCs express Notch receptors (Stump et al. 2002; Basak et al. 2012), the transcription factor Rbpj (Imayoshi et al. 2010) and express Notch target genes *Hes5* and *Hes1* (Hatakeyama et al. 2004). Notch receptor expression reduces along the neurogenic lineage, while Rbpj expression remains unaffected. The progenitors of NSCs express the ligands JAGGED and DLL (Stump et al. 2002)

that become progressively postmitotic and, eventually, differentiate into neurons, astrocytes and oligodendrocytes (Ihrie and Alvarez-Buylla 2011; Gage 2000; Rolando et al. 2016; Lugert et al. 2010; Bonaguidi et al. 2012). Active NSCs may also divide symmetrically to generate two NSCs which is common during early stages of brain development but remains a contentious issue in the adult brain (Bonaguidi et al. 2011). These two types of NSCs can be found in both adult neurogenic niches – the SVZ of the lateral ventricle (LV) and the SGZ of the DG (Fig. 1).

3.1 Neurogenesis in the SVZ

NSCs in the SVZ are found in the subependymal layer between the LV and the striatum. A single layer of ependymal cells separates the SVZ from the cerebrospinal fluid in the LV (Ihrie and Alvarez-Buylla 2011). New neurons originating in the SVZ migrate along the RMS to the OB. Under physiological conditions, the OB is provided continuously with new interneurons from the SVZ which terminally differentiate and integrate into the local circuitry (Lois 1996).

The NSCs in the SVZ of the LV project apical sensory cilium through the ependyma into the CSF and radially to blood vessels to obtain systemic inputs (Merkle et al. 2007; Doetsch et al. 1999; Mirzadeh et al. 2008). Notch signaling components are broadly expressed in the postnatal and adult brain (Stump et al. 2002; Basak et al. 2012; Nyfeler et al. 2005) (Fig.1). It has been shown that SVZ NSCs express Notch receptors (Basak et al. 2012). The NSCs are provided with ligands with their niche. The ependymal cells and astrocytes of the SVZ express Jagged1 and the transient amplifying progenitors (TAPs) express Dll1 (Stump et al. 2002; Nyfeler et al. 2005; Basak et al. 2012; Furutachi et al. 2013). In addition, the BVs underlying the SVZ express Dll ligands and Jagged1 (Temple 2001; Ottone et al. 2014). These juxtacrine signals play a pivotal role in maintenance of NSCs in a quiescent and undifferentiated state (Ottone et al. 2014; Nyfeler et al. 2005). Likely, direct interactions between NSCs and surrounding cells balance the populations of NSCs and TAPs in the niche. Both NSCs and TAPs present and secrete a vast array of proteins involved in regulating neurogenesis including Notch receptors and Notch ligand Jagged (Aguirre et al. 2010; Hermann et al. 2014; Drago et al. 2013; Nyfeler et al. 2005; Basak et al. 2012).

3.2 Role of Notch Components in the Adult SVZ

In the SVZ Notch signaling is regulated via the Notch ligands Jagged1 and Dll1, whereby JAGGED is the predominantly expressed ligand in the adult brain (Stump et al. 2002). The loss of

Table 2 Notch in Adult Neurogenesis; Effects of Notch signaling manipulation in the postnatal and adult brain

Postnatal/Adult		
Notch1	cKO: Block of NSCs self renewal	Ables et al. (2010)
		Basak et al. (2012)
Notch2		
Rbpj	cKO: Depletion and exhaustion of NSCs	Imayoshi et al. (2010)
		Ehm et al. (2010)
Hes1		
Hes5	Expressed in NSCs and astrocytes	Basak and Taylor (2007)
		Lugert et al. (2010)
Jagged	cKO: Block of NSCs self renewal	Nyfeler et al. (2005)
Dll1	cKO: Loss of quiescent NSCs	Kawaguchi et al. (2013)
Dll4	GOF: Increased NSC proliferation and survival	Androutsellis-Theotokis et al. (2006)

Jagged1 from the postnatal brain results in a block of NSCs self-renewal, similar to the phenotypes observed by a Notch loss of function (Nyfeler et al. 2005; Ottone et al. 2014). DLL1 ligand is expressed in the brain albeit at lower levels (Stump et al. 2002). The loss of *Dll1* by conditional gene inactivation induced a loss of quiescent NSCs (Kawaguchi et al. 2013) and the infusion of DLL4 into the SVZ, stimulated NSCs proliferation and increased survival of their progeny (Androutsellis-Theotokis et al. 2006). Conversely, DLL3 seems not be involved in NSCs maintenance, but rather in directing neurons along their differentiation path (Dunwoodie 2009). Interestingly, canonical Notch signaling leads to the transcriptional repression of some Notch ligand genes particularly *Dll1* (Shimojo et al. 2011). Therefore, Notch signaling is autoregulatory within the niche with ligand expression being repressed in cells that receive an activating signal through Notch. The Notch target genes mediate the Notch phenotypes and have been described to be essential in neuronal differentiation, survival and plasticity (Lutolf et al. 2002; Breunig et al. 2007; Alberi et al. 2011) (Table 2).

The role of Notch signaling in NSCs of the adult SVZ becomes evident upon conditional ablation of *Rbpj*. RBPJ-deficient NSCs are no longer maintained, leading to an initial activation of the NSC pool and an expansion of the progenitor population (Imayoshi et al. 2010). This causes a depletion and exhaustion of the quiescent and active NSCs pools in the SVZ and an eventual loss of neurogenesis (Imayoshi et al. 2010). Interestingly, when *Notch1* was deleted from the same NSC populations, only active NSCs were affected and quiescent NSCs were spared (Basak et al. 2012) suggesting the involvement of other Notch receptors in the maintenance of the quiescent NSC pool (Table 2). The Notch induced transcription factors HES1 and HES5 regulate *Ascl1* expression during embryogenesis (Kageyama et al. 2007). The proneural factor ASCL1 is involved in neuronal fate commitment as well as neural differentiation and NSC proliferation (Castro et al. 2011). ASCL1 mediates the transition of quiescence NSCs to an active state (Andersen et al. 2014). In order to control this important step, NSCs tightly regulate the levels of ASCL1 by degradation via the E3 ubiquitin protein ligase HUWE1 to return NSCs to a quiescent state (Urban et al. 2016).

3.3 Neurogenesis in the DG of the Hippocampus

The DG of the hippocampus is part of the limbic system and plays a key role in memory consolidation and spatial navigation. Neurogenesis in the DG is found in the SGZ of adult rodents, primates as well as humans (Spalding et al. 2013) and ongoing neurogenesis in the adult SGZ has been proposed to be important in certain forms of learning and memory (Zhao et al. 2008).

The nomenclature of the cells within the neurogenic lineage in the SGZ is different to that in the SVZ. Type-1 cells are the NSCs and can be divided into radial quiescent, radial active, horizontal quiescent and horizontal active depending upon their morphology and mitotic activity (Lugert et al. 2010). Type-1 cells give-rise to Type-2 cells, which can be subdivided into Type-2a (early progenitors) and Type-2b (late progenitors) cells. The Type-2 cells are mitotic intermediate progenitors (IPs) of the DG neurogenic lineage and give-rise to Type-3 cells, which exit the cell cycle and are fate-committed neuroblasts (Ehninger and Kempermann 2008). Newly generated neurons in the hippocampus integrate into established networks, making neurogenesis a unique form of neuronal plasticity.

Just as in the SVZ, the NSCs in the SGZ are provided with external stimuli via their local niche and the vasculature. In the SGZ, the NSCs are positioned close to their progeny and the endothelial cells of blood vessels. The NSCs receive signals from the neuronal circuitry of the DG and even form synapse-like structures with interneurons that regulate their proliferative activity by the release of the neurotransmitter γ-Aminobutyric acid (Giachino et al. 2014a; Song et al. 2012). However, major differences exist between the two adult neurogenic niches. Particularly, NSC progeny in the SGZ, the neuroblasts and newborn neurons, do not migrate out of the niche area, settle within a few cell diameters of the NSCs and thereby are in an ideal position to directly control neurogenic activity within the niche.

3.4 Role of Notch Components in Adult DG Neurogenesis

Notch signaling also plays a primary role in the control of neurogenesis in the SGZ of the DG (Ables et al. 2011; Ables et al. 2010; Ehm et al. 2010; Lugert et al. 2010; Ming and Song 2011; Pierfelice et al. 2011; Breunig et al. 2007). Notch receptors are expressed throughout the DG including on the NSCs and progenitors in the SGZ (Stump et al. 2002). Active Notch signaling is prominent in both radial and horizontal NSCs (Type-1 cells) but is absent from the IPs (Type-2 cells) and immature neurons (Type-3 cell) (Breunig et al. 2007; Lugert et al. 2010; Lugert et al. 2012). The transcription of the Notch target *Hes5* efficiently discriminates the NSCs from other cells including proliferative committed progenitors in the DG (Lugert et al. 2010; Lugert et al. 2012; Ehm et al. 2010). The Notch ligand JAGGED1 is preferentially expressed by IPCs and neurons in the DG, although its expression has also been found in radial glia-like stem cells (Lavado et al. 2010; Lavado and Oliver 2014). *Hes5* is a target of Notch signaling in the central nervous system and a good indicator of Notch activity (Basak and Taylor 2007). Using reporter mice where *GFP* is driven by the regulatory elements of the *Hes5* gene (*Hes5::GFP*), both radial and horizontal Type-1 cells were found to express *Hes5* in the adult DG (Basak and Taylor 2007; Lugert et al. 2010) (Table 2).

Genetic ablation experiments indicate that Notch and Notch signaling pathway components are important in regulation of adult DG neurogenesis. Conditional deletion of *Notch1* or overexpression of activated NOTCH1 has a prominent effect on neurogenesis in the adult DG (Breunig et al. 2007; Ables et al. 2010). Loss of *Notch1* results in a loss of Type-1 cells and abolition of neurogenesis as well as a reduction in mitotic progenitors (Ables et al. 2010; Breunig et al. 2007). Conversely, activation of Notch1 signaling by expression of the NICD increased radial NSCs and resulted in the generation of glial cells at the expense of neurons (Breunig et al. 2007). Interestingly, and similar to the role of Notch1 in the SVZ, genetic ablation of *Notch1* caused a loss of active NSCs but did not reduce the quiescent radial cells (Ables et al. 2010). However, Notch1 also plays roles in DG neurogenesis at later stages in the lineage and newborn *Notch1* knockout neurons showed impaired dendritic arborisation probably via non-canonical pathway (Breunig et al. 2007) (Table 2).

Conditional gene inactivation of *Rbpj* caused depletion of NSCs, leading to a transient burst in proliferation and production of IPs and neuroblasts (Ehm et al. 2010; Lugert et al. 2010). The activation of NSCs by loss of Rbpj accelerated the exhaustion of the NSC reservoir leading to the impaired adult DG neurogenesis (Ehm et al. 2010). Postnatal deletion of *Jagged1* results in a smaller dentate gyrus and a transient increase in the SGZ neurogenesis accompanied by exhaustion of NSCs, which mimics the phenotype of *Rbpj* conditional knockout (Lavado and Oliver 2014) (Table 2).

Although there are no data showing the effects of deletion of *Hes* genes in the adult DG,

Hes5::GFP expression revealed heterogeneity and behavioral differences between the different Type-1 cells in the DG in response to pathophysiological stimuli (Lugert et al. 2010). Lineage marking and tracing revealed that radial NSCs are more quiescent but are activated by physical exercise whereas horizontal NSCs are more active and respond to epileptic seizures (Lugert et al. 2010; Jessberger and Parent 2015). Hence, Notch signaling is not only a central regulator of NSCs activity in the DG but is also a common signaling pathway shared by different stem cell populations in the same neurogenic niche.

4 Conclusion

Notch signaling is essential in the maintenance of NSCs in the developing and adult brain. The mediators of this maintenance are Notch targets of the Hes and Hey families. These bHLH transcription factors suppress the expression of proneural genes in a cyclic manner and thus keep NSCs in an undifferentiated state. Using canonical Notch signaling targets including *Hes5*, one is able to distinguish NSCs and their progeny. *Hes5* is expressed almost exclusively in the developing nervous system in the embryo and in the adult brain. However, despite expression by many cells in the adult brain, conditional deletion of Rbpj and Notch1 produces distinct phenotypes in the adult murine brain niches suggesting that there is still much to learn about the role of Notch receptors and signaling components in the control of NSC fate(s) and activity(ies) (Basak et al. 2012; Imayoshi et al. 2010) (Table 1 and Table 2). Thus, it is of interest to understand whether distinct Notch homologues are the key to NSCs heterogeneity in the adult brain or whether other factors, including crosstalk between Notch signaling and other pathways maintain distinct NSCs pools. Given the importance of Notch signaling in the control of NSCs and neurogenesis, it is essential to elucidate the canonical, epigenetic, non-canonical and crosstalk functions in the pathway. These regulatory networks could then form the basis for potential future clinical therapies for regeneration or rejuvenation of the brain.

Acknowledgments We thank the members of the Taylor lab for critical reading of the manuscript and for helpful discussions. This work was supported by the Swiss National Science Foundation (VT) and the SystemsX.ch, NeuroStemX project (VT) and the University of Basel (VT) and the Forschungsfond of the University of Basel (AE).

References

Ables JL, Decarolis NA, Johnson MA, Rivera PD, Gao Z, Cooper DC, Radtke F, Hsieh J, Eisch AJ (2010) Notch1 is required for maintenance of the reservoir of adult hippocampal stem cells. J Neurosci Off J Soc Neurosci 30(31):10484–10492. https://doi.org/10.1523/JNEUROSCI.4721-09.2010

Ables JL, Breunig JJ, Eisch AJ, Rakic P (2011) Not(ch) just development: Notch signalling in the adult brain. Nat Rev Neurosci 12(5):269–283. https://doi.org/10.1038/nrn3024

Aguirre A, Rubio ME, Gallo V (2010) Notch and EGFR pathway interaction regulates neural stem cell number and self-renewal. Nature 467(7313):323–327. https://doi.org/10.1038/nature09347

Alberi L, Liu S, Wang Y, Badie R, Smith-Hicks C, Wu J, Pierfelice TJ, Abazyan B, Mattson MP, Kuhl D, Pletnikov M, Worley PF, Gaiano N (2011) Activity-induced Notch signaling in neurons requires arc/Arg3.1 and is essential for synaptic plasticity in hippocampal networks. Neuron 69(3):437–444. https://doi.org/10.1016/j.neuron.2011.01.004

Alunni A, Bally-Cuif L (2016) A comparative view of regenerative neurogenesis in vertebrates. Development 143(5):741–753. https://doi.org/10.1242/dev.122796

Andersen J, Urban N, Achimastou A, Ito A, Simic M, Ullom K, Martynoga B, Lebel M, Goritz C, Frisen J, Nakafuku M, Guillemot F (2014) A transcriptional mechanism integrating inputs from extracellular signals to activate hippocampal stem cells. Neuron 83(5):1085–1097. https://doi.org/10.1016/j.neuron.2014.08.004

Androutsellis-Theotokis A, Leker RR, Soldner F, Hoeppner DJ, Ravin R, Poser SW, Rueger MA, Bae SK, Kittappa R, McKay RD (2006) Notch signalling regulates stem cell numbers in vitro and in vivo. Nature 442(7104):823–826. https://doi.org/10.1038/nature04940

Artavanis-Tsakonas S, Rand MD, Lake RJ (1999) Notch signaling: cell fate control and signal integration in development. Science 284(5415):770–776

Basak O, Taylor V (2007) Identification of self-replicating multipotent progenitors in the embryonic nervous system by high Notch activity and Hes5 expression. Eur J Neurosci 25(4):1006–1022. https://doi.org/10.1111/j.1460-9568.2007.05370.x

Basak O, Giachino C, Fiorini E, Macdonald HR, Taylor V (2012) Neurogenic subventricular zone stem/progenitor cells are Notch1-dependent in their active but not quiescent state. J Neurosci Off J Soc Neurosci 32(16):5654–5666. https://doi.org/10.1523/JNEUROSCI.0455-12.2012

Bhatt S, Diaz R, Trainor PA (2013) Signals and switches in Mammalian neural crest cell differentiation. Cold Spring Harb Perspect Biol 5(2). https://doi.org/10.1101/cshperspect.a008326

Bonaguidi MA, Wheeler MA, Shapiro JS, Stadel RP, Sun GJ, Ming GL, Song H (2011) In vivo clonal analysis reveals self-renewing and multipotent adult neural stem cell characteristics. Cell 145(7):1142–1155. https://doi.org/10.1016/j.cell.2011.05.024

Bonaguidi MA, Song J, Ming GL, Song H (2012) A unifying hypothesis on mammalian neural stem cell properties in the adult hippocampus. Curr Opin Neurobiol 22(5):754–761. https://doi.org/10.1016/j.conb.2012.03.013

Breunig JJ, Silbereis J, Vaccarino FM, Sestan N, Rakic P (2007) Notch regulates cell fate and dendrite morphology of newborn neurons in the postnatal dentate gyrus. Proc Natl Acad Sci U S A 104(51):20558–20563. https://doi.org/10.1073/pnas.0710156104

Castro DS, Martynoga B, Parras C, Ramesh V, Pacary E, Johnston C, Drechsel D, Lebel-Potter M, Garcia LG, Hunt C, Dolle D, Bithell A, Ettwiller L, Buckley N, Guillemot F (2011) A novel function of the proneural factor Ascl1 in progenitor proliferation identified by genome-wide characterization of its targets. Genes Dev 25(9):930–945. https://doi.org/10.1101/gad.627811

Codega P, Silva-Vargas V, Paul A, Maldonado-Soto AR, Deleo AM, Pastrana E, Doetsch F (2014) Prospective identification and purification of quiescent adult neural stem cells from their in vivo niche. Neuron 82(3):545–559. https://doi.org/10.1016/j.neuron.2014.02.039

Conlon RA, Reaume AG, Rossant J (1995) Notch1 is required for the coordinate segmentation of somites. Development 121(5):1533–1545

de la Pompa JL, Wakeham A, Correia KM, Samper E, Brown S, Aguilera RJ, Nakano T, Honjo T, Mak TW, Rossant J, Conlon RA (1997) Conservation of the Notch signalling pathway in mammalian neurogenesis. Development 124(6):1139–1148

Doetsch F, Caille I, Lim DA, Garcia-Verdugo JM, Alvarez-Buylla A (1999) Subventricular zone astrocytes are neural stem cells in the adult mammalian brain. Cell 97(6):703–716

Drago D, Cossetti C, Iraci N, Gaude E, Musco G, Bachi A, Pluchino S (2013) The stem cell secretome and its role in brain repair. Biochimie 95(12):2271–2285. https://doi.org/10.1016/j.biochi.2013.06.020

Dunwoodie SL (2009) The role of Notch in patterning the human vertebral column. Curr Opin Genet Dev 19(4):329–337. https://doi.org/10.1016/j.gde.2009.06.005

Ehm O, Goritz C, Covic M, Schaffner I, Schwarz TJ, Karaca E, Kempkes B, Kremmer E, Pfrieger FW, Espinosa L, Bigas A, Giachino C, Taylor V, Frisen J, Lie DC (2010) RBPJkappa-dependent signaling is essential for long-term maintenance of neural stem cells in the adult hippocampus. J Neurosci Off J Soc Neurosci 30(41):13794–13807. https://doi.org/10.1523/JNEUROSCI.1567-10.2010

Ehninger D, Kempermann G (2008) Neurogenesis in the adult hippocampus. Cell Tissue Res 331(1):243–250. https://doi.org/10.1007/s00441-007-0478-3

Franco SJ, Muller U (2013) Shaping our minds: stem and progenitor cell diversity in the mammalian neocortex. Neuron 77(1):19–34. https://doi.org/10.1016/j.neuron.2012.12.022

Franklin JL, Berechid BE, Cutting FB, Presente A, Chambers CB, Foltz DR, Ferreira A, Nye JS (1999) Autonomous and non-autonomous regulation of mammalian neurite development by Notch1 and Delta1. Curr Biol 9(24):1448–1457

Fuentealba LC, Rompani SB, Parraguez JI, Obernier K, Romero R, Cepko CL, Alvarez-Buylla A (2015) Embryonic origin of postnatal neural stem cells. Cell 161(7):1644–1655. https://doi.org/10.1016/j.cell.2015.05.041

Furutachi S, Matsumoto A, Nakayama KI, Gotoh Y (2013) p57 controls adult neural stem cell quiescence and modulates the pace of lifelong neurogenesis. EMBO J 32(7):970–981. https://doi.org/10.1038/emboj.2013.50

Furutachi S, Miya H, Watanabe T, Kawai H, Yamasaki N, Harada Y, Imayoshi I, Nelson M, Nakayama KI, Hirabayashi Y, Gotoh Y (2015) Slowly dividing neural progenitors are an embryonic origin of adult neural stem cells. Nat Neurosci 18(5):657–665. https://doi.org/10.1038/nn.3989

Gage FH (2000) Mammalian neural stem cells. Science 287(5457):1433–1438. https://doi.org/10.1126/science.287.5457.1433

Gaiano N, Nye JS, Fishell G (2000) Radial glial identity is promoted by Notch1 signaling in the murine forebrain. Neuron 26(2):395–404

Giachino C, Barz M, Tchorz JS, Tome M, Gassmann M, Bischofberger J, Bettler B, Taylor V (2014a) GABA suppresses neurogenesis in the adult hippocampus through GABAB receptors. Development 141(1):83–90. https://doi.org/10.1242/dev.102608

Giachino C, Basak O, Lugert S, Knuckles P, Obernier K, Fiorelli R, Frank S, Raineteau O, Alvarez-Buylla A, Taylor V (2014b) Molecular diversity subdivides the adult forebrain neural stem cell population. Stem Cells 32(1):70–84. https://doi.org/10.1002/stem.1520

Gotz M, Huttner WB (2005) The cell biology of neurogenesis. Nat Rev Mol Cell Biol 6(10):777–788. https://doi.org/10.1038/nrm1739

Greig LC, Woodworth MB, Galazo MJ, Padmanabhan H, Macklis JD (2013) Molecular logic of neocortical projection neuron specification, development and diversity. Nat Rev Neurosci 14(11):755–769. https://doi.org/10.1038/nrn3586

Guo C, Eckler MJ, McKenna WL, McKinsey GL, Rubenstein JL, Chen B (2013) Fezf2 expression identifies a multipotent progenitor for neocortical projection neurons, astrocytes, and oligodendrocytes. Neuron 80(5):1167–1174. https://doi.org/10.1016/j.neuron.2013.09.037

Hamada Y, Kadokawa Y, Okabe M, Ikawa M, Coleman JR, Tsujimoto Y (1999) Mutation in ankyrin repeats of the mouse Notch2 gene induces early embryonic lethality. Development 126(15):3415–3424

Hatakeyama J, Tomita K, Inoue T, Kageyama R (2001) Roles of homeobox and bHLH genes in specification of a retinal cell type. Development 128(8):1313–1322

Hatakeyama J, Bessho Y, Katoh K, Ookawara S, Fujioka M, Guillemot F, Kageyama R (2004) Hes genes regulate size, shape and histogenesis of the nervous system by control of the timing of neural stem cell differentiation. Development 131(22):5539–5550. https://doi.org/10.1242/dev.01436

Hermann DM, Peruzzotti-Jametti L, Schlechter J, Bernstock JD, Doeppner TR, Pluchino S (2014) Neural precursor cells in the ischemic brain–integration, cellular crosstalk, and consequences for stroke recovery. Front Cell Neurosci 8:291. https://doi.org/10.3389/fncel.2014.00291

Hirata H, Yoshiura S, Ohtsuka T, Bessho Y, Harada T, Yoshikawa K, Kageyama R (2002) Oscillatory expression of the bHLH factor Hes1 regulated by a negative feedback loop. Science 298(5594):840–843. https://doi.org/10.1126/science.1074560

Hitoshi S, Alexson T, Tropepe V, Donoviel D, Elia AJ, Nye JS, Conlon RA, Mak TW, Bernstein A, van der Kooy D (2002) Notch pathway molecules are essential for the maintenance, but not the generation, of mammalian neural stem cells. Genes Dev 16(7):846–858. https://doi.org/10.1101/gad.975202

Homem CC, Knoblich JA (2012) Drosophila neuroblasts: a model for stem cell biology. Development 139(23):4297–4310. https://doi.org/10.1242/dev.080515

Honjo T (1996) The shortest path from the surface to the nucleus: RBP-J kappa/Su(H) transcription factor. Genes Cells Devot Mol Cell Mech 1(1):1–9

Ihrie RA, Alvarez-Buylla A (2011) Lake-front property: a unique germinal niche by the lateral ventricles of the adult brain. Neuron 70(4):674–686. https://doi.org/10.1016/j.neuron.2011.05.004

Imayoshi I, Sakamoto M, Yamaguchi M, Mori K, Kageyama R (2010) Essential roles of Notch signaling in maintenance of neural stem cells in developing and adult brains. J Neurosci Off J Soc Neurosci 30(9):3489–3498. https://doi.org/10.1523/JNEUROSCI.4987-09.2010

Jessberger S, Parent JM (2015) Epilepsy and Adult Neurogenesis. Cold Spring Harb Perspect Biol 7(12). https://doi.org/10.1101/cshperspect.a020677

Kageyama R, Ohtsuka T, Kobayashi T (2007) The Hes gene family: repressors and oscillators that orchestrate embryogenesis. Development 134(7):1243–1251. https://doi.org/10.1242/dev.000786

Kawaguchi D, Furutachi S, Kawai H, Hozumi K, Gotoh Y (2013) Dll1 maintains quiescence of adult neural stem cells and segregates asymmetrically during mitosis. Nat Commun 4:1880. https://doi.org/10.1038/ncomms2895

Kazanis I, Lathia J, Moss L, ffrench-Constant C (2008) The neural stem cell microenvironment. In: StemBook. Cambridge, MA. doi:https://doi.org/10.3824/stembook.1.15.1

Lavado A, Oliver G (2014) Jagged1 is necessary for postnatal and adult neurogenesis in the dentate gyrus. Dev Biol 388(1):11–21. https://doi.org/10.1016/j.ydbio.2014.02.004

Lavado A, Lagutin OV, Chow LM, Baker SJ, Oliver G (2010) Prox1 is required for granule cell maturation and intermediate progenitor maintenance during brain neurogenesis. PLoS Biol 8(8). https://doi.org/10.1371/journal.pbio.1000460

Lindsell CE, Boulter J, diSibio G, Gossler A, Weinmaster G (1996) Expression patterns of Jagged, Delta1, Notch1, Notch2, and Notch3 genes identify ligand-receptor pairs that may function in neural development. Mol Cell Neurosci 8(1):14–27. https://doi.org/10.1006/mcne.1996.0040

Liu XS, Chopp M, Zhang RL, Tao T, Wang XL, Kassis H, Hozeska-Solgot A, Zhang L, Chen C, Zhang ZG (2011) MicroRNA profiling in subventricular zone after stroke: MiR-124a regulates proliferation of neural progenitor cells through Notch signaling pathway. PLoS One 6(8):e23461. https://doi.org/10.1371/journal.pone.0023461

Lois C (1996) Chain migration of neuronal precursors. Science 271:978–981

Louvi A, Artavanis-Tsakonas S (2006) Notch signalling in vertebrate neural development. Nat Rev Neurosci 7(2):93–102. https://doi.org/10.1038/nrn1847

Lugert S, Basak O, Knuckles P, Haussler U, Fabel K, Gotz M, Haas CA, Kempermann G, Taylor V, Giachino C (2010) Quiescent and active hippocampal neural stem cells with distinct morphologies respond selectively to physiological and pathological stimuli and aging. Cell Stem Cell 6(5):445–456. https://doi.org/10.1016/j.stem.2010.03.017

Lugert S, Vogt M, Tchorz JS, Muller M, Giachino C, Taylor V (2012) Homeostatic neurogenesis in the adult hippocampus does not involve amplification of Ascl1(high) intermediate progenitors. Nat Commun 3:670. https://doi.org/10.1038/ncomms1670

Lupo G, Harris WA, Lewis KE (2006) Mechanisms of ventral patterning in the vertebrate nervous system. Nat Rev Neurosci 7(2):103–114. https://doi.org/10.1038/nrn1843

Lutolf S, Radtke F, Aguet M, Suter U, Taylor V (2002) Notch1 is required for neuronal and glial differentiation in the cerebellum. Development 129(2):373–385

Marin O (2013) Cellular and molecular mechanisms controlling the migration of neocortical interneurons. Eur J Neurosci 38(1):2019–2029. https://doi.org/10.1111/ejn.12225

Masamizu Y, Ohtsuka T, Takashima Y, Nagahara H, Takenaka Y, Yoshikawa K, Okamura H, Kageyama R (2006) Real-time imaging of the somite segmentation clock: revelation of unstable oscillators in the individual presomitic mesoderm cells. Proc Natl Acad Sci U S A 103(5):1313–1318. https://doi.org/10.1073/pnas.0508658103

Mason HA, Rakowiecki SM, Gridley T, Fishell G (2006) Loss of notch activity in the developing central nervous system leads to increased cell death. Dev Neurosci 28(1–2):49–57. https://doi.org/10.1159/000090752

Merkle FT, Mirzadeh Z, Alvarez-Buylla A (2007) Mosaic organization of neural stem cells in the adult brain. Science 317(5836):381–384. https://doi.org/10.1126/science.1144914

Ming GL, Song H (2011) Adult neurogenesis in the mammalian brain: significant answers and significant questions. Neuron 70(4):687–702. https://doi.org/10.1016/j.neuron.2011.05.001

Mirzadeh Z, Merkle FT, Soriano-Navarro M, Garcia-Verdugo JM, Alvarez-Buylla A (2008) Neural stem cells confer unique pinwheel architecture to the ventricular surface in neurogenic regions of the adult brain. Cell Stem Cell 3(3):265–278. https://doi.org/10.1016/j.stem.2008.07.004

Nicola Z, Fabel K, Kempermann G (2015) Development of the adult neurogenic niche in the hippocampus of mice. Front Neuroanat 9:53. https://doi.org/10.3389/fnana.2015.00053

Noctor SC, Martinez-Cerdeno V, Ivic L, Kriegstein AR (2004) Cortical neurons arise in symmetric and asymmetric division zones and migrate through specific phases. Nat Neurosci 7(2):136–144. https://doi.org/10.1038/nn1172

Noctor SC, Martinez-Cerdeno V, Kriegstein AR (2007) Contribution of intermediate progenitor cells to cortical histogenesis. Arch Neurol 64(5):639–642. https://doi.org/10.1001/archneur.64.5.639

Nyfeler Y, Kirch RD, Mantei N, Leone DP, Radtke F, Suter U, Taylor V (2005) Jagged1 signals in the postnatal subventricular zone are required for neural stem cell self-renewal. EMBO J 24:3504–3515

Ohtsuka T, Ishibashi M, Gradwohl G, Nakanishi S, Guillemot F, Kageyama R (1999) Hes1 and Hes5 as notch effectors in mammalian neuronal differentiation. EMBO J 18(8):2196–2207. https://doi.org/10.1093/emboj/18.8.2196

Ohtsuka T, Sakamoto M, Guillemot F, Kageyama R (2001) Roles of the basic helix-loop-helix genes Hes1 and Hes5 in expansion of neural stem cells of the developing brain. J Biol Chem 276(32):30467–30474. https://doi.org/10.1074/jbc.M102420200

Ottone C, Krusche B, Whitby A, Clements M, Quadrato G, Pitulescu ME, Adams RH, Parrinello S (2014) Direct cell-cell contact with the vascular niche maintains quiescent neural stem cells. Nat Cell Biol 16(11):1045–1056. https://doi.org/10.1038/ncb3045

Pierfelice T, Alberi L, Gaiano N (2011) Notch in the vertebrate nervous system: an old dog with new tricks. Neuron 69(5):840–855. https://doi.org/10.1016/j.neuron.2011.02.031

Rolando C, Taylor V (2014) Neural stem cell of the hippocampus: development, physiology regulation, and dysfunction in disease. Curr Top Dev Biol 107:183–206. https://doi.org/10.1016/B978-0-12-416022-4.00007-X

Rolando C, Erni A, Grison A, Beattie R, Engler A, Gokhale PJ, Milo M, Wegleiter T, Jessberger S, Taylor V (2016) Multipotency of adult hippocampal NSCs in vivo is restricted by Drosha/NFIB. Cell Stem Cell 19(5):653–662. https://doi.org/10.1016/j.stem.2016.07.003

Sauka-Spengler T, Bronner M (2010) Snapshot: neural crest. Cell 143(3):486–486. e481. https://doi.org/10.1016/j.cell.2010.10.025

Shimojo H, Ohtsuka T, Kageyama R (2008) Oscillations in notch signaling regulate maintenance of neural progenitors. Neuron 58(1):52–64. https://doi.org/10.1016/j.neuron.2008.02.014

Shimojo H, Ohtsuka T, Kageyama R (2011) Dynamic expression of notch signaling genes in neural stem/progenitor cells. Front Neurosci 5:78. https://doi.org/10.3389/fnins.2011.00078

Song J, Zhong C, Bonaguidi MA, Sun GJ, Hsu D, Gu Y, Meletis K, Huang ZJ, Ge S, Enikolopov G, Deisseroth K, Luscher B, Christian KM, Ming GL, Song H (2012) Neuronal circuitry mechanism regulating adult quiescent neural stem-cell fate decision. Nature 489(7414):150–154. https://doi.org/10.1038/nature11306

Spalding KL, Bergmann O, Alkass K, Bernard S, Salehpour M, Huttner HB, Bostrom E, Westerlund I, Vial C, Buchholz BA, Possnert G, Mash DC, Druid H, Frisen J (2013) Dynamics of hippocampal neurogenesis in adult humans. Cell 153(6):1219–1227. https://doi.org/10.1016/j.cell.2013.05.002

Stump G, Durrer A, Klein AL, Lutolf S, Suter U, Taylor V (2002) Notch1 and its ligands Delta-like and Jagged are expressed and active in distinct cell populations in the postnatal mouse brain. Mech Dev 114(1–2):153–159

Tam PP, Loebel DA (2007) Gene function in mouse embryogenesis: get set for gastrulation. Nat Rev Genet 8(5):368–381. https://doi.org/10.1038/nrg2084

Telley L, Govindan S, Prados J, Stevant I, Nef S, Dermitzakis E, Dayer A, Jabaudon D (2016) Sequential transcriptional waves direct the differentiation of newborn neurons in the mouse neocortex. Science 351(6280):1443–1446. https://doi.org/10.1126/science.aad8361

Temple S (2001) The development of neural stem cells. Nature 414:112–117

Urban N, van den Berg DL, Forget A, Andersen J, Demmers JA, Hunt C, Ayrault O, Guillemot F (2016) Return to quiescence of mouse neural stem cells by degradation of a proactivation protein. Science 353(6296):292–295. https://doi.org/10.1126/science.aaf4802

Weinmaster G, Roberts VJ, Lemke G (1991) A homolog of Drosophila Notch expressed during mammalian development. Development 113(1):199–205

Weller M, Krautler N, Mantei N, Suter U, Taylor V (2006) Jagged1 ablation results in cerebellar granule cell migration defects and depletion of Bergmann glia. Dev Neurosci 28(1–2):70–80. https://doi.org/10.1159/000090754

Xue Y, Gao X, Lindsell CE, Norton CR, Chang B, Hicks C, Gendron-Maguire M, Rand EB, Weinmaster G, Gridley T (1999) Embryonic lethality and vascular defects in mice lacking the Notch ligand Jagged1. Hum Mol Genet 8(5):723–730

Zhao C, Deng W, Gage FH (2008) Mechanisms and functional implications of adult neurogenesis. Cell 132(4):645–660. https://doi.org/10.1016/j.cell.2008.01.033

Notch and Stem Cells

Anna Bigas and Cristina Porcheri

Abstract

The Notch pathway is crucial in the regulation of stem cells biology. Notch-mediated signalling controls several aspects of tissue homeostasis in both embryonic and adult tissues, balancing stem cells maintenance and differentiation. Although the major elements of the pathway are well conserved throughout evolution, its fine regulation varies among different systems. In this review, we are focusing at the differences and analogies of Notch activity in different animal models, comparing stem cells of various tissues in both adulthood and development. We summarize the major mode of action of the Notch-pathway in dependency to the type of ligand, cross-talk control and transcriptional regulation adopted by stem cells to preserve their undifferentiation status or complete their maturation.

Keywords

Notch · Self-renewing · Differentiation · Stem cells

Abbreviations

Dll	Delta-like
Jag	Jagged
RBPJ	Recombination signal binding protein for immunoglobulin kappa J region
Ngn2	Neurogenin2
ES	Embryonic stem cells
LIF	Leukemia inhibitory factor
BMP	Bone morphogenetic protein
NSCs	Neural stem cells
SOP	Sensory organ precursor
GFAP	Glial fibrillary acidic protein
EGF	Epidermal growth factor
HSCs	Hematopoietic stem cells
EMPs	Erythroid-myeloid progenitors
Hes1	Hairy and enhancer of split 1
Hey2	Hairy/enhancer-of-split related with YRPW motif 2
TAD	Transactivation domain
AGM	Aorta-gonad-mesonephros
BM	Bone marrow
DN1/DN3	Double negative (thymocytes)
TCR	T-cell receptor
Th1	Lymphocyte T-helper1
Th2	Lymphocyte T-helper2

A. Bigas (✉)
Program in Cancer Research, Institut Hospital del Mar d'Investigacions Mèdiques (IMIM), Barcelona, Spain

Centro de Investigación Biomédica en Red de Cáncer (CIBERONC), Barcelona, Spain
e-mail: abigas@imim.es

C. Porcheri
Program in Cancer Research, Institut Hospital del Mar d'Investigacions Mèdiques (IMIM), Barcelona, Spain

T. Borggrefe, B. D. Giaimo (eds.), *Molecular Mechanisms of Notch Signaling*, Advances in Experimental Medicine and Biology 1066, https://doi.org/10.1007/978-3-319-89512-3_12

T-ALL	T-cell acute lymphoblastic leukemia
B-CLL	Chronic lymphocytic leukemia of the B-cell lineage
LGR5	Leucine rich repeat containing G protein-coupled receptor 5
IFE	Interfollicular epidermis
HF	Hair follicles
SG	Sebaceous glands
ROCK2	Rho-associated protein kinase 2
RECK	Reversion-inducing Cysteine-rich Protein with Kazal Motifs
ADAM	A disintegrin and metalloproteinase
NRARP	NOTCH-regulated ankyrin repeat-containing protein
MyoD	Myogenic differentiation antigen
TRAF6	TNF-receptor-associated factor 6
POGLUT1	Protein O-glucosyltransferase 1
PTEN	Phosphatase and tensin homologue
PAX3/7	Paired homeobox transcription factors
BRG1	Brahma-related gene-1
APC	Adenomatous polyposis coli
NLK	Nemo-like kinase
GDE2	Glycerophosphodiester phosphodiesterase 2

1 Mechanisms of Notch Signaling Relevant for Stem Cell Biology

Stem cells are characterized by their ability to self-renew and differentiate into a variety of specialized cells. Their functionality is at the basis of tissue homeostasis and organ function and many pathways cooperate to steer fate determination of a single stem cell. The interplay between internal and external signals is finely regulated to support tissue integrity and avoid stem cell depletion over time.

Notch signalling is central for somatic stem cell homeostasis although its function is specific for different tissues (Koch et al. 2013). Differences exist between Notch function in stem cell specification (mainly occurring in the embryo) and stem cell maintenance or self-renewal (most important in the adult organism).

The canonical Notch cascade is activated upon juxtacrine cell-to-cell interaction, resulting in changes of gene expression and specific fate acquisition. When NOTCH receptors are activated by the interaction with specific ligands, the intracellular portion of NOTCH is cleaved and translocated to the nucleus where it becomes part of a transcriptional regulation complex orchestrating gene expression (Bray 2006; Fortini 2009; Hori et al. 2013). In the trans-activation system, Notch signal can induce lateral inhibition or lateral induction by regulating fate determination in adjacent cells (Fig. 1). *Lateral inhibition* is classically used to prevent spread of cell differentiation by restricting Notch signal, when one cell blocks neighbouring ones from expressing the ligand (Heitzler and Simpson 1991). In other contexts, Notch promotes production of ligand resulting in a cooperative differentiation into the same fate (Daudet and Lewis 2005).

Other types of interaction between ligands and receptors can modulate Notch activation in a cell-autonomous manner, as in the case of *cis-inhibition*. In this type of regulation, receptors and ligands are exposed on the surface of the same cell, causing inhibition of the pathway upon interaction.

Concurrently with these types of signalling, the intensity and extent of Notch signal can be calibrated through feedback circuits that create temporary and/or *oscillatory* expression patterns of its effectors. This is the case of the hairy and enhancer of split family of factors that regulate their own transcription, producing timely controlled fluctuations between increased and decreased levels, which in turn affect the transcription of other targets (Fig. 2).

Some of these mechanisms cooperate in stem cells to produce a dynamic regulation of the Notch activity, varying output in time and strength, and finally determining stem cells maintenance or progression into lineage specification.

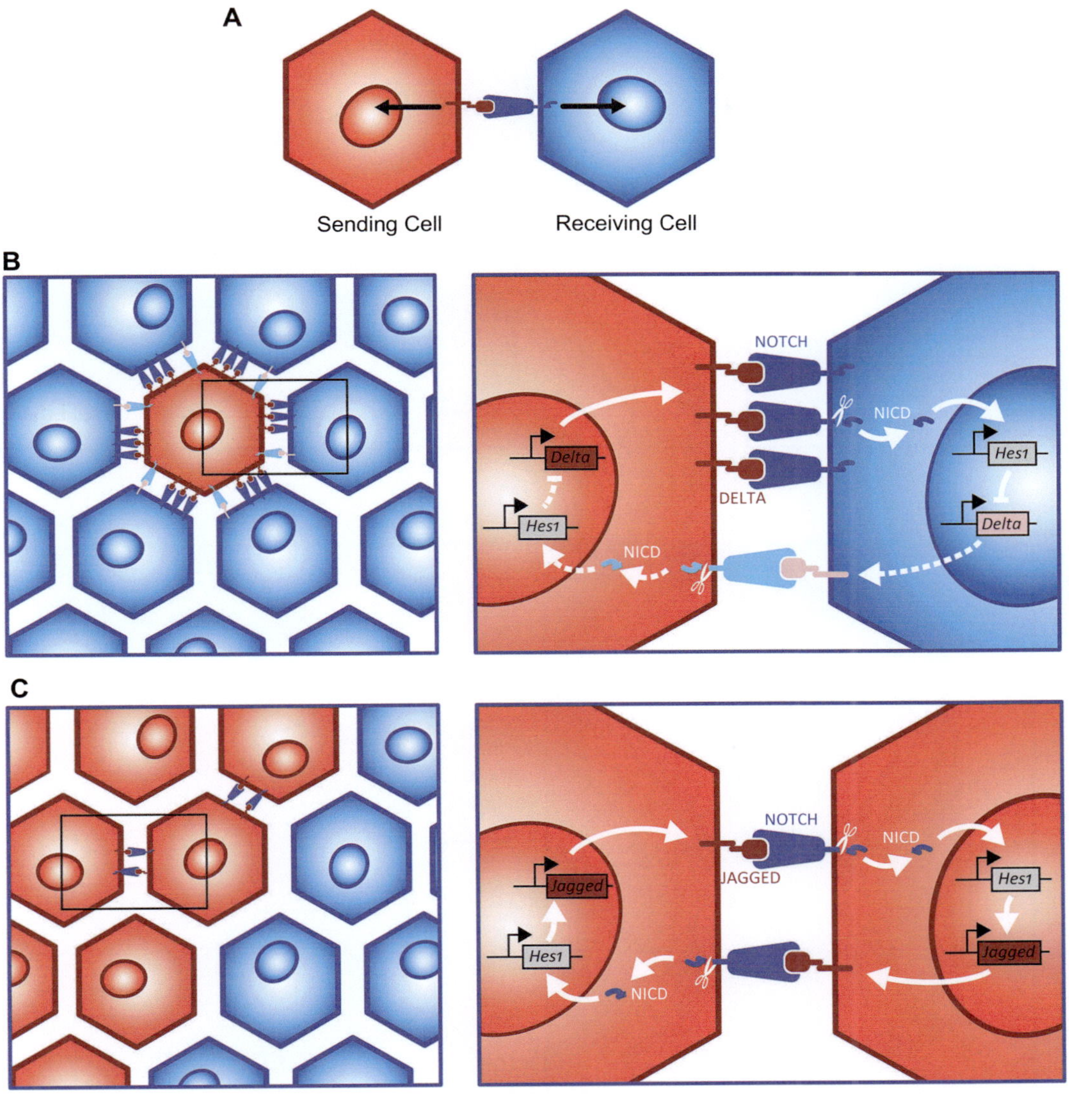

Fig. 1 Notch regulates intercellular communication. (**a**) Trans-activation mode of action upon Notch-ligand interaction. (**b**) Schematic representation of lateral inhibition. Cells expressing NOTCH receptors induce *Hes1* transcription (blue cell), which in turn inhibits expression of Notch-ligands. Low levels of ligand expression are reflected in lower induction of *Hes1* in the neighbouring cell (red cell, right panel). Overview of the resulting tissue structure (left panel). (**c**) Schematic representation of lateral induction. Cells expressing high Notch also express high levels of ligands, strengthening the loop for acquisition of similar fates (right panel). Overview of the resulting tissue organization (left panel)

1.1 Lateral Inhibition

In stem cells, an important mechanism of Notch-dependent fate determination is *lateral inhibition*. Lateral inhibition stabilizes and amplifies small initial differences in Notch levels until they are determining for fate acquisition (Lewis 1998). For example, in *Drosophila* neuroblast precursors, activation of Notch can repress the production of the Notch ligand *Delta* (*Dl*) through Enhancer of Split repressors. Thus, cells experiencing high levels of Notch activity also face a

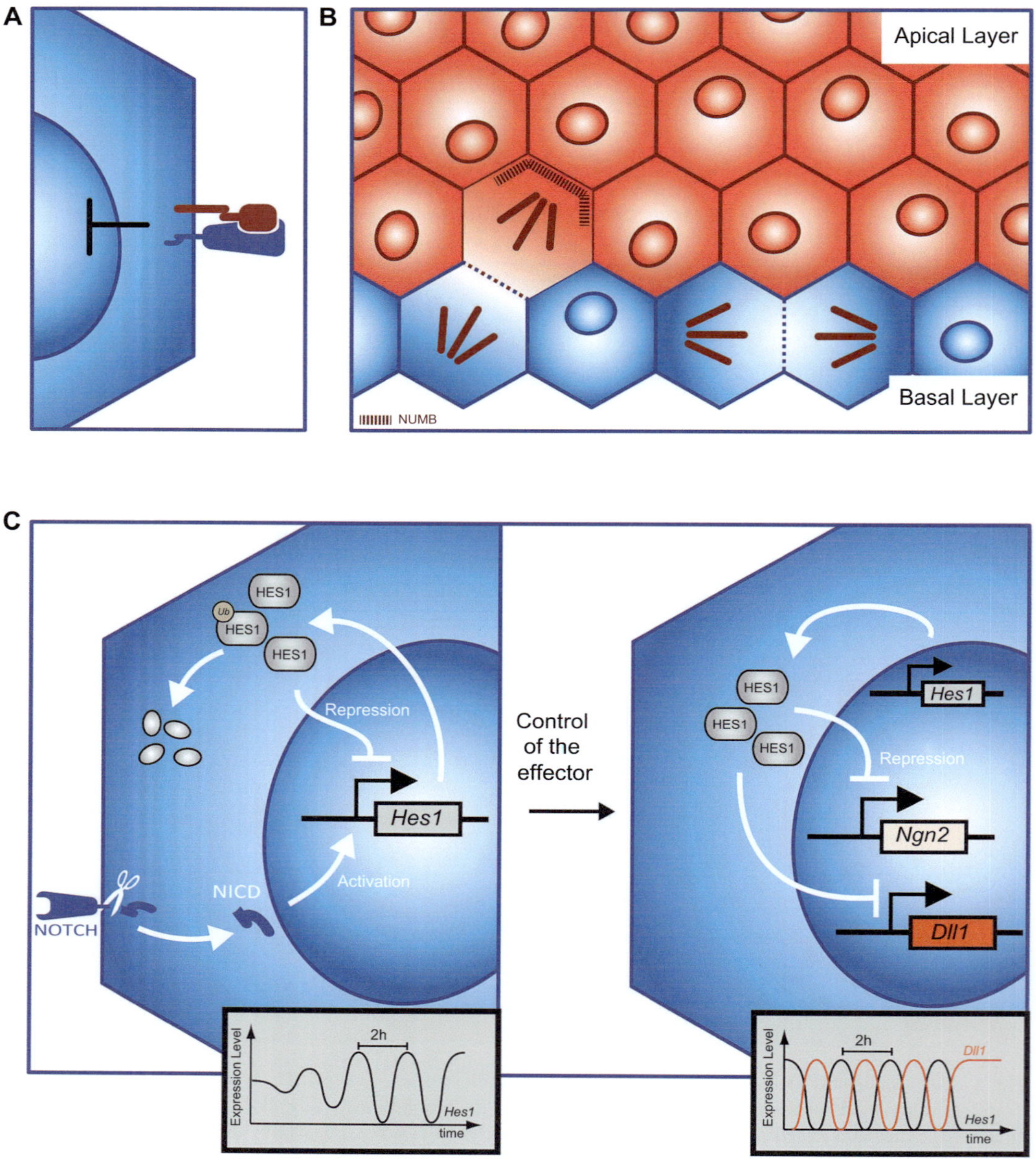

Fig. 2 Notch regulates intracellular events. (**a**) Cis-inhibition upon receptor-ligand interaction on the same cell. (**b**) Schematic representation of differential NUMB-mediated segregation of NOTCH during asymmetric cell division. (**c**) Oscillatory events drive the regulation of fate effectors. Initial events of *Hes1* oscillatory expression (left panel), are reflected in established oscillatory behaviour of fate-determining genes (right panel) as in the example of embryonic neural stem cells

reduction of the ligand and, consequently, a cell with more ligand forces adjacent cells to produce less [(Simpson 1990; Louvi and Artavanis-Tsakonas 2006; Artavanis-Tsakonas et al. 1999), Fig. 1].

The initial difference, either determined or stochastically generated, is therefore rapidly amplified by other signals activating Notch which in turn deregulates its ligand in one cell while inducing its expression on the neighbouring ones.

This system guarantees tight control of the number of cells acquiring a specific fate and significantly contributes to the organization of tissue architecture. Blockage of Notch signalling results in extra cells adopting a specific fate, while an excess of Notch interferes with this differentiation (Heitzler and Simpson 1991).

Lateral inhibition plays a central role during embryonic neurogenesis where cells expressing high levels of Dl ligand acquire a neuronal fate, while Notch-expressing cells block proneural genes and keep their stem cell properties (Artavanis-Tsakonas et al. 1991; Louvi and Artavanis-Tsakonas 2006). Similarly, in the mammalian intestine stem cell, niche-forming cells provide a constant source of ligands to the Notch-expressing intestinal stem cells preserving their undifferentiated character (Riccio et al. 2008; Pellegrinet et al. 2011; Sancho et al. 2013).

Notch-ligand interactions are therefore essential regulators of stem cell maintenance. Cells initially equipotent undergo specification through finely tuned activation and inhibition, orchestrated by their neighbouring cells.

1.2 Lateral Induction

Notch also modulates a mechanism defined as "*lateral induction*" where positive signals lead to specialisation of the juxtaposed cell by Notch. In this case, the tissue displays homogeneous clusters of Notch activity, in contrast with the salt-and-pepper distribution representative of lateral inhibition (Fig. 1) (Owen et al. 1999; Owen et al. 2000; Wearing and Sherratt 2000; Lai 2004). A classic example of Notch inductive signalling is the boundary formation at the *Drosophila* wing margin, a line of cells that controls the outgrowth of the wing. Notch activity turns on the transcription of the coactivator vestigial in the presumptive wing margin. The final effect is the production of sharp boundaries of gene expression (Kim et al. 1996).

During ear neurogenesis, lateral inhibition and lateral induction cooperate in driving prosensory specification or hair cell determination, respectively. Notch activity is low during prosensory stages but increases during hair cell determination. Expression pattern and functional studies suggest that the possible outcome of Notch activity depends from the type of ligand-initiating signalling. JAG1 mainly drives lateral induction while DLL1 promotes lateral inhibition. JAG1 acts as a competitor of DLL1 for the NOTCH1 interaction, leading to lateral induction and finally facilitating differentiation into hair cells (Brooker et al. 2006; Petrovic et al. 2014).

1.3 Cis-Inhibition

NOTCH receptors can also directly interact with the ligands present on the same cell. Gain and loss of function studies have demonstrated that these ligands recognize the receptor via their extracellular domain and cis-inhibit (in a cell autonomous manner) the ability to receive Notch signals coming from adjacent cells [(del Alamo et al. 2011), Fig. 2]. Mathematical models attempting to explain the function of cis-interactions show that a sharp, switch-like response, is induced from cis-DELTA (Sprinzak et al. 2010).

Cis-inhibition was initially described in the context of dorso-ventral boundary formation in the *Drosophila* wing imaginal disc (de Celis and Bray 1997) and since then, many examples have been described during development of both *Drosophila* and other organisms (Bray 1998; Yaron and Sprinzak 2012; Bray 2016).

In mammalian organisms however, cis-inhibition has been elusive. Evidence for cis-inhibition has been obtained in culture experimental systems of mammalian cells (Sprinzak et al. 2010) but conclusive *in vivo* evidences are still lacking. A putative role for cis-inhibition was proposed in epidermal stem cell maintenance. In both human and mouse cells, the DLL1 ligand is expressed in clusters of uncommitted progenitors within the basal layer of the interfollicular epidermis, while NOTCH receptors are expressed in all epidermal layers (Favier et al. 2000). In these conditions, NOTCH-DLL1 cis-interaction in the stem cells of the clusters maintains their undifferentiated state while, outside the clusters, differentiation proceeds (Lowell et al. 2000; Estrach et al. 2008). Although in this context Notch is involved in regulating epidermal differentiation, the mechanism supporting this function remains to be clarified.

1.4 Oscillatory Expression of Effectors

At a transcriptional level, the effects of Notch activity are tightly regulated mainly through positive and negative feedback loops. Specifically, *Hes* transcriptional repressors are Notch targets which downregulate themselves and other NOTCH targets, producing an oscillatory signalling in the transcriptional activity. The relevance of these oscillations in stem cells is particularly evident during embryonic neurogenesis, when *Hes1* expression dynamically changes from high to low (Shimojo et al. 2008). When HES1 protein levels are kept constant by overexpression, control of differentiation is lost. In mouse telencephalon development, HES1 oscillation controls the levels of proneuronal factors [such as *Delta-like 1*(*Dll1*) and *Neurogenin2* (*Ngn2*)]: when HES1 protein levels are high, *Dll1* and *Ngn2* are downregulated, blocking acquisition of the neuronal fate [(Shimojo et al. 2008; Shimojo et al. 2016), Fig. 2]. Oscillation of NOTCH-targets was also described for other stem cells such as embryonic stem cells, where low levels of *Hes1* lead to acquisition of a neuronal fate and cells carrying high levels of *Hes1* enter a mesodermal program (Kobayashi et al. 2009).

Somitogenesis is another process that highly depends on the oscillation of Notch signalling components. Oscillating gene expression involves an auto-inhibitory activity of the Notch-activated bHLH repressors HES1 and HES7, which turn their own transcription off (Kageyama et al. 2012).

Taken together, these findings show that cycling Notch-dependent factors orchestrate differentiation of stem cells and play an essential role in tissue morphogenesis. The Notch-driven regulation of genetic expression is therefore more complex than a mere on-off system, suggesting that stem cells are exposed to an elaborate series of specific signals regulated in intensity and time.

1.5 Asymmetric Segregation

Maintenance of tissue homeostasis, cell number and cell identity requires fine regulation in time and space. Stem cells carry the potential to regenerate a whole tissue, although increasing their number through multiple cellular divisions amplifies the risk of genetic instability and cell death. To preserve stem cells from depletion and genomic alterations, self-renewal can occur through asymmetric cell division (Li 2013). In this scenario, specific fate determinants segregate differentially between the two daughter cells, giving rise to a stem cell and a committed progenitor [(Neumuller and Knoblich 2009; Knoblich 2001), Fig. 2]. Progenitors can amplify through a series of symmetric divisions, while the newly generated stem cell enters a quiescent or semi-quiescent state.

Asymmetric distribution of Notch upon division provides a crucial element for cell fate determination and stem cell maintenance. Asymmetric Notch distribution has been widely associated with the activity of Numb, which is involved in the internalisation of the receptor and thus is responsible for reducing the number of Notch molecules on the surface membrane of one particular daughter cell (Rhyu et al. 1994; Knoblich 2001; Knoblich et al. 1995).

Orientation of cell division can also influence distribution of fate determinants, as in the case of the developing cortex, where a vertical mitotic spindle leads to asymmetric division and parallel positioning results in symmetric mitosis. During cortical neurogenesis, stem cells of the basal layer divide asymmetrically as a result to an acquired polarity. When a neuroepithelial progenitor divides with a vertical spindle, only the apical daughter cell inherits NUMB, while in the case of horizontal divisions NUMB segregates equally between the two daughter cells. Numb regulates membrane exposure of Notch potentially by direct binding with its intracellular tail and promoting its endocytosis (Furthauer and

Gonzalez-Gaitan 2009b, a). Therefore, when the polarity regulator complex sequesters NUMB to the apical surface, it consequently increases the number of NOTCH molecules on the basal surface (Zhong et al. 1996). Upon division, the basal cells will contain high Notch activity responsible for maintaining a stem cell character while the apical cell will have lower levels of Notch and can proceed toward differentiation (Cayouette and Raff 2002).

Mouse intestinal stem cells predominantly undergo symmetric division (Lopez-Garcia et al. 2010). Importantly, upon exposure to inflammation, putative stem cells start to divide asymmetrically in a process that controls Notch signaling through NUMB and *miR34a.* NUMB expression increases with differentiation while Notch remains higher in the stem cell compartment, in line with its role in self-renewal. NUMB is more present in the basal cell of the asymmetric division, segregated together with another Notch controller *miR-34a*, a specific *Notch* (and *Numb*) binding microRNA (Bu et al. 2013). Numb regulates NOTCH in a graded, continuous way, while *miR-34a* action is more abrupt and depends on reaching a specific threshold level. The added effects lead to a stronger blockage of the Notch pathway where in absence of NUMB, *Notch* levels are more sensitive to *miR-34a* action, producing a more robust switch (Bu et al. 2016).

Asymmetric division also occurs during skin formation in embryonic development and it is essential for preserving skin morphology in the adult. Skin differentiation appears concomitantly with vertical orientation of mitotic spindle during progenitor division. This allows differential distribution of Notch receptors between basal and suprabasal layers and lays the basis for epidermal stratification (Williams et al. 2011). Alteration of the asymmetric division machinery leads to improper compartmentalization of NOTCH and disrupts tissue architecture. In contrast with what observed in other systems, NUMB plays a marginal role in asymmetric division of the epidermis, as its disruption leads only to minor differentiation defects (Williams et al. 2011).

As these examples show, asymmetric cell division is a well-understood model for stem cell self-renewal (Knoblich 2008). Nevertheless, recent imaging studies combined with modelling show that asymmetric divisions occur much less frequently than previously thought and stem cells from different tissues often divide symmetrically (Knoblich 2008; Snippert et al. 2010).

2 Pluripotent Stem Cells: Embryonic Stem Cells

Embryonic stem cells (ES) are an excellent model to study pluripotency and the molecular mechanisms driving differentiation. Under appropriate experimental conditions they can differentiate into three-dimensional clusters known as embryoid bodies containing different embryonic tissues. Undifferentiated mouse ES cells express the receptors *Notch1* and *Notch2* and their ligands *Jag1*, *Jag2*, *Dll1* and *Dll3* (Lowell et al. 2006). Genetic ablation of *rbpj*, *Notch1* or *Notch2* allow normal progression of embryonic development until the formation of the three germ layers (Oka et al. 1995; Huppert et al. 2005), suggesting that Notch is dispensable for ES homeostasis in the initial stages of embryonic life. In contrast, Notch signalling plays an essential role in directing differentiation. Activation of Notch in mesodermal cells blocks cardiogenesis, hematopoiesis and endothelial differentiation and ablation of *rbpj* induces acquisition of the cardiomyocyte fate (Nemir et al. 2006; Schroeder et al. 2006; Schroeder et al. 2003). Additionally, induction of Notch pathway leads to neuroectodermal differentiation in the absence of self-renewal factors [such as leukemia inhibitory factor (LIF) or Bone morphogenetic protein 4 (BMP4)] while blocking non-ectodermal fate. Similarly, pharmacological or genetic blockage of the Notch signalling compromises acquisition of the neuronal fate (Lowell et al. 2006).

In human ES cells, Notch activation is not required to maintain pluripotent undifferentiated cells but similar to mouse ES cells, it plays an essential role for their differentiation into endodermal, mesodermal and ectodermal cell layers (Noggle et al. 2006). A peak of *Dll1* expression in the first 48 h of embryoid bodies formation produces a time-restricted wave of Notch activity. This pulse of Notch activation correlates with the decrease of *Oct4* expression, allowing undifferentiated cells to specify the three germ layers (Yu et al. 2008).

In conclusion, ES cells use Notch activation to regulate the initial stages of tissue formation, while no evidence exists for a putative role in pluripotency maintenance or self-renewal.

3 Notch in Somatic Stem Cells: Generation and Maintenance

Notch is crucial in the specification and maintenance of different types of somatic stem cells during both embryonic life and adulthood. In this section, we will focus on the essential role of Notch in regulating neural, hematopoietic, intestinal, epidermal and muscular stem cells.

3.1 Neural Stem Cells

The first indication that Notch is involved in the control of neural differentiation was obtained in the *Drosophila* system (Artavanis-Tsakonas et al. 1999). In this organism, embryonic ectoderm uses Notch signalling to generate cellular diversity (Gaiano and Fishell 2002) that results in a patched-pattern of neural progenitors intermingled in the predominant epithelial layer [see lateral inhibition section and (Bertet et al. 2014; Gaiano and Fishell 2002)]. Similarly, in mammals, the vast majority of neurons and glial cells that forms the structural basis of intellectual function are produced during a restricted period of embryonic development (Gotz and Huttner 2005; Caviness et al. Caviness. et al. 2009). In cortical neurogenesis, undifferentiated progenitors line the wall of ventricular cavities, where they undergo either symmetric or asymmetric division. Symmetric division predominates the earliest phases of cortical neurogenesis, amplifying the pool of undifferentiated progenitors and maintaining their identity. Later on, asymmetric division gives rise to a population of neuroblasts that in turn migrate along radial glial fascicles to reach the outer rim of the cortical plate and complete their differentiation into post-mitotic neurons (Haydar et al. 2003; Kosodo et al. 2004; Gotz and Huttner 2005; Hansen et al. 2010). Asymmetric division causes a differential distribution of fate determinants in between the two daughter cells. Maintenance of neural stem cells (NSCs) depends on differential segregation of Notch regulators and NOTCH receptors, most likely achieved through the NUMB protein, selectively localized in the apical membrane of dividing stem cells (Zhong et al. 1996; Chenn and McConnell 1995). In *Drosophila* Sensory Organ Precursor (SOP) determination, the mechanistic details of the Notch-Numb interaction have been widely elucidated. NUMB regulates the endocytosis of different components, such as the NOTCH receptor, by interacting with clathrin-coated pits (Berdnik et al. 2002; Hutterer and Knoblich 2005; Knoblich et al. 1995). Numb-mediated endocytosis is important for Notch inactivation in one of the asymmetrically dividing cells. Interestingly, inheritance of NOTCH-DELTA containing endosomes in the other cell will determine the activation of Notch (Coumailleau et al. 2009; Furthauer and Gonzalez-Gaitan 2009a). Hence, Numb directly influences Notch-mediated cell-to-cell interactions, which in turn mediate lateral inhibition and participate in fate determination. In zebrafish, neural progenitors also divide asymmetrically, unequally distributing endocytosed ligands of the Delta family in between the two daughter cells (Kressmann et al. 2015; Dong et al. 2012). In chick neural progenitors, differential segregation disposes the inheritance of the Notch-regulator MINDBOMB1 (MIB1) by the prospective neuron (Tozer et al. 2017). Similar regulations have been described in mammalian neurogenesis, with highly polarized progenitor cells undergoing asymmetric division and retaining NUMB on their apical or apical-lateral side of neural progenitor cells (Zhong and Chia 2008). MIB1 is also part of the endocytosis machinery in mammals, essential in the maintenance of the radial glia progenitors, as its mutation results in a complete loss of Notch and premature differentiation of neurons (Yoon et al. 2008). Thus, regulation of

Notch during stem cell division is well conserved throughout species.

Since the initial phases of telencephalon formation, Notch activity is associated with a blockage of differentiation, preserving the pool of stem/progenitors cells in the developing brain. NOTCH1 receptor confined to the basal membrane controls neuronal fate in undifferentiated progenitors (Mizutani et al. 2007; Del Bene et al. 2008). Specifically, NOTCH1 regulates ventral-basal gradient of Sonic hedgehog (SHH) in the telencephalon by regulating subcellular localization of PATCHED receptor and consequently changing the identity of ventral progenitors (Kong et al. 2015).

Genetic ablation of either *Notch1* or its effector *rbpj* induces premature differentiation into neurons. Gain-of function experiments overexpressing a dominant-active form of *Notch* or target genes of the *Hes* family confirmed the antineuronal function of NOTCH (Yoon and Gaiano 2005).

During embryonic development the regulation of Notch-pathway components follows dynamic patterns of active and inactive states. The Notch effectors HES1 and HES7 are expressed in an oscillatory manner and control expression of proneural factors such as *Achaete-scute homolog 1* (*Ascl1/Mash1*) and *Ngn2* (Shimojo et al. 2008; Imayoshi et al. 2013). The *Dll1* ligand is also regulated in a similar manner as shown by live-imaging analysis. A forced steady expression of *Dll1* promotes neurogenesis and inhibits proliferation of neural progenitors (Shimojo et al. 2016).

Later in development, NSCs give rise to glial progenitors that can further differentiate into oligodendrocytes or astrocytes. Notch regulates glial fate decision by blocking expression of oligodendrocytic-specific genes and directly activating transcription of astrocytic effectors, such as Glial fibrillary acidic protein (GFAP) (Ishibashi et al. 1994; Sakamoto et al. 2003; Ohtsuka et al. 1999; Gaiano et al. 2000; Chambers et al. 2001). Thus, Notch pathway selectively expands the astroglia pool, while the acquisition of oligodendrocyte lineage is inhibited (Gaiano et al. 2000; Chambers et al. 2001; Lutolf et al. 2002).

Together, these findings show that Notch is crucial at different time points of the nervous system development. First, Notch is essential to maintain NSCs undifferentiation and block acquisition of neuronal fate. Later, Notch induces differentiation of astrocytes instead of oligodendrocytic fate acquisition (Fig. 3).

Adult NSCs remain in special hubs of the central nervous system (CNS) partially retaining their ability of self-renewal and generation of all neural fates. Two main regions of the brain function as neural stem cells niches: the subgranular zone (SGZ) of the dentate gyrus in the hippocampus and the subventricular zone (SVZ), lining the walls of the lateral ventricles (Doetsch et al. 1999; Gage 2000; Gage et al. 1995). Notch-pathway components are broadly expressed in the whole brain and specifically found in NSCs of the SGZ and SVZ (Stump et al. 2002; Irvin et al. 2004; Givogri et al. 2006; Ehm et al. 2010). Conditional ablation of *rbpj* in adult NSCs results in an increased amount of the transit amplifying population and differentiated neurons at the expenses of the stem cell pool, thus recapitulating the Notch function observed in embryonic neurogenesis (Ehm et al. 2010).

As for other somatic stem cells, adult NSCs remain in a semiquiescent state and proliferate only sporadically. Notch is specifically relevant for actively cycling NSCs in the SVZ as *Notch1*-deletion has no effects on the quiescent subpopulation of NSCs but ablate the cycling ones (Basak et al. 2012). In the SGZ, both quiescent and cycling NSCs experience active NOTCH as shown by tracing of the HES5$^+$ cells in the dentate gyrus (Lugert et al. 2010). Genetic deletion of *Notch1* in the SGZ reduces the number of NSCs and lead to a consequent reduction of transit amplifying and newly generated neurons (Ables et al. 2010).

How Notch mediates regulation of the quiescence state in the adult niche is still unclear but the presence of its ligand JAG1 on ependymal cells and niche astrocytes suggests that Notch signalling is a crucial connector between stem cells and their microenvironment. JAG1 is indeed responsible for NOTCH1 activation inside the stem cell compartment, thus main-

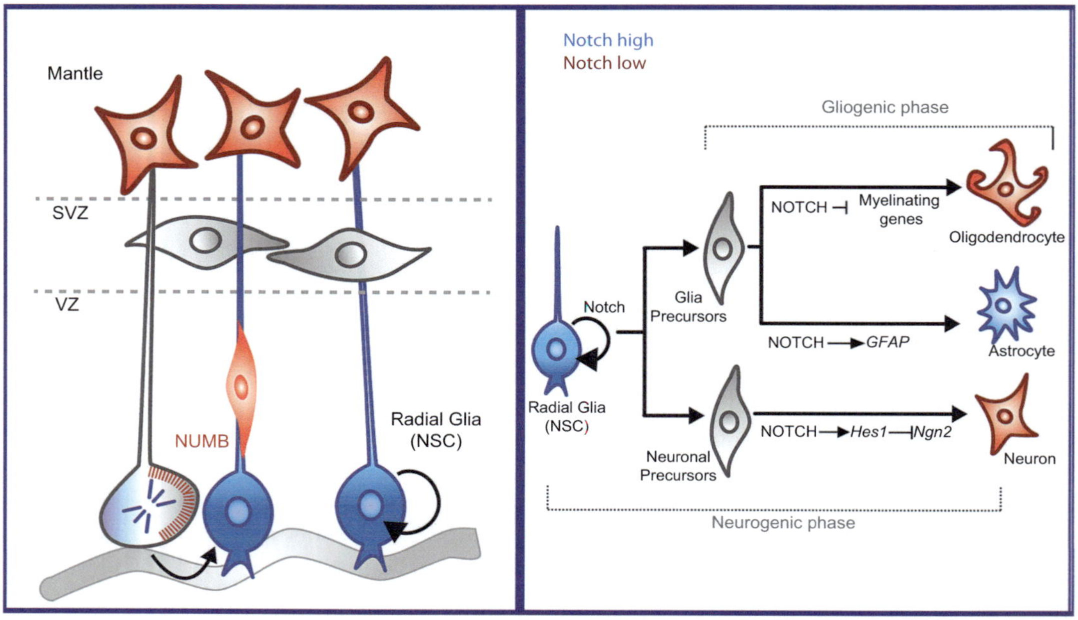

Fig. 3 Notch regulation of neural stem cells. Organization of a neural stem cell niche in the embryonic telencephalon (left panel). Blue colour shows cells with high Notch activity; orange colour represents low Notch activity. Asymmetric division preserves Notch activity in the progeny of stem cells, while Notch is downregulated in differentiated neurons. Importance of Notch in neural fate acquisition (right panel). In glial precursors high Notch drives acquisition of astrocytic fate while high activity of Notch blocks oligodendrocyte differentiation. Neuroblasts complete their differentiation into functional neurons only in absence of Notch. SVZ: Subventricular Zone; VZ: Ventricular Zone

taining their undifferentiated state (Nyfeler et al. 2005), in contrast to DLL4 ligand that promotes generation of neural precursors as shown in the adult rat brain (Androutsellis-Theotokis et al. 2006). *Notch1* and NOTCH ligands are expressed by niche components such as astrocytes (Magnusson et al. 2014; Ferron et al. 2011; Benner et al. 2013), and blockage of Notch in astrocytes outside the canonical neurogenic niches triggers ectopic neurogenesis (Magnusson et al. 2014).

Notch-dependent control of transcription depends on the formation of a ternary complex formed by Notch intracellular domain, CSL and a member of the Mastermind family. In fish neurogenesis, the Nemo-like kinase (NLK) interferes with the formation of the ternary complex of transcription (Ishitani et al. 2010) while in mammals the details of Notch-dependent transcriptional control are still lacking. Growth factors and morphogens regulate changes in Notch activity in spatial and temporal coordinated manner (Pierfelice et al. 2011).

During development, several factors [such as BOTCH, SLIT and Glycerophosphodiester Phosphodiesterase 2 (GDE2)] block Notch activities in undifferentiated cells to trigger differentiation. Botch inhibits cleavage of NOTCH maintaining the receptor in an immature inactive full length form. Blockage of Botch induces retention of progenitors into the ventricular and subventricular regions, while overexpression of it promotes their migration into the more external layer to complete differentiation (Chi et al. 2012). The Slit/ Roundabout homolog (ROBO) signalling activates *Hes1* effector in telencephalic progenitors and deletion of *Robo* receptors or their ligands ablate mitosis in the ventricular wall (Borrell et al. 2012). Finally, GDE2 inactivates Notch specifically in spinal cord progenitors to allow formation of motor neurons through the release of the protease inhibitor Reversion-inducing Cysteine-rich Protein with Kazal Motifs (RECK). Release of RECK disinhibits the disintegrin and metalloprotease (ADAM) which in turn is responsible for DLL1 shedding, finally leading to inactivation of the Notch signalling (Park et al. 2013).

Close to birth, the telencephalic expression of the ubiquitin complex member FBW7 reduces the levels of Notch inducing stem cell differentiation (Hoeck et al. 2010).

In the adult tissue, repression of *Hes5* is needed for progression of neurogenesis. The B-cell lymphoma 6 protein (BCL6) proneurogenic factor alters the transcriptional complex composition on the *Hes5* promoter by excluding MASTERMIND1 and recruiting SIRT. In this manner, the *Hes5* activity is silenced despite the presence of active Notch signalling (Tiberi et al. 2012). SOX21 is a member of the Sox family of transcription factors directly acting on the Notch-effector *Hes5*. In the SGZ, SOX21 represses *Hes5* gene expression balancing out Notch effect (Matsuda et al. 2012). Finally, growth factors present in the niche cooperate in the control of Notch, as it is the case for EGF signalling on transit amplifying progenitors, resulting in *Numb* expression and Notch inhibition on NSCs (Aguirre et al. 2010).

3.2 Hematopoietic Stem Cells

Vertebrate haematopoiesis originates in sequential waves during development starting in the yolk sac, producing erythrocytes and macrophages (primitive hematopoiesis), followed by the production of definitive-like erythroid-myeloid progenitors [EMPs, (Medvinsky et al. 2011; Dzierzak and Speck 2008; Palis et al. 1999)]. Next, the hematopoietic stem activity is found in the aorta of the Aorta-gonad-mesonephros (AGM) region where definitive hematopoietic stem cells (HSCs) are produced and finally migrate to other tissues to be amplified (fetal liver or caudal hematopoietic tissues). Soon after that, other extraembryonic tissues become hemogenic in mammalian embryos such as the placenta and the vitelline and umbilical arteries, also producing a definitive type of hematopoiesis.

Primitive hematopoietic cells and endothelium are thought to derive from a common precursor, the hemangioblast (Huber et al. 2004). Although Notch is heavily involved in the regulation of vascular fate, absence of Notch signal is not required for the first extraembryonic waves of hematopoiesis. However, Notch inhibition leads to an increased number of primitive erythrocytes as suggested by the phenotype of *rbpj*-deficient mice and confirmed by gain of function experiments (Robert-Moreno et al. 2007; Lee et al. 2009).

Hematopoietic production is truly definitive with the generation of hematopoietic stem cells, capable of long-term self-maintenance and generation of all blood cell types. In vertebrates, definitive HSCs mainly originate from the ventral wall of the aortic endothelium in the AGM (Dieterlen-Lievre and Martin 1981; Medvinsky et al. 1993; Ciau-Uitz et al. 2000; de Jong and Zon 2005; Ivanovs et al. 2014; Dzierzak and Speck 2008; Lee et al. 2009). Endothelial cells in the aortic wall undergo an endothelial-to-hematopoietic transition ending into the formation of clusters with hematopoietic potential. Once circulation has been properly established between extraembryonic tissues and embryo, HSCs are detected in the different extraembryonic tissues (Inman and Downs 2007; Gekas et al. 2005; Ottersbach and Dzierzak 2005). The complete molecular program of emerging and maturing of HSCs is not entirely characterized, although it is well established that the Notch signalling is required for these processes (Kanz et al. 2016; Butko et al. 2016).

The expression of Notch family members in the embryonic aorta is dynamic in time and space (Oh et al. 2013). NOTCH1 and NOTCH4 receptors are both present in the aortic endothelium at the time of hematopoietic clusters formation (E9.5-E11.5) and while the NOTCH1 receptor is indispensable for arterial specification, lack of NOTCH4 does not perturb normal arteriogenesis (Robert-Moreno et al. 2008; Krebs et al. 2000; Krebs et al. 2004). Subsequently, NOTCH2 has also been detected

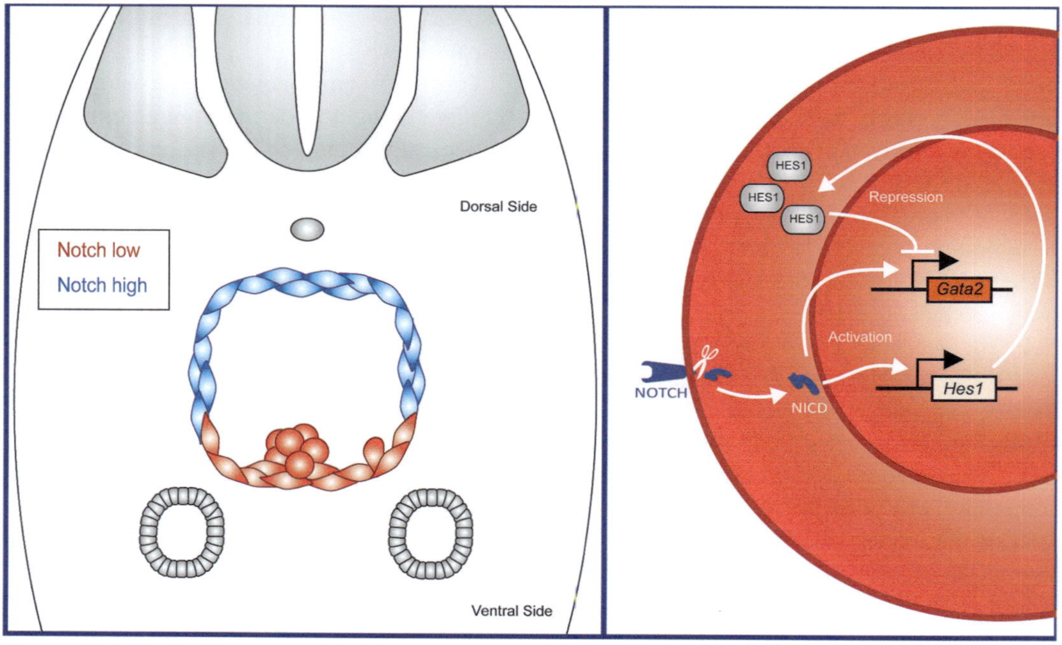

Fig. 4 Notch activity in hematopoietic stem cells. Organization of the hematopoietic niche in the embryonic aorta-gonad-mesonephros (left panel). High Notch is needed to preserve the endothelial program, while low levels of Notch in the ventral part of the aortic endothelium allow endothelial to hematopoietic transition and formation of hematopoietic clusters. Molecular mechanism of hematopoietic control inside the cluster (right panel). Activation of NOTCH induces the expression of *Hes1* gene which in turn blocks the expression of *Gata2*. Additionally, NOTCH directly induces *Gata2* producing an incoherent feed-forward loop

by flow cytometry in hematopoietic progenitors (Hadland et al. 2015). In the AGM, Notch regulates classical target genes such as *Hes1* and *Hey2*, as well as other tissue-specific genes such as *Gata2* and *EphrinB2* (Davis and Turner 2001; Guiu et al. 2013; Robert-Moreno et al. 2005; Iso et al. 2003) (Fig. 4). There is also evidence that *Runx1* is downstream of NOTCH but likely controlled by GATA2 (Robert-Moreno et al. 2005; Nakagawa et al. 2006; Richard et al. 2013).

Analysis of Notch ligands in the aortic endothelium has been extremely informative to elucidate Notch function. *Dll4*, *Jag1* and *Jag2* ligands are expressed in different areas of the AGM endothelium and hematopoietic clusters (Robert-Moreno et al. 2008; Gama-Norton et al. 2015). While *Jag2*-mutants do not show major alterations in the aortic endothelium, ablation of *Dll4* or *Jag1* exerts a major effect on how Notch steers the decision between endothelial, hematopoietic and arterial fate. *Dll4* mutants are embryonic lethal and display broad vascular defects before HSCs emergence, masking the direct effect on hematopoietic development. In contrast, overexpression of *Dll4* activates arterial program in the venous endothelium, indicating that this ligand is the main instructor for arterial specification (Krebs et al. 2004; Duarte et al. 2004).

Several *Notch* mutant cells fail to contribute to hematopoiesis but also fail to activate the arterial program, thus precluding the study of Notch function in HSCs emergence (Hadland et al. 2004; Kumano et al. 2003; Robert-Moreno et al. 2005). However, *Jag1*- or *Hes1/5*-mutant embryos maintain the expression of arterial markers while containing reduced numbers of hematopoietic progenitors (Robert-Moreno et al. 2005; Robert-Moreno et al. 2008; Guiu et al. 2013), indicating that the hematopoietic program can be indeed uncoupled from the Notch-driven arterial program. As hematopoietic development proceeds, hemogenic precursors become less dependent or totally independent of Notch signalling. Specifically, recent studies exclusively position Notch requirement in the formation of the first HSC precursors in the AGM but not thereafter (Souilhol et al. 2016).

Induction of expression of Notch pathway elements in HSC precursors is not well-understood but SOX17 is one of the positive transcription factors upstream of *Notch1* (Clarke et al. 2013) and pro-inflammatory signals have been shown to induce *Jag1* expression and promote NOTCH activity (He et al. 2015; Espin-Palazon et al. 2014). The fact that HSCs have to turn-off *Notch* at different points of development, implies that several Notch inhibition mechanisms are also active at the end of the process (Zhang et al. 2015; Lizama et al. 2015). Recent studies have revealed new information about the ligands specific function in the AGM, indicating that JAG1 is able to interfere with the DLL4 signal, inhibiting the endothelial program (Gama-Norton et al. 2015). In addition, HSC precursors were found to have low Notch activity while the endothelial/arterial precursors had high Notch activity, suggesting that fate acquisition is dependent on Notch strength. Although mechanistic details are missing, these results indicate that JAG1 ligand is responsible for the low Notch signal strength required for the activation of the hematopoietic program and HSCs formation, while at the same time blocking endothelial fate acquisition. In the absence of JAG1, DLL4 induces high strength of Notch activity resulting in endothelial fate acquisition (Gama-Norton et al. 2015).

As embryonic development proceeds, the aorta environment becomes less permissive for HSCs generation and the progenitor/HSC pool relocates into the fetal liver (Morrison et al. 1995). Here, a massive production of hematopoietic cells and HSCs occurs. Importantly, HSCs amplification in the fetal liver also depends on low levels of NOTCH1 activity as observed in mutant *Notch1* embryos containing a deletion of the trans-activation domain (TAD) that leads to attenuated NOTCH1 activity. This activity was sufficient to bypass the lethality of previous knock-out models and allowed hematopoietic development up to the fetal liver stage. Mutant HSCs were successfully generated and migrated into the fetal liver but their expansion capacity was severely impaired (Gerhardt et al. 2014).

Close to birth, a final embryonic wave of migration locates the HSCs population to the bone marrow (BM) where they remain for the rest of lifespan (Dzierzak and Speck 2008). In the adult BM niche, endosteal and vascular cells play essential roles in maintaining functional HSCs. In the endosteum, parathormone-dependent activation increases the number of *Jag1*-expressing osteoblasts, which in turn leads to an increased number of HSCs. As this increment was eliminated by γ-secretase inhibition, it was hypothesized that Notch regulation in the adult niche controls the homeostasis of HSCs, signalling through JAG1 ligand (Calvi et al. 2003). In the vascular niche, conditional deletion of *Jag1* in endothelial cells resulted in a rapid loss of the HSCs pool, indicating that Notch is involved in the balance between quiescence and self-renewal of long term HSCs (Poulos et al. 2013). The role of Notch as stem cell regulator in the adult BM has been supported by *in vitro* and functional assays, where HSCs were cocultured with feeder cells expressing *Jag1*, *Jag2* or *Dll1*, showing maintained and even enhanced self-renewal (Karanu et al. 2000; Varnum-Finney et al. 1998; Ohishi et al. 2002; Butler et al. 2010). In contrast with these findings, conditional loss of function of *Notch1* and *Notch2* receptor or *rbpj* did not show alteration at HSCs levels and more studies are needed to clarify the role of Notch in adult HSCs (Radtke et al. 1999; Saito et al. 2003; Maillard et al. 2008).

Although the specific role of Notch in HSCs is still under debate, Notch signalling is unequivocally involved at different stages of T-lymphocyte differentiation. In the thymus, early activation of NOTCH1 determines the switch between B- and T-fate differentiation, when interacting with the DLL4 ligand (Wilson et al. 2001; Radtke et al. 2004). Notch is required for immature T-cells to progress through double negative (DN1 to DN3) up to the point of checking pre-TCR rearrangements (β-selection). When T cells progress to double positive CD4⁺CD8⁺, they become insensitive to Notch signalling. However, mature T cells can respond again to Notch signalling during their maturation in Th1 and Th2 cells (Osborne and Minter 2007) as well as during antigen-mediated immune responses in secondary lymphoid tissues to protect cells from TCR-induced apoptosis (Jehn et al. 1999). Recent results also implicate DLL4-NOTCH1 as the initial signal in the bone marrow that leads T cell precursors to the thymus (Yu et al. 2015).

In contrast, immature B-lymphocytes cannot progress further in their lineage differentiation when Notch signal stays active and, only upon relocation into peripheral organs, interaction of NOTCH2 with DLL1 ligands induces transitional B-cells (Saito et al. 2003; Radtke et al. 1999). In myeloid cells, Notch signal also needs to be downmodulated for differentiation to proceed. There are many different cell-autonomous and non-cell autonomous hematopoietic phenotypes observed in hematopoietic Notch-deficient mice which underlie the complex but important role of Notch in the maintenance of hematopoietic homeostasis. In this sense, lack of signalling can give rise to myeloproliferative diseases (Lobry et al. 2014; Wang et al. 2016) while aberrant activation of Notch in hematopoietic progenitors gives rise to acute lymphoblastic leukemia of the T-cell lineage (T-ALL) or chronic lymphocytic leukemia of the B-cell lineage (B-CLL).

3.3 Intestinal Stem Cells

The intestine is a highly proliferative tissue in mammals, facing regeneration every 4–5 days. This ability to regenerate is permitted by the intestinal stem cells which are able to self-renew and give rise to a population of fast-cycling transit amplifying progenitors that differentiate into secretory and absorptive lineages. Stem cells in the intestinal niche are located at the bottom of the crypt in direct contact with transit amplifying cells and Paneth cells (Barker et al. 2007; van der Flier and Clevers 2009; Sato et al. 2011).

Expression of *Notch1* and *Notch2* is found in the adult intestinal epithelium and lineage-tracing studies showed that both receptors are expressed in the stem cells, although the specific role for each one remains to be fully characterized (Sander and Powell 2004) (Fig. 5). Expression of one receptor is sufficient to maintain the control of normal proliferation but both receptors are necessary when the system is challenged [i.e. crypt regeneration upon irradiation, (Riccio et al. 2008; Carulli et al. 2015; Gifford et al. 2016)].

The NOTCH receptor signals in the intestinal stem cells by interaction with DLL1 and DLL4 ligands expressed in the adjacent Paneth cells. Although genetic alteration of single ligands has minor effect on intestinal epithelium, ablation of both *Dll1* and *Dll4* converts proliferating progenitors into goblet cells, suggesting a role for Notch in the maintenance of undifferentiation (Pellegrinet et al. 2011). *Dll1* is additionally expressed above the Paneth/stem cell zone in few undifferentiated secretory precursors that could convert into stem cells under specific stress conditions (van Es et al. 2012).

Activation of the Notch pathway is also essential to control stem cell proliferation and survival as its inhibition leads to a dramatic decrease of the stem cell compartment Leucine-rich repeat-containing G-protein coupled receptor 5 (*Lgr5*) and B lymphoma Mo-MLV insertion region 1 homolog (*Bmi1*), and decreased expression of proliferation markers, such as Ki67 (VanDussen et al. 2012; Lopez-Arribillaga et al. 2015). Upstream control of Notch remains to be completely elucidated, although recent reports indicate that thyroid hormone activates *Notch1*, *Dll1*, *Dll4* and *Hes1* expression via its nuclear receptor TRA1 (Sirakov et al. 2015).

Fate-tracing experiments demonstrated that *Notch1*-expressing cells (Fre et al. 2011) and cells that experienced NOTCH1 activation (Vooijs et al. 2007) are able to regenerate the whole crypt. The main role of Notch in the intestine is the correct specification of the enterocytic lineage versus the secretory lineage. Thus, generation of enterocytes depends entirely on the NOTCH1-HES1-mediated repression of the transcription factor *Atoh1* that is a master regulator of the secretory lineages including goblet, endocrine or Paneth cells (Yang et al. 2001; Shroyer

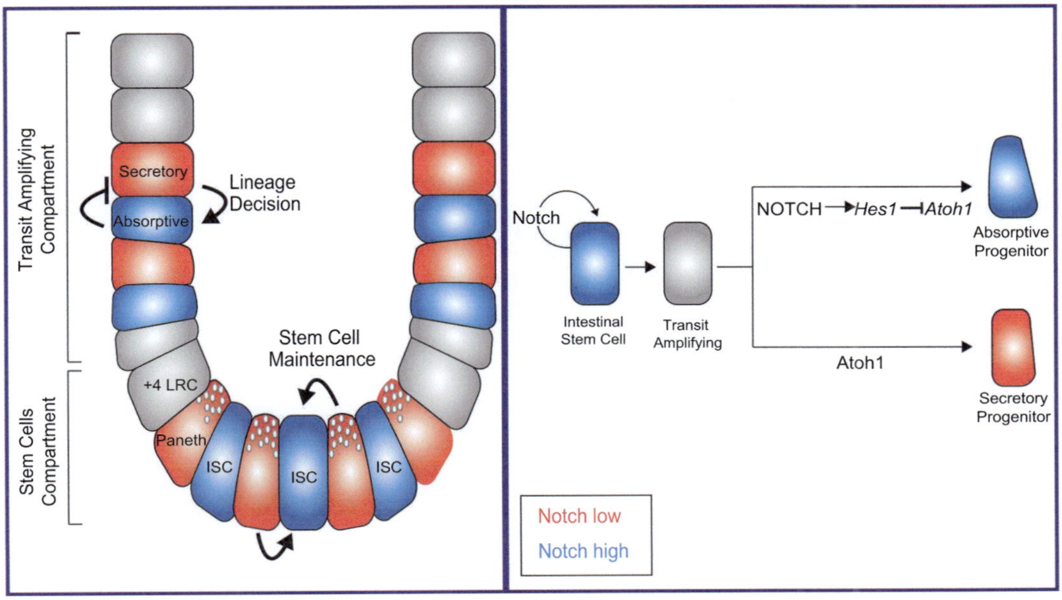

Fig. 5 Notch activity in intestinal stem cells. Schematic representation of an adult intestinal stem cell niche (left panel). Intestinal stem cells (blue) with high Notch activity are intermingled with Paneth cells (orange) expressing low levels of Notch. Notch regulates both self-maintenance and differentiation in the adult intestinal niche (right panel). High levels of Notch allow self-renewal of intestinal stem cells. Levels of Notch are also important in fate decisions: high levels of NOTCH block the expression of *Atoh1* and induce acquisition of an absorptive fate; low levels of NOTCH allow differentiation into a secretory cell

et al. 2007; VanDussen and Samuelson 2010; Li et al. 2012; van Es et al. 2010). Ablation of *Notch1* or inhibition of Notch signalling causes forced differentiation of the intestinal progenitors into the secretory lineages, thus resulting in disruption of the intestine architecture (Fre et al. 2005; Stanger et al. 2005; Pellegrinet et al. 2011; Riccio et al. 2008; VanDussen et al. 2012; Sasaki et al. 2016). In parallel to inducing transcription of lineage-specific factors, *Atoh1* reinforces its own expression and directly regulates *Dll1* and *Dll4*, thus enabling lateral inhibition on adjacent cells (Kim et al. 2014). Chromatin accessibility plays a role in regulation of intestinal Notch, as shown by specific mutation on the chromatin remodelling subunit Brahma-related gene-1 (*Brg1*). *Brg1*-mutant mice display abnormal crypt and increased stem cells loss, together with a dramatic downregulation of *Notch1* (Takada et al. 2016).

During intestinal development *Notch1* is also expressed in the intestinal mesenchyme. Constitutive Notch signal leads to reduced basal intestinal proliferation and alters morphogenesis of the villi (Moriyama et al. 2006; Stanger et al. 2005). As the observed phenotype was very similar to the one induced by Wnt-signalling disruption (Pinto et al. 2003), a crosstalk between Notch and Wnt regulating intestinal stem cells was established in normal and tumour intestinal cells (Fre et al. 2009; Rodilla et al. 2009; Tian et al. 2015). In fact, while the proliferation effects could not be rescued without Wnt activation, differentiation to goblet cells upon Notch activation was independent of Wnt signalling.

Balancing the levels of Wnt and Notch also reflects in the control of stem cell kinetics (Hirata et al. 2013; Pin et al. 2012). High levels of Notch activity in progenitors with active β-CATENIN correlates with slow cell cycle and inhibition of Notch turns slow-dividing cells into fast-proliferating progenitors (Hirata et al. 2013). Moreover, using other combinations of Notch and Wnt composite mutants, it was further determined the collaborative function of both pathways in intestinal stem cells and their importance at the transcriptional level of stem cell genes, such as *Lgr5*, *Olfm4*, *Ascl2* and *Bmi1* (Lopez-Arribillaga et al. 2015).

In *Drosophila* intestine, the GATAe transcription factor regulates stemness in the adult midgut (Okumura et al. 2016). Similarly, GATA4 and GATA6 play an essential role in the formation of villi architecture during mouse intestinal development. Expression of the *Dll1* ligand and the Notch target *Olm4* were reduced in *Gata4/6* mutant, together with changes in the expression of markers for enterocytes and the number of goblet cells. Chromatin analysis revealed that *Dll1* gene contains GATA binding sites recognized by GATA4 in the small intestine, suggesting a direct regulation of the Notch ligand transcription (Walker et al. 2014).

Because of its basic activity in intestinal stem cell maintenance and differentiation, alteration of the Notch pathway has been described in pathological conditions. Notch activity is dysregulated in chronic inflammatory diseases, such as colitis and Crohn's syndrome, and mounting evidences associate Notch-signalling activation with colon rectal cancer (Noah and Shroyer 2013; Ahmed et al. 2012; Guilmeau 2012; Kazanjian and Shroyer 2011). Components of the Notch pathway such as *Notch1*, *Jag1*, *Jag2* and *Hes1* factors, are highly expressed in human colon adenocarcinomas and colon rectal cancer models (Reedijk et al. 2008; Guilmeau 2012). Functional implications of Notch in colorectal cancer were shown in animal models that combined strong activation of Notch with mutations in adenomatous polyposis coli (Apc), inhibitor of β-CATENIN hyperactivation (Fre et al. 2009). Mice overexpressing *Notch1* develop additional tumours and have worse survival compared to littermates that only carry the *Apc* mutation. Furthermore, mutation in *Hes1* reduces proliferation and induces differentiation in existing Apc-tumours with minor effect on the adjacent mucosa, suggesting Hes1 as a suitable target for future therapies aiming to block tumoural growth (Ueo et al. 2012).

3.4 Epidermal Stem Cells

Epidermis is one of the main barriers against mechanical stress and traumas in the mammalian body. Epidermal cells face a continuous regeneration supported by the stem cells in the basal layer. The stem cells are anchored through integrins to the basal lamina and can give rise to three terminal differentiated population: the interfollicular epidermis (IFE), hair follicles (HF) and sebaceous glands [SG, (Blanpain and Fuchs 2006)]. Signalling coming from cell-to-cell and cell-to-extracellular matrix interaction play a crucial role in preserving epidermal homeostasis, balancing between self-renewal and induction of differentiation (Fuchs 2008; Jensen et al. 1999).

Epidermal stratification is strictly regulated by two mechanisms: delamination and asymmetric division. Basal cells progressing outward into a more differentiated fate lose their attachment downregulating their level of integrins. Gain of function studies showed that the active NOTCH-RBPJ pathway downregulates basal features and allow progression of differentiation (Estrach et al. 2007; Blanpain et al. 2006). In this context, NOTCH mediates downregulation of the Rho-associated protein kinase2 (ROCK2) in keratinocytes, reducing integrin expression and promoting differentiation (Lefort et al. 2007).

Expression of all Notch receptors (*Notch1*, *Notch2*, *Notch3* and *Notch4*) have been detected in the IFE, clustering specifically in the suprabasal region, while their expression in the HF is limited to the base of the follicle where proliferation and differentiation occur (Blanpain et al. 2006; Kopan and Weintraub 1993; Nickoloff et al. 2002; Pan et al. 2004). The first suprabasal layer, the spinous layer, is formed as a consequence of asymmetric division of stem cells resident in the basal layer. Basal cells self-renew dividing asymmetrically and giving rise to a transit amplifying cell and a basal progenitor. Transit amplifying cells are responsible for the amplification of the keratinocyte pool until their migration toward the most external stratum corneum. Finally, shedding cells of the stratum corneum

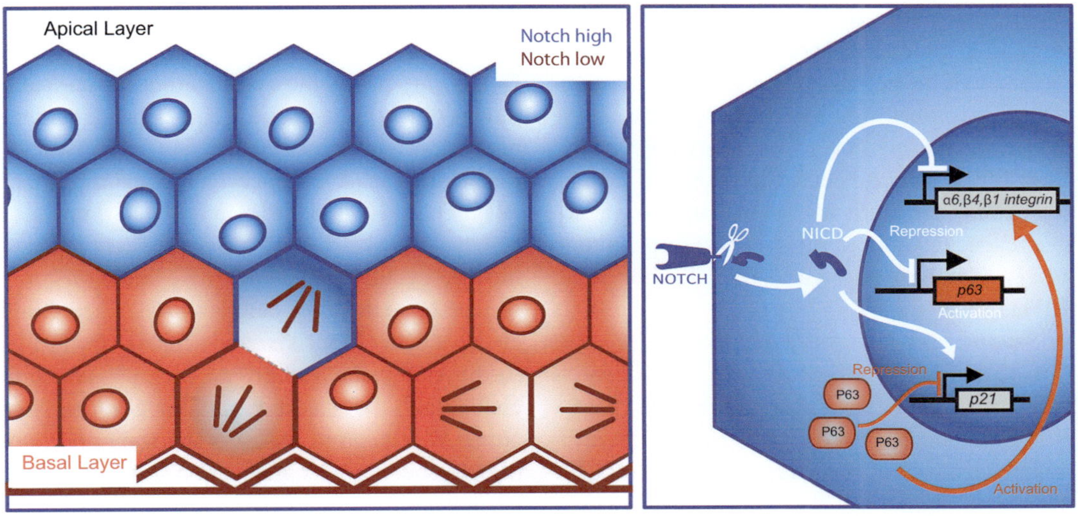

Fig. 6 Notch activity in epidermal stem cells. Notch is mainly active in the apical region of the epidermis (left panel). Molecular mechanisms involving Notch regulation of differentiation (right panel). Notch activity induces cell cycle progression through *p21* expression and blocks *p63* and integrins expression. In turn, P63 has opposite effects and is responsible to keep epidermal stem cells undifferentiated in the basal layer, blocking *p21* and inducing integrins expression

are continuously replaced by newly generated keratinocytes (Fuchs 2008). This process regulates Notch availability between daughter cells, as blockage of the machinery involved in the asymmetric cell division it also blocks Notch activation and leads to a decreased number of spinous cells formed (Moriyama et al. 2006; Williams et al. 2011).

The major Notch ligands found in epidermis are *Dll1*, *Jag1* and *Jag2* (Estrach et al. 2006; Lee et al. 2007; Watt et al. 2008). Jag ligands are complementarily expressed in the adult epidermis, with *Jag1* mainly present in the suprabasal layer and *Jag2* in basal layer beneath (Estrach et al. 2006). Notch signalling controls keratinocytes differentiation primarily via RBPJ, as shown by *in vitro* experiments. Differentiation of keratinocytes involves a Notch-regulated exit from cell cycle where RBPJ recognition of the *p21* promoter results in its induction of expression. In parallel, NOTCH1 represses *p63*, which in turn regulates *Hes1* expression in mutual antagonism to control epidermal homeostasis (Rangarajan et al. 2001; Tadeu and Horsley 2013) (Fig. 6). Additionally, HES1 is involved in the induction of the spinous layer but has no effect on basal cell features (Moriyama et al. 2006; Mammucari et al. 2005). Depletion of *Jag1* in the IFE induces thickening of the epidermal layer and abnormalities in HF maturation similarly to ablation of *Notch1* (Knoblich 2001; Estrach et al. 2006).

While the Jag ligands are the cardinal mediators for Notch activity in adult epidermis, the Dll1 ligand plays its major role during embryonic development. Deletion of *Dll1* in the embryo has effects on IFE, with increased proliferation and altered differentiation, and in embryonic HF formation, delaying the first wave of hair growth (Lowell et al. 2000; Estrach et al. 2008; Estrach et al. 2007). Similarly, deletion of *Notch1* in developing skin produces disruption of layer organization and lack of sebocytes, while deletion of *Notch 2, 3* and *4* does not produce major defects (Watt et al. 2008; Nicolas et al. 2003; Pan et al. 2004).

3.5 Muscle Stem Cells

In the embryo, muscles are formed from a population of progenitors in the dermomyotome somitic compartment. The initial population of stem/progenitor cells starts to express myogenic factors, exit the cell cycle and fuse into multinucleated myotubes. Some stem cells relocate at the side of the muscle fibres where they persist until adult life as satellite cells. Satellite cells display stem cells characteristics, such as their regenerative capacities upon injury or transplantation assays (Sambasivan et al. 2011; Lepper et al. 2011).

Satellite cells in the adult express *Notch1*, *Notch2* and *Notch3*, together with Notch ligands *Dll1* and *Jag1* (Relaix and Marcelle 2009). NOTCH1 and NOTCH2 act as potent transcriptional activators of their target genes via RBPJ-dependent signalling; by contrast, NOTCH3 is distinct from NOTCH1 and acts as a *Notch1* repressor by activating *Nrarp*, a negative regulator of Notch signalling (Kitamoto and Hanaoka 2010; Brack and Rando 2012).

In early experiments with mammalian C2C12 muscle cell lines, it was already demonstrated that forced activation of NOTCH1 inhibited their differentiation into myofibers by interfering with the master myogenic factor of differentiation myogenic differentiation antigen (*MyoD*) (Kopan et al. 1994). Since then, more physiological experiments have established an important role for Notch in the maintenance of the muscle tissue.

Notch plays a role in myogenesis starting from the initial embryonic phases. Loss of function experiments showed that ablation of the Notch effector *rbpj* or *Dll1*-ligand in dermomyotomes results in the exhaustion of the initial progenitor pool and completely inhibits muscle formation (Vasyutina et al. 2007; Schuster-Gossler et al. 2007; Czajkowski et al. 2014). Muscle formation relies on a proliferating population of progenitor cells that express the *paired homeobox transcription factors 3* and *7* (*Pax3* and *Pax7*) (Buckingham and Relaix 2007). Multipotent PAX3$^+$ cells in the somites give rise to both vasculature and skeletal muscle. The balance between those two fate choices is Notch dependent as overexpression of active Notch favours the endothelial differentiation altering the balance of expression between *Pax3* and *forkhead box C2* (*Foxc2*) (Mayeuf-Louchart et al. 2014). Additionally, constitutive activation of Notch in myogenic progenitors specifically blocks differentiation in a RBPJ-dependent manner (Mourikis et al. 2012a) but induces self-renewing of muscle stem cells upregulating *Pax7* (Wen et al. 2012). Specifically, Notch regulates cell cycle in skeletal muscle progenitors repressing the cyclin-dependent kinase inhibitors *p21* and *p57*, controlling cellular quiescence (Zalc et al. 2014; Mourikis and Tajbakhsh 2014). Additionally, satellite cells are miss positioned in conditional mutants for *rbpj*, suggesting that Notch is regulating not only the stem cell pool directly but also its interaction with the niche (Brohl et al. 2012).

Notch remains an essential controller of muscle stem cells during adult life and several factors cooperate with Notch to regulate differentiation and self-renewal (Bjornson et al. 2012; Mourikis et al. 2012b). Notch regulates fate decisions in satellite cells in dependency to the type of ligand it is exposed to. In skeletal myoblasts, interaction with DLL4 ligand but not DLL1, downregulates myogenic genes and activate the pericyte programme (Cappellari et al. 2013).

Similarly to the embryonic development, the crosstalk between Pax7 and Notch balances self-renewal and differentiation in adult satellite cells. Loss of *Pax7* dramatically reduces the number of satellite cells, which can be reverted by constitutive activation of intracellular NOTCH1. In addition, undifferentiated cells lacking *Pax7* but with active NOTCH1 are unable to upregulate *MyoD* and acquire adipogenic features (Pasut et al. 2016).

Notch activity can be regulated by levels of oxygen in the stem cell niche. Myoblasts maintained in hypoxic culture conditions downregulate *MyoD* and low oxygen favours upregulation of *Pax7* and self-renewal. Mechanistically, this is explained by hypoxia-dependent activation of the Notch signalling, which downregulates *miR-1* and *miR-206* via HES/HEY proteins consequently increasing the levels of *Pax7* (Liu et al. 2012). Additionally, establishment of a quiescent pool of satellite stem cells depends on sex hormones during juvenile life. Androgens and oestrogens activate NOTCH via induction of *Mib1* and force exit from cell cycle (Kim et al. 2016). Finally, upstream regulation of *Notch* in progenitor cells is crucial for muscle homeostasis. Skeletal muscle cells express the transcription factor *Prospero Homeobox 1(Prox1)* on differentiation. Silencing of *Prox1* induces activation of *Notch1* and consequent inhibition of myoblasts differentiation (Kivela et al. 2016). An additional upstream control of Notch in satellite cells is provided by PTEN (phosphatase and tensin homologue). Ablation of *Pten* increases RAC-alpha serine/threonine-protein kinase (AKT) phosphorylation which in turn suppresses Notch signalling and triggers premature differentiation, depleting the stem cell pool (Yue et al. 2017). Therefore the various factors regulating quiescence and differentiation of satellite cells converge in a Notch-dependent mechanism.

Notch1 receptor expressed in quiescent satellite cells of the adult muscles is downregulated only in case of injury, when stem cells exit quiescence to start regeneration (Bjornson et al. 2012; Wen et al. 2012). Regeneration depends on an adequate number of satellite cells, which proliferation is controlled by the Notch pathway (Mourikis and Tajbakhsh 2014). Upregulation of *Notch* is improved by the inhibition of the TNF-receptor-associated factor 6 (TRAF6) and results in improved regeneration of myofibers (Hindi et al. 2012).

Muscular disorders, such as the Duchenne muscular dystrophy, have a link with alterations of the Notch pathway. In the canine model of the disease, low levels of JAG1 ligand correlate with a more severe condition and functional assays in zebrafish reveal that upregulation of *Jag1* can dramatically improve the phenotype (Vieira et al. 2015). Similarly, mutation of the protein O-glucosyltransferase 1 (*Poglut1*), involved in Notch post-translational modifications, impairs muscle development, causing dystrophy and loss of satellite cells (Servian-Morilla et al. 2016).

In agreement with the role of Notch as stem-cell keeper, the levels of Notch in satellite cells is dramatically reduced in aged muscles, coherent with the exhaustion of satellite cells and reduced efficiency in muscle regeneration (Conboy et al. 2003).

In conclusion, Notch signalling is central in muscle physiology, protecting stem cells from depletion and playing a major role in cell-to-cell communication between progenitors and their surrounding microenvironment (Koch et al. 2013).

4 Conclusions

Various tissues during embryonic and adult life rely on the Notch pathway to control stem cell activity. Notch receptors are specifically present on tissues and additionally regulated by other modifiers (glycosylation or ubiquitination). Furthermore, the type and position of ligands impose another level of control to the pathway and determine the intensity of Notch activity. The same type of interaction in a different context can result in different outcomes, such as induction or inhibition of fate determination. Disrupting the Notch pathway has dramatic but different consequences in many stem cell systems, which underline the extreme necessity of specific regulation to preserve homeostasis and functionality of mammalian tissues. Further studies are still required to fully understand the role of Notch in the specification and maintenance of every tissue.

Acknowledgements We would like to thank L. Espinosa for comments and discussion on the manuscript and H.B.F. Pohl for the help with the graphics. This work was supported by Spanish Ministry of Economy and Innovation (MINECO) (SAF2013-40922-R and SAF2016-75613-R). CP is a recipient of Juan de la Cierva fellowship (FJCI2014-19870).

The authors declare no competing financial interests.

References

Ables JL, Decarolis NA, Johnson MA, Rivera PD, Gao Z, Cooper DC, Radtke F, Hsieh J, Eisch AJ (2010) Notch1 is required for maintenance of the reservoir of adult hippocampal stem cells. J Neurosci 30(31):10484–10492. https://doi.org/10.1523/JNEUROSCI.4721-09.2010

Aguirre A, Rubio ME, Gallo V (2010) Notch and EGFR pathway interaction regulates neural stem cell number and self-renewal. Nature 467(7313):323–327. https://doi.org/10.1038/nature09347

Ahmed I, Chandrakesan P, Tawfik O, Xia L, Anant S, Umar S (2012) Critical roles of Notch and Wnt/beta-catenin pathways in the regulation of hyperplasia and/or colitis in response to bacterial infection. Infect Immun 80(9):3107–3121. https://doi.org/10.1128/IAI.00236-12

Androutsellis-Theotokis A, Leker RR, Soldner F, Hoeppner DJ, Ravin R, Poser SW, Rueger MA, Bae SK, Kittappa R, McKay RD (2006) Notch signalling regulates stem cell numbers in vitro and in vivo. Nature 442(7104):823–826. https://doi.org/10.1038/nature04940

Artavanis-Tsakonas S, Delidakis C, Fehon RG (1991) The Notch locus and the cell biology of neuroblast segregation. Annu Rev Cell Biol 7:427–452. https://doi.org/10.1146/annurev.cb.07.110191.002235

Artavanis-Tsakonas S, Rand MD, Lake RJ (1999) Notch signaling: cell fate control and signal integration in development. Science 284(5415):770–776

Barker N, van Es JH, Kuipers J, Kujala P, van den Born M, Cozijnsen M, Haegebarth A, Korving J, Begthel H, Peters PJ, Clevers H (2007) Identification of stem cells in small intestine and colon by marker gene Lgr5. Nature 449(7165):1003–1007. https://doi.org/10.1038/nature06196

Basak O, Giachino C, Fiorini E, Macdonald HR, Taylor V (2012) Neurogenic subventricular zone stem/progenitor cells are Notch1-dependent in their active but not quiescent state. J Neurosci 32(16):5654–5666. https://doi.org/10.1523/JNEUROSCI.0455-12.2012

Benner EJ, Luciano D, Jo R, Abdi K, Paez-Gonzalez P, Sheng H, Warner DS, Liu C, Eroglu C, Kuo CT (2013) Protective astrogenesis from the SVZ niche after injury is controlled by Notch modulator Thbs4. Nature 497(7449):369–373. https://doi.org/10.1038/nature12069

Berdnik D, Torok T, Gonzalez-Gaitan M, Knoblich JA (2002) The endocytic protein alpha-Adaptin is required for numb-mediated asymmetric cell division in Drosophila. Dev Cell 3(2):221–231

Bertet C, Li X, Erclik T, Cavey M, Wells B, Desplan C (2014) Temporal patterning of neuroblasts controls Notch-mediated cell survival through regulation of Hid or Reaper. Cell 158(5):1173–1186. https://doi.org/10.1016/j.cell.2014.07.045

Bjornson CR, Cheung TH, Liu L, Tripathi PV, Steeper KM, Rando TA (2012) Notch signaling is necessary to maintain quiescence in adult muscle stem cells. Stem Cells 30(2):232–242. https://doi.org/10.1002/stem.773

Blanpain C, Fuchs E (2006) Epidermal stem cells of the skin. Annu Rev Cell Dev Biol 22:339–373. https://doi.org/10.1146/annurev.cellbio.22.010305.104357

Blanpain C, Lowry WE, Pasolli HA, Fuchs E (2006) Canonical notch signaling functions as a commitment switch in the epidermal lineage. Genes Dev 20(21):3022–3035. https://doi.org/10.1101/gad.1477606

Borrell V, Cardenas A, Ciceri G, Galceran J, Flames N, Pla R, Nobrega-Pereira S, Garcia-Frigola C, Peregrin S, Zhao Z, Ma L, Tessier-Lavigne M, Marin O (2012) Slit/Robo signaling modulates the proliferation of central nervous system progenitors. Neuron 76(2):338–352. https://doi.org/10.1016/j.neuron.2012.08.003

Brack AS, Rando TA (2012) Tissue-specific stem cells: lessons from the skeletal muscle satellite cell. Cell Stem Cell 10(5):504–514. https://doi.org/10.1016/j.stem.2012.04.001

Bray S (1998) A Notch affair. Cell 93(4):499–503

Bray SJ (2006) Notch signalling: a simple pathway becomes complex. Nat Rev Mol Cell Biol 7(9):678–689. https://doi.org/10.1038/nrm2009

Bray SJ (2016) Notch signalling in context. Nat Rev Mol Cell Biol 17(11):722–735. https://doi.org/10.1038/nrm.2016.94

Brohl D, Vasyutina E, Czajkowski MT, Griger J, Rassek C, Rahn HP, Purfurst B, Wende H, Birchmeier C (2012) Colonization of the satellite cell niche by skeletal muscle progenitor cells depends on Notch signals. Dev Cell 23(3):469–481. https://doi.org/10.1016/j.devcel.2012.07.014

Brooker R, Hozumi K, Lewis J (2006) Notch ligands with contrasting functions: Jagged1 and Delta1 in the mouse inner ear. Development 133(7):1277–1286. https://doi.org/10.1242/dev.02284

Bu P, Chen KY, Chen JH, Wang L, Walters J, Shin YJ, Goerger JP, Sun J, Witherspoon M, Rakhilin N, Li J, Yang H, Milsom J, Lee S, Zipfel W, Jin MM, Gumus ZH, Lipkin SM, Shen X (2013) A microRNA miR-34a-regulated bimodal switch targets Notch in colon cancer stem cells. Cell Stem Cell 12(5):602–615. https://doi.org/10.1016/j.stem.2013.03.002

Bu P, Wang L, Chen KY, Srinivasan T, Murthy PK, Tung KL, Varanko AK, Chen HJ, Ai Y, King S, Lipkin SM, Shen X (2016) A miR-34a-Numb feedforward loop triggered by inflammation regulates asymmetric stem cell division in intestine and colon cancer. Cell Stem Cell 18(2):189–202. https://doi.org/10.1016/j.stem.2016.01.006

Buckingham M, Relaix F (2007) The role of Pax genes in the development of tissues and organs: Pax3 and Pax7 regulate muscle progenitor cell functions. Annu Rev Cell Dev Biol 23:645–673. https://doi.org/10.1146/annurev.cellbio.23.090506.123438

Butko E, Pouget C, Traver D (2016) Complex regulation of HSC emergence by the Notch signaling pathway. Dev Biol 409(1):129–138. https://doi.org/10.1016/j.ydbio.2015.11.008

Butler JM, Nolan DJ, Vertes EL, Varnum-Finney B, Kobayashi H, Hooper AT, Seandel M, Shido K, White IA, Kobayashi M, Witte L, May C, Shawber C, Kimura Y, Kitajewski J, Rosenwaks Z, Bernstein ID, Rafii S (2010) Endothelial cells are essential for the self-renewal and repopulation of Notch-dependent hematopoietic stem cells. Cell Stem Cell 6(3):251–264. https://doi.org/10.1016/j.stem.2010.02.001

Calvi LM, Adams GB, Weibrecht KW, Weber JM, Olson DP, Knight MC, Martin RP, Schipani E, Divieti P, Bringhurst FR, Milner LA, Kronenberg HM, Scadden

DT (2003) Osteoblastic cells regulate the haematopoietic stem cell niche. Nature 425(6960):841–846. https://doi.org/10.1038/nature02040

Cappellari O, Benedetti S, Innocenzi A, Tedesco FS, Moreno-Fortuny A, Ugarte G, Lampugnani MG, Messina G, Cossu G (2013) Dll4 and PDGF-BB convert committed skeletal myoblasts to pericytes without erasing their myogenic memory. Dev Cell 24(6):586–599. https://doi.org/10.1016/j.devcel.2013.01.022

Carulli AJ, Keeley TM, Demitrack ES, Chung J, Maillard I, Samuelson LC (2015) Notch receptor regulation of intestinal stem cell homeostasis and crypt regeneration. Dev Biol 402(1):98–108. https://doi.org/10.1016/j.ydbio.2015.03.012

Caviness VS Jr, Nowakowski RS, Bhide PG (2009) Neocortical neurogenesis: morphogenetic gradients and beyond. Trends Neurosci 32(8):443–450. https://doi.org/10.1016/j.tins.2009.05.003

Cayouette M, Raff M (2002) Asymmetric segregation of Numb: a mechanism for neural specification from Drosophila to mammals. Nat Neurosci 5(12):1265–1269. https://doi.org/10.1038/nn1202-1265

Chambers CB, Peng Y, Nguyen H, Gaiano N, Fishell G, Nye JS (2001) Spatiotemporal selectivity of response to Notch1 signals in mammalian forebrain precursors. Development 128(5):689–702

Chenn A, McConnell SK (1995) Cleavage orientation and the asymmetric inheritance of Notch1 immunoreactivity in mammalian neurogenesis. Cell 82(4):631–641

Chi Z, Zhang J, Tokunaga A, Harraz MM, Byrne ST, Dolinko A, Xu J, Blackshaw S, Gaiano N, Dawson TM, Dawson VL (2012) Botch promotes neurogenesis by antagonizing Notch. Dev Cell 22(4):707–720. https://doi.org/10.1016/j.devcel.2012.02.011

Ciau-Uitz A, Walmsley M, Patient R (2000) Distinct origins of adult and embryonic blood in Xenopus. Cell 102(6):787–796

Clarke RL, Yzaguirre AD, Yashiro-Ohtani Y, Bondue A, Blanpain C, Pear WS, Speck NA, Keller G (2013) The expression of Sox17 identifies and regulates haemogenic endothelium. Nat Cell Biol 15(5):502–510. https://doi.org/10.1038/ncb2724

Conboy IM, Conboy MJ, Smythe GM, Rando TA (2003) Notch-mediated restoration of regenerative potential to aged muscle. Science 302(5650):1575–1577. https://doi.org/10.1126/science.1087573

Coumailleau F, Furthauer M, Knoblich JA, Gonzalez-Gaitan M (2009) Directional Delta and Notch trafficking in Sara endosomes during asymmetric cell division. Nature 458(7241):1051–1055. https://doi.org/10.1038/nature07854

Czajkowski MT, Rassek C, Lenhard DC, Brohl D, Birchmeier C (2014) Divergent and conserved roles of Dll1 signaling in development of craniofacial and trunk muscle. Dev Biol 395(2):307–316. https://doi.org/10.1016/j.ydbio.2014.09.005

Daudet N, Lewis J (2005) Two contrasting roles for Notch activity in chick inner ear development: specification of prosensory patches and lateral inhibition of hair-cell differentiation. Development 132(3):541–551. https://doi.org/10.1242/dev.01589

Davis RL, Turner DL (2001) Vertebrate hairy and Enhancer of split related proteins: transcriptional repressors regulating cellular differentiation and embryonic patterning. Oncogene 20(58):8342–8357. https://doi.org/10.1038/sj.onc.1205094

de Celis JF, Bray S (1997) Feed-back mechanisms affecting Notch activation at the dorsoventral boundary in the Drosophila wing. Development 124(17):3241–3251

de Jong JL, Zon LI (2005) Use of the zebrafish system to study primitive and definitive hematopoiesis. Annu Rev Genet 39:481–501. https://doi.org/10.1146/annurev.genet.39.073003.095931

del Alamo D, Rouault H, Schweisguth F (2011) Mechanism and significance of cis-inhibition in Notch signalling. Curr Biol 21(1):R40–R47. https://doi.org/10.1016/j.cub.2010.10.034

Del Bene F, Wehman AM, Link BA, Baier H (2008) Regulation of neurogenesis by interkinetic nuclear migration through an apical-basal notch gradient. Cell 134(6):1055–1065. https://doi.org/10.1016/j.cell.2008.07.017

Dieterlen-Lievre F, Martin C (1981) Diffuse intraembryonic hemopoiesis in normal and chimeric avian development. Dev Biol 88(1):180–191

Doetsch F, Caille I, Lim DA, Garcia-Verdugo JM, Alvarez-Buylla A (1999) Subventricular zone astrocytes are neural stem cells in the adult mammalian brain. Cell 97(6):703–716

Dong Z, Yang N, Yeo SY, Chitnis A, Guo S (2012) Intralineage directional Notch signaling regulates self-renewal and differentiation of asymmetrically dividing radial glia. Neuron 74(1):65–78. https://doi.org/10.1016/j.neuron.2012.01.031

Duarte A, Hirashima M, Benedito R, Trindade A, Diniz P, Bekman E, Costa L, Henrique D, Rossant J (2004) Dosage-sensitive requirement for mouse Dll4 in artery development. Genes Dev 18(20):2474–2478. https://doi.org/10.1101/gad.1239004

Dzierzak E, Speck NA (2008) Of lineage and legacy: the development of mammalian hematopoietic stem cells. Nat Immunol 9(2):129–136. https://doi.org/10.1038/ni1560

Ehm O, Goritz C, Covic M, Schaffner I, Schwarz TJ, Karaca E, Kempkes B, Kremmer E, Pfrieger FW, Espinosa L, Bigas A, Giachino C, Taylor V, Frisen J, Lie DC (2010) RBPJkappa-dependent signaling is essential for long-term maintenance of neural stem cells in the adult hippocampus. J Neurosci 30(41):13794–13807. https://doi.org/10.1523/JNEUROSCI.1567-10.2010

van Es JH, de Geest N, van de Born M, Clevers H, Hassan BA (2010) Intestinal stem cells lacking the Math1 tumour suppressor are refractory to Notch inhibitors. Nat Commun 1:18. https://doi.org/10.1038/ncomms1017

Espin-Palazon R, Stachura DL, Campbell CA, Garcia-Moreno D, Del Cid N, Kim AD, Candel S, Meseguer J, Mulero V, Traver D (2014) Proinflammatory signaling regulates hematopoietic stem cell emergence. Cell 159(5):1070–1085. https://doi.org/10.1016/j.cell.2014.10.031

Estrach S, Ambler CA, Lo Celso C, Hozumi K, Watt FM (2006) Jagged 1 is a beta-catenin target gene required for ectopic hair follicle formation in adult epidermis. Development 133(22):4427–4438. https://doi.org/10.1242/dev.02644

Estrach S, Legg J, Watt FM (2007) Syntenin mediates Delta1-induced cohesiveness of epidermal stem cells in culture. J Cell Sci 120(Pt 16):2944–2952. https://doi.org/10.1242/jcs.016253

Estrach S, Cordes R, Hozumi K, Gossler A, Watt FM (2008) Role of the Notch ligand Delta1 in embryonic and adult mouse epidermis. J Invest Dermatol 128(4):825–832. https://doi.org/10.1038/sj.jid.5701113

Favier B, Fliniaux I, Thelu J, Viallet JP, Demarchez M, Jahoda CA, Dhouailly D (2000) Localisation of members of the notch system and the differentiation of vibrissa hair follicles: receptors, ligands, and fringe modulators. Dev Dyn 218(3):426–437. https://doi.org/10.1002/1097-0177(200007)218:3<426::AID-DVDY1004>3.0.CO;2-4

Ferron SR, Charalambous M, Radford E, McEwen K, Wildner H, Hind E, Morante-Redolat JM, Laborda J, Guillemot F, Bauer SR, Farinas I, Ferguson-Smith AC (2011) Postnatal loss of Dlk1 imprinting in stem cells and niche astrocytes regulates neurogenesis. Nature 475(7356):381–385. https://doi.org/10.1038/nature10229

Fortini ME (2009) Notch signaling: the core pathway and its posttranslational regulation. Dev Cell 16(5):633–647. https://doi.org/10.1016/j.devcel.2009.03.010

Fre S, Huyghe M, Mourikis P, Robine S, Louvard D, Artavanis-Tsakonas S (2005) Notch signals control the fate of immature progenitor cells in the intestine. Nature 435(7044):964–968. https://doi.org/10.1038/nature03589

Fre S, Pallavi SK, Huyghe M, Lae M, Janssen KP, Robine S, Artavanis-Tsakonas S, Louvard D (2009) Notch and Wnt signals cooperatively control cell proliferation and tumorigenesis in the intestine. Proc Natl Acad Sci U S A 106(15):6309–6314. https://doi.org/10.1073/pnas.0900427106

Fre S, Hannezo E, Sale S, Huyghe M, Lafkas D, Kissel H, Louvi A, Greve J, Louvard D, Artavanis-Tsakonas S (2011) Notch lineages and activity in intestinal stem cells determined by a new set of knock-in mice. PLoS One 6(10):e25785. https://doi.org/10.1371/journal.pone.0025785

Fuchs E (2008) Skin stem cells: rising to the surface. J Cell Biol 180(2):273–284. https://doi.org/10.1083/jcb.200708185

Furthauer M, Gonzalez-Gaitan M (2009a) Endocytic regulation of notch signalling during development. Traffic 10(7):792–802. https://doi.org/10.1111/j.1600-0854.2009.00914.x

Furthauer M, Gonzalez-Gaitan M (2009b) Endocytosis, asymmetric cell division, stem cells and cancer: unus pro omnibus, omnes pro uno. Mol Oncol 3(4):339–353. https://doi.org/10.1016/j.molonc.2009.05.006

Gage FH (2000) Mammalian neural stem cells. Science 287(5457):1433–1438

Gage FH, Coates PW, Palmer TD, Kuhn HG, Fisher LJ, Suhonen JO, Peterson DA, Suhr ST, Ray J (1995) Survival and differentiation of adult neuronal progenitor cells transplanted to the adult brain. Proc Natl Acad Sci U S A 92(25):11879–11883

Gaiano N, Fishell G (2002) The role of notch in promoting glial and neural stem cell fates. Annu Rev Neurosci 25:471–490. https://doi.org/10.1146/annurev.neuro.25.030702.130823

Gaiano N, Nye JS, Fishell G (2000) Radial glial identity is promoted by Notch1 signaling in the murine forebrain. Neuron 26(2):395–404

Gama-Norton L, Ferrando E, Ruiz-Herguido C, Liu Z, Guiu J, Islam AB, Lee SU, Yan M, Guidos CJ, Lopez-Bigas N, Maeda T, Espinosa L, Kopan R, Bigas A (2015) Notch signal strength controls cell fate in the haemogenic endothelium. Nat Commun 6:8510. https://doi.org/10.1038/ncomms9510

Gekas C, Dieterlen-Lievre F, Orkin SH, Mikkola HK (2005) The placenta is a niche for hematopoietic stem cells. Dev Cell 8(3):365–375. https://doi.org/10.1016/j.devcel.2004.12.016

Gerhardt DM, Pajcini KV, D'Altri T, Tu L, Jain R, Xu L, Chen MJ, Rentschler S, Shestova O, Wertheim GB, Tobias JW, Kluk M, Wood AW, Aster JC, Gimotty PA, Epstein JA, Speck N, Bigas A, Pear WS (2014) The Notch1 transcriptional activation domain is required for development and reveals a novel role for Notch1 signaling in fetal hematopoietic stem cells. Genes Dev 28(6):576–593. https://doi.org/10.1101/gad.227496.113

Gifford GB, Demitrack ES, Keeley TM, Tam A, La Cunza N, Dedhia PH, Spence JR, Simeone DM, Saotome I, Louvi A, Siebel CW, Samuelson LC (2016) Notch1 and Notch2 receptors regulate mouse and human gastric antral epithelial cell homoeostasis. Gut. https://doi.org/10.1136/gutjnl-2015-310811

Givogri MI, de Planell M, Galbiati F, Superchi D, Gritti A, Vescovi A, de Vellis J, Bongarzone ER (2006) Notch signaling in astrocytes and neuroblasts of the adult subventricular zone in health and after cortical injury. Dev Neurosci 28(1–2):81–91. https://doi.org/10.1159/000090755

Gotz M, Huttner WB (2005) The cell biology of neurogenesis. Nat Rev Mol Cell Biol 6(10):777–788. https://doi.org/10.1038/nrm1739

Guilmeau S (2012) Notch signaling and intestinal cancer. Adv Exp Med Biol 727:272–288. https://doi.org/10.1007/978-1-4614-0899-4_21

Guiu J, Shimizu R, D'Altri T, Fraser ST, Hatakeyama J, Bresnick EH, Kageyama R, Dzierzak E, Yamamoto M, Espinosa L, Bigas A (2013) Hes repressors are essential regulators of hematopoietic stem cell development downstream of Notch signaling. J Exp Med 210(1):71–84. https://doi.org/10.1084/jem.20120993

Hadland BK, Huppert SS, Kanungo J, Xue Y, Jiang R, Gridley T, Conlon RA, Cheng AM, Kopan R, Longmore GD (2004) A requirement for Notch1 distinguishes 2 phases of definitive hematopoiesis during development. Blood 104(10):3097–3105. https://doi.org/10.1182/blood-2004-03-1224

Hadland BK, Varnum-Finney B, Poulos MG, Moon RT, Butler JM, Rafii S, Bernstein ID (2015) Endothelium and NOTCH specify and amplify aorta-gonad-mesonephros-derived hematopoietic stem cells. J Clin Invest 125(5):2032–2045. https://doi.org/10.1172/JCI80137

Hansen DV, Lui JH, Parker PR, Kriegstein AR (2010) Neurogenic radial glia in the outer subventricular zone of human neocortex. Nature 464(7288):554–561. https://doi.org/10.1038/nature08845

Haydar TF, Jr. Ang E, Rakic P (2003) Mitotic spindle rotation and mode of cell division in the developing telencephalon. Proc Natl Acad Sci U S A 100(5):2890–2895. https://doi.org/10.1073/pnas.0437969100

He Q, Zhang C, Wang L, Zhang P, Ma D, Lv J, Liu F (2015) Inflammatory signaling regulates hematopoietic stem and progenitor cell emergence in vertebrates. Blood 125(7):1098–1106. https://doi.org/10.1182/blood-2014-09-601542

Heitzler P, Simpson P (1991) The choice of cell fate in the epidermis of Drosophila. Cell 64(6):1083–1092

Hindi SM, Paul PK, Dahiya S, Mishra V, Bhatnagar S, Kuang S, Choi Y, Kumar A (2012) Reciprocal interaction between TRAF6 and notch signaling regulates adult myofiber regeneration upon injury. Mol Cell Biol 32(23):4833–4845. https://doi.org/10.1128/MCB.00717-12

Hirata A, Utikal J, Yamashita S, Aoki H, Watanabe A, Yamamoto T, Okano H, Bardeesy N, Kunisada T, Ushijima T, Hara A, Jaenisch R, Hochedlinger K, Yamada Y (2013) Dose-dependent roles for canonical Wnt signalling in de novo crypt formation and cell cycle properties of the colonic epithelium. Development 140(1):66–75. https://doi.org/10.1242/dev.084103

Hoeck JD, Jandke A, Blake SM, Nye E, Spencer-Dene B, Brandner S, Behrens A (2010) Fbw7 controls neural stem cell differentiation and progenitor apoptosis via Notch and c-Jun. Nat Neurosci 13(11):1365–1372. https://doi.org/10.1038/nn.2644

Hori K, Sen A, Artavanis-Tsakonas S (2013) Notch signaling at a glance. J Cell Sci 126(Pt 10):2135–2140. https://doi.org/10.1242/jcs.127308

Huber TL, Kouskoff V, Fehling HJ, Palis J, Keller G (2004) Haemangioblast commitment is initiated in the primitive streak of the mouse embryo. Nature 432(7017):625–630. https://doi.org/10.1038/nature03122

Huppert SS, Ilagan MX, De Strooper B, Kopan R (2005) Analysis of Notch function in presomitic mesoderm suggests a gamma-secretase-independent role for presenilins in somite differentiation. Dev Cell 8(5):677–688. https://doi.org/10.1016/j.devcel.2005.02.019

Hutterer A, Knoblich JA (2005) Numb and alpha-Adaptin regulate Sanpodo endocytosis to specify cell fate in Drosophila external sensory organs. EMBO Rep 6(9):836–842. https://doi.org/10.1038/sj.embor.7400500

Imayoshi I, Isomura A, Harima Y, Kawaguchi K, Kori H, Miyachi H, Fujiwara T, Ishidate F, Kageyama R (2013) Oscillatory control of factors determining multipotency and fate in mouse neural progenitors. Science 342(6163):1203–1208. https://doi.org/10.1126/science.1242366

Inman KE, Downs KM (2007) The murine allantois: emerging paradigms in development of the mammalian umbilical cord and its relation to the fetus. Genesis 45(5):237–258. https://doi.org/10.1002/dvg.20281

Irvin DK, Nakano I, Paucar A, Kornblum HI (2004) Patterns of Jagged1, Jagged2, Delta-like 1 and Delta-like 3 expression during late embryonic and postnatal brain development suggest multiple functional roles in progenitors and differentiated cells. J Neurosci Res 75(3):330–343. https://doi.org/10.1002/jnr.10843

Ishibashi M, Moriyoshi K, Sasai Y, Shiota K, Nakanishi S, Kageyama R (1994) Persistent expression of helix-loop-helix factor HES-1 prevents mammalian neural differentiation in the central nervous system. EMBO J 13(8):1799–1805

Ishitani T, Hirao T, Suzuki M, Isoda M, Ishitani S, Harigaya K, Kitagawa M, Matsumoto K, Itoh M (2010) Nemo-like kinase suppresses Notch signalling by interfering with formation of the Notch active transcriptional complex. Nat Cell Biol 12(3):278–285. https://doi.org/10.1038/ncb2028

Iso T, Kedes L, Hamamori Y (2003) HES and HERP families: multiple effectors of the Notch signaling pathway. J Cell Physiol 194(3):237–255. https://doi.org/10.1002/jcp.10208

Ivanovs A, Rybtsov S, Anderson RA, Turner ML, Medvinsky A (2014) Identification of the niche and phenotype of the first human hematopoietic stem cells. Stem Cell Reports 2(4):449–456. https://doi.org/10.1016/j.stemcr.2014.02.004

Jehn BM, Bielke W, Pear WS, Osborne BA (1999) Cutting edge: protective effects of notch-1 on TCR-induced apoptosis. J Immunol 162(2):635–638

Jensen UB, Lowell S, Watt FM (1999) The spatial relationship between stem cells and their progeny in the basal layer of human epidermis: a new view based on whole-mount labelling and lineage analysis. Development 126(11):2409–2418

Kageyama R, Niwa Y, Isomura A, Gonzalez A, Harima Y (2012) Oscillatory gene expression and somitogenesis. Wiley Interdiscip Rev Dev Biol 1(5):629–641. https://doi.org/10.1002/wdev.46

Kanz D, Konantz M, Alghisi E, North TE, Lengerke C (2016) Endothelial-to-hematopoietic transition: Notch-ing vessels into blood. Ann N Y Acad Sci 1370(1):97–108. https://doi.org/10.1111/nyas.13030

Karanu FN, Murdoch B, Gallacher L, Wu DM, Koremoto M, Sakano S, Bhatia M (2000) The notch ligand jagged-1 represents a novel growth factor of human hematopoietic stem cells. J Exp Med 192(9):1365–1372

Kazanjian A, Shroyer NF (2011) NOTCH Signaling and ATOH1 in Colorectal Cancers. Curr Colorectal Cancer Rep 7(2):121–127. https://doi.org/10.1007/s11888-011-0090-5

Kim J, Sebring A, Esch JJ, Kraus ME, Vorwerk K, Magee J, Carroll SB (1996) Integration of positional signals and regulation of wing formation and identity by Drosophila vestigial gene. Nature 382(6587):133–138. https://doi.org/10.1038/382133a0

Kim TH, Li F, Ferreiro-Neira I, Ho LL, Luyten A, Nalapareddy K, Long H, Verzi M, Shivdasani RA (2014) Broadly permissive intestinal chromatin underlies lateral inhibition and cell plasticity. Nature 506(7489):511–515. https://doi.org/10.1038/nature12903

Kim JH, Han GC, Seo JY, Park I, Park W, Jeong HW, Lee SH, Bae SH, Seong J, Yum MK, Hann SH, Kwon YG, Seo D, Choi MH, Kong YY (2016) Sex hormones establish a reserve pool of adult muscle stem cells. Nat Cell Biol 18(9):930–940. https://doi.org/10.1038/ncb3401

Kitamoto T, Hanaoka K (2010) Notch3 null mutation in mice causes muscle hyperplasia by repetitive muscle regeneration. Stem Cells 28(12):2205–2216. https://doi.org/10.1002/stem.547

Kivela R, Salmela I, Nguyen YH, Petrova TV, Koistinen HA, Wiener Z, Alitalo K (2016) The transcription factor Prox1 is essential for satellite cell differentiation and muscle fibre-type regulation. Nat Commun 7:13124. https://doi.org/10.1038/ncomms13124

Knoblich JA (2001) Asymmetric cell division during animal development. Nat Rev Mol Cell Biol 2(1):11–20. https://doi.org/10.1038/35048085

Knoblich JA (2008) Mechanisms of asymmetric stem cell division. Cell 132(4):583–597. https://doi.org/10.1016/j.cell.2008.02.007

Knoblich JA, Jan LY, Jan YN (1995) Asymmetric segregation of Numb and Prospero during cell division. Nature 377(6550):624–627. https://doi.org/10.1038/377624a0

Kobayashi T, Mizuno H, Imayoshi I, Furusawa C, Shirahige K, Kageyama R (2009) The cyclic gene Hes1 contributes to diverse differentiation responses of embryonic stem cells. Genes Dev 23(16):1870–1875. https://doi.org/10.1101/gad.1823109

Koch U, Lehal R, Radtke F (2013) Stem cells living with a Notch. Development 140(4):689–704. https://doi.org/10.1242/dev.080614

Kong JH, Yang L, Dessaud E, Chuang K, Moore DM, Rohatgi R, Briscoe J, Novitch BG (2015) Notch activity modulates the responsiveness of neural progenitors to sonic hedgehog signaling. Dev Cell 33(4):373–387. https://doi.org/10.1016/j.devcel.2015.03.005

Kopan R, Weintraub H (1993) Mouse notch: expression in hair follicles correlates with cell fate determination. J Cell Biol 121(3):631–641

Kopan R, Nye JS, Weintraub H (1994) The intracellular domain of mouse Notch: a constitutively activated repressor of myogenesis directed at the basic helix-loop-helix region of MyoD. Development 120(9):2385–2396

Kosodo Y, Roper K, Haubensak W, Marzesco AM, Corbeil D, Huttner WB (2004) Asymmetric distribution of the apical plasma membrane during neurogenic divisions of mammalian neuroepithelial cells. EMBO J 23(11):2314–2324. https://doi.org/10.1038/sj.emboj.7600223

Krebs LT, Xue Y, Norton CR, Shutter JR, Maguire M, Sundberg JP, Gallahan D, Closson V, Kitajewski J, Callahan R, Smith GH, Stark KL, Gridley T (2000) Notch signaling is essential for vascular morphogenesis in mice. Genes Dev 14(11):1343–1352

Krebs LT, Shutter JR, Tanigaki K, Honjo T, Stark KL, Gridley T (2004) Haploinsufficient lethality and formation of arteriovenous malformations in Notch pathway mutants. Genes Dev 18(20):2469–2473. https://doi.org/10.1101/gad.1239204

Kressmann S, Campos C, Castanon I, Furthauer M, Gonzalez-Gaitan M (2015) Directional Notch trafficking in Sara endosomes during asymmetric cell division in the spinal cord. Nat Cell Biol 17(3):333–339. https://doi.org/10.1038/ncb3119

Kumano K, Chiba S, Kunisato A, Sata M, Saito T, Nakagami-Yamaguchi E, Yamaguchi T, Masuda S, Shimizu K, Takahashi T, Ogawa S, Hamada Y, Hirai H (2003) Notch1 but not Notch2 is essential for generating hematopoietic stem cells from endothelial cells. Immunity 18(5):699–711

Lai EC (2004) Notch signaling: control of cell communication and cell fate. Development 131(5):965–973. https://doi.org/10.1242/dev.01074

Lee J, Basak JM, Demehri S, Kopan R (2007) Bi-compartmental communication contributes to the opposite proliferative behavior of Notch1-deficient hair follicle and epidermal keratinocytes. Development 134(15):2795–2806. https://doi.org/10.1242/dev.02868

Lee CY, Vogeli KM, Kim SH, Chong SW, Jiang YJ, Stainier DY, Jin SW (2009) Notch signaling functions as a cell-fate switch between the endothelial and hematopoietic lineages. Curr Biol 19(19):1616–1622. https://doi.org/10.1016/j.cub.2009.07.073

Lefort K, Mandinova A, Ostano P, Kolev V, Calpini V, Kolfschoten I, Devgan V, Lieb J, Raffoul W, Hohl D, Neel V, Garlick J, Chiorino G, Dotto GP (2007) Notch1 is a p53 target gene involved in human keratinocyte tumor suppression through negative regulation of ROCK1/2 and MRCKalpha kinases. Genes Dev 21(5):562–577. https://doi.org/10.1101/gad.1484707

Lepper C, Partridge TA, Fan CM (2011) An absolute requirement for Pax7-positive satellite cells in acute injury-induced skeletal muscle regeneration. Development 138(17):3639–3646. https://doi.org/10.1242/dev.067595

Lewis J (1998) Notch signalling and the control of cell fate choices in vertebrates. Semin Cell Dev Biol 9(6):583–589. https://doi.org/10.1006/scdb.1998.0266

Li R (2013) The art of choreographing asymmetric cell division. Dev Cell 25(5):439–450. https://doi.org/10.1016/j.devcel.2013.05.003

Li HJ, Kapoor A, Giel-Moloney M, Rindi G, Leiter AB (2012) Notch signaling differentially regulates the cell fate of early endocrine precursor cells and their maturing descendants in the mouse pancreas and intestine. Dev Biol 371(2):156–169. https://doi.org/10.1016/j.ydbio.2012.08.023

Liu W, Wen Y, Bi P, Lai X, Liu XS, Liu X, Kuang S (2012) Hypoxia promotes satellite cell self-renewal and enhances the efficiency of myoblast transplantation. Development 139(16):2857–2865. https://doi.org/10.1242/dev.079665

Lizama CO, Hawkins JS, Schmitt CE, Bos FL, Zape JP, Cautivo KM, Borges Pinto H, Rhyner AM, Yu H, Donohoe ME, Wythe JD, Zovein AC (2015) Repression of arterial genes in hemogenic endothelium is sufficient for haematopoietic fate acquisition. Nat Commun 6:7739. https://doi.org/10.1038/ncomms8739

Lobry C, Oh P, Mansour MR, Look AT, Aifantis I (2014) Notch signaling: switching an oncogene to a tumor suppressor. Blood 123(16):2451–2459. https://doi.org/10.1182/blood-2013-08-355818

Lopez-Arribillaga E, Rodilla V, Pellegrinet L, Guiu J, Iglesias M, Roman AC, Gutarra S, Gonzalez S, Munoz-Canoves P, Fernandez-Salguero P, Radtke F, Bigas A, Espinosa L (2015) Bmi1 regulates murine intestinal stem cell proliferation and self-renewal downstream of Notch. Development 142(1):41–50. https://doi.org/10.1242/dev.107714

Lopez-Garcia C, Klein AM, Simons BD, Winton DJ (2010) Intestinal stem cell replacement follows a pattern of neutral drift. Science 330(6005):822–825. https://doi.org/10.1126/science.1196236

Louvi A, Artavanis-Tsakonas S (2006) Notch signalling in vertebrate neural development. Nat Rev Neurosci 7(2):93–102. https://doi.org/10.1038/nrn1847

Lowell S, Jones P, Le Roux I, Dunne J, Watt FM (2000) Stimulation of human epidermal differentiation by delta-notch signalling at the boundaries of stem-cell clusters. Curr Biol 10(9):491–500

Lowell S, Benchoua A, Heavey B, Smith AG (2006) Notch promotes neural lineage entry by pluripotent embryonic stem cells. PLoS Biol 4(5):e121. https://doi.org/10.1371/journal.pbio.0040121

Lugert S, Basak O, Knuckles P, Haussler U, Fabel K, Gotz M, Haas CA, Kempermann G, Taylor V, Giachino C (2010) Quiescent and active hippocampal neural stem cells with distinct morphologies respond selectively to physiological and pathological stimuli and aging. Cell Stem Cell 6(5):445–456. https://doi.org/10.1016/j.stem.2010.03.017

Lutolf S, Radtke F, Aguet M, Suter U, Taylor V (2002) Notch1 is required for neuronal and glial differentiation in the cerebellum. Development 129(2):373–385

Magnusson JP, Goritz C, Tatarishvili J, Dias DO, Smith EM, Lindvall O, Kokaia Z, Frisen J (2014) A latent neurogenic program in astrocytes regulated by Notch signaling in the mouse. Science 346(6206):237–241. https://doi.org/10.1126/science.346.6206.237

Maillard I, Koch U, Dumortier A, Shestova O, Xu L, Sai H, Pross SE, Aster JC, Bhandoola A, Radtke F, Pear WS (2008) Canonical notch signaling is dispensable for the maintenance of adult hematopoietic stem cells. Cell Stem Cell 2(4):356–366. https://doi.org/10.1016/j.stem.2008.02.011

Mammucari C, Tommasi di Vignano A, Sharov AA, Neilson J, Havrda MC, Roop DR, Botchkarev VA, Crabtree GR, Dotto GP (2005) Integration of Notch 1 and calcineurin/NFAT signaling pathways in keratinocyte growth and differentiation control. Dev Cell 8(5):665–676. https://doi.org/10.1016/j.devcel.2005.02.016

Matsuda S, Kuwako K, Okano HJ, Tsutsumi S, Aburatani H, Saga Y, Matsuzaki Y, Akaike A, Sugimoto H, Okano H (2012) Sox21 promotes hippocampal adult neurogenesis via the transcriptional repression of the Hes5 gene. J Neurosci 32(36):12543–12557. https://doi.org/10.1523/JNEUROSCI.5803-11.2012

Mayeuf-Louchart A, Lagha M, Danckaert A, Rocancourt D, Relaix F, Vincent SD, Buckingham M (2014) Notch regulation of myogenic versus endothelial fates of cells that migrate from the somite to the limb. Proc Natl Acad Sci U S A 111(24):8844–8849. https://doi.org/10.1073/pnas.1407606111

Medvinsky AL, Samoylina NL, Muller AM, Dzierzak EA (1993) An early pre-liver intra-embryonic source of CFU-S in the developing mouse. Nature 364(6432):64–67. https://doi.org/10.1038/364064a0

Medvinsky A, Rybtsov S, Taoudi S (2011) Embryonic origin of the adult hematopoietic system: advances and questions. Development 138(6):1017–1031. https://doi.org/10.1242/dev.040998

Mizutani K, Yoon K, Dang L, Tokunaga A, Gaiano N (2007) Differential Notch signalling distinguishes neural stem cells from intermediate progenitors. Nature 449(7160):351–355. https://doi.org/10.1038/nature06090

Moriyama M, Osawa M, Mak SS, Ohtsuka T, Yamamoto N, Han H, Delmas V, Kageyama R, Beermann F, Larue L, Nishikawa S (2006) Notch signaling via Hes1 transcription factor maintains survival of melanoblasts and melanocyte stem cells. J Cell Biol 173(3):333–339. https://doi.org/10.1083/jcb.200509084

Morrison SJ, Hemmati HD, Wandycz AM, Weissman IL (1995) The purification and characterization of fetal liver hematopoietic stem cells. Proc Natl Acad Sci U S A 92(22):10302–10306

Mourikis P, Tajbakhsh S (2014) Distinct contextual roles for Notch signalling in skeletal muscle stem cells. BMC Dev Biol 14:2. https://doi.org/10.1186/1471-213X-14-2

Mourikis P, Gopalakrishnan S, Sambasivan R, Tajbakhsh S (2012a) Cell-autonomous Notch activity maintains the temporal specification potential of skeletal muscle stem cells. Development 139(24):4536–4548. https://doi.org/10.1242/dev.084756

Mourikis P, Sambasivan R, Castel D, Rocheteau P, Bizzarro V, Tajbakhsh S (2012b) A critical requirement for notch signaling in maintenance of the quiescent skeletal muscle stem cell state. Stem Cells 30(2):243–252. https://doi.org/10.1002/stem.775

Nakagawa M, Ichikawa M, Kumano K, Goyama S, Kawazu M, Asai T, Ogawa S, Kurokawa M, Chiba S (2006) AML1/Runx1 rescues Notch1-null mutation-induced deficiency of para-aortic splanchnopleural hematopoiesis. Blood 108(10):3329–3334. https://doi.org/10.1182/blood-2006-04-019570

Nemir M, Croquelois A, Pedrazzini T, Radtke F (2006) Induction of cardiogenesis in embryonic stem cells via downregulation of Notch1 signaling. Circ Res 98(12):1471–1478. https://doi.org/10.1161/01.RES.0000226497.52052.2a

Neumuller RA, Knoblich JA (2009) Dividing cellular asymmetry: asymmetric cell division and its implications for stem cells and cancer. Genes Dev 23(23):2675–2699. https://doi.org/10.1101/gad.1850809

Nickoloff BJ, Qin JZ, Chaturvedi V, Denning MF, Bonish B, Miele L (2002) Jagged-1 mediated activation of notch signaling induces complete maturation of human keratinocytes through NF-kappaB and PPARgamma. Cell Death Differ 9(8):842–855. https://doi.org/10.1038/sj.cdd.4401036

Nicolas M, Wolfer A, Raj K, Kummer JA, Mill P, van Noort M, Hui CC, Clevers H, Dotto GP, Radtke F (2003) Notch1 functions as a tumor suppressor in mouse skin. Nat Genet 33(3):416–421. https://doi.org/10.1038/ng1099

Noah TK, Shroyer NF (2013) Notch in the intestine: regulation of homeostasis and pathogenesis. Annu Rev Physiol 75:263–288. https://doi.org/10.1146/annurev-physiol-030212-183741

Noggle SA, Weiler D, Condie BG (2006) Notch signaling is inactive but inducible in human embryonic stem cells. Stem Cells 24(7):1646–1653. https://doi.org/10.1634/stemcells.2005-0314

Nyfeler Y, Kirch RD, Mantei N, Leone DP, Radtke F, Suter U, Taylor V (2005) Jagged1 signals in the postnatal subventricular zone are required for neural stem cell self-renewal. EMBO J 24(19):3504–3515. https://doi.org/10.1038/sj.emboj.7600816

Oh P, Lobry C, Gao J, Tikhonova A, Loizou E, Manent J, van Handel B, Ibrahim S, Greve J, Mikkola H, Artavanis-Tsakonas S, Aifantis I (2013) In vivo mapping of notch pathway activity in normal and stress hematopoiesis. Cell Stem Cell 13(2):190–204. https://doi.org/10.1016/j.stem.2013.05.015

Ohishi K, Varnum-Finney B, Bernstein ID (2002) Delta-1 enhances marrow and thymus repopulating ability of human CD34(+)CD38(-) cord blood cells. J Clin Invest 110(8):1165–1174. https://doi.org/10.1172/JCI16167

Ohtsuka T, Ishibashi M, Gradwohl G, Nakanishi S, Guillemot F, Kageyama R (1999) Hes1 and Hes5 as notch effectors in mammalian neuronal differentiation. EMBO J 18(8):2196–2207. https://doi.org/10.1093/emboj/18.8.2196

Oka C, Nakano T, Wakeham A, de la Pompa JL, Mori C, Sakai T, Okazaki S, Kawaichi M, Shiota K, Mak TW, Honjo T (1995) Disruption of the mouse RBP-J kappa gene results in early embryonic death. Development 121(10):3291–3301

Okumura T, Takeda K, Kuchiki M, Akaishi M, Taniguchi K, Adachi-Yamada T (2016) GATAe regulates intestinal stem cell maintenance and differentiation in Drosophila adult midgut. Dev Biol 410(1):24–35. https://doi.org/10.1016/j.ydbio.2015.12.017

Osborne BA, Minter LM (2007) Notch signalling during peripheral T-cell activation and differentiation. Nat Rev Immunol 7(1):64–75. https://doi.org/10.1038/nri1998

Ottersbach K, Dzierzak E (2005) The murine placenta contains hematopoietic stem cells within the vascular labyrinth region. Dev Cell 8(3):377–387. https://doi.org/10.1016/j.devcel.2005.02.001

Owen AM, Herrod NJ, Menon DK, Clark JC, Downey SP, Carpenter TA, Minhas PS, Turkheimer FE, Williams EJ, Robbins TW, Sahakian BJ, Petrides M, Pickard JD (1999) Redefining the functional organization of working memory processes within human lateral prefrontal cortex. Eur J Neurosci 11(2):567–574

Owen MR, Sherratt JA, Wearing HJ (2000) Lateral induction by juxtacrine signaling is a new mechanism for pattern formation. Dev Biol 217(1):54–61. https://doi.org/10.1006/dbio.1999.9536

Palis J, Robertson S, Kennedy M, Wall C, Keller G (1999) Development of erythroid and myeloid progenitors in the yolk sac and embryo proper of the mouse. Development 126(22):5073–5084

Pan Y, Lin MH, Tian X, Cheng HT, Gridley T, Shen J, Kopan R (2004) gamma-secretase functions through Notch signaling to maintain skin appendages but is not required for their patterning or initial morphogenesis. Dev Cell 7(5):731–743. https://doi.org/10.1016/j.devcel.2004.09.014

Park S, Lee C, Sabharwal P, Zhang M, Meyers CL, Sockanathan S (2013) GDE2 promotes neurogenesis by glycosylphosphatidylinositol-anchor cleavage of RECK. Science 339(6117):324–328. https://doi.org/10.1126/science.1231921

Pasut A, Chang NC, Rodriguez UG, Faulkes S, Yin H, Lacaria M, Ming H, Rudnicki MA (2016) Notch Signaling Rescues Loss of Satellite Cells Lacking Pax7 and Promotes Brown Adipogenic Differentiation. Cell Rep 16(2):333–343. https://doi.org/10.1016/j.celrep.2016.06.001

Pellegrinet L, Rodilla V, Liu Z, Chen S, Koch U, Espinosa L, Kaestner KH, Kopan R, Lewis J, Radtke F (2011) Dll1- and dll4-mediated notch signaling are required for homeostasis of intestinal stem cells. Gastroenterology 140(4):1230–1240, e1231–1237. doi:https://doi.org/10.1053/j.gastro.2011.01.005

Petrovic J, Formosa-Jordan P, Luna-Escalante JC, Abello G, Ibanes M, Neves J, Giraldez F (2014) Ligand-dependent Notch signaling strength orchestrates lateral induction and lateral inhibition in the developing inner ear. Development 141(11):2313–2324. https://doi.org/10.1242/dev.108100

Pierfelice T, Alberi L, Gaiano N (2011) Notch in the vertebrate nervous system: an old dog with new tricks. Neuron 69(5):840–855. https://doi.org/10.1016/j.neuron.2011.02.031

Pin C, Watson AJ, Carding SR (2012) Modelling the spatio-temporal cell dynamics reveals novel insights on cell differentiation and proliferation in the small intestinal crypt. PLoS One 7(5):e37115. https://doi.org/10.1371/journal.pone.0037115

Pinto D, Gregorieff A, Begthel H, Clevers H (2003) Canonical Wnt signals are essential for homeostasis of the intestinal epithelium. Genes Dev 17(14):1709–1713. https://doi.org/10.1101/gad.267103

Poulos MG, Guo P, Kofler NM, Pinho S, Gutkin MC, Tikhonova A, Aifantis I, Frenette PS, Kitajewski J, Rafii S, Butler JM (2013) Endothelial Jagged-1 is necessary for homeostatic and regenerative hematopoiesis. Cell Rep 4(5):1022–1034. https://doi.org/10.1016/j.celrep.2013.07.048

Radtke F, Wilson A, Stark G, Bauer M, van Meerwijk J, MacDonald HR, Aguet M (1999) Deficient T cell fate specification in mice with an induced inactivation of Notch1. Immunity 10(5):547–558

Radtke F, Wilson A, Mancini SJ, MacDonald HR (2004) Notch regulation of lymphocyte development and function. Nat Immunol 5(3):247–253. https://doi.org/10.1038/ni1045

Rangarajan A, Talora C, Okuyama R, Nicolas M, Mammucari C, Oh H, Aster JC, Krishna S, Metzger D, Chambon P, Miele L, Aguet M, Radtke F, Dotto GP (2001) Notch signaling is a direct determinant of keratinocyte growth arrest and entry into differentiation. EMBO J 20(13):3427–3436. https://doi.org/10.1093/emboj/20.13.3427

Reedijk M, Odorcic S, Zhang H, Chetty R, Tennert C, Dickson BC, Lockwood G, Gallinger S, Egan SE (2008) Activation of Notch signaling in human colon adenocarcinoma. Int J Oncol 33(6):1223–1229

Relaix F, Marcelle C (2009) Muscle stem cells. Curr Opin Cell Biol 21(6):748–753. https://doi.org/10.1016/j.ceb.2009.10.002

Rhyu MS, Jan LY, Jan YN (1994) Asymmetric distribution of numb protein during division of the sensory organ precursor cell confers distinct fates to daughter cells. Cell 76(3):477–491

Riccio O, van Gijn ME, Bezdek AC, Pellegrinet L, van Es JH, Zimber-Strobl U, Strobl LJ, Honjo T, Clevers H, Radtke F (2008) Loss of intestinal crypt progenitor cells owing to inactivation of both Notch1 and Notch2 is accompanied by derepression of CDK inhibitors p27Kip1 and p57Kip2. EMBO Rep 9(4):377–383. https://doi.org/10.1038/embor.2008.7

Richard C, Drevon C, Canto PY, Villain G, Bollerot K, Lempereur A, Teillet MA, Vincent C, Rossello Castillo C, Torres M, Piwarzyk E, Speck NA, Souyri M, Jaffredo T (2013) Endothelio-mesenchymal interaction controls runx1 expression and modulates the notch pathway to initiate aortic hematopoiesis. Dev Cell 24(6):600–611. https://doi.org/10.1016/j.devcel.2013.02.011

Robert-Moreno A, Espinosa L, de la Pompa JL, Bigas A (2005) RBPjkappa-dependent Notch function regulates Gata2 and is essential for the formation of intra-embryonic hematopoietic cells. Development 132(5):1117–1126. https://doi.org/10.1242/dev.01660

Robert-Moreno A, Espinosa L, Sanchez MJ, de la Pompa JL, Bigas A (2007) The notch pathway positively regulates programmed cell death during erythroid differentiation. Leukemia 21(7):1496–1503. https://doi.org/10.1038/sj.leu.2404705

Robert-Moreno A, Guiu J, Ruiz-Herguido C, Lopez ME, Ingles-Esteve J, Riera L, Tipping A, Enver T, Dzierzak E, Gridley T, Espinosa L, Bigas A (2008) Impaired embryonic haematopoiesis yet normal arterial development in the absence of the Notch ligand Jagged1. EMBO J 27(13):1886–1895. https://doi.org/10.1038/emboj.2008.113

Rodilla V, Villanueva A, Obrador-Hevia A, Robert-Moreno A, Fernandez-Majada V, Grilli A, Lopez-Bigas N, Bellora N, Alba MM, Torres F, Dunach M, Sanjuan X, Gonzalez S, Gridley T, Capella G, Bigas A, Espinosa L (2009) Jagged1 is the pathological link between Wnt and Notch pathways in colorectal cancer. Proc Natl Acad Sci U S A 106(15):6315–6320. https://doi.org/10.1073/pnas.0813221106

Saito T, Chiba S, Ichikawa M, Kunisato A, Asai T, Shimizu K, Yamaguchi T, Yamamoto G, Seo S, Kumano K, Nakagami-Yamaguchi E, Hamada Y, Aizawa S, Hirai H (2003) Notch2 is preferentially expressed in mature B cells and indispensable for marginal zone B lineage development. Immunity 18(5):675–685

Sakamoto M, Hirata H, Ohtsuka T, Bessho Y, Kageyama R (2003) The basic helix-loop-helix genes Hesr1/Hey1 and Hesr2/Hey2 regulate maintenance of neural precursor cells in the brain. J Biol Chem 278(45):44808–44815. https://doi.org/10.1074/jbc.M300448200

Sambasivan R, Yao R, Kissenpfennig A, Van Wittenberghe L, Paldi A, Gayraud-Morel B, Guenou H, Malissen B, Tajbakhsh S, Galy A (2011) Pax7-expressing satellite cells are indispensable for adult skeletal muscle regeneration. Development 138(17):3647–3656. https://doi.org/10.1242/dev.067587

Sancho R, Blake SM, Tendeng C, Clurman BE, Lewis J, Behrens A (2013) Fbw7 repression by hes5 creates a feedback loop that modulates Notch-mediated intestinal and neural stem cell fate decisions. PLoS Biol 11(6):e1001586. https://doi.org/10.1371/journal.pbio.1001586

Sander GR, Powell BC (2004) Expression of notch receptors and ligands in the adult gut. J Histochem Cytochem 52(4):509–516. https://doi.org/10.1177/002215540405200409

Sasaki N, Sachs N, Wiebrands K, Ellenbroek SI, Fumagalli A, Lyubimova A, Begthel H, van den Born M, van Es JH, Karthaus WR, Li VS, Lopez-Iglesias C, Peters PJ, van Rheenen J, van Oudenaarden A, Clevers H (2016) Reg4+ deep crypt secretory cells function as epithelial niche for Lgr5+ stem cells in colon. Proc Natl Acad Sci U S A 113(37):E5399–E5407. https://doi.org/10.1073/pnas.1607327113

Sato T, van Es JH, Snippert HJ, Stange DE, Vries RG, van den Born M, Barker N, Shroyer NF, van de Wetering M, Clevers H (2011) Paneth cells constitute the niche for Lgr5 stem cells in intestinal crypts. Nature 469(7330):415–418. https://doi.org/10.1038/nature09637

Schroeder T, Kohlhof H, Rieber N, Just U (2003) Notch signaling induces multilineage myeloid differentiation and up-regulates PU.1 expression. J Immunol 170(11):5538–5548

Schroeder T, Meier-Stiegen F, Schwanbeck R, Eilken H, Nishikawa S, Hasler R, Schreiber S, Bornkamm GW, Just U (2006) Activated Notch1 alters differentiation of embryonic stem cells into mesodermal cell lineages at multiple stages of development. Mech Dev 123(7):570–579. https://doi.org/10.1016/j.mod.2006.05.002

Schuster-Gossler K, Cordes R, Gossler A (2007) Premature myogenic differentiation and depletion of progenitor cells cause severe muscle hypotrophy in Delta1 mutants. Proc Natl Acad Sci U S A 104(2):537–542. https://doi.org/10.1073/pnas.0608281104

Servian-Morilla E, Takeuchi H, Lee TV, Clarimon J, Mavillard F, Area-Gomez E, Rivas E, Nieto-Gonzalez JL, Rivero MC, Cabrera-Serrano M, Gomez-Sanchez L, Martinez-Lopez JA, Estrada B, Marquez C, Morgado Y, Suarez-Calvet X, Pita G, Bigot A, Gallardo E, Fernandez-Chacon R, Hirano M, Haltiwanger RS, Jafar-Nejad H, Paradas C (2016) A POGLUT1 mutation causes a muscular dystrophy with reduced Notch signaling and satellite cell loss. EMBO Mol Med 8(11):1289–1309. https://doi.org/10.15252/emmm.201505815

Shimojo H, Ohtsuka T, Kageyama R (2008) Oscillations in notch signaling regulate maintenance of neural progenitors. Neuron 58(1):52–64. https://doi.org/10.1016/j.neuron.2008.02.014

Shimojo H, Isomura A, Ohtsuka T, Kori H, Miyachi H, Kageyama R (2016) Oscillatory control of Delta-like1 in cell interactions regulates dynamic gene expression and tissue morphogenesis. Genes Dev 30(1):102–116. https://doi.org/10.1101/gad.270785.115

Shroyer NF, Helmrath MA, Wang VY, Antalffy B, Henning SJ, Zoghbi HY (2007) Intestine-specific ablation of mouse atonal homolog 1 (Math1) reveals a role in cellular homeostasis. Gastroenterology 132(7):2478–2488. https://doi.org/10.1053/j.gastro.2007.03.047

Simpson P (1990) Lateral inhibition and the development of the sensory bristles of the adult peripheral nervous system of Drosophila. Development 109(3):509–519

Sirakov M, Boussouar A, Kress E, Frau C, Lone IN, Nadjar J, Angelov D, Plateroti M (2015) The thyroid hormone nuclear receptor TRalpha1 controls the Notch signaling pathway and cell fate in murine intestine. Development 142(16):2764–2774. https://doi.org/10.1242/dev.121962

Snippert HJ, van der Flier LG, Sato T, van Es JH, van den Born M, Kroon-Veenboer C, Barker N, Klein AM, van Rheenen J, Simons BD, Clevers H (2010) Intestinal crypt homeostasis results from neutral competition between symmetrically dividing Lgr5 stem cells. Cell 143(1):134–144. https://doi.org/10.1016/j.cell.2010.09.016

Souilhol C, Lendinez JG, Rybtsov S, Murphy F, Wilson H, Hills D, Batsivari A, Binagui-Casas A, McGarvey AC, MacDonald HR, Kageyama R, Siebel C, Zhao S, Medvinsky A (2016) Developing HSCs become Notch independent by the end of maturation in the AGM region. Blood 128(12):1567–1577. https://doi.org/10.1182/blood-2016-03-708164

Sprinzak D, Lakhanpal A, Lebon L, Santat LA, Fontes ME, Anderson GA, Garcia-Ojalvo J, Elowitz MB (2010) Cis-interactions between Notch and Delta generate mutually exclusive signalling states. Nature 465(7294):86–90. https://doi.org/10.1038/nature08959

Stanger BZ, Datar R, Murtaugh LC, Melton DA (2005) Direct regulation of intestinal fate by Notch. Proc Natl Acad Sci U S A 102(35):12443–12448. https://doi.org/10.1073/pnas.0505690102

Stump G, Durrer A, Klein AL, Lutolf S, Suter U, Taylor V (2002) Notch1 and its ligands Delta-like and Jagged are expressed and active in distinct cell populations in the postnatal mouse brain. Mech Dev 114(1–2):153–159

Tadeu AM, Horsley V (2013) Notch signaling represses p63 expression in the developing surface ectoderm. Development 140(18):3777–3786. https://doi.org/10.1242/dev.093948

Takada Y, Fukuda A, Chiba T, Seno H (2016) Brg1 plays an essential role in development and homeostasis of the duodenum through regulation of Notch signaling. Development 143(19):3532–3539. https://doi.org/10.1242/dev.141549

Tian H, Biehs B, Chiu C, Siebel CW, Wu Y, Costa M, de Sauvage FJ, Klein OD (2015) Opposing activities of Notch and Wnt signaling regulate intestinal stem cells and gut homeostasis. Cell Rep 11(1):33–42. https://doi.org/10.1016/j.celrep.2015.03.007

Tiberi L, van den Ameele J, Dimidschstein J, Piccirilli J, Gall D, Herpoel A, Bilheu A, Bonnefont J, Iacovino M, Kyba M, Bouschet T, Vanderhaeghen P (2012) BCL6 controls neurogenesis through Sirt1-dependent epigenetic repression of selective Notch targets. Nat Neurosci 15(12):1627–1635. https://doi.org/10.1038/nn.3264

Tozer S, Baek C, Fischer E, Goiame R, Morin X (2017) Differential routing of Mindbomb1 via centriolar satellites regulates asymmetric divisions of neural progenitors. Neuron 93(3):542–551. e544. https://doi.org/10.1016/j.neuron.2016.12.042

Ueo T, Imayoshi I, Kobayashi T, Ohtsuka T, Seno H, Nakase H, Chiba T, Kageyama R (2012) The role of Hes genes in intestinal development, homeostasis and tumor formation. Development 139(6):1071–1082. https://doi.org/10.1242/dev.069070

van der Flier LG, Clevers H (2009) Stem cells, self-renewal, and differentiation in the intestinal epithelium. Annu Rev Physiol 71:241–260. https://doi.org/10.1146/annurev.physiol.010908.163145

van Es JH, Sato T, van de Wetering M, Lyubimova A, Nee AN, Gregorieff A, Sasaki N, Zeinstra L, van den Born M, Korving J, Martens AC, Barker N, van Oudenaarden A, Clevers H (2012) Dll1+ secretory progenitor cells revert to stem cells upon crypt damage. Nat Cell Biol 14(10):1099–1104. https://doi.org/10.1038/ncb2581

VanDussen KL, Samuelson LC (2010) Mouse atonal homolog 1 directs intestinal progenitors to secretory cell rather than absorptive cell fate. Dev Biol 346(2):215–223. https://doi.org/10.1016/j.ydbio.2010.07.026

VanDussen KL, Carulli AJ, Keeley TM, Patel SR, Puthoff BJ, Magness ST, Tran IT, Maillard I, Siebel C, Kolterud A, Grosse AS, Gumucio DL, Ernst SA, Tsai YH, Dempsey PJ, Samuelson LC (2012) Notch signaling modulates proliferation and differentiation of intestinal crypt base columnar stem cells. Development 139(3):488–497. https://doi.org/10.1242/dev.070763

Varnum-Finney B, Purton LE, Yu M, Brashem-Stein C, Flowers D, Staats S, Moore KA, Le Roux I, Mann R, Gray G, Artavanis-Tsakonas S, Bernstein ID (1998) The Notch ligand, Jagged-1, influences the development of primitive hematopoietic precursor cells. Blood 91(11):4084–4091

Vasyutina E, Lenhard DC, Wende H, Erdmann B, Epstein JA, Birchmeier C (2007) RBP-J (Rbpsuh) is essential to maintain muscle progenitor cells and to generate satellite cells. Proc Natl Acad Sci U S A 104(11):4443–4448. https://doi.org/10.1073/pnas.0610647104

Vieira NM, Elvers I, Alexander MS, Moreira YB, Eran A, Gomes JP, Marshall JL, Karlsson EK, Verjovski-Almeida S, Lindblad-Toh K, Kunkel LM, Zatz M (2015) Jagged 1 rescues the duchenne muscular dystrophy phenotype. Cell 163(5):1204–1213. https://doi.org/10.1016/j.cell.2015.10.049

Vooijs M, Ong CT, Hadland B, Huppert S, Liu Z, Korving J, van den Born M, Stappenbeck T, Wu Y, Clevers H, Kopan R (2007) Mapping the consequence of Notch1 proteolysis in vivo with NIP-CRE. Development 134(3):535–544. https://doi.org/10.1242/dev.02733

Walker EM, Thompson CA, Battle MA (2014) GATA4 and GATA6 regulate intestinal epithelial cytodifferentiation during development. Dev Biol 392(2):283–294. https://doi.org/10.1016/j.ydbio.2014.05.017

Wang W, Zimmerman G, Huang X, Yu S, Myers J, Wang Y, Moreton S, Nthale J, Awadallah A, Beck R, Xin W, Wald D, Huang AY, Zhou L (2016) Aberrant Notch signaling in the bone marrow microenvironment of acute lymphoid leukemia suppresses osteoblast-mediated support of hematopoietic niche function. Cancer Res 76(6):1641–1652. https://doi.org/10.1158/0008-5472.CAN-15-2092

Watt FM, Estrach S, Ambler CA (2008) Epidermal Notch signalling: differentiation, cancer and adhesion. Curr Opin Cell Biol 20(2):171–179. https://doi.org/10.1016/j.ceb.2008.01.010

Wearing HJ, Sherratt JA (2000) Keratinocyte growth factor signalling: a mathematical model of dermal-epidermal interaction in epidermal wound healing. Math Biosci 165(1):41–62

Wen Y, Bi P, Liu W, Asakura A, Keller C, Kuang S (2012) Constitutive Notch activation upregulates Pax7 and promotes the self-renewal of skeletal muscle satellite cells. Mol Cell Biol 32(12):2300–2311. https://doi.org/10.1128/MCB.06753-11

Williams SE, Beronja S, Pasolli HA, Fuchs E (2011) Asymmetric cell divisions promote Notch-dependent epidermal differentiation. Nature 470(7334):353–358. https://doi.org/10.1038/nature09793

Wilson A, MacDonald HR, Radtke F (2001) Notch 1-deficient common lymphoid precursors adopt a B cell fate in the thymus. J Exp Med 194(7):1003–1012

Yang Q, Bermingham NA, Finegold MJ, Zoghbi HY (2001) Requirement of Math1 for secretory cell lineage commitment in the mouse intestine. Science 294(5549):2155–2158. https://doi.org/10.1126/science.1065718

Yaron A, Sprinzak D (2012) The cis side of juxtacrine signaling: a new role in the development of the nervous system. Trends Neurosci 35(4):230–239. https://doi.org/10.1016/j.tins.2011.12.003

Yoon K, Gaiano N (2005) Notch signaling in the mammalian central nervous system: insights from mouse mutants. Nat Neurosci 8(6):709–715. https://doi.org/10.1038/nn1475

Yoon KJ, Koo BK, Im SK, Jeong HW, Ghim J, Kwon MC, Moon JS, Miyata T, Kong YY (2008) Mind bomb 1-expressing intermediate progenitors generate notch signaling to maintain radial glial cells. Neuron 58(4):519–531. https://doi.org/10.1016/j.neuron.2008.03.018

Yu X, Zou J, Ye Z, Hammond H, Chen G, Tokunaga A, Mali P, Li YM, Civin C, Gaiano N, Cheng L (2008) Notch signaling activation in human embryonic stem cells is required for embryonic, but not trophoblastic, lineage commitment. Cell Stem Cell 2(5):461–471. https://doi.org/10.1016/j.stem.2008.03.001

Yu VW, Saez B, Cook C, Lotinun S, Pardo-Saganta A, Wang YH, Lymperi S, Ferraro F, Raaijmakers MH, Wu JY, Zhou L, Rajagopal J, Kronenberg HM, Baron R, Scadden DT (2015) Specific bone cells produce DLL4 to generate thymus-seeding progenitors from bone marrow. J Exp Med 212(5):759–774. https://doi.org/10.1084/jem.20141843

Yue F, Bi P, Wang C, Shan T, Nie Y, Ratliff TL, Gavin TP, Kuang S (2017) Pten is necessary for the quiescence and maintenance of adult muscle stem cells. Nat Commun 8:14328. https://doi.org/10.1038/ncomms14328

Zalc A, Hayashi S, Aurade F, Brohl D, Chang T, Mademtzoglou D, Mourikis P, Yao Z, Cao Y, Birchmeier C, Relaix F (2014) Antagonistic regulation of p57kip2 by Hes/Hey downstream of Notch signaling and muscle regulatory factors regulates skeletal muscle growth arrest. Development 141(14):2780–2790. https://doi.org/10.1242/dev.110155

Zhang P, He Q, Chen D, Liu W, Wang L, Zhang C, Ma D, Li W, Liu B, Liu F (2015) G protein-coupled receptor 183 facilitates endothelial-to-hematopoietic transition via Notch1 inhibition. Cell Res 25(10):1093–1107. https://doi.org/10.1038/cr.2015.109

Zhong W, Chia W (2008) Neurogenesis and asymmetric cell division. Curr Opin Neurobiol 18(1):4–11. https://doi.org/10.1016/j.conb.2008.05.002

Zhong W, Feder JN, Jiang MM, Jan LY, Jan YN (1996) Asymmetric localization of a mammalian numb homolog during mouse cortical neurogenesis. Neuron 17(1):43–53

Oscillatory Control of Notch Signaling in Development

Ryoichiro Kageyama, Hiromi Shimojo, and Akihiro Isomura

Abstract

The Notch effectors Hes1 and Hes7 and the Notch ligand Delta-like1 (Dll1) are expressed in an oscillatory manner during neurogenesis and somitogenesis. These two biological events exhibit different types of oscillations: anti-/out-of-phase oscillation in neural stem cells during neurogenesis and in-phase oscillation in presomitic mesoderm (PSM) cells during somitogenesis. Accelerated or delayed Dll1 expression by shortening or elongating the size of the Dll1 gene, respectively, dampens or quenches Dll1 oscillation at intermediate levels, a phenomenon known as "amplitude/oscillation death" of coupled oscillators. Under this condition, both Hes1 oscillation in neural stem cells and Hes7 oscillation in PSM cells are also dampened. As a result, maintenance of neural stem cells is impaired, leading to microcephaly, while somite segmentation is impaired, leading to severe fusion of somites and their derivatives, such as vertebrae and ribs. Thus, the appropriate timing of Dll1 expression is critical for the oscillatory expression in Notch signaling and normal processes of neurogenesis and somitogenesis. Optogenetic analysis indicated that Dll1 oscillations transfer the oscillatory information between neighboring cells, which may induce anti-/out-of-phase and in-phase oscillations depending on the delay in signaling transmission. These oscillatory dynamics can be described in a unified manner by mathematical modeling.

Keywords

Oscillatory expression · Neural stem cell · Presomitic mesoderm · Somite segmentation clock

R. Kageyama (✉)
Institute for Frontier Life and Medical Sciences, Kyoto University, Kyoto, Japan

World Premier International Research Initiative–Institute for Integrated Cell-Material Sciences (WPI-iCeMS), Kyoto University, Kyoto, Japan

Kyoto University Graduate School of Medicine, Kyoto, Japan

Kyoto University Graduate School of Biostudies, Kyoto, Japan
e-mail: rkageyam@infront.kyoto-u.ac.jp

H. Shimojo
Institute for Frontier Life and Medical Sciences, Kyoto University, Kyoto, Japan

A. Isomura
Institute for Frontier Life and Medical Sciences, Kyoto University, Kyoto, Japan

Japan Science and Technology Agency, PRESTO, Saitama, Japan

T. Borggrefe, B. D. Giaimo (eds.), *Molecular Mechanisms of Notch Signaling*, Advances in Experimental Medicine and Biology 1066, https://doi.org/10.1007/978-3-319-89512-3_13

Abbreviations

Dll1	Delta-like1
hGAVPO	humanized and optimized factor consisting of Gal4 DNA-binding domain, *Neurospora crassa* photoreceptor Vivid, and p65 activation domain
Notch	Notch intracellular domain
PSM	Presomitic mesoderm

1 Introduction

Notch ligand–expressing cells activate Notch signaling in their neighboring cells (Artavanis-Tsakonas et al. 1999; Fortini 2009; Gaiano and Fishell 2002; Kopan and Ilagan 2009; Pierfelice et al. 2011; Aster et al. 2016; Bray 2016). Upon activation of Notch signaling, the transmembrane receptor Notch is processed, releasing the Notch intracellular domain (NICD). NICD subsequently migrates into the nucleus and forms a complex with the DNA-binding protein RBPJ and the transcriptional co-activator Mastermind, which induces the expression of the transcriptional repressors Hes1 and Hes7. Hes1 and Hes7 then repress the expression of Notch ligands such as Delta-like1 (Dll1) (Fig. 1a). Thus, Notch ligand–expressing cells make their neighbors Notch ligand-negative. It was previously thought that this regulation is unidirectional, generating two different cell types, Notch ligand–positive and –negative cells. However, recent studies revealed that the expressions of Dll1 and the Notch effectors Hes1 and Hes7 are oscillatory in various cell types, and that this regulation is dynamic and bidirectional between neighboring cells (Fig. 1a) (Kageyama et al. 2008). In this chapter, we discuss the significance of oscillatory control of Notch signaling in two different developmental settings, neurogenesis and somitogenesis.

2 Oscillatory Expression in Neurogenesis

In the developing nervous system, neural stem cells change their competence over time (Alvarez-Buylla et al. 2001; Fishell and Kriegstein 2003; Fujita 2003; Götz and Huttner 2005; Miller and Gauthier 2007). They initially undergo a symmetric cell division to expand in cell number and subsequently enter the neurogenic stage. During the neurogenic stage, neural stem cells give rise first to deep layer neurons and later to superficial layer neurons. Finally, neural stem cells stop neurogenesis and differentiate into astrocytes. Thus, it is very important to maintain neural stem cells until the last stage of development to generate a full diversity of cell types. During the neurogenic stage, neural stem cells undergo an asymmetric cell division which generates two different daughter cells, one initiating neuronal differentiation and the other remaining a neural stem cell (Fig. 1b). In the former daughter cell, proneural genes such as *Ascl1* and *Neurog2* not only induce neuronal differentiation but also upregulate the expression of Notch ligands such as Dll1. Dll1 activates Notch signaling in the latter daughter cell. Activation of Notch1 releases NICD, which induces Hes1 expression (Jarriault et al. 1995; Ohtsuka et al. 1999). Hes1 represses the expression of proneural genes and Notch ligand genes, thereby inhibiting neuronal differentiation and maintaining neural stem cells (Fig. 1b) (Ishibashi et al. 1994; Chen et al. 1997). In the absence of Hes1, both daughter cells express proneural genes and differentiate into neurons. As a result, neural stem cells are prematurely exhausted without making a sufficient number and a full diversity of cells, leading to microcephaly or anencephaly in *Hes1*-null mice (Ishibashi et al. 1995; Ohtsuka et al. 1999; Hatakeyama et al. 2004). Thus, the Dll1-Notch-Hes1 pathway is essential for maintenance of neural stem cells and development of the nervous system.

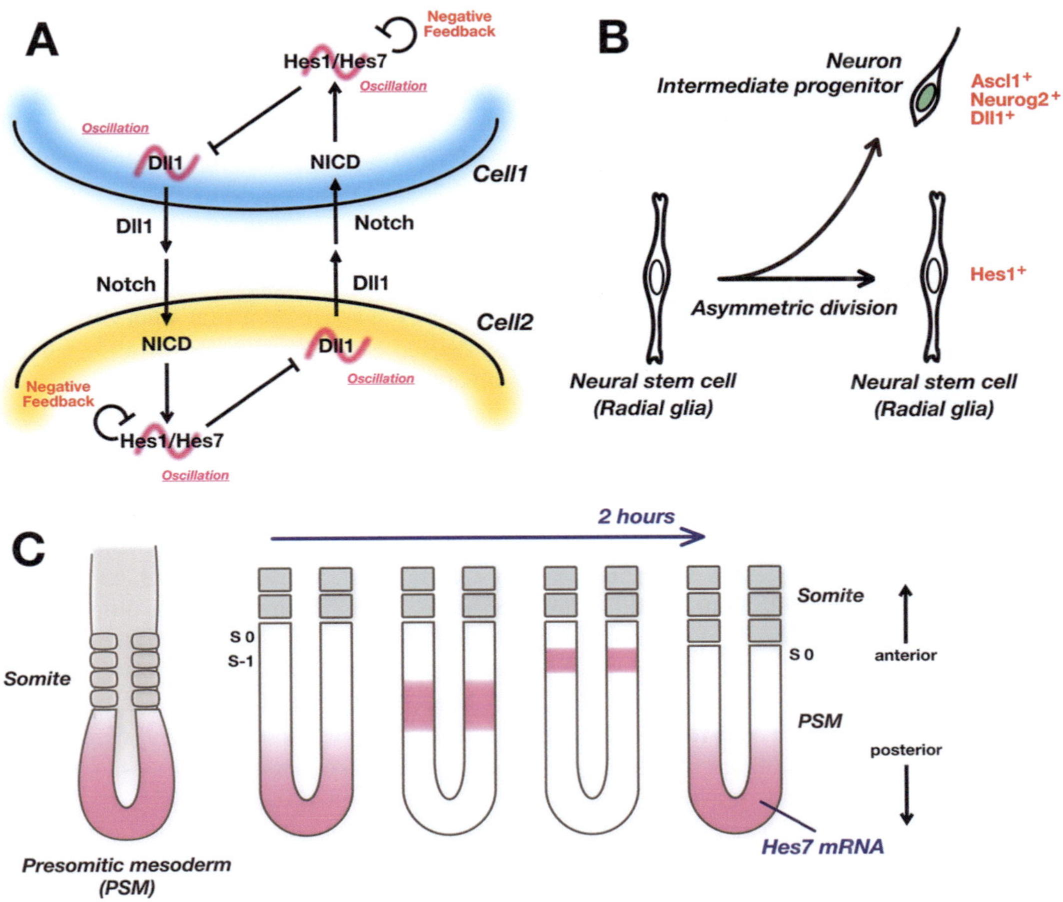

Fig. 1 The oscillatory networks of Notch signaling in neurogenesis and somitogenesis. (**a**) The Dll1-Notch-Hes1/Hes7 pathway. Dll1 activates Notch signaling in a neighboring cell. The activation of Notch signaling liberates the NICD from Notch receptor and induces Hes1/Hes7 expression, which represses Dll1 expression. Hes1/Hes7 expression oscillates by negative feedback and their oscillations drive Dll1 oscillation. This regulation is bidirectional between neighboring cells. (**b**) Asymmetric cell division of neural stem cells. One daughter cell expresses Ascl1/Neurog2, which induce Dll1 expression, as consequence this cell differentiates into a neuron or an intermediate progenitor. Dll1 activates Notch signaling in the other daughter cell which expresses Hes1 and this cell remains a neural stem cell. (**c**) Hes7 expression patterns in the presomitic mesoderm (PSM). Hes7 expression is initiated in the posterior region (Phase 1), propagates anteriorly (Phase 2) and reaches the S-1 region (Phase 3). After the disappearance of Hes7 expression in the S-1 region, a new segmentation occurs between the S-1 and S0 regions, thereby forming a bilateral pair of somites (the rightmost panel). Now, the S0 region becomes a new somite, while the S-1 region becomes the new S0 region

While the Dll1-Notch-Hes1 pathway is important for maintenance of neural stem cells, it has an inherent problem: activation of Notch signaling requires Dll1 expression in neighboring neurons. Then, how are neural stem cells maintained at an early stage of development before neurons are born? Is Notch signaling active or not? In situ hybridization and immunostaining analyses showed that *Hes1*, *Ascl1*, and *Dll1* are expressed in a salt-and-pepper pattern (variable levels of expression are mixed) by neural stem cells before differentiating neurons appear (Bettenhausen et al. 1995; Hatakeyama et al. 2004; Kageyama et al. 2008). Live imaging analyses using *Hes1*, *Ascl1*, and *Dll1* promoter-driven luciferase reporters showed that the expression of these genes oscillate in neural stem cells (Masamizu et al. 2006; Shimojo et al. 2008; Imayoshi et al.

2013; Shimojo et al. 2016). Because these oscillations are out of synchrony between neighboring cells, snapshots of their expression revealed by in situ hybridization and immunostaining analyses showed salt-and-pepper patterns. These results suggest that Notch signaling is active not only during the neurogenic stage but also before the neurogenic stage begins.

The oscillatory expression in neural stem cells is driven by the Hes1 oscillator (Fig. 2a) (Hirata et al. 2002). Hes1 binds to its own promoter and represses its own expression but, when *Hes1* promoter is repressed, both *Hes1* mRNA and Hes1 protein disappear rapidly, because of their extremely poor stability. When Hes1 protein disappears, *Hes1* promoter is reactivated, resulting in a next round of expression. In this way, Hes1 expression autonomously oscillates with a period of about 2–3 h by this negative feedback. Hes1 oscillation periodically represses Ascl1, Neurog2, and Dll1 expression, thereby inducing Ascl1, Neurog2, and Dll1 oscillations in neural stem cells (Fig. 2b and c). It is likely that these oscillations may be important for mutual activation of Notch signaling among neural stem cells. The current model is as follows (Fig. 2b): when Hes1 expression is low, Ascl1, Neurog2 and Dll1 expressions become high in a neural stem cell (Neural stem cell 1 in the left side of Fig. 2b). This high Dll1 expression activates Notch signaling in a neighboring neural stem cell, where Hes1 expression becomes high, thereby repressing Ascl1, Neurog2, and Dll1 (Neural stem cell 2 in the left side of Fig. 2b). Because of oscillatory expression, high Hes1 expression becomes low 1 h later in the latter neural stem cell, which then expresses high levels of Ascl1, Neurog2, and Dll1 (Neural stem cell 2 in the right side of Fig. 2b). This high Dll1 expression next activates Notch signaling in the former neural stem cell (Neural stem cell 1 in the right side of Fig. 2b). Thus, it is likely that this kind of oscillatory expression of Dll1 mutually activates Notch signaling in neighboring neural stem cells without any help of differentiating neurons. Indeed, when Dll1 and Hes1 oscillations are dampened, maintenance and proliferation of neural stem cells are impaired, resulting in microcephaly (see below in paragraph 4).

In differentiating neurons, Hes1 expression disappears while Ascl1, Neurog2, and Dll1 expression becomes sustained (Fig. 2c). In differentiating astrocytes, Hes1 expression is sustained while Ascl1, Neurog2, and Dll1 expression disappears. Similarly, Olig2, which regulates oligodendrocyte development, is expressed in an oscillatory manner by neural stem cells but its expression becomes sustained in differentiating oligodendrocytes. Thus, multiple fate determination factors like Hes1, Ascl1/Neurog2 and Olig2 are expressed in an oscillatory manner by neural stem cells but one of them is selected and expressed in a sustained manner by differentiating cells, suggesting that the expression dynamics of fate determination factors are different between neural stem cells and differentiating cells (Shimojo et al. 2008; Imayoshi et al. 2013; Imayoshi and Kageyama 2014).

It has been reported that three cell fate determination factors, Ascl1, Hes1 and Olig2, which promote differentiation of neurons, astrocytes, and oligodendrocytes, respectively, play an important role in maintenance and proliferation of neural stem cells, suggesting that these factors have opposing functions (Ohtsuka et al. 2001; Castro et al. 2011). Because the expression dynamics are different (oscillatory versus sustained) between neural stem cells and differentiating cells, different expression dynamics could be involved in opposing functions. To examine this hypothesis, we employed a light-switchable system for gene expression, using the protein consisting of Gal4 DNA-binding domain, *Neurospora crassa* photoreceptor Vivid, and p65 activation domain, called GAVP (Wang et al. 2012). Two mutations were further introduced to reduce the background in the dark (optimized GAVP, called GAVPO), and its codon usage was humanized to increase the expression efficiency (hGAVPO) (Wang et al. 2012; Imayoshi et al. 2013). Upon blue light illumination, Vivid is activated and forms a homodimer. The dimer form of hGAVPO binds to UAS sequences via its Gal4 DNA-binding domain and activates downstream gene expression via its p65 activation domain. In

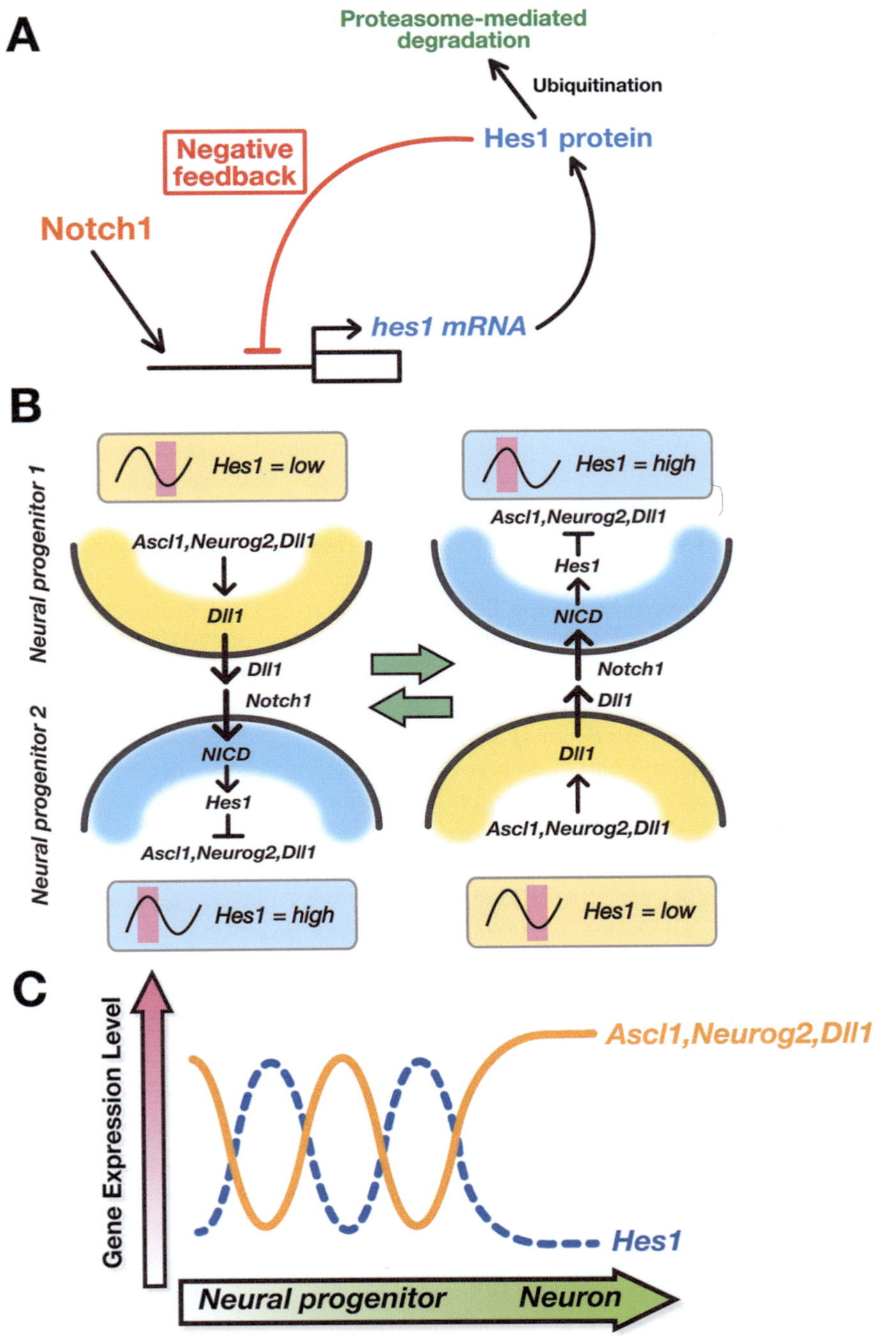

Fig. 2 *Hes1* oscillation in neural stem cells. (**a**) Hes1 expression oscillates by negative feedback. Hes1 binds to its own promoter and represses its own expression but, when *Hes1* promoter is repressed, both *Hes1* mRNA and Hes1 protein disappear rapidly, because of their extremely low stability. When Hes1 protein disappears, Notch1 signaling activates *Hes1* promoter. (**b**) Mutual activation of Notch1 signaling between neighboring neural stem cells by Dll1 oscillations. Hes1 oscillations drive the oscillatory expression of Ascl1, Neurog2 and Dll1 in neural stem cells. Dll1 oscillations may lead to mutual activation of Notch signaling among neural stem cells. (**c**) In neural stem cells, Hes1 oscillations drive Ascl1, Neurog2, and Dll1 oscillations. In differentiating neurons, Hes1 expression disappears while Ascl1, Neurog2, and Dll1 expression becomes sustained

the dark, hGAVPO dissociates and the downstream gene expression is switched off. This system enables to induce oscillatory versus sustained gene expression by blue light. It was found that while sustained expression of Ascl1 induces neuronal differentiation, oscillatory expression of Ascl1 activates proliferation of neural stem cells, suggesting that the expression dynamics are important for the Ascl1 activities (Imayoshi et al. 2013). Thus, it is likely that cell fate determination factors, such as Ascl1, exert opposing functions depending on their oscillatory versus sustained expression patterns.

3 Oscillatory Expression in Somitogenesis

Somites are metameric structures, which later give rise to vertebrae, ribs, skeletal muscles and subcutaneous tissues (Pourquié 2011). A bilateral pair of somites is formed by segmentation of the anterior parts of the presomitic mesoderm (PSM), which is located in the posterior part of embryos (Fig. 1c). This process is called somite segmentation, which is repeated every 2 h in mouse embryos, and this periodic event is controlled by the somite segmentation clock gene *Hes7* (Bessho et al. 2001; Sparrow et al. 2012). Hes7 expression starts from the posterior region of the PSM (Phase 1, Fig. 1c) and then propagates into the anterior region of the PSM (Phases 2 and 3, Fig. 1c). This dynamic Hes7 expression is caused by oscillatory expression in individual PSM cells, and each cycle of Hes7 oscillations leads to formation of a pair of somites (Fig. 1c). In the absence of Hes7, all somites are severely fused as well as the vertebrae and ribs (Bessho et al. 2001). When steady expression of Hes7 is induced, again all somites are severely fused (Takashima et al. 2011). Thus, both loss of expression and steady expression of Hes7 lead to somite fusion, suggesting that the oscillatory expression of Hes7 is required for somite segmentation.

Hes7 oscillation is controlled by negative feedback, just like Hes1 oscillation in neural stem cells (Bessho et al. 2003). It has been shown that negative feedback with delayed timing is essential for Hes7 oscillation: deletion of all introns from *Hes7*, which has three introns, accelerates Hes7 protein expression because the time required to transcribe and remove the intronic sequences by splicing is not necessary, and this accelerated negative feedback leads to steady (non-oscillatory) Hes7 expression and severe somite fusion (Takashima et al. 2011). Interestingly, deletion of only two introns (leaving one intron) from the Hes7 gene moderately accelerates Hes7 protein expression and this moderate acceleration increases the tempo of Hes7 oscillation. As a result, the tempo of the segmentation clock is accelerated, forming more somites and vertebrae, although Hes7 oscillation is later dampened (Harima et al. 2013). These two different phenotypes depending on the extent of acceleration in negative feedback were successfully simulated by mathematical modeling (Takashima et al. 2011; Harima et al. 2013). These data together support the notion that *Hes7* is the central gene of the somite segmentation clock.

While oscillation occurs in phase between neighboring PSM cells along the mediolateral axis, the oscillation phase is delayed in the anterior compared to the posterior PSM, thereby generating wave-like propagation from the posterior to anterior direction (Fig. 1c). How such phase relationship between neighboring PSM cells is controlled remains to be analyzed. It has been shown that the oscillatory expression becomes unstable and out of phase when the cells are dissociated, suggesting that cell-cell communication is important for stable synchronized oscillation (Maroto et al. 2005; Masamizu et al. 2006). Indeed, dissociated PSM cells soon exhibited synchronized oscillations when they were reaggregated (Tsiairis and Aulehla 2016). It has been shown that Notch signaling is required for the synchronization of oscillatory expression. In zebrafish embryos lacking genes encoding the components of the Notch pathway or treated with γ–secretase inhibitors, which inhibit Notch signaling, oscillatory gene expression is desynchronized between neighboring PSM cells, forming salt-and-pepper expression patterns (Jiang et al. 2000; Riedel-Kruse et al. 2007; Özbudak and

Lewis 2008; Delaune et al. 2012). Wash-out of γ–secretase inhibitors reactivates Notch signaling and synchronization recovers rapidly (Riedel-Kruse et al. 2007). These results indicate that Notch signaling regulates synchronization between neighboring PSM cells, which makes the in-phase oscillatory expression resistant to perturbation such as mitosis and cell movement (Horikawa et al. 2006). In zebrafish, expression of the Notch ligand DeltaC protein oscillates under the control of *her* oscillations (Giudicelli et al. 2007), and DeltaC oscillation is likely to drive synchronization by periodic activation of Notch signaling (Mara et al. 2007; Özbudak et al., 2008). In mice, the expression of Dll1 as well as Lunatic fringe (Lfng, also known as b1,3-N-acetylglucoseaminyl-transferase), which modulates the Notch and Dll1 activities (Panin et al. 2002; Okubo et al. 2012), oscillates and both Dll1 and Lfng genes are important for synchronized oscillations (Hrabe de Angelis et al. 1997; Evrard et al. 1998; Zhang and Gridley 1998; Maruhashi et al. 2005; Niwa et al. 2011; Bone et al. 2014; Shimojo et al. 2016). These results together suggest that Notch ligand oscillations may be key to synchronized oscillations in the somite segmentation clock.

4 Significance of Dll1 Oscillation in Development

To reveal the significance of Dll1 oscillation in neurogenesis and somitogenesis, steady or non-oscillatory Dll1 expression was induced without changing the average expression levels. Mathematical simulation suggested that accelerated or delayed Dll1 expression dampens or quenches Dll1 oscillation, leading to non-oscillatory Dll1 expression (see Fig. 5) (Shimojo et al. 2016). Thus, to accelerate or delay the timing of Dll1 expression, the size of the Dll1 gene was shortened or elongated, respectively. To shorten the Dll1 gene, all introns were removed from the Dll1 locus (*Dll1* type 1 mutant mice) and this mutation was found to accelerate Dll1 expression (Shimojo et al. 2016). To elongate the Dll1 gene, an extra sequence was inserted (*Dll1* type 2 mutant mice), and this mutation was found to delay Dll1 expression (Shimojo et al. 2016).

In neural stem cells of both *Dll1* type 1 and type 2 homozygous-mutant mice (Fig. 3a), Dll1 oscillation was severely dampened or mostly quenched (Fig. 3b), although the average expression levels remained almost unaffected compared to wild type controls. Furthermore, Hes1 oscillation was also severely dampened in the mutant neural stem cells (Fig. 3c), although the expression levels were not much changed compared to the wild type. In these mutant mice, neural stem cells were not maintained properly and started neuronal differentiation prematurely, resulting in a smaller brain (Fig. 3d) (Shimojo et al. 2016). Furthermore, optogenetic induction of Dll1 oscillation enhanced the maintenance of neural stem cells. Thus, steady or non-oscillatory Dll1 is able to activate Hes1 expression in neighboring cells but it is not sufficient for the maintenance and proliferation of neural stem cells.

In the PSM of both *Dll1* type 1 and type 2 homozygous-mutant mice (Fig. 3e), Dll1 oscillation was severely dampened or mostly quenched (Fig. 3f), although the expression levels were not much changed, compared to the wild type. Both types of *Dll1* mutant mice exhibited severe segmentation defects: somites and their derivatives such as the vertebrae and ribs were severely fused (Fig. 3i) (Shimojo et al. 2016). Interestingly, in these mutants, Hes7 oscillation was also dampened at intermediate levels (Fig. 3g) and steady Hes7 expression repressed Lfng expression (Fig. 3h). Steady or non-oscillatory Dll1 seemed to be able to activate Notch signaling, because Hes7 expression levels were not much changed in the mutants compared to the wild type, suggesting that non-oscillatory Dll1 is still functional for the activation of gene expression in neighboring PSM cells. However, non-oscillatory Dll1 is definitely non-functional in regards to the segmentation clock, indicating that the normal timing of Dll1 expression is critical for Dll1-Hes7-Lfng oscillatory networks, and that Dll1 oscillation is essential for the segmentation clock.

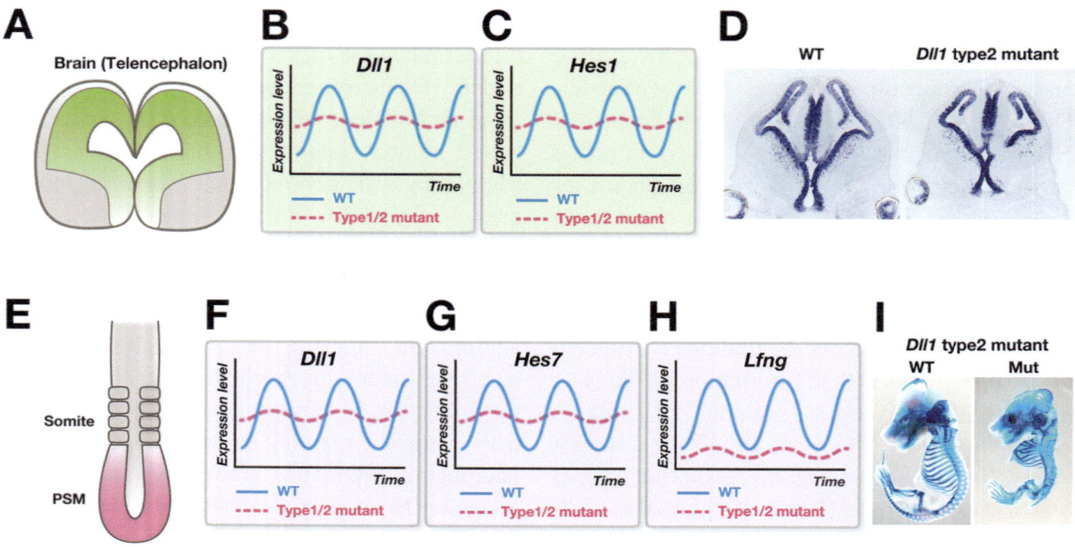

Fig. 3 The phenotypes of *Dll1* type 1 and type 2 homozygous mutant mice in which the size of *Dll1* gene is shortened and lengthened, respectively. In *Dll1* type 1 and type 2 homozygous mutant mice, Dll1 expression is accelerated and delayed, respectively (Shimojo et al. 2016). (**a**) Schema of the developing brain. The ventricular zone, which contains neural stem cells, is shaded with green. (**b**, **c**) Dll1 and Hes1 expression dynamics and brain development of wild type and *Dll1* type 1 and type 2 mutant mice. *Dll1* (**b**) and *Hes1* (**c**) expression oscillates dynamically in wild type (WT, blue lines) neural stem cells but these oscillations are dampened in both *Dll1* type 1 and type 2 mutant neural stem cells (red broken lines). (**d**) The size of the *Dll1* type 2 mutant brain is smaller compared to that of the wild type (WT) brain. (**e**) Schema of the somites and the presomitic mesoderm. (**f–h**) *Dll1*, *Hes7* and *Lfng* expression dynamics and segmentation in wild type and *Dll1* type 1 and type 2 mutant mice. *Dll1* (**f**), *Hes7* (**g**) and *Lfng* (**h**) expression oscillates dynamically in the wild type (WT, blue lines) PSM but these oscillations are dampened in both *Dll1* type 1 and type 2 mutant PSM (red broken lines). (**i**) Vertebrae and ribs are normally segmented in a wild type (WT) mouse but are severely fused in a *Dll1* type 2 mutant mouse. Adapted from (Shimojo et al. 2016)

5 Transmission of the Oscillatory Information Via Dll1 Oscillation

Dll1 expression oscillates in an anti-phase/out-of-phase manner between neighboring neural stem cells, while it oscillates in an in-phase manner between neighboring PSM cells. Furthermore, accelerated and delayed Dll1 expression dampens or quenches oscillations in both neural stem cells and PSM cells, raising the possibility that Dll1 oscillation regulates both the anti-phase/out-of-phase and in-phase synchronized oscillations. However, it was not known whether Dll1 can convey the oscillatory information to neighboring cells and whether Dll1 oscillations are sufficient to entrain oscillatory expression at a population level. Recently, this notion was directly tested by optogenetics using hGAVPO (Isomura et al. 2017): photo-sensitive sender cells carry an optogenetic perturbation module for Dll1 induction (Fig. 4a, Optogenetic perturbation) while photo-insensitive receiver cells have the native Hes1 oscillator and the *Hes1* reporter (Fig. 4a, Native oscillator and Reporter, respectively). These two cell lines were co-cultured, and Hes1 expression was monitored after blue light illumination. When the cells were exposed to sustained light illumination, photo-sensitive sender cells steadily expressed Dll1. Under this condition, the receiver cells showed an asynchronous oscillatory pattern of *Hes1* expression (Fig. 4b, Sustained light). By contrast, when synchronized Dll1 oscillations

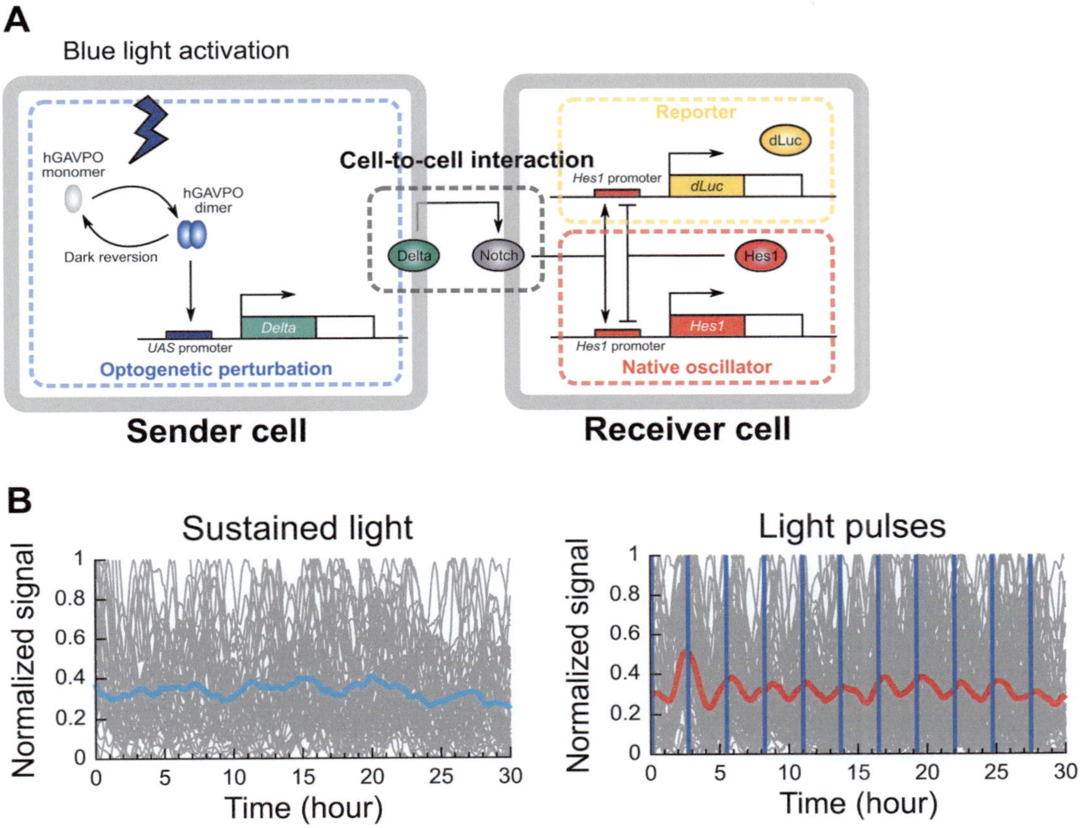

Fig. 4 Cell-cell transmission of the oscillatory information via Dll1 oscillations. (**a**) Schematic representation of the genetic network comprising cell-to-cell interaction of the Notch signaling pathway connected to an optogenetic perturbation module in a sender cell and the native Hes1 oscillator and a Hes1 reporter in a receiver cell. Sender cells express Dll1 (Delta) by blue light stimulation. (**b**) Sustained light illumination induced steady Dll1 expression and failed to entrain *Hes1* oscillation in receiver cells. By contrast, light pulses (blue vertical lines) induced periodic Dll1 expression in sender cells and entrained *Hes1* oscillation in receiver cells at the population level. Grey lines indicate Hes1 expression in individual cells while blue and red lines show the average of Hes1 expression at the population level. Adapted from (Isomura et al. 2017)

with 2.5 h periodicity were induced in photo-sensitive sender cells by periodic blue light illumination (2-min duration with 2.5 h intervals), receiver cells also exhibited synchronized Hes1 oscillation with the same periodicity as the external perturbation (Fig. 4b, Light pulses). These data show that the temporal information of the stimulus was transferred from the photo-sensitive sender cells to the photo-insensitive receiver cells.

A single-cell time-series analysis demonstrated that the single-cell genetic oscillators were responsible for initiating phase modulation depending on the timing of external perturbation in surrounding cells and that periodic inputs of Notch signaling entrain intrinsic oscillations by frequency tuning and phase shifting at the single-cell level (Isomura et al. 2017). This result indicates that single-cell genetic oscillators can transmit and decode dynamic information through Dll1-mediated multicellular interactions, thereby synchronizing the population of oscillators.

6 Mathematical Simulation of Oscillation/Amplitude Death

The above results together indicate that Dll1 transmits the oscillatory information to neighboring cells to control neurogenesis and somitogenesis. During these processes, Dll1 expression oscillates out of phase in neural stem cells and in phase in PSM cells and these different types of oscillations can be described in a unified manner. Mathematical modeling suggests that coupling delays between cells are very important for such different types of oscillations (Ramana Reddy et al. 1998; Shimojo et al. 2016). In both neural stem cells and PSM cells, Hes1/Hes7 expression oscillates by delayed negative feedback and Hes1/Hes7 oscillations drive Dll1 oscillation (Fig. 5a). Dll1 activates Hes1/Hes7 expression in neighboring cells with some delay (Fig. 5a).

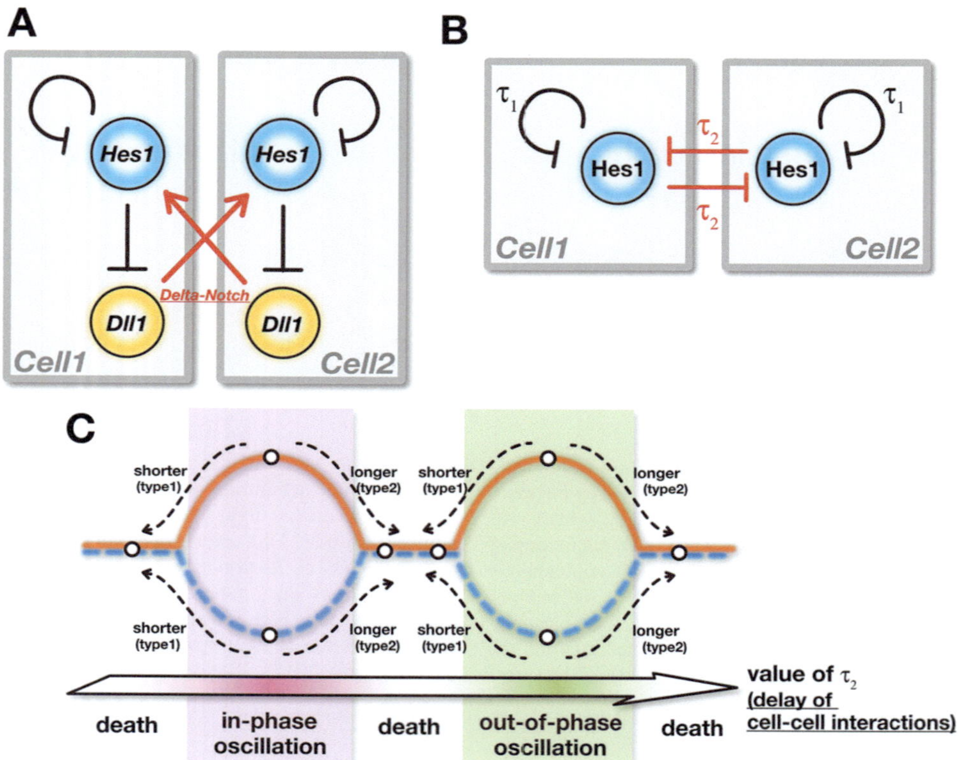

Fig. 5 Simulations for different oscillatory dynamics by mathematical modeling. (**a**) The Dll1-Hes1 regulatory pathway between cells. Hes1 forms negative feedback. Hes1 represses Dll1 in the same cell while Dll1 induces Hes1 in a neighboring cell. (**b**) The regulation shown in (**a**) can be simplified as double negative feedback loops. Hes1 represses its own expression with the delay τ_1 within the same cell while repressing Hes1 expression in a neighboring cell with the delay τ_2. (**c**) In-phase and anti-phase/out-of-phase oscillations and amplitude/oscillation death of coupled oscillators. The orange line and blue broken line represent the highest and lowest levels of Hes1/Hes7 expression, respectively. The distance between these two lines indicates the amplitude of oscillation. According to the mathematical modeling (Shimojo et al. 2016), Hes1/Hes7 expression oscillates in phase, as in PSM cells (pink shaded area), or out of phase, as in neural stem cells (green shaded area), depending on the τ_2 values (the horizontal axis). When Dll1 expression is accelerated or delayed, τ_2 values should be decreased or increased, respectively. When τ_2 values are decreased or increased, both in-phase and out-of-phase oscillations would be dampened (lower amplitudes indicated by broken arrows) or quenched (non-shaded area), a phenomenon known as "amplitude death" or "oscillation death" of coupled oscillators. These data suggest that the timing of Dll1 expression is very important for the dynamics of coupled oscillators. Adapted from (Shimojo et al. 2016)

This indicates that Hes1/Hes7 also repress Hes1/Hes7 in a neighboring cell via repression of Dll1. Thus, this regulation can be simplified as double-negative feedback loops where Hes1/Hes7 repress their own expression with a delay τ_1 within a cell, while repressing Hes1/Hes7 expression in a neighboring cell with a delay τ_2 (Fig. 5b). According to the mathematical modeling (Shimojo et al. 2016), both in-phase (pink shaded area in Fig. 5c) and out-of-phase (green shaded area in Fig. 5c) oscillations can be induced by simply changing τ_2 values (horizontal axis in Fig. 5c). This result suggests that the delay in Delta-Notch signaling transmission between cells is very important for oscillatory dynamics.

Another issue predicted from the mathematical modeling is that the timing of Dll1 expression is very important for oscillatory expression. Accelerated or delayed Dll1 expression may decrease or increase τ_2, respectively. In Fig. 5c, the orange and blue lines represent the highest and lowest levels of Hes1/Hes7 expression, respectively and thus the distance between these two lines represents the amplitude of oscillatory expression. According to the mathematical modeling, both decreased and increased τ_2 values would decrease the amplitude and therefore dampen the in-phase and out-of-phase oscillations (Fig. 5c, broken arrows). Furthermore, these oscillations would be quenched if τ_2 values are further decreased or increased (non-shaded areas in Fig. 5c). This is the transition state between in-phase and out-of-phase oscillations, a phenomenon known as "amplitude death" or "oscillation death" of coupled oscillators (Fig. 5c) (Ramana Reddy et al. 1998). Both Hes1 and Hes7 oscillations were indeed severely dampened or almost quenched at intermediate levels when Dll1 expression was accelerated (decreased τ_2) or delayed (increased τ_2) (see Fig. 3) (Shimojo et al. 2016). These data indicate that intercellular-coupled oscillators change their gene expression dynamics, depending on the delay in cell-cell interactions and that the precise timing for the activation of Notch signaling is important for these dynamic gene expression. However, this mathematical modeling still needs more validation and adjustment. For example, it remains to be determined whether τ_2 values are different between neural stem cells and PSM cells and further analyses will increase our understanding of the mechanism of how these dynamics of gene expression are controlled.

7 Conclusions

The oscillatory expression driven by the oscillator genes *Hes1* and *Hes7* is important for various developmental processes such as neurogenesis and somitogenesis. During these processes, the oscillatory information seems to be transmitted from cell to cell via Dll1 to coordinate the gene expression at the tissue level. Mathematical modeling suggests that in-phase and anti-phase/out-of-phase oscillations, occurring during somitogenesis and neurogenesis respectively, critically depend on the delay in Dll1-mediated cell-cell interactions and that both increase and decrease of this delay would dampen or quench the oscillatory expression. Indeed, both increased and decreased delays in Dll1-mediated cell-cell interactions dampen or quench the oscillations and impair the processes of neurogenesis and somitogenesis. Thus, Dll1-mediated cell-cell transmission of the oscillatory information at proper timings is critical to the coordinated gene expression in tissue morphogenesis.

Acknowledgements This work was supported by Core Research for Evolutional Science and Technology (R.K.), Grant-in-Aid for Scientific Research on Innovative Areas (MEXT 16H06480 to R.K., MEXT 24116705 to H.S., and MEXT 26119708 to A.I.), Scientific Research (A) (JSPS 24240049) (R.K.), Young Scientists (A) (JSPS 15H05326) (A.I.) and Young Scientists (B) (JSPS 24700354) (H.S.) and Platform for Dynamic Approaches to Living System from the Ministry of Education, Culture, Sports, Science and Technology, Japan.

References

Alvarez-Buylla A, Garcia-Verdugo JM, Tramontin AD (2001) A unified hypothesis on the lineage of neural stem cells. Nat Rev Neurosci 2:287–293

Artavanis-Tsakonas S, Rand MD, Lake RJ (1999) Notch signaling: cell fate control and signal integration in development. Science 284:770–776

Aster JC, Pear WS, Blacklow SC (2016) The varied roles of notch in cancer. Annu Rev Pathol 12:245–275

Bessho Y, Sakata R, Komatsu S, Shiota K, Yamada S, Kageyama R (2001) Dynamic expression and essential functions of *Hes7* in somite segmentation. Genes Dev 15:2642–2647

Bessho Y, Hirata H, Masamizu Y, Kageyama R (2003) Periodic repression by the bHLH factor Hes7 is an essential mechanism for the somite segmentation clock. Genes Dev 17:1451–1456

Bettenhausen B, Hrabe de Angelis M, Simon D, Guenet JL, Gossler A (1995) Transient and restricted expression during mouse embryogenesis of *Dll1*, a murine gene closely related to *Drosophila Delta*. Development 121:2407–2418

Bone RA, Bailey CSL, Wiedermann G, Ferjentsik Z, Appleton PL, Murray PJ, Maroto M, Dale JK (2014) Spatiotemporal oscillations of Notch1, Dll1 and NICD are coordinated across the mouse PSM. Development 141:4806–4816

Bray SJ (2016) Notch signalling in context. Nat Rev Mol Cell Biol 17:722–735

Castro DS, Martynoga B, Parras C, Ramesh V, Pacary E, Johnston C, Drechsel D, Lebel-Potter M, Garcia LG, Hunt C et al (2011) A novel function of the proneural factor Ascl1 in progenitor proliferation identified by genome-wide characterization of its targets. Genes Dev 25:930–945

Chen H, Thiagalingam A, Chopra H, Borges MW, Feder JN, Nelkin BD, Baylin SB, Ball DW (1997) Conservation of the *Drosophila* lateral inhibition pathway in human lung cancer: a hairy-related protein (HES-1) directly represses achaete-scute homolog-1 expression. Proc Natl Acad Sci U S A 94:5355–5360

Delaune EA, François P, Shih NP, Amacher SL (2012) Single-cell-resolution imaging of the impact of notch signaing and mitosis on segmentation clock dynamics. Dev Cell 23:995–1005

Evrard YA, Lun Y, Aulehla A, Gan L, Johnson RL (1998) *lunatic fringe* is an essential mediator of somite segmentation and patterning. Nature 394:377–381

Fishell G, Kriegstein AR (2003) Neurons from radial glia: the consequences of asymmetric inheritance. Curr Opin Neurobiol 13:34–41

Fortini ME (2009) Notch signaling: the core pathway and its posttranslational regulation. Dev Cell 16:633–647

Fujita S (2003) The discovery of the matrix cell, the identification of the multipotent neural stem cell and the development of the central nervous system. Cell Struct Funct 28:205–228

Gaiano N, Fishell G (2002) The role of notch in promoting glial and neural stem cell fates. Annu Rev Neurosci 25:471–490

Giudicelli F, Ozbudak EM, Wright GJ, Lewis J (2007) Setting the tempo in development: an investigation of the zebrafish somite clock mechanism. PLoS Biol 5:e150

Götz M, Huttner WB (2005) The cell biology of neurogenesis. Nat Rev Mol Cell Biol 6:777–788

Harima Y, Takashima Y, Ueda Y, Ohtsuka T, Kageyama R (2013) Accelerating the tempo of the segmentation clock by reducing the number of introns in the *Hes7* gene. Cell Rep 3:1–7

Hatakeyama J, Bessho Y, Katoh K, Ookawara S, Fujioka M, Guillemot F, Kageyama R (2004) Hes genes regulate size, shape and histogenesis of the nervous system by control of the timing of neural stem cell differentiation. Development 131:5539–5550

Hirata H, Yoshiura S, Ohtsuka T, Bessho Y, Harada T, Yoshikawa K, Kageyama R (2002) Oscillatory expression of the bHLH factor Hes1 regulated by a negative feedback loop. Science 298:840–843

Horikawa K, Ishimatsu K, Yoshimoto E, Kondo S, Takeda H (2006) Noise-resistant and synchronized oscillation of the segmentation clock. Nature 441:719–723

Hrabe de Angelis M, McIntyre J II, Gossler A (1997) Maintenance of somite borders in mice requires the Delta homologue Dll1. Nature 386:717–721

Imayoshi I, Kageyama R (2014) bHLH factors in self-renewal, multipotency, and fate choice of neural progenitor cells. Neuron 82:9–23

Imayoshi I, Isomura A, Harima Y, Kawaguchi K, Kori H, Miyachi H, Fujiwara TK, Ishidate F, Kageyama R (2013) Oscillatory control of factors determining multipotency and fate in mouse neural progenitors. Science 342:1203–1208

Ishibashi M, Moriyoshi K, Sasai Y, Shiota K, Nakanishi S, Kageyama R (1994) Persistent expression of helix-loop-helix factor HES-1 prevents mammalian neural differentiation in the central nervous system. EMBO J 13:1799–1805

Ishibashi M, Ang SL, Shiota K, Nakanishi S, Kageyama R, Guillemot F (1995) Targeted disruption of mammalian hairy and enhancer of split homolog-1 (HES-1) leads to up-regulation of neural helix-loop-helix factors, premature neurogenesis, and severe neural tube defects. Genes Dev 9:3136–3148

Isomura A, Ogushi F, Kori H, Kageyama R (2017) Optogenetic perturbation and bioluminescence imaging to analyze cell-to-cell transfer of oscillatory information. Genes Dev 31:524–535

Jarriault S, Brou C, Logeat F, Schroeter EH, Kopan R, Israel A (1995) Signalling downstream of activated mammalian notch. Nature 377:355–358

Jiang Y-J, Aerne BL, Smithers L, Haddon C, Ish-Horowicz D, Lewis J (2000) Notch signaling and the synchronization of the somite segmentation clock. Nature 408:475–479

Kageyama R, Ohtsuka T, Shimojo H, Imayoshi I (2008) Dynamic notch signaling in neural progenitor cells and a revised view of lateral inhibition. Nat Neurosci 11:1247–1251

Kopan R, Ilagan MXG (2009) The canonical notch signaling pathway: unfolding the activation mechanism. Cell 137:216–233

Mara A, Schroeder J, Chalouni C, Holley SA (2007) Priming, initiation and synchronization of the segmentation clock by *deltaD* and *deltaC*. Nat Cell Biol 9:523–530

Maroto M, Dale JK, Dequéant M-L, Petit A-C, Pourquié O (2005) Synchronised cycling gene oscillations in presomitic mesoderm cells require cell-cell contact. Int J Dev Biol 49:309–315

Maruhashi M, Van de Putte T, Huylebroeck D, Kondoh H, Higashi Y (2005) Involvement of SIP1 in positioning of somite boundaries in the mouse embryo. Dev Dyn 234:332–338

Masamizu Y, Ohtsuka T, Takashima Y, Nagahara H, Takenaka Y, Yoshikawa K, Okamura H, Kageyama R (2006) Real-time imaging of the somite segmentation clock: revelation of unstable oscillators in the individual presomitic mesoderm cells. Proc Natl Acad Sci U S A 103:1313–1318

Miller ED, Gauthier AS (2007) Timing is everything: making neurons versus glia in the developing cortex. Neuron 54:357–369

Niwa Y, Shimojo H, Isomura A, González A, Miyachi H, Kageyama R (2011) Different types of oscillations in notch and Fgf signaling regulate the spatiotemporal periodicity of somitogenesis. Genes Dev 25:1115–1120

Ohtsuka T, Ishibashi M, Gradwohl G, Nakanishi S, Guillemot F, Kageyama R (1999) *Hes1* and *Hes5* as notch effectors in mammalian neuronal differentiation. EMBO J 18:2196–2207

Ohtsuka T, Sakamoto M, Guillemot F, Kageyama R (2001) Roles of the basic helix-loop-helix genes *Hes1* and *Hes5* in expansion of neural stem cells of the developing brain. J Biol Chem 276:30467–30474

Okubo Y, Sugawara T, Abe-Koduka N, Kanno J, Kimura A, Saga Y (2012) Lfng regulates the synchronized oscillation of the mouse segmentation clock via trans-repression of notch signalling. Nat Commun 3:1141

Özbudak EM, Lewis J (2008) Notch signaling synchronizes the zebrafish segmentation clock but is not needed to create somite boundaries. PLoS Genet 4:e15

Panin VM, Shao L, Lei L, Moloney DJ, Irvine KD, Haltiwanger RS (2002) Notch ligands are substrates for protein *O*-fucosyltransferase-1 and fringe. J Biol Chem 277:29945–29952

Pierfelice T, Alberi L, Gaiano N (2011) Notch in the vertebrate nervous system: an old dog with new tricks. Neuron 69:840–855

Pourquié O (2011) Vertebrate segmentation: from cyclic gene networks to scoliosis. Cell 145:650–663

Ramana Reddy DV, Sen A, Johnston GL (1998) Experimental evidence of time delay induced death in coupled limit cycle oscillators. Phys Rev Lett 80:5109–5112

Riedel-Kruse IH, Müller C, Oates AC (2007) Synchrony dynamics during initiation, failure, and rescue of the segmentation clock. Science 317:1911–1915

Shimojo H, Ohtsuka T, Kageyama R (2008) Oscillations in notch signaling regulate maintenance of neural progenitors. Neuron 58:52–64

Shimojo H, Isomura A, Ohtsuka T, Kori H, Miyachi H, Kageyama R (2016) Oscillatory control of Delta-like1 in cell interactions regulate dynamic gene expression and tissue morphogenesis. Genes Dev 30:102–116

Sparrow DB, Chapman G, Smith AJ, Mattar MZ, Major JA, O'Reilly VC, Saga Y, Zackai EH, Dormans JP, Alman BA, McGregor L, Kageyama R, Kusumi K, Dunwoodie SL (2012) A mechanism for gene-environment interaction in the etiology of congenital scoliosis. Cell 149:295–306

Takashima Y, Ohtsuka T, González A, Miyachi H, Kageyama R (2011) Intronic delay is essential for oscillatory expression in the segmentation clock. Proc Natl Acad Sci U S A 108:3300–3305

Tsiairis CD, Aulehla A (2016) Self-organization of embryonic genetic oscillators into spatiotemporal wave patterns. Cell 164:656–667

Wang X, Chen X, Yang Y (2012) Spatiotemporal control of gene expression by a light-switchable transgene system. Nat Methods 9:266–269

Zhang N, Gridley T (1998) Defects in somite formation in lunatic fringe-deficient mice. Nature 394:374–377

CSL-Associated Corepressor and Coactivator Complexes

Franz Oswald and Rhett A. Kovall

Abstract

The highly conserved Notch signal transduction pathway orchestrates fundamental cellular processes including, differentiation, proliferation, and apoptosis during embryonic development and in the adult organism. Dysregulated Notch signaling underlies the etiology of a variety of human diseases, such as certain types of cancers, developmental disorders and cardiovascular disease. Ligand binding induces proteolytic cleavage of the Notch receptor and nuclear translocation of the Notch intracellular domain (NICD), which forms a ternary complex with the transcription factor CSL and the coactivator MAML to upregulate transcription of Notch target genes. The DNA-binding protein CSL is the centrepiece of transcriptional regulation in the Notch pathway, acting as a molecular hub for interactions with either corepressors or coactivators to repress or activate, respectively, transcription. Here we review previous structure-function studies of CSL-associated coregulator complexes and discuss the molecular insights gleaned from this research. We discuss the functional consequences of both activating and repressing binding partners using the same interaction platforms on CSL. We also emphasize that although there has been a significant uptick in structural information over the past decade, it is still under debate how the molecular switch from repression to activation mediated by CSL occurs at Notch target genes and whether it will be possible to manipulate these transcription complexes therapeutically in the future.

Keywords

Notch · CSL · Structure analysis · RAM domain · Coactivator complex · Corepressor complex · DNA-binding · Transcription

Abbreviations

CBF1	C-promoter Binding Factor 1
LAG-1	abnormal cell LINeage-12 (Lin-12) And abnormal Germ line proliferation phenotype-1 (Glp-1)
RBP-J	Recombination Signal-Binding Protein for immunoglobin kappa J region
Su(H)	Suppressor of Hairless

F. Oswald (✉)
University Medical Center Ulm, Center for Internal Medicine, Department of Internal Medicine I, University of Ulm, Ulm, Germany
e-mail: franz.oswald@uni-ulm.de

R. A. Kovall (✉)
Department of Molecular Genetics, Biochemistry and Microbiology, University of Cincinnati College of Medicine, Cincinnati, OH, USA
e-mail: kovallra@ucmail.uc.edu

T. Borggrefe, B. D. Giaimo (eds.), *Molecular Mechanisms of Notch Signaling*, Advances in Experimental Medicine and Biology 1066, https://doi.org/10.1007/978-3-319-89512-3_14

CBP/CREBBP	C-Adenosine Mono Phosphate Responsive Element (cAMP-RE)-Binding protein (CREB)-Binding Protein; KAT3A
EP300	E1A Binding Protein P300, KAT3B
PCAF	P300/CBP-Associated Factor; KAT2B
GCN5	General Control Of AmiNo Acid Synthesis Protein 5-Like 2; KAT2A
CDK8	Cyclin-Dependent Kinase 8
SCF	S-Phase Kinase Associated Protein1/Cullin/F-Box Protein
SEL10	Suppressor and/or Enhancer of abnormal cell LINeage-12 (Lin-12)-10
FBWX7	F-Box and WD Repeat Domain containing 7
SIRT-1	Sirtuin-1
CARM1	Coactivator-Associated Arginine Methyltransferase1
PRMT4	Protein Arginine N-MethylTransferase 4
CTBP	C-Terminal Binding Protein
CTIP	CTBP Interacting Protein
KYOT2/FHL1	Four and a Half LIM domains 1
NCoR	Nuclear Receptor CoRepressor
SMRT	Silencing Mediator For Retinoid And Thyroid Hormone Receptors
SHARP	SMRT/HDAC1-Associated Repressor Protein
SPEN	SPlit ENds family transcriptional repressor
LID	Little Imaginal Disks
KDM5A	Lysine(K) Demethylase 5A
CIR	Corepressor Interacting with RBPJ
SKIP	Sloan-KetterIng-retroviral oncogene (SKI) -Interacting Protein
L3MBTL3	Lethal(3)Malignant Brain Tumor-Like Protein 3
RITA1	RBPJ Interacting and Tubulin Associated 1
EBNA2	Epstein-Barr Virus Nuclear Antigen 2
NFAT	Nuclear Factor of Activated T-cells
NF-κB1	Nuclear Factor κB1
POFUT1	Protein O-Fucosyltransferase 1
Fringe	Beta-1,3-N-Acetylglucosaminyltransferase

1 Introduction

The Notch signaling pathway is evolutionary conserved in metazoan organisms and represents a short-range cell-to-cell communication mechanism. A fly mutant with *"notches"* in its wing ends served as an eponym for the gene responsible for this particular phenotype (Morgan 1917). In 1985 the *Notch* gene was first cloned in *Drosophila melanogaster* and was found to encode a putative type I transmembrane protein with an extracellular region, a single transmembrane domain, and an intracellular region (Wharton et al. 1985). Further studies in *Drosophila* showed that the NOTCH protein serves as a receptor for two specific ligands, SERRATE and DELTA, which are also type I transmembrane proteins (Struhl and Adachi 1998; Artavanis-Tsakonas et al. 1999). During embryonic development and in the adult organism, Notch signaling affects and regulates stem cell maintenance, cell fate decisions, and cell lineage identity, as well as cell proliferation, differentiation and apoptosis (Borggrefe and Oswald 2009). These different outcomes of Notch signaling seem to be highly dependent on cellular context (Bray 2016). Although Notch signaling has pleiotropic functions, the pathway itself, which is devoid of second messengers and enzyme cascades, is mechanistically very simple.

Five ligands (JAGGED 1 and 2, DELTA-LIKE 1, 3 and 4) and four NOTCH receptors (Notch1–4) are present in mammals (Bray 2006; Kovall et al. 2017). The Notch receptor contains multiple epidermal growth factor (EGF)-like repeats (36 EGF repeats in mammalian NOTCH1), and three LNR (LIN1–2/Notch) repeats, which are located within the so-called Negative Regulatory Region (NRR) in the extracellular domain. The intracellular part of the Notch receptor contains the RAM (RBPJ-associated molecule) domain and seven ankyrin (ANK) repeats, which are followed by a trans-activation domain (TAD) and a PEST [rich in proline (P), glutamic acid (E), serine (S) and threonine (T) residues] domain at its carboxy terminus.

NOTCH receptors undergo multiple cleavage events and post-translational modifications during their maturation and in response to ligand binding (Fig. 1). The first cleavage event (S1) is ligand independent and occurs in the trans-golgi network by a furin-like convertase (Logeat et al. 1998) (Fig. 1A). S1 results in two protein fragments, the Notch extracellular domain (NECD) and an intracellular fragment that contains the transmembrane domain, which are non-covalently held together and presented as a heterodimer at the cell surface. The extracellular domains of NOTCH receptors are also modified by O-linked glycosylation and fucosylation, which can modulate specific ligand-receptor interactions, thereby affecting signaling outcome (Takeuchi and Haltiwanger 2010, 2014; Rana and Haltiwanger 2011). These modifications within the EGF repeats are catalyzed by protein O-fucosyltransferase 1 (POFUT1) and the fringe glycosyl-transferases RADICAL FRINGE (RFNG), LUNATIC FRINGE (LFNG) and MANIC FRINGE (MFNG) (Okajima et al. 2003; Moloney et al. 2000; Bruckner et al. 2000) (Fig. 1A). After ligand binding, a mechanical pulling force is thought to expose a second cleavage site (S2) in the NRR due to conformational changes that occur within the LNR domain (Fig. 1B) (Gordon et al. 2015). This ligand dependent cleavage step is catalyzed by members of the ADAM (A Disintegrin And Metalloprotease) metalloproteases family, ADAM10 and ADAM17 (Struhl and Greenwald 1999; Brou et al. 2000; Bozkulak and Weinmaster 2009). Subsequently, the remaining transmembrane NOTCH fragment, also called Notch extracellular truncation (NEXT), undergoes a final cleavage step (S3), which occurs within the cellular membrane and is catalyzed by the γ-secretase complex (Mumm et al. 2000). Cleavage at the S3 site releases the Notch intracellular domain (NICD) from the cell membrane (Schroeter et al. 1998) and subsequently NICD translocates to the nucleus to activate transcription of Notch target genes (Fig. 1C) (Struhl and Adachi 1998).

NICD does not bind to DNA itself but rather interacts with the DNA binding transcription factor CSL [for CBF1/RBPJ (C-promoter Binding Factor1/ Recombination Binding Protein Jk), Su(H) (Suppressor of Hairless), and Lag-1] and the transcriptional coactivator MASTERMIND-LIKE (MAML) to form a DNA-bound transactivation complex (Nam et al. 2006; Wilson and Kovall 2006; Kopan and Ilagan 2009; Kovall and Blacklow 2010). The CSL-NICD-MAML transactivation complex recruits histone modifying coactivators, like CREBBP/EP300 (CREB Binding Protein/E1A Binding Protein P300) or PCAF (P300/CBP-associated factor, *aka* KAT2B) and GCN5 (General control of amino acid synthesis protein 5, *aka* KAT2A), together with chromatin remodeling complexes to activate transcription (Fig. 1D) (Kurooka and Honjo 2000; Oswald et al. 2001; Wallberg et al. 2002; Kadam and Emerson 2003). NICD is a short-lived protein, as its PEST domain is phosphorylated by CYCLINC/CDK8 (Fryer et al. 2004), resulting in its ubiquitilation by the SCF/SEL10/FBXW7 E3 ubiquitin ligase complex, leading to its degradation by the proteasome (Fig. 1E). A number of additional post-translational modifications regulate the activity and stability of NICD, *e.g.* deacetylation by SIRT-1 (silent mating type information regulation 2 homolog, *aka* SIRTUIN1) (Guarani et al. 2011) and methylation by CARM1 (Coactivator Associated Arginine Methyltransferase 1)/PRMT4 (Protein Arginine N-Methyltransferase 4) (Hein et al. 2015), which regulate the amplitude and duration of the Notch response (Wu et al. 2001; Tsunematsu et al. 2004).

In the absence of an active Notch signal CSL acts as a transcriptional repressor (Fig. 1F) (Dou et al. 1994). In *Drosophila*, the CSL ortholog Su(H) recruits the HAIRLESS/CtBP (C-terminal Binding Protein)/GROUCHO corepressor complex (Morel et al. 2001; Barolo et al. 2002). In vertebrates, RBPJ directly interacts with corepressor components KYOT2/FHL1 (Taniguchi et al. 1998), SHARP (SMRT/HDAC1-associated

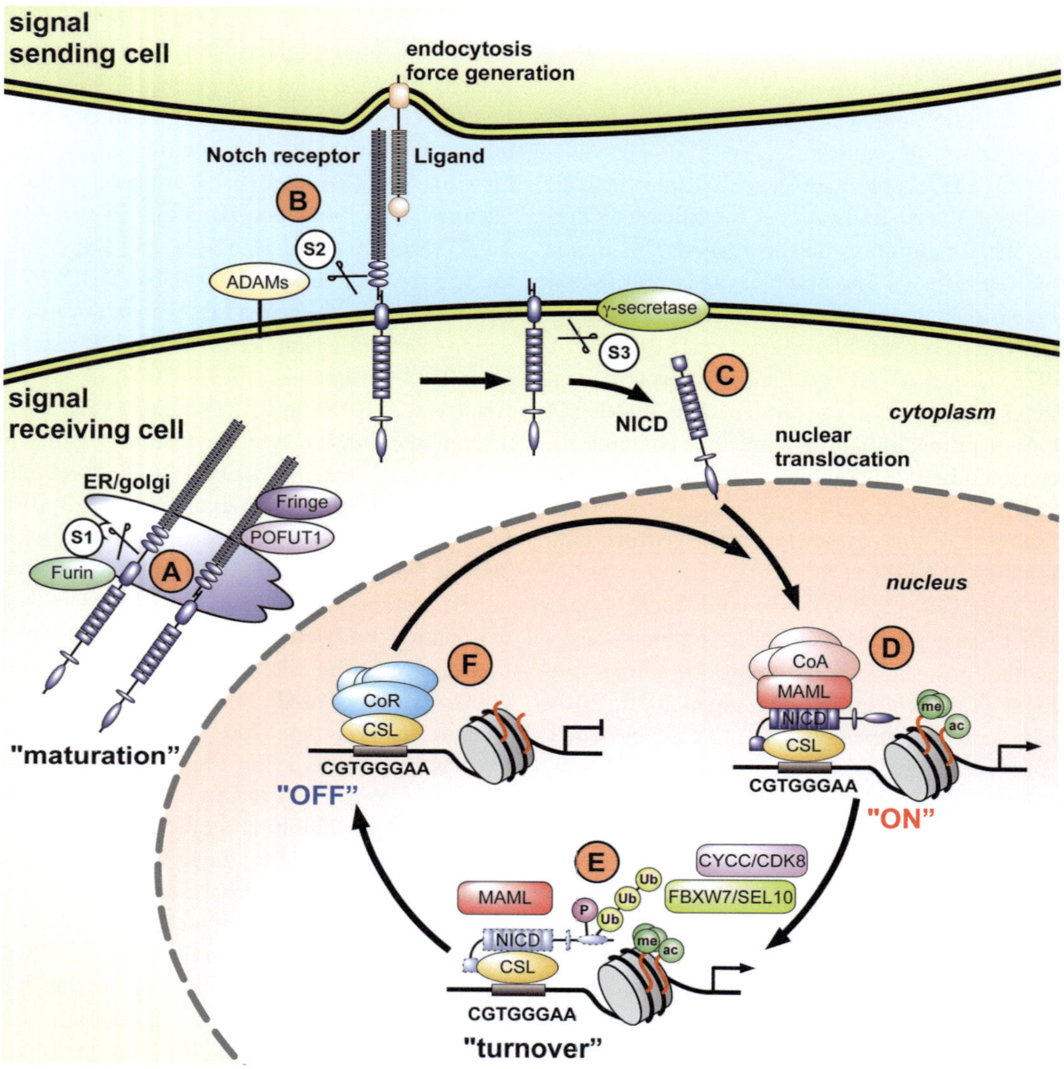

Fig. 1 Schematic representation of the major molecular events during Notch signaling: (**A**) Posttranslational modifications of the Notch receptor during maturation in the trans-golgi network. The Notch receptor precursor protein is cleaved by a furin convertase (S1). Protein fragments are non-covalently linked together as a heterodimer. Additional modifications are catalyzed by fringe glycosyltransferases (FRINGE) and protein O-fucosyltransferase 1 (POFUT1). (**B**) Notch receptors and ligands are single transmembrane spanning proteins. Ligand binding and its endocytosis generate a mechanical pulling force to expose the second cleavage site (S2) and processing by ADAM family metalloproteases. (**C**) A further cleavage step (S3) is catalyzed by a gamma-secretase containing complex, releasing the Notch Intracellular Domain (NICD) that translocates to the nucleus. (**D**) Nuclear NICD interacts with the DNA-binding protein CSL and recruits a coactivator complex composed of Mastermind (MAML) and additional chromatin modifying factors to activate transcription of Notch target genes ("ON"). (**E**) Phosphorylation by the mediator subunit CYCLINC/CDK8 and subsequent ubiquitylation by the FBXW7/SEL10 containing E3 ubiquitin ligase complex lead to rapid degradation of NICD by the proteasome ("turnover"). (**E**) In the absence of activated Notch signaling, CSL recruits various corepressor complexes to downregulate transcription of Notch target genes

repressor protein)/SPEN (Split Ends), also called MINT (Msx2-interacting nuclear target protein) (Oswald et al. 2002; Kuroda et al. 2003), L3MBTL3 [Lethal(3)Malignant Brain Tumor-Like Protein 3] (Xu et al. 2017), the H3K4 demethylase KDM5A [Lysine (K)-Specific Demethylase 5A)/LID (Little imaginal discs), (Moshkin et al. 2009; Liefke et al. 2010) or other cofactors like CIR (CBF1-Associated Corepressor) (Hsieh et al. 1999) and SKIP (Ski-interacting protein) (Zhou et al. 2000). These direct RBPJ binding partners recruit further corepressors, such as CtIP (CtBP interacting protein)/CtBP (Oswald et al. 2005), NCoR(Nuclear receptor corepressor1)/SMRT (silencing mediator for retinoid or thyroid-hormone receptors) (Zhou and Hayward 2001; Oswald et al. 2016), histone modifying enzymes (Xu et al. 2017; Hsieh et al. 1999; Olave et al. 1998) or Polycomb complex components (Qin et al. 2004; Qin et al. 2005) to silence Notch target genes. Therefore, CSL has dual roles within the Notch signaling pathway, acting either as an activator or repressor of transcription, depending on the status of Notch activity. As CSL plays a pivotal role in the regulation of transcription of Notch target genes, here we review the X-ray structures of CSL-mediated transcription complexes and what has been learned from these structural studies.

2 Overall Fold of Transcription Factor CSL

CSL proteins are DNA binding proteins that recognize the consensus sequence –C/tGTGGGAA– (Del Bianco et al. 2010; Meng et al. 2005; Tun et al. 1994) and regulate transcriptional activation and repression of Notch target genes by interacting with coactivators and corepressors, respectively. As originally shown in the X-ray structure of LAG-1 bound to DNA (Fig. 2A) (Kovall and Hendrickson 2004), all CSL proteins contain a conserved structural core that is largely composed of β-strands and consists of three domains: NTD (N-terminal domain), BTD (β-trefoil domain), and CTD (C-terminal domain). Additionally, CSL proteins from different organisms contain poorly conserved N- and C-terminal extensions of the structural core that appear unstructured by secondary-structure/disorder prediction algorithms. In general, the function of these regions is not well understood, but in certain orthologs the N-terminal regions appear to play a role in DNA binding and cooperative interactions with other transcription factors (Prevorovsky et al. 2011; Neves et al. 2007).

The NTD and CTD have immunoglobin type folds, whereas the BTD has a β-trefoil fold, similar to fibroblast growth factors and interleukin-1 (Kovall and Hendrickson 2004). The BTD of CSL has an atypical β-trefoil fold, as it is missing two of the canonical 12 β-strands that compose the classic β-trefoil fold. This results in a large exposed hydrophobic pocket on the surface of CSL, which is the binding site for many of the coregulators that interact with CSL (*see below*), including the RAM domain of NOTCH (Wilson and Kovall 2006; Friedmann et al. 2008), FHL1 (Four and a half LIM domains protein 1) (*aka* KyoT2) (Collins et al. 2014), RITA1 (RBPJ-interacting and tubulin-associated protein 1) (Tabaja et al. 2017), EBNA2 (Epstein–Barr virus nuclear antigen 2) (Johnson et al. 2010), and SPEN (*aka* MINT or SHARP) (VanderWielen et al. 2011). CSL proteins share some structurally similarity to the Rel Homology Domain (RHD) proteins, such as the transcription factors NF-κB1 (Nuclear Factor-κB1) and NFAT (Nuclear factor of activated T-cells) (Kovall and Hendrickson 2004). The NTD and CTD of CSL structurally align with RHD-N and RHD-C domains, respectively. However, the overall fold of CSL is distinct from other RHD members in that the BTD lies between the RHD-N and RHD-C domains of CSL, whereas typical RHD proteins have a RHD-N immediately followed by a RHD-C domain. Moreover, RHD proteins typically bind DNA as homodimers or heterodimers, whereas CSL proteins bind DNA as monomers. The NTD and BTD of CSL form a continuous electropositive surface in which to interact with DNA (Fig. 2) (Kovall and Hendrickson 2004). Much like other RHD proteins, the NTD of CSL inserts a β-hairpin loop within the major groove of DNA to make both specific and nonspecific contacts,

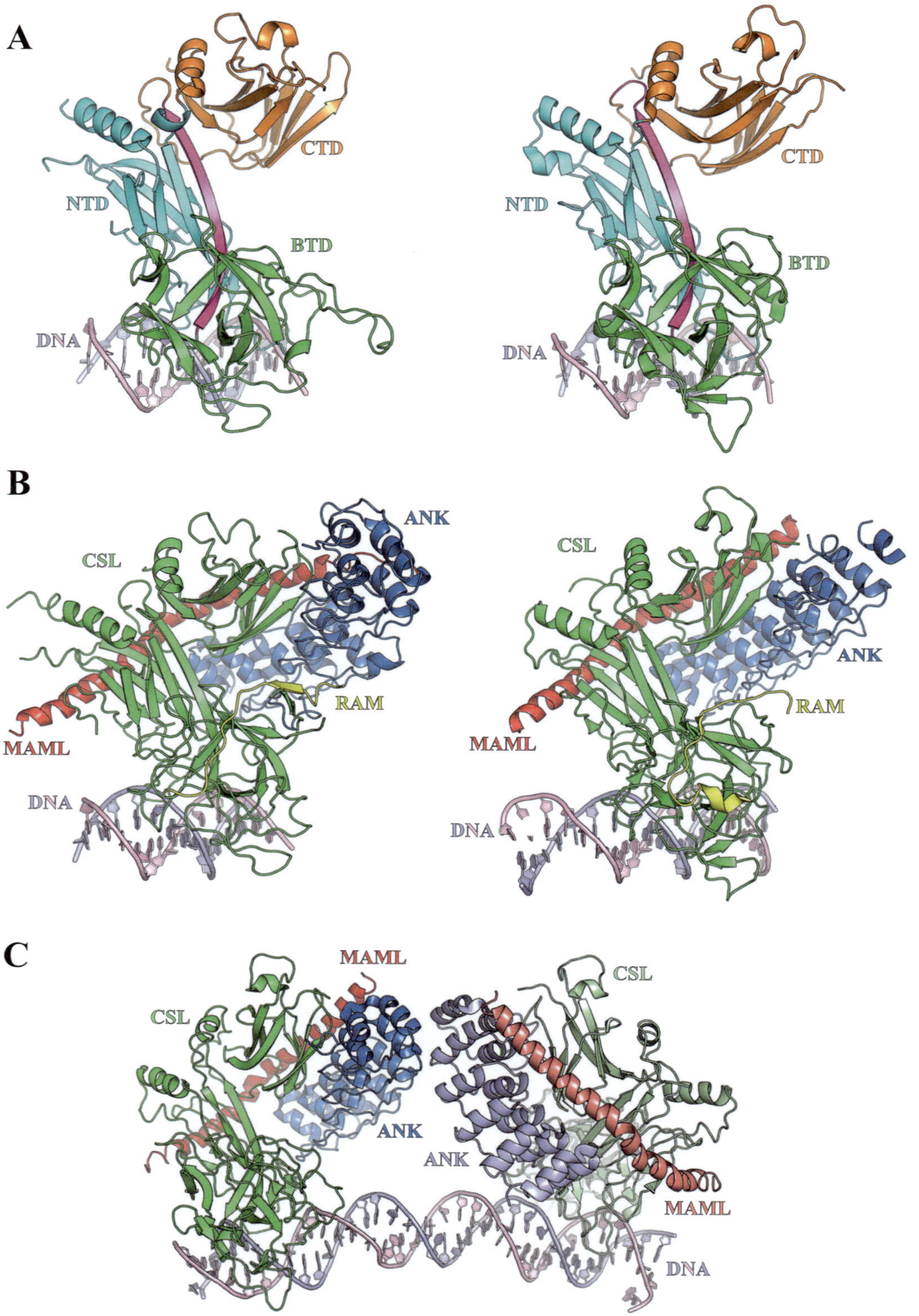

Fig. 2 X-ray structures of unbound CSL and CSL-NICD-MAML ternary complexes bound to DNA: (**A**) *Left,* ribbon diagram of LAG-1 bound to DNA (PDBID: 1TTU) and *right,* ribbon diagram of RBPJ bound to DNA(PDBID: 3IAG). The NTD, BTD, and CTD are colored cyan, green, and orange respectively. A β-strand that

largely recognizing the second half of its consensus binding site (–GGGAA–). The BTD also contributes to DNA binding, in which a β-hairpin loop inserts into the minor groove of DNA, making both specific and nonspecific contacts to the first base steps in the consensus binding site (–CG–) (Fig. 2).

3 The CSL-NICD-MAML Activation Complex

An obligatory step to activate transcription of target genes in response to a Notch signal is the formation of the ternary complex composed of CSL, NICD, and a member of the MAML family of transcriptional coactivators (MAML1–3 in mammals). The activation complex structures of the *Caenorhabditis elegans* and human orthologous proteins have been determined (Fig. 2B) (Nam et al. 2006; Wilson and Kovall 2006), and demonstrate that the RAM domain and ANK repeats of NICD bind the BTD and CTD, respectively, of CSL. MAML, which adopts a short bent α-helical conformation in the complex, forms a tripartite interaction with ANK, and the CTD and NTD of CSL (Fig. 2B). Similar to the RHD-C domains in other proteins, the CTD of CSL functions as a protein-protein interaction domain, binding MAML and NICD in the activation complex, as well as the corepressors SPEN and HAIRLESS detailed below (VanderWielen et al. 2011; Yuan et al. 2016). MAML coactivators are relatively large proteins (~1000 residues) that also interact with CBP/EP300 and the CDK8 module of the Mediator complex to activate transcription (Oswald et al. 2001; Wallberg et al. 2002; Fryer et al. 2004), but only require a small N-terminal domain to form a complex with NICD and CSL (Fig. 2B) (Nam et al. 2006; Wilson and Kovall 2006; Nam et al. 2003). Interestingly, constructs that only correspond to this N-terminal region are termed DN-MAML (dominant-negative MAML), and expressed in cells, these constructs are potent inhibitors of Notch signaling due to the ability of DN-MAML to form ternary complexes with CSL-NICD, but are unable to recruit CBP/EP300 and CDK8 to activate transcription (Weng et al. 2003).

The RAM domain of NICD binds in an extended conformation across the BTD of CSL in a manner that blankets the exposed hydrophobic surface on the BTD (Figs. 2B and 3A) (Wilson and Kovall 2006; Choi et al. 2012). The RAM domains of all NOTCH receptors (NOTCH1–4 in mammals), as well as a number of other coregulators that bind BTD, have a conserved hydrophobic tetrapeptide motif (φWφP), where φ is any nonpolar amino acid. In addition to the φWφP motif, RAM domains have other conserved motifs that are important for interacting with BTD, including an N-terminal basic region, and –HG– and –GF– dipeptide motifs (Johnson et al. 2010; Lubman et al. 2007). Interestingly, other coregulators that bind BTD similarly to RAM share some, but not all of these other motifs conserved in RAM. Prior to interacting with CSL, RAM is a random coil in solution (Nam et al. 2003; Bertagna et al. 2008). While RAM is ~100 residues in length, only ~20 N-terminal residues are required for interacting with the BTD of CSL (Wilson and Kovall 2006; Friedmann et al. 2008; Choi et al. 2012). The remaining ~80 residues between the RAM domain and ANK repeats of NICD were not resolved in the X-structure of the activation complex (Fig. 2B). However, this intervening region appears to be important for formation of the ternary complex, because (1) statistical models suggest that the length of RAM has been tuned through evolution to optimize the interactions between ANK and CTD (Bertagna et al. 2008), and (2) mutation of

Fig. 2 (continued) makes hydrogen bonding interactions with all three domains is colored magenta. The DNA is colored light pink and light blue. (**B**) Ribbon diagrams of CSL-NICD-MAML ternary complexes bound to DNA for Notch components from *Caenorhabditis elegans* (*left*, PDBID: 2FO1) and humans (*right*, PDBID: 3V79). CSL and MAML are colored green and red, respectively; the ANK and RAM domains of NICD are colored blue and yellow, respectively; and the DNA is colored light pink and light blue. (**C**) Ribbon diagram of dimeric CSL-NICD-MAML complexes bound to SPS element. Coloring is the same as (**B**)

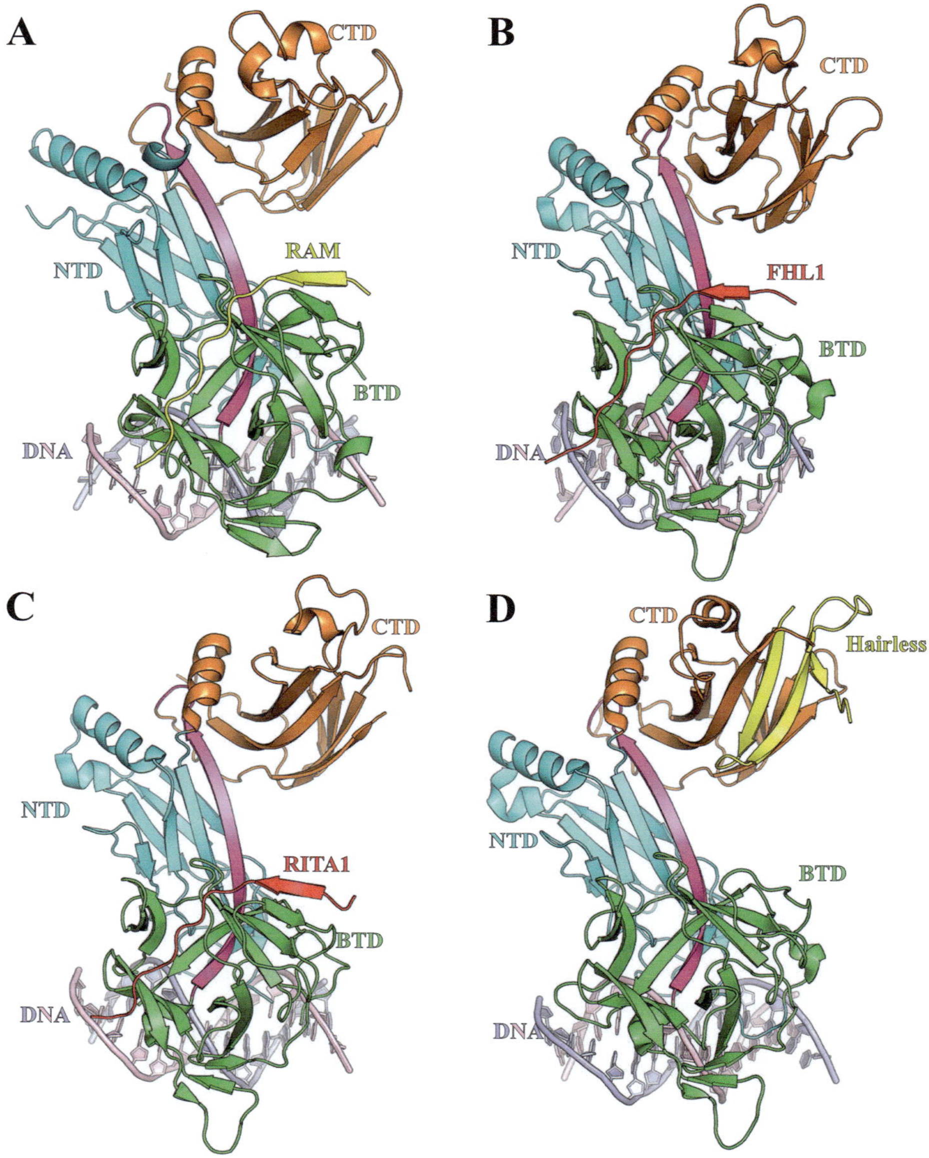

Fig. 3 X-ray structures of CSL-coregulator complexes: (**A**) Ribbon diagram of the RAM domain of NICD bound to LAG-1 and DNA. Coloring is the same as Fig. 2. (**B**) Ribbon diagram of the RBPJ-FHL1 complex bound to DNA. RBPJ-DNA coloring is the same as Fig. 2 and FHL1 is colored red. (**C**) Ribbon diagram of the RBPJ-RITA1-DNA complex. RBPJ-DNA coloring is the same as Fig. 2 and RITA1 is colored red. (**D**) Ribbon diagram of the Su(H)-HAIRLESS-DNA corepressor complex. Su(H)-DNA is colored the same as Fig. 2 and HAIRLESS is colored yellow

sequence specific elements within this intervening region of RAM adversely affect cellular reporter assays, suggesting that this region also contributes to proper transcriptional activation by NICD (Sherry et al. 2015).

There are seven ankyrin repeats within the ANK domain of NICD, as well as an N-terminal capping repeat (Fig. 2B) (Wilson and Kovall 2006). The folding of the terminal repeats is coupled to forming a complex with CSL and MAML

(Choi et al. 2012). There are several structures of the isolated ANK repeats of NICD (Nam et al. 2006; Zweifel et al. 2003), which overlay with a high degree of correspondence with the ANK repeats from the ternary complex structures, suggesting that formation of the CSL-NICD-MAML ternary complex does not induce any large conformational changes within ANK. There was a large rigid body shift observed in the domains of CSL when comparing the unbound structure with the activation complex, such that CSL assumed a more compact conformation with its BTD and CTD moving closer together (Wilson and Kovall 2006). However, these domain movements were only observed in the ternary complex structure with the *Caenorhabditis elegans* orthologous proteins. Whether this conformational change is organism specific or a general property of the activation complex remains to be determined.

In vitro studies using purified recombinant proteins have analyzed the interactions that consititute the CSL-NICD-MAML ternary complex and suggest that its assembly occurs in a stepwise manner (Kovall and Blacklow 2010). The RAM domain of NOTCH was originally identified in a yeast two-hybrid screen for RBPJ binding partners (Tamura et al. 1995) and subsequently shown to form a high affinity (K_d ~10 nM) interaction with the BTD of CSL (Friedmann et al. 2008; Lubman et al. 2007; Del Bianco et al. 2008). In the absence of MAML, the ANK repeats of NOTCH bind weakly to the CTD of CSL (Friedmann et al. 2008; Lubman et al. 2007; Del Bianco et al. 2008). Interestingly, the affinity of ANK for CTD seems to vary in different organisms – in mammals and nematodes the ANK-CTD interaction is very weak and technically difficult to detect (Friedmann et al. 2008; Lubman et al. 2007; Del Bianco et al. 2008), whereas the affinity of ANK for CTD in flies is stronger and binds with ~0.5 μM affinity (Contreras et al. 2015). Why the strength of ANK-CTD interactions varies in different organisms is unclear, but in the case for *Drosophila*, perhaps this is due to competition with the corepressor HAIRLESS, which also binds the CTD with high affinity (K_d ~1 nM) (Maier et al. 2011). MAML does not interact with CSL or NICD individually, but binds to the preformed CSL-NICD binary complex, rigidifying and stabilizing the ternary complex (Nam et al. 2003; Choi et al. 2012). To date, there are no studies that have quantitated the affinity of MAML for CSL-NICD. Taken together, these studies suggest that the high affinity RAM interaction for BTD targets NICD to CSL in the nucleus. The intrisically disordered region of RAM ideally positions ANK to interact with the CTD, and subsequently, MAML binds a groove formed by the CTD and ANK (Kovall and Blacklow 2010).

4 CSL-DNA Binding

CSL proteins bind the consensus sequence –C/tGTGGGAA– with a modest affinity of ~100 nM (Friedmann and Kovall 2010), although some known *in vivo* sites that deviate from the consensus bind considerably weaker (K_d ~1 μM) (Torella et al. 2014). The residues in CSL that contact DNA are absolutely conserved and comparative binding/structural studies of the mouse, worm, and fly orthologs suggest that all CSL proteins bind DNA in a similar manner with similar affinities (Kovall and Hendrickson 2004; Friedmann and Kovall 2010). This is in contrast to the protein-protein interactions that CSL makes with coregulators, e.g. the RAM domain of NICD, in which the affinities for complex formation can vary significantly (>10 fold) (Friedmann et al. 2008; Contreras et al. 2015). As mentioned above, the NTD of CSL interacts specifically with the major groove of DNA, whereas the BTD makes specific contacts in the minor groove (Kovall and Blacklow 2010). All CSL structures to date show very similar major groove contacts made by the NTD; however, in some CSL structures there is variability in how the BTD contacts the minor groove of DNA (Friedmann et al. 2008; Yuan et al. 2016; Friedmann and Kovall 2010). Specifically, a β-hairpin loop in BTD can assume several different conformations to make seemingly equivalent specific and nonspecific interactions with DNA. This may suggest that the BTD can assume different conformations to interact with DNA depending on the nearby base pairs it contacts, which is consistent with the variability observed in the consensus sequence for CSL (Del Bianco et al. 2010; Meng et al. 2005; Tun et al. 1994).

In addition to binding monomeric DNA binding sites, in some metazoans CSL can also bind dimeric sites, which are known as SPS [Su(H) Paired Sites or Sequence Paired Sites] (Bailey and Posakony 1995). SPS are composed of two CSL binding sites arranged in a head-to-head arrangement with 15–19 base pairs separating the two sites (Nam et al. 2007). A typical SPS contains two CSL consensus-binding sites; however, cryptic paired sites have also been identified, in which one of the DNA binding sites significantly deviates from the consensus and is unable to support binding of monomeric CSL complexes (Arnett et al. 2010). When the Notch pathway is activated, two CSL-NICD-MAML can bind an SPS in a cooperative manner, whereby modest interactions between the ANK repeats of the two NICD molecules mediate the cooperativity (Fig. 2C). Interestingly, mutations that abrogate the cooperative interactions between ANK molecules affect transcription from Notch target genes that contain an SPS, but have no effect on targets that only contain monomeric sites (Arnett et al. 2010).

Classical models of Notch transcriptional regulation posit that CSL is constitutively bound to DNA, and corepressors and coactivators are exchanged on the DNA (Kao et al. 1998; Hsieh and Hayward 1995). However, more recent studies cast serious doubt on this model and suggest that the exchange of CSL-mediated corepressor and coactivator complexes is a much more dynamic process, and likely occurs in the nucleoplasm rather than while CSL is bound to DNA (Castel et al. 2013; Krejci and Bray 2007). Previous genome wide studies have shown that when the Notch pathway is activated CSL binds more sites at target genes than when the pathway is inactive (Castel et al. 2013; Krejci and Bray 2007; Hass et al. 2016). Although the molecular basis of this observation is unknown, generally speaking, two possibilities exist: (1) the affinity of CSL for DNA increases when bound to NICD and MAML and/or other general transcription factors; and (2) increased Notch activity or the activity of other transcription factors, *e.g.* pioneer factors, change the local chromatin environment, making it more accessible for CSL to bind. *In vitro* studies have shown that neither the affinity of CSL for DNA changes when it is bound to coregulators, such as NICD, FHL1, HAIRLESS, SPEN or RITA1, nor does the specificity of CSL change when bound to NICD and MAML (Del Bianco et al. 2010; Friedmann et al. 2008; Collins et al. 2014; Tabaja et al. 2017; VanderWielen et al. 2011; Maier et al. 2011). Albeit these *in vitro* studies have used only constructs that correspond to the structural cores of CSL, NICD, and MAML, and have not been performed with full-length proteins. Thus, it is an open question in the field as to what is the molecular basis that underlies the observed increase in CSL binding genome wide when Notch is active in cells.

5 CSL as a Repressor

Without a doubt CSL, in conjunction with NOTCH and MAML, plays an essential role in the upregulation of transcription from all Notch target genes in all organisms; however, its role as a transcriptional repressor is a bit more enigmatic and may have different roles in different organisms. In the model organism *D. melanogaster*, there is overwhelming genetic, cellular/biochemical, and structural evidence that Su(H) (the fly CSL ortholog), when in complex with the corepressor HAIRLESS, functions as a transcriptional repressor (Brockmann et al. 2014; Maier 2006). In other organisms, such as mammals and nematodes, the function of CSL as a repressor is not as clear. There is compelling biochemical, cellular, and structural data that RBPJ (the mammalian CSL ortholog) interacts with the corepressors FHL1, RITA1, SPEN, and L3MBTL3 (Taniguchi et al. 1998; Oswald et al. 2002; Kuroda et al. 2003; Xu et al. 2017; Tabaja et al. 2017; VanderWielen et al. 2011; Wacker et al. 2011). However, there is not a preponderance of genetic data supporting the function of RBPJ as a repressor. Nonetheless, there are several cellular and genetic studies that suggest loss of RBPJ results in upregulation of transcription at some Notch target genes (Castel et al. 2013; Hu et al. 2012; Surendran et al. 2010). Interestingly, loss of RBPJ has been shown *in vivo* to promote tumorigenesis (Kulic et al. 2015), suggesting its role as a repressor may be important for tumor suppressor functions. While its role as a transcriptional repressor in the Notch pathway remains to be completely elucidated, the emerging picture seems to suggest that CSL is required for activation of all target genes, but its role as a repressor is important for a subset of target genes.

HAIRLESS is the major antagonist of Notch signaling in *Drosophila* and binds Su(H) with high affinity via a relative short peptide-like sequence (Yuan et al. 2016; Maier et al. 2011). HAIRLESS also interacts with the corepressors CtBP (C-terminal Binding Protein) and GROUCHO in order to function as a transcriptional repressor (Morel et al. 2001; Barolo et al. 2002; Nagel et al. 2005). Consistent with previous studies, HAIRLESS binds the CTD of Su(H) (Fig. 3D) (Yuan et al. 2016; Maier et al. 2011). Unexpectedly, HAIRLESS binding induces a large conformational change in the CTD, whereby HAIRLESS wedges itself between the two β-sheets that compose the Ig fold of the CTD (Yuan et al. 2016). This results in HAIRLESS primarily interacting with residues that form the hydrophobic core of the CTD rather than surface exposed residues (Fig. 3D). This large structural change is incompatible with NICD and MAML binding (Yuan et al. 2016). In future studies, it will be interesting to see whether other coregulators, such as SPEN, interact with this conserved binding pocket on the CTD.

Two other corepressors, FHL1 and RITA1, interact with RBPJ via a peptide-like sequence that resembles the RAM of NICD (Fig. 3B, C) (Taniguchi et al. 1998; Collins et al. 2014; Tabaja et al. 2017; Wacker et al. 2011). FHL1 proteins are characterized by N-terminal LIM (**LIN**11, **I**SL-1 & **M**EC-3) domains, which are protein-protein interaction motifs thought to interact with PRC (Polycomb Repressive Complex), and a C-terminal sequence that binds the BTD of RBPJ (Fig. 3B) (Qin et al. 2004; Qin et al. 2005). FHL1 binds RBPJ with high affinity and has a hydrophobic tetrapeptide sequence similar to RAM (Collins et al. 2014). However, FHL1 does not contain the other motifs in RAM, e.g. N-terminal basic residues, and –HG– and –GF–, required for high affinity binding of RBPJ. RITA1 also contains a hydrophobic tetrapeptide motif that is essential for its interaction with RBPJ and is also missing the other motifs in RAM that are required for high affinity interactions with RBPJ (Fig. 3C) (Tabaja et al. 2017; Wacker et al. 2011). In contrast to FHL1, RITA1 only binds RBPJ with moderate affinity (~1uM K_d) (Tabaja et al. 2017). Additionally, RITA1 has other functional domains, such as nuclear import and export sequences, and a C-terminal domain that interacts with tubulin, and interestingly, RITA1 appears to have Notch independent functions outside the nucleus (Wacker et al. 2011; Steinhauser et al. 2016).

6 Coregulator Competition

An open question in the field is whether corepressors and coactivators compete for binding to CSL in the nucleus, or alternatively, are there different pools of CSL-mediated transcription complexes in the nucleus that are then recruited to different Notch target genes. As mentioned previously, the classical model of Notch signaling proposes that CSL is constitutively bound to DNA, and in the absence of a Notch signal, DNA bound CSL-corepressor complexes actively repress transcription from Notch target genes; when Notch becomes activated in the cell, NICD translocates to the nucleus directly binding CSL, recruiting MAML and simultaneously displacing corepressors, thereby activating transcription at these sites. Numerous *in vitro* studies have shown that corepressors and coactivators can compete for binding to CSL. In pulldown assays from cellular extracts it has been shown that overexpression of one coregulator can displace the binding of another coregulator to CSL (Xu et al. 2017). For example, overexpression of NICD in cells can outcompete SHARP/SPEN for binding to CSL (Oswald et al. 2002; Kuroda et al. 2003).

Similarly, with purified recombinant proteins it has been shown that coregulators can compete for binding to CSL (Johnson et al. 2010; VanderWielen et al. 2011). Another example is the competitive binding of NICD and HAIRLESS for Su(H) (Maier et al. 2011). In this case, NICD is very effective at competing off HAIRLESS bound to Su(H) even in the absence of MAM. Similar experiments performed with the mammalian proteins RBPJ, MAML, NICD, and SPEN demonstrate that MAML is required for NICD to effectively compete off SPEN binding to RBPJ (VanderWielen et al. 2011). While it has been shown *in vitro* that corepressors and coactivators can compete for binding to CSL, it is not clear whether this actually occurs in cells under normal physiological conditions. Put another way, does every NICD molecule have to compete with a CSL bound corepressor in order to activate transcription or are their free molecules of CSL in the nucleus that NICD can easily access, and therefore corepressor displacement is an *in vitro* artifact? At the present time it is unclear whether one or both of these mechanisms are functioning in cells. Certainly, future studies that quantitate the number of CSL, corepressor, and NICD molecules within the cell, coupled with the known *in vitro* affinities of these complexes, will then begin to allow for a clearer picture of whether coregulators compete for CSL binding or not.

7 Modulation of CSL-Mediated Transcription Complexes

Given that numerous corepressors and coactivators bind to the BTD of CSL raises the question as to whether small molecules or biologic reagents can be identified that inhibit the binding of one, or some, coregulators, but not inhibit interactions with all coregulators. On the face of it this seems to be an arduous task because of the structurally similar manner, in which many coregulators bind to the nonpolar surface on the BTD of CSL. However, there is some experimental data that suggests it may be possible to identify reagents that selectively inhibit one coregulator, sparing the binding of others. A number of years ago, the Kempkes laboratory, using a yeast two-hybrid screen, identified mutations in RBPJ that selectively inhibited binding to the RAM domain of NICD or the viral coactivator EBNA2, but not to both (Fuchs et al. 2001). Interestingly, these subtle mutations lie right in the middle of the RAM binding site on the BTD. More recently, these binding results were confirmed by the Barrick laboratory using purified recombinant proteins and isothermal titration calorimetry (Johnson et al. 2010). Moreover, RBPJ binding data from the Kovall laboratory is consistent with the Kempkes results, i.e. in some cases mutations in the BTD can have drastically different impacts on the binding of different coregulators (Xu et al. 2017; Collins et al. 2014; Tabaja et al. 2017; Yuan et al. 2012). Taken together, these results raise the exciting prospect that it may be possible to identify selective reagents that affect either the repression or activation function of CSL, but not both, which could have biomedical applications for human diseases that are characterized by either insufficient or overactive Notch signaling.

8 Summary, Concluding Remarks and Open Questions

Progress made over the past decade has provided amazing insights into the molecular structures of the transcriptional components of the Notch signaling pathway. Available structures that contain CSL transcription complexes are summarized in Table 1. Structural studies of CSL-associated coactivator and corepressor complexes from different species have revealed the intriguing evolutionary conservation of these molecular interactions and mechanism, albeit with some species-specific differences. We now know that many corepressors interact with CSL by "mimicking" the RAM domain of NICD and its interactions with the BTD of CSL; however, there appear to be significant differences associated with their affinities and specificities for CSL. This has led to an understanding as to why there is competitive binding of NICD and KyoT2 or

Table 1 Available CSL complex structure data (**P**rotein **D**ata **B**ank, PDB)

PDB-ID	Complex	Species	Reference
1TTU	CSL bound to DNA	*C. elegans*	Kovall et al. (2004)
2FO1	Activator complex bound to DNA[a]	*C. elegans*	Wilson et al. (2006)
2F8X	Activator complex bound to DNA[b]	*H. sapiens*	Nam et al. (2006)
3BRD	CSL-RAM bound to DNA	*C. elegans*	Friedmann et al. (2008)
3BRF	CSL-RAM bound to DNA	*C. elegans*	Friedmann et al. (2008)
3BRG	CSL bound to DNA	*M. musculus*	Friedmann et al. (2008)
3NBN	Activator complex dimer bound to DNA	*H. sapiens*	Arnett et al. (2010)
3V79	Activator complex bound to DNA[a]	*H. sapiens*	Choi et al. (2012)
3IAG	CSL bound to DNA	*M. musculus*	Friedmann et al. (2010)
4J2X	Repressor complex bound to DNA[c]	*M. musculus*	Collins et al. (2014)
5E24	Repressor complex bound to DNA[d]	*D. melanogaster*	Yuan et al. (2016)
5EG6	Repressor complex bound to DNA[e]	*M. musculus, H. sapiens*	Tabaja et al. (2017)

[a](CSL/ANK/RAM/MAML)
[b](CSL/ANK/MAML)
[c](CSL/KYOT2)
[d](Su[H]/HAIRLESS)
[e](CSL/RITA1)

RITA1 for CSL. Future studies that seek to elucidate the structures of CSL complexes like CSL-SHARP and CSL-L3MBTL3 will provide additional molecular insights into how CSL functions as a repressor and will further refine our knowledge of these transcription factor-switching mechanisms. Despite this progress, there are still a lot of open questions in the field, for example: (I) Do the CSL-associated coactivator and corepressor complexes exchange on DNA or are there pre-existing complexes in the nucleoplasm or is it some combination of both mechanisms? (II) Are CSL-corepressor complexes gene-, binding site- and cell type-specific, and if so, how are these specificities regulated? (III) Does CSL DNA-binding affinity change when complexed with NICD or corepressors? And finally, (IV) will it be possible to manipulate CSL specific cofactor binding with small molecules or biologics in order to modulate the Notch response for clinical applications in the future?

Acknowledgments We want to thank Bernd Baumann for critical reading of the manuscript. Research in the F.O. laboratory is supported by the DFG (SFB1074/A3) and the BMBF (Federal Ministry of Education and Research, research nucleus SyStAR). Research in the R.A.K. laboratory is supported by the NIH (CA178974), NSF (MCB-1715822), and the Bankhead-Coley Cancer Research Program.

References

Arnett KL, Hass M, McArthur DG, Ilagan MX, Aster JC, Kopan R, Blacklow SC (2010) Structural and mechanistic insights into cooperative assembly of dimeric Notch transcription complexes. Nat Struct Mol Biol 17(11):1312–1317. https://doi.org/10.1038/nsmb.1938

Artavanis-Tsakonas S, Rand MD, Lake RJ (1999) Notch signaling: cell fate control and signal integration in development. Science 284(5415):770–776 https://doi.org/10.1126/science.284.5415.770

Bailey AM, Posakony JW (1995) Suppressor of hairless directly activates transcription of enhancer of split complex genes in response to Notch receptor activity. Genes Dev 9(21):2609–2622 https://doi.org/10.1101/gad.9.21.2609

Barolo S, Stone T, Bang AG, Posakony JW (2002) Default repression and Notch signaling: Hairless acts as an adaptor to recruit the corepressors Groucho and dCtBP to Suppressor of Hairless. Genes Dev 16(15):1964–1976. https://doi.org/10.1101/gad.987402

Bertagna A, Toptygin D, Brand L, Barrick D (2008) The effects of conformational heterogeneity on the binding of the Notch intracellular domain to effector proteins: a case of biologically tuned disorder. Biochem Soc Trans 36(Pt 2):157–166. https://doi.org/10.1042/BST0360157

Borggrefe T, Oswald F (2009) The Notch signaling pathway: transcriptional regulation at Notch target genes. Cell Mol Life Sci 66(10):1631–1646 https://doi.org/10.1007/s00018-009-8668-7

Bozkulak EC, Weinmaster G (2009) Selective use of ADAM10 and ADAM17 in activation of Notch1 signaling. Mol Cell Biol 29(21):5679–5695. https://doi.org/10.1128/mcb.00406-09

Bray SJ (2006) Notch signalling: a simple pathway becomes complex. Nat Rev Mol Cell Biol 7(9):678–689. https://doi.org/10.1038/nrm2009

Bray SJ (2016) Notch signalling in context. Nat Rev Mol Cell Biol 17(11):722–735. https://doi.org/10.1038/nrm.2016.94

Brockmann B, Mastel H, Oswald F, Maier D (2014) Analysis of the interaction between human RITA and Drosophila Suppressor of Hairless. Hereditas 151(6):209–219. https://doi.org/10.1111/hrd2.00074

Brou C, Logeat F, Gupta N, Bessia C, LeBail O, Doedens JR, Cumano A, Roux P, Black RA, Israel A (2000) A novel proteolytic cleavage involved in Notch signaling: the role of the disintegrin-metalloprotease TACE. Mol Cell 5(2):207–216 https://doi.org/10.1016/S1097-2765(00)80417-7

Bruckner K, Perez L, Clausen H, Cohen S (2000) Glycosyltransferase activity of Fringe modulates Notch-Delta interactions. Nature 406(6794):411–415. https://doi.org/10.1038/35019075

Castel D, Mourikis P, Bartels SJ, Brinkman AB, Tajbakhsh S, Stunnenberg HG (2013) Dynamic binding of RBPJ is determined by Notch signaling status. Genes Dev 27(9):1059–1071. https://doi.org/10.1101/gad.211912.112

Choi SH, Wales TE, Nam Y, O'Donovan DJ, Sliz P, Engen JR, Blacklow SC (2012) Conformational locking upon cooperative assembly of notch transcription complexes. Structure 20(2):340–349. https://doi.org/10.1016/j.str.2011.12.011

Collins KJ, Yuan Z, Kovall RA (2014) Structure and function of the CSL-KyoT2 corepressor complex: a negative regulator of Notch signaling. Structure 22(1):70–81. https://doi.org/10.1016/j.str.2013.10.010

Contreras AN, Yuan Z, Kovall RA (2015) Thermodynamic binding analysis of Notch transcription complexes from Drosophila melanogaster. Protein Sci Publ Protein Soc 24(5):812–822. https://doi.org/10.1002/pro.2652

Del Bianco C, Aster JC, Blacklow SC (2008) Mutational and energetic studies of Notch 1 transcription complexes. J Mol Biol 376(1):131–140 https://doi.org/10.1016/j.jmb.2007.11.061

Del Bianco C, Vedenko A, Choi SH, Berger MF, Shokri L, Bulyk ML, Blacklow SC (2010) Notch and MAML-1 complexation do not detectably alter the dna binding specificity of the transcription factor CSL. PLoS One 5(11):e15034. https://doi.org/10.1371/journal.pone.0015034

Dou S, Zeng X, Cortes P, Erdjument-Bromage H, Tempst P, Honjo T, Vales LD (1994) The recombination signal sequence-binding protein RBP-2N functions as a transcriptional repressor. Mol Cell Biol 14(5):3310–3319 https://doi.org/10.1128/MCB.14.5.3310

Friedmann DR, Kovall RA (2010) Thermodynamic and structural insights into CSL-DNA complexes. Protein Sci Publ Protein Soc 19(1):34–46. https://doi.org/10.1002/pro.280

Friedmann DR, Wilson JJ, Kovall RA (2008) RAM-induced allostery facilitates assembly of a notch pathway active transcription complex. J Biol Chem 283(21):14781–14791. https://doi.org/10.1074/jbc.M709501200

Fryer CJ, White JB, Jones KA (2004) Mastermind recruits CycC:CDK8 to phosphorylate the Notch ICD and coordinate activation with turnover. Mol Cell 16(4):509–520. https://doi.org/10.1016/j.molcel.2004.10.014

Fuchs KP, Bommer G, Dumont E, Christoph B, Vidal M, Kremmer E, Kempkes B (2001) Mutational analysis of the J recombination signal sequence binding protein (RBPJ)/ Epstein-Barr virus nuclear antigen 2 (EBNA2) and RBP-J/Notch interaction. Eur J Biochem 268(17):4639–4646 https://doi.org/10.1046/j.1432-1327.2001.02387.x

Gordon WR, Zimmerman B, He L, Miles LJ, Huang J, Tiyanont K, McArthur DG, Aster JC, Perrimon N, Loparo JJ, Blacklow SC (2015) Mechanical Allostery: evidence for a force requirement in the proteolytic activation of Notch. Dev Cell 33(6):729–736. https://doi.org/10.1016/j.devcel.2015.05.004

Guarani V, Deflorian G, Franco CA, Kruger M, Phng LK, Bentley K, Toussaint L, Dequiedt F, Mostoslavsky R, Schmidt MH, Zimmermann B, Brandes RP, Mione M, Westphal CH, Braun T, Zeiher AM, Gerhardt H, Dimmeler S, Potente M (2011) Acetylation-dependent regulation of endothelial Notch signalling by the SIRT1 deacetylase. Nature 473(7346):234–238. https://doi.org/10.1038/nature09917

Hass MR, Liow HH, Chen X, Sharma A, Inoue YU, Inoue T, Reeb A, Martens A, Fulbright M, Raju S, Stevens M, Boyle S, Park JS, Weirauch MT, Brent MR, Kopan R (2016) SpDamID: marking DNA bound by protein complexes identifies Notch-dimer responsive enhancers. Mol Cell 64(1):213. https://doi.org/10.1016/j.molcel.2016.09.035

Hein K, Mittler G, Cizelsky W, Kuhl M, Ferrante F, Liefke R, Berger IM, Just S, Strang JE, Kestler HA, Oswald F, Borggrefe T (2015) Site-specific methylation of Notch1 controls the amplitude and duration of the Notch1 response. Sci Signal 8(369):ra30. https://doi.org/10.1126/scisignal.2005892

Hsieh JJ, Hayward SD (1995) Masking of the CBF1/RBPJ kappa transcriptional repression domain by Epstein-Barr virus EBNA2. Science 268(5210):560–563 https://doi.org/10.1126/science.7725102

Hsieh JJ, Zhou S, Chen L, Young DB, Hayward SD (1999) CIR, a corepressor linking the DNA binding factor CBF1 to the histone deacetylase complex. Proc Natl Acad Sci U S A 96(1):23–28 https://doi.org/10.1073/pnas.96.1.23

Hu B, Castillo E, Harewood L, Ostano P, Reymond A, Dummer R, Raffoul W, Hoetzenecker W, Hofbauer GF, Dotto GP (2012) Multifocal epithelial tumors and field cancerization from loss of mesenchymal CSL signaling. Cell 149(6):1207–1220. https://doi.org/10.1016/j.cell.2012.03.048

Johnson SE, Ilagan MX, Kopan R, Barrick D (2010) Thermodynamic analysis of the CSL x Notch interaction: distribution of binding energy of the Notch RAM region to the CSL beta-trefoil domain and the mode of competition with the viral transactivator EBNA2. J Biol Chem 285(9):6681–6692. https://doi.org/10.1074/jbc.M109.019968

Kadam S, Emerson BM (2003) Transcriptional specificity of human SWI/SNF BRG1 and BRM chromatin remodeling complexes. Mol Cell 11(2):377–389 https://doi.org/10.1016/S1097-2765(03)00034-0

Kao HY, Ordentlich P, Koyano-Nakagawa N, Tang Z, Downes M, Kintner CR, Evans RM, Kadesch T (1998) A histone deacetylase corepressor complex regulates the Notch signal transduction pathway. Genes Dev 12(15):2269–2277 https://doi.org/10.1101/gad.12.15.2269

Kopan R, Ilagan MX (2009) The canonical Notch signaling pathway: unfolding the activation mechanism. Cell 137(2):216–233. https://doi.org/10.1016/j.cell.2009.03.045

Kovall RA, Blacklow SC (2010) Mechanistic insights into Notch receptor signaling from structural and biochemical studies. Curr Top Dev Biol 92:31–71. https://doi.org/10.1016/s0070-2153(10)92002-4

Kovall RA, Hendrickson WA (2004) Crystal structure of the nuclear effector of Notch signaling, CSL, bound to DNA. EMBO J 23(17):3441–3451. https://doi.org/10.1038/sj.emboj.7600349

Kovall RA, Gebelein B, Sprinzak D, Kopan R (2017) The canonical notch signaling pathway: structural and biochemical insights into shape, sugar, and force. Dev Cell 41(3):228–241. https://doi.org/10.1016/j.devcel.2017.04.001

Krejci A, Bray S (2007) Notch activation stimulates transient and selective binding of Su(H)/CSL to target enhancers. Genes Dev 21(11):1322–1327 https://doi.org/10.1101/gad.424607

Kulic I, Robertson G, Chang L, Baker JH, Lockwood WW, Mok W, Fuller M, Fournier M, Wong N, Chou V, Robinson MD, Chun HJ, Gilks B, Kempkes B, Thomson TA, Hirst M, Minchinton AI, Lam WL, Jones S, Marra M, Karsan A (2015) Loss of the Notch effector RBPJ promotes tumorigenesis. J Exp Med 212(1):37–52. https://doi.org/10.1084/jem.20121192

Kuroda K, Han H, Tani S, Tanigaki K, Tun T, Furukawa T, Taniguchi Y, Kurooka H, Hamada Y, Toyokuni S, Honjo T (2003) Regulation of marginal zone B cell development by MINT, a suppressor of Notch/RBP-J signaling pathway. Immunity 18(2):301–312 https://doi.org/10.1016/S1074-7613(03)00029-3

Kurooka H, Honjo T (2000) Functional interaction between the mouse notch1 intracellular region and histone acetyltransferases PCAF and GCN5. J Biol Chem 275(22):17211–17220. https://doi.org/10.1074/jbc.M000909200

Liefke R, Oswald F, Alvarado C, Ferres-Marco D, Mittler G, Rodriguez P, Dominguez M, Borggrefe T (2010) Histone demethylase KDM5A is an integral part of the core Notch- RBP-J repressor complex. Genes Dev 24(6):590–601 https://doi.org/10.1101/gad.563210

Logeat F, Bessia C, Brou C, LeBail O, Jarriault S, Seidah NG, Israel A (1998) The Notch1 receptor is cleaved constitutively by a furin-like convertase. Proc Natl Acad Sci U S A 95(14):8108–8112 https://doi.org/10.1073/pnas.95.14.8108

Lubman OY, Ilagan MX, Kopan R, Barrick D (2007) Quantitative dissection of the Notch:CSL interaction: insights into the Notch-mediated transcriptional switch. J Mol Biol 365(3):577–589. https://doi.org/10.1016/j.jmb.2006.09.071

Maier D (2006) Hairless: the ignored antagonist of the Notch signalling pathway. Hereditas 143(2006):212–221. https://doi.org/10.1111/j.2007.0018-0661.01971.x

Maier D, Kurth P, Schulz A, Russell A, Yuan Z, Gruber K, Kovall RA, Preiss A (2011) Structural and functional analysis of the repressor complex in the Notch signaling pathway of Drosophila melanogaster. Mol Biol Cell 22(17):3242–3252 https://doi.org/10.1091/mbc.E11-05-0420

Meng X, Brodsky MH, Wolfe SA (2005) A bacterial one-hybrid system for determining the DNA-binding specificity of transcription factors. Nat Biotechnol 23(8):988–994 https://doi.org/10.1038/nbt1120

Moloney DJ, Panin VM, Johnston SH, Chen J, Shao L, Wilson R, Wang Y, Stanley P, Irvine KD, Haltiwanger RS, Vogt TF (2000) Fringe is a glycosyltransferase that modifies Notch. Nature 406(6794):369–375. https://doi.org/10.1038/35019000

Morel V, Lecourtois M, Massiani O, Maier D, Preiss A, Schweisguth F (2001) Transcriptional repression by suppressor of hairless involves the binding of a hairless-dCtBP complex in Drosophila. Curr Biol 11(10):789–792 https://doi.org/10.1016/S0960-9822(01)00224-X

Morgan TH (1917) The theory of the gene. Am Nat 51:513–544 https://doi.org/10.1086/279629

Moshkin YM, Kan TW, Goodfellow H, Bezstarosti K, Maeda RK, Pilyugin M, Karch F, Bray SJ, Demmers JA, Verrijzer CP (2009) Histone chaperones ASF1 and NAP1 differentially modulate removal of active histone marks by LID-RPD3 complexes during NOTCH silencing. Mol Cell 35(6):782–793. https://doi.org/10.1016/j.molcel.2009.07.020

Mumm JS, Schroeter EH, Saxena MT, Griesemer A, Tian X, Pan DJ, Ray WJ, Kopan R (2000) A ligand-induced extracellular cleavage regulates gamma-secretase-like proteolytic activation of Notch1. Mol Cell 5(2):197–206 https://doi.org/10.1016/S1097-2765(00)80416-5

Nagel AC, Krejci A, Tenin G, Bravo-Patino A, Bray S, Maier D, Preiss A (2005) Hairless-mediated repression of notch target genes requires the combined activity of Groucho and CtBP corepressors. Mol Cell Biol 25(23):10433-10441. https://doi.org/10.1128/MCB.25.23.10433-10441.2005

Nam Y, Weng AP, Aster JC, Blacklow SC (2003) Structural requirements for assembly of the CSL. intracellular Notch1. Mastermind-like 1 transcriptional activation complex. J Biol Chem 278(23):21232–21239. https://doi.org/10.1074/jbc.M301567200

Nam Y, Sliz P, Song L, Aster JC, Blacklow SC (2006) Structural basis for cooperativity in recruitment of

MAML coactivators to Notch transcription complexes. Cell 124(5):973–983. https://doi.org/10.1016/j.cell.2005.12.037

Nam Y, Sliz P, Pear WS, Aster JC, Blacklow SC (2007) Cooperative assembly of higher-order Notch complexes functions as a switch to induce transcription. Proc Natl Acad Sci U S A 104(7):2103–2108. doi:0611092104 [pii]. https://doi.org/10.1073/pnas.0611092104

Neves A, English K, Priess JR (2007) Notch-GATA synergy promotes endoderm-specific expression of ref-1 in C. elegans. Development 134(24):4459–4468 https://doi.org/10.1242/dev.008680

Okajima T, Xu A, Irvine KD (2003) Modulation of notch-ligand binding by protein O-fucosyltransferase 1 and fringe. J Biol Chem 278(43):42340–42345. https://doi.org/10.1074/jbc.M308687200

Olave I, Reinberg D, Vales LD (1998) The mammalian transcriptional repressor RBP (CBF1) targets TFIID and TFIIA to prevent activated transcription. Genes Dev 12(11):1621–1637 https://doi.org/10.1101/gad.12.11.1621

Oswald F, Tauber B, Dobner T, Bourteele S, Kostezka U, Adler G, Liptay S, Schmid RM (2001) p300 acts as a transcriptional coactivator for mammalian Notch-1. Mol Cell Biol 21(22):7761–7774. https://doi.org/10.1128/mcb.21.22.7761-7774.2001

Oswald F, Kostezka U, Astrahantseff K, Bourteele S, Dillinger K, Zechner U, Ludwig L, Wilda M, Hameister H, Knochel W, Liptay S, Schmid RM (2002) SHARP is a novel component of the Notch/RBP-Jkappa signalling pathway. EMBO J 21(20):5417–5426 https://doi.org/10.1093/emboj/cdf549

Oswald F, Winkler M, Cao Y, Astrahantseff K, Bourteele S, Knochel W, Borggrefe T (2005) RBP-Jkappa/SHARP recruits CtIP/CtBP corepressors to silence Notch target genes. Mol Cell Biol 25(23):10379–10390 https://doi.org/10.1128/MCB.25.23.10379-10390.2005

Oswald F, Rodriguez P, Giaimo BD, Antonello ZA, Mira L, Mittler G, Thiel VN, Collins KJ, Tabaja N, Cizelsky W, Rothe M, Kuhl SJ, Kuhl M, Ferrante F, Hein K, Kovall RA, Dominguez M, Borggrefe T (2016) A phospho-dependent mechanism involving NCoR and KMT2D controls a permissive chromatin state at Notch target genes. Nucleic Acids Res 44(10):4703–4720 https://doi.org/10.1093/nar/gkw105

Prevorovsky M, Atkinson SR, Ptackova M, McLean JR, Gould K, Folk P, Puta F, Bahler J (2011) N-termini of fungal CSL transcription factors are disordered, enriched in regulatory motifs and inhibit DNA binding in fission yeast. PLoS One 6(8):e23650. https://doi.org/10.1371/journal.pone.0023650

Qin H, Wang J, Liang Y, Taniguchi Y, Tanigaki K, Han H (2004) RING1 inhibits transactivation of RBP-J by Notch through interaction with LIM protein KyoT2. Nucleic Acids Res 32(4):1492–1501. https://doi.org/10.1093/nar/gkh295

Qin H, Du D, Zhu Y, Li J, Feng L, Liang Y, Han H (2005) The PcG protein HPC2 inhibits RBP-J-mediated transcription by interacting with LIM protein KyoT2. FEBS Lett 579(5):1220–1226. https://doi.org/10.1016/j.febslet.2005.01.022

Rana NA, Haltiwanger RS (2011) Fringe benefits: functional and structural impacts of O-glycosylation on the extracellular domain of Notch receptors. Curr Opin Struct Biol 21(5):583–589. https://doi.org/10.1016/j.sbi.2011.08.008

Schroeter EH, Kisslinger JA, Kopan R (1998) Notch-1 signalling requires ligand-induced proteolytic release of intracellular domain. Nature 393(6683):382–386. https://doi.org/10.1038/30756

Sherry KP, Johnson SE, Hatem CL, Majumdar A, Barrick D (2015) Effects of linker length and transient secondary structure elements in the intrinsically disordered notch RAM region on notch signaling. J Mol Biol 427(22):3587–3597. https://doi.org/10.1016/j.jmb.2015.09.001

Steinhauser K, Kloble P, Kreis NN, Ritter A, Friemel A, Roth S, Reichel JM, Michaelis J, Rieger MA, Louwen F, Oswald F, Yuan J (2016) Deficiency of RITA results in multiple mitotic defects by affecting microtubule dynamics. Oncogene. https://doi.org/10.1038/onc.2016.372

Struhl G, Adachi A (1998) Nuclear access and action of notch in vivo. Cell 93(4):649–660 https://doi.org/10.1016/S0092-8674(00)81193-9

Struhl G, Greenwald I (1999) Presenilin is required for activity and nuclear access of Notch in Drosophila. Nature 398(6727):522–525. https://doi.org/10.1038/19091

Surendran K, Boyle S, Barak H, Kim M, Stomberski C, McCright B, Kopan R (2010) The contribution of Notch1 to nephron segmentation in the developing kidney is revealed in a sensitized Notch2 background and can be augmented by reducing Mint dosage. Dev Biol 337(2):386–395. https://doi.org/10.1016/j.ydbio.2009.11.017

Tabaja N, Yuan Z, Oswald F, Kovall RA (2017) Structure-function analysis of RBP-J-interacting and tubulin-associated (RITA) reveals regions critical for repression of Notch target genes. J Biol Chem 292(25):10549–10563. https://doi.org/10.1074/jbc.M117.791707

Takeuchi H, Haltiwanger RS (2010) Role of glycosylation of Notch in development. Semin Cell Dev Biol 21(6):638–645. https://doi.org/10.1016/j.semcdb.2010.03.003

Takeuchi H, Haltiwanger RS (2014) Significance of glycosylation in Notch signaling. Biochem Biophys Res Commun 453(2):235–242. https://doi.org/10.1016/j.bbrc.2014.05.115

Tamura K, Taniguchi Y, Minoguchi S, Sakai T, Tun T, Furukawa T, Honjo T (1995) Physical interaction between a novel domain of the receptor Notch and the transcription factor RBP-J kappa/Su(H). Curr Biol 5(12):1416–1423 https://doi.org/10.1016/S0960-9822(95)00279-X

Taniguchi Y, Furukawa T, Tun T, Han H, Honjo T (1998) LIM protein KyoT2 negatively regulates transcription by association with the RBP-J DNA-binding protein.

Mol Cell Biol 18(1):644–654 https://doi.org/10.1128/MCB.18.1.644

Torella R, Li J, Kinrade E, Cerda-Moya G, Contreras AN, Foy R, Stojnic R, Glen RC, Kovall RA, Adryan B, Bray SJ (2014) A combination of computational and experimental approaches identifies DNA sequence constraints associated with target site binding specificity of the transcription factor CSL. Nucleic Acids Res 42(16):10550–10563. https://doi.org/10.1093/nar/gku730

Tsunematsu R, Nakayama K, Oike Y, Nishiyama M, Ishida N, Hatakeyama S, Bessho Y, Kageyama R, Suda T, Nakayama KI (2004) Mouse Fbw7/Sel-10/Cdc4 is required for notch degradation during vascular development. J Biol Chem 279(10):9417–9423. https://doi.org/10.1074/jbc.M312337200

Tun T, Hamaguchi Y, Matsunami N, Furukawa T, Honjo T, Kawaichi M (1994) Recognition sequence of a highly conserved DNA binding protein RBP-J kappa. Nucleic Acids Res 22(6):965–971

VanderWielen BD, Yuan Z, Friedmann DR, Kovall RA (2011) Transcriptional repression in the Notch pathway: thermodynamic characterization of CSL-MINT (Msx2-interacting nuclear target protein) complexes. J Biol Chem 286(17):14892–14902. https://doi.org/10.1074/jbc.M110.181156

Wacker SA, Alvarado C, von Wichert G, Knippschild U, Wiedenmann J, Clauss K, Nienhaus GU, Hameister H, Baumann B, Borggrefe T, Knochel W, Oswald F (2011) RITA, a novel modulator of Notch signalling, acts via nuclear export of RBP-J. EMBO J 30(1):43–56 https://doi.org/10.1038/emboj.2010.289

Wallberg AE, Pedersen K, Lendahl U, Roeder RG (2002) p300 and PCAF act cooperatively to mediate transcriptional activation from chromatin templates by notch intracellular domains in vitro. Mol Cell Biol 22(22):7812–7819 https://doi.org/10.1128/MCB.22.22.7812-7819.2002

Weng AP, Nam Y, Wolfe MS, Pear WS, Griffin JD, Blacklow SC, Aster JC (2003) Growth suppression of pre-T acute lymphoblastic leukemia cells by inhibition of notch signaling. Mol Cell Biol 23(2):655–664 https://doi.org/10.1128/MCB.23.2.655-664.2003

Wharton KA, Johansen KM, Xu T, Artavanis-Tsakonas S (1985) Nucleotide sequence from the neurogenic locus notch implies a gene product that shares homology with proteins containing EGF-like repeats. Cell 43(3 Pt 2):567–581

Wilson JJ, Kovall RA (2006) Crystal structure of the CSL-Notch-Mastermind ternary complex bound to DNA. Cell 124(5):985–996. https://doi.org/10.1016/j.cell.2006.01.035

Wu G, Lyapina S, Das I, Li J, Gurney M, Pauley A, Chui I, Deshaies RJ, Kitajewski J (2001) SEL-10 is an inhibitor of notch signaling that targets notch for ubiquitin-mediated protein degradation. Mol Cell Biol 21(21):7403–7415. https://doi.org/10.1128/mcb.21.21.7403-7415.2001

Xu T, Park SS, Giaimo BD, Hall D, Ferrante F, Ho DM, Hori K, Anhezini L, Ertl I, Bartkuhn M, Zhang H, Milon E, Ha K, Conlon KP, Kuick R, Govindarajoo B, Zhang Y, Sun Y, Dou Y, Basrur V, Elenitoba-Johnson KS, Nesvizhskii AI, Ceron J, Lee CY, Borggrefe T, Kovall RA, Rual JF (2017) RBPJ/CBF1 interacts with L3MBTL3/MBT1 to promote repression of Notch signaling via histone demethylase KDM1A/LSD1. EMBO J 36(21):3232–3249. https://doi.org/10.15252/embj.201796525

Yuan Z, Friedmann DR, VanderWielen BD, Collins KJ, Kovall RA (2012) Characterization of CSL (CBF-1, Su(H), Lag-1) mutants reveals differences in signaling mediated by Notch1 and Notch2. J Biol Chem 287(42):34904–34916. doi:M112.403287 [pii]. https://doi.org/10.1074/jbc.M112.403287

Yuan Z, Praxenthaler H, Tabaja N, Torella R, Preiss A, Maier D, Kovall RA (2016) Structure and function of the Su(H)-Hairless repressor complex, the major antagonist of notch signaling in Drosophila melanogaster. PLoS Biol 14(7):e1002509. https://doi.org/10.1371/journal.pbio.1002509

Zhou S, Hayward SD (2001) Nuclear localization of CBF1 is regulated by interactions with the SMRT corepressor complex. Mol Cell Biol 21(18):6222–6232 https://doi.org/10.1128/MCB.21.18.6222-6232.2001

Zhou S, Fujimuro M, Hsieh JJ, Chen L, Miyamoto A, Weinmaster G, Hayward SD (2000) SKIP, a CBF1-associated protein, interacts with the ankyrin repeat domain of NotchIC to facilitate NotchIC function. Mol Cell Biol 20(7):2400–2410 https://doi.org/10.1128/MCB.20.7.2400-2410.2000

Zweifel ME, Leahy DJ, Hughson FM, Barrick D (2003) Structure and stability of the ankyrin domain of the Drosophila Notch receptor. Protein Sci Publ Protein Soc 12(11):2622–2632. https://doi.org/10.1110/ps.03279003

Part IV

Disease Links and Therapeutics

Notch and Senescence

Matthew Hoare and Masashi Narita

Abstract

Cellular senescence, previously thought of as an autonomous tumour suppressor mechanism, is emerging as a phenotype and effector present throughout the life of an organism from embryogenesis to senile decline. Senescent cells have powerful non-autonomous effects upon multiple players within their microenvironment mainly through their secretory phenotype. How senescent cells co-ordinate numerous, sometimes functionally contrasting outputs through their secretome had previously been unclear. The Notch pathway, originally identified for its involvement in *Drosophila* wing development, has more recently been found to underpin diverse effects in human cancer. Here we discuss recent findings that suggest that Notch is intimately involved in the development of senescence and how it acts to co-ordinate the composition and functional effects of the senescence secretome. We also highlight the complex physical and functional interplay between Notch and p53, critical to both senescence and cancer. Understanding the interplay between Notch, p53 and senescence could allow us develop the therapeutics of the future for cancer and ageing.

Keywords

NOTCH · Senescence · SASP · Secretome · TGF-beta · Interleukins · Immune surveillance · RAS

Abbreviations

bHLH	Basic helix-loop-helix transcription factor
BRAF	B-Raf proto-oncogene
BRD4	bromodomain containing 4
C/EBPβ	CCAAT/enhancer binding protein beta
CCL	C-C motif chemokine ligand
CDK	cyclin-dependent kinase
CXCL	chemokine (C-X-C motif) ligand
CXCR	C-X-C motif chemokine receptor
DLL	delta-like ligand
ECM	extracellular matrix
EMT	epithelial-mesenchymal transition
FBXW7	F-box and WD repeat domain containing 7
G-CSF	granulocyte colony stimulating factor

M. Hoare (✉)
Cancer Research UK Cambridge Institute, University of Cambridge, Cambridge, UK

Department of Medicine, Addenbrooke's Hospital, University of Cambridge, Cambridge, UK
e-mail: matthew.hoare@cruk.cam.ac.uk

M. Narita (✉)
Cancer Research UK Cambridge Institute, University of Cambridge, Cambridge, UK
e-mail: masashi.narita@cruk.cam.ac.uk

T. Borggrefe, B. D. Giaimo (eds.), *Molecular Mechanisms of Notch Signaling*, Advances in Experimental Medicine and Biology 1066, https://doi.org/10.1007/978-3-319-89512-3_15

H3K27ac	histone 3 acetylated at lysine 27
HES	hairy and enhancer of split
HEY	hairy/enhancer-of-split related with YRPW motif
IL	Interleukin
JAG1	Jagged-1
JAK	Janus kinase
MAML1	mastermind like transcriptional coactivator 1
MEF	mouse embryonic fibroblast
MYC	myelocytomatosis proto-oncogene
NK cells	natural killer cells
PD-1	programmed cell death 1
PTEN	phosphatase and tensin homolog
RAS	rat sarcoma virus oncogene
RBPJ	recombination signal binding protein for immunoglobulin kappa J region
RelA	v-rel reticuloendotheliosis viral oncogene homolog A
SA β-GAL	senescence-associated beta-galactosidase
SASP	senescence-associated secretory phenotype
SHH	sonic hedgehog
SMAD	Mothers against decapentaplegic
STAT	signal transducer and activator of transcription
T-ALL	T-cell acute lymphoblastic leukaemia
TGFβ1	transforming growth factor beta 1

1 Cellular Senescence

Somatic cells have a variety of tumour suppressor mechanisms to prevent cellular damage leading to transformation into cancer. Amongst these there is increasing recognition that cellular senescence not only plays a crucial role in the pathogenesis of cancer and the cancer microenvironment, but also more generally in wound healing and embryological development. When cells enter cellular senescence they undergo long-term stable exit from the cell cycle, but can remain viable and metabolically active for a prolonged period. Cellular senescence was originally identified from cultured cells demonstrating a finite *in vitro* growth capacity. Subsequent work linked this proliferative arrest to progressive attrition of telomere length and the development of a telomere-derived DNA damage signal (Muñoz-Espín and Serrano 2014).

The identification that activation of oncogenic RAS in primary human cells could also lead to the development of senescence linked this process to tumour suppression (Fig. 1a) (Serrano et al. 1997). Senescence was then found to underpin the suppression of human cancers, such as arresting BRAF-expressing cells in senescence and thereby preventing melanoma development (Michaloglou et al. 2005). Since this time, many genetic lesions leading to oncogene expression or loss of tumour suppressor activity have been found to drive cellular senescence. Evidence of senescence has been described in a variety of human pre-neoplastic lesions suggesting that the acquisition of tumorigenic mutations is actively repressed *in vivo* by cellular senescence (Collado et al. 2005; Collado and Serrano 2010).

Senescence has also been shown to underpin the successful response of some tumours to chemotherapy. In the murine Eμ-Myc model of lymphoma, treatment with chemotherapy induces senescence within the tumour and leads to tumour regression. Abrogation of senescence in these mice prevents the response to treatment and leads to a worsening of survival (Schmitt et al. 2002). Further, in established murine liver cancer with inactivation of senescence, re-establishment of senescence can lead to complete resolution of the tumour (Xue et al. 2007). Therefore, not only

Fig. 1 (continued) DAPI-dense foci of heterochromatin called senescence-associated heterochromatic foci (SAHF). Senescent cells have an expansion of their lysosomal compartment and express galactosidase activity at a non-optimal pH, termed senescence-associated beta-galactosidase (SA β-GAL). Senescent cells are highly secretory and produce a range of cytokines, growth factors and matrix-modifying enzymes termed the senescence-associated secretory phenotype (SASP); shown here are RAS-senescent IMR90s expressing the chemokine IL8. The most fundamental characteristic of senescent cells is their lack of proliferation, even upon growth factor or oncogenic stimulation, demonstrated here by lack of colony forming ability compared to control cells

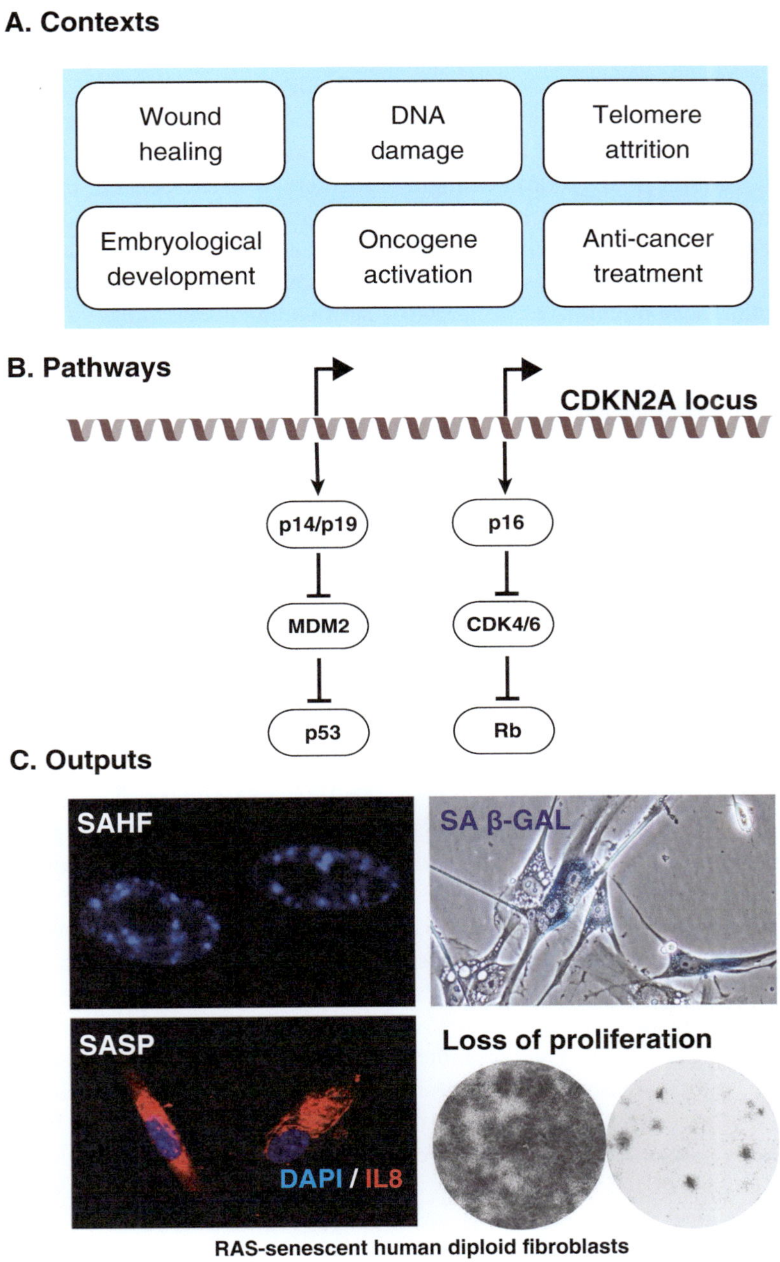

Fig. 1 Cellular senescence is a highly conserved cellular pathway involved in diverse biological settings. (**a**) Whilst originally identified in the context of telomere attrition and then suppression of oncogene-induced transformation, cellular senescence is now recognized to occur in contexts as diverse as embryological development, wound healing and the response to anti-cancer therapies. (**b**) Cellular senescence is underpinned by two cellular pathways driven by gene-products of the CDKN2A locus. Both p14^{ARF} (p19 in mice) and p16^{INK4A} are expressed from this locus and result in p53 and Rb-dependent cellular responses, respectively. There is enormous interest in the recently developed CDK4/6 inhibitors, such as palbociclib, which can restore a tumour suppressive senescence response in cancer cells that have intact Rb signalling. (**c**) The presence of senescence within a cell is inferred by a panel of markers and downstream effector functions, none of which are truly specific for senescence. The chromatin of RAS-senescent IMR90 human diploid fibroblasts undergoes a significant architectural change to form

does senescence resist the development of cancer, but may also underpin the successful response to cancer treatment.

Senescent cells accumulate in otherwise healthy organisms with progressive ageing (van Deursen 2014; Herbig et al. 2006). Utilising genetic labeling from the promoter of p16, a cyclin-dependent kinase (CDK) inhibitor and putative senescence marker, several studies have found differential accumulation of senescent cells within diverse organs. In wild-type mice, the number of senescent cells progressively increases with age, but is heterogeneous across otherwise genetically identical litter-mates (Burd et al. 2013). Further, the level of senescence within an organ does not predict the development of tumours. Targeted clearance of these senescent cells increases the healthy lifespan of both wild-type and prematurely aged mice through reduction of both tumorigenesis and age-related pathologies (Baker et al. 2016; Baker et al. 2011). This is, perhaps, paradoxical, but as we shall see later, senescence can have contrasting oncogenic and tumour suppressive effects. Importantly, even when the senescent cells are deleted late in life, when age-related pathologies have already developed, this prevents the progression of those pathologies, holding the promise of future therapies that arrest or even reverse age-related decline. These findings have reinforced the notion that senescence demonstrates antagonistic pleiotropy: a process that defends the organism and promotes reproductive fitness by repressing tumorigenesis early in life, but is deleterious in later life through declining organ function and age-related illness (Giaimo 2012).

More recent studies have broadened our understanding of senescence as a developmental mechanism underpinning both healing and organogenesis. Senescent cells can be found in skin wounds in mice (Fig. 1a). These cells are actively involved in the appropriate restoration of homeostasis as deletion of senescent cells in these mice delays the healing of the wound (Demaria et al. 2014). Senescence can also be detected during embryogenesis in the developing inner ear and urinary tract of mice where it is crucial to appropriate development of these organs (Muñoz-Espín et al. 2013; Storer et al. 2013). This form of senescence shares all the features of other models of senescence (see later section on markers of senescence), other than a DNA-damage signal. Importantly, these developmental senescent cells express a typical pro-inflammatory secretome that is crucial in the regulation of the surrounding tissue. Loss of senescence during development leads to developmental abnormalities. Therefore, more than simply a tumour suppressor mechanism, senescence seems to be a highly conserved developmental pathway, intrinsic to a range of cellular behaviours, that can function in a stress-responsive mode to resist transformation.

Critical to the development of senescence are two major pathways (Fig. 1b) frequently mutated in human cancer: the p53-p21 (Serrano et al. 1997; Brugarolas et al. 1995) and p16-Rb pathways (Alcorta et al. 1996). Viral oncoproteins, that can drive the development of human cancers, are known to inactivate these pathways: the SV40 large T antigen is able to bind both of these factors leading to their inactivation and subsequent senescence bypass; similarly, the adenoviral E1A protein inhibits their function and promotes tumour formation (Shay et al. 1991). Activation of p53 and Rb in senescence seems to rely, in large part, on the activity of two proteins, $p16^{INK4A}$ and $p14^{ARF}$, expressed from the *CDKN2A* locus. p16 inhibits the CDK4/6-dependent inactivation of Rb, whereas p14 (p19 in mice) interferes with the ability of MDM2 to inhibit p53. Ectopic expression of p16 can induce a senescent phenotype in cancer cell lines (Dai and Enders 2000) and this relationship has more recently become a potential therapeutic target with the development of CDK4 inhibitors, such as palbociclib. In fact, these drugs mimic the effect of p16 by preventing CDK4-dependent Rb phosphorylation and thereby repressing E2F-target genes, crucial for cell cycle progression. Chronic CDK4 inhibitor treatment is able to drive senescence in cancer cells that have lost both p53 and p16, but only when Rb remains intact (Anders et al. 2011). This raises the possibility of using such drugs to restore an appropriate senescence response in cancer, even when some endogenous tumour suppressors have been lost (Yoshida et al. 2016).

p53, on the other hand has a multitude of effects in senescence in a range of cellular pathways (Johmura and Nakanishi 2016). This seems to be in part related to a distinct set of chromatin binding sites and therefore a distinct transcriptional programme compared to that seen in the acute activation of p53 in acute cellular stress (Kirschner et al. 2015). In the autonomous aspects of senescence, p53 promotes growth arrest by upregulating p21 expression that acts in turn to inhibit CDK2-dependent Rb phosphorylation. Significantly, p53 seems to repress some of the non-autonomous activities of senescence (Rodier et al. 2009) that, as we shall discuss later, underpin much of the functionality of the senescent cell.

1.1 Markers of Senescence

A number of putative defining characteristics of senescence have been identified (Fig. 1c). However, none are truly sensitive or specific, which has hampered efforts to understand the role of senescence in human disease. Due to their lack of specificity, the presence of senescence is normally inferred by the simultaneous presence of several of these markers. In fact, senescent cells are defined by a combination of several of the following features: a lack of proliferation; activation of p53-p21 and p16-Rb pathways (Alcorta et al. 1996); formation of senescence-associated heterochromatic foci (SAHF), at least in oncogene-induced senescence (Narita et al. 2003); a persistent DNA-damage response (DDR) (d'Adda di Fagagna et al. 2003); expression of a lysosomal enzyme termed senescence-associated beta-galactosidase (SA β-GAL) (Dimri et al. 1995); and the secretion of a range of cytokines, chemokines and extracellular matrix (ECM) modifying factors termed the senescence-associated secretory phenotype (SASP) (Fig. 1c) (Coppé et al. 2010; Salama et al. 2014).

1.2 The Senescence-Associated Secretory Phenotype

It is through the SASP that senescent cells exert significant effects upon their surrounding environment. Most previous studies have focused on the secretome of cells undergoing oncogene-induced senescence (OIS) or DNA-damage-induced senescence (DDIS) and demonstrated that the typical SASP consists of proinflammatory cytokines such as interleukin-1α (IL1α) and IL6, and chemokines such as IL8 and C-C motif chemokine ligand 2 (CCL2) (Coppé et al. 2010). Transcriptionally the SASP is positively regulated by the transcription factors v-rel reticuloendotheliosis viral oncogene homolog A (RelA / p65 (an NF-κB family member)) (Chien et al. 2011), CCAAT/enhancer binding protein beta (C/EBPβ) (Kuilman et al. 2008) (possibly in a tight positive feedback loop with IL1α (Orjalo et al. 2009)) and the chromatin binding factor bromodomain protein 4 (BRD4) which dynamically binds to super-enhancers, related to many SASP genes (Tasdemir et al. 2016). Chien et al. identified RelA through an unbiased proteomic screen of RAS-senescent chromatin. Subsequent functional investigation found that loss of RelA, during *in vitro* senescence, failed to bypass senescence but did prevent the senescence-associated upregulation of IL1α, IL6 and IL8 (Chien et al. 2011). *In vivo*, loss of RelA leads to a failure to develop treatment-induced senescence and relapse after chemotherapy (Chien et al. 2011).

C/EBPβ was identified as a SASP regulator through a search for putative transcriptional regulators of the prototypic SASP component IL6 in BRAF-induced senescence (Kuilman et al. 2008). Endogenous C/EBPβ binds to the core promoter of IL6 during OIS and ectopic C/EBPβ drives IL6 expression; loss of either IL6 or C/EBPβ can bypass BRAF-induced-senescence in primary human cells. Ectopic expression of C/EBPβ induces senescence in both primary human cells (Kuilman et al. 2008) and transformed breast cancer cell lines (Atwood and Sealy 2010). However, crucial to the activation of C/EBPβ in response to the RAS/MAPK pathway is the activity of the cell-cycle inhibitor p19 (p14 in humans). RAS/MAPK activation in transformed cells lacking p19 fails to drive C/EBPβ expression and stimulates proliferation rather than senescence. Restoration of p19, leading to an upregulation of C/EBPβ, or ectopic C/EBPβ expression re-establishes the senescence response to RAS/MAPK activation (Sebastian and Johnson 2009). Therefore, C/EBPβ sits downstream of p19 in the development of RAS-senescence.

The identification of BRD4 as a major regulator of the SASP has emerged from analysis of the changing epigenetic landscape of RAS-senescent cells that must underpin the simultaneous repression of cell-cycle-related genes and the activation of secretory-related genes (Tasdemir et al. 2016). In other biological contexts where significant functional reprogramming occurs, this is underpinned by changes at genetic regulatory elements termed enhancers, marked by acetylation of histone 3 at lysine 27 (H3K27ac). Amongst these enhancer elements, those marked by long stretches of H3K27ac are termed super-enhancers. Analysis found significant remodeling of super-enhancers in the context of senescence adjacent to genes encoding SASP components. As a putative H3K27ac binding partner, increased chromatin binding of BRD4 was confirmed at these loci and its inhibition, either genetically or pharmacologically, leads to abrogation of the proinflammatory SASP and reduction in SASP signalling to immunocytes both *in vitro* and *in vivo* (Tasdemir et al. 2016).

The secretome is also significantly modulated at the post-translational level through the inflammasome (Acosta et al. 2013), p38 MAPK (Freund et al. 2011), mTOR (Herranz et al. 2015; Laberge et al. 2015) and autophagy pathways (Young et al. 2009; Narita et al. 2011). The SASP has been found is nearly all forms of senescence thus-far described, other than senescence induced by overexpression of p16 (Coppé et al. 2011) and mostly relies on a persistent DNA-damage signal (Rodier et al. 2009).

Functionally the SASP is important due to the diverse downstream effects that senescent cells can exert on multiple players within the microenvironment. Firstly the secretome can act in an autocrine manner to reinforce the senescent phenotype. Here signalling from C/EBPβ or through C-X-C motif chemokine receptor 2 (CXCR2) leads to senescence, whereas loss of these factors results in senescence bypass (Kuilman et al. 2008; Acosta et al. 2008). Secondly the SASP can enforce a paracrine senescence upon surrounding normal cells, through the secretion of transforming growth factor beta 1 (TGFβ1) and IL1β, potentially providing a means of controlling transformation in the context of a cancerisation field effect (Acosta et al. 2013; Hubackova et al. 2012). Thirdly, the SASP from senescent cells has been demonstrated to have significantly pro-oncogenic effects upon certain cell types. In *Drosophila* Ras-induced imaginal epithelial senescence drives proliferation of neighbouring epithelial cells through the SASP (Nakamura et al. 2014). Similarly, senescent human fibroblasts are able to drive the growth of co-cultured pre-malignant and fully transformed human cell lines, in addition to promoting their growth in xenografts (Krtolica et al. 2001). This effect is, at least partially, dependent upon NF-κB, as metformin-induced loss of NF-κB signalling prevents the senescence-driven growth of adjacent prostatic cancer cell lines (Moiseeva 2013). Furthermore, senescent cells in co-culture can promote the epithelial-mesenchymal transition (EMT), with enhanced invasiveness, in the target cell population (Coppé et al. 2008). Lastly the SASP has been shown to have significant effects upon components of the immune system.

One of the critical findings of senescence in most model systems is that senescent cells are able to trigger their own immune-mediated destruction. Through the pro-inflammatory SASP, senescent cells recruit diverse members of the immune system leading to targeted killing and subsequent clearance, in a process termed senescence surveillance. In mouse models of NRAS-induced hepatocyte senescence, the NRAS-expressing cells secrete pro-inflammatory cytokines and are progressively removed from the liver (Kang et al. 2011). CD4$^+$ T-lymphocytes are necessary for this process, as their deletion allows persistence of senescent cells and subsequent liver cancer development. Other groups have demonstrated that differing immunocytes are responsible for senescence surveillance in different contexts, such as macrophages and NK (natural killer) cells (Krizhanovsky et al. 2008; Lujambio et al. 2013). Not only are senescent cells responsible for immune cell recruitment, but also for re-programming and controlling the downstream function of the recruited immunocytes (Eggert et al. 2016). Senescent hepatic stellate cells secrete a SASP that is able to modulate the polarisation, secretome and function of recruited macrophages (Lujambio et al. 2013). Curiously, this SASP-dependent modulation of monocyte maturation and function can be antagonised *in vivo* by fully transformed cells within the same environment, through mechanisms that are not fully understood at present (Eggert et al. 2016).

Therefore, there is a range of different downstream functions of the SASP, some with clearly contrasting effects on different target cell populations: senescent cells are able to direct development, control wound healing, resist transformation and control the composition and function of parts of the immune system. Until recently it remained unclear how a senescent cell could coordinate these different effects through a SASP of singular or static composition. However, recent data point to a role for the Notch pathway in the dynamic control of both SASP composition and its net functional output.

2 Notch

2.1 Notch Signalling Pathway

In order to respond to cues from neighbouring cells or the microenvironment, a variety of different signalling pathways have evolved to sense and direct cellular behaviour. Among these, Notch has emerged as a critical pathway in a variety of different cellular contexts. The role of Notch in development was originally identified from spontaneous mutations in *Drosophila*, where haploinsufficiency leads to an obvious notch in the wing edge (Greenwald 2012). Since this fortuitous discovery, comparative genomics has demonstrated that components of the Notch-signalling pathway are highly conserved in bilateria through evolution, suggesting that this signalling pathway arose around 550 million years ago. Subsequent work has demonstrated the significant role that Notch plays not only in directing cell-fate decisions during development, where loss of function of Notch receptors or ligands leads to dysgenesis of the vasculature, biliary tree and nervous system, but also in the development and progression of cancer.

In all organisms where Notch is described, the pathway is notable for the simplicity of the components involved in the core signalling pathway and downstream transduction. Indeed, given the simplicity of the core components, lack of enzymatic amplification steps and the multiplicity of downstream functional outcomes that Notch has been linked with, there must be a significant role for the diverse set of non-core components that have been described to modulate Notch signalling. In its simplest form the Notch pathway consists of a single-pass transmembrane Notch receptor which, when bound to a canonical ligand on an adjacent cell, undergoes a conformational change and subsequent proteolytic cleavage by the transmembrane metalloproteinase ADAM17 (Kopan and Ilagan 2009). This results in a membrane-tethered intermediate form that is susceptible to further cleavage by γ-secretase, a multi-molecular complex responsible for cleavage of a range of membrane-bound substrates including Notch receptors. This second cleavage releases the Notch intracellular domain (NICD) from the inner envelope of the plasma membrane to traverse the cytoplasm and enter the nucleus. Within the nucleus the NICD binds to the highly conserved DNA binding protein recombination signal binding protein for immunoglobulin kappa J region (RBPJ [CBF1 / LAG-1 / Su(H)]) displacing transcriptional repressors and recruiting transcriptional activators, such as mastermind-like 1 (MAML1). Upon binding to NICD, RBPJ is converted to a transcriptional activator, recruiting the acetyl-transferase p300 and initiating the downstream Notch transcriptional program (Kopan and Ilagan 2009). Notch is able to regulate a range of genes, including the hairy and enhancer of split (HES) and hairy/enhancer-of-split related with YRPW motif (HEY) family of transcription factors, MYC and Cyclin D3. In addition to being a critical transcriptional co-activator, MAML1 controls the half-life of the NICD through regulating its phosphorylation by CDK8 (Fryer et al. 2002; Fryer et al. 2004). This phosphorylation renders the NICD susceptible to ubiquitination by F-box and WD repeat domain containing 7 (FBXW7) (O'Neil et al. 2007) and subsequent degradation, thereby limiting the duration of signalling.

Within mammals there are four separate Notch genes, all of which are able to liberate a distinct intracellular domain and drive distinct downstream signalling events, despite all binding to RBPJ. Similarly, in *Drosophila* there are two (Delta and Serrate), but in mammals at least five (Delta-like (DLL) 1; DLL3; DLL4; Jagged1 and Jagged 2), canonical Notch ligands. These ligands have differing binding affinities for the different Notch receptors and drive distinct downstream functions (Bray 2016). Control of the affinity of the different ligands for the Notch receptors is in part controlled by the Fringe-mediated post-translational glycosylation of the receptors prior to their trafficking to the plasma membrane (Bray 2016). However, the basis for distinct functional outcomes from a pathway involving multiple receptors and ligands, but a single DNA binding protein remains unclear.

2.2 Notch in Cancer

Alterations in the Notch-signalling pathway have been linked to the development and progression of cancer. The earliest suggestion that Notch could be oncogenic came with the identification of a rare chromosomal translocation between chromosomes 7 and 9 leading to the constitutive expression of the NOTCH1 intracellular domain (N1ICD) in lymphocytes in human T-cell acute lymphoblastic leukaemia (T-ALL) (Ellisen et al. 1991). Subsequent studies revealed that the majority of patients with T-ALL had activating mutations due to indels of the NOTCH1 gene rather than major structural rearrangements (Weng et al. 2004). These mutations either led to ligand-independent receptor cleavage or enhanced the stability of the NICD once liberated from the membrane (Ranganathan et al. 2011). The same studies found that a significant minority of patients without NOTCH1 mutations had mutations of FBXW7, leading to increased stability of the N1ICD (O'Neil et al. 2007).

With increasing knowledge of the genetic structure of diverse cancer types it has become apparent that NOTCH is frequently mutated or that the Notch signalling pathway is activated in several human cancers, other than T-ALL. Activating mutations or increased downstream signalling have been described in many solid organ malignancies such as breast cancer, hepatocellular carcinoma, cholangiocarcinoma, colorectal carcinoma and melanoma (Ranganathan et al. 2011; Aster et al. 2017). The precise pathways that Notch regulates to drive cancer remain unclear as, in addition to driving its own transcriptional program, Notch also exerts significant cross-talk to diverse other cellular pathways such as Wnt-β-catenin, RAS-MAPK and others. Candidate pathway approaches suggest that Notch exerts its actions through transcriptional regulation of both cell-cycle and apoptosis-related genes; in particular, Cyclin D1 and D3 are direct transcriptional targets of N1ICD and drive cell cycle progression (Ronchini and Capobianco 2001).

Conversely it has become apparent that Notch and downstream signalling can be tumour suppressive in some circumstances. Sequencing has demonstrated that a significant number of patients with bladder cancer (Rampias et al. 2014), head and neck squamous cell cancer (Agrawal et al. 2011), skin cancer (Nicolas et al. 2003) and small-cell lung carcinoma (George et al. 2015) have inactivating mutations of NOTCH1 suggesting a tumour suppressive role in these cancers (Nowell and Radtke 2017). Notch has also been demonstrated to be tumour suppressive in mouse models. Inducible knock-out of Notch1 leads to the spontaneous development of basal-cell carcinoma of the skin and accelerated tumour development after chemical carcinogenesis, partly due to loss of Notch1-mediated repression of the sonic hedgehog (Shh) pathway, previously implicated in basal cell carcinoma of the skin (Nicolas et al. 2003; Demehri et al. 2009). Similarly, mesenchymal-specific loss of RBPJ, the DNA-binding protein for Notch, leads to the development of skin inflammation and subsequent tumour generation (Hu et al. 2012).

In some tissue types modulation of different Notch receptors has been demonstrated to have opposing effects upon tumorigenesis. Human pancreatic carcinoma is ubiquitously associated with the expression of oncogenic KRAS, in addition to other genetic lesions such as loss of the p53 encoding gene TP53. Mouse models with pancreas-specific expression of $KRas^{G12D}$ lead to the development of the pre-neoplastic lesion pancreatic intraepithelial neoplasia (PanIN). In mice with combined pancreas-specific expression of $KRas^{G12D}$ and loss of Notch1, there is an increased number and more advanced grade of PanIN compared to mice with $KRas^{G12D}$ alone (Hanlon et al. 2010; Mazur et al. 2010). However, in similar mice with pancreas-specific loss of Notch2, survival is longer with reduced development of PanIN, but the mice develop late, highly anaplastic pancreatic carcinoma (Mazur et al. 2010). Therefore, in the same tissue, different Notch receptors can have complex and potentially opposing effects upon tissue differentiation and tumorigenesis.

It is clear that Notch receptors can be either oncogenic or tumour suppressive in different tissues, depending on context. The molecular basis for this duality of function in different cancer types is currently unknown, but one possibility is the emerging role for Notch and downstream signalling in the autonomous and non-autonomous functions of senescence.

2.3 Notch in Senescence

The association of Notch signalling with cellular senescence is a relatively recent finding and several studies have identified different NOTCH receptors in different senescence model systems. Replicative senescence is associated with the upregulation of NOTCH1 in both normal human prostatic cells and oesophageal keratinocytes (Bhatia et al. 2008; Kagawa et al. 2015). Similarly, all of the NOTCH receptors are upregulated during *in vitro* culture of human endothelial cells (Venkatesh et al. 2011) and upregulated in murine endothelium overlying atherosclerosis, thought to have features of senescence (Liu et al. 2012). Manipulation of downstream Notch function is also able to modulate these senescent phenotypes. Pharmacological inhibition of Notch signalling by the γ-secretase inhibitor DAPT is able to increase *in vitro* replicative lifespan and reduce features of replicative senescence in oesophageal keratinocytes, such as SA β-GAL and p16 expression (Kagawa et al. 2015). Several Notch receptors are upregulated in other forms of senescence beyond replicative senescence. There is increased NOTCH3 expression in several forms of stress-induced senescence including replicative senescence, DDIS and oxidative stress-induced senescence in human fibroblasts (Cui et al. 2013). In this context, knockdown of NOTCH3 is able to delay the onset of proliferation arrest and reduces features of senescence, such as SA β-GAL and p21 expression (Cui et al. 2013).

2.4 Multiple Notch Receptors Can Drive a Senescent Phenotype

In addition to modulating the senescent phenotype induced by diverse other stressors, several Notch receptors are able to drive senescent phenotypes independently of other stimuli. Overexpression of NOTCH3 induces both a proliferative arrest by up-regulating the cyclin-dependent kinase inhibitor p21, through direct binding to its core promoter and a senescent phenotype (Cui et al. 2013). Interruption of downstream NOTCH3 signalling, through expression of a dominant negative MAML1 (dnMAML1), or knockdown of p21 are able to partially rescue this NOTCH3-induced senescent phenotype.

Cancers may bypass this NOTCH-induced senescence (NIS) through reduction of NOTCH receptor expression. Expression of NOTCH3 is significantly down-regulated in human breast cancer compared to normal breast tissue and is correlated with the level of p21 expression in the same tumour. Ectopic NOTCH3 expression, in breast cancer cell lines with low endogenous NOTCH3 expression, is able to drive the cells into NIS, suggesting that some degree of senescence-bypass could be associated with reduction of NOTCH3 signalling or that selection for clones with low Notch signalling could occur in human breast cancer (Cui et al. 2013).

Similarly to NOTCH3, several studies have shown that ectopic NOTCH1 also drives a senescent phenotype with reduced proliferation, increased SA β-GAL expression and upregulation of the CDK4/6 inhibitor p16, with subsequent loss of Rb phosphorylation. This NIS also requires NOTCH-mediated transcription, as it can be rescued by inhibition through co-expression of dnMAML1 (Kagawa et al. 2015; Hoare et al. 2016). Importantly, the cells remain arrested in NOTCH-induced senescence, even after removal of ectopic N1ICD, a cardinal feature of senescence (Hoare et al. 2016). This confirms that this phenotype is true senescence and not simply quiescence, that can be induced through expression of HES1 (Sang et al. 2008). In this context NOTCH1-induced senescence seems to be dependent upon the p16-Rb signalling axis as knockdown of p16, but not p14, is able to rescue this NIS phenotype (Kagawa et al. 2015). Whether there is genuine specificity of NIS for the p16-Rb or p53-p21 pathways induced by signalling from the different Notch receptors or whether these different reports are describing a common, conserved NIS phenotype remains unclear.

The role of RBPJ in NIS is more controversial. In one context, N1ICD-mediated senescence can be rescued by concurrent shRNA-mediated knockdown of RBPJ, with reduced p16 expression and continued proliferation (Kagawa et al. 2015). However, other studies have suggested that loss of RBPJ, in the absence of ectopic N1ICD expression can also lead to the development of senescence (Procopio et al. 2015). Murine dermal fibroblasts with loss of Rbpj or human fibroblasts with shRNA-mediated knockdown of RBPJ have increased expression of p15, p16, p21 and SA β-GAL (Procopio et al. 2015). Further, RBPJ can directly bind to DNA at enhancer elements upstream of both p16 and p21 genes (Procopio et al. 2015). This apparent discrepancy can be explained by the dual role that RBPJ plays, dependent upon the presence of the NICD. In the Notch-inactive state RBPJ acts as a transcriptional repressor of multiple genes through constitutive DNA binding (Wang et al. 2011). Upon Notch-activation, binding of the NICD leads to conversion of RBPJ to a transcriptional activator. Therefore, loss of RBPJ in the study by Procopio and colleagues, in the absence of NICD, removes the transcriptional repression on these genes and drives a similar senescent phenotype to NICD-mediated conversion to a transcriptional activator. Consistent with this, ectopic expression of N1ICD in their models leads to a similar phenotype to RBPJ loss (Procopio et al. 2015).

At present there is no evidence that NIS is specific for NOTCH1 or 3. Indeed, in cultured endothelial cells, ectopic expression of the NICDs from NOTCH1, 2 or 4, but not the NOTCH-target genes HEY1 or HEY2 are able to drive a similar senescent phenotype with reduced proliferation, increased expression of SA β-GAL and upregulation of both p16 and p21 (Venkatesh et al. 2011; Liu et al. 2012). Functionally this is associated with increased endothelial permeability (Venkatesh et al. 2011) and increased endothelial pro-inflammatory cytokine expression (Liu et al. 2012) *in vitro*. Therefore, seemingly all of the NOTCH receptors can trigger a senescence response in certain contexts.

This NOTCH-induced senescent phenotype does not involve the HEY family of transcription factors (Venkatesh et al. 2011). Indeed, other studies have found that HES1, a canonical NOTCH-target gene, is important in resisting irreversible cell-cycle exit associated with prolonged expression of p21 in fibroblasts and thereby controlling the cell-fate decision between quiescence and senescence. In this context,

4 days of expression of p21 are sufficient to lead to irreversible senescence. However, restoration of HES1 expression permitted cell cycle entry and proliferation even after long periods of proliferative arrest (Sang et al. 2008). This function of HES1 was also found in the context of OIS, where HES1 was able to resist the entry into RAS-induced senescence and prolong cellular proliferation.

It is interesting to speculate that cMyc could represent a plausible intermediary underpinning NIS. Myc is known to be a direct transcriptional target of NOTCH1 (Palomero et al. 2006; Weng et al. 2006). Chronic activation of cMyc has also been shown to drive a form of senescence, when the cyclin-dependent kinase CDK2 is lost or inactivated (Campaner et al. 2010). In the Eu-Myc mouse model of lymphoma, cellular senescence has been demonstrated to underpin the response to chemotherapy and is critical to an improvement in survival (Schmitt et al. 2002); Eu-Myc mice with biallelic loss of CDK2 have spontaneous development of senescence within lymphoid tissue and significantly improved prognosis compared to CDK2 heterozygotes (Campaner et al. 2010). Notch is also known to repress the expression of CDK2 (Qi et al. 2003), suggesting a model where NOTCH expression could simultaneously up and down-regulate cMyc and CDK2 respectively, driving senescence; this remains to be tested.

In addition to senescence driven by dysregulated signalling from the activated forms of the Notch receptors, abrogation of FBXW7 and interruption of normal degradation of the NICD has also been linked to senescence. Disruption of FBXW7 promotes endogenous Notch signalling and is able to lead to a proliferative arrest and senescence-like phenotype (Ishikawa et al. 2008). Loss of FBXW7 in mouse embryonic fibroblasts (MEFs) leads to a specific retention of N1ICD, amongst other FBXW7 targets, and significant up-regulation of a range of Notch-target genes. The growth arrest could be rescued by inhibition of Notch signalling by DAPT or subsequent loss of p53 function. This suggests that loss of FBXW7 leads to prolonged and upregulated Notch1 signalling that is able to drive a senescent phenotype.

These effects of Notch signalling upon senescence can be recapitulated in mouse model systems. Specific expression of N1ICD in murine endothelial cells is associated with reduced angiogenesis and increased *ex vivo* SA β-GAL expression in cultured aortic tissue, compared to tissue from control mice (Venkatesh et al. 2011). Ectopic expression of N1ICD specifically in the renal tubules after renal injury, not only prolongs the resolution of injury, but also leads to increased markers of tubular senescence with upregulation of both p16 and p21 (Sörensen-Zender et al. 2014). In these mice, treatment with the Notch inhibitor DAPT leads to a significant reduction in the level of both p16 and p21, suggesting either enhanced clearance or reduced development of senescence in the kidneys.

Therefore, there is abundant evidence that sustained Notch activation from increased activity or impaired degradation of several Notch family members is able to drive a senescent phenotype, including *in vivo* senescence. However, the basis for Notch acting as a tumour suppressor to drive senescence or as an oncogene leading to malignancy, such as T-ALL remains elusive.

2.5 Notch Regulates the SASP

There has been indirect evidence of a link between NOTCH and non-autonomous signaling previously; loss of Notch1 in mouse skin is associated with increased influx of immune cells, suggesting a role for Notch in suppression of inflammatory signaling (Demehri et al. 2009). Mechanistically Notch signalling has a complex relationship with the secretome and TGFβ signalling in particular. There seems to be a strong positive feedback loop between NOTCH and TGFβ1. Treatment of mesenchymal cells with TGFβ1 leads to the upregulation of the Notch ligand JAG1, through the canonical TGFβ1 targets Mothers against decapentaplegic 3 (SMAD3) (Kurpinski et al. 2010) and SMAD4 (Sethi et al. 2011), whereas in epithelial cells the TGFβ1-JAG1 pathway can drive an EMT through Notch

(Zavadil et al. 2004). Indeed, there is some evidence that the NICD is able to physically bind to both SMAD9 (Yatim et al. 2012) and SMAD3; the latter interaction has been demonstrated to enhance downstream Notch signalling (Blokzijl et al. 2003) Certainly this signalling axis seems to underpin the proliferative arrest (Niimi et al. 2007) and Notch-induced senescence of cells in response to TGFβ1 treatment (Kagawa et al. 2015). Co-operative signalling through these two pathways seems to be critical for induction of p21 (Niimi et al. 2007). Both TGFβ1-mediated growth arrest and SA β-GAL expression are rescued through concurrent treatment with DAPT or knockdown of NOTCH1. Therefore, NOTCH signalling seems to be a downstream effector of non-autonomous signalling through TGFβ1. Indeed, in addition to blindly augmenting TGFβ-signalling, activation of Notch seems to be able to shape the response to TGFβ. Notch is able to modulate the relative expression levels of the different SMAD proteins, promoting SMAD3 in particular and altering downstream signalling networks from TGFβ (Fu et al. 2009).

From our own work we have established that N1ICD is sufficient to induce expression of both TGFβ1 and TGFβ3, cell surface expression of the latency-associated peptide, cleaved from the TGFβ1 propeptide and to induce downstream TGFβ-signalling with increased chromatin-bound SMAD3. In addition, blockade of TGFβ1 signalling through antagonists of TGFβR1 or expression of a dominant negative form of SMAD4 is able to partially rescue the NIS phenotype (Hoare et al. 2016).

However, the relationship between Notch and a more widespread role in control of the composition of the secretome and thereby net functional non-autonomous output of a cell was much less clear. NOTCH1 was identified as significantly upregulated in an unbiased plasma membrane proteomic screen looking for senescence-associated cell surface proteins. Validation confirmed that NOTCH1 was upregulated in several forms of senescence and in different cell types. Despite being progressively upregulated through the transition to RAS-induced senescence (RIS), NOTCH1 is only functionally active during the transition to senescence, with loss of downstream signalling when cells are fully senescent. Through pharmacological and genetic pathway manipulation during senescence it was possible to identify that NOTCH1 is able to drive expression of several TGFβ-family members at the same time as repressing the typical pro-inflammatory SASP of RIS cells. Transcriptional profiling confirmed that RAS and NOTCH1 co-regulated the secretome towards two polar opposite secretory phenotypes. In the case of RAS, the secretome consisted of proinflammatory cytokines such as IL1α, IL1β and ECM-degrading matrix metalloproteases (MMP), whereas the NOTCH-driven secretome consisted of several TGFβ-family members, collagens and extracellular matrix components such as fibronectin. Importantly, when co-expressed, N1ICD is dominant over RAS in determining the secretome composition. Therefore, the level of NOTCH signalling acts as a rheostat upon the secretome composition and net functional output of cells undergoing senescence. The burst of NOTCH1 signalling during the transition to senescence is able to direct a pro-fibrotic and immunosuppressive SASP, prior to subsequent loss of Notch signalling and secretome switch to an anti-fibrotic and proinflammatory SASP (Hoare et al. 2016). We found that Notch was functionally active during *in vivo* RAS-induced hepatocyte senescence, where autonomous expression of Notch1 was increased. Utilising hydrodynamic tail-vein delivery of an oncogenic NRAS-containing transposon, we were able to induce RAS-senescence of hepatocytes (Kang et al. 2011; Hoare et al. 2016). Co-delivery of dnMAML1 with RAS promoted recruitment of T-lymphocytes to the liver, associated with enhanced clearance of the RAS-senescent hepatocytes, presumably related to an enhanced pro-inflammatory SASP. A role for Notch in suppressing inflammation would be consistent with previous studies of Notch in other contexts. In mice with mesenchymal-specific loss of Notch signalling spontaneous inflammation of the skin was observed, with upregulation of a diverse array of inflammatory mediators and MMPs, ultimately driving the development of tumours (Hu et al. 2012).

Mechanistically NOTCH1 inhibits the pro-inflammatory SASP through repression of both expression and chromatin binding of the transcription factor C/EBPβ. C/EBPβ is thought to act in concert with the NF-κB component RelA in transcriptionally regulating the SASP (Chien et al. 2011; Kuilman et al. 2008). C/EBPβ has been demonstrated to be a critical factor in the generation of the SASP, with loss of C/EBPβ causing loss of pro-inflammatory cytokine expression and senescence bypass (Kuilman et al. 2008). Importantly, we could not find any significant effect of NOTCH1 activation upon expression or chromatin binding of any NF-κB family member in the context of senescence, despite previous evidence of a link between NOTCH and NF-κB (Oakley et al. 2003). N1ICD was able to reduce the binding of C/EBPβ to enhancer elements upstream of the *IL1A* locus, as well as previously identified binding sites in the core promoters of *IL6* and *IL8* (Hoare et al. 2016). Therefore, our data place NOTCH1 amongst the master regulators of the senescence secretome. In particular, NOTCH1 appears to be upstream of IL1α, critically important for the regulation of various inflammatory cytokines including IL6 and IL8. The precise mechanism by which NOTCH1 is able to repress C/EBPβ remains unclear, including whether this repression is direct. Previous studies have identified that the canonical NOTCH1-target HES1 is able to transcriptionally repress C/EBPα (De Obaldia et al. 2013), but whether a similar mechanism operates for C/EBPβ is unknown.

Some evidence for a SASP with an evolving composition and downstream functionality has been found before. In skin wound healing senescent myofibroblasts are important to normal wound repair and their loss prolongs the time to restoration of the wound (Demaria et al. 2014). In this context the SASP directs the operation of the healing microenvironment, before ultimately directing the immune-mediated destruction of the senescent cell, necessitating a time-dependent switch between non-autonomous signalling modules with contrasting downstream functionalities. Indeed, many studies have demonstrated that the onset of senescence is associated with a reduction in tissue fibrosis, potentially relating to a switch from a pro-fibrotic to pro-inflammatory secretome (Krizhanovsky et al. 2008; Jun and Lau 2010).

Therapeutically, there is much interest in the concept of manipulating the composition of the SASP to enhance passage into senescence or clearance of senescent cells to prevent the development of cancer. In the context of PTEN-loss associated senescence in the prostate, the typical SASP is immunosuppressive, leading to recruitment of myeloid-derived suppressor cells and preventing immune-mediated clearance of the senescent cells (Toso et al. 2014). These cells also show evidence of signalling through the Jak2/Stat3 pathway leading to expression of chemokine (C-X-C motif) ligand 2 (Cxcl2) and granulocyte colony-stimulating factor (G-CSF). Upon genetic deletion of Stat3 or pharmacological inhibition of Jak2, the SASP of the senescent prostatic epithelium shows significant reductions in Cxcl2 and G-CSF, associated with enhanced immune-cell infiltration and reduction in subsequent tumour development (Toso et al. 2014). In the case of Notch, genetic inhibition through dnMAML1 is able to enhance the clearance of RAS-senescent hepatocytes from the mouse liver, associated with enhanced recruitment of $CD3^+$ T-lymphocytes. Therefore, modulation of SASP composition, in order to promote chemotaxis and senescence surveillance, can be demonstrated through two different pathways in two distinct models of senescence. Therefore, the combination of SASP modulation to enhance recruitment and immune-checkpoint blockade, such as anti-programmed cell death 1 (PD-1) therapy, to enhance immune activation could be a rational combination in the treatment of neoplastic and pre-neoplastic lesions.

2.6 Notch and p53

TP53 is one of the most commonly mutated genes in human cancers. It also plays a critical role in both senescence and the SASP. There is increasing evidence that Notch and p53 have a complex relationship with regulation of both fac-

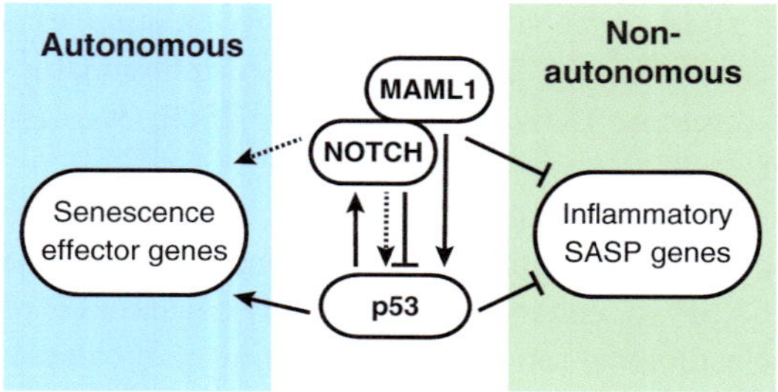

Fig. 2 NOTCH and p53 are involved in complex reciprocal regulation, but drive coordinated outputs in senescence. NOTCH and p53 have been demonstrated to reciprocally regulate each other, including through physical binding, leading to stimulation or inhibition in a context-dependent manner. However, in the context of senescence, both drive common outputs with an autonomous cell-cycle arrest and subsequent senescence, in addition to inhibiting the pro-inflammatory senescence secretome and therefore coordinately modulating the non-autonomous functionality of senescent cells

tors by the other (Fig. Fig. 2) (Dotto 2009). In *Drosophila*, Notch is a direct target of p53 activity and mediates p53-dependent cell behaviour between apoptosis and proliferation (Simón et al. 2014). Similarly, in both mammalian keratinocytes and epithelial cells NOTCH has been demonstrated to be a direct p53 transcriptional target, where p53 up-regulates NOTCH1 expression (Yugawa et al. 2007; Lefort et al. 2007). UV irradiation of the skin leads to upregulation of both Notch1 expression and activity, in a p53-dependent manner (Mandinova et al. 2008), where Notch acts to repress UV-damage induced apoptosis.

Previous studies of Notch-mediated regulation of p53 have again revealed a duality of Notch function between activation and repression of p53. In the context of haematological malignancy, Notch is known to repress p53 function, potentially through regulation of the p53-regulatory protein MDM2 (Beverly et al. 2005) or through direct physical interaction with p53 itself (Kim et al. 2007). Indeed p53 has been demonstrated to bind to both RBPJ (Procopio et al. 2015) and MAML1 (Zhao et al. 2007; Yun et al. 2015) in different contexts. Through this direct interaction, the N1ICD is able to repress the expression of p53-target genes such as p21, in a dose-dependent manner (Kim et al. 2007). In some tumour cell lines, NOTCH1 activity is responsible for repression of p53-dependent apoptosis, through reducing the stability of the p53 protein (Licciulli et al. 2013).

However, other studies have shown that activated Notch-signalling can positively regulate p53 function. The canonical Notch targets HES1 and HEY1 have been shown to positively regulate p53 activity, through negative regulation of MDM2 (Huang et al. 2004). The Notch-dependent cell-cycle arrest attendant with FBXW7 loss can be rescued by knockout of p53 (Ishikawa et al. 2008).

Despite their complex reciprocal regulation, in the context of senescence, Notch and p53 signalling seem to drive coordinated endpoints with autonomous cell cycle arrest and cellular senescence. Similarly, loss of the constitutive repressive RBPJ activity is also able to drive this phenotype (Procopio et al. 2015). Physically RBPJ binds both to the p53 promoter (Boggs et al. 2009) and to p53 itself and reduces its transcriptional activity (Procopio et al. 2015); bait oligonucleotides containing the promoter sequence of the canonical p53-target p21 were able to pull down both p53 and RBPJ, suggesting that these two factors are physically and functionally linked in the same gene space, related to the senescence program. Indeed increasing levels of ectopic RBPJ lead to a dose-dependent reduction of p53 transcriptional activity (Procopio et al. 2015), suggesting a functional interaction between these factors. Paradoxically, increasing levels of the NOTCH co-activator MAML1 are able to increase p53-directed gene transcription (Zhao et al. 2007; Yun et al. 2015), potentially though enhancing its stability and activation (Yun et al. 2015).

However, whereas NOTCH, RBPJ and p53 coordinately regulate the autonomous features of senescence, their roles in the regulation of non-autonomous functionality, and the SASP in particular, are sometimes contrasting. NOTCH1 (Kagawa et al. 2015; Hoare et al. 2016) and p53 (Coppé et al. 2008) both function to promote cellular senescence, whilst repressing the pro-inflammatory SASP. Constitutive RBPJ functions to repress both senescence and the SASP

(Procopio et al. 2015), suggesting that NOTCH converts RBPJ to an activated state at senescence-associated genes, but not at genes regulating the SASP or potentially that NOTCH1-mediated regulation of SASP or C/EBPβ could be independent of RBPJ. This remains to be directly tested, but intriguing evidence suggests that this RBPJ-independent, non-canonical function of NOTCH1 could occur in the regulation of IL6, when p53 is lost (Jin et al. 2012).

Basal-type breast cancer is associated with increased Notch signalling and expression of IL6. Ectopic expression of N1ICD or activation of endogenous NOTCH through JAG1 leads to upregulation of IL6, but only in basal-type breast cancer cell lines that express mutated p53 (Jin et al. 2012). Further, expression of a dominant negative RBPJ reduces the expression of canonical Notch-target genes, but has no effect upon N1ICD-regulated IL6 expression. Conversely, N1ICD lacking the RBPJ-binding domain or cytoplasmically-retained N1ICD are both able to up-regulate IL6, but had no effect upon expression of HES and HEY proteins. This effect on IL6 expression could be reversed by co-expression of wild-type p53. The precise mechanism of this interaction remains elusive and whether this occurs with other secreted factors or in contexts outside of breast cancer remains to be evaluated. However, the findings reinforce that various components of the Notch-signalling pathway could have divergent effects upon autonomous senescence and the senescence-secretome and that some of this functionality may not require nuclear localisation or the apparatus of the canonical Notch pathway.

2.7 Notch-Mediated Juxtacrine Signalling

In addition to regulating the non-autonomous behaviour of senescent cells through the secretome, Notch also regulates signalling to the microenvironment through cell-contact dependent pathways. Studies of embryological development have identified two modes of Notch-dependent signalling through a tissue: lateral inhibition and lateral induction (Artavanis-Tsakonas et al. 1999). In the former, activated Notch signalling represses the expression of Notch ligands within the same cell leading to a reduction in signal transmitted to neighbouring cells. Thereby, there is a differentiation between Notch active and neighbouring Notch-inactive cells. This mode of signalling has been commonly described as a mode of differentiating cell fate decisions at the level of cells and boundary formation at the level of tissues (Boni et al. 2008; Guo et al. 1996; Lim et al. 2015).

The second mode, termed lateral induction, describes how Notch-signalling drives autonomous expression of Notch ligands leading to increased transmission of a Notch-signal to neighbouring cells. In this situation both signal sending and receiving cells will be Notch-active. This mode allows for co-ordination of cell fate and a spatial expansion of coordinated Notch-signalling across a tissue (Hartman et al. 2010; Petrovic et al. 2014). The cellular decision to induce or repress Notch ligand expression seems to involve the strength of the Notch signal and therefore likely the balance and post-translational modification of Notch ligands on neighbouring cells (Petrovic et al. 2014).

We identified that ectopic N1ICD was able to specifically induce the expression of JAG1 amongst the other Notch ligands. This up-regulation of JAG1 transmits a Notch signal to surrounding cells leading to non-autonomous transmission of Notch-induced senescence with upregulation of p16 and reduced proliferation in the signal-receiving cells (Hoare et al. 2016). This form of senescence could be rescued through knockdown of JAG1 expression in the sending cell, inhibiting Notch signalling with dnMAML1 or with DAPT in the receiving cell. Therefore, this represented clear evidence of *in vitro* N1ICD-mediated lateral induction of NOTCH signalling and NOTCH-induced senescence through JAG1. In the mouse liver there was also evidence of both lateral induction of Hes1 and p21 expression from RAS-senescent hepatocytes, suggesting that RAS-induced senescence is associated with the transmission of a cell-contact dependent lateral induction of Notch signalling.

Previously non-autonomous signalling in senescence was thought to involve paracrine, secreted factors alone. The finding of Notch-mediated cell-contact dependent pathways adds complexity to senescence signalling to other players in the microenvironment. It will be interesting to see, not only the effects of this Notch-mediated signalling pathway upon surrounding parenchymal cells, but also upon members of the immune system, where Notch is known to play a profound role in regulating cellular differentiation (Backer et al. 2014).

3 Conclusions

It is becoming clear that senescence, far from a simple tumour suppressor mechanism, is a highly conserved pathway that is utilised in a variety of physiological and pathophysiological contexts throughout the life-cycle from embryogenesis to age-related decline. Fundamental to our understanding of the role of senescence will be to understand how its non-autonomous functionality is regulated and the net output or signal to the various players within the microenvironment is delivered. This output must be dynamically regulated to deliver behaviours as diverse as inner ear development and co-ordination of skin wound healing. We are only just beginning to understand some of the players that control this process. Notch activity is able to modulate both the net secretory output of the senescent cell as well as a cell-contact dependent form of lateral induction, previously thought of as a developmental patterning program.

We do not understand the many contradictions and dualities that have been described to occur with Notch signalling: how is activation of this pathway oncogenic in one context but tumour suppressive in another?

The ultimate prizes for understanding how senescent cells arise, function and then are cleared will be therapies that may target preneoplastic lesions before they develop into cancer and also treatments for non-cancerous age-related pathologies where senescent cells underpin the decline in function with age.

Acknowledgements MH and MN are supported by an MRC Research Grant (MR/R010013/1). MH is supported by a CRUK Clinician Scientist Fellowship (C52489/A19924). MN is supported by a Cancer Research UK Cambridge Centre Core Grant (C14303/A17197).

References

Acosta JC, O'Loghlen A, Banito A et al (2008) Chemokine signaling via the CXCR2 receptor reinforces senescence. Cell 133:1006–1018. https://doi.org/10.1016/j.cell.2008.03.038

Acosta JC, Banito A, Wuestefeld T et al (2013) A complex secretory program orchestrated by the inflammasome controls paracrine senescence. Nat Cell Biol 15:978–990. https://doi.org/10.1038/ncb2784

Agrawal N, Frederick MJ, Pickering CR et al (2011) Exome sequencing of head and neck squamous cell carcinoma reveals inactivating mutations in NOTCH1. Science 333:1154–1157. https://doi.org/10.1126/science.1206923

Alcorta DA, Xiong Y, Phelps D et al (1996) Involvement of the cyclin-dependent kinase inhibitor p16 (INK4a) in replicative senescence of normal human fibroblasts. Proc Natl Acad Sci U S A 93:13742–13747

Anders L, Ke N, Hydbring P et al (2011) A systematic screen for CDK4/6 substrates links FOXM1 phosphorylation to senescence suppression in Cancer cells. Cancer Cell 20:620–634. https://doi.org/10.1016/j.ccr.2011.10.001

Artavanis-Tsakonas S, Rand MD, Lake RJ (1999) Notch signaling: cell fate control and signal integration in development. Science 284:770–776

Aster JC, Pear WS, Blacklow SC (2017) The varied roles of notch in Cancer. Annu Rev Pathol 12:245–275. https://doi.org/10.1146/annurev-pathol-052016-100127

Atwood AA, Sealy L (2010) Regulation of C/EBPbeta1 by Ras in mammary epithelial cells and the role of C/EBPbeta1 in oncogene-induced senescence. Oncogene 29:6004–6015. https://doi.org/10.1038/onc.2010.336

Backer RA, Helbig C, Gentek R et al (2014) A central role for Notch in effector CD8+ T cell differentiation. Nat Immunol 15:1143–1151. https://doi.org/10.1038/ni.3027

Baker DJ, Wijshake T, Tchkonia T et al (2011) Clearance of p16Ink4a-positive senescent cells delays ageing-associated disorders. Nature 479:232–236. https://doi.org/10.1038/nature10600

Baker DJ, Childs BG, Durik M et al (2016) Naturally occurring p16(Ink4a)-positive cells shorten healthy lifespan. Nature 530:184–189. https://doi.org/10.1038/nature16932

Beverly LJ, Felsher DW, Capobianco AJ (2005) Suppression of p53 by Notch in lymphomagenesis: implications for initiation and regression. Cancer Res 65:7159–7168. https://doi.org/10.1158/0008-5472.CAN-05-1664

Bhatia B, Multani AS, Patrawala L et al (2008) Evidence that senescent human prostate epithelial cells enhance tumorigenicity: cell fusion as a potential mechanism and inhibition by p16INK4a and hTERT. Int J Cancer 122:1483–1495. https://doi.org/10.1002/ijc.23222

Blokzijl A, Dahlqvist C, Reissmann E et al (2003) Cross-talk between the Notch and TGF-beta signaling pathways mediated by interaction of the Notch intracellular domain with Smad3. J Cell Biol 163:723–728. https://doi.org/10.1083/jcb.200305112

Boggs K, Henderson B, Reisman D (2009) RBP-Jkappa binds to and represses transcription of the p53 tumor suppressor gene. Cell Biol Int 33:318–324. https://doi.org/10.1016/j.cellbi.2008.12.005

Boni A, Urbanek K, Nascimbene A et al (2008) Notch1 regulates the fate of cardiac progenitor cells. Proc Natl Acad Sci 105:15529–15534. https://doi.org/10.1073/pnas.0808357105

Bray SJ (2016) Notch signalling in context. Nat Rev Mol Cell Biol. https://doi.org/10.1038/nrm.2016.94

Brugarolas J, Chandrasekaran C, Gordon JI et al (1995) Radiation-induced cell cycle arrest compromised by p21 deficiency. Nature 377:552–557. https://doi.org/10.1038/377552a0

Burd CE, Sorrentino JA, Clark KS et al (2013) Monitoring tumorigenesis and senescence in vivo with a p16(INK4a)-luciferase model. Cell 152:340–351. https://doi.org/10.1016/j.cell.2012.12.010

Campaner S, Doni M, Hydbring P et al (2010) Cdk2 suppresses cellular senescence induced by the c-myc oncogene. Nat Cell Biol 12(54):1–14. https://doi.org/10.1038/ncb2004

Chien Y, Scuoppo C, Wang X et al (2011) Control of the senescence-associated secretory phenotype by NF-κB promotes senescence and enhances chemosensitivity. Genes Dev 25:2125–2136. https://doi.org/10.1101/gad.17276711

Collado M, Serrano M (2010) Senescence in tumours: evidence from mice and humans. Nat Rev Cancer 10:51–57. https://doi.org/10.1038/nrc2772

Collado M, Gil J, Efeyan A et al (2005) Tumour biology: senescence in premalignant tumours. Nature 436:642. https://doi.org/10.1038/436642a

Coppé J-P, Patil CK, Rodier F et al (2008) Senescence-associated secretory phenotypes reveal cell-nonautonomous functions of oncogenic RAS and the p53 tumor suppressor. PLoS Biol 6:2853–2868. https://doi.org/10.1371/journal.pbio.0060301

Coppé J-P, Desprez P-Y, Krtolica A, Campisi J (2010) The senescence-associated secretory phenotype: the dark side of tumor suppression. Annu Rev Pathol Mech Dis 5:99–118. https://doi.org/10.1146/annurev-pathol-121808-102144

Coppé J-P, Rodier F, Patil CK et al (2011) Tumor suppressor and aging biomarker p16(INK4a) induces cellular senescence without the associated inflammatory secretory phenotype. J Biol Chem 286:36396–36403. https://doi.org/10.1074/jbc.M111.257071

Cui H, Kong Y, Xu M, Zhang H (2013) Notch3 functions as a tumor suppressor by controlling cellular senescence. Cancer Res 73:3451–3459. https://doi.org/10.1158/0008-5472.CAN-12-3902

d'Adda di Fagagna F, Reaper PM, Clay-Farrace L et al (2003) A DNA damage checkpoint response in telomere-initiated senescence. Nature 426:194–198. https://doi.org/10.1038/nature02118

Dai CY, Enders GH (2000) p16 INK4a can initiate an autonomous senescence program. Oncogene 19:1613–1622. https://doi.org/10.1038/sj.onc.1203438

De Obaldia ME, Bell JJ, Wang X et al (2013) T cell development requires constraint of the myeloid regulator C/EBP-α by the Notch target and transcriptional repressor Hes1. Nat Publ Group 14:1277–1284. https://doi.org/10.1038/ni.2760

Demaria M, Ohtani N, Youssef SA et al (2014) An essential role for senescent cells in optimal wound healing through secretion of PDGF-AA. Dev Cell 31:722–733. https://doi.org/10.1016/j.devcel.2014.11.012

Demehri S, Turkoz A, Kopan R (2009) Epidermal Notch1 loss promotes skin tumorigenesis by impacting the stromal microenvironment. Cancer Cell 16:55–66. https://doi.org/10.1016/j.ccr.2009.05.016

Dimri GP, Lee X, Basile G et al (1995) A biomarker that identifies senescent human cells in culture and in aging skin in vivo. Proc Natl Acad Sci U S A 92:9363–9367

Dotto GP (2009) Crosstalk of Notch with p53 and p63 in cancer growth control. Nat Rev Cancer 9:587–595. https://doi.org/10.1038/nrc2675

Eggert T, Wolter K, Ji J et al (2016) Distinct functions of senescence-associated immune responses in liver tumor surveillance and tumor progression. Cancer Cell 30:533–547. https://doi.org/10.1016/j.ccell.2016.09.003

Ellisen LW, Bird J, West DC et al (1991) TAN-1, the human homolog of the Drosophila notch gene, is broken by chromosomal translocations in T lymphoblastic neoplasms. Cell 66:649–661

Freund A, Patil CK, Campisi J (2011) p38MAPK is a novel DNA damage response-independent regulator of the senescence-associated secretory phenotype. EMBO J 30:1536–1548. https://doi.org/10.1038/emboj.2011.69

Fryer CJ, Lamar E, Turbachova I et al (2002) Mastermind mediates chromatin-specific transcription and turnover of the Notch enhancer complex. Genes Dev 16:1397–1411. https://doi.org/10.1101/gad.991602

Fryer CJ, White JB, Jones KA (2004) Mastermind recruits CycC:CDK8 to phosphorylate the Notch ICD and coordinate activation with turnover. Mol Cell 16:509–520. https://doi.org/10.1016/j.molcel.2004.10.014

Fu Y, Chang A, Chang L et al (2009) Differential regulation of transforming growth factor beta signaling pathways by Notch in human endothelial cells. J Biol Chem 284:19452–19462. https://doi.org/10.1074/jbc.M109.011833

George J, Lim JS, Jang SJ et al (2015) Comprehensive genomic profiles of small cell lung cancer. Nature 524:47–53. https://doi.org/10.1038/nature14664

Giaimo S, d'Adda di Fagagna F (2012) Is cellular senescence an example of antagonistic pleiotropy? Aging Cell 11:378–383. https://doi.org/10.1111/j.1474-9726.2012.00807.x

Greenwald I (2012) Notch and the awesome power of genetics. Genetics 191:655–669. https://doi.org/10.1534/genetics.112.141812

Guo M, Jan LY, Jan YN (1996) Control of daughter cell fates during asymmetric division: interaction of Numb and Notch. Neuron 17:27–41

Hanlon L, Avila JL, Demarest RM et al (2010) Notch1 functions as a tumor suppressor in a model of K-ras-induced pancreatic ductal adenocarcinoma. Cancer Res 70:4280–4286. https://doi.org/10.1158/0008-5472.CAN-09-4645

Hartman BH, Reh TA, Bermingham-McDonogh O (2010) Notch signaling specifies prosensory domains via lateral induction in the developing mammalian inner ear. Proc Natl Acad Sci 107:15792–15797. https://doi.org/10.1073/pnas.1002827107

Herbig U, Ferreira M, Condel L et al (2006) Cellular senescence in aging primates. Science 311:1257. https://doi.org/10.1126/science.1122446

Herranz N, Gallage S, Mellone M et al (2015) mTOR regulates MAPKAPK2 translation to control the senescence-associated secretory phenotype. Nat Cell Biol 17:1205–1217. https://doi.org/10.1038/ncb3225

Hoare M, Ito Y, Kang T-W et al (2016) NOTCH1 mediates a switch between two distinct secretomes during senescence. Nat Cell Biol 18:979–992. https://doi.org/10.1038/ncb3397

Hu B, Castillo E, Harewood L et al (2012) Multifocal epithelial tumors and field cancerization from loss of mesenchymal CSL signaling. Cell 149:1207–1220. https://doi.org/10.1016/j.cell.2012.03.048

Huang Q, Raya A, DeJesus P et al (2004) Identification of p53 regulators by genome-wide functional analysis. Proc Natl Acad Sci U S A 101:3456–3461. https://doi.org/10.1073/pnas.0308562100

Hubackova S, Krejcikova K, Bartek J, Hodny Z (2012) IL1- and TGFβ-Nox4 signaling, oxidative stress and DNA damage response are shared features of replicative, oncogene-induced, and drug-induced paracrine 'Bystander senescence. Aging (Albany NY) 4:932–951

Ishikawa Y, Onoyama I, Nakayama KI, Nakayama K (2008) Notch-dependent cell cycle arrest and apoptosis in mouse embryonic fibroblasts lacking Fbxw7. Oncogene 27:6164–6174. https://doi.org/10.1038/onc.2008.216

Jin S, Mutvei AP, Chivukula IV et al (2012) Non-canonical Notch signaling activates IL-6/JAK/STAT signaling in breast tumor cells and is controlled by p53 and IKKα/IKKβ. Oncogene. https://doi.org/10.1038/onc.2012.517

Johmura Y, Nakanishi M (2016) Multiple facets of p53 in senescence induction and maintenance. Cancer Sci 107:1550–1555. https://doi.org/10.1111/cas.13060

Jun J-I, Lau LF (2010) The matricellular protein CCN1 induces fibroblast senescence and restricts fibrosis in cutaneous wound healing. Nat Cell Biol 12:676–685. https://doi.org/10.1038/ncb2070

Kagawa S, Natsuizaka M, Whelan KA et al (2015) Cellular senescence checkpoint function determines differential Notch1-dependent oncogenic and tumor-suppressor activities. Oncogene 34:2347–2359. https://doi.org/10.1038/onc.2014.169

Kang T-W, Yevsa T, Woller N et al (2011) Senescence surveillance of pre-malignant hepatocytes limits liver cancer development. Nature 479:547–551. https://doi.org/10.1038/nature10599

Kim SB, Chae GW, Lee J et al (2007) Activated Notch1 interacts with p53 to inhibit its phosphorylation and transactivation. Cell Death Differ 14:982–991. https://doi.org/10.1038/sj.cdd.4402083

Kirschner K, Samarajiwa SA, Cairns JM et al (2015) Phenotype specific analyses reveal distinct regulatory mechanism for chronically activated p53. PLoS Genet 11:e1005053. https://doi.org/10.1371/journal.pgen.1005053

Kopan R, Ilagan MXG (2009) The canonical Notch signaling pathway: unfolding the activation mechanism. Cell 137:216–233. https://doi.org/10.1016/j.cell.2009.03.045

Krizhanovsky V, Yon M, Dickins RA et al (2008) Senescence of activated stellate cells limits liver fibrosis. Cell 134:657–667. https://doi.org/10.1016/j.cell.2008.06.049

Krtolica A, Parrinello S, Lockett S et al (2001) Senescent fibroblasts promote epithelial cell growth and tumorigenesis: a link between cancer and aging. Proc Natl Acad Sci U S A 98:12072–12077. https://doi.org/10.1073/pnas.211053698

Kuilman T, Michaloglou C, Vredeveld LCW et al (2008) Oncogene-induced senescence relayed by an interleukin-dependent inflammatory network. Cell 133:1019–1031. https://doi.org/10.1016/j.cell.2008.03.039

Kurpinski K, Lam H, Chu J et al (2010) Transforming growth factor-beta and notch signaling mediate stem cell differentiation into smooth muscle cells. Stem Cells 28:734–742. https://doi.org/10.1002/stem.319

Laberge R-M, Sun Y, Orjalo AV et al (2015) MTOR regulates the pro-tumorigenic senescence-associated secretory phenotype by promoting IL1A translation. Nat Cell Biol 17:1049–1061. https://doi.org/10.1038/ncb3195

Lefort K, Mandinova A, Ostano P et al (2007) Notch1 is a p53 target gene involved in human keratinocyte tumor suppression through negative regulation of ROCK1/2 and MRCKalpha kinases. Genes Dev 21:562–577. https://doi.org/10.1101/gad.1484707

Licciulli S, Avila JL, Hanlon L et al (2013) Notch1 is required for Kras-induced lung adenocarcinoma and controls tumor cell survival via p53. Cancer Res 73:5974–5984. https://doi.org/10.1158/0008-5472.CAN-13-1384

Lim KJ, Brandt WD, Heth JA et al (2015) Lateral inhibition of Notch signaling in neoplastic cells. Oncotarget 6:1666–1677

Liu Z-J, Tan Y, Beecham GW et al (2012) Notch activation induces endothelial cell senescence and pro-inflammatory response: implication of Notch signaling in atherosclerosis. Atherosclerosis 225:296–303. https://doi.org/10.1016/j.atherosclerosis.2012.04.010

Lujambio A, Akkari L, Simon J et al (2013) Non-cell-autonomous tumor suppression by p53. Cell 153:449–460. https://doi.org/10.1016/j.cell.2013.03.020

Mandinova A, Lefort K, Tommasi di Vignano A et al (2008) The FoxO3a gene is a key negative target of canonical Notch signalling in the keratinocyte UVB response. EMBO J 27:1243–1254. https://doi.org/10.1038/emboj.2008.45

Mazur PK, Einwächter H, Lee M et al (2010) Notch2 is required for progression of pancreatic intraepithelial neoplasia and development of pancreatic ductal adenocarcinoma. Proc Natl Acad Sci 107:13438–13443. https://doi.org/10.1073/pnas.1002423107

Michaloglou C, Vredeveld LCW, Soengas MS et al (2005) BRAFE600-associated senescence-like cell cycle arrest of human naevi. Nature 436:720–724. https://doi.org/10.1038/nature03890

Moiseeva O (2013) Metformin inhibits the senescence-associated secretory phenotype by interfering with IKK/NF-κB activation. Aging Cell 12:489–498. https://doi.org/10.1111/acel.12075

Muñoz-Espín D, Serrano M (2014) Cellular senescence: from physiology to pathology. Nat Rev Mol Cell Biol 15:482–496. https://doi.org/10.1038/nrm3823

Muñoz-Espín D, Cañamero M, Maraver A et al (2013) Programmed cell senescence during mammalian embryonic development. Cell 155:1104–1118. https://doi.org/10.1016/j.cell.2013.10.019

Nakamura M, Ohsawa S, Igaki T (2014) Mitochondrial defects trigger proliferation of neighbouring cells via a senescence-associated secretory phenotype in Drosophila. Nat Commun 5:5264. https://doi.org/10.1038/ncomms6264

Narita M, Nuñez S, Heard E et al (2003) Rb-mediated heterochromatin formation and silencing of E2F target genes during cellular senescence. Cell 113:703–716

Narita M, Young ARJ, Arakawa S et al (2011) Spatial coupling of mTOR and autophagy augments secretory phenotypes. Science 332:966–970. https://doi.org/10.1126/science.1205407

Nicolas M, Wolfer A, Raj K et al (2003) Notch1 functions as a tumor suppressor in mouse skin. Nat Genet 33:416–421. https://doi.org/10.1038/ng1099

Niimi H, Pardali K, Vanlandewijck M et al (2007) Notch signaling is necessary for epithelial growth arrest by TGF-beta. J Cell Biol 176:695–707. https://doi.org/10.1083/jcb.200612129

Nowell CS, Radtke F (2017) Notch as a tumour suppressor. Nat Rev Cancer 17:145–159. https://doi.org/10.1038/nrc.2016.145

Oakley F, Mann J, Ruddell R-G et al (2003) Basal expression of IkappaBalpha is controlled by the mammalian transcriptional repressor RBP-J (CBF1) and its activator Notch1. J Biol Chem 278:24359–24370. https://doi.org/10.1074/jbc.M211051200

O'Neil J, Grim J, Strack P et al (2007) FBW7 mutations in leukemic cells mediate NOTCH pathway activation and resistance to gamma-secretase inhibitors. J Exp Med 204:1813–1824. https://doi.org/10.1084/jem.20070876

Orjalo AV, Bhaumik D, Gengler BK et al (2009) Cell surface-bound IL-1alpha is an upstream regulator of the senescence-associated IL-6/IL-8 cytokine network. Proc Natl Acad Sci U S A 106:17031–17036. https://doi.org/10.1073/pnas.0905299106

Palomero T, Lim WK, Odom DT et al (2006) NOTCH1 directly regulates c-MYC and activates a feed-forward-loop transcriptional network promoting leukemic cell growth. Proc Natl Acad Sci U S A 103:18261–18266. https://doi.org/10.1073/pnas.0606108103

Petrovic J, Formosa-Jordan P, Luna-Escalante JC et al (2014) Ligand-dependent Notch signaling strength orchestrates lateral induction and lateral inhibition in the developing inner ear. Development 141:2313–2324. https://doi.org/10.1242/dev.108100

Procopio M-G, Laszlo C, Labban Al D et al (2015) Combined CSL and p53 downregulation promotes cancer-associated fibroblast activation. Nat Cell Biol 17:1193–1204. https://doi.org/10.1038/ncb3228

Qi R, An H, Yu Y et al (2003) Notch1 signaling inhibits growth of human hepatocellular carcinoma through induction of cell cycle arrest and apoptosis. Cancer Res 63:8323–8329

Rampias T, Vgenopoulou P, Avgeris M et al (2014) A new tumor suppressor role for the Notch pathway in bladder cancer. Nat Med. https://doi.org/10.1038/nm.3678

Ranganathan P, Weaver KL, Capobianco AJ (2011) Notch signalling in solid tumours: a little bit of everything but not all the time. Nat Rev Cancer 11:338–351. https://doi.org/10.1038/nrc3035

Rodier F, Coppé J-P, Patil CK et al (2009) Persistent DNA damage signalling triggers senescence-associated inflammatory cytokine secretion. Nat Cell Biol 11:973–979. https://doi.org/10.1038/ncb1909

Ronchini C, Capobianco AJ (2001) Induction of cyclin D1 transcription and CDK2 activity by Notchic: implication for cell cycle disruption in transformation by Notchic. Mol Cell Biol 21:5925–5934. https://doi.org/10.1128/MCB.21.17.5925-5934.2001

Salama R, Sadaie M, Hoare M, Narita M (2014) Cellular senescence and its effector programs. Genes Dev 28:99–114. https://doi.org/10.1101/gad.235184.113

Sang L, Coller HA, Roberts JM (2008) Control of the reversibility of cellular quiescence by the transcriptional repressor HES1. Science 321:1095–1100. https://doi.org/10.1126/science.1155998

Schmitt CA, Fridman JS, Yang M et al (2002) A senescence program controlled by p53 and p16INK4a contributes to the outcome of cancer therapy. Cell 109:335–346

Sebastian T, Johnson PF (2009) RasV12-mediated down-regulation of CCAAT/enhancer binding protein beta in immortalized fibroblasts requires loss of p19Arf and facilitates bypass of oncogene-induced senescence. Cancer Res 69:2588–2598. https://doi.org/10.1158/0008-5472.CAN-08-2312

Serrano M, Lin AW, McCurrach ME et al (1997) Oncogenic ras provokes premature cell senescence associated with accumulation of p53 and p16INK4a. Cell 88:593–602

Sethi N, Dai X, Winter CG, Kang Y (2011) Tumor-derived JAGGED1 promotes osteolytic bone metastasis of breast cancer by engaging notch signaling in bone cells. Cancer Cell 19:192–205. https://doi.org/10.1016/j.ccr.2010.12.022

Shay JW, Pereira-Smith OM, Wright WE (1991) A role for both RB and p53 in the regulation of human cellular senescence. Exp Cell Res 196:33–39

Simón R, Aparicio R, Housden BE et al (2014) Drosophila p53 controls Notch expression and balances apoptosis and proliferation. Apoptosis 19:1430–1443. https://doi.org/10.1007/s10495-014-1000-5

Sörensen-Zender I, Rong S, Susnik N et al (2014) Renal tubular Notch signaling triggers a prosenescent state after acute kidney injury. Am J Physiol Renal Physiol 306:F907–F915. https://doi.org/10.1152/ajprenal.00030.2014

Storer M, Mas A, Robert-Moreno A et al (2013) Senescence is a developmental mechanism that contributes to embryonic growth and patterning. Cell. https://doi.org/10.1016/j.cell.2013.10.041

Tasdemir N, Banito A, Roe J-S et al (2016) BRD4 connects enhancer remodeling to senescence immune surveillance. Cancer Discov 6:612–629. https://doi.org/10.1158/2159-8290.CD-16-0217

Toso A, Revandkar A, Di Mitri D et al (2014) Enhancing chemotherapy efficacy in pten-deficient prostate tumors by activating the senescence-associated antitumor immunity. Cell Rep 9:75–89. https://doi.org/10.1016/j.celrep.2014.08.044

van Deursen JM (2014) The role of senescent cells in ageing. Nature 509:439–446. https://doi.org/10.1038/nature13193

Venkatesh D, Fredette N, Rostama B et al (2011) RhoA-mediated signaling in Notch-induced senescence-like growth arrest and endothelial barrier dysfunction. Arterioscler Thromb Vasc Biol 31:876–882. https://doi.org/10.1161/ATVBAHA.110.221945

Wang H, Zou J, Zhao B et al (2011) Genome-wide analysis reveals conserved and divergent features of Notch1/RBPJ binding in human and murine T-lymphoblastic leukemia cells. Proc Natl Acad Sci U S A 108:14908–14913. https://doi.org/10.1073/pnas.1109023108

Weng AP, Ferrando AA, Lee W et al (2004) Activating mutations of NOTCH1 in human T cell acute lymphoblastic leukemia. Science 306:269–271. https://doi.org/10.1126/science.1102160

Weng AP, Millholland JM, Yashiro-Ohtani Y et al (2006) c-Myc is an important direct target of Notch1 in T-cell acute lymphoblastic leukemia/lymphoma. Genes Dev 20:2096–2109. https://doi.org/10.1101/gad.1450406

Xue W, Zender L, Miething C et al (2007) Senescence and tumour clearance is triggered by p53 restoration in murine liver carcinomas. Nature 445:656–660. https://doi.org/10.1038/nature05529

Yatim A, Benne C, Sobhian B et al (2012) NOTCH1 nuclear Interactome reveals key regulators of its transcriptional activity and oncogenic function. Mol Cell. https://doi.org/10.1016/j.molcel.2012.08.022

Yoshida A, Lee EK, Diehl JA (2016) Induction of therapeutic senescence in Vemurafenib-resistant melanoma by extended inhibition of CDK4/6. Cancer Res 76:2990–3002. https://doi.org/10.1158/0008-5472.CAN-15-2931

Young ARJ, Narita M, Ferreira M et al (2009) Autophagy mediates the mitotic senescence transition. Genes Dev 23:798–803. https://doi.org/10.1101/gad.519709

Yugawa T, Handa K, Narisawa-Saito M et al (2007) Regulation of Notch1 gene expression by p53 in epithelial cells. Mol Cell Biol 27:3732–3742. https://doi.org/10.1128/MCB.02119-06

Yun J, Espinoza I, Pannuti A et al (2015) p53 modulates notch signaling in MCF-7 breast Cancer cells by associating with the notch transcriptional complex via MAML1. J Cell Physiol 230:3115–3127. https://doi.org/10.1002/jcp.25052

Zavadil J, Cermak L, Soto-Nieves N, Böttinger EP (2004) Integration of TGF-beta/Smad and Jagged1/Notch signalling in epithelial-to-mesenchymal transition. EMBO J 23:1155–1165. https://doi.org/10.1038/sj.emboj.7600069

Zhao Y, Katzman RB, Delmolino LM et al (2007) The notch regulator MAML1 interacts with p53 and functions as a coactivator. J Biol Chem 282:11969–11981. https://doi.org/10.1074/jbc.M608974200

Control of Blood Vessel Formation by Notch Signaling

Fabian Tetzlaff and Andreas Fischer

Abstract

Blood vessels span throughout the body to nourish tissue cells and to provide gateways for immune surveillance. Endothelial cells that line capillaries have the remarkable capacity to be quiescent for years but to switch rapidly into the activated state once new blood vessels need to be formed. In addition, endothelial cells generate niches for progenitor and tumor cells and provide organ-specific paracrine (angiocrine) factors that control organ development and regeneration, maintenance of homeostasis and tumor progression. Recent data indicate a pivotal role for blood vessels in responding to metabolic changes and that endothelial cell metabolism is a novel regulator of angiogenesis. The Notch pathway is the central signaling mode that cooperates with VEGF, WNT, BMP, TGF-β, angiopoietin signaling and cell metabolism to orchestrate angiogenesis, tip/stalk cell selection and arteriovenous specification. Here, we summarize the current knowledge and implications regarding the complex roles of Notch signaling during physiological and tumor angiogenesis, the dynamic nature of tip/stalk cell selection in the nascent vessel sprout and arteriovenous differentiation. Furthermore, we shed light on recent work on endothelial cell metabolism, perfusion-independent angiocrine functions of endothelial cells in organ-specific vascular beds and how manipulation of Notch signaling may be used to target the tumor vasculature.

Keywords

Angiogenesis · Notch signaling · Arteriovenous differentiation · Tumor angiogenesis · Angiocrine signaling · Endothelial metabolism · Endothelial cells

Abbreviations

ADAM	A disintegrin and metalloprotease
ALK	Activin receptor-like kinase
BMP	Bone morphogenetic protein

F. Tetzlaff
Vascular Signaling and Cancer (A270), German Cancer Research Center (DKFZ), Heidelberg, Germany

Vascular Biology, CBTM, Medical Faculty Mannheim, Heidelberg University, Mannheim, Germany

A. Fischer (✉)
Vascular Signaling and Cancer (A270), German Cancer Research Center (DKFZ), Heidelberg, Germany

Vascular Biology, CBTM, Medical Faculty Mannheim, Heidelberg University, Mannheim, Germany

Department of Medicine I and Clinical Chemistry, Heidelberg University Hospital, Heidelberg, Germany
e-mail: a.fischer@dkfz.de

T. Borggrefe, B. D. Giaimo (eds.), *Molecular Mechanisms of Notch Signaling*, Advances in Experimental Medicine and Biology 1066, https://doi.org/10.1007/978-3-319-89512-3_16

CADASIL	Cerebral Autosomal dominant arteriopathy with subcortical infarcts and leukoencephalopathy
DLL	Delta-like
EC	Endothelial cell
FGF	Fibroblast growth factor
FOX	Forkhead box protein
HES	Hairy and enhancer of split
HEY	Hairy/enhancer-of-split related with YRPW motif
IL	Interleukin
NICD	Notch intracellular domain
NRARP	Notch-regulated ankyrin repeat-containing protein
NRP	Neuropilin
PFKFB3	6-Phosphofructo-2-kinase
PI3K	Phosphatidylinositol 4,5-Bisphosphate 3-kinase
RBPJ	Recombining binding protein suppressor of hairless
SHH	Sonic hedgehog
SMAD	Mothers against decapentaplegic
TGF	Transforming growth factor
VEGF	Vascular endothelial growth factor
VEGFR	Vascular endothelial growth factor receptor

1 Introduction

The vasculature comprises one of the largest organs in mammals. Blood vessels nourish all tissues in the body and provide gateways for immune surveillance. In addition, vascular cells provide organ-specific paracrine factors, also termed angiocrine factors, which instruct the behavior of neighboring cells. Angiocrine signaling is essential for the maintenance of homeostasis and metabolism, stem cell differentiation, organ regeneration and tumor progression (Rafii et al. 2016). The importance of the vasculature becomes apparent by studying vascular dysfunction, which is the major contributor to human mortality. Abnormalities in vessel functionality are causative for heart infarction, stroke, neurodegenerative diseases, dementia, diabetic complications and obesity-associated disorders, while excessive blood vessel formation is a hallmark of cancer, chronic inflammation and eye diseases such as wet macular degeneration (Folkman 2007). Drugs that inhibit blood vessel growth have recently become first-line therapies for certain eye and tumor diseases (Carmeliet and Jain 2011) .

Blood vessels are formed by endothelial cells (ECs), which provide an anti-thrombotic surface, and by mural cells (vascular smooth muscle cells and pericytes). In mature vessels, ECs are in a quiescent state, divide rarely and form barriers between blood and surrounding tissues. ECs have the remarkable capacity to switch between the quiescent and the activated state during injuries, hypoxia, inflammation or tissue growth, when the formation of new blood vessels is required (Potente et al. 2011).

The *de novo* formation of blood vessels from mesodermal-derived endothelial precursor cells is called vasculogenesis (Risau and Flamme 1995). It occurs predominantly during early development to generate a primordial vascular plexus and the first large vessels such as the dorsal aorta. The vascular plexus is further remodeled and new vessels are formed from the pre-existing ones in a process called angiogenesis (Herbert and Stainier 2011). Similarly to new branches growing on a tree, angiogenesis occurs primarily by sprouting of new branches from existing microvessels. Angiogenesis occurs throughout life as capillaries grow and regress accordingly to functional demands. For example, physical exercise stimulates angiogenesis in skeletal muscle (Hellsten and Hoier 2014) and expansion of adipose tissue is also associated with the formation of new blood vessels (Cao 2010). Intussusception (vessel splitting) is another way of generating new vessels. During this process blood vessels develop transluminal tissue pillars which subsequently fuse resulting in new vascular entities (Makanya et al. 2009). Once the new vessels establish nutrient and oxygen supplies that meet the metabolic tissue demand, the ECs will turn quiescent (Risau 1997).

Notch signaling is of utmost importance for vessel morphogenesis and function. Based on a series of previously published excellent review articles (Blanco and Gerhardt 2013; Carmeliet and Jain 2011; Eilken and Adams 2010; Gridley 2010; Siekmann et al. 2013; Potente et al. 2011), this chapter will summarize the current view about Notch signaling in the vasculature with a focus on vessel sprouting, arteriovenous differentiation, EC metabolism and tumor angiogenesis. We will also highlight recent work showing the tight interconnections of the Notch pathway with other core signaling pathways and its roles for organ-specific angiocrine signaling.

2 Notch Signaling in Endothelial Cells

Canonical Notch signaling requires the interaction of membrane-bound Notch ligands on the signal-sending cell with Notch receptors on the signal-receiving cell to trigger proteolytic cleavages of the Notch receptors. γ-secretase releases the active Notch intracellular domain (NICD) from the cell membrane, which translocates to the nucleus, binds to the transcription factor Rbpj [also known as CSL, CBF1, Su(H) or Lag2] and activates gene expression (Kopan and Ilagan 2009). In principle, expression of the Notch ligands Dll1, Dll4, Jag1 and Jag2 and the Notch receptors Notch1 and Notch4 on ECs has been reported (Hofmann and Luisa Iruela-Arispe 2007). However, one needs to keep in mind that the individual endothelial expression patterns are quite variable in different vascular beds (e.g. Notch signaling is much higher in arterial than venous ECs), and are depending on the developmental state. Compared to the normal, quiescent vasculature in tissues of the adult, the expression of Notch ligands is typically stronger in tumor blood vessels (Patel et al. 2005; Lu et al. 2007; Jubb et al. 2012; Gale et al. 2004; Mailhos et al. 2001; Scehnet et al. 2007). Prototypical Notch1 target genes in ECs are *Hey1*, *Hey2*, *Hes1*, *Nrarp*, *EphrinB2*, but also the Notch ligand-encoding gene *Dll4* (Dou et al. 2008; Fischer et al. 2004; Taylor et al. 2002; Liu et al. 2006; Krebs et al. 2001; Phng et al. 2009; Lawson et al. 2002; Ridgway et al. 2006; Lobov et al. 2007; Iso et al. 2006; Patel et al. 2005). The latter is quite unusual and suggests a positive Dll4-Notch1 feedback loop in ECs (Diez et al. 2007; Lanner et al. 2013). Notch ligands are also cleaved by the γ-secretase and their intracellular domain enters the nucleus. However, no functional role for a potential "Notch reverse signaling" during angiogenesis could be detected (Liebler et al. 2012; Redeker et al. 2013).

Gene targeting studies in mice revealed that deletion of *Dll4* (Duarte et al. 2004; Krebs et al. 2004; Gale et al. 2004), *Jag1* (Xue et al. 1999), *Notch1* (Huppert et al. 2000; Krebs et al. 2000; Limbourg et al. 2005), *Notch1/Notch4* (Krebs et al. 2000), the Notch S2 cleavage enzyme *Adam10* (Glomski et al. 2011), components of the γ-secretase complex (Herreman et al. 1999; Li et al. 2003), *Rbpj* (Krebs et al. 2004), *Hey1/Hey2* (Kokubo et al. 2005; Fischer et al. 2004), or a constitutive endothelium-specific expression of activated alleles for *Notch1* (Krebs et al. 2010) or *Notch4* (Uyttendaele et al. 2001) lead to embryonic lethality with severe vascular remodeling abnormalities and defects in arteriovenous specification. Besides embryonic development, Notch signaling coordinates vascular remodeling also in the adult (Limbourg et al. 2007; Takeshita et al. 2007). Interestingly, the loss of a single *Dll4* allele already results in severe angiogenesis defects (Duarte et al. 2004; Gale et al. 2004; Krebs et al. 2004). *Dll4* and *Vegf-a* belong to the very few genes, of which heterozygosity results in a lethal embryonic phenotype.

One could assume that endothelial Notch ligands act in a redundant manner. However, it was shown that they play distinct roles in blood vessel morphogenesis and do not act redundantly (Preuße et al. 2015). Expression of Dll1 on ECs begins later than that of Dll4 during fetal mouse development. While Dll4 is needed to establish arterial cell fate (*see below*), Dll1 is required for maintenance of arterial cell fate (Sörensen et al. 2009). On the other hand, Jag1 can even antagonize Dll4/Notch1 signaling in ECs during tip/stalk cell selection depending on the glycosylation pattern of Notch1 receptor (Benedito et al. 2009).

3 Sprouting Angiogenesis

The outgrowth of a new vessel branch is stimulated by proangiogenic growth factors, which are released during hypoxia, inflammation, nutrient starvation or from oncogene-transformed cells. These shift the balance between proangiogenic (e.g. VEGF, FGF) and antiangiogenic (e.g. endostatin, angiostatin, tumstatin, soluble VEGFR1) factors towards a proangiogenic outcome, an event termed the "angiogenic switch" (Folkman 1995; Folkman 2007). The most important proangiogenic protein is vascular endothelial growth factor (VEGF-A; hereafter called VEGF). The complex signaling biology of VEGF family members [VEGF-A, -B, -C, –D, –E and placenta growth factor (PlGF)] and VEGF-A splice isoforms has been reviewed elsewhere (Simons et al. 2016). Deletion of *Vegf* or its receptors in mice leads to embryonic death as consequence of abnormal vascular development (Fong et al. 1995; Dumont et al. 1998; Shalaby et al. 1995; Carmeliet et al. 1996; Ferrara et al. 1996). In the postnatal mouse retina, a Vegf gradient is generated by the already existing astrocyte network that serves as a guiding scaffold for the developing blood vessels (Ruhrberg et al. 2002; Gerhardt et al. 2003).

Angiogenesis is induced by VEGF, which signals through VEGFR2 and VEGFR3 to activate quiescent ECs. Activated ECs protrude filopodia, secrete matrix metalloproteinases to degrade the basement membrane and become invasive (Arroyo and Iruela-Arispe 2010). The breakdown of basement membrane is in particular mediated by EC podosome rosettes (Seano et al. 2014). Podosomes are specialized actin-based structures that degrade extracellular matrix and promote invasive cell migration (Murphy and Courtneidge 2011). The formation of EC podosomes is controlled by VEGF and Notch signaling (Spuul et al. 2016). Furthermore, stimulated ECs release angiopoietin-2 leading to detachment of pericytes. This further allows ECs to invade the surrounding tissue (Augustin et al. 2009). During invasion ECs usually remain connected to the vessel network (Blanco and Gerhardt 2013).

The nascent sprout contains two different cell phenotypes: tip and stalk cells (Fig. 1). The leading tip cell is characterized by its position, its long and dynamic filopodia and its pro-invasive and migratory behavior (Gerhardt et al. 2003), but also its highly glycolytic metabolic activity (De Bock et al. 2013). Similar to axonal growth cones, tip cells integrate attractive and repellent guidance cues (e.g. Semaphorin, Netrin, VEGF or Slit proteins) to define the route in which the new sprout grows (Adams and Eichmann 2010). Guidance is facilitated by actin-rich filopodia on the tip cells, whose formation is driven by VEGF via RhoGTPase signaling. Interestingly, filopodia are not absolutely necessary for migration of ECs as lamellipodia can partially compensate for their function (Phng et al. 2013). It was reported that there can be two cells that extend filopodia and have significant overlap in space and time at the tip of angiogenic sprouts (Pelton et al. 2014). This surprising observation challenges the model of a single EC at the sprout tip. The trailing stalk cells are proliferative, less migratory than tip cells and form the nascent vascular lumen (Gerhardt et al. 2003). Furthermore, tip and stalk cells possess distinct gene expression profiles (e.g. higher expression of *Dll4*, *Vegfr2*, *Vegfr3*, *Pdgfb*, *Unc5b*, *Cxcr4*, *Nidogen-2*, *Esm1*, *Angiopoietin-2*, *Apelin* in tip cells) (Del Toro et al. 2010; Blanco and Gerhardt 2013). For cell proliferation, stalk cells have to generate biomass (nucleotides, protein, lipids). Therefore, cell metabolism differs between tip and stalk cells (see 3.4). Stalk cells produce extracellular matrix and recruit pericytes that attach to the new vessel sprout (Fig. 1). ECs in new vessel loops that are well covered by mural cells and have again become quiescent were named "phalanx cells" (Mazzone et al. 2009).

3.1 VEGF and Notch Signaling Control Tip/Stalk Cell Selection

The ability of ECs to lead a nascent sprout is strongly dependent on their VEGF receptor expression profile and their competence to

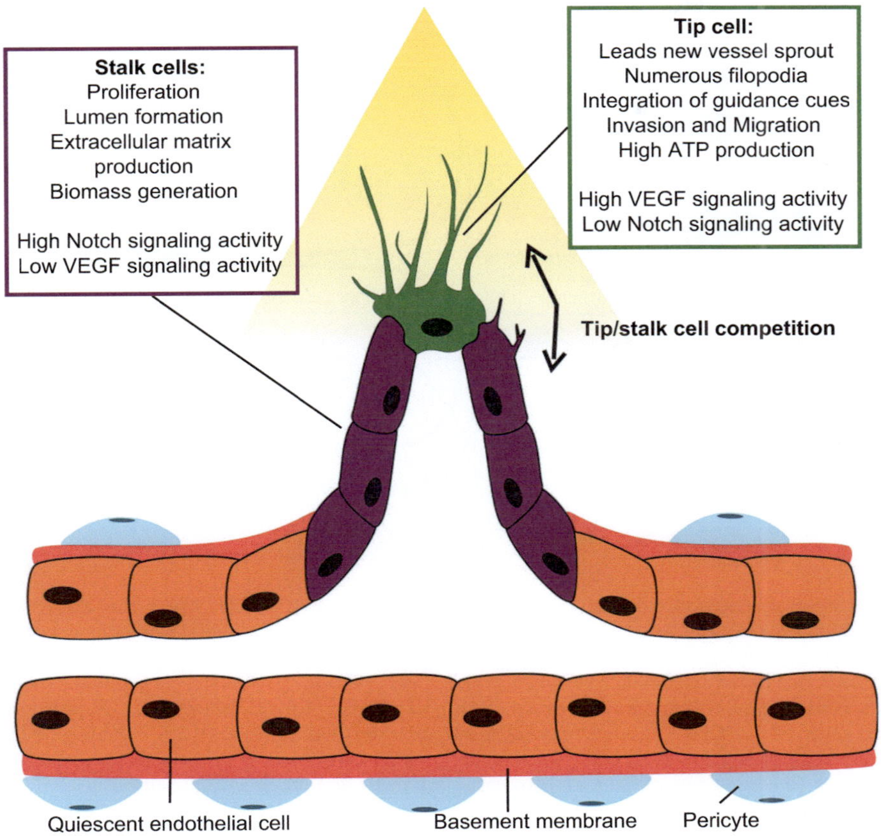

Fig. 1 Model of tip/stalk cell phenotypes. The leading tip cell protrudes many filopodia and guides the new vessel sprout towards the VEGF gradient. Tip cells are highly invasive and migratory and require high ATP amounts, which are predominantly generated by glycolysis. The trailing stalk cells proliferate and form a new vessel lumen. The newly formed vessel sprout gets covered by extracellular matrix proteins and by pericytes. However, this is a dynamic process and stalk cells battle for the tip position to take over the lead

respond to VEGF (Jakobsson et al. 2010; Gerhardt et al. 2003). While tip cells are characterized by high expression levels of Vegfr2 and also Vegfr3 (Tammela et al. 2008; Tammela et al. 2011; Zarkada et al. 2015; Blanco and Gerhardt 2013), the role of the Vegfr1, which acts as a VEGF trap, is less clear (Siekmann et al. 2013). In zebrafish, Notch-driven Vegfr1 expression acts as a negative regulator of tip cell differentiation (Krueger et al. 2011) and neuronal-derived soluble Vegfr1 is critical for guiding the direction of vessel growth (Wild et al. 2017).

VEGF signaling acts upstream of the Notch pathway and induces Dll4 expression (Lawson et al. 2002; Ridgway et al. 2006; Lobov et al. 2007; Patel et al. 2005). It has been suggested that Vegf acts via the PI3K pathway activating the Forkhead family transcription factors Foxc1 and Foxc2, which then bind to a *Dll4* enhancer element, or alternatively via the disassembly of a repressor complex at the *Dll4* promoter (Seo et al. 2006; Hayashi and Kume 2008). Subsequently, Dll4 binds and signals to Notch1 receptors on adjacent ECs. The Notch-induced transcription factors Hey1 and Hey2 decrease expression of Vegfr2/3 and thereby reduce responsiveness to VEGF. Such cells will most likely behave as stalk cells (Blanco and Gerhardt 2013). Therefore the nascent sprout is guided by a tip cell with high Dll4 expression and low Notch signaling activity followed by stalk cells with high Notch signaling output (Fig. 2).

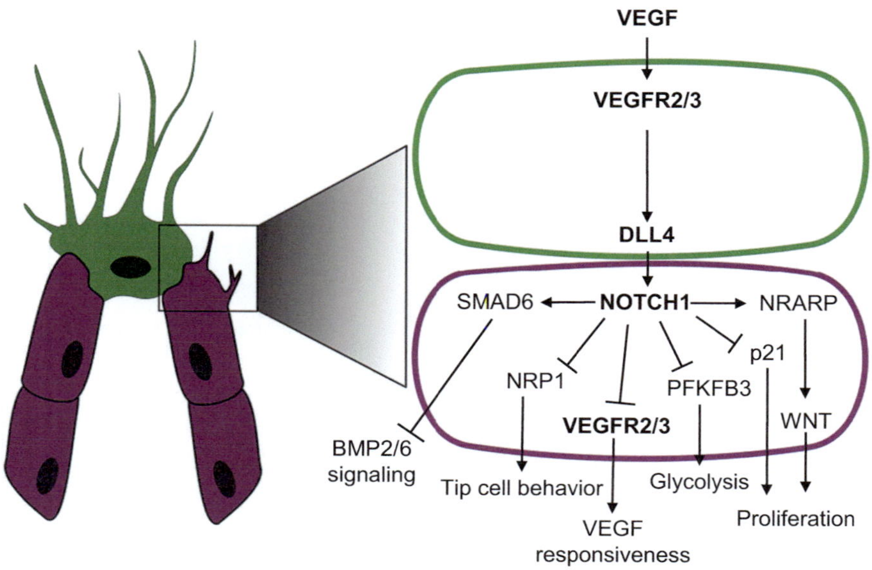

Fig. 2 Core signaling pathways during tip/stalk cell selection. VEGF induces tip cell behavior and expression of the Notch ligand DLL4. This leads to NOTCH1 activation in adjacent cells which adopt the stalk cell phenotype. In stalk cells, Notch signaling represses expression of tip cell-enriched genes like VEGFR2/3 and thereby suppress responsiveness to the pro-angiogenic VEGF. Notch inhibits its expression of PFKFB3, an activator of glycolysis, which is required to adopt the tip cell phenotype. Moreover, Notch inhibits proliferation via inhibition of p21 but this is counteracted via WNT signaling since stalk cells need to proliferate. In addition, Notch activates expression of the inhibitory SMAD6 proteins to counteract pro-angiogenic BMP2/6 signaling. Notch inhibits NRP1 expression, which suppresses the stalk cell phenotype by limiting SMAD2/3 activation

Studies with genetic or pharmacologic inhibition of Notch signaling underlined the importance of this pathway during sprouting angiogenesis and tip/stalk cell selection. Notch inhibition leads to the formation of excessive tip cell numbers and vessel branches, a process called hypersprouting (Noguera-Troise et al. 2007; Ridgway et al. 2006; Hellström et al. 2007; Lobov et al. 2007; Siekmann and Lawson 2007; Suchting et al. 2007; Sainson et al. 2005; Leslie et al. 2007). Accordingly, ECs with low Notch signaling activity dominate at the tip cell position, whereas Notch-active ECs are mostly excluded (Jakobsson et al. 2010; Hellström et al. 2007; Siekmann and Lawson 2007; Benedito et al. 2009).

Dll4/Notch1 is the most important ligand and receptor pair in coordinating angiogenesis. However, the situation is more complex. For example, stalk cells also express few Dll4 ligands on their membrane and this could potentially lead to signaling back to Notch receptors on tip cells. This is antagonized by the Notch ligand Jag1, which is strongly expressed on stalk cells (Hofmann and Luisa Iruela-Arispe 2007; Benedito et al. 2009) and inhibits Dll4/Notch1 signaling. Thereby, Jag1 antagonizes signaling from the stalk back to the tip cell (Benedito et al. 2009) and it may also prevent Notch overactivation in the stalk cell plexus.

3.2 Crosstalk Between Notch and Other Signaling Pathways to Control Tip/Stalk Cell Selection

Numerous additional molecules influence tip or stalk cell fate selection through interactions with Notch signaling. In brief, WNT/β-catenin signaling promotes transcription of *Dll4* by binding to an enhancer element (Corada et al. 2010) or through protein interaction of β-catenin with Rbpj (Yamamizu et al. 2010). Furthermore,

WNT signaling induces expression of the transcription factor Sox17, which can activate Notch signaling and promote expression of tip cell-enriched genes (Lee et al. 2014; Corada et al. 2013). On the other hand, Sox17 expression is repressed by Notch signaling in stalk cells (Lee et al. 2014). It was demonstrated, that the mRNA level of *Sox17* is not altered by Notch whereas the protein level of Sox17 is. This shows that Sox17 is post-transcriptionally regulated by the Notch pathway. Taken together this indicates that through a negative feedback loop, hypersprouting is prevented. Similarly, Notch and WNT signaling are linked via Nrarp to control the stability of new vessels. Notch induces Nrarp expression, which in turn limits Notch signaling and promotes WNT signaling in stalk cells (Phng et al. 2009).

The competence of ECs to become a tip cell is also influenced by bone morphogenetic proteins (BMPs) and TGF-β signaling. Bmp9 signals through Alk1 in stalk cells to induce Smad1/5/8 phosphorylation. These Smads synergize with activated Notch receptors to induce expression of Notch targets Hey1 and Hey2, which inhibit VEGF receptor expression (Larrivée et al. 2012; Moya et al. 2012). This is further promoted by Smad1/5-mediated induction of Id proteins which augment Hes1 protein levels (Moya et al. 2012). However, the roles of BMP signaling for tip/stalk selection and angiogenesis are not fully defined yet and still controversial. Very recently, it was reported that Notch promotes expression of the inhibitory Smad6 protein and thereby limits the responsiveness of stalk cells towards the proangiogenic Bmp2 and Bmp6 (Mouillesseaux et al. 2016). Lastly, it was reported that the stalk cell phenotype has to be actively repressed to allow tip cell formation. Neuropilin-1 (Nrp1) plays a key role in suppressing the stalk cell phenotype through limiting Smad2/3 activation. Nrp1 promotes tip cell behavior and the formation of filopodia (Fantin et al. 2013; Fantin et al. 2015). Notch downregulates Nrp1 expression and thus promotes stalk cell behavior (Aspalter et al. 2015).

The Notch-dependent acquisition of the stalk cell phenotype also requires the phosphatase Pten (Serra et al. 2015). Furthermore, Dll4 expression in tip cells is regulated via laminin/integrin signaling (Stenzel et al. 2011). Besides crosstalk of Notch signaling with other signaling pathways, direct protein-protein interactions influence tip-stalk-cell selection. Synaptojanin-2-binding protein (Synj2bp) stabilizes Delta-like protein expression in stalk cells to allow continuous Notch signaling within the stalk cell plexus and to prevent formation of ectopic vessel branches (Adam et al. 2013).

3.3 The Dynamic Nature of Tip/Stalk Cell Differentiation

EC tip and stalk cell specification does not represent permanent cell fate decisions but rather dynamic fluctuations in cell phenotypes (Blanco and Gerhardt 2013). The Gerhardt laboratory has shown that stalk cells compete in a highly dynamic manner for the tip position leading to frequent exchange of the tip cells (Jakobsson et al. 2010). Such EC shuffling occurs every few hours (Ubezio et al. 2016). Mechanistically, the VEGF-Dll4/Notch feedback system drives the competition for the tip/stalk cell selection. This is facilitated by the oscillatory output strength of Notch signaling (Kageyama et al. 2007). As such, the expression of Dll4 fluctuates in individual ECs within sprouting vessels (Ubezio et al. 2016). Therefore, one can assume that concomitantly the levels of Vegfrs, Dll4 and Notch target genes change constantly as ECs interact with each other. As a result, the competence of acting as a tip cell changes constantly, certain stalk cells are relieved from tip cell inhibition and overtake the lead position (Blanco and Gerhardt 2013). This leads to a dynamic position shuffle in the growing sprout.

The tip cell competence concept is further strengthened by the finding that the continual flux in Notch signaling output strength in individual ECs results in differential VE-cadherin turnover to generate spatial differentials in cell-cell adhesions and polarized junctional protrusions. These permanent switches between active and inactive cell junctions allow EC rearrangements during sprout elongation (Bentley et al. 2014).

3.4 Control of Angiogenesis by Metabolism

The vasculature contributes to systemic metabolism control. On the one hand the endothelium controls the shuttling of nutrients from blood to tissue cells in an organ-specific manner (Robciuc et al. 2016; Jais et al. 2016; Hagberg et al. 2010; Corvera and Gealekman 2014) and therefore plays a critical, but poorly understood role, for organ homeostasis. On the other hand, metabolism controls angiogenesis. For example, the expansion of adipose tissue requires angiogenesis, which is stimulated by proangiogenic factors released from adipocytes (Corvera and Gealekman 2014). ECs contain metabolic sensors and their effectors (Sirtuins, mTOR, Pgc1α, Lkb1, Ampk, Foxos and Sirt1) (Potente and Carmeliet 2017) and respond to alteration in nutrient supply. To understand how cellular metabolism affects angiogenesis, one needs to consider how ECs generate ATP. Research from the Carmeliet laboratory revealed that ECs are very glycolytic and produce the majority of ATP by metabolizing glucose into lactate rather than by oxidative phosphorylation, even if plenty of oxygen is available (De Bock et al. 2013). As such, ECs behave similar to cancer cells, which consume high amount of glucose for aerobic glycolysis (Schulze and Harris 2012). Although much less ATP is gained compared to oxidative phosphorylation, glycolysis has the advantage of generating ATP in a very rapid manner and glycolysis allows energy production in hypoxic areas, into which angiogenic ECs need to migrate (Potente and Carmeliet 2017).

Activated ECs require in particular high glycolytic flux for migration and invasion (De Bock et al. 2013; Cruys et al. 2016). This is facilitated by VEGF and hypoxia signaling that together increase the uptake and breakdown of glucose by up-regulating glucose transporter type-1 and glycolytic enzymes, such as 6-Phosphofructo-2-kinase (Pfkfb3) and lactate dehydrogenase-A (Yeh et al. 2008; Peters et al. 2009; Nakazawa et al. 2016; De Bock et al. 2013). Even in ECs with constitutive Notch1 signaling, which are genetically determined to become stalk cells, enhanced glycolysis by Pfkfb3 activation induces tip cell behavior (De Bock et al. 2013). This indicates that EC metabolism can exert control over genetic circuits (Potente and Carmeliet 2017).

In stalk cells, Notch signaling reduces but not eliminates the expression of Pfkfb3 and Pfkfbp3-driven glycolysis, as it is also essential for stalk cells (De Bock et al. 2013). Moreover, stalk cells must synthesize all cellular components (e.g. nucleotides, proteins and lipids) for cell division and cell growth. Therefore, ECs also break down fatty acids to generate carbons for the *de novo* nucleotide synthesis and not only for energy production (Schoors et al. 2015).

It will have to be determined how exactly the metabolic status influences the EC genetic program and vice versa. Fluctuations of Notch and VEGF signaling outputs alter glycolysis rates and ATP production in ECs and thereby change the fitness of ECs to battle for the tip position (Spuul et al. 2016; Potente and Carmeliet 2017; De Bock et al. 2013). The energy status also controls the activity of Foxo1 by Sirt1 and the latter inhibits Notch signaling through deacetylation of the Nicd1 resulting in increased angiogenesis (Guarani et al. 2011). Latest research showed that Foxo1 is an essential regulator of vascular growth by coupling metabolic and proliferative activities in ECs via inhibition of Myc, which fuels glycolysis and mitochondrial metabolism (Wilhelm et al. 2016). In addition, the Notch signaling activity in ECs is influenced by plasma glucose levels (Yoon et al. 2014) and by the presence of certain pro-inflammatory fatty acids (Briot et al. 2015). Taken together, these reports show that Notch signaling integrates angiogenic signaling with the metabolic status.

3.5 Anastomosis of Vessel Sprouts and Remodeling of the New Vessel Network

Newly formed sprouts need to connect with other sprouts or existing vessels to generate a new circulatory loop. Anastomosis is a complex process that has not yet been fully resolved (Betz et al. 2016). Tip cells contact other tip cells to initiate fusing of two sprouts (Isogai et al. 2003), which is supported by tissue-resident macrophages (Tammela et al. 2011; Fantin et al. 2010; Outtz et al. 2011). Anastomosis requires the formation of new VE-Cadherin-containing EC junctions to consolidate the connection (Bentley et al. 2014). Such junctions are essential for EC polarization and lumen formation. After formation of a patent lumen, blood flow contributes to stabilize the new vascular loop. Increasing oxygen tension decreases VEGF production and helps to switch the activated EC status into a quiescent one. Further vessel maturation includes production of extracellular matrix, recruitment of mural cells, remodeling into a hierarchical network and the pruning of excessive vessel branches (Potente et al. 2011). Notch signaling is critically involved in the recruitment and the tight interactions of ECs with pericytes and smooth muscle cells (Fouillade et al. 2012). Further research is required to elucidate the detailed mechanisms of how Notch signaling is involved in vessel pruning.

4 Arteriovenous Differentiation

After the assembly of the first primitive vessels in the embryo or in a growing tissue of the adult (e.g. muscle or adipose tissue) a rapid differentiation into a hierarchically organized network of arteries, capillaries, veins and lymphatic vessels occurs. The specification of lymphatics has been reviewed elsewhere (Yang and Oliver 2014). Arteries transport blood away from the heart towards the capillaries. As such, arterial vessels are subjected to high blood pressure and pulsatile shear stress, whereas veins face low-pressure gradients can contain valves to prevent backflow and are more distensible than arteries (Corada et al. 2014).

Several studies indicated that vascular progenitor cells, which form the first large vessels in the embryo, are already committed for arterial or venous cell fate (Quillien et al. 2014; Kohli et al. 2013). On the other hand, it was shown that venous-fated EphB4-positive ECs migrate away from arterial-fated EphrinB2-positive ECs in mixed vessels to establish the first artery and vein (Lindskog et al. 2014; Herbert et al. 2009). Subsequently, new branches sprout out of the first arteries and veins. Time-lapse movies of zebrafish embryos demonstrated that vessel sprouts can disconnect from the originating vein and reconnect with the adjacent artery (Betz et al. 2016). Also tip cells from venous sprouts can migrate backwards and incorporate into newly formed arteries in mice and fish (Xu et al. 2014). This suggests that the arteriovenous cell fate is not terminally defined in the early stage of development.

4.1 Arterial Differentiation

Vascular remodeling can occur in absence of blood flow and is largely determined by genetic factors whereby the VEGF and Notch pathways play key roles. Arterial and venous ECs possess specific molecular identities such as EphrinB2 expression exclusively in arterial and EphB4 exclusively in venous beds (Wang et al. 1998). Notch pathway components are expressed at much higher levels in arterial than venous ECs (Villa et al. 2001; Claxton and Fruttiger 2004) and are major players during embryonic arterial differentiation (Gridley 2010; Swift and Weinstein 2009). This was demonstrated by gene targeting approaches in mouse and zebrafish, which revealed that disruption of the Notch pathway does not only lead to impaired vessel sprouting but also to poorly formed arterial vessels, loss of arterial markers (e.g. EphrinB2, Hey2, Cxcr4, Cx40, Nrp1) and/or ectopic expression of venous markers (e.g. EphB4, COUP-TFII (Nr2f2), Nrp2) (Lawson et al. 2001; Zhong et al. 2001; Zhong et al. 2000; Fischer et al. 2004; Lawson et al. 2002; Duarte et al. 2004; Krebs et al. 2004; Sörensen et al. 2009).

Dll4-mediated Notch signaling induces expression of arterial-specific genes (Kim et al. 2008; Iso et al. 2006) and suppresses the expression of the master regulator of venous specification, COUP-TFII (Swift et al. 2014). Dll1 plays a distinct role. Dll1 is expressed selectively on fetal mouse arteries and is not required for the establishment but for the maintenance of arterial identity and VEGF receptor expression (Sörensen et al. 2009). It should be taken into account that blood pressure, blood flow dynamics and hypoxia are also important for the proper differentiation and the maintenance of arteriovenous identity (Le Noble et al. 2005; Lanner et al. 2013; Diez et al. 2007).

Once the circulatory system is formed and fully functional, the arteriovenous fate needs to be actively maintained to prevent the formation of arteriovenous shunts. Arteriovenous malformations in the brain are an important cause of intracerebral hemorrhage in young adults (Lawton et al. 2015). Increased NOTCH1 activity has been observed in human arteriovenous malformations (Murphy et al. 2009; Zhuge et al. 2009). Based on gene targeting approaches, Notch signaling appears to be involved in its pathogenesis. Interestingly, both endothelial-specific inhibition and over-activation of Notch signaling can lead to the formation of arteriovenous malformations at least in certain vascular beds (Trindade et al. 2008; Carlson et al. 2005; Miniati et al. 2010; Murphy et al. 2012; Murphy et al. 2014; Murphy et al. 2008; Gale et al. 2004; Murphy et al. 2009).

Besides maintaining arterio-venous identity, Notch signaling is required to maintain integrity of vascular smooth muscle cells. Neomorphic mutations in *NOTCH3*, which often lead to unequal numbers of cysteine residues in the extracellular domain, cause cerebral autosomal dominant arteriopathy with subcortical infarcts and leukoencephalopathy (CADASIL). This leads to degeneration of vascular smooth muscle cells in small-sized arteries, changes in brain blood perfusion that cause migraine attacks, stroke and dementia. Gene targeting experiments have shown that mice carrying a CADASIL-causing *Notch3* point mutation displayed attenuated myogenic responses and reduced caliber of brain arteries as well as impaired cerebrovascular autoregulation and functional hyperemia (Chabriat et al. 2009; Joutel et al. 2010).

4.2 Venous Specification

It was previously believed that venous differentiation is the default differentiation pathway in the absence of Notch activation. However, mouse knockout studies revealed a pivotal role for the transcription factor COUP-TFII, which is exclusively expressed in venous and lymphatic ECs to establish venous fate (You et al. 2005). Interestingly, Notch signaling suppresses COUP-TFII expression, most likely via Hey transcriptional repressors, and thereby allows arterial fate specification (Swift et al. 2014). In summary, it appears likely that Notch and COUP-TFII repress each other to allow the establishment of the arterial or venous gene expression programs, respectively.

4.3 Upstream Regulators of Notch During Arteriovenous Differentiation

It still remains unclear what mechanisms act upstream of Notch signaling in early phases of arteriovenous differentiation. Hypoxia might play an important role by inducing DLL4 expression (Diez et al. 2007; Patel et al. 2005). In zebrafish, Shh and Vegf-a act upstream of Notch to promote arterial differentiation. Alternatively, Shh might promote arterial differentiation independently of VEGF signaling via the calcitonin receptor-like receptor (Wilkinson et al. 2012). In mammals, neurons or glial cells release VEGF to support arterial differentiation. VEGF signaling via Erk induces transcription of Dll4 and arterial-specific genes (Deng et al. 2013; Ren et al. 2010). However, VEGF signaling can also induce Pi3k activity, which has an opposite effect on arterial morphogenesis (Hong et al. 2008; Ren et al. 2010), indicating that other factors are needed to fine-tune VEGF signaling branches. Neuropilin-1, which is more abundantly expressed on arterial than venous ECs, could be one of these factors as it promotes Vegfr2 trafficking and Erk signaling (Lanahan et al. 2013).

Besides VEGF signaling, the expression of Dll4 during arterial differentiation is also promoted by SoxF transcription factors (Corada et al. 2013; Sacilotto et al. 2013), WNT/β-catenin signaling (Corada et al. 2010; Yamamizu et al. 2010), angiopoietin-1 (Zhang et al. 2011) and the transcription factors Foxc1 and Foxc2 (Seo et al. 2006; Hayashi and Kume 2008). Lastly, it should be noted that also blood flow dynamics induce the expression of Notch pathway components and other arterial-specific genes in cultured ECs (Lehoux and Jones 2016) and endothelial cells in mice (Ramasamy et al. 2016). Furthermore, studies using cultured cells have shown that such physiologic forces can sensitize the negative regulatory region of Notch1 to ADAM-mediated cleavage (Gordon et al. 2015). As such, a large amount of genetic and environmental factors promote EC Notch signaling to enable and maintain arterial morphogenesis.

5 Organ-Specific Vascular Beds and Angiocrine Signaling

A major challenge for the research field will be the analysis of organ-specific vascular beds. Blood vessel anatomy and function differs dramatically between organs and even within the same organ (e.g. the fenestrated endothelium in kidney glomeruli vs. the continuous endothelium in peritubular capillaries). The tightness of vessels is adapted to the organ-specific requirements with e.g. tight EC connections in the central nervous system and gaps (fenestrations) in the sinusoidal endothelium of liver, endocrine organs or bone marrow (Aird 2007). Little is known so far regarding how these differences are established during development and maintained throughout life. This is, however, of utmost importance. For example, treatment of mice with tyrosine kinase inhibitors targeting VEGF receptors led to pronounced regression of fenestrated capillaries, that are typically present in endocrine organs and that under normal conditions express high levels of Vegfr2/3 (Kamba et al. 2006). Similar data were obtained in pancreatic islets by a genetic approach (Lammert et al. 2003), indicating that VEGF acts as a survival factor for fenestrated capillaries in endocrine organs.

Several angiocrine functions have recently been described in which ECs control organ development and regeneration by secreting e.g. growth factors or by providing niches and cell surface molecules for hematopoietic stem cells or tumor cells (Rafii et al. 2016). Here we focus on such examples in which the Notch pathway is critically involved.

Work from the Adams laboratory gave fascinating insights on how blood vessels orchestrate the formation, function and remodeling of bone (Kusumbe et al. 2014). In contrast to other organs, active Notch signaling in bone ECs promotes blood vessel growth. Furthermore, Notch regulates the angiocrine release of Noggin, which is involved in bone growth, mineralization and chrondrocyte maturation (Ramasamy et al. 2014). It is known that many diseases lead to impaired skeletal blood flow. Interestingly, flow-responsive genes induce endothelial Notch signaling in bone. Therefore, impaired blood flow hampers osteogenesis and rejuvenation of bone through impaired EC Notch signaling and decreased angiogenesis (Ramasamy et al. 2016).

In the liver, Notch1 is important to maintain quiescence and morphology of the specialized sinusoidal vasculature. Disruption of *Notch1* using the rather tissue-unspecific Mx-Cre line led to de-differentiation of sinusoidal ECs, vascular remodeling, detachment of mural cells and intussusceptive angiogenesis (Dill et al. 2012; Dimova et al. 2013). In the bone marrow, Jag1 expression on ECs is important for hematopoietic stem cell differentiation (Poulos et al. 2013) and niche-forming vessels can be restored by activation of EC Notch signaling (Kusumbe et al. 2016). In the lung, Jag1 expressed on pulmonary capillary ECs induces Notch signaling in perivascular fibroblasts and thereby enhances lung fibrosis (Cao et al. 2016).

Lastly, we want to emphasize that aside from their role in angiogenesis, tumor ECs possess additional roles. ECs within a solid tumor mass are in close contact with tumor cells and many immune cells and their released cytokines. As such, tumor ECs often do not form tight barriers any more, exhibit altered gene expression programs and also actively alter the behavior of adjacent cells in the tumor microenvironment. In this regard, ECs can provide several membrane-bound and secreted factors that promote tumor progression (Butler et al. 2010). Notch ligands of the Delta-like and Jagged families are frequently present on tumor ECs and can promote Notch signaling in adjacent tumor cells. This increased aggressiveness of lymphoma cells (Cao et al. 2014), promotes the cancer stem cell phenotype (Lu et al. 2013; Zhu et al. 2011), increases tumor cell survival (Pedrosa et al. 2015) and facilitates metastasis (Sonoshita et al. 2011). Interestingly, Notch ligands can also be secreted by tumor cells via exosomes and be incorporated in EC membranes at distant sites to either activate or inhibit Notch signaling (Sharghi-Namini et al. 2014; Sheldon et al. 2010). Furthermore, Notch activation in ECs can be driven by inflammation and this in turn contributes to increased expression of leukocyte adhesion molecules (Liu et al. 2012; Verginelli et al. 2014). Work from our group showed that sustained NOTCH1 activation in ECs leads to senescence, expression of adhesion molecules and weakening of cell junctions that promote transmigration and homing of circulating tumor cells (Wieland et al. 2017).

6 Tumor Angiogenesis and Notch Targeting Agents

Angiogenesis is a hallmark of cancer (Hanahan and Weinberg 2011). The growth of small tumor cell clumps into a clinically relevant tumor is only possible by the induction of blood vessel growth into the tumor mass. Tumor vessels have an abnormal structure and often function poorly. The endothelial lining contains gaps and disorganized cell-cell junction integrity. Also the coverage with pericytes is frequently impaired making vessels leaky. This increases interstitial pressure what impairs the transport of nutrients and drugs towards tumor cells. Moreover, vascular leakiness facilitates intravasation of tumor cells and dissemination (Goel et al. 2011). The tumor vasculature lacks a strict hierarchical structure, arteriovenous identity is poorly defined, vessels have irregular lumen sizes, are often tortuous shaped and thin-walled. Both hyper-vascularized and poorly vascularized tumor areas accompany tumor heterogeneity. Irregular vessel branches, shunts, blind-ended branches, weak vessel contractility and irregular lumen sizes together lead to abnormal and very heterogeneous perfusion rates. Irregular perfusion impairs oxygen, nutrient and drug delivery, thereby limiting the efficiency of chemotherapy and radiation. Impaired perfusion causes aggravation, as hypoxic tumor areas secrete even higher amounts of proangiogenic factors leading to the formation of even more chaotic vessel structures with increased permeability (Carmeliet and Jain 2011; Potente et al. 2011).

VEGF targeting substances are in clinical use but show limited efficiency (Carmeliet and Jain 2011; Potente et al. 2011). Anti-VEGF drugs inhibit the formation of new vessel sprouts and also induce regression of pre-existing tumor vessels, in particular immature vessels. It is assumed that the mode of action is not starving the tumor to death but rather to normalize the tumor vasculature by regression of immature vessels and maturation of the remaining ones. The normalized tumor vasculature is better perfused and enables better delivery of cytotoxic agents to tumor cells (Goel et al. 2011). It is assumed that many initially sensitive tumors develop resistance against VEGF-targeting drugs by secretion of other proangiogenic proteins (e.g. FGF2, PDFG, PlGF, IL-8, ANG2) and by other means of vessel formation (e.g. cooption of already existing vessels) (Bergers and Hanahan 2008). This indicates that better combination therapies are required to target the tumor vasculature.

Besides VEGF, Notch signaling is an interesting target. As in physiological angiogenesis, Notch signaling is involved in tumor angiogenesis (Noguera-Troise et al. 2006; Ridgway et al. 2006; Lobov et al. 2007). However, the pathological high VEGF concentrations may disrupt oscillatory Notch signaling outputs and thereby impair the formation of proper cell junctions and promote vessel expansion (Bentley et al. 2014; Ubezio et al. 2016). Dll4 and Jag1 are

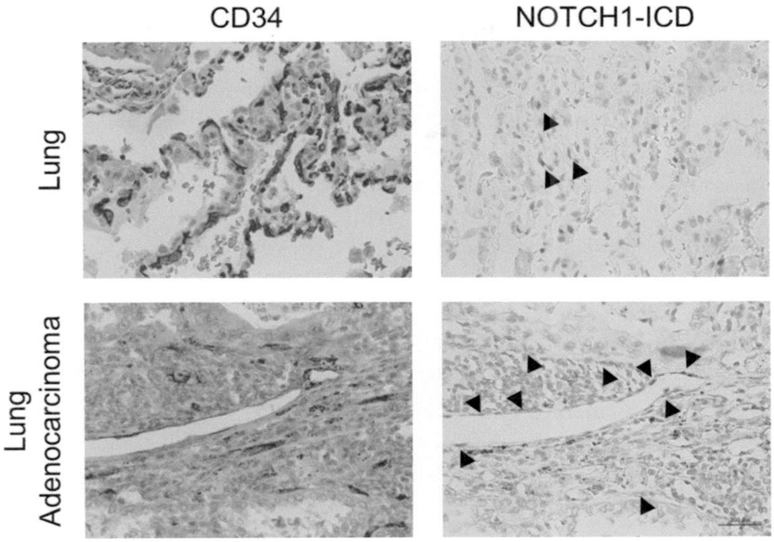

Fig. 3 Notch signaling is active in blood vessels of adult and tumor blood vessels. Sections of normal lung and lung adenocarcinoma were stained against the endothelial marker CD34 (brown color) or the cleaved NOTCH1 receptor (NOTCH1-ICD). Cell nuclei were counterstained with hematoxylin (blue color). Magnification 400-fold

abundantly expressed on tumor vessels (Patel et al. 2005; Lu et al. 2007; Jubb et al. 2012; Gale et al. 2004; Mailhos et al. 2001; Scehnet et al. 2007) and tumor vessels often exhibit strong Notch1 activity (Fig. 3). By computational modeling, it was suggested that the higher production levels of the antagonistic ligand Jag1 give rise to a hybrid tip/stalk phenotype that leads to poorly perfused vessels (Boareto et al. 2015).

Manipulation of EC Notch signaling appears to be an attractive target to interfere with tumor progression. Notch signaling is often hyperactive in cancer cells (in particular in the cancer stem cells) and acts as an oncogene in many tumor entities. Therefore, Notch inhibition could target tumor cells and tumor vessels simultaneously. Many academic groups and pharmaceutical companies have developed Notch inhibiting substances and several ones are in phase I/II trials (Andersson and Lendahl 2014). In rodent models, blockade of Dll4, Notch1 or γ-secretase leads to a non-productive hypersprouting phenotype resulting in central tumor necrosis (Noguera-Troise et al. 2006; Ridgway et al. 2006; Scehnet et al. 2007). This may sound paradoxical, but the excessive vessel branches generate such a chaotic network that dramatically diminishes tumor perfusion. Whether this can also be achieved in human cancer patients is not clear yet. γ-secretase inhibitors, which block the activity of all four Notch receptors, have quite profound adverse effects (e.g. gastrointestinal toxicity) in clinical trials (Andersson and Lendahl 2014) but neutralizing antibodies against individual Notch receptors might be able to overcome this (Wu et al. 2010). In addition, antibodies targeting individual Notch ligands have also been developed (Andersson and Lendahl 2014). Nevertheless, DLL4-neutralizing antibody can also cause severe adverse effects (Yan et al. 2010), e.g. development of congestive heart failure was observed in clinical phase I studies (Chiorean et al. 2015; Falchook et al. 2015; Smith et al. 2014). It will be important to study the underlying mechanisms to overcome this problem.

As outlined above, it appears to be more reasonable to induce tumor vessel normalization instead of tumor vessel regression. A novel approach to achieve this might be targeting EC metabolism. ECs are highly glycolytic and high rates of glucose breakdown are instrumental for adopting the tip cell phenotype during sprouting. A rather mild inhibition of glycolysis can be achieved by targeting its activator Pfkfb3. In mouse cancer models, Pfkfb3 inhibition tightened the vascular barrier, improved adhesion of pericytes and reduced the pro-inflammatory phenotype of tumor ECs that facilitates metastasis (Cantelmo et al. 2016). Another option to normalize the tumor vasculature could be the activation of EC Notch signaling. As shown by genetic approaches in mice, this reduced tumor angiogenesis, but increased vessel diameter and improved perfusion and oxygenation (Li et al. 2007). Notch activation could also help to reduce glycolysis in ECs as Notch signaling reduces expression of Pfkfb3 (De Bock et al. 2013). While Notch-inhibiting substances are in clinical trials, we still lack fully validated drugs to activate Notch signaling in a therapeutic manner. The Kitajewski laboratory has generated soluble Notch1 extracellular domain proteins fused to IgG-Fc (Notch decoys) that bind and inhibit selectively either the stimulatory Delta-like or the inhibitory Jagged ligands (Funahashi et al. 2008; Kangsamaksin et al. 2015). These Notch decoys inhibit sprouting angiogenesis and also target pericytes in the vessel wall (Klose et al. 2015; Funahashi et al. 2008; Kangsamaksin et al. 2015). Future experiments will determine whether Notch-activating substances can be used successfully in combination with chemotherapy to better target tumor cells.

7 Perspectives

In recent years there has been a significant progress in the understanding of Notch signaling during sprouting angiogenesis. However, much remains to be learned. As the tip/stalk cell selection is tightly dependent on subtle fluctuations in Notch signal output strengths, it is necessary to determine multiple genetic and environmental factors, such as hemodynamics and metabolites that fine-tune ligand expression and localization at the cell surface, receptor glycosylation, NICD protein stability, nuclear NICD complex formation and the dynamic control of Notch target gene expression.

Inducible tissue-specific transgene models and therapeutic antibodies will be key to determine how VEGF and Notch signaling are involved in organ-specific angiogenesis, maintenance of EC quiescence, as well as barrier and transport functions throughout life. There is already solid evidence that VEGF does not only control blood vessel formation, but also acts as a survival factor for ECs (Domigan et al. 2015) and non-vascular cells (Mackenzie and Ruhrberg 2012). Similar to this, basal Dll4/Notch activity has been detected in quiescent ECs (Zhang et al. 2011) and is important to maintain vascular integrity and function (Liu et al. 2011; Yan et al. 2010). Lastly, angiocrine functions of ECs have attracted enormous attention (Rafii et al. 2016). It will be fascinating to see how ECs control the function of parenchymal cells through the secretion of signaling molecules or through providing membrane-bound factors that orchestrate the behavior of its neighboring cells in organ-specific vascular beds.

Acknowledgments We thank all members of the Fischer laboratory for critical discussion and apologize to all colleagues whose work could not be cited in this chapter. Work in our laboratory on these topics has been supported by the Deutsche Forschungsgemeinschaft (DFG-FI1568/3-1; FI1568/5-1; SFB-TR23), the Helmholtz Society and the German Cancer Research Center Heidelberg.

References

Adam MG et al (2013) Synaptojanin-2 binding protein stabilizes the Notch ligands DLL1 and DLL4 and inhibits sprouting angiogenesis. Circ Res 113(11):1206–1218

Adams RH, Eichmann A (2010) Axon guidance molecules in vascular patterning. Cold Spring Harb Perspect Biol 2(5):a001875

Aird WC (2007) Phenotypic heterogeneity of the endothelium: I. Structure, function, and mechanisms. Circ Res 100(2):158–173

Andersson ER, Lendahl U (2014) Therapeutic modulation of Notch signalling--are we there yet? Nat Rev Drug Discov 13(5):357–378

Arroyo AG, Iruela-Arispe ML (2010) Extracellular matrix, inflammation, and the angiogenic response. Cardiovasc Res 86(2):226–235

Aspalter IM et al (2015) Alk1 and Alk5 inhibition by Nrp1 controls vascular sprouting downstream of Notch. Nat Commun 6:7264

Augustin HG et al (2009) Control of vascular morphogenesis and homeostasis through the angiopoietin - Tie system. Nat Rev Mol Cell Biol 10(3):165–177

Benedito R et al (2009) The notch ligands Dll4 and Jagged1 have opposing effects on angiogenesis. Cell 137(6):1124–1135

Bentley K et al (2014) The role of differential VE-cadherin dynamics in cell rearrangement during angiogenesis. Nat Cell Biol 16(4):309–321

Bergers G, Hanahan D (2008) Modes of resistance to anti-angiogenic therapy. Nat Rev Cancer 8(8):592–603

Betz C et al (2016) Cell behaviors and dynamics during angiogenesis. Development 143(13):2249–2260
Blanco R, Gerhardt H (2013) VEGF and Notch in tip and stalk cell selection. Cold Spring Harb Perspect Med 3(1):a006569
Boareto M et al (2015) Jagged mediates differences in normal and tumor angiogenesis by affecting tip-stalk fate decision. PNAS 112(29):E3836–E3844
Briot A et al (2015) Endothelial NOTCH1 is suppressed by circulating lipids and antagonizes inflammation during atherosclerosis. J Exp Med 212(12):2147–2163
Butler JM, Kobayashi H, Rafii S (2010) Instructive role of the vascular niche in promoting tumour growth and tissue repair by angiocrine factors. Nat Rev Cancer 10(2):138–146
Cantelmo AR et al (2016) Inhibition of the glycolytic activator PFKFB3 in endothelium induces tumor vessel normalization, impairs metastasis, and improves chemotherapy. Cancer Cell 30(6):968–985
Cao Y (2010) Adipose tissue angiogenesis as a therapeutic target for obesity and metabolic diseases. Nat Rev Drug Discov 9(2):107–115
Cao Z et al (2014) Angiocrine factors deployed by tumor vascular niche induce B cell lymphoma invasiveness and chemoresistance. Cancer Cell 25(3):350–365
Cao Z et al (2016) Targeting of the pulmonary capillary vascular niche promotes lung alveolar repair and ameliorates fibrosis. Nat Med 22(2):154–162
Carlson TR et al (2005) Endothelial expression of constitutively active Notch4 elicits reversible arteriovenous malformations in adult mice. PNAS 102(28):9884–9889
Carmeliet P, Jain RK (2011) Molecular mechanisms and clinical applications of angiogenesis. Nature 473(7347):298–307
Carmeliet P et al (1996) Abnormal blood vessel development and lethality in embryos lacking a single VEGF allele. Nature 380(6573):435–439
Chabriat H et al (2009) CADASIL. Lancet Neurol 8(7):643–653
Chiorean EG et al (2015) A phase I first-in-human study of Enoticumab (REGN421), a fully human delta-like ligand 4 (Dll4) monoclonal antibody in patients with advanced solid tumors. Clin Cancer Res 21(12):2695–2703
Claxton S, Fruttiger M (2004) Periodic Delta-like 4 expression in developing retinal arteries. Gene Expr Patterns 5(1):123–127
Corada M et al (2010) The Wnt/β-catenin pathway modulates vascular remodeling and specification by upregulating Dll4/notch signaling. Dev Cell 18(6):938–949
Corada M et al (2013) Sox17 is indispensable for acquisition and maintenance of arterial identity. Nat Commun 4:2609
Corada M, Morini MF, Dejana E (2014) Signaling pathways in the specification of arteries and veins. Arterioscler Thromb Vasc Biol 34(11):2372–2377
Corvera S, Gealekman O (2014) Adipose tissue angiogenesis: impact on obesity and type-2 diabetes. Biochim Biophys Acta Mol basis Dis 1842(3):463–472
Cruys B et al (2016) Glycolytic regulation of cell rearrangement in angiogenesis. Nat Commun 7:12240
De Bock K et al (2013) Role of PFKFB3-driven glycolysis in vessel sprouting. Cell 154(3):651–663
Del Toro R et al (2010) Identification and functional analysis of endothelial tip cell-enriched genes. Blood 116(19):4025–4033
Deng Y et al (2013) Endothelial RAF1 / ERK activation regulates arterial morphogenesis Endothelial RAF1 / ERK activation regulates arterial morphogenesis. Blood 121(19):3988–3997
Diez H et al (2007) Hypoxia-mediated activation of Dll4-Notch-Hey2 signaling in endothelial progenitor cells and adoption of arterial cell fate. Exp Cell Res 313(1):1–9
Dill MT et al (2012) Disruption of Notch1 induces vascular remodeling, intussusceptive angiogenesis, and angiosarcomas in livers of mice. Gastroenterology 142(4):967–977.e2
Dimova I et al (2013) Inhibition of Notch signaling induces extensive intussusceptive neo-angiogenesis by recruitment of mononuclear cells. Angiogenesis 16(4):921–937
Domigan CK et al (2015) Autocrine VEGF maintains endothelial survival through regulation of metabolism and autophagy. J Cell Sci 128(12):2236–2248
Dou G-R et al (2008) RBP-J, the transcription factor downstream of Notch receptors, is essential for the maintenance of vascular homeostasis in adult mice. FASEB J 22(5):1606–1617
Duarte A et al (2004) Dosage-sensitive requirement for mouse Dll4 in artery development. Genes Dev 18(20):2474–2478
Dumont DJ et al (1998) Cardiovascular failure in mouse embryos deficient in VEGF receptor-3. Science 282(5390):946–949
Eilken HM, Adams RH (2010) Dynamics of endothelial cell behavior in sprouting angiogenesis. Curr Opin Cell Biol 22(5):617–625
Falchook G et al (2015) Phase 1 study of MEDI0639 in patients with advanced solid tumors. J Clin Oncol 33(suppl):abstr 3024
Fantin A et al (2010) Tissue macrophages act as cellular chaperones for vascular anastomosis downstream of VEGF-mediated endothelial tip cell induction. Blood 116(5):829–840
Fantin A et al (2013) NRP1 acts cell autonomously in endothelium to promote tip cell function during sprouting angiogenesis. Blood 121(12):2352–2362
Fantin A et al (2015) NRP1 regulates CDC42 activation to promote Filopodia formation in endothelial tip cells. Cell Rep 11(10):1577–1590
Ferrara N et al (1996) Heterozygous embryonic lethality induced by targeted inactivation of the VEGF gene. Nature 380(6573):439–442
Fischer A et al (2004) The Notch target genes Hey1 and Hey2 are required for embryonic vascular development. Genes Dev 18(8):901–911

Folkman J (1995) Angiogenesis in cancer, vascular, rheumatoid and other disease. Nat Med 1(1):27–31

Folkman J (2007) Is angiogenesis an organizing principle in biology and medicine? J Pediatr Surg 42(1):1–11

Fong G-HH et al (1995) Role of the Flt-1 receptor tyrosine kinase in regulating the assembly of vascular endothelium. Nature 376(6535):66–70

Fouillade C et al (2012) Notch signalling in smooth muscle cells during development and disease. Cardiovasc Res 95(2):138–146

Funahashi Y et al (2008) A notch1 ectodomain construct inhibits endothelial notch signaling, tumor growth, and angiogenesis. Cancer Res 68(12):4727–4735

Gale NW et al (2004) Haploinsufficiency of delta-like 4 ligand results in embryonic lethality due to major defects in arterial and vascular development. PNAS 101(45):15949–15954

Gerhardt H et al (2003) VEGF guides angiogenic sprouting utilizing endothelial tip cell filopodia. J Cell Biol 161(6):1163–1177

Glomski K et al (2011) Deletion of Adam10 in endothelial cells leads to defects in organ-specific vascular structures. Blood 118(4):1163–1174

Goel S et al (2011) Normalization of the vasculature for treatment of cancer and other diseases. Physiol Rev 91(3):1071–1121

Gordon WR et al (2015) Mechanical Allostery: evidence for a force requirement in the proteolytic activation of notch. Dev Cell 33(6):729–736

Gridley T (2010) Notch signaling in the vasculature. Curr Top Dev Biol 92:277–309

Guarani V et al (2011) Acetylation-dependent regulation of endothelial Notch signalling by the SIRT1 deacetylase. Nature 473(7346):234–238

Hagberg CE et al (2010) Vascular endothelial growth factor B controls endothelial fatty acid uptake. Nature 464(7290):917–921

Hanahan D, Weinberg RA (2011) Hallmarks of cancer: the next generation. Cell 144(5):646–674

Hayashi H, Kume T (2008) Foxc transcription factors directly regulate DLL4 and hey2 expression by interacting with the VEGF-notch signaling pathways in endothelial cells. PLoS One 3(6):1–9

Hellsten Y, Hoier B (2014) Capillary growth in human skeletal muscle: physiological factors and the balance between pro-angiogenic and angiostatic factors. Biochem Soc Trans 42(6):1616–1622

Hellström M et al (2007) Dll4 signalling through Notch1 regulates formation of tip cells during angiogenesis. Nature 1(3):133–136

Herbert SP, Stainier DYR (2011) Molecular control of endothelial cell behaviour during blood vessel morphogenesis. Nat Rev Mol Cell Biol 12(9):551–564

Herbert SP et al (2009) Arterial-venous segregation by selective cell sprouting: an alternative mode of blood vessel formation. Science 326(5950):294–298

Herreman A et al (1999) Presenilin 2 deficiency causes a mild pulmonary phenotype and no changes in amyloid precursor protein processing but enhances the embryonic lethal phenotype of presenilin 1 deficiency. PNAS 96(21):11872–11877

Hofmann JJ, Luisa Iruela-Arispe M (2007) Notch expression patterns in the retina: an eye on receptor-ligand distribution during angiogenesis. Gene Expr Patterns 7(4):461–470

Hong CC, Kume T, Peterson RT (2008) Role of crosstalk between phosphatidylinositol 3-kinase and extracellular signal-regulated kinase/mitogen-activated protein kinase pathways in artery-vein specification. Circ Res 103(6):573–579

Huppert SS et al (2000) Embryonic lethality in mice homozygous for a processing-deficient allele of Notch1. Nature 405(June):966–970

Iso T et al (2006) Dll4-selective Notch signaling induces ephrinB2 gene expression in endothelial cells. Biochem Biophys Res Commun 341(3):708–714

Isogai S et al (2003) Angiogenic network formation in the developing vertebrate trunk. Development 130(21):5281–5290

Jais A et al (2016) Myeloid-cell-derived VEGF maintains brain glucose uptake and limits cognitive impairment in obesity. Cell 166(5):1338–1340

Jakobsson L et al (2010) Endothelial cells dynamically compete for the tip cell position during angiogenic sprouting. Nat Cell Biol 12(10):943–953

Joutel A et al (2010) Cerebrovascular dysfunction and microcirculation rarefaction precede white matter lesions in a mouse genetic model of cerebral ischemic small vessel disease. J Clin Investig 120(2):433–445

Jubb AM et al (2012) Expression of vascular Notch ligands Delta-like 4 and Jagged-1 in glioblastoma. Histopathology 60(5):740–747

Kageyama R, Masamizu Y, Niwa Y (2007) Oscillator mechanism of Notch pathway in the segmentation clock. Dev Dyn 236(6):1403–1409

Kamba T et al (2006) VEGF-dependent plasticity of fenestrated capillaries in the normal adult microvasculature. Am J Physiol Heart Circ Physiol 290(2):H560–H576

Kangsamaksin T et al (2015) NOTCH decoys that selectively block DLL/NOTCH or JAG/NOTCH disrupt angiogenesis by unique mechanisms to inhibit tumor growth. Cancer Discov 5(2):182–197

Kim YH et al (2008) Artery and vein size is balanced by Notch and ephrin B2/EphB4 during angiogenesis. Development 135(22):3755–3764

Klose R et al (2015) Soluble Notch ligand and receptor peptides act antagonistically during angiogenesis. Cardiovasc Res 107(1):153–163

Kohli V et al (2013) Arterial and venous progenitors of the major axial vessels originate at distinct locations. Dev Cell 25(2):196–206

Kokubo H et al (2005) Mouse hesr1 and hesr2 genes are redundantly required to mediate Notch signaling in the developing cardiovascular system. Dev Biol 278(2):301–309

Kopan R, Ilagan MXG (2009) The canonical notch signaling pathway: unfolding the activation mechanism. Cell 137(2):216–233

Krebs LT et al (2000) Notch signaling is essential for vascular morphogenesis in mice. Genes Dev 14(11):1343–1352

Krebs LT et al (2001) The Nrarp gene encodes an ankyrin-repeat protein that is transcriptionally regulated by the notch signaling pathway. Dev Biol 238(1):110–119

Krebs LT et al (2004) Haploinsufficient lethality and formation of arteriovenous malformations in Notch pathway mutants. Genes Dev 18(20):2469–2473

Krebs LT et al (2010) Notch1 activation in mice causes arteriovenous malformations phenocopied by EphrinB2 and EphB4 mutants. Genesis 48(3):146–150

Krueger J et al (2011) Flt1 acts as a negative regulator of tip cell formation and branching morphogenesis in the zebrafish embryo. Development 138(10):2111–2120

Kusumbe AP, Ramasamy SK, Adams RH (2014) Coupling of angiogenesis and osteogenesis by a specific vessel subtype in bone. Nature 507(7492):323–328

Kusumbe AP et al (2016) Age-dependent modulation of vascular niches for haematopoietic stem cells. Nature 532(7599):380–384

Lammert E et al (2003) Role of VEGF-A in vascularization of pancreatic islets. Curr Biol 13(12):1070–1074

Lanahan A et al (2013) The Neuropilin 1 cytoplasmic domain is required for VEGF-A-dependent Arteriogenesis. Dev Cell 25(2):156–168

Lanner F et al (2013) Hypoxia-induced arterial differentiation requires adrenomedullin and notch signaling. Stem Cells Dev 22(9):1360–1369

Larrivée B et al (2012) ALK1 signaling inhibits angiogenesis by cooperating with the notch pathway. Dev Cell 22(3):489–500

Lawson ND et al (2001) Notch signaling is required for arterial-venous differentiation during embryonic vascular development. Development 128(19):3675–3683

Lawson ND, Vogel AM, Weinstein BM (2002) Sonic hedgehog and vascular endothelial growth factor act upstream of the Notch pathway during arterial endothelial differentiation. Dev Cell 3(1):127–136

Lawton MT et al (2015) Brain arteriovenous malformations. Nat Rev Dis Prim 1(Table 1):15008

Le Noble F et al (2005) Control of arterial branching morphogenesis in embryogenesis: go with the flow. Cardiovasc Res 65(3):619–628

Lee SH et al (2014) Notch pathway targets proangiogenic regulator Sox17 to restrict angiogenesis. Circ Res 115(2):215–226

Lehoux S, Jones EA (2016) Shear stress, arterial identity and atherosclerosis. Thromb Haemost 115(3):467–473

Leslie JD et al (2007) Endothelial signalling by the Notch ligand Delta-like 4 restricts angiogenesis. Development 134(5):839–844

Li T et al (2003) Nicastrin is required for assembly of Presenilin/γ-Secretase complexes to mediate notch signaling and for processing and trafficking of β-amyloid precursor protein in mammals. J Neurosci 23(8):3272–3277

Li JL et al (2007) Delta-like 4 Notch ligand regulates tumor angiogenesis, improves tumor vascular function, and promotes tumor growth in vivo. Cancer Res 67(23):11244–11253

Liebler SS et al (2012) No evidence for a functional role of bi-directional notch signaling during angiogenesis. PLoS One 7(12):1–10

Limbourg FP et al (2005) Essential role of endothelial Notch1 in angiogenesis. Circulation 111(14):1826–1832

Limbourg A et al (2007) Notch ligand Delta-like 1 is essential for postnatal arteriogenesis. Circ Res 100:363–371

Lindskog H et al (2014) Molecular identification of venous progenitors in the dorsal aorta reveals an aortic origin for the cardinal vein in mammals. Development 141(5):1120–1128

Liu ZJ et al (2006) Inhibition of endothelial cell proliferation by Notch1 signaling is mediated by repressing MAPK and PI3K/Akt pathways and requires MAML1. FASEB J 20(7):1009–1011

Liu Z et al (2011) Notch1 loss of heterozygosity causes vascular tumors and lethal hemorrhage in mice. J Clin Investig 121(2):800–808

Liu Z-J et al (2012) Notch activation induces endothelial cell senescence and pro-inflammatory response: implication of Notch signaling in atherosclerosis. Atherosclerosis 225(2):296–303

Lobov IB et al (2007) Delta-like ligand 4 (Dll4) is induced by VEGF as a negative regulator of angiogenic sprouting. PNAS 104(9):3219–3224

Lu C et al (2007) Gene alterations identified by expression profiling in tumor-associated endothelial cells from invasive ovarian carcinoma. Cancer Res 67(4):1757–1768

Lu J et al (2013) Endothelial cells promote the colorectal Cancer stem cell phenotype through a soluble form of Jagged-1. Cancer Cell 23(2):171–185

Mackenzie F, Ruhrberg C (2012) Diverse roles for VEGF-A in the nervous system. Development 139(8):1371–1380

Mailhos C et al (2001) Delta4, an endothelial specific notch ligand expressed at sites of physiological and tumor angiogenesis. Differ Res Biol Divers 69(2–3):135–144

Makanya AN, Hlushchuk R, Djonov VG (2009) Intussusceptive angiogenesis and its role in vascular morphogenesis, patterning, and remodeling. Angiogenesis 12(2):113–123

Mazzone M et al (2009) Heterozygous deficiency of PHD2 restores tumor oxygenation and inhibits metastasis via endothelial normalization. Cell 136(5):839–851

Miniati D et al (2010) Constitutively active endothelial Notch4 causes lung arteriovenous shunts in mice. Am J Physiol Lung Cell Mol Physiol 298(2):L169–L177

Mouillesseaux KP et al (2016) Notch regulates BMP responsiveness and lateral branching in vessel networks via SMAD6. Nat Commun 7:13247

Moya IM et al (2012) Stalk cell phenotype depends on integration of notch and Smad1/5 signaling cascades. Dev Cell 22(3):501–514

Murphy DA, Courtneidge SA (2011) The "ins" and "outs" of podosomes and invadopodia: characteristics, formation and function. Nat Rev Mol Cell Biol 12(7):413–426

Murphy P a et al (2008) Endothelial Notch4 signaling induces hallmarks of brain arteriovenous malformations in mice. PNAS 105(31):10901–10906

Murphy P a et al (2009) Endothelial notch signaling is upregulated in human brain arteriovenous malformations and a mouse model of the disease. Lab Investig 89(9):971–982

Murphy PA et al (2012) Notch4 normalization reduces blood vessel size in arteriovenous malformations. Sci Transl Med 4(117):117ra8

Murphy PA et al (2014) Constitutively active Notch4 receptor elicits brain arteriovenous malformations through enlargement of capillary-like vessels. PNAS 111(50):18007–18012

Nakazawa MS, Keith B, Simon MC (2016) Oxygen availability and metabolic adaptations. Nat Rev Cancer 16(10):663–673

Noguera-Troise I et al (2006) Blockade of Dll4 inhibits tumour growth by promoting non-productive angiogenesis. Nature 444(7122):1032–1037

Noguera-Troise I et al (2007) Blockade of Dll4 inhibits tumour growth by promoting non-productive angiogenesis. Novartis Found Symp 283:106–120

Outtz HH et al (2011) Notch1 controls macrophage recruitment and Notch signaling is activated at sites of endothelial cell anastomosis during retinal angiogenesis in mice. Blood 118(12):3436–3439

Patel NS et al (2005) Up-regulation of delta-like 4 ligand in human tumor vasculature and the role of basal expression in endothelial cell function. Cancer Res 65(19):8690–8697

Pedrosa A-R et al (2015) Endothelial Jagged1 promotes solid tumor growth through both pro-angiogenic and angiocrine functions. Oncotarget 6(27):24404–24423

Pelton JC et al (2014) Multiple endothelial cells constitute the tip of developing blood vessels and polarize to promote lumen formation. Development 141(21):4121–4126

Peters K et al (2009) Changes in human endothelial cell energy metabolic capacities during in vitro cultivation. The role of "aerobic glycolysis" and proliferation. Cell Physiol Biochem 24(5–6):483–492

Phng LK et al (2009) Nrarp coordinates endothelial notch and Wnt signaling to control vessel density in angiogenesis. Dev Cell 16(1):70–82

Phng L-K, Stanchi F, Gerhardt H (2013) Filopodia are dispensable for endothelial tip cell guidance. Development 140(19):4031–4040

Potente M, Carmeliet P (2017) The link between angiogenesis and endothelial metabolism. Annu Rev Physiol 79(1). p.annurev-physiol-021115-105134

Potente M, Gerhardt H, Carmeliet P (2011) Basic and therapeutic aspects of angiogenesis. Cell 146(6):873–887

Poulos MG et al (2013) Endothelial Jagged-1 is necessary for homeostatic and regenerative hematopoiesis. Cell Rep 4(5):1022–1034

Preuße K et al (2015) Context-dependent functional divergence of the notch ligands DLL1 and DLL4 in vivo. PLoS Genet 11(6):e1005328

Quillien A et al (2014) Distinct Notch signaling outputs pattern the developing arterial system. Development 141(7):1544–1552

Rafii S, Butler JM, Ding B-S (2016) Angiocrine functions of organ-specific endothelial cells. Nature 529(7586):316–325

Ramasamy SK et al (2014) Endothelial Notch activity promotes angiogenesis and osteogenesis in bone. Nature 507(7492):376–380

Ramasamy SK et al (2016) Blood flow controls bone vascular function and osteogenesis. Nat Commun 7:13601

Redeker C et al (2013) Normal development in mice overexpressing the intracellular domain of DLL1 argues against reverse signaling by DLL1 in vivo. PLoS One 8(10):e79050

Ren B et al (2010) ERK1/2-Akt1 crosstalk regulates arteriogenesis in mice and zebrafish. J Clin Investig 120(4):1217–1228

Ridgway J et al (2006) Inhibition of Dll4 signalling inhibits tumour growth by deregulating angiogenesis. Nature 444(7122):1083–1087

Risau W (1997) Mechanisms of angiogenesis. Nature 386(6626):671–674

Risau W, Flamme I (1995) Vasculogenesis. Annu Rev Cell Dev Biol 11:73–91

Robciuc MR et al (2016) VEGFB/VEGFR1-induced expansion of adipose vasculature counteracts obesity and related metabolic complications. Cell Metab 23(4):712–724

Ruhrberg C et al (2002) Spatially restricted patterning cues provided by heparin-binding VEGF-A control blood vessel branching morphogenesis. Genes Dev 16(20):2684–2698

Sacilotto N et al (2013) Analysis of Dll4 regulation reveals a combinatorial role for Sox and Notch in arterial development. PNAS 110(29):11893–11898

Sainson RC et al (2005) Cell-autonomous notch signaling regulates endothelial cell branching and proliferation during vascular tubulogenesis. FASEB J 19(8):1027–1029

Scehnet JS et al (2007) Inhibition of Dll4-mediated signaling induces proliferation of immature vessels and results in poor tissue perfusion. Blood 109(11):4753–4760

Schoors S et al (2015) Fatty acid carbon is essential for dNTP synthesis in endothelial cells. Nature 520(7546):192–197

Schulze A, Harris AL (2012) How cancer metabolism is tuned for proliferation and vulnerable to disruption. Nature 491(7424):364–373

Seano G et al (2014) Endothelial podosome rosettes regulate vascular branching in tumour angiogenesis. Nat Cell Biol 16(10):931–941

Seo S et al (2006) The forkhead transcription factors, Foxc1 and Foxc2, are required for arterial specification and lymphatic sprouting during vascular development. Dev Biol 294(2):458–470

Serra H et al (2015) PTEN mediates Notch-dependent stalk cell arrest in angiogenesis. Nat Commun 6:7935

Shalaby F et al (1995) Failure of blood-island formation and vasculogenesis in Flk-1-deficient mice. Nature 376:62–66

Sharghi-Namini S et al (2014) Dll4-containing exosomes induce capillary sprout retraction in a 3D microenvironment. Sci Rep 4:4031

Sheldon H et al (2010) New mechanism for Notch signaling to endothelium at a distance by delta-like 4 incorporation into exosomes. Blood 116(13):2385–2394

Siekmann AF, Lawson ND (2007) Notch signalling limits angiogenic cell behaviour in developing zebrafish arteries. Nature 445(7129):781–784

Siekmann AF, Affolter M, Belting HG (2013) The tip cell concept 10 years after: new players tune in for a common theme. Exp Cell Res 319(9):1255–1263

Simons M, Gordon E, Claesson-Welsh L (2016) Mechanisms and regulation of endothelial VEGF receptor signalling. Nat Rev Mol Cell Biol 10(10):611–625

Smith DC et al (2014) A phase I dose escalation and expansion study of the anticancer stem cell agent Demcizumab (anti-DLL4) in patients with previously treated solid tumors. Clin Cancer Res 20(24):6295–6303

Sonoshita M et al (2011) Suppression of colon cancer metastasis by Aes through inhibition of Notch signaling. Cancer Cell 19(1):125–137

Sörensen I, Adams RH, Gossler A (2009) DLL1-mediated Notch activation regulates endothelial identity in mouse fetal arteries. Blood 113(22):5680–5688

Spuul P et al (2016) VEGF-A/notch-induced Podosomes Proteolyse basement membrane collagen-IV during retinal sprouting angiogenesis. Cell Rep 17(2):484–500

Stenzel D et al (2011) Endothelial basement membrane limits tip cell formation by inducing Dll4/Notch signalling in vivo. EMBO Rep 12(11):1135–1143

Suchting S et al (2007) The Notch ligand Delta-like 4 negatively regulates endothelial tip cell formation and vessel branching. PNAS 104(9):3225–3230

Swift MR, Weinstein BM (2009) Arterial-venous specification during development. Circ Res 104(5):576–588

Swift MR et al (2014) SoxF factors and Notch regulate nr2f2 gene expression during venous differentiation in zebrafish. Dev Biol 390(2):116–125

Takeshita K et al (2007) Critical role of endothelial Notch1 signaling in postnatal angiogenesis. Circ Res 100(1):70–78

Tammela T et al (2008) Blocking VEGFR-3 suppresses angiogenic sprouting and vascular network formation. Nature 454(7204):656–660

Tammela T et al (2011) VEGFR-3 controls tip to stalk conversion at vessel fusion sites by reinforcing Notch signalling. Nat Cell Biol 13(10):1202–1213

Taylor KL, Henderson AM, Hughes CCW (2002) Notch activation during endothelial cell network formation in vitro targets the basic HLH transcription factor HESR-1 and Downregulates VEGFR-2/KDR expression. Microvasc Res 64(3):372–383

Trindade A et al (2008) Overexpression of delta-like 4 induces arterialization and attenuates vessel formation in developing mouse embryos. Blood 112(5):1720–1729

Ubezio B et al (2016) Synchronization of endothelial Dll4-Notch dynamics switch blood vessels from branching to expansion. eLife 5:e12167

Uyttendaele H et al (2001) Vascular patterning defects associated with expression of activated Notch4 in embryonic endothelium. PNAS 98(10):5643–5648

Verginelli F et al (2014) Activation of an endothelial Notch1-Jagged1 circuit induces VCAM1 expression, an effect amplified by interleukin-1β. Oncotarget 6(41):43216–43229

Villa N et al (2001) Vascular expression of Notch pathway receptors and ligands is restricted to arterial vessels. Mech Dev 108(1–2):161–164

Wang HU, Chen ZF, Anderson DJ (1998) Molecular distinction and angiogenic interaction between embryonic arteries and veins revealed by ephrin-B2 and its receptor Eph-B4. Cell 93(5):741–753

Wieland E et al (2017) Endothelial Notch1 activity facilitates metastasis. Cancer Cell 31(3):355–367

Wild R et al (2017) Neuronal sFlt1 and Vegfaa determine venous sprouting and spinal cord vascularization. Nat Commun 10(8)

Wilhelm K et al (2016) FOXO1 couples metabolic activity and growth state in the vascular endothelium. Nature 529(7585):216–220

Wilkinson RN et al (2012) Hedgehog signaling via a calcitonin receptor-like receptor can induce arterial differentiation independently of VEGF signaling in zebrafish. Blood 120(2):477–488

Wu Y et al (2010) Therapeutic antibody targeting of individual Notch receptors. Nature 464(7291):1052–1057

Xu C et al (2014) Arteries are formed by vein-derived endothelial tip cells. Nat Commun 5:5758

Xue Y et al (1999) Embryonic lethality and vascular defects in mice lacking the Notch ligand Jagged1. Hum Mol Genet 8(5):723–730

Yamamizu K et al (2010) Convergence of Notch and β-catenin signaling induces arterial fate in vascular progenitors. J Cell Biol 189(2):325–338

Yan M et al (2010) Chronic DLL4 blockade induces vascular neoplasms. Nature 463(7282):E6–E7

Yang Y, Oliver G (2014) Development of the mammalian lymphatic vasculature. J Clin Investig 124(3):888–897

Yeh WL, Lin CJ, Fu WM (2008) Enhancement of glucose transporter expression of brain endothelial cells by vascular endothelial growth factor derived from glioma exposed to hypoxia. Mol Pharmacol 73(1):170–177

Yoon CH et al (2014) High glucose-induced jagged 1 in endothelial cells disturbs notch signaling for angiogenesis: a novel mechanism of diabetic vasculopathy. J Mol Cell Cardiol 69:52–66

You L-R et al (2005) Suppression of Notch signalling by the COUP-TFII transcription factor regulates vein identity. Nature 435(7038):98–104

Zarkada G et al (2015) VEGFR3 does not sustain retinal angiogenesis without VEGFR2. PNAS 112(3):761–766

Zhang J et al (2011) Angiopoietin-1/Tie2 signal augments basal Notch signal controlling vascular quiescence by inducing Delta-like 4 expression through AKT-mediated activation of β-catenin. J Biol Chem 286(10):8055–8066

Zhong TP et al (2000) Gridlock, an HLH gene required for assembly of the aorta in zebrafish. Science 287(5459):1820–1824

Zhong TP et al (2001) Gridlock signalling pathway fashions the first embryonic artery. Nature 414(6860):216–220

Zhu TS et al (2011) Endothelial cells create a stem cell niche in glioblastoma by providing NOTCH ligands that nurture self-renewal of cancer stem-like cells. Cancer Res 71(18):6061–6072

Zhuge Q et al (2009) Notch-1 signalling is activated in brain arteriovenous malformations in humans. Brain 132(12):3231–3241

Notch and T Cell Function – A Complex Tale

Jyothi Vijayaraghavan and Barbara A. Osborne

Abstract

Notch drives critical decisions in a multitude of developmental decisions in many invertebrate and vertebrate organisms including flies, worms, fish, mice and humans. Therefore, it is not surprising that Notch family members also play a key role in cell fate choices in the vertebrate immune system. This review highlights the critical function of Notch in the development of mature T lymphocytes from hematopoietic precursors and describes the role of Notch in mature T cell activation, proliferation and differentiation.

Keywords

Notch · T cell · CD4$^+$ · CD8$^+$ · Th1 · Th2 · Th17 · Tregs

J. Vijayaraghavan
Department of Veterinary & Animal Sciences, Program in Animal Biotechnology and Biomedical Sciences, University of Massachusetts, Amherst, MA, USA

B. A. Osborne (✉)
Department of Veterinary & Animal Sciences, Program in Animal Biotechnology and Biomedical Sciences, University of Massachusetts, Amherst, MA, USA

Program in Molecular & Cellular Biology, University of Massachusetts, Amherst, MA, USA
e-mail: osborne@vasci.umass.edu

Abbreviations

CTL	Cytotoxic T lymphocytes.
Dll	Delta-like
GSI, γ	Secretase inhibitor
IFNχ	Interferon gamma
IL17	Interleukin 17
IL2	Interleukin 2
IL4	Interleukin 4
Jag	Jagged
NICD	Notch intracellular domain
RBPJ	Recombination signal binding protein for immunoglobulin kappa J region
Tregs	T regulatory cells

1 Introduction

Notch is a protein that is highly conserved throughout evolution and the signaling pathway regulated by Notch performs critical functions in many invertebrates as well as vertebrates. The 300 kilodalton Notch protein is processed in the trans-Golgi by a furin protease resulting in the appearance, on the cell surface, of the Notch heterodimer. This heterodimer consists of an extracellular domain (NECD) that is non-covalently associated with a transmembrane bound peptide referred to as Notch-Tm. Canonical Notch signaling in all species studied to date involves interaction of the NECD with a ligand. In mam-

T. Borggrefe, B. D. Giaimo (eds.), *Molecular Mechanisms of Notch Signaling*, Advances in Experimental Medicine and Biology 1066, https://doi.org/10.1007/978-3-319-89512-3_17

mals, four Notch receptors (Notch1, 2, 3 or 4) can interact with five ligands, Delta-like (Dll) 1, 2 or 4 or Jagged (Jag) 1 or 2. Following interaction with ligand, the NECD is forcefully "ripped", from the cell surface of the Notch-bearing cell and endocytosed by the ligand-bearing cell. This exposes a site on Notch-Tm called NRR (negative regulatory region; see also "The Molecular Mechanism of Notch Activation" by Lovendahl/Blacklow/Gordon) making it susceptible to cleavage by an ADAM protease. In most instances, either ADAM 10 or ADAM 17 carry out this cleavage. Following ADAM cleavage, a conformational change occurs in the Notch-Tm, rendering it a substrate for cleavage by the intramembranous protease γ-secretase, resulting in the release of the intracellular domain of Notch (NICD). NICD rapidly translocates to the nucleus, displacing co-repressors bound to the DNA binding protein RBPJ (recombination signal binding protein for immunoglobulin kappa J region), and recruiting co-activators such as Mastermind-like and p300 and initiating a transcriptional program (reviewed in Bray 2016; see also "CSL-Associated Corepressor and Coactivator Complexes" by Oswald/Kovall).

In addition to interaction with RBPJ to initiate canonical signaling, Notch can interact with a variety of other intracellular proteins and participate in non-canonical signaling pathways ("Mechanisms of non-canonical signaling in health and disease: Diversity to take therapy up a Notch?"). In lymphocytes, Notch can interact with such diverse proteins as AKT, mTOR, NF-κB, mitofusin and CARMA1 to name a few (Perumalsamy et al. 2009, 2010; Shin et al. 2006, 2014). In many instances, these "non-canonical" interactions influence Notch function (reviewed in Ayaz and Osborne 2014). For example, Notch interaction with CARMA1 is required for the activation of the IKK complex (Shin et al. 2014) and Notch interaction with mitochondrial proteins such as mitofusion is an important component of Notch mediated survival signals (Perumalsamy et al. 2010).

Notch can also initiate signaling independent of interaction with ligands. The best evidence for ligand independent Notch activation comes from work conducted in *Drosophila melanogaster*, where genetic studies demonstrate that, in some situations, the Notch heterodimer is endocytosed and activated through interaction with Sima, the fly homologue of HIF-1α [Hypoxia inducible factor 1 alpha (Hori et al. 2011; Mukherjee et al. 2011)]. In mammalian cells, ligand-independent activation of Notch may be induced using Ca^{2+} chelators such as EDTA (Rand et al. 2000). Ca^{2+} interaction with the NRR of the Notch-Tm ensures proper folding of the protein and removal of Ca^{2+} disrupts folding and renders Notch-Tm susceptible to cleavage by ADAMs (van Tetering et al. 2009). Whether ligand independent activation of mammalian Notch occurs *in vivo* remains to be determined. However, as discussed below, it is possible that ligand-independent Notch activation may occur in mature T cells.

Notch signaling is important in many cells of the immune system but perhaps the best characterized effects of Notch in the immune system are in early T cell development and mature T cell function (reviewed in Amsen et al. 2015; Shah and Zúñiga-Pflücker 2014; Rothenberg et al. 2016). Indeed, some of the earliest examples of Notch function in mammals comes from studies conducted in the hematopoietic system. Once it was apparent from the report from Ellisen et al. (1991) that activated Notch is aberrantly expressed in T-ALL, many groups focused on the role of Notch signaling in normal T cell development. Before we consider these studies, it is useful to briefly review the important events during T cell development.

2 Notch in T Cell Development

All cells of the immune system are derived from a multi-potent hematopoietic stem cell (HSC). The HSC can differentiate into a common lymphoid progenitor (CLP) that can give rise to either T or B cells depending upon the surrounding environment. In case the CLP migrates to the thymus, this cell progresses through a differentiation program that eventually results in the production of mature $CD4^+$ and $CD8^+$ T cells. In studies pioneered by Zuniga-Pflucker and colleagues, it is apparent that a key feature of the thymic environment driving this developmental process is Notch/ligand interactions where a Notch-bearing early T cell encounters ligands (likely Dll4) at several points in the developmental process (Schmitt and Zúñiga-Pflücker 2002; Ciofani et al. 2004). The development of a mature T cell from the CLP is a complex process but some of the key steps in this progression are readily recognized by expression of easily identified cell surface markers. During early maturation events, T cell precursors lack both CD4 and CD8 and are called DN (for double negative) T cells. These DN T cells undergo a series of events whereby they acquire various cell surface markers that are detected by flow cytometry. The DN T cell proceeds developmentally through at least four documented stages called DN1, 2, 3 and 4. During the progression through DN1 to DN4, T-cell receptor (TCR) rearrangements occur. Immature thymocytes can mature to express αβ or γδ TCRs however, the majority of mature T cells produced in the thymus express an αβ TCR. In the cortico-medullary junction in the thymus, where cortico-epithelial cells express Dll4, DN1 cells inevitably are driven to become mature αβ T cells. Thus, the interaction between Notch1 on DN1 cells and Dll4 on cortico-epithelial cells is critical in driving the DN1 cell towards assuming a mature T cell phenotype (Schmitt and Zúñiga-Pflücker 2002; reviewed in Shah and Zúñiga-Pflücker 2014).

In addition to the role of Notch signaling in specifying a cell lineage fate in DN1 cells, Notch is also important in other steps as the DN1 cell begins its progress to become a mature T cell [(reviewed in Shah and Zúñiga-Pflücker 2014; Rothenberg et al. 2016), see Fig. 1]. DN1 cells require Notch signaling to become a DN2 cell. The DN2 cell then begins to rearrange the genes encoding TCR. As mentioned above, TCRs come in two distinct varieties, αβ or γδ. Both TCRs require rearrangement of gene segments to produce a functional TCR and the majority of early T cells become what are termed αβ T cells. A first step in αβ gene rearrangement is the rearrangement of the β chain. The β chain associates in the cytosol of the DN2 cell with an invariant preTCRα chain and is displayed on the cell surface as a dimer with TCRβ associated with preTCRα. The appearance of this heterodimer on the cell surface is a signal to begin the DNA rearrangements necessary to produce a functional mature TCRα chain. Notch signaling is a

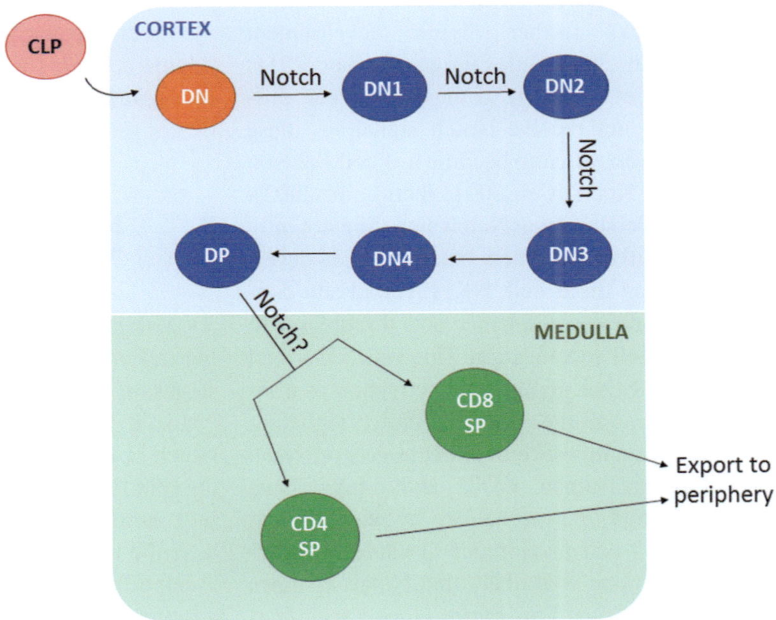

Fig. 1 Notch in early T cell development. CLP - Common Lymphoid Progenitor, DN – Double Negative, DP- Double Positive, SP – Single Positive

key component of this stage of thymic development as it is required for the transition from DN1 to DN2 (Schmitt et al. 2004). At the next stage of development, DN3, a process called β selection occurs whereby preTCR/CD3 signaling drives proliferation as well as inhibits apoptosis allowing progression to the DN4 stage of development and the acquisition of CD4 and CD8. At this point, these immature cells are termed DP (double positive) T cells because they express both CD4 and CD8. The DP cell now must undergo several selective processes. Because TCRs only recognize antigen presented to the cell by self-MHC (major histocompatibility complex), any DP cell that has a TCR that does not recognize self-MHC is deleted or more specifically allowed to die by neglect. The cell receives no stimulatory signals and hence dies. However, in a process called positive selection, if the TCR recognizes self-MHC, this T cell is allowed to survive and mature. Finally, any cell that carries a TCR that strongly recognizes both self-MHC plus self-antigen is negatively selected or instructed to undergo apoptosis. Negative selection ensures that self-reactive T cells, T cells that can cause havoc when mature and functional, are deleted in the thymus.

Notch is critical at the early stages of T cell development up until DN3, failure to encounter DLL4 blocks further thymic development (reviewed in Shah and Zúñiga-Pflücker 2014; Rothenberg et al. 2016). If the early T cell progenitors do not receive Notch signaling, these cells may even turn and become a B cell (Wilson et al. 2001; Koch et al. 2001; Izon et al. 2002). We now understand that Notch signaling not only induces T cell development but also blocks development along the B cell, NK, myeloid and dendritic cell lineages and hence acts a repressor to promote T cell specification. One well-described outcome of Notch signaling is activation of transcription through the canonical Notch signaling pathway. Although several direct targets of Notch including preT-alpha, CD25 and c-myc, have been identified, the mechanism by which Notch drives early T cell development is not fully understood. Therefore, it is likely that Notch assumes many distinct functions during T cell specification and the commitment to a T cell lineage. Indeed, Notch signals in early T cell precursors enhance cell proliferation but are not essential for viability, while at the DN3 stage Notch signaling is essential for survival (Ciofani and Zúñiga-Pflücker 2005). Thus, while we have a detailed understanding of the requirement for Notch signaling in early T cell development, the precise mechanisms that Notch uses to effect T cell specification are unknown.

The influence of Notch on later processes of T cell development are less well-delineated. As described above, DN4 cells acquire the cell surface markers, CD4 and CD8. Early experiments using a truncated version of NICD supported a role for Notch in CD4 versus CD8 lineage decisions with Notch1 overexpression driving DP thymocytes to a CD8 lineage and reducing the number of CD4 single positive (SP) T cells (Robey et al. 1996). However, these findings are controversial because other experiments employing targeted deletion of Notch1 at this point in developing T cells did not observe an effect on CD4 or CD8 lineage decisions (Wolfer et al. 2001). More recently, using thymocytes from mice with targeted deletion of presenilin 1/2, the enzymatic subunit of γ-secretase, a role for Notch1 in CD4 versus CD8 lineages is again supported (Laky et al. 2015). However, because γ-secretase has over 100 identified substrates (Golde et al. 2013) and many of these substrates are expressed in T cells, caution in the interpretation of these experiments is suggested.

3 Notch Activity in Peripheral T Cells

As described above, it is now evident that signaling through Notch plays crucial roles at various stages of T cell development. In more recent years, it has become increasingly evident that Notch is also involved in the activation and differentiation of naïve $CD4^+$ and $CD8^+$ SP T cells into various subsets in the periphery. Here, we describe how this pathway regulates such varied T cell differentiation programs, whether it acts as a molecular switch in peripheral T cell function and address some of the controversies in this emerging field.

3.1 Notch in T Cell Activation and Proliferation

For a T cell to mount an immune response against an infectious challenge, it needs to be activated. This involves interaction of the TCR with its cognate antigenic peptide presented on the surface of an antigen presenting cell (APC), bound to an MHC Class I (interacts with $CD8^+$ T cells) or II (interacts with $CD4^+$ T cells) molecule. This signal is further augmented by co-stimulatory molecules resulting in full activation of a T cell, subsequent IL2 (Interleukin 2) production, ultimately leading to T cell proliferation. The activation process is complex involving multiple intracellular signaling events and to add to this complexity, recent studies have found that Notch proteins can affect the activation and subsequent proliferation of naïve $CD4^+$ and $CD8^+$ SP T cells.

In 2003, studies by two independent groups revealed previously unrecognized roles for Notch proteins in T cells. Adler and colleagues demonstrated that $CD4^+$ T cell stimulation with anti-CD3 and anti-CD28 not only increases the expression of all four Notch genes but also induces Notch1 activation (Adler et al. 2003). Furthermore, pharmacological blockage of Notch1 activation inhibits T cell proliferation *in vitro,* which is associated with a decrease in CD25 expression and IL2 production. In agreement with the above data, Palaga and colleagues revealed that activation of both $CD4^+$ and $CD8^+$ SP T cells results in Notch activation and subsequent upregulation of the Notch protein (Palaga et al. 2003). They also reported a decrease in T cell activation, proliferation and IFNγ (Interferon gamma) production upon GSI (γ-Secretase Inhibitor) mediated inhibition of Notch in peripheral T cells. A subsequent study by Benson and colleagues demonstrated that Notch1 is upregulated and colocalizes with CD4 on the cell surface following *in vitro* activation of $CD4^+$ T cells (Benson et al. 2005). However, in this study, pharmacological blockage of Notch signaling does not affect proliferation but attenuates cytokine production. This group also showed that upon transfection of a constitutively active form of Notch1, $CD4^+$ T cells fail to proliferate but exhibit enhanced secretion of cytokines on stimulation. In another report, Rutz and colleagues documented that distinct Notch ligands differentially affect early T cell activation and proliferation, where Dll1 and Jag1 inhibit proliferation and the expression of early activation markers – CD69 and CD25, while Dll4 has the opposite effect (Rutz et al. 2005). The binding capacity of the Notch ligands to resting and activated T cells also differ considerably, with Dll4 showing the strongest binding, followed by Dll1 and Jagged1 (Rutz et al. 2005).

Almost a decade later, several studies revealed novel roles for Notch in T cell activation and proliferation. A 2013 study suggested that Notch can directly regulate PD1 (Programmed death 1) expression in activated $CD8^+$ T cells (Mathieu et al. 2013). Following anti-CD3/CD28 stimulation of a co-culture of purified $CD8^+$ T cells and APCs or bulk splenocytes, the authors observed that PD-1 expression was significantly reduced in $CD8^+$ T cells when Notch signaling was blocked using the GSI DAPT. These results are interpreted to suggest that prolonged activation of Notch signaling during chronic infection, due to continued antigen presentation by APCs expressing Notch ligands, may lead to Notch-induced expression of PD-1, thereby regulating the immune response. It is important to note that these experiments, like many of the studies reviewed here, interpret results obtained using GSI as an effect of Notch. γ-secretase substrates number over 100 and the use of GSIs to block Notch activity must be verified by targeted deletion of Notch in the cell in question. Others have used dominant negative forms of mammalian Mastermind (DN-MAML) which is a more direct approach to inhibiting the canonical Notch signaling pathway,

The following year, our laboratory showed that Notch affects activation, proliferation and differentiation of $CD4^+$ T cells in a non-canonical fashion (Dongre et al. 2014). Notch signaling that occurs independent of its canonical partner –

RBPJ, is termed non-canonical Notch signaling. In experiments using conditional Notch1 and RBPJ knockouts, we demonstrated that CD4+ T cell activation and proliferation is impaired in the absence of Notch1 but remains unaffected when RBPJ is deleted. This non-canonical role of Notch in regulating peripheral T cell function is not only novel but may also explain some of the differential effects of Notch. Another group tested the ability of Dll4-bearing APCs to drive CD4+ T cell priming and found that Dll4-deficient APCs less efficiently promote activation, metabolism, proliferation and IL2 secretion of CD4+ T cells (Laky et al. 2015). Furthermore, they documented that APCs can fine tune the antigen sensitivity of CD4+ T cells via Dll4-induced Notch signaling, where Dll4-Notch interaction through PI3K (Phosphoinositide-3-kinase) signaling allows naïve CD4+ T cells to respond to low doses of antigen. This Dll4-induced effect of Notch signaling on T cell activation agrees with the work of Rutz and colleagues (Rutz et al. 2005).

The available data suggest to us that Notch can promote or inhibit T cell activation and proliferation based on environmental cues and the presence or absence of different Notch ligands. Thus, as suggested by others, individual ligands may have differing biological effects and this may be influenced by environmental cues. Lending further credence to this idea are studies showing that distinct Notch ligands can induce differential effects in a particular cell, for instance during human lymphoid differentiation (Jaleco et al. 2001) or T lineage commitment (Lehar et al. 2005). However, because each group uses unique experimental approaches it is difficult to reach an overarching consensus. Differences in the cell populations studied (purified T cells versus T cells in the presence of APCs and other cells), the pharmacological inhibitors used and activation of T cells in the presence or absence of ligands, clearly indicate that more work needs to be done in this direction to obtain a clearer picture of how individual ligands influence Notch in T cell function.

3.2 Notch in CD4+ T Cell Differentiation

CD4+ T cells are multifaceted and therefore an integral part of the immune system. Among other functions, they can orchestrate an immune response against a wide range of pathogens and can also regulate these responses, thereby preventing autoimmune disorders. How does a CD4+ T cell manage to perform such diverse functions? Depending on the cytokine milieu during TCR activation, naïve CD4+ T cells can differentiate into several lineages of T helper (Th) lymphocytes, including Th1, Th2, Th17 and regulatory T (Tregs), that are defined by their function and cytokine production (see Fig. 2). Notch has been found to be important in the differentiation of most Th cells however, whether it acts as a molecular switch or plays a more subtle context-dependent role in Th differentiation remains to be determined.

3.2.1 Notch in Th1 and Th2 Differentiation

Th1 and Th2 cells express T-bet (Th1) and Gata3 (Th2) as the driving differentiation factors and produce IFNγ and IL4 (Interleukin 4) as signature cytokines, respectively. While Th1 cells fight intracellular viruses and bacteria, Th2 cells direct immunity against extracellular helminthic infections and play a role in allergies. IL12 and IL4 are believed to be the major inducers of Th1 and Th2 differentiation, respectively; however, other pathways have been shown to be involved as well (Skokos and Nussenzweig 2007).

The first evidence of a role for Notch in CD4+ T cell differentiation came in 2003, when Maekawa and colleagues showed that the Dll1-Notch3 interaction induces differentiation towards the Th1 lineage (Maekawa et al. 2003). Dll1-Fc stimulation of CD4+ T cells not only substantially increased the number of IFNγ secreting cells over IL4 producing cells, but also induced the expression of T-bet. These results were further strengthened when *in vivo* administration of

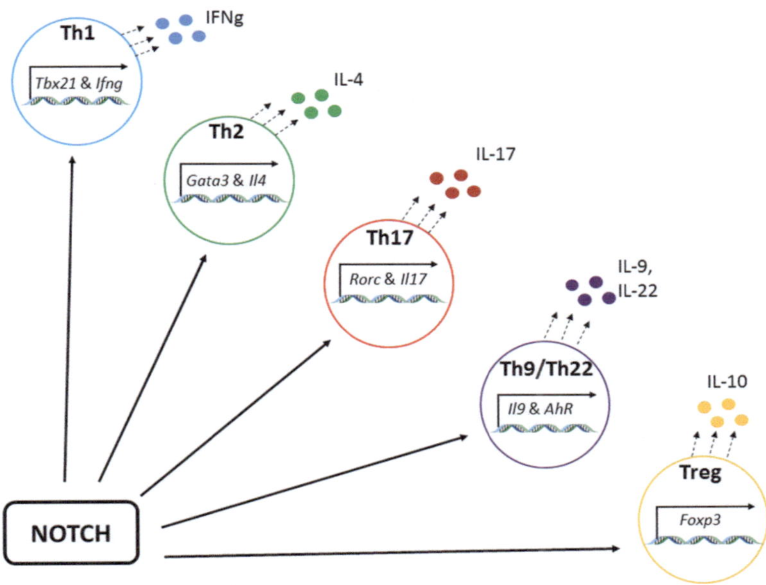

Fig. 2 Notch in CD4^{+} T cell differentiation. Notch can drive CD4^{+} T cell differentiation into most subtypes by regulating the master transcription factors. However, whether Notch does so via its canonical or non-canonical partners requires further study

Dll1-Fc resulted in a Th1 response against *Leishmania major* (*L. major*) infection in normally susceptible BALB/C mice. Moreover, retroviral overexpression of NICD3 in CD4^{+} T cells increased IFNγ secretion while decreasing IL4 production and this skewing towards the Th1 phenotype was found to be dependent on Dll1-Notch3 interaction. In a contrasting report, using Notch1$^{fl/fl}$x CD4-Cre mice where the peripheral T cells are deficient in Notch1, Tacchini-Cottier and colleagues showed that Notch1 is dispensable for Th1 and Th2 differentiation *in vitro* (Tacchini-Cottier et al. 2004). Moreover, *L. major* infection in Notch1$^{-/-}$ CD4-Cre mice resulted in a protective Th1 response characterized by high IFNγ levels and low IL4 levels similar to resistant C57BL/6 mice, indicating that Notch1 is not critical for Th1 differentiation. Challenging these results, Minter and colleagues demonstrated that GSI-mediated inhibition of Notch signaling attenuates polarization towards Th1 by preventing *Tbx21* upregulation, the gene encoding T-bet (Minter et al. 2005). Further, GSI treatment of CD4^{+} T cells reduces the levels of Notch, *Tbx21* and IFNγ on Th1 polarization while IL4 production remains unaffected in polarized Th2 cells. *In vivo*, administration of GSI to mice with experimental autoimmune encephalomyelitis (EAE), a classical Th1 mediated model of multiple sclerosis, significantly reduced the symptoms of EAE. The authors further showed that Notch1 directly regulates *Tbx21* expression by forming a Notch1/RBPJ complex on the *Tbx21* promoter. In contrast to the work of Tacchini-Cottier and colleagues, these results point towards a T cell intrinsic mechanism for Notch1 in Th1 differentiation.

Studies in the subsequent years by both groups resolved some of the controversies regarding the role of Notch in Th1 differentiation. Using mice with T cell specific ablation of both Notch1 and Notch2 on a C57Bl/6 – *L. major*-resistant genetic background, the Tacchini-Cottier laboratory showed that lack of both these receptors renders the mice susceptible to *L. major* infection while mice lacking either Notch1 or Notch2 develop a protective Th1 response (Auderset et al. 2012). Their data point towards a redundant role for Notch1 and Notch2 in driving a Th1 response. Further, in 2013, the Minter laboratory reported that NICD1 is increased in T cells from mice with aplastic anemia, a Th1-mediated disease, and that blocking Notch attenuates the disease (Roderick et al. 2013). In support of their earlier results, they show that NICD1 is bound to the *TBX21* promoter in PBMCs (Peripheral blood mononuclear cells) from patients with untreated aplastic anemia. These results highlight a strong role for Notch in regulating Th1-mediated responses.

Several subsequent studies provide a clearer view of Notch in regulating the Th1 differentiation program. A study exploring how dendritic cells induce a Th1 response upon Toll-like receptor (TLR) ligation in the absence of the major inducing cytokine IL12, showed that the Notch ligand Dll4 is involved in this process, implicating Notch signaling in IL12-independent Th1 differentiation (Skokos and Nussenzweig 2007). This suggests that Notch and IL12 are redundant and that this redundancy may explain some of the discrepancies in the contribution of Notch to Th1 differentiation. Another report showed that over-expression of NICD3 in CD4⁺ T cells during differentiation led to strong IL10 production in Notch-transduced Th1 cells (Rutz et al. 2008). IL10 is an anti-inflammatory cytokine that is involved in controlling immune responses. In this study, Notch signaling was found to be responsible for inducing IL10 production in a STAT4 (Signal transducer and activator of transcription 4) dependent manner converting a pro-inflammatory Th1 response into a regulatory one, thus providing novel opportunities to use this pathway to attenuate Th1-mediated immune disorders. In addition to T cell activation and proliferation (described earlier), our laboratory has also shown that differentiation into the Th1 lineage, although Notch1 dependent, is independent of signaling through its canonical partner, RBPJ (Dongre et al. 2014).

The data described so far suggest that Notch regulates differentiation into Th1 but not Th2 cell fate. However, there is enough evidence to implicate Notch in Th2 differentiation as well. An early study by Amsen and colleagues documented that APCs that express the Notch ligand Dll1 induce a Th1 fate whereas Jagged1 expression potentiates differentiation into Th2 (Amsen et al. 2004). Additionally, the authors report that differentiation into the Th2 lineage requires an intact canonical Notch pathway, which induces Gata3 expression and directly regulates the *Il4* gene but this mechanism is independent of STAT6. They also show that retroviral expression of both NICD1 and NICD2 in CD4⁺ T cells promotes IL4 production independent of STAT6. In a subsequent study, the same group highlighted that direct regulation of Gata3 by Notch is required to generate optimal Th2 responses (Amsen et al. 2007). These results were confirmed by another group in the same year (Fang et al. 2007). Together, their data reveal that Notch in conjunction with RBPJ binds to the *Gata3* promoter to induce IL4 production, promoting the Th2 phenotype. Furthermore, Amsen and co-authors go on to show that in the absence of Gata3, Notch turns from being an inducer of Th2 to a strong Th1 inducer, indicating that Gata3 acts as a molecular switch in Notch-induced Th differentiation. A separate study demonstrated that signaling through Notch controls the initial IL4 expression by regulating the IL4 enhancer – conserved noncoding sequence-2 (CNS-2) in memory phenotype CD4⁺ T cells and Natural Killer T (NKT) cells (Tanaka et al. 2006). Their data demonstrate that loss of Th2 development in RBPJ deficient mice is due to the lack of initial IL4 production by CNS-2-regulated T cells, suggesting that Notch/RBPJ-mediated control of initial IL4 production may direct whether naïve CD4⁺ T cells can adopt a Th2 phenotype. In total, the studies described above clearly demonstrate that Notch, through Gata-3, regulates IL4 expression and this can influence Th2 development. Therefore, it is possible that extrinsic Notch regulation of IL4 production in another cell provides IL4 to a developing Th2 cell. This interpretation is supported by the fact that *in vitro* T helper polarization to Th1 requires Notch while Notch is dispensible for Th2 polarization (Minter et al. 2005; Dongre et al. 2014).

3.2.2 Notch in Th17 Differentiation

Apart from Th1 and Th2, several other subsets of Th cells have been discovered and Notch has been shown to be involved in immune responses through those lineages as well. Th17 cells mount defenses against extracellular fungi and bacteria and are important modulators of several autoimmune disorders. These cells express RORγt (RAR-related orphan receptor gamma t) as their master transcriptional regulator, produce IL17A and IL17F as major cytokines and are induced by TGFβ (Transforming growth factor beta) and

IL6. Through experiments using TCR transgenic cells (DO11.10), a 2009 study revealed that Dll-4 enhances IL17 production in the presence of TGFβ and IL6, while inhibition of Notch signals fail to do so even under skewing conditions (Mukherjee et al. 2009). They further showed that RBPJ, the canonical partner of Notch, directly interacts with the RORγt and IL17 promoter to regulate IL17 production in response to Dll4. Strengthening these observations, Keerthivasan and colleagues reported that Notch inhibition, using GSI or Notch1 siRNA, reduces IL17 production during mouse and human Th17 polarization (Keerthivasan et al. 2011). Additionally, GSI administration ameliorates EAE symptoms and dampens the Th17-mediated response in this model. This group also found that Notch1 directly binds to both IL17 and RORγt promoters, implying a direct regulation of Th17 differentiation by Notch1.

3.2.3 Notch in the Differentiation of Other Th Subsets

Th9 cells, another class of Th cells, produce IL9 and are generated under the influence of IL4 and TGFβ. The transcriptional regulation of this subset and whether they act as immune response mediators or sustain inflammation is still not clear. Shedding light on these questions, Elyaman and co-authors showed that Notch1 and Smad3 together bind to the *Il9* promoter and activate IL9 production (Elyaman et al. 2012). Moreover, using an EAE model, they showed that Jag2-induced IL9 production can alleviate or exacerbate EAE symptoms based on whether the mice are pretreated or treated with anti-Jag2 monoclonal antibody at the same time when EAE is induced. This suggests that IL9 producing cells can play dual roles in the immune system, depending on the timing of the co-stimulation and the cytokine microenvironment. IL22 is a cytokine that can be produced by Th1, Th17 cells as well as some other cells. Its production is induced by IL6 and driven by the expression of the aryl hydrocarbon receptor (AhR). Notch was found to be involved in the regulation of IL22 production as well by inducing the production of stimulators of AhR (Alam et al. 2010).

There is no dispute that Notch is important in the differentiation of CD4$^+$ T cells into multiple lineages. Evidence from ligand data suggests that the Delta family of ligands promotes Th1 and Th17 responses, whereas the Jagged family may be important for Th2 and Th9 differentiation. Although limited, there exists adequate evidence pointing towards a role for Notch in regulating differentiation towards Th17 and Th9 lineages as well. Despite the conflicting views on Notch's role in Th1 and Th2, the direct regulation of lineage regulators *Tbx21* and *Gata3* by Notch clearly show that Notch can play crucial roles in both Th1 and Th2 differentiation. Since signaling through the TCR activates Notch, it is possible that Notch acts as a "co-receptor" and cooperates with environmental signals to drive differentiation pathways. Additionally, the interplay between Notch/RBPJ and chromatin is an important feature of Notch signaling (see "CSL-Associated Corepressor and Coactivator Complexes" by Oswald/Kovall; Oswald et al. 2016). In light of ample evidence of epigenetic regulation of T helper lineages (reviewed by Zhu et al. 2010), it is tempting to speculate that the state of the chromatin near Notch target genes likely contributes to T helper lineage decisions. Nevertheless, questions as to which Notch receptors are involved and whether the effects of Notch are cell intrinsic or extrinsic remain unresolved.

3.3 Notch in CD8$^+$ T Cell Differentiation

To add to the already long list of functions for Notch, studies suggest that Notch is also involved in regulating CD8$^+$ T cell responses (see Fig. 3). CD8$^+$ T cells or cytotoxic T lymphocytes (CTLs) are involved in killing tumor cells or virally-infected cells. Data implicating Notch in CD8$^+$ T cell differentiation was provided by the Yasumoto group who showed that retroviral expression of Dll1 on bone marrow-derived dendritic cells (BMDC) enhanced the differentiation of CD8$^+$ cells into CTLs, whereas lack of Notch2 in peripheral CD8$^+$ T cells failed to induce this dif-

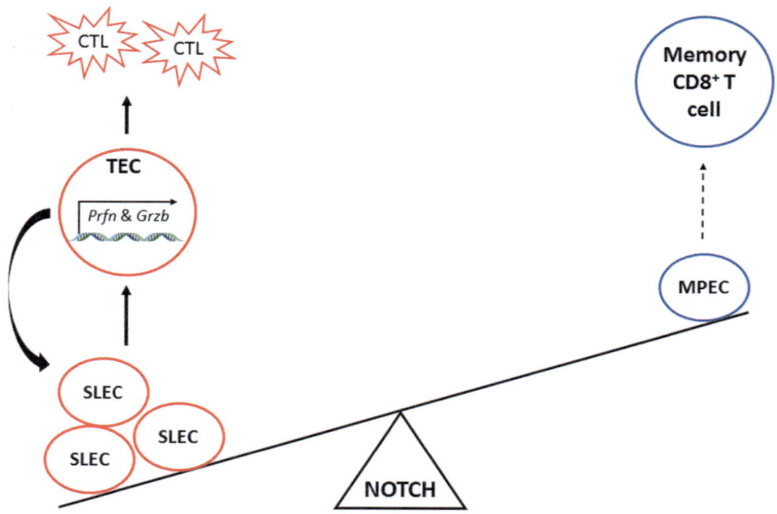

Fig. 3 Notch in CD8⁺ T cell differentiation. CTL – Cytotoxic T Lymphocytes, TEC – Terminal Effector cells, SLEC – Short-lived Effector cells, MPEC – Memory Precursor Effector cells

ferentiation *in vitro* and *in vivo* (Maekawa et al. 2008). Further, Notch2 in a complex with CREB1 was found to directly control the transcription of the gene encoding granzyme B (CTL effector molecule), independent of Eomes – the key regulator of granzyme B and perforin. Using mice that lack Notch2 in CD8$^+$ T cells, the same group then went on to show that signaling through Notch2 is essential for antitumor CTL responses *in vivo* (Sugimoto et al. 2010). Consistent with these findings, data from our lab demonstrate that both GSI-mediated inhibition of Notch and genetic reduction of Notch1 decrease the mRNA and protein levels of cytolytic effectors - perforin and granzyme B in CD8$^+$ T cells (Cho et al. 2009). This could be the result of direct binding of Notch1 to the promoters of Eomes, perforin and granzyme B, thereby linking Notch signaling to the regulation of these CTLs effector molecules. This effect of Notch held true for human CD8$^+$ T cells as well (Kuijk et al. 2013). In another study on the role of Notch in antitumor responses, the authors demonstrated that Notch expression was reduced in T cells from tumors in mice (Sierra et al. 2014). Moreover, transgenic expression of NICD1 in antigen-specific CD8$^+$ T cells increased granzyme B levels and resulted in higher cytotoxic effects, suggesting a strong potential for Notch in enhancing the efficacy of T cell based immunotherapies.

Notch was also found to regulate the choice between terminal effector cells (TEC) or memory precursor cell (MPC) fates in CD8$^+$ T cells (Backer et al. 2014). Here, Amsen and colleagues describe that signaling through Notch promotes differentiation towards TECs and feeds back into the TEC promoting pathways giving rise to fully protective TECs. Similarly, Mathieu and colleagues document that Notch is crucial for the formation of short lived effector cells (SLECs) but is dispensable for the generation of memory precursor effector cells [MPECs, (Mathieu et al. 2015)]. Their data also suggest a context-dependent role for Notch during CD8$^+$ T cell response, where Notch is required for maximal IFNγ production and only selectively required for IL2 and TNFα (Tumor necrosis factor alpha) production after *Listeria monocytogenes* infection and vaccination with dendritic cells. Therefore, the current evidence clearly point towards a crucial role for Notch in immune responses through CD8$^+$ T cells, implicating Notch as a strong candidate for immunotherapy in cancer.

3.4 Notch in Regulatory T Cells

Regulatory T cells or Tregs, as the name suggests, are a subset of CD4$^+$ T cells that can suppress an immune response. They are defined by the expression of their master transcriptional factor FoxP3. Tregs that are derived from the thymus are termed naturally occurring Tregs or nTregs. Tregs can also be induced *in vitro* from naïve $CD4^+$ T cells in the presence of TGFβ and these are called induced Tregs or iTregs. Another emerging category of regulatory T cells are $CD8^+$ suppressor T cells. Although, these are less explored than $CD4^+$ Tregs, multiple populations have been described based on the expression of several markers but only a small number of $CD8^+$ Tregs express FoxP3 (Tang et al. 2005; Dinesh et al. 2010).

The first indication of a role for Notch in inhibiting an immune response came with a study reporting that overexpression of Notch ligand Serrate1 (Jag1) on APCs leads to differentiation of antigen-specific $CD4^+$ T cells into regulatory cells (Hoyne et al. 2000). The authors demonstrated that these regulatory cells can inhibit primary and secondary immune responses and can also transfer this antigen-specific tolerance to recipient mice. In the following years, reports from several groups further strengthened the role of Notch signaling in Treg development. In 2003, two studies by the same group revealed that co-culture of Epstein-Barr virus lymphoblastoid B cells (EBV-LCL) overexpressing Jag1 with T cells induces the generation of human Tregs that can inhibit proliferative and cytotoxic immune responses towards a specific antigen (Vigouroux et al. 2003) or alloantigen (Yvon et al. 2003). Furthermore, both studies showed that this inhibition of immune response is transferable, since the Notch-induced Tregs could also inhibit immune responses of fresh T cells that have not been exposed to Jag1. Evidence of additional involvement of the Notch receptors in Treg function was provided by the Screpanti laboratory, who showed that the presence of constitutively active NICD3 in the T cells of transgenic mice prevents the development of experimental autoimmune diabetes (Anastasi et al. 2003). Failure to develop disease was associated with an enhanced number of $CD4^+$ $CD25^+$ Tregs and increased expression of the Treg specific cytokine, IL10. Work from our group in collaboration with colleagues concur with the above findings. We have shown that *in vitro* GSI inhibition of signaling through Notch blocks TGFβ-induced expression of FoxP3 and its target genes (Samon et al. 2008). Lending *in vivo* support to this finding, GSI administration to C57BL/6 mice reduced FoxP3 expression resulting in symptoms reminiscent of a disease involving dysregulation of TGFβ and Tregs. Our chromatin immunoprecipitation (ChIP) data further suggest that Notch1 directly regulates FoxP3 expression cooperatively with TGFβ signaling. This result was corroborated by a subsequent study where the authors report that NICD binds to the *Foxp3* promoter in Tregs in a complex with RBPJ (Ou-Yang et al. 2009). On similar lines, the Screpanti group demonstrated that Notch in conjunction with PKC-theta and NFκB controls FoxP3 expression, thereby regulating Tregs generation (Barbarulo et al. 2011). A novel study aimed at generating iTregs *in vitro,* demonstrated that Dll1-mediated Notch signaling efficiently converts human memory $CD4^+$ T cells into iTregs (Mota et al. 2014). Their data further suggest that Notch signaling through Dll1 plays a dual role in promoting iTreg development - by directly regulating FoxP3 expression and interacting with the TGFβ pathway. Therefore, it is evident that Notch channels signals through multiple partners to promote the development of Tregs. Hinting at a role for Notch in $CD8^+$ Tregs, another study showed that pretreatment of alloantigen bearing cells with Dll1 inhibits responses to subsequent exposure of the same antigen, resulting in prolongation of graft survival in a mouse model of cardiac allograft (Wong et al. 2003). Their data further suggest that this inhibition of graft rejection is because Notch ligation on $CD8^+$ T cells enhances their IL10 production, altering their differentiation potential from a T1-type response to an inhibitory one.

The role of Notch in Treg development and function, however, is not without controversy. Evidence opposing the abovementioned findings was provided by Bassil and colleagues, where neutralization of Dll4 using a blocking antibody during the induction phase of EAE alleviated EAE symptoms by drastically increasing the $CD4^+FoxP3^+$ Tregs population in the periphery and the CNS (Central nervous system) (Bassil et al. 2011). Additionally, the authors reported that Dll4-induced Notch signaling inhibits Tregs development by regulating the phosphorylation of STAT5 (Signal transducer and activator of transcription 5), a key regulator of FoxP3 expression. Adding to an already complex view of Notch in Tregs, a recent study revealed that Tregs-specific deletion of components of Notch signaling augmented Tregs-mediated suppression of Th1 response, whereas NICD1 overexpression reversed this effect (Charbonnier et al. 2015). Their data also suggest roles for both canonical and non-canonical Notch pathways in the dysregulation of Tregs.

The current view on the role of Notch in Tregs is divided. While there is more evidence indicating that Notch signaling promotes Tregs generation than inhibiting development of this population of cells, additional studies are warranted before a consensus can be reached on the matter. Furthermore, numerous studies suggest that Notch ligands are critical with Serrate and Jag inducing Tregs generation, while signaling through Dll ligands appear to have the opposite effect. Therefore, as suggested earlier, it is possible that the opposing evidence on the role of Notch in Tregs function could be the result of signaling through different Notch ligands. Nonetheless, further experiments are needed to test if this idea is indeed true.

4 Notch and Diseases Mediated by T Lymphocytes

Understanding how Notch influences T cell function is important because of the well-defined role it plays in the development of a variety of T cell related diseases. Indeed, the first report of a mammalian Notch homologue was as a translocation in T-ALL (Ellisen et al. 1991) demonstrating a key role for Notch in T cell malignancy. Over the ensuing two decades, Notch has been implicated in many cancers, including those of the immune system (reviewed in Chiang et al. 2016). Perhaps not surprisingly, due to its role in T cell activation, Notch is also known to influence a variety of autoimmune diseases. More than a decade ago, our lab described a role for Notch in mediating EAE, a disease known to require Th1 responses (Minter et al. 2005). Roderick et al. (2013) demonstrated a key contribution of Notch in the development of bone marrow failure, another autoimmune disease mediated by Th1 cells. There also is increasing evidence that Notch may contribute to several other autoimmune conditions (reviewed in Kuksin and Minter 2015). Additionally, data from the Maillard lab (Tran et al. 2013) demonstrate that targeting Notch with blocking antibodies in a mouse model of graft versus host disease (GVHD) may ameliorate GVHD. These data are particularly important since this group used Notch blocking antibodies to abrogate disease, a therapy that may be clinically useful in the near term. Although Notch is implicated in many diseases, including those of the immune system, blockade of Notch in a clinical setting is fraught with potential problems because of the requirement for Notch signaling in a vast array of cells and tissues. Acute blockade using antibodies may possibly alleviate the clinical complications observed with gamma secretase inhibitors.

5 Concluding Remarks

Notwithstanding the contradictory views on how Notch affects T cell activation, differentiation and function, it is beyond dispute that signaling through Notch is critical for T cell function. The available data suggest more and more that Notch plays a highly versatile and context-dependent role in relaying signals downstream and modifying outcomes based on its immediate environmental cues. However, there is still much to be learned to obtain a complete picture and fully understand the implications of Notch-based immunotherapies. Considering the pleiotropic effects of signaling through Notch, the use of consistent experimental approaches and in-depth analysis of their functions are crucial to reach a consensus regarding how this signaling pathway controls so many aspects of T cell-mediated immune responses. However, designing experiments to study the downstream effects of Notch signaling can be very tricky. As mentioned earlier, results from experiments using GSIs need to be interpreted cautiously considering that they have multiple substrates and Notch is only one of them. Knockout experiments are difficult since some Notch receptors (Notch1 and Notch2) are critical during development and therefore blocking signaling through them *in vivo* can cause embryonic lethality. Although, one can get around this issue using conditional knockouts, there is an additional problem of compensation by other Notch receptors when one receptor is knocked out *in vivo*. Further, a Notch loss-of-function phenotype can be mimicked using dominant negative Mastermind-like protein 1 (dnMAML1) that will prevent the binding of wild-type MAML1 to Notch and RBPJ, thus preventing target gene expression downstream of Notch. However, this construct does not account for signaling via non-canonical partners of Notch nor does it take into consideration the effects of MAML1 on other unrelated signaling pathways. These complexities call for careful and detailed design of experiments and cautious analysis and interpretation of results to fully understand T-cell mediated responses regulated by Notch and to develop Notch-based therapies to treat immune disorders.

Acknowledgments The authors thank the members of the Osborne and Minter labs for lively discussions and the National Multiple Sclerosis Society (RG 5151A1/2) and the NIH (CA166009) for support.

References

Adler SH, Chiffoleau E, Xu L, Dalton NM, Burg JM, Well AD, Wolfe MS, Turka LA, Pear WS (2003) Notch signaling augments T cell responsiveness by enhancing CD25 expression. J Immunol 171(6):2896–2903

Alam MS, Maekawa Y, Kitamura A, Tanigaki K, Yoshimoto T, Kishihara K, Yasutomo K (2010) Notch signaling drives IL-22 secretion in CD4+ T cells by stimulating the aryl hydrocarbon receptor. Proc Natl Acad Sci U S A 107(13):5943–5948

Amsen D, Blander JM, Lee GR, Tanigaki K, Honjo T, Flavell RA (2004) Instruction of distinct CD4 T helper cell fates by different notch ligands on antigen-presenting cells. Cell 117(4):515–526

Amsen D, Antov A, Jankovic D, Sher A, Radtke F, Souabni A, Busslinger M, McCright B, Gridley T, Flavell RA (2007) Direct regulation of Gata3 expression determines the T helper differentiation potential of notch. Immunity 27(1):89–99

Amsen D, Helbig C, Backer RA (2015) Notch in T cell differentiation: all things considered. Trends Immunol 36(12):802–814

Anastasi E, Campese AF, Bellavia D, Bulotta A, Balestri A, Pascucci M, Checquolo S, Gradini R, Lendahl U, Frati L, Gulino A, Di Mario U, Screpanti I (2003) Expression of activated Notch3 in transgenic mice enhances generation of T regulatory cells and protects against experimental autoimmune diabetes. J Immunol 171(9):4504–4511

Auderset F, Schuster S, Coutaz M, Koch U, Desgranges F, Merck E, MacDonald HR, Radtke F, Tacchini-Cottier F (2012) Redundant Notch1 and Notch2 signaling is necessary for IFNgamma secretion by T helper 1 cells during infection with Leishmania major. PLoS Pathog 8(3):e1002560

Ayaz F, Osborne BA (2014) Non-canonical notch signaling in cancer and immunity. Front Oncol 4(345):1–6

Backer RA, Helbig C, Gentek R, Kent A, Laidlaw BJ, Dominguez CX, de Souza YS, van Trierum SE, van Beek R, Rimmelzwaan GF, ten Brinke A, Willemsen AM, van Kampen AH, Kaech SM, Blander JM, van Gisbergen K, Amsen D (2014) A central role for notch in effector CD8(+) T cell differentiation. Nat Immunol 15(12):1143–1151

Barbarulo A, Grazioli P, Campese AF, Bellavia D, Di Mario G, Pelullo M, Ciuffetta A, Colantoni S, Vacca A, Frati L, Gulino A, Felli MP, Screpanti I (2011) Notch3 and canonical NF-kappaB signaling pathways cooperatively regulate Foxp3 transcription. J Immunol 186(11):6199–6206

Bassil R, Zhu B, Lahoud Y, Riella LV, Yagita H, Elyaman W, Khoury SJ (2011) Notch ligand delta-like 4 blockade alleviates experimental autoimmune encephalomyelitis by promoting regulatory T cell development. J Immunol 187(5):2322–2328

Benson RA, Adamson K, Corsin-Jimenez M, Marley JV, Wahl KA, Lamb JR, Howie SE (2005) Notch1 co-localizes with CD4 on activated T cells and notch signaling is required for IL-10 production. Eur J Immunol 35(3):859–869

Bray SJ (2016) Notch signalling in context. Nat Rev Mol Cell Biol 17(11):722–735

Charbonnier LM, Wang S, Georgiev P, Sefik E, Chatila TA (2015) Control of peripheral tolerance by regulatory T cell-intrinsic notch signaling. Nat Immunol 16(11):1162–1173

Chiang MY, Radojcic V, Maillard I (2016) Oncogenic notch signaling in T-cell and B-cell lymphoproliferative disorders. Curr Opin Hematol 23(4):362–370

Cho OH, Shin HM, Miele L, Golde TE, Fauq A, Minter LM, Osborne BA (2009) Notch regulates cytolytic effector function in CD8+ T cells. J Immunol 182(6):3380–3389

Ciofani M, Zúñiga-Pflücker JC (2005) Notch promotes survival of pre-T cells at the beta-selection checkpoint by regulating cellular metabolism. Nat Immunol 6(9):881–888

Ciofani M, Schmitt TM, Ciofani A, Michie AM, Cuburu N, Aublin A, Maryanski JL, Zúñiga-Pflücker JC (2004) Obligatory role for cooperative signaling by pre-TCR and notch during thymocyte differentiation. J Immunol 172(9):5230–5239

Dinesh RK, Skaggs BJ, La Cava A, Hahn BH, Singh RP (2010) CD8+ Tregs in lupus, autoimmunity, and beyond. Autoimmun Rev 9(8):560–568

Dongre A, Surampudi L, Lawlor RG, Fauq AH, Miele L, Golde TE, Minter LM, Osborne BA (2014) Non-canonical notch signaling drives activation and differentiation of peripheral CD4(+) T cells. Front Immunol 5:54

Ellisen LW, Bird J, West DC, Soreng AL, Reynolds TC, Smith SD, Sklar J (1991) TAN-1, the human homolog of the Drosophila notch gene, is broken by chromosomal translocations in T lymphoblastic neoplasms. Cell 66(4):649–661

Elyaman W, Bassil R, Bradshaw EM, Orent W, Lahoud Y, Zhu B, Radtke F, Yagita H, Khoury SJ (2012) Notch receptors and Smad3 signaling cooperate in the induction of interleukin-9-producing T cells. Immunity 36(4):623–634

Fang TC, Yashiro-Ohtani Y, Del Bianco C, Knoblock DM, Blacklow SC, Pear WS (2007) Notch directly regulates Gata3 expression during T helper 2 cell differentiation. Immunity 27(1):100–110

Golde TE, Koo EH, Felsenstein KM, Osborne BA, Miele L (2013) γ-Secretase inhibitors and modulators. Biochim Biophys Acta 1828(12):2898–2907

Hori K, Sen A, Kirchhausen T, Artavanis-Tsakonas S (2011) Synergy between theESCRT-III complex and Deltex defines a ligand-independent notch signal. J Cell Biol 195(6):1005–1015

Hoyne GF, Le Roux I, Corsin-Jimenez M, Tan K, Dunne J, Forsyth LM, Dallman MJ, Owen MJ, Ish-Horowicz D, Lamb JR (2000) Serrate1-induced notch signalling regulates the decision between immunity and tolerance made by peripheral CD4(+) T cells. Int Immunol 12(2):177–185

Izon DJ, Aster JC, He Y, Weng A, Karnell FG, Patriub V, Xu L, Bakkour S, Rodriguez C, Allman D, Pear WS (2002) Deltex1 redirects lymphoid progenitors to the B cell lineage by antagonizing Notch1. Immunity 16(2):231–243

Jaleco AC, Neves H, Hooijberg E, Gameiro P, Clode N, Haury M, Henrique D, Parreira L (2001) Differential effects of notch ligands Delta-1 and Jagged-1 in human lymphoid differentiation. J Exp Med 194(7):991–1002

Keerthivasan S, Suleiman R, Lawlor R, Roderick J, Bates T, Minter L, Anguita J, Juncadella I, Nickoloff BJ, Le Poole IC, Miele L, Osborne BA (2011) Notch signaling regulates mouse and human Th17 differentiation. J Immunol 187(2):692–701

Koch U, Lacombe TA, Holland D, Bowman JL, Cohen BL, Egan SE, Guidos CJ (2001) Subversion of the T/B lineage decision in the thymus by lunatic fringe-mediated inhibition of Notch-1. Immunity 15(2):225–236

Kuijk LM, Verstege MI, Rekers NV, Bruijns SC, Hooijberg E, Roep BO, de Gruijl TD, van Kooyk Y, Unger WW (2013) Notch controls generation and function of human effector CD8+ T cells. Blood 121(14):2638–2646

Kuksin CA, Minter LM (2015) The link between autoimmunity and lymphoma: does NOTCH signaling play a contributing role? Front Oncol 5:51. https://doi.org/10.3389/fonc.2015.00051

Laky K, Evans S, Perez-Diez A, Fowlkes BJ (2015) Notch signaling regulates antigen sensitivity of naive CD4+ T cells by tuning co-stimulation. Immunity 42(1):80–94

Lehar SM, Dooley J, Farr AG, Bevan MJ (2005) Notch ligands Delta 1 and Jagged1 transmit distinct signals to T-cell precursors. Blood 105(4):1440–1447

Maekawa Y, Tsukumo S, Chiba S, Hirai H, Hayashi Y, Okada H, Kishihara K, Yasutomo K (2003) Delta1-Notch3 interactions bias the functional differentiation of activated CD4+ T cells. Immunity 19(4):549–559

Maekawa Y, Minato Y, Ishifune C, Kurihara T, Kitamura A, Kojima H, Yagita H, Sakata-Yanagimoto M, Saito T, Taniuchi I, Chiba S, Sone S, Yasutomo K (2008) Notch2 integrates signaling by the transcription factors RBP-J and CREB1 to promote T cell cytotoxicity. Nat Immunol 9(10):1140–1147

Mathieu M, Cotta-Grand N, Daudelin JF, Thebault P, Labrecque N (2013) Notch signaling regulates PD-1 expression during CD8(+) T-cell activation. Immunol Cell Biol 91(1):82–88

Mathieu M, Duval F, Daudelin JF, Labrecque N (2015) The notch signaling pathway controls short-lived effector CD8+ T cell differentiation but is dispensable for memory generation. J Immunol 194(12):5654–5662

Minter LM, Turley DM, Das P, Shin HM, Joshi I, Lawlor RG, Cho OH, Palaga T, Gattipati S, Telfer JC, Kostura L, Fauq AH, Simpson K, Such KA, Miele L, Golde TE, Miller SD, Osborne BA (2005) Inhibitors of gamma-secretase block in vivo and in vitro T helper type 1 polarization by preventing notch upregulation of Tbx21. Nat Immunol 6(7):680–688

Mota C, Nunes-Silva V, Pires AR, Matoso P, Victorino RM, Sousa AE, Caramalho I (2014) Delta-like 1-mediated notch signaling enhances the in vitro conversion of human memory CD4 T cells into FOXP3-expressing regulatory T cells. J Immunol 193(12):5854–5862

Mukherjee S, Schaller MA, Neupane R, Kunkel SL, Lukacs NW (2009) Regulation of T cell activation by notch ligand, DLL4, promotes IL-17 production and Rorc activation. J Immunol 182(12):7381–7388

Mukherjee T, Kim WS, Mandal L, Banerjee U (2011) Interaction between notch and Hif-alpha in development and survival of Drosophila blood cells. Science 332(6034):1210–1213

Oswald F, Rodriguez P, Giaimo BD, Antonello ZA, Mira L, Mittler G, Thiel VN, Collins KJ, Tabaja N, Cizelsky W, Rothe M, Kühl SJ, Kühl M, Ferrante F, Hein K, Kovall RA, Dominguez M, Borggrefe T (2016) A phospho-dependent mechanism involving NCoR and KMT2D controls a permissive chromatin state at Notch target genes. Nucleic Acids Res 44(10):4703–4720

Ou-Yang HF, Zhang HW, Wu CG, Zhang P, Zhang J, Li JC, Hou LH, He F, Ti XY, Song LQ, Zhang SZ, Feng L, Qi HW, Han H (2009) Notch signaling regulates the FOXP3 promoter through RBP-J- and Hes1-dependent mechanisms. Mol Cell Biochem 320(1–2):109–114

Palaga T, Miele L, Golde TE, Osborne BA (2003) TCR-mediated notch signaling regulates proliferation and IFN-gamma production in peripheral T cells. J Immunol 171(6):3019–3024

Perumalsamy LR, Nagala M, Banerjee P, Sarin A (2009) A hierarchical cascade activated by non-canonical notch signaling and the mTOR-Rictor complex regulates neglect-induced death in mammalian cells. Cell Death Differ 16(6):879–889

Perumalsamy LR, Nagala M, Sarin A (2010) Notch-activated signaling cascade interacts with mitochondrial remodeling proteins to regulate cell survival. Proc Natl Acad Sci U S A 107(15):6882–6887

Rand MD, Grimm LM, Artavanis-Tsakonas S, Patriub V, Blacklow SC, Sklar J, Aster JC (2000) Calcium depletion dissociates and activates heterodimeric notch receptors. Mol Cell Biol 20(5):1825–1835

Robey E, Chang D, Itano A, Cado D, Alexander H, Lans D, Weinmaster G, Salmon P (1996) An activated form of notch influences the choice between CD4 and CD8 T cell lineages. Cell 87(3):483–492

Roderick JE, Gonzalez-Perez G, Kuksin CA, Dongre A, Roberts ER, Srinivasan J, Andrzejewski C Jr, Fauq AH, Golde TE, Miele L, Minter LM (2013) Therapeutic targeting of NOTCH signaling ameliorates immune-mediated bone marrow failure of aplastic anemia. J Exp Med 210(7):1311–1329

Rothenberg EV, Ungerbäck J, Champhekar A (2016) Forging T-lymphocyte identity: intersecting networks of transcriptional control. Adv Immunol 129:109–174

Rutz S, Mordmuller B, Sakano S, Scheffold A (2005) Notch ligands Delta-like1, Delta-like4 and Jagged1 differentially regulate activation of peripheral T helper cells. Eur J Immunol 35(8):2443–2451

Rutz S, Janke M, Kassner N, Hohnstein T, Krueger M, Scheffold A (2008) Notch regulates IL-10 production by T helper 1 cells. Proc Natl Acad Sci U S A 105(9):3497–3502

Sade H, Krishna S, Sarin A (2004) The anti-apoptotic effect of Notch-1 requires p56lck-dependent, Akt/PKB-mediated signaling in T cells. J Biol Chem 279(4):2937–2944

Samon JB, Champhekar A, Minter LM, Telfer JC, Miele L, Fauq A, Das P, Golde TE, Osborne BA (2008) Notch1 and TGFbeta1 cooperatively regulate Foxp3 expression and the maintenance of peripheral regulatory T cells. Blood 112(5):1813–1821

Schmitt TM, Zúñiga-Pflücker JC (2002) Induction of T cell development from hematopoietic progenitor cells by delta-like-1 in vitro. Immunity 17(6):749–756

Schmitt TM, Ciofani M, Petrie HT, Zúñiga-Pflücker JC (2004) Maintenance of T cell specification and differentiation requires recurrent notch receptor-ligand interactions. J Exp Med 200(4):469–479

Shah DK, Zúñiga-Pflücker JC (2014) An overview of the intrathymic intricacies of T cell development. J Immunol 192(9):4017–4023

Shin HM, Minter LM, Cho OH, Gottipati S, Fauq AH, Golde TE, Sonenshein GE, Osborne BA (2006) Notch1 augments NF-kappaB activity by facilitating its nuclear retention. EMBO J 25(1):129–138

Shin HM, Tilahun ME, Cho OH, Chandiran K, Kuksin CA, Keerthivasan S, Fauq AH, Golde TE, Miele L, Thome M, Osborne BA, Minter LM (2014) NOTCH1 can initiate NF-κB activation via cytosolic interactions with components of the T cell signalosome. Front Immunol 5:249

Sierra RA, Thevenot P, Raber PL, Cui Y, Parsons C, Ochoa AC, Trillo-Tinoco J, Del Valle L, Rodriguez PC (2014) Rescue of notch-1 signaling in antigen-specific CD8+ T cells overcomes tumor-induced T-cell suppression and enhances immunotherapy in cancer. Cancer Immunol Res 2(8):800–811

Skokos D, Nussenzweig MC (2007) CD8- DCs induce IL-12-independent Th1 differentiation through Delta 4 notch-like ligand in response to bacterial LPS. J Exp Med 204(7):1525–1531

Sugimoto K, Maekawa Y, Kitamura A, Nishida J, Koyanagi A, Yagita H, Chiba S, Shimada M, Yasutomo K (2010) Notch2 signaling is required for potent antitumor immunity in vivo. J Immunol 184(9):4673–4678

Tacchini-Cottier F, Allenbach C, Otten LA, Radtke F (2004) Notch1 expression on T cells is not required for CD4+ T helper differentiation. Eur J Immunol 34(6):1588–1596

Tanaka S, Tsukada J, Suzuki W, Hayashi K, Tanigaki K, Tsuji M, Inoue H, Honjo T, Kubo M (2006) The interleukin-4 enhancer CNS-2 is regulated by notch signals and controls initial expression in NKT cells and memory-type CD4 T cells. Immunity 24(6):689–701

Tang XL, Smith TR, Kumar V (2005) Specific control of immunity by regulatory CD8 T cells. Cell Mol Immunol 2(1):11–19

Tran IT, Sandy AR, Carulli AJ, Ebens C, Chung J, Shan GT, Radojcic V, Friedman A, Gridley T, Shelton A, Reddy P, Samuelson LC, Yan M, Siebel CW, Maillard I (2013) Blockade of individual notch ligands and receptors controls graft-versus-host disease. J Clin Investig 123(4):1590–1604 van Tetering G, van Diest P, Verlaan I, van der Wall E, Kopan R, Vooijs M (2009) Metalloprotease ADAM10 is required for Notch1 site 2 cleavage. J Biol Chem 284(45):31018–31027

Vigouroux S, Yvon E, Wagner HJ, Biagi E, Dotti G, Sili U, Lira C, Rooney CM, Brenner MK (2003) Induction of antigen-specific regulatory T cells following overexpression of a notch ligand by human B lymphocytes. J Virol 77(20):10872–10880

Wilson A, MacDonald HR, Radtke F (2001) Notch 1-deficient common lymphoid precursors adopt a B cell fate in the thymus. J Exp Med 194(7):1003–1012

Wolfer A, Bakker T, Wilson A, Nicolas M, Ioannidis V, Littman DR, Wilson CB, Held W, Robson MacDonald H, Radtke F (2001) Inactivation of Notch1 in immature thymocytes does not perturb CD4 or CD8 T cell development. Nat Immunol 2(3):235–241

Wong KK, Carpenter MJ, Young LL, Walker SJ, McKenzie G, Rust AJ, Ward G, Packwood L, Wahl K, Dalriviere L, Hoyne G, Gibbs P, Champion BR, Lamb JR, Dallman MJ (2003) Notch ligation by Delta1 inhibits peripheral immune responses to transplantation antigens by a CD8+ cell-dependent mechanism. J Clin Invest 112(11):1741–1750

Yvon ES, Vigouroux S, Rousseau RF, Biagi E, Amrolia P, Dotti G, Wagner HJ, Brenner MK (2003) Overexpression of the notch ligand, Jagged-1, induces alloantigen-specific human regulatory T cells. Blood 102(10):3815–3821

Zhu J, Yamane H, Paul WE (2010) Differentiation of effector CD4 T cell populations. Annu Rev Immunol 28:445–489

Notch in Leukemia

Anna C. McCarter, Qing Wang, and Mark Chiang

Abstract

Notch is commonly activated in lymphoid malignancies through ligand-independent and ligand-dependent mechanisms. In T-cell acute lymphoblastic leukemia/lymphoma (T-ALL), ligand-independent activation predominates. Negative Regulatory Region (NRR) mutations trigger supraphysiological Notch1 activation by exposing the S2 site to proteolytic cleavage in the absence of ligand. Subsequently, cleavage at the S3 site generates the activated form of Notch, intracellular Notch (ICN). In contrast to T-ALL, in mature lymphoid neoplasms such as chronic lymphocytic leukemia (CLL), the S2 cleavage site is exposed through ligand-receptor interactions. Thus, agents that disrupt ligand-receptor interactions might be useful for treating these malignancies. Notch activation can be enhanced by mutations that delete the C-terminal proline (P), glutamic acid (E), serine (S), and threonine (T) (PEST) domain. These mutations do not activate the Notch pathway per se, but rather impair degradation of ICN. In this chapter, we review the mechanisms of Notch activation and the importance of Notch for the genesis and maintenance of lymphoid malignancies. Unfortunately, targeting the Notch pathway with pan-Notch inhibitors in clinical trials has proven challenging. These clinical trials have encountered dose-limiting on-target toxicities and primary resistance. Strategies to overcome these challenges have emerged from the identification and improved understanding of direct oncogenic Notch target genes. Other strategies have arisen from new insights into the "nuclear context" that selectively directs Notch functions in lymphoid cancers. This nuclear context is created by factors that co-bind ICN at cell-type specific transcriptional regulatory elements. Disrupting the functions of these proteins or inhibiting downstream oncogenic pathways might combat cancer without the intolerable side effects of pan-Notch inhibition.

Keywords

Notch · T-cell acute lymphoblastic leukemia · Chronic lymphocytic leukemia · MYC · AKT

A. C. McCarter
Cell and Molecular Biology Program, University of Michigan, Ann Arbor, MI, USA
e-mail: gormleya@med.umich.edu

Q. Wang · M. Chiang (✉)
Division of Hematology-Oncology, Department of Internal Medicine, University of Michigan, Ann Arbor, MI, USA
e-mail: qingwa@med.umich.edu; markchia@umich.edu

T. Borggrefe, B. D. Giaimo (eds.), *Molecular Mechanisms of Notch Signaling*, Advances in Experimental Medicine and Biology 1066, https://doi.org/10.1007/978-3-319-89512-3_18

Abbreviations

ADAM10 A Disintegrin And Metalloproteinase Domain-Containing Protein 10
AF4p12 ALL1-Fused Gene From Chromosome 4p12 Protein
AITL Angioimmunoblastic T-Cell Lymphomas
AKT AK Mouse Transforming
AMKL Acute Megakaryocytic Leukemia
AML Acute Myeloid Leukemia
AML1 Acute Myeloid Leukemia 1
APL Acute Promyelocytic Leukemia
APH1A Anterior Pharynx Defective 1 Homolog A
BBC3 BCL-2 Binding Component 3
BCL2 B-Cell CLL/Lymphoma 2
BCL6 B-Cell CLL/Lymphoma 6
BCL11B B-Cell CLL/Lymphoma 11B
BCR B-Cell Receptor
BHLH Basic Helix-Loop-Helix
BLK B Lymphocyte Kinase
BLNK B-cell Linker Protein
BM Bone Marrow
BPTES Bis-2-(5-Phenylacetamido-1,2,4-Thiadiazol-2-Yl) Ethyl Sulfide
BRD4 Bromodomain Containing 4
CARM1 Coactivator-associated Arginine Methyltransferase 1
CAS9 CRISPR Associated Protein 9
CBFβ Core Binding Factor β
CCNC Cyclin C
CCND1 Cyclin D1
CCND3 Cyclin D3
CCNE1 Cyclin E1
CCR7 C-C Chemokine Receptor Type 7
CDK3 Cyclin-dependent kinase 3
CDK4 Cyclin-dependent kinase 4
CDK6 Cyclin-dependent kinase 6
CDK8 Cyclin-dependent kinase 8
CDK19 Cyclin-dependent kinase 19
CDKN1B Cyclin Dependent Kinase Inhibitor 1B
CDKN2 Cyclin Dependent Kinase Inhibitor 2D
CHD4 Chromodomain Helicase DNA Binding Protein 4
ChIP Chromatin Immuno-precipitation
CLL Chronic Lymphocytic Leukemia
CR2 Complement C3d Receptor 2
CRISPR Clustered Regularly Interspaced Short Palindromic Repeats
CXCR4 C-X-C Chemokine Receptor Type 4
CYLD Ubiquitin Carboxyl-Terminal Hydrolase
DDX5 DEAD-box RNA Helicase 5
DEPTOR DEP Domain Containing MTOR Interacting Protein
DLL1 Delta-Like 1
DLL4 Delta-Like 4
DLX5 Distal-Less Homeobox 5
DN-MAML Dominant-Negative Mastermind
DNA Deoxyribonucleic Acid
DTX1 Deltex E3 Ubiquitin Ligase 1
E2A E2A Immunoglobulin Enhancer-Binding Factor E12/E47
EBNA2 Epstein–Barr Nuclear Antigen 2
EBV Epstein–Barr Virus
EIF2A Eukaryotic Translation Initiation Factor 2A
ERK Extracellular Signal-regulated Kinase
ETO Eight-Twenty One
ETS1 E26 Avian Leukemia Oncogene 1
ETP-ALL Early T-cell Precursor Acute Lymphoblastic Leukemia

FBXW7	F-Box and WD Repeat-Containing Protein 7	LEF1	Lymphoid Enhancer Binding Factor 1
FDA	Food and Drug Administration	LIC	Leukemia-Initiating Cells
		LMO2	Lim Domain Only 2
FL	Follicular Lymphoma	LN	Lymph Node
G-CSF	Granulocyte-Colony Stimulating Factor	LNR	Lin12/Notch Repeats
		LUNAR1	Leukemia-Associated Non-Coding IGF1R Activator RNA 1
GABPA	GA Binding Protein Transcription Factor Alpha Subunit	LSC	Leukemia Stem Cells
GATA3	GATA Binding Protein 3	LSD1	Lysine (K)-Specific Demethylase 1A
GATAD2B	GATA Zinc Finger Domain Containing 2B	LYN	Lck/Yes-Related Novel Protein Tyrosine Kinase
GEMM	Genetically Engineered Mouse Model Of T-ALL	MAL	Megakaryocytic Acute Leukemia
GI	Gastrointestinal		
GOF	Gain-Of-Function	MAML	Mastermind-Like
GSI	γ-Secretase Inhibitor	MCL	Mantle Cell Lymphoma
HD	Heterodimerization Domain	MCL1	Myeloid Leukemia Cell Differentiation Protein 1
HDAC1	Histone Deacetylase 1		
HEB	E2A/Hela E Box-Binding	MTOR	Mechanistic Target Of Rapamycin
HES1	Hairy And Enhancer Of Split 1	MYC	Myelocytomatosis proto-oncogene
HSC	Hematopoietic Stem Cell		
HSP90	Heat Shock Protein 90	N-ME	Notch-MYC enhancer
ICN1	Intracellular Notch1	NDME	Notch-dependent MYC enhancer
ID3	Inhibitor Of DNA Binding 3 HLH Protein	NEMO	NF-κB Essential Modulator
IGF1R	Insulin Like Growth Factor 1 Receptor	NF-κB	Nuclear Factor-κB
		NFκB2	Nuclear Factor Of Kappa Light Polypeptide Gene Enhancer In B Cells 2
IHC	Immunohistochemistry		
IKK	Inhibitor Of Kappa-B Kinase	NOS	Not Otherwise Specified
IKKα	Inhibitor Of Kappa-B Kinase Subunit Alpha	NRR	Negative Regulatory Region
IKKβ	Inhibitor Of Kappa-B Kinase Subunit Beta	NURD	Nucleosome Remodeling Deacetylase
IKZF1	Ikaros Family Zinc Finger 1	OTT	One-Twenty Two
IL7R	Interleukin 7 Receptor	p70S6K	Ribosomal protein S6 kinase beta-1
INK4	Inhibitor of Cyclin-Dependent Kinase 4	PB1	Polybromo 1
IRF4	Interferon Regulatory Factor 4	PBAF	Polybromo-Associated BRG1- Or HBRM-Associated Factors
KRAS	Kirsten Rat Sarcoma Viral Oncogene Homolog	PCAF	P300/CBP-Associated Factor
JME	Juxtamembrane Extracellular	PCR	Polymerase Chain Reaction
JUN	Ju-nana (Japanese number 17)	PDX	Patient-Derived Xenograft

PEST	Proline (P), Glutamic Acid (E), Serine (S), And Threonine (T)
PHF8	PHD Finger Protein 8
PI3K	Phosphatidylinositol 4,5-Bisphosphate 3-Kinase
PIAS	Protein Inhibitor of Activated STAT
PIK3CD	Phosphatidylinositol-4,5-Bisphosphate 3-Kinase Catalytic Subunit Delta
PIP3	Phosphatidylinositol(3,4,5)-trisphosphate P_3
PLCγ	Phospholipase Cγ
PML	Promyelocytic Leukemia Locus Gene
POFUT1	Protein O-Fucosyltrans-ferase 1
PRL2	Phosphatase of Regene-rating Liver
PTCRA invariant	preTα chain of the pre-T cell receptor
PTEN	Phosphatase And Tensin Homolog
QRT-PCR	Quantitative Real Time PCR
RARA	Retinoic Acid Receptor-Alpha
RAS	Rat Sarcoma virus oncogene
RB	Retinoblastoma Protein
RBBP4	RB Binding Protein 4, Chromatin Remodeling Factor
RBPJ	Recombination Signal Binding Protein For Immunoglobulin Kappa J Region
RELB	Relaxed B Proto-Oncogene
RNA	Ribonucleic Acid
RNAPII	RNA Polymerase II
RNF40	Ring Finger Protein 40
RUNX1	Runt Related Transcription Factor 1
SAHM	MAM-like Stapled Peptides
SERCA	Sarco/Endoplasmic Reticulum Calcium ATPase
SLL	Small Lymphocytic Lymphoma
SKP2	S-Phase Kinase Associated Protein 2
STUB1	STIP1 Homology And U-Box Containing Protein 1
SYK	Spleen Tyrosine Kinase
T-ALL	T-cell Acute Lymphoblastic Leukemia
TAL1	T-cell Acute Lymphoblastic Leukemia 1
TCF1	Transcription Factor 1
TCRB	T-cell Receptor β
TP53	Tumor Protein P53
TPR	Tetratricopeptide Repeat
TRIB2	Tribbles Pseudokinase 2
TSS	Transcriptional Start Site
UTR	Untranslated Region
ZFP36L1	Zinc Finger Protein C3H Type 36-Like 1
ZFP36L2	Zinc Finger Protein C3H Type 36-Like 2
ZMIZ1	Zinc Finger MIZ-Type Containing 1

1 General

Human *NOTCH1* was first recognized as an oncogene based on chromosomal translocations generating a constitutively active *NOTCH1* allele in T-cell acute lymphoblastic leukemia/lymphoma [T-ALL, (Ellisen et al. 1991)]. T-ALL is an aggressive lymphoid malignancy that arises from immature thymocytes in the thymus or the bone marrow (BM). These rare translocations were the initial clue that subsequently led others to identify much more common gain-of-function *NOTCH1* mutations (~60% of cases) in T-ALL (Weng et al. 2004). Subsequently, gain-of-function Notch mutations were identified in various types of mature lymphoid neoplasms. These cancers arise from mature B and T cells, which circulate in lymph nodes (LNs), spleen and other organs as part of the adaptive immune system. In contrast to its predominantly oncogenic role in lymphoid neoplasms, Notch might act primarily as a tumor suppressor in myeloid neoplasms. These cancers arise from developing or mature myeloid cells, such as monocytes, which are important for the innate immune system. In this chapter, we review the current understanding of Notch signaling in blood cancers. Because of the size of this topic and because Notch has tissue-dependent pleotropic functions, we will focus on experiments conducted in hematological cancer cells. We will also focus on T-ALL as the most insightful information has been generated for this cancer. We will be mindful of extending these insights to other hematological cancers given the context dependence of Notch signaling within the hematopoietic system.

2 Mechanisms of Notch Activation in Lymphoid Neoplasms

The most compelling evidence for Notch activation in cancer is the identification of acquired, recurrent, gain-of-function mutations in human Notch genes. In the absence of mutation, strong evidence can also be obtained through antibodies that recognize neoepitopes created by γ-secretase cleavage of NOTCH1 (Kluk et al. 2013) or NOTCH3 (Bernasconi-Elias et al. 2016), even if showing cleavage of NOTCH can be technically challenging. These antibodies have been useful to show NOTCH1 or NOTCH3 activation in the absence of mutations. *NOTCH2* and *NOTCH4* mutations that trigger cleavage are very rare (*NOTCH2*) or have not yet been detected (*NOTCH4*) in hematological malignancies. Antibodies recognizing cleaved NOTCH2 and NOTCH4 have not yet been developed. Thus, it has been difficult to test the possibility that NOTCH2 or NOTCH4 is cleaved (and thus activated) in cases where mutations are absent. Notch activation can also be inferred by inhibition mediated by γ-secretase inhibitors (GSI). However, these studies have to be interpreted carefully due to potential off-target effects.

2.1 Mutational Mechanisms of Notch Activation in T-ALL

After the initial discovery of recurrent *NOTCH1* mutations in human pediatric T-ALL (Weng et al. 2004), other large studies confirmed these findings and extended them to adult T-ALL [(Asnafi et al. 2009; Trinquand et al. 2013; Mansour et al. 2009; Jenkinson et al. 2013; Clappier et al. 2010; Kox et al. 2010) (Zhang et al. 2012; Van Vlierberghe et al. 2011; Neumann et al. 2013), Table 1]. To date, *NOTCH1* remains one of the most prevalent oncogenes across almost all T-ALL oncogenomic subtypes (Weng et al. 2004). The exception is early T-cell precursor ALL (ETP-ALL) where *NOTCH1* mutations are relatively less frequent than in conventional T-ALL (~11-38% of ETP-ALLs versus ~50–62% of conventional T-ALL). ETP-ALL is a primitive form of T-ALL with stem cell and/or myeloid features (Coustan-Smith et al. 2009).

NOTCH1 mutations in T-ALL cluster in two hotspots that dysregulate pathway activation through distinct mechanisms (Fig. 1). The first hotspot consists of single amino acid substitutions and in-frame insertions in the extracellular negative regulatory region (NRR mutations) and in-frame insertions in the juxtamembrane extracellular region (JME mutations). The NRR consists of the Lin12/Notch repeats (LNR, Exon 25) and the heterodimerization domain (HD, Exons 26 and 27). The NRR locks the Notch receptor in the inactive conformation (Sanchez-Irizarry et al. 2004), burying the S2 cleavage site within a hydrophobic pocket, thus autoinhibiting S2 cleavage by a disintegrin and metalloproteinase domain-containing protein 10 [ADAM10, (Gordon et al. 2007; Gordon et al. 2009)]. LNR mutations (e.g. H1545P) disengage the LNR clamp from the HD domain possibly through loss of calcium binding [(Gordon et al. 2009), Fig. 2A]. HD class 1A mutations (e.g. L1601P) separate the extracellular domain from the transmembrane domain (Malecki et al. 2006) while HD class 1B mutations (e.g. L1594P) cause conformational changes. HD class 2 mutations (e.g.

Table 1 *NOTCH1/NOTCH2* Gain-of-function mutations in lymphoid neoplasms

Lymphoid neoplasm	Gene	Frequency%	Combined Frequency%[a]	GOF Notch Mutation	References
T-ALL (pediatric)	*NOTCH1* *FBXW7*	50-62% 14-18%	54-65%	NRR PEST	Weng (2004); Jenkinson (2013); Clappier (2010); Kox (2010)
T-ALL (adult)	*NOTCH1* *FBXW7*	60-62% 18-24%	67-74%	NRR PEST	Asnafi (2009); Trinquand (2013); Mansour (2009)
ETP-ALL (pediatric)	*NOTCH1* *FBXW7*	11% 5%	13%	NRR PEST	Zhang (2012)
ETP-ALL (adult)	*NOTCH1* *FBXW7*	15-38% 3%	15-41%	NRR PEST	Van Vlierberghe (2011); Neumann (2013)
CLL/SLL	*NOTCH1* *FBXW7*	5-22%[b] 3%		PEST	Di Ianni (2009); Puente (2011); Puente (2015); Baliakas (2015); Balatti (2012); Sportoletti (2010); Fabbri (2011); Jeromin (2014); Rossi (2012a); Wang (2011b); Nadeu (2016); Rasi (2016)
Mantle cell lymphoma	*NOTCH1* *NOTCH2*	5-12% 5%		NRR (rare) PEST	Kridel (2012); Bea (2013)
Marginal zone lymphoma	*NOTCH2*	5-25%		NRR (rare) PEST	Kiel (2012); Rossi (2012b); Troen (2008)
Diffuse large B-cell lymphoma	*NOTCH1* *NOTCH2*	4-8% 20%		PEST	Lohr (2012); Arcaini (2015)
Splenic diffuse red pulp small B-cell lymphoma	*NOTCH1*	11%		PEST	Martinez (2016)
Adult T-cell leukemia/ lymphoma	*NOTCH1*	4-33%		PEST	Pancewicz (2010); Shimizu (2007)
Follicular lymphoma	*NOTCH1* *NOTCH2*	rare rare		PEST	Karube (2014); Krysiak (2016)

T-ALL, T-cell acute lymphoblastic leukemia/lymphoma; ETP-ALL, early T-cell precursor acute lymphoblastic leukemia; CLL, Chronic Lymphocytic Leukemia; SLL, Small lymphocytic lymphoma; GOF, gain-of-function; NRR, negative regulatory region; PEST, proline, glutamate, serine, and threonine; FBXW7, F-Box and WD repeat domain containing 7

[a]Combined frequency of either *NOTCH1* or *FBXW7* mutations; [b]1.5% in Ballatti (2012), the rest of studies were >5%

A1721_V1722InsARLGSLNIPYKIEA) or JME alleles (e.g. A1739_A1740InsQAVEPPPPA QLHFMYVA) insert short peptides that separate the NRR from the S2 site and/or membrane (Sulis et al. 2008; Malecki et al. 2006). All these ligand-independent changes, induced by mutations, expose the S2 cleavage site, leading to generation of the transcriptionally active form of Notch1, the intracellular Notch1 (ICN1).

The second hotspot of *NOTCH1* mutations consists of nonsense mutations or short insertion/deletions dispersed throughout the C-terminal PEST domain (Exon 34) (Fig. 2B). These mutations truncate the C-terminus through premature STOP codons. PEST truncations remove "degron" sites that are phosphorylated by serine/threonine kinases, such as the cyclin C- cyclin-dependent kinase 8 (CDK8)/cyclin-dependent kinase 19 (CDK19)/cyclin-dependent kinase 3 (CDK3) complexes and unknown kinases that phosphorylate a WSSSSP motif (Chiang et al. 2006; Fryer et al. 2004; Li et al. 2014). Phosphorylated degrons are recognized and

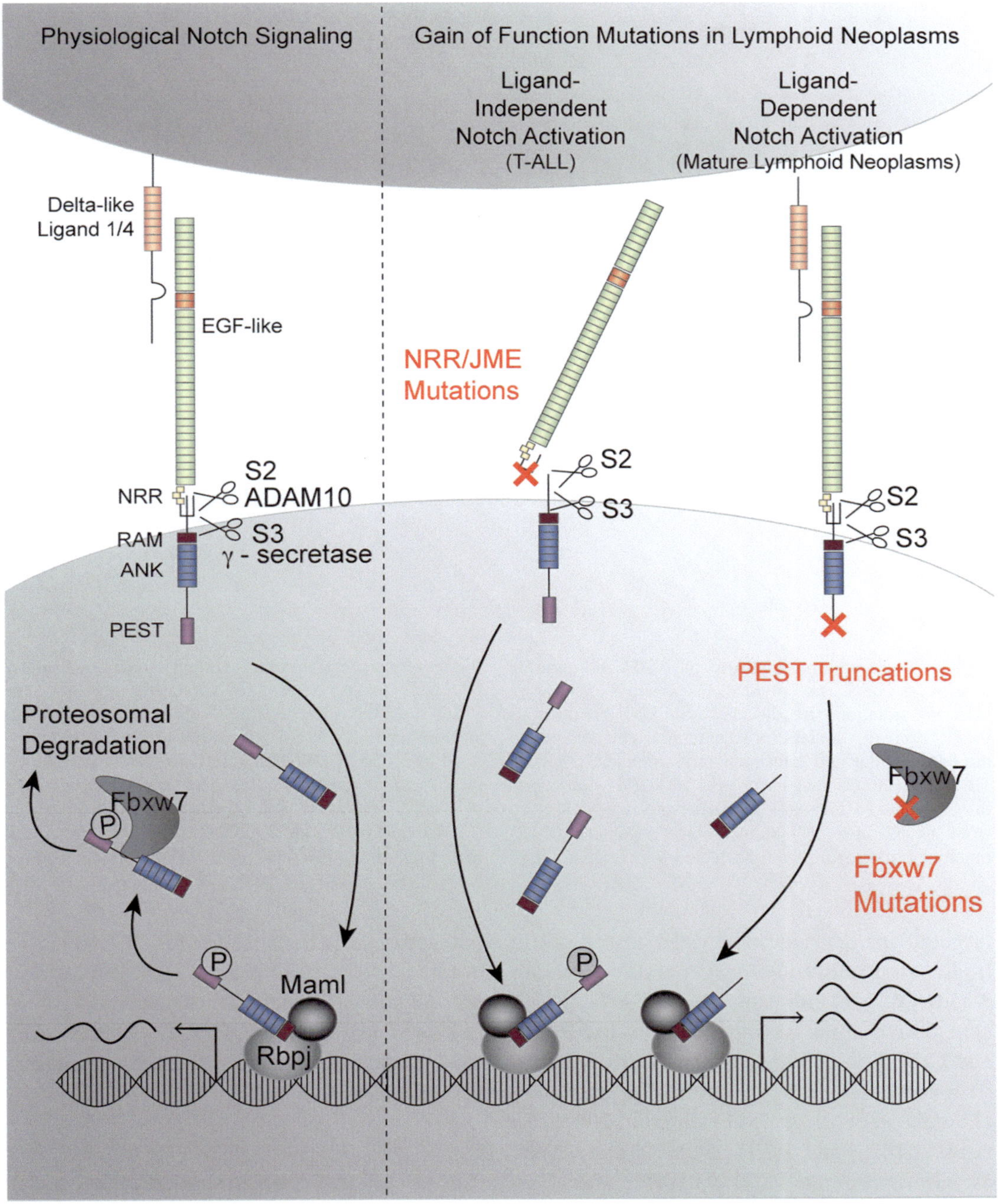

Fig. 1 Oncogenic Notch signaling in lymphoid malignancies. (left) Physiological Notch signaling. (right) Ligand-independent cleavage triggered by NRR NOTCH1 mutations in T-ALL and ligand-dependent cleavage of NOTCH1/2 in mature lymphoid neoplasms. Activation can be enhanced by PEST/FBXW7 mutations which increase half-life of intracellular Notch (ICN)1/2

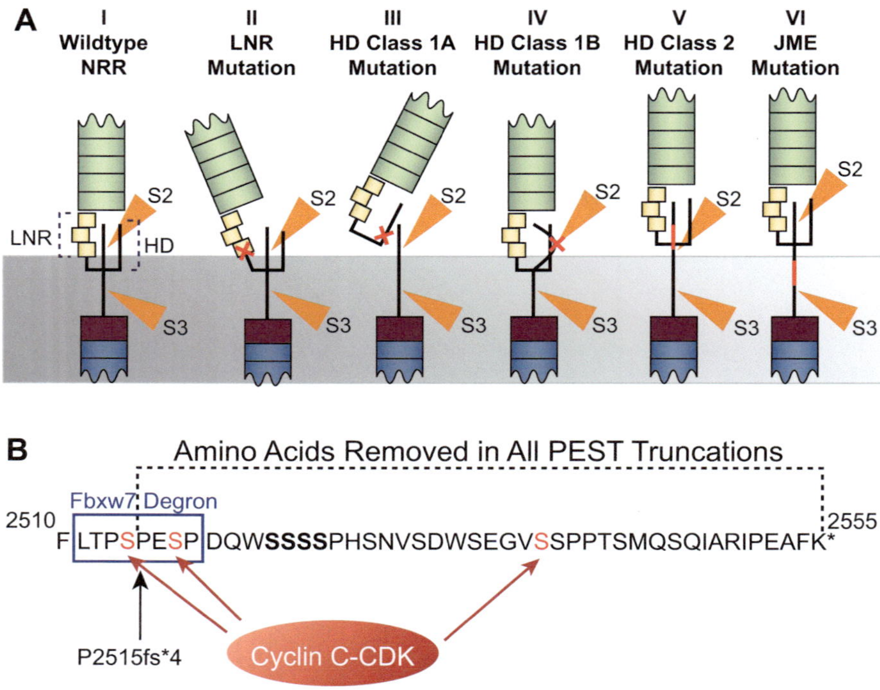

Fig. 2 Mutational mechanisms of NOTCH1 gain-of-function **A** The S2 proteolytic cleavage site is exposed by (1) LNR mutations that disengage the LNR clamp from the HD domain; (2) class 1A HD mutations that dissociate the extracellular and transmembrane domains; (3) class 1B HD mutations that disturb the NOTCH1 conformation; (4) class 2 HD mutations that separate the NRR away from the S2 site; (5) juxtamembrane expansion (JME) mutations that separate the NRR and S2 site away from the membrane. **B** C-terminal protein sequence of human NOTCH1 (amino acids 2510-2555) that are lost upon PEST mutations. The most C-terminal truncation occurs at amino acid 2516. S (*red*), serine residues phosphorylated by CCNC-CDK; **SSSS**, serine residues phosphorylated by unknown kinase; P2515fs*4 represents ~70-80% of all PEST mutations in CLL and results in a frameshift (fs) starting at Proline #2515 with a shift frame of 4 including the stop codon (*); FBXW7 degron motif (*box*)

targeted for degradation by the E3 ubiquitin ligase f-box and WD repeat containing protein 7 (FBXW7) (Thompson et al. 2007; O'Neil et al. 2007b). Inactivating mutations in *FBXW7* and *CCNC* (which encodes cyclin C) occur in ~14–24% and ~9% of patients, respectively (Thompson et al. 2007; O'Neil et al. 2007b; Li et al. 2014). Thus, PEST, FBXW7 and CCNC mutations prolong ICN1 half-life. The overlapping roles of PEST and FBXW7 mutations might explain why they are mutually exclusive. In contrast to NRR/JME mutations, PEST/FBXW7 mutations do not trigger ligand-independent Notch activation.

In ~20% of patients, NRR and PEST/FBXW7 mutations occur in cis, which cooperatively augments Notch signals (Weng et al. 2004). Altogether, *NOTCH1* and/or *FBXW7* mutations occur in ~54–74% of patients. In the absence of rat sarcoma virus oncogene (RAS) and phosphatase and tensin homolog (PTEN) mutations, they are associated with favorable prognosis (Trinquand et al. 2013).

2.2 Ligand-Dependent Notch Activation in T-ALL

2.2.1 Evidence for Ligand-Dependent Notch Signaling

Ligand interactions with unmutated receptors may trigger Notch activation in T-ALL. Mouse studies predict that circulating human T-ALL blasts will encounter the NOTCH1 ligand Delta-like 4 (DLL4) on thymic epithelial cells (Koch et al. 2008; Hozumi et al. 2008); fibroblastic reticular cells and follicular dendritic cells in LNs (Fasnacht et al. 2014; Chung et al. 2017); endothelial cells (Ramasamy et al. 2014); and bone-producing osteocalcin-expressing cells in the BM (Yu et al. 2015). Although NRR-mutated Notch receptors activate Notch signaling constitutively, they also respond to ligands based on *in vitro* cell-based reporter assays (Malecki et al. 2006). Accordingly, patient-derived xenografts (PDX) with NRR-mutated or wildtype NOTCH1 receptors respond to ligand stimulation in proliferation and leukemia initiating assays (Armstrong et al. 2009). ~23% of *NOTCH1* mutations are PEST mutations without an NRR mutation in cis (Weng et al. 2004). PEST mutations do not activate the Notch pathway per se. Thus, these receptors would require ligand to trigger activation (Malecki et al. 2006; Weng et al. 2004). NRR mutations that trigger cleavage of NOTCH3 have not been observed. Nevertheless, cleaved NOTCH3 was identified in 2 of 40 primary T-ALL samples and in 12 of 24 PDX samples (Bernasconi-Elias et al. 2016). Presumably, cleavage of NOTCH3 had been triggered by ligand. If ligands are important, then Notch inhibitors might have antileukemic activity even when *NOTCH1* mutations are absent or subclonal. Accordingly, PDX samples with wildtype receptors were inhibited by an anti-Dll4 antibody in mice (Minuzzo et al. 2015). In a clinical trial, patients with wildtype receptors responded to GSI (Zweidler-McKay et al. 2014). These data suggest that ligand-receptor interactions might be important in T-ALL.

2.2.2 Evidence Against Ligand-Dependent Notch Signaling

In ~43% of human T-ALL BM specimens, ICN1 cannot be detected by immunohistochemistry [IHC, (Kluk et al. 2013)]. The authors concluded that these samples were unrelated to NOTCH1 activation. However, we note that this study did not perform *NOTCH1* mutational analysis. In other words, the presence or absence of ICN1 for each sample was not linked to the presence or absence of a *NOTCH1* mutation. If some ICN1-negative samples had PEST mutations, we hypothesize that Notch might have been activated by ligand in a different niche (e.g. thymus, spleen, or LN), or in an ancestral clone, or was activated by ligand at low levels below the sensitivity of the technique. Conversely, if some ICN1$^+$ samples lacked NRR mutations, we hypothesize that ligand was triggering cleavage.

2.3 Ligand-Dependent Notch Activation in Mature Lymphoid Neoplasms

2.3.1 Mature B-cell Neoplasms

Gain-of-function *NOTCH1* mutations were identified in chronic lymphocytic leukemia [CLL, ~5–22%, (Di Ianni et al. 2009; Puente et al. 2011; Puente et al. 2015; Baliakas et al. 2015; Balatti et al. 2012; Sportoletti et al. 2010; Fabbri et al. 2011; Jeromin et al. 2014; Rossi et al. 2012a; Nadeu et al. 2016; Rasi et al. 2016)], mantle cell lymphoma [MCL, ~5–12%, (Kridel et al. 2012; Bea et al. 2013)], diffuse large B-cell lymphoma [~4-8%, (Lohr et al. 2012; Arcaini et al. 2015)], follicular lymphoma [FL, rare, (Karube et al. 2014; Krysiak et al. 2016)] and in splenic diffuse red pulp small B-cell lymphoma [~11%, (Martinez et al. 2016)]. *NOTCH2* mutations have been identified in MCL [~5%, (Bea et al. 2013)], diffuse large B-cell lymphoma [~20%, (Arcaini et al. 2015)], FL [rare, (Karube et al. 2014; Krysiak et al. 2016)] and splenic marginal zone lymphoma [~5–25%, (Kiel et al. 2012; Rossi et al. 2012b; Troen et al. 2008)]. It is possible that conventional sequencing methods underestimate the prevalence of subclonal Notch activation. Accordingly, a digital droplet polymerase chain reaction (PCR) based assay found a mutation rate of 53% in 88 CLL patients (Minervini et al. 2016). Further, Western blotting paired with immunofluorescence detects ICN1 expression in 51% of CLL blood samples with unmutated *NOTCH1* alleles (Fabbri et al. 2017). IHC studies of CLL lymph nodes are even more sensitive, detecting ICN1 expression in 85-90% of unmutated samples (Kluk et al. 2013; Onaindia et al. 2015). *NOTCH1* mutations in CLL are associated with aggressive features, such as shorter survival, chemotherapy resistance, and transformation to a higher-grade malignancy (Balatti et al. 2012; Fabbri et al. 2011).

The nature of Notch mutations in B-cell neoplasms is different than those in T-ALL. NRR mutations are rare. Instead, mutations almost exclusively target the PEST domain. Thus, Notch activation is predicted to rely on interactions with ligands expressed on stromal cells. Accordingly, culturing CLL cells *in vitro* causes rapid downregulation of ICN1 and Notch target genes [e.g. myelocytomatosis proto-oncogene (*MYC*), deltex e3 ubiquitin ligase 1 (*DTX1*), and hairy and enhancer of split 1 (*HES1*), (Arruga et al. 2014)], but culturing them on stromal cells that express Notch ligands reactivates the Notch pathway (Jitschin et al. 2015; Arruga et al. 2014; Fabbri et al. 2017). In contrast to T-ALL, PEST mutations are not dispersed throughout the PEST domain in fact, ~70–80% of the mutations in CLL are the same – p.P2515fs*4 (c.7544-7545delCT). In contrast to T-ALL, non-coding mutations in the *NOTCH1* 3'untranslated region (UTR) occur in ~2% of CLL patients (Puente et al. 2015; Larrayoz et al. 2016). These mutations introduce a new splice acceptor site in the 3'UTR, which excises the PEST-coding sequences. Similar to T-ALL, PEST truncations improve ICN1 protein stability (Arruga et al. 2014) and *FBWX7* mutations are mutually exclusive of PEST mutations. However, the prevalence of *FBWX7* mutations (~2.5%) might be lower than in T-ALL (Wang et al. 2011b; Jeromin et al. 2014). Definitive proof for the importance of ligand for B-cell lymphoma will require *in vivo* studies using antibodies that block ligand-receptor interactions.

2.3.2 Candidate Sources of Notch Ligands

For NOTCH1, human B-lymphoma cells might encounter the same sources of DLL4 ligands as described for T-ALL (Sect. 2.2.). For NOTCH2, candidate ligands are DLL1-expressing fibroblastic cells in secondary lymphoid organs, such as the spleen. These cells stimulate the Notch2-dependent development of marginal zone B cells and Esam$^+$ dendritic cells in mice (Fasnacht et al. 2014). Other possible sources are endothelial cells, which upregulate Jagged-1 in response to human and murine B-cell lymphomas (Dai et al. 2014). In PDX models of MCL, expression of Notch target genes was higher in the spleen compared to BM or blood (Ryan et al. 2016). In CLL patients, expression of Notch target genes was higher in LN compared to BM or blood (Arruga et al. 2014). An IHC study described 13 cases of CLL in which the tumor cells extended across the LN capsule. In 12 of these cases, ICN1 was present in the LN but dropped in the cells outside of it (Kluk et al. 2013). These data suggest that secondary lymphoid organs, not the BM and blood, contain the major cellular sources of Notch ligands in B-cell neoplasms.

2.3.3 Mature T-Cell Neoplasms

In contrast to T-ALL, Notch activation in mature T-cell neoplasms is uncommon. *NOTCH1* mutations occur in ~4-33% of a rare T-cell lymphoma type called human T-cell leukemia virus type 1-associated adult T-cell leukemia/lymphoma (Pancewicz et al. 2010; Shimizu et al. 2007). However, mutations have not been reported for other mature T-cell lymphomas. The prevalence of NOTCH1 activation in T-cell lymphoma might be underestimated by mutation rates (Kluk et al. 2013). Accordingly, IHC study for ICN1 showed that subsets of neoplastic cells express ICN1 in ~86% of angioimmunoblastic T-cell lymphomas (AITL), ~42% of peripheral T-cell lymphomas not otherwise specified (NOS), ~17% of cutaneous T-cell lymphomas and ~14% of anaplastic large cell lymphomas. Lymphoma cells outside the LN capsule show diminished ICN1, suggesting a role for ligands in the LN. Further, murine studies show that Dll4-expressing fibroblastic cells in the LNs drive the Notch1/2-dependent development of T follicular helper cells (Fasnacht et al. 2014). These cells are believed to be the cell-of-origin for AITL. Thus, ICN1 might be important for lymphoma initiation. The significance of ICN1 for lymphoma maintenance is unclear. The presence of ICN1 might simply reflect an activated T-cell state since ICN1 is normally generated for cytokine, differentiation and other immune responses ("Notch and T Cell Function – A Complex Tale"). Further, only subsets of lymphoma cells express ICN1. Thus, even if a subset of the cells were found to be dependent on Notch signals, the other cells would appear to be independent.

3 Role and Regulation of Notch Signaling in T-ALL

3.1 The Role of Notch Activation in T-ALL

3.1.1 Notch in T-ALL Initiation

NOTCH1 mutations can occur as an initiating event based on the detection of NRR mutations in peripheral blood cells at birth (Eguchi-Ishimae et al. 2008). These mutations occur prior to the acquisition of other oncogenic events and well in advance of the clinical appearance of childhood T-ALL. Further, the t(7;9) translocation occurs during normal T-cell development when the T-cell receptor β gene (*TCRB*) recombines at the pre-T-cell stage. Aberrant recombination creates the t(7;9) translocation, which juxtaposes the deoxyribonucleic acid (DNA) sequence encoding part of the transmembrane NOTCH1 subunit with the *TCRB* regulatory sequences. This rearrangement generates either ICN1 (Ellisen et al. 1991) or a membrane-tethered version of ICN1 (Palomero et al. 2006a). ICN1 is a potent initiator of T-ALL in mouse models (Pear et al. 1996; Aster et al. 2000; Deftos et al. 2000); however, t(7;9) is extremely rare compared to NRR mutations. Like ICN1 alleles, NRR mutant alleles can transactivate Notch reporters *in vitro* and induce ectopic T-cell development in the BM from hematopoietic stem cells (HSCs) in mouse models (Chiang et al. 2008). However, NRR alleles have much weaker effects than ICN1 alleles and are weak initiators of T-ALL. The NRR mutations are eventually lost from the HSCs and T-cell progenitor compartments because of excessive T-cell commitment (Chiang et al. 2013). Thus, Notch1 mutations can initiate T-ALL only in limited contexts.

3.1.2 Notch in T-ALL Progression

The role of Notch during T-ALL progression is based on the finding that *NOTCH1* mutations can be subclonal or occur at relapse but not at diagnosis (Mansour et al. 2007). By this measure, Notch activation frequently occurs during progression. A large sequencing study found that *NOTCH1* mutations are subclonal in ~44% of the cases with *NOTCH1* mutations (Liu et al. 2017). However, these data seem at odds with an IHC study showing that all cases of T-ALL (N=14) were diffusely positive for ICN1 [>80% of cells, (Kluk et al. 2013)]. Thus, *NOTCH1* mutations but not NOTCH1 activation might be subclonal. To reconcile these data, it is possible that intratumoral heterogeneity can occur with some cells activating NOTCH1 through ligand-receptor interactions while other cells are also activating NOTCH1 through mutations. This possibility would be consistent with observations of several transgenic mouse lines that ectopically express oncogenes associated with human T-ALL (Aster et al. 2007). In these mice, the transgenes initiate T-ALL in the thymus where Notch1 is activated by Dll4 expressed on stromal cells. Subsequently, Notch signals are raised during disease progression by the spontaneous acquisition of *Notch1* mutations. In one of these models, developing T-ALL cells were forced to express a dominant-negative mastermind (DN-MAML) transgene that blocked Notch signals. In response, the cells deleted DN-MAML rather than finding alternatives for Notch signals (Chiang et al. 2016). Thus, Notch activation can be essential for progression.

3.1.3 Notch in T-ALL Maintenance

In established Notch1-activated T-ALL cells, Notch withdrawal frequently induces G1/S arrest and sometimes apoptosis (Weng et al. 2004; O'Neil et al. 2006; Weng et al. 2003). Notch inhibition also downregulates glycolysis and glutaminolysis and increases autophagy (Herranz et al. 2015). The target genes that are responsible for these effects are discussed in Sect. 3.2. Would Notch inhibition have curative potential by targeting leukemia stem cells (LSCs)? LSCs are the subset of cells responsible for propagating the cancer. LSCs are measured by transplanting leukemia cells at limiting dilution into recipient mice and estimating the number of leukemia-initiating cells (LICs) based on the fraction of mice that develop leukemia. Notch activation drives the development of LICs in mouse models of Notch-induced T-ALL (Chiang et al. 2013; Li et al. 2008). Inhibiting Notch signals with GSI decreases LICs in patient-derived xenograft models (Armstrong et al. 2009) and a T-cell acute lymphoblastic leukemia (Tal1)/Lim domain only 2 (Lmo2)-induced mouse model of T-ALL (Tatarek et al. 2011). DN-MAML also reduces LICs in the Kirsten rat sarcoma viral oncogene homolog $(\text{Kras})^{\text{G12D}}$-induced mouse model of T-ALL (Chiang et al. 2016). These data suggest that Notch inhibitors could contribute to curative therapy by targeting LSCs.

3.2 Oncogenic Direct Notch Target Genes and Pathways in T-ALL

Identifying direct Notch target genes with oncogenic functions is important given the on-target toxicities and primary resistance seen with pan-Notch inhibitors. Inhibiting important Notch target proteins rather than Notch itself might avoid toxicities and bypass resistance. We are mindful that Notch regulates target genes in a context-dependent manner, even within the confines of T-ALL transcriptomes. Of note, Notch likely regulates several important targets only indirectly, such as C-C chemokine receptor type 7 (*CCR7*) (Buonamici et al. 2009), immunoglobulin enhancer-binding factor E12/E47 protein (E2A, encoded by *TCF3*) (Nie et al. 2003), phosphatase of regenerating liver (*PRL2*) (Dong et al. 2014; Dong et al. 2012; Kobayashi et al. 2016) and tumor protein p53 (*TP53*) (Beverly et al. 2005).

To date, the direct NOTCH1 target genes that have been shown to maintain T-ALL proliferation are *MYC, HES1*, interleukin 7 receptor (*IL7R*), insulin like growth factor 1 Receptor (*IGF1R*), leukemia-associated non-coding IGF1R activator ribonucleic acid 1 (*LUNAR1*), cyclin D3 (*CCND3*), and nuclear factor of kappa light polypeptide gene enhancer in B cells 2 (*NFκB2*)/relaxed B proto-oncogene (*RELB*). Less well understood are DEP domain containing MTOR interacting protein (*DEPTOR*) (Hu et al. 2016), tribbles pseudokinase 2 (*TRIB2*) (Sanda et al. 2012; Wouters et al. 2007) and *NOTCH3* (Bellavia et al. 2000; Wang et al. 2011a). The evidence for each of these targets is summarized in Table 2. To show that these target genes were direct, earlier studies used chromatin immunoprecipitation (ChIP) assays at promoter regions. However, the promoter ChIP assay has been supplanted by ChIP-Seq, which has shown that ICN1 dynamically binds proximal and/or distal enhancers of target genes with greater frequency and often-greater affinity than the promoter region (Margolin et al. 2009; Wang et al. 2014).

3.2.1 PI3K/AKT/mTOR

Notch1 was initially believed to activate the phosphatidylinositol 4,5-bisphosphate 3-kinase (PI3K)/Ak mouse transforming (AKT)/ mechanistic target of rapamycin (MTOR) pathway by inducing *MYC* (Chan et al. 2007). Subsequent reports showed that several other direct Notch1 target genes converge on this pathway – *HES1* (Palomero et al. 2007), *IGF1R* (Medyouf et al. 2011), *LUNAR1* (Trimarchi et al. 2014), *IL7R* (Gonzalez-Garcia et al. 2009), invariant preTα chain of the pre-T cell receptor (*PTCRA*) (Reizis and Leder 2002) and possibly *DEPTOR* (Hu et al. 2016). We are mindful that these target genes have roles in PI3K/AKT-independent pathways but have grouped them here to emphasize the importance of PI3K/AKT as a downstream oncogenic mediator of converging Notch signals (Palomero et al. 2007; Cullion et al. 2009; Chiarini et al. 2009; Avellino et al. 2005; Herranz et al. 2015).

3.2.1.1 Pten Inactivation Through Hes1

HES1, a basic helix-loop-helix (bHLH) transcription factor, induces the PI3K/AKT pathway by suppressing the transcription of *PTEN* (Palomero et al. 2007). PTEN blocks PI3K signaling by dephosphorylating phosphatidylinositol (3,4,5)-trisphosphate P_3 (PIP3) which inactivates AKT (Palomero et al. 2007). Notch inhibition with GSI has broad transcriptional effects on metabolic genes including those

Table 2 Oncogenic Direct NOTCH1 Target Genes in T-ALL[a] (For references, see text – Sect. 3.2)

	BOX1: Assays testing direct regulation by NOTCH1				**BOX2**: Assays testing importance for Notch-activated T-ALL maintenance			
Direct NOTCH1 target gene	Location of major ICN1/ RBPJ peak(s)[b]	GSI Wash-out Assay[c]	Luciferase Reporter Assay[d]	Induces T-ALL in GEMM	Human T-ALL cell lines (*in vitro*)	Rescues GSI-treated T-ALL cell lines (*in vitro*)[e]	Notch-activated GEMM (*in vivo*)	Primary human T-ALL[f]
MYC	+1.4Mb	x	x	x	x	x	x	both
HES1	promoter region	x	x	x	x		x	*in vitro*
IL7R	+27Kb; +43Kb	x	x	g	x	x		*in vitro*
IGF1R	intron 20 enhancer	x	x		x		x	*in vitro*
LUNAR1	intron 20, *IGFR1R*		x		x			
CCND3	intron 1 enhancer				x	x	x	*in vivo*
NFkB2	exon 14 enhancer				x		x	
RELB	promoter region				x		x	
DEPTOR	promoter region[h]				x			
TRIB2	intron 2 enhancer				x			
NOTCH3	intron 1 enhancer	x		x	x[i]			

T-ALL, T-cell acute lymphoblastic leukemia/lymphoma; GEMM, Genetically engineered mouse model of T-ALL; GSI, γ-secretase inhibitor; ICN1, Intracellular Notch1

[a]defined as genes directly regulated by ICN1 binding (**BOX1**) and for which inhibition of encoded proteins (e.g. genetically or with pharmacological inhibitors, **BOX2**) reduces Notch-activated T-ALL proliferation; [b]Peak locations in CUTLL1 extracted from GSM1252936 (Wang et al. 2014) relative to transcriptional start site (except for *DEPTOR*[h]); [c]details in (Bailis et al. 2014), CUTLL1 expression data extracted from (Wang et al. 2011a); [d]using genomic fragment containing the ICN1 peak(s) with adequate controls; [e]full or partial rescue of GSI-mediated growth inhibition when target gene is ectopically expressed; [f]*in vitro*, *in vivo* or both; [g]mature lymphoma; [h]by ChIP-qPCR not ChIP-Seq (Hu et al., 2016); [i]only the *NOTCH3*-mutated TALL-1 cell line (Bernasconi-Elias et al. 2016).

important for nucleic acid and amino acid biosynthesis, protein translation and ribosome biosynthesis (Herranz et al. 2015). However, *Pten* inactivation largely reverses these effects in mouse models of Notch1-induced T-ALL (Herranz et al. 2015).

3.2.1.2 Cell Surface Receptors that Activate PI3K

Direct Notch target genes that encode receptors presumably activate AKT through canonical receptor-ligand interactions [*IL7R* and *IGF1R* (Johnson et al. 2008)] or through tonic signals [(*PTCRA* (Sade et al. 2004)]. NOTCH1 binds a pair of distal enhancer sites 3' of the *IL7R* gene (Wang et al. 2014). Enforced expression of IL7R rescues withdrawal of Notch signals in a human T-ALL cell line (Gonzalez-Garcia et al. 2009). High levels of IL7-IL7R interactions can also drive peripheral B and T-cell lymphomas in mice, although not T-ALL (Rich et al. 1993; Abraham et al. 2005). IL7 is required to grow primary human T-ALL cells in culture (Barata et al. 2004; Armstrong et al. 2009). The importance of IL7R is further highlighted by the discovery of ~10% of patients with activating *IL7R* mutations

(Zenatti et al. 2011; Shochat et al. 2011). NOTCH1 directly binds an intronic enhancer in the human *IGF1R* gene to induce its transcription (Medyouf et al. 2011). Genetic inactivation of mouse *Igf1r* inhibited LIC activity in a mouse model of Notch1-induced T-ALL (Medyouf et al. 2011). Antibodies or drugs that block IGF1R signals reduced the proliferation of human T-ALL cell lines and patient-derived xenograft cells. Interestingly, the intronic enhancer that regulates *IGF1R* also regulates *LUNAR1*, which encodes an enhancer-like long noncoding ribonucleic acid (RNA) that induces *IGF1R* transcription through long-range interactions (Trimarchi et al. 2014). Silencing *LUNAR1* inhibited the proliferation of human T-ALL cell lines *in vitro* and *in vivo*. Thus, NOTCH1 induces *IGF1R* directly and indirectly through LUNAR1. Pre-T cell receptor signals are important for leukemia initiation in mouse models of Notch-induced T-ALL (Allman et al. 2001; Campese et al. 2006; Bellavia et al. 2002). However, the relevance of these signals for human T-ALL maintenance is unclear.

3.2.2 Myc

3.2.2.1 Myc in T-ALL Initiation and Progression

MYC is a bHLH transcription factor regulating diverse target genes important for proliferation and metabolism (Palomero et al. 2006b). ICN1 binds strongly to the Notch-dependent and T-cell specific *Myc* enhancer ("N-ME" or "NDME") located at +1.4 MB downstream of the transcriptional start site [TSS, (Herranz et al. 2014; Yashiro-Ohtani et al. 2014)]. Notch may also stabilize Myc protein indirectly through activation of the PI3K/AKT pathway (Bonnet et al. 2011; Palomero et al. 2007). Ectopic expression of Myc drives T-ALL initiation in mice (Felsher and Bishop 1999) and zebrafish (Langenau et al. 2003). These leukemias lack *Notch1* mutations or cleaved ICN1. Retroviral insertional mutagenesis in mice showed mutual exclusivity of *Notch1* and *Myc* insertions in virtually all murine T-ALL tumors (Sharma et al. 2006; Uren et al. 2008). Similarly, human T-ALLs with *MYC* translocations are typically devoid of *NOTCH1* mutations (La Starza et al. 2014). Thus, Myc can substitute for ICN1 for leukemia initiation. Myc is also required for initiation as its genetic inactivation prevents the development of Notch1-induced murine T-ALL (Herranz et al. 2014; Li et al. 2008). In the $Kras^{G12D}$-induced mouse model of T-ALL, spontaneous *Notch1* mutations are acquired in >90% of tumors (Chiang et al. 2008). However, enforced expression of *Myc* relieves the selective pressure for Notch activation, leading to T-ALLs that lack *Notch1* mutations or cleaved ICN1 (Chiang et al. 2016). Thus, the Myc protein can replace activated Notch in T-ALL progression. In contrast, enforced expression of other Notch targets, such as *Hes1* (Dudley et al. 2009) or *Akt1* in the transgenic $Kras^{G12D}$ mouse model (Chiang et al. 2016), leads to T-ALLs that retain *Notch1* mutations or cleaved ICN1. In this regard, *MYC* appears to hold a unique role among Notch target genes during initiation and progression. However, the role of *MYC* during maintenance seems somewhat less prominent (next section).

3.2.2.2 Myc in T-ALL Maintenance

Genetic inactivation of the Notch-driven *Myc*-enhancer in established murine Notch1-induced T-ALLs causes tumor regression (Herranz et al. 2014). Similarly, *MYC* transcription can be reduced through genetic or pharmacological inhibition of bromodomain containing 4 (BRD4). This inhibits proliferation and LIC activity in mouse models of Notch-induced T-ALL and patient-derived xenografts (King et al. 2013; Roderick et al. 2014; Knoechel et al. 2014; Loosveld et al. 2014). In contrast to its unique role during initiation and progression of T-ALL, Myc does not appear to be the dominant contribution of Notch1 during maintenance. For example, enforced expression of *MYC* only rescues a subset of human T-ALL cell lines from withdrawal of Notch signals (Weng et al. 2006; Palomero et al. 2006b). Similarly, enforced expression of *Myc* only rescues a subset of transgenic $Kras^{G12D}$-induced murine T-ALL cell lines from Notch withdrawal (Chiang, unpublished observations). Overexpression of Myc in an ICN1-initiated mouse model cannot maintain T-ALL when ICN1 is withdrawn (Demarest et al. 2011). Finally, a patient whose blasts overexpressed MYC independently of Notch nevertheless achieved a complete remission upon GSI treatment (Knoechel et al. 2015). Thus, Myc seems to be a dominant contributor for Notch-driven T-ALL maintenance in only some contexts.

3.2.3 NF-κB

The nuclear factor-κB (NF-κB) pathway can be blocked using bortezomib (Vilimas et al. 2007) or more specifically with a peptide inhibitor that disrupts the binding of NF-κB essential modulator (NEMO) to inhibitor of kappa-B kinase subunit beta (Ikkβ) (Espinosa et al. 2010). Both agents inhibit the proliferation of Notch-activated human T-ALL cell lines and maintenance of Notch1-induced T-ALL mouse models. NOTCH1 activates the NF-κB pathway by directly binding the *NFκB2* and *RELB* loci and inducing their transcription (Wang et al. 2014; Vilimas et al. 2007). NOTCH1 can also activate the NF-κB pathway indirectly by inducing HES1 protein expression, which represses the expression of ubiquitin carboxyl-terminal hydrolase (CYLD), which antagonizes inhibitor of kappa-B kinase (IKK). Elevated IKK then degrades IκBα, activating the NF-κB pathway (Espinosa et al. 2010). In activated murine T cells, ICN1 promotes the nuclear retention of NF-κB factors, possibly by directly binding and competing for inhibitor of nuclear factor kappa-B kinase subunit alpha (IκBα) (Shin et al. 2006). Thus, Notch promotes the NF-κB pathway through multiple mechanisms.

3.2.4 Hes1

ICN1 binds directly to the promoter region of *HES1* in human T-ALL cell lines (Wang et al. 2014). Enforced expression of Hes1 induces T-ALL in mouse models (Dudley et al. 2009) and conversely, its genetic inactivation in mouse models of Notch-induced T-ALL impairs leukemia initiation and maintenance (Wendorff et al. 2010; Schnell et al. 2015). While HES1 promotes the PI3K/AKT (Sect. 3.2.1) and NF-κB (Sect. 3.2.3) pathways, it suppresses transcription of BCL-2 binding component 3 [*BBC3,* which encodes the Puma pro-apoptotic protein (Schnell et al. 2015)]. Given the diverse mechanisms by which HES1 drives T-ALL proliferation, there has been considerable interest in finding ways to therapeutically target HES1. Investigators identified drugs that phenocopy the gene expression changes induced by genetic deletion of *Hes1* (Schnell et al. 2015). In detail, perhexiline, an inhibitor of mitochondrial carnitine palmitoyltransferase-1, induces antileukemic effects on primary human T-ALL cells and in a mouse model of Notch-induced T-ALL (Schnell et al. 2015). Perhexiline is better tolerated than GSI in clinical studies and it is being used to treat cardiac disease. While this is promising, the target of this drug within the Hes1 pathway remains unknown.

3.2.5 Cell Cycle

Notch inhibition induces G1/S cycle arrest (Weng et al. 2004; Weng et al. 2003). In fact, it is known that Notch drives the G1/S transition by binding enhancers in the *CCND3* locus (Wang et al. 2014) and inducing its transcription (Joshi et al. 2009) as well as by regulating Myc (Bretones et al. 2015). Additionally, Notch indirectly induces cyclin-dependent kinase 4 (*CDK4*) and cyclin-dependent kinase 6 (*CDK6*) and represses transcription of cyclin-dependent kinase inhibitors cyclin dependent kinase inhibitor 2d (*CDKN2D*) and cyclin dependent kinase inhibitor 1B (*CDKN1B*) presumably through indirect mechanisms (Rao et al. 2009). Inactivation of *Ccnd3* impairs initiation and maintenance of Notch1-induced T-ALL in mouse models (Sicinska et al. 2003; Choi et al. 2012) while enforced expression of CDK4 or CDK6, in combination with CCND3, partially rescues human T-ALL cell lines from Notch inhibition (Joshi et al. 2009). CDK4/6 inhibitors block the proliferation of human T-ALL cell lines and primary cells both *in vitro* and *in vivo* (Choi et al. 2012; Sawai et al. 2012; Rao et al. 2009; Pikman et al. 2016). Finally, Myc and Notch directly induce the transcription of S-phase kinase associated protein 2 (Skp2) in T-ALL and *Skp2* encodes an ubiquitin ligase that promotes cell cycle progression by targeting the Cdk inhibitor p27Kip1 for degradation

(Dohda et al. 2007). Thus, Notch promotes cell cycle progression through multiple mechanisms.

3.3 Regulation of Notch1 Expression in T-ALL

It is important to understand the pathways and factors that regulate Notch1 expression in order to devise new strategies to downregulate the Notch pathway for therapeutic purposes. Since the regulators of Notch expression are frequently tissue-specific, such strategies might have less toxicity compared to pan-Notch inhibitors. However, only a few studies have explored this possibility.

3.3.1 Direct Regulators of Notch1 Transcription in T-ALL

Notch1 directly auto-regulates its own transcription in conjunction with E2A transcription factors in murine thymocytes and T-ALL cell lines (Yashiro-Ohtani et al. 2009). Ectopic overexpression of inhibitor of DNA binding 3 HLH protein (Id3), an inhibitor of E2A, reduced *Notch1* transcripts and inhibited the growth of murine T-ALL cell lines (Yashiro-Ohtani et al. 2009). Targeting E2A might seem attractive as E2A-deficient mice are generally healthy with defects limited to lymphoid development (Bain et al. 1994; Bain et al. 1997). However, chronic inhibition might be problematic as E2A-deficient mice frequently develop T-ALL. The homeobox transcription factor distal-less homeobox 5 (Dlx5) binds an enhancer downstream of *Notch1* and an intronic enhancer in *Notch3* (Tan et al. 2017), inducing their transcription and driving T-ALL in transgenic mice. Targeting Dlx5 might be challenging as Dlx5-deficient mice develop multiple craniofacial abnormalities and do not survive post birth (Depew et al. 1999).

3.3.2 Post-Transcriptional Regulators of Notch1 Expression in T-ALL

RNA binding proteins zinc finger protein C3H type 36-like 1 (ZFP36L1) and zinc finger protein C3H type 36-like 1 (ZFP36L2) bind conserved AU-rich regulatory elements in the 3′ UTR of Notch1 mRNA, impairing its stability and translation (Hodson et al. 2010). Mice deficient for these proteins develop T-ALL associated with Notch1 overexpression (Hodson et al. 2010). Similarly, MiR-101 binds the 3' UTR of *NOTCH1* and reduces NOTCH1 protein levels when overexpressed at high levels (Qian et al. 2016). Since MiR-101 levels are relatively repressed in human T-ALL samples, this might help promote NOTCH1 overexpression (Qian et al. 2016). Given that enhancing Notch-depleting mechanisms such as the Ccnc/Fbxw7 pathway (Sect. 2.1) might be challenging as a therapeutic strategy, one alternative approach would be to block the mechanisms that process or protect functional Notch receptors. For example, thapsigargin, a chemical inhibitor of sarco/endoplasmic reticulum calcium ATPase (SERCA), blocks the normal processing of wildtype and mutated Notch receptors in the endoplasmic reticulum and Golgi compartments, leading to depletion of ICN1, which then induces growth arrest of human T-ALL cell lines both *in vitro* and *in vivo* (Roti et al. 2013). In another example, the chaperone protein heat shock protein 90 (Hsp90) protects Notch proteins from E3 ligase STIP1 homology and U-box containing protein 1 (Stub1)-dependent degradation (Wang et al. 2017). Inhibitors of SERCA or HSP90 had anti-leukemic effects in human T-ALL xenografts without major toxicity (Wang et al. 2017; Roti et al. 2013) and similar effects would also be predicted for inhibitors of protein O-fucosyltransferase 1 (POFUT1), which is important for glycosylation of Notch receptors, as genetic inactivation of *POFUT1* impaired leukemia-associated NOTCH1 mutant signaling (McMillan et al. 2017). Since SERCA and HSP90 inhibitors did not cause the major toxicities associated with pan-Notch inhibition, mutated Notch receptors might have an increased dependence compared to their wildtype counterparts on post-translational regulators.

3.4 Downstream Regulation of Notch1 Signals in T-ALL

In the nucleus, ICN1 engages mastermind-like protein (Maml) and recombination signal binding protein for immunoglobulin kappa J region (Rbpj) to form the core Notch transcriptional complex. Additional regulatory proteins bind the core complex to amplify Notch activity. Mass spectrometry screens of proteins pulled down by an anti-MAML1 antibody in human T-ALL cells (KOPT-K1) or streptavidin bead pulldown of biotin-tagged ICN1 in murine T-ALL cells (Beko) identified proteins that bind the core complex, such as DEAD-box RNA helicase 5 (Ddx5/p68), an ATP-dependent RNA helicase (Lin et al. 2013; Jung et al. 2013). Knockdown of Ddx5 reduced Notch target gene transcription and proliferation of T-ALL cell lines both *in vitro* and *in vivo* (Lin et al. 2013; Jung et al. 2013). Another mass spectrometry study in a human T-ALL cell line (SUPT1) used an epitope-tagged ICN1 as bait (Yatim et al. 2012) and identified candidate interacting proteins such as histone demethylases [e.g. lysine (K)-specific demethylase 1A (LSD1) and PHD finger protein 8 (PHF8)], chromatin remodeling complexes [e.g. cohesin complex components and polybromo-associated BRG1- or HBRM-associated factors (PBAF) complex components – BRG1 and polybromo 1 (PB1)], co-activators [e.g. ALL1-fused gene from chromosome 4p12 protein (AF4p12)], histone acetylases [e.g. p300 and p300/CBP-associated factor (PCAF)], histone ubiquitinases [e.g. Bre1 subunit ring finger protein 40 (RNF40)] and transcription factors [e.g. B-cell CLL/lymphoma 11B (BCL11b), E2A/Hela E box-binding (HEB), runt related transcription factor 1 (RUNX1), NOTCH1, NOTCH2, and NOTCH3, (Yatim et al. 2012)]. Apart from PBAF, these Notch-interacting proteins were previously discovered by other investigators in other cell types [for more comprehensive review see (Borggrefe and Liefke 2012)]. It is unclear whether these proteins bind NOTCH1 directly but it is clear that BRG1, PB1, LSD1 and PHF8 are recruited by ICN1 to Notch binding sites. These cofactors catalyze chromatin changes, such as H3K9me1/2 demethylation and H3K27me2 demethylation and their genetic silencing leads to downregulation of oncogenic target genes (*IL7R, MYC,* and *HES1*) and slower proliferation of human T-ALL cell lines both *in vitro* and *in vivo* (Yatim et al. 2012). Coactivator-associated arginine methyltransferase 1 (CARM1) binds directly to ICN1, is recruited by the Notch complex to enhancers, methylates five arginine residues within the Notch transcriptional activation domain, and drives transcriptional activity at enhancers. Inhibiting the function of these cofactors, perhaps by disrupting their interaction with ICN1, might be an effective therapeutic strategy with eventually fewer side effects compared to pan-Notch inhibition.

3.4.1 Context-Dependent Regulation – Cooperativity

The Notch complex is not a "pioneer factor" that can activate enhancers by itself but requires cell type-specific nuclear contexts to transactivate and implement diverse functions (Bray 2016). ChIP-Seq analyses have identified several transcriptional regulators that might contribute to the nuclear context that promotes ICN1 activity in T-cell lineages. These factors frequently bind adjacent to ICN1 at enhancers and include GA binding protein transcription factor alpha Subunit (Gabpa), E26 avian leukemia oncogene 1 (Ets1), Runx1, Heb, E47, Tal1, GATA binding protein 3 (Gata3), zinc finger MIZ-type containing 1 (Zmiz1) and Ikaros (Palii et al. 2011; Sanda et al. 2012; Wang et al. 2011a; Wei et al. 2011; Pinnell et al. 2015; Geimer Le Lay et al. 2014; Wang et al. 2014). Like Notch1, several of them are important for normal T-cell development based on murine knockout studies (Yu et al. 2010; Muthusamy et al. 1995; Bories et al. 1995; Barton et al. 1998; Egawa et al. 2007; Xu et al. 2013; Wang et al. 1996; Garcia-Ojeda et al. 2013; Scripture-Adams et al. 2014; Hattori et al. 1996; Hosoya et al. 2009; Pinnell et al. 2015). Thus, the combinatorial action of "ICN1-adjacent" transcriptional regulators might create the favorable nuclear context that promotes ICN1 activity in T-ALL cells.

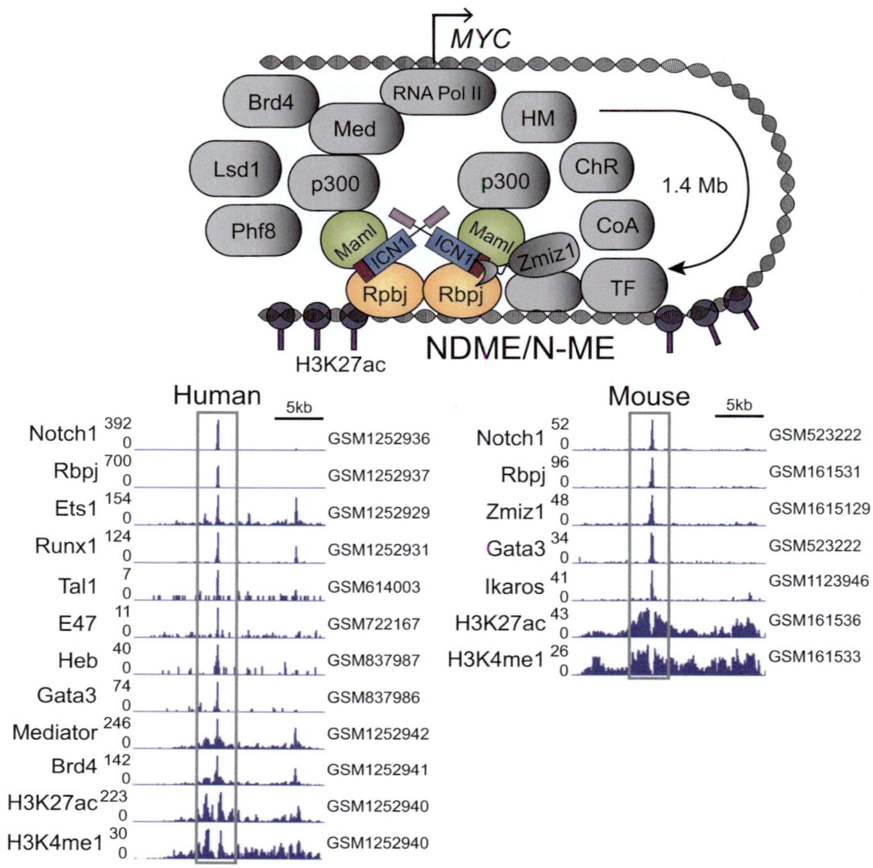

Fig. 3 Context-dependent regulation of the T-cell specific Notch-dependent *MYC* enhancer (NDME or N-ME). Transcription factors (TF), chromatin remodelers (ChR), histone modifiers (HM) and coactivators (CoA) bind to a +1.4 MB *MYC* enhancer in a large complex around the core ICN1/RBPJ. These factors might create a favorable nuclear context in T-ALL cells, at this specific enhancer, that allows ICN1/RBPJ to transactivate the *MYC* promoter through long-range interactions

3.4.1.1 The Notch-Dependent MYC Enhancer (NDME) in T-ALL

The prototypical, leukemia-relevant example of context-dependent regulation is the Notch-dependent Myc-enhancer NDME [also known as N-ME, (Sect3.2.2)]. This enhancer is active exclusively in T cells. In fact, it is inactive in non-T cells even if ICN is ectopically overexpressed (Herranz et al. 2014; Yashiro-Ohtani et al. 2014). What are the cooperating transcription factors than make this enhancer responsive to Notch signals? Several transcription factors bind the NDME, for example Ets1, Runx1, Heb, E47, Gabpa, Tal1, Zmiz1, Gata3 and Ikaros [(Palii et al. 2011; Sanda et al. 2012; Wang et al. 2011a; Pinnell et al. 2015; Wang et al. 2014), Fig. 3]. In theory, one could disrupt NDME functions in order to target Myc specifically in T cells probably avoiding the intolerable effects of pan-Notch inhibition. To show proof-of-principle, tissue-wide deletion of the 1.1 Kb region encompassing the N-ME in mice blocks *Myc* transcription and proliferation of T-ALL cells but has no effect on other cell types (Herranz et al. 2014). More precise clustered regularly interspaced short palindromic repeats (CRISPR)/CRISPR associated protein 9 (Cas9)-mediated excision of the NDME in a human T-ALL cell line reduced H3K27 acetylation, *MYC* transcription and cell proliferation (Ohtani et al. 2015). These studies suggest that targeting the transcriptional regulators that activate Notch responsive elements might be effective and safer than pan-Notch inhibition.

3.4.1.2 Targeting the Nuclear Context of ICN1 in T-ALL

How to target Notch responsive elements without pan-Notch inhibition? One option is to inhibit histone modifiers such as BRD4, a reader of acetylated histones that binds the NDME [Sect. 3.2.2, (Wang et al. 2014; Yashiro-Ohtani et al. 2014)] or, alternatively, to target protein-protein interactions between ICN1 and its cofactors. We showed that the protein inhibitor of activated STAT (PIAS)-like coactivator Zmiz1 directly binds ICN1 through a tetratricopeptide repeat (TPR) domain (Pinnell et al. 2015). Zmiz1 recruits ICN1/Rbpj to the NDME resulting in elevated H3K27ac and *Myc* transcription (Pinnell et al. 2015). In T-ALL, Zmiz1 regulates ~43% of Notch target genes and binds ~75% of overlapping ICN1/Rbpj sites, especially those enriched for Ets, Runx, Tal1/E2A and transcription factor 1 (Tcf1)/lymphoid enhancer binding factor 1 (Lef1) motifs (Pinnell et al. 2015). Accordingly, Zmiz1 also binds Ets1, perhaps forming a "bridge" that connects ICN1 to Ets1 (Helbig and Amsen 2015). Genetic inactivation of *Zmiz1* or disruption of the Notch1-Zmiz1 interaction using a dominant-negative TPR inhibitor, slowed leukemic proliferation or prolonged survival in Notch-induced T-ALL mouse models without significant toxicities (Pinnell et al. 2015). Thus, targeting context-dependent direct cofactors of ICN1 might selectively disable the oncogenic functions of Notch while sparing its essential normal functions.

3.4.1.3 Notch Dimerization

In T-ALL cells, Notch1 homo-dimerizes selectively at sequence-paired sites that regulate *Hes1*, *Ptcra* and *Myc* (Bailey and Posakony 1995; Nam et al. 2007; Liu et al. 2010; Yashiro-Ohtani et al. 2014). Dimerization is important for murine Notch-induced T-ALL leukemogenesis and proliferation in part, through inducing Myc (Liu et al. 2010). Since the residues that mediate dimerization are conserved in Notch2 and Notch3, it is possible that also Notch2 and Notch3 dimerize, as suggested by mass spectrometry data showing that NOTCH2 and NOTCH3 bind ICN1 (Yatim et al. 2012). In theory, disrupting the dimerization might be useful for avoiding the toxicities of pan-Notch inhibition. However, in contrast to the murine disease, in the human disease dimerization-defective ICN1 mutations can induce human *MYC* transcription and rescue the proliferation of human T-ALL cell lines upon withdrawal of Notch signals (Yashiro-Ohtani et al. 2014). Thus, in contrast to the murine disease, dimerization appears to be unimportant for the human disease.

3.4.1.4 AF4p12

AF54p12 is a coactivator that is recruited by ICN1 to Notch1 binding sites (Yatim et al. 2012). By recruiting RNA polymerase II (RNAPII), AF54p12 drives the transcription of oncogenic genes like *IL7R* and *HES1* (Yatim et al. 2012). It also appears to be selective for some Notch target genes as it does not regulate or recruit RNAPII to the *DTX1* locus. However, the mechanism for its context dependence is unclear.

3.4.2 Context-Dependent Regulation – Antagonism

While some transcriptional regulators strengthen Notch signals, others weaken them. As discussed in Sect. 2.1, CCNC/CDK associates with ICN1 leading to its degradation. ICN1 might also associate with transcriptional repressors such as retinoblasoma binding protein 4, chromatin remodeling factor (RBBP4), chromodomain helicase DNA binding protein 4 (CHD4), GATA zinc finger domain containing 2B (GATAD2B), and histone deacetylase 1 (HDAC1) which are components of the nucleosome remodeling deacetylase (NuRD) complex (Yatim et al. 2012). It is unclear whether these repressors are selective for certain target genes or cell types.

3.4.2.1 Ikaros

Ikaros, encoded by Ikaros family zinc finger 1 (*Ikzf1*) gene, is a transcriptional repressor that binds ~60% of ICN1/Rbpj sites (including the

NDME, Fig. 20.3) and suppresses a subset of Notch target genes in murine thymocytes and T-ALL (Geimer Le Lay et al. 2014; Chari and Winandy 2008). Ikaros orchestrates this regulation by competing with ICN1/Rbpj for shared binding sites (Beverly and Capobianco 2003; Kleinmann et al. 2008). Alternatively, Ikaros binds next to Rbpj and interferes with ICN1 function possibly through protein-protein interactions (Geimer Le Lay et al. 2014; Kathrein et al. 2008). The low frequency of inactivating *IKZF1* mutations (<5%) in human T-ALL seems to argue against IKAROS having broad functional significance (Marcais et al. 2010). However, ICN1 represses *IKZF1* transcription in human T-ALL (Witkowski et al. 2015) and induces dominant-negative isoforms of Ikaros in mouse models through alternative splicing (Bellavia et al. 2007). It must be noted that repression of Pten might also induce dominant-negative Ikaros isoforms (Yuan et al. 2017) and that restoring Ikaros levels induces tumor regression (Witkowski et al. 2015). Thus, it is intriguing to conceptualize therapeutic strategies that enhance Ikaros function, perhaps by interfering with its degradation (Song et al. 2015a; Song et al. 2015b).

4 Role of Notch in Mature Lymphoid Neoplasms

Notch is generally thought to play an oncogenic role in mature lymphoid neoplasms because of the identification of Notch gain-of-function mutations. However, Notch can also play a tumor suppressor role in lymphoma as it does in several solid cancers [reviewed by (Nowell and Radtke 2017)], Accordingly, Notch2 and other Notch pathway components are silenced due to direct transcriptional repression by Bcl6 (B-cell CLL/lymphoma 6) in normal follicular B cells and follicular lymphoma (FL) cells (Valls et al. 2017). Inhibiting Bcl6 restores Notch2 functions, thereby reducing FL viability. This tumor suppressive function might reflect normal B-cell biology since Notch2 suppresses the follicular B-cell phenotype and promotes the marginal zone B-cell phenotype in mice (Pillai and Cariappa 2009). Accordingly, gain-of-function *NOTCH2* mutations are rare in FLs but common in marginal zone lymphomas (Sect. 2.3). The role of Notch as oncogene in mature lymphoid cancers has only been preliminarily tested in animal models. Inactivation of *Notch2* in B cells slows the development of CLL-like disease in interferon regulatory factor 4 (IRF4) knock-in mice (Shukla et al. 2016). This might indicate a role for Notch during initiation of CLL-like disease. However, the relevance to human disease remains unclear as the mutated receptor in human CLL is *NOTCH1* not *NOTCH2*. Notch activation might also play a role during progression of human CLL as *NOTCH1* mutations in CLL are subclonal in ~35% of mutated cases (Nadeu et al. 2016). Notch1 is physiologically activated in human naïve and memory B cells, which are considered to be the cell-of-origin for CLL (Fabbri et al. 2017). Thus, CLL lymphomagenesis might be analogous to how T-ALL develops in the thymic microenvironment. We hypothesize that Notch signals are initially activated through ligand-receptor interactions and are then intensified as the disease progresses through mutations.

4.1 Notch Target Genes and Pathways in Mature Lymphoid Neoplasms

In contrast to the wealth of information in T-ALL, little is known about the oncogenic Notch target genes that drive mature lymphoid neoplasms. As a starting point, investigators have tested the possibility that genes relevant in T-ALL are also important in mature lymphoid neoplasms. This analysis unveiled that the CUTLL1 T-ALL cell line and the Rec-1 MCL B-cell line share 21 target genes that are downregulated by GSI (Stoeck et al. 2014). These genes include T-ALL drivers like *MYC* and *HES1*, which were also confirmed as Notch targets in other B-cell lines and primary B-cell lymphoma cells (Jitschin et al. 2015; Arruga et al. 2014; Fabbri et al. 2017). An integrated ChIP-Seq/RNA-Seq study of a CLL cell line that overexpresses tagged ICN1 identified *CCND3* as a direct Notch target and additionally identified novel targets not previously shown to be directly regulated by Notch in T-ALL, as among them B-cell CLL/lymphoma 2 (*BCL2*), C-X-C chemokine receptor type 4 (*CXCR4*) and myeloid leukemia cell differentiation protein 1 (*MCL1*) (Fabbri et al. 2017).

4.1.1 Myc

The NDME, which is active in T-ALL (Sect. 3.2.2), is inactive in mature B-cell lymphoma cells. In contrast, a separate B-cell specific Notch-dependent enhancer located at -0.5 Mb relatively to the TSS is active (Fabbri et al. 2017; Ryan et al. 2015). This enhancer is duplicated in about 4% of CLL patients (Fabbri et al. 2017) and is likely to be the same enhancer, located at -0.428 Mb, that was previously described to bind the Notch-like and Rbpj binding protein Epstein–Barr nuclear antigen 2 (EBNA2) in human Epstein-Barr virus (EBV)-transformed B cells (Zhao et al. 2011).

4.1.2 B-cell Receptor Signaling

B-lymphoma cells require constitutive activation of their B-cell receptors (BCR) for survival and inhibitors that target downstream BCR signals (e.g. ibrutinib or idelalisib) are Food and Drug Administration (FDA)-approved and emerging as front-line therapies. Preclinical data suggests that Notch and BCR signaling can cooperate during activation of normal B cells (Thomas et al. 2007). Accordingly, *NOTCH1* mutations might be associated with the expression of specific BCR subsets in CLL (Rossi et al. 2013). Further, in CLL cells that overexpress tagged ICN1, Notch1 directly induces the transcription of positive regulators of BCR signaling, such as *LYN* (Lck/yes-related novel protein tyrosine kinase), B lymphocyte kinase (*BLK*), spleen tyrosine kinase (*SYK*), complement C3d receptor 2 (*CR2*), phosphatidylinositol-4,5-bisphosphate 3-kinase catalytic subunit delta (*PIK3CD*), B-cell linker protein (*BLNK*) and members of the Ras and NF-κB pathways (Fabbri et al. 2017). Primary CLL and MCL cells stimulated *in vitro* for activation of both Notch and BCR signaling showed increased BCR signal transduction, such as phosphorylation of SYK and phospholipase Cγ (PLCγ) (Ryan et al. 2016). More extensive validation of these target genes as direct Notch drivers awaits future investigation.

5 Therapeutic Targeting of the Notch Signaling Pathway

5.1 Pan-Notch Inhibition

5.1.1 γ-secretase Inhibitors

γ-secretase inhibitors (GSI) prevent proteolysis of all four Notch receptors thus blocking all Notch signals (mutant or wildtype) with the exception of the rare t(7;9) translocation that removes the S3 cleavage site. Unfortunately, GSIs have been disappointing in clinical trials. More than a dozen clinical trials tested GSIs in patients with mostly solid cancers showed responses seen in less than 5% of patients [reviewed in (Andersson and Lendahl 2014)]. These studies were generally hampered by dose-limiting gastrointestinal (GI) toxicity, which was attributed to on-target effects of pan-Notch inhibition on the intestinal epithelium (VanDussen et al. 2012; van Es et al. 2005; Riccio et al. 2008; Milano et al. 2004). The first clinical trial testing GSIs (in particular MK-0752) in relapsed/refractory T-ALL was halted due to excessive diarrhea (Deangelo et al. 2009). However, 1 out of 7 patients had a 45% reduction in mediastinal mass (Deangelo et al. 2009). In a phase I trial of 8 relapsed/refractory adult T-ALL patients treated with the GSI PF-03084014, one patient achieved a complete remission for ~3 months (Papayannidis et al. 2015). In a preliminary report of an active Phase I trial, 25 relapsed/refractory pediatric T-ALL patients were treated with the GSI BMS-906024 with or without corticosteroids (Zweidler-McKay et al. 2014). The response rate was particularly encouraging at 32% of patients, perhaps reflecting the synergy between GSI and corticosteroids (Sect. 5.3.2). One of these patients was relapse-free for >19 months (Knoechel et al. 2015).

5.1.1.1 Ineffective Inhibition of Notch Signals or Primary Resistance?

Why have there been modest response rates of GSI in clinical trials, including the three mentioned above in T-ALL? One possibility is primary resistance, which will be discussed in Sect. 5.3. A second possibility is ineffective inhibition due to on-target toxicities. Because of these toxicities, GSI must be given intermittently, such as once or three times a week instead of daily. While intermittent dosing is more tolerable, it fails to achieve continuous suppression of Notch signaling given the short half-life of the GSI molecules (Krop et al. 2012; Cullion et al. 2009). Intermittent dosing is predicted to be less effective for tumors with PEST or FBXW7 mutations. Since these mutations prolong the half-life of ICN1, the level of ICN1 after each dose of GSI administration decreases more slowly than in cells without these mutations. Unfortunately, there is no robust and comprehensive set of biomarkers to quantitatively determine whether sufficient GSI has been given to achieve cancer-killing levels. Preclinical studies in genetically engineered mouse models of T-ALL give clues to whether ineffective inhibition of Notch signals or primary resistance is the greater clinical problem (Cullion et al. 2009; Rakowski et al. 2011). In these studies, an intermittent dosing schedule was used (three daily doses of GSI followed by 4 days off). Both studies showed disappointingly transient *in vivo* responses to GSI. However, tumor cells harvested from mice that succumbed to T-ALL despite treatment were highly sensitive to GSI inhibition *ex vivo*. Thus, the inability to raise Notch inhibition to cancer-killing levels might be an important clinical problem.

5.1.2 Other Pan-Notch Inhibitors

Besides GSI, other strategies that block all Notch signals are being developed and tested at the preclinical stage (Andersson and Lendahl 2014; Tosello and Ferrando 2013; Roti and Stegmaier 2014). These strategies include antibodies that target the γ-secretase complex (e.g. Nicastrin), inhibitors of ADAM protease, SERCA and HSP90 (Sect. 3.3) as well as MAM-like stapled peptides (SAHM) that disrupt the Notch transcriptional complex (Moellering et al. 2009). Since these strategies inhibit both normal and mutant Notch signals, it is possible that they might run into the same challenges that have plagued GSIs in clinical trials. However, there might be a therapeutic window not seen with GSI. For example, mice treated with SERCA inhibitors, HSP90 inhibitors or SAHM peptides did not develop the gastrointestinal toxicities of pan-Notch inhibition at doses that were effective in inhibiting growth of T-ALL xenografts (Moellering et al. 2009; Roti et al. 2013). The reduced toxicity could indicate antileukemic, Notch-independent effects of these drugs or that mutated Notch proteins might be more dependent on the Notch pathway machinery than the wild-type Notch proteins performing essential physiological functions.

5.2 Selective Notch Inhibition

Antibodies that target specific receptors or ligands would be predicted to have less toxicity than pan-Notch inhibition. Biotech companies have developed blocking antibodies that clamp and stabilize the NRR even in the presence of a Class I NRR mutation (Wu et al. 2010; Aste-Amezaga et al. 2010; Agnusdei et al. 2014). In mouse studies, the GI toxicities of Notch1 antibodies were lower compared to GSIs or combined Notch1/Notch2 antibodies (Wu et al. 2010). However, the cost of lowered toxicity might be a reduced efficacy. Accordingly, signaling by Class II NRR and JME mutations are resistant to Notch1 antibodies (Wu et al. 2010; Aste-Amezaga et al. 2010). Further, the inhibitory effect of Notch1 antibodies in a variety of *in vitro* signaling assays is frequently inferior to GSI, including assays of human T-ALL cell line proliferation (Wu et al. 2010; Aste-Amezaga et al. 2010). Selective Notch inhibition is also being tested with antibodies that bind Notch ligands. Such antibodies are predicted to be less effective in cancers driven by NRR mutations (e.g. T-ALL) but more effective in cancers driven by ligands (e.g. CLL).

5.3 Resistance to Anti-Notch Agents

5.3.1 Mechanisms of Resistance

About two-thirds of human Notch-activated T-ALL cell lines are resistant to GSI (Weng et al. 2004) as well as about one-third of patient-derived xenografts tested *ex vivo* (Roderick et al. 2014). Why do T-ALL cells develop resistance to Notch inhibitors without ever encountering Notch inhibitors? Investigators have speculated that Notch activation is an efficient mechanism for developing tumors to access multiple growth-promoting pathways simultaneously (Sect. 3.2). It is possible that tumor cells undergo further selection to acquire additional mechanisms that amplify these pathways. In this way, the tumors become less dependent on the original Notch signals. Accordingly, the major mechanisms of resistance appear to be through Notch-independent activation of two Notch-driven pathways: PI3K/AKT and MYC. It is important to note that neither pathway is sufficient to confer resistance in all contexts; however, both might be sufficient. Accordingly, combined expression of *Akt1* and *Myc* (but neither one alone) confers GSI resistance in 8 out of 8 $Kras^{G12D}$-induced murine T-ALL cell lines (Chiang, unpublished observations).

5.3.1.1 Pten/Akt

Inactivating mutations or deletions of *PTEN* occur in 10-30% of T-ALL patients (Palomero et al. 2007; Mendes et al. 2014; Gutierrez et al. 2009; Zuurbier et al. 2012). Inactivation of *PTEN* was invariably associated with GSI resistance of T-ALL cell lines (Palomero et al. 2007) however, a subsequent study identified four cell lines with *PTEN*-inactivating mutations that retained sensitivity to GSI (Zuurbier et al. 2012). Thus, PTEN loss might promote resistance to Notch inhibitors but is not sufficient in all contexts. Accordingly, induction of genetic deletion of *Pten* confers *in vivo* GSI resistance to established murine Notch-induced T-ALL (Herranz et al. 2015). However, T-ALL cell lines generated by transducing activated Notch1 into Pten-deficient murine hematopoietic progenitors, which are then transplanted into mice, retain high sensitivity to GSI (Medyouf et al. 2010). To reconcile these data, it has been suggested that PI3K/AKT activation induces GSI resistance only when it is a late event during pathogenesis (Mendes et al. 2016). Consistent with this idea, we observed that transduction of *Akt1* into $Kras^{G12D}$-expressing murine hematopoietic progenitors (which eventually develop into tumors when transplanted into recipient mice) does not confer GSI resistance (Chiang et al. 2016). In contrast, transduction of *Akt1* into some established $Kras^{G12D}$-induced murine T-ALL cell lines confers GSI resistance (Chiang, unpublished observations). Similarly, CRISPR-Cas9-mediated *PTEN* inactivation in some human cell lines confers GSI resistance (Herranz et al. 2015). Thus, it is possible that PTEN inactivation/AKT activation can confer GSI resistance in the context of a late event.

5.3.1.2 Myc

In contrast to Pten loss, Myc activation seems to confer GSI resistance more often in the context of an early rather than late event. For example, enforced expression of *Myc* induces T-ALL in animal models that are GSI resistant (Sect. 3.2.2) (Felsher and Bishop 1999; Langenau et al. 2003; Chiang et al. 2016). However, enforced expression of *MYC* in established human T-ALL cell lines confers GSI resistance in only limited contexts (Sect. 3.2.2) (Weng et al. 2006; Palomero et al. 2006b). In contrast to PI3K/AKT, which is frequently activated through mutations (Gutierrez et al. 2009), genetic lesions that induce MYC occur in only ~10% of human T-ALLs through t(8;14) translocation (Lange et al. 1992) or focal duplications of the N-ME (Herranz et al. 2014). However, additional pathways that activate MYC likely remain to be discovered. For example, when GSI-sensitive human T-ALL cell lines are chronically treated with low doses of GSI, these cells develop GSI resistance by activating a Notch-independent, BRD4-dependent *MYC* enhancer (Knoechel et al. 2014). FBXW7 targets MYC protein for degradation, so it is not surprising that MYC is upregulated by *FBXW7* mutations (Welcker et al. 2004; Yada et al. 2004; Thompson et al. 2007; O'Neil et al. 2007a). In

T-ALL cell lines, *FBXW7* mutations correlate with GSI resistance (Thompson et al. 2007; O'Neil et al. 2007a). Although suggestive, this study does not definitively implicate MYC as the resistance mechanism since FBXW7 degrades other oncoproteins such as c-ju-nana (JUN), MCL1, MTOR, and cyclin E1 (CCNE1) (Mao et al. 2008; Koepp et al. 2001; Strohmaier et al. 2001; Wei et al. 2005; Wertz et al. 2011; Inuzuka et al. 2011).

5.3.2 Agents That Increase Sensitivity to Notch Inhibitors

Given their relatively modest effects in clinical trials, GSIs are being tested in combination with other agents. Many of the agents discussed below inhibit known growth-promoting pathways downstream of Notch signals. The ability of these agents to enhance the effects of GSI is consistent with the notion that tumor cells have deepened their "addiction" to downstream Notch pathways by acquiring Notch-independent mechanisms to activate these pathways.

5.3.2.1 GSI and Glucocorticoids

Activation of Notch1 confers resistance to glucocorticoids in murine cell lines and thymocytes (Deftos et al. 1998). Accordingly, inhibiting Notch promotes glucocorticoid sensitivity of human T-ALL cell lines (Real et al. 2009). Notch confers resistance in part through inducing HES1 protein, which directly suppresses the transcription of the glucocorticoid receptor (Real et al. 2009). Notch also activates AKT, which phosphorylates the glucocorticoid receptor preventing its nuclear translocation (Piovan et al. 2013). Conversely, glucocorticoids also promote GSI sensitivity. In fact, glucocorticoids scored frequently in a library screen for compounds that promote GSI-mediated inhibition of a human T-ALL cell line (Gutierrez et al. 2014). Glucocorticoids can also protect mice from the GI toxicities induced by pan-Notch inhibition (Real et al. 2009). These data provide rationale for combined therapies with GSI and glucocorticoids. Accordingly, preliminary results of a clinical trial testing GSI in combination with glucocorticoids showed relatively high efficacy and low toxicity (Zweidler-McKay et al. 2014) (Sect. 5.1.1).

5.3.2.2 GSI and PI3K/MTOR Inhibitors

In theory, MTOR inhibitors like rapamycin could help treating GSI-resistant T-ALLs by activating the PI3K pathway (Sect. 5.3.1). Combining rapamycin with GSI synergistically enhances apoptosis and growth inhibition of Tal1/inhibitor of cyclin-dependent kinase 4 $(Ink4)^{+/-}$-induced murine T-ALLs and human T-ALL cell lines both *in vitro* and *in vivo* (Cullion et al. 2009; Chan et al. 2007). Combining the dual-specificity of PI3K/MTOR inhibitor PI-103 with GSI had additive antileukemic effects on $Kras^{G12D}$-induced murine T-ALL cell lines (Dail et al. 2010). Phenothiazines (e.g. perphenazine) target PI3K and ribosomal protein S6 kinase beta-1 (p70S6K) by binding and activating the tumor suppressor human protein phosphatase 2A. Combining perphenazine with GSI synergistically inhibits T-ALL proliferation (Gutierrez et al. 2014). However, this study is challenging to interpret since phenothiazines inhibit other targets like extracellular signal-regulated kinase (ERK). ERK inhibitors also enhance the anti-leukemic effects of GSI on $Kras^{G12D}$-induced murine T-ALL cell lines (Dail et al. 2010).

5.3.2.3 GSI and Inhibitors of Metabolism and Protein Synthesis

Since Notch has diverse metabolic effects (Herranz et al. 2015), it might seem unlikely that Notch-activated T-ALLs would be particularly dependent upon a single metabolic pathway. However, inhibiting glutaminolysis with the glutaminase inhibitor bis-2-(5-phenylacetamido-1,2,4-thiadiazol-2-yl)ethyl sulfide (BPTES) induced apoptosis of human T-ALL cell lines and PDX cells both *in vitro* and *in vivo* (Herranz et al. 2015). Further, combining BPTES with GSI had synergistic antileukemic effects. In a subsequent study, the same group identified associations between gene expression signatures given by various drugs and by GSI (Sanchez-Martin et al. 2017). Several of these drugs had strong synergistic effects in combination with GSI – withaferin A, rapamycin, vorinostat, parthenolide, wortmannin, astmizole, trifluoperazine and trichostatin A (Sanchez-Martin et al. 2017). Withaferin A is a steroidal lactone with multiple cell-type specific effects. In T-ALL cells, it inhibits translation by targeting eukaryotic translation initiation factor 2a (EIF2A). Combining GSI with withaferin A or another translation inhibitor (e.g. silvestrol) had synergistic *in vitro* growth inhibitory effects on Notch-activated human T-ALL cell lines. Withaferin A also had synergistic *in vivo* effects on Notch-induced murine T-ALL and PDXs (Sanchez-Martin et al. 2017).

5.3.2.4 GSI and Cyclin/CDK Inhibitors

Since GSI inhibits cell cycle progression (Sect. 3.2.5), there is rationale for combining GSI with cell cycle inhibitors. Accordingly, combining GSI with a cyclin D1 (CCND1)/CDK4 inhibitor (6-substituted indolocarbazole) enhanced retinoblastoma protein (RB) hypophosphorylation, G1/S arrest and apoptosis of Notch-activated human T-ALL cell lines (Rao et al. 2009). Another study found that combining GSI with the CDK4/6 inhibitor LEE011 synergistically inhibited proliferation of Notch1-activated T-ALL cell lines (Pikman et al. 2016). However, in both studies, these inhibitors had no effect on GSI-resistant T-ALL cell lines.

6 Role of Notch in Myeloid Neoplasms

6.1 Myeloproliferative Disease

In contrast to the incontrovertible role of Notch as an oncogene in several types of lymphoid cancers, the role of Notch in myeloid cancers is less clear. Notch has been proposed to be a tumor suppressor. This seems reasonable as Notch activation inhibits myeloid differentiation in human and mouse models both *in vivo* and *in vitro* (Chiang et al. 2008; de Pooter et al. 2006; Stier et al. 2002; Carlesso et al. 1999). Conversely, loss of Notch function in some mouse models leads to myeloid hyperplasia or myeloproliferative disease. These mouse models include conditional loss of *Ncstn* (which encodes Nicastrin) (Klinakis et al. 2011); combined loss of *Notch1/Notch2* (Klinakis et al. 2011; Dumortier et al. 2010); loss of *Pofut1* (Zhou et al. 2008; Yao et al. 2011); and loss of *Adam10* (Yoda et al. 2011). The myeloid phenotype appears to be cell-autonomous and cell non-autonomous (Yao et al. 2011; Yoda et al. 2011). One cell non-autonomous mechanism appears to be through epithelial loss of Notch signals (Dumortier et al. 2010). Notch-deprived epithelial cells secrete thymic stromal lymphopoietin, which simulates the secretion of high levels of granulocyte colony stimulating factor (G-CSF), a myeloid growth factor. Finally, ~12% of patients with chronic myelomonocytic leukemia harbor inactivating mutations in various Notch pathway components such as *NCSTN, MAML1, NOTCH2,* and anterior pharynx defective 1 homolog A (*APH1A*) (Klinakis et al. 2011).

6.2 Acute myeloid Leukemia

The role of Notch in acute myeloid leukemia (AML) appears context-dependent. Mutations that cause loss of Notch function have not been identified in AML. However, gene expression profiling of some primary human AML samples shows that Notch signals are silenced despite high levels of Notch2 expression (Lobry et al. 2013; Kannan et al. 2013). Reactivation of the Notch pathway with ICN in mouse models of AML or treating primary human AML samples with a Notch agonist peptide induced differentiation, growth arrest and apoptosis (Lobry et al. 2013; Kannan et al. 2013). Thus, Notch might be a tumor suppressor in AML in some contexts. In other contexts, Notch might have oncogenic functions. For example, a rare subset of AML called acute megakaryocytic leukemia (AMKL) contains one-twenty two-megakaryocytic acute leukemia (*OTT-MAL*) fusion oncogenes. OTT-MAL binds Rbpj independently of Notch and activates Notch signals in a mouse model of AMKL (Mercher et al. 2009). A second subset of AML is defined by the expression of the fusion oncogene, acute myeloid leukemia 1-eight-twenty one (*AML1-ETO*). These patients upregulate JAGGED1 and a subset of Notch target genes

(Alcalay et al. 2003). AML1-ETO binds regulatory sites within Notch target gene loci and induces transcription by interfering with the repressive function of ETO on RBPJ and requires its interaction with core binding factor β (CBFβ) (Salat et al. 2008; Thiel et al. 2017). A third subset of AML called acute promyelocytic leukemia (APL) is defined by the presence of fusion oncogenes involving the retinoic acid receptor-alpha (*RARA*) gene, such as promyelocytic leukemia locus gene (*PML-RARA*). Gene expression profiling of APL patient samples shows high transcripts of JAGGED1 and Notch target genes (Grieselhuber et al. 2013; Alcalay et al. 2003). Further, enforced expression of *PML-RARA* in a transgenic mouse model generated AMLs with similar Notch-activated profiles as the APL patients. Treating these leukemias *in vitro* or *in vivo* with GSI or DN-MAML reduced leukemic growth of a subset of the murine tumors (Grieselhuber et al. 2013). Thus, like FL (Sect. 4), AML might be heterogeneous with regard to the role of Notch as oncogene or tumor suppressor.

7 Summary

Notch activation is emerging as a major driver in an expanding range of lymphoid neoplasms. Notch activation can be ligand-independent or ligand-dependent. While the role of Notch is best understood in T-ALL, emerging data in other hematological neoplasms shows similarities but also important differences from T-ALL. In T-ALL, Notch activation promotes oncogenesis by efficiently providing the developing tumor cells with access to multiple tumor-promoting pathways, in particular PI3K/AKT and MYC. The relative importance of each pathway is highly context-dependent. Additional Notch-independent genetic alterations can be acquired that deepen the reliance of tumor cells on these downstream pathways. In some contexts, the tumor cells are weaned off their original dependence on Notch activation. In other contexts, tumor cells retain their dependence. These cells must also rely on the enzymes, chaperones, transcriptional regulators and other factors that facilitate the generation and/or function of ICN1. Thus, when patients present to their oncologists, their tumors can have two major weaknesses --- (1) a dependence on growth-promoting pathways downstream of Notch and (2) a synthetic lethal dependence on the Notch pathway machinery. Better understanding of these weaknesses, particularly their context dependence, could reveal ways to overcome the twin challenges of on-target toxicity and primary resistance that pan-Notch inhibitors have faced in clinical trials.

8 Future Directions

It would be helpful to develop a robust set of biomarkers to quantitate the level of Notch inhibition in tumor cells. With these biomarkers in hand, clinical investigators could determine whether Notch inhibition is high enough to expect anti-leukemic effects based on preclinical studies. For example, one might envision sorting T-ALL cells from the blood or CLL cells from lymph node aspirates and immediately processing them for a quantitative immunofluorescence assay for ICN1 or a quantitative real time PCR (qRT-PCR) panel of several oncogenic Notch1 target genes. The level of Notch inhibition based on these biomarkers would be highly informative when poor responses are seen in clinical trials. On one hand, these biomarkers might indicate that Notch inhibition is effective and sustained. If so, then finding strategies to overcome primary resistance becomes the top priority. On the other hand, these biomarkers might show that Notch inhibition cannot be raised and sustained at cancer-killing levels because of dose-limiting toxicities. If so, then finding safer strategies to target Notch assumes top priority.

For mature lymphoid cancers, the challenges are more fundamental than the ones for T-ALL. For example, Notch is generally activated in only a subset of cells. Thus, the question that needs to be addressed is whether targeting a mere subset could have therapeutic value. Does this subset have elite status such as exclusive access to stem cell potential or resistance to chemotherapy? Mature T cells normally activate Notch upon interaction with antigen presenting cells to generate the appropriate cytokine and differentiation response for immunity. Thus, it is unclear whether the cleaved ICN1 seen in T-cell lymphoma samples has pro-oncogenic effects or is simply an unrelated consequence of T-cell activation. Finally, mature lymphoid cancers are predicted to rely on ligand-receptor interactions. If this is correct, then antibodies that block specific ligands or receptors could be effective and have less toxicity than pan-Notch inhibitors. Thus, the top priority is to confirm, using *in vivo* studies, that these cancers are indeed ligand-dependent.

Acknowledgements We thank members of the Maillard and Chiang laboratories for helpful discussions and review of the manuscript. A.M. was supported by the NIH Ruth L. Kirschstein National Research Service Award (F30CA228228), the NIH Cellular and Molecular Biology Training Grant (T32-GM007315), the MSTP Training Grant (T32-GM007863-37), and the Rackham Graduate Student Research Grant. Q.W. and M.C. were supported by the NIH (R01CA196604). M.C. was also supported by the Rally Foundation for Childhood Cancer Research and the Alex's Lemonade Stand Foundation. The authors apologize to those whose work could not be mentioned owing to space limitations.

References

Abraham N, Ma MC, Snow JW, Miners MJ, Herndier BG, Goldsmith MA (2005) Haploinsufficiency identifies STAT5 as a modifier of IL-7-induced lymphomas. Oncogene 24(33):5252–5257. doi:1208726 [pii]. https://doi.org/10.1038/sj.onc.1208726

Agnusdei V, Minuzzo S, Frasson C, Grassi A, Axelrod F, Satyal S, Gurney A, Hoey T, Seganfreddo E, Basso G, Valtorta S, Moresco RM, Amadori A, Indraccolo S (2014) Therapeutic antibody targeting of Notch1 in T-acute lymphoblastic leukemia xenografts. Leukemia 28(2):278–288. https://doi.org/10.1038/leu.2013.183

Alcalay M, Meani N, Gelmetti V, Fantozzi A, Fagioli M, Orleth A, Riganelli D, Sebastiani C, Cappelli E, Casciari C, Sciurpi MT, Mariano AR, Minardi SP, Luzi L, Muller H, Di Fiore PP, Frosina G, Pelicci PG (2003) Acute myeloid leukemia fusion proteins deregulate genes involved in stem cell maintenance and DNA repair. J Clin Invest 112(11):1751–1761

Allman D, Karnell FG, Punt JA, Bakkour S, Xu L, Myung PS, Koretzky GA, Pui JC, Aster JC, Pear WS (2001) Separation of Notch1 promoted lineage commitment and expansion/transformation in developing T cells. J Exp Med 194(1):99–106

Andersson ER, Lendahl U (2014) Therapeutic modulation of Notch signalling--are we there yet? Nat Rev Drug Discov 13(5):357–378. https://doi.org/10.1038/nrd4252

Arcaini L, Rossi D, Lucioni M, Nicola M, Bruscaggin A, Fiaccadori V, Riboni R, Ramponi A, Ferretti VV, Cresta S, Casaluci GM, Bonfichi M, Gotti M, Merli M, Maffi A, Arra M, Varettoni M, Rattotti S, Morello L, Guerrera ML, Sciarra R, Gaidano G, Cazzola M, Paulli M (2015) The NOTCH pathway is recurrently mutated in diffuse large B-cell lymphoma associated with hepatitis C virus infection. Haematologica 100(2):246–252. https://doi.org/10.3324/haematol.2014.116855

Armstrong F, Brunet de la Grange P, Gerby B, Rouyez MC, Calvo J, Fontenay M, Boissel N, Dombret H, Baruchel A, Landman-Parker J, Romeo PH, Ballerini P, Pflumio F (2009) NOTCH is a key regulator of human T-cell acute leukemia initiating cell activity. Blood 113(8):1730–1740. doi:blood-2008-02-138172 [pii]. https://doi.org/10.1182/blood-2008-02-138172

Arruga F, Gizdic B, Serra S, Vaisitti T, Ciardullo C, Coscia M, Laurenti L, D'Arena G, Jaksic O, Inghirami G, Rossi D, Gaidano G, Deaglio S (2014) Functional impact of NOTCH1 mutations in chronic lymphocytic leukemia. Leukemia 28(5):1060–1070. https://doi.org/10.1038/leu.2013.319

Asnafi V, Buzyn A, Le Noir S, Baleydier F, Simon A, Beldjord K, Reman O, Witz F, Fagot T, Tavernier E, Turlure P, Leguay T, Huguet F, Vernant JP, Daniel F, Bene MC, Ifrah N, Thomas X, Dombret H, Macintyre E (2009) NOTCH1/FBXW7 mutation identifies a large subgroup with favorable outcome in adult T-cell acute lymphoblastic leukemia (T-ALL): a Group for Research on Adult Acute Lymphoblastic Leukemia (GRAALL) study. Blood 113(17):3918–3924. doi:blood-2008-10-184069 [pii]. https://doi.org/10.1182/blood-2008-10-184069

Aste-Amezaga M, Zhang N, Lineberger JE, Arnold BA, Toner TJ, Gu M, Huang L, Vitelli S, Vo KT, Haytko P, Zhao JZ, Baleydier F, L'Heureux S, Wang H, Gordon WR, Thoryk E, Andrawes MB, Tiyanont K, Stegmaier K, Roti G, Ross KN, Franlin LL, Wang F, Chastain M, Bett AJ, Audoly LP, Aster JC, Blacklow SC, Huber HE (2010) Characterization of Notch1 antibodies that inhibit signaling of both normal and mutated Notch1 receptors. PLoS One 5(2):e9094. https://doi.org/10.1371/journal.pone.0009094

Aster JC, Xu L, Karnell FG, Patriub V, Pui JC, Pear WS (2000) Essential roles for ankyrin repeat and transactivation domains in induction of T-cell leukemia by Notch1. Mol Cell Biol 20(20):7505–7515

Aster JC, Pear WS, Blacklow SC (2007) Notch signaling in Leukemia. Annu Rev Pathol 3:587–613

Avellino R, Romano S, Parasole R, Bisogni R, Lamberti A, Poggi V, Venuta S, Romano MF (2005) Rapamycin stimulates apoptosis of childhood acute lymphoblastic leukemia cells. Blood 106(4):1400–1406. https://doi.org/10.1182/blood-2005-03-0929

Bailey AM, Posakony JW (1995) Suppressor of hairless directly activates transcription of enhancer of split complex genes in response to Notch receptor activity. Genes Dev 9(21):2609–2622

Bain G, Maandag EC, Izon DJ, Amsen D, Kruisbeek AM, Weintraub BC, Krop I, Schlissel MS, Feeney AJ, van Roon M et al (1994) E2A proteins are required for proper B cell development and initiation of immunoglobulin gene rearrangements [see comments]. Cell 79(5):885–892

Bain G, Engel I, Robanus Maandag EC, te Riele HP, Voland JR, Sharp LL, Chun J, Huey B, Pinkel D, Murre C (1997) E2A deficiency leads to abnormalities in alphabeta T-cell development and to rapid development of T-cell lymphomas. Mol Cell Biol 17(8):4782–4791

Balatti V, Bottoni A, Palamarchuk A, Alder H, Rassenti LZ, Kipps TJ, Pekarsky Y, Croce CM (2012) NOTCH1 mutations in CLL associated with trisomy 12. Blood 119(2):329–331. https://doi.org/10.1182/blood-2011-10-386144

Baliakas P, Hadzidimitriou A, Sutton LA, Rossi D, Minga E, Villamor N, Larrayoz M, Kminkova J, Agathangelidis A, Davis Z, Tausch E, Stalika E, Kantorova B, Mansouri L, Scarfo L, Cortese D, Navrkalova V, Rose-Zerilli MJ, Smedby KE, Juliusson G, Anagnostopoulos A, Makris AM, Navarro A, Delgado J, Oscier D, Belessi C, Stilgenbauer S, Ghia P, Pospisilova S, Gaidano G, Campo E, Strefford JC, Stamatopoulos K, Rosenquist R, European Research Initiative on CLL (2015) Recurrent mutations refine prognosis in chronic lymphocytic leukemia. Leukemia 29(2):329–336. https://doi.org/10.1038/leu.2014.196

Barata JT, Keenan TD, Silva A, Nadler LM, Boussiotis VA, Cardoso AA (2004) Common gamma chain-signaling cytokines promote proliferation of T-cell acute lymphoblastic leukemia. Haematologica 89(12):1459–1467

Barton K, Muthusamy N, Fischer C, Ting CN, Walunas TL, Lanier LL, Leiden JM (1998) The Ets-1 transcription factor is required for the development of natural killer cells in mice. Immunity 9(4):555–563. doi:S1074-7613(00)80638-X [pii]

Bea S, Valdes-Mas R, Navarro A, Salaverria I, Martin-Garcia D, Jares P, Gine E, Pinyol M, Royo C, Nadeu F, Conde L, Juan M, Clot G, Vizan P, Di Croce L, Puente DA, Lopez-Guerra M, Moros A, Roue G, Aymerich M, Villamor N, Colomo L, Martinez A, Valera A, Martin-Subero JI, Amador V, Hernandez L, Rozman M, Enjuanes A, Forcada P, Muntanola A, Hartmann EM, Calasanz MJ, Rosenwald A, Ott G, Hernandez-Rivas JM, Klapper W, Siebert R, Wiestner A, Wilson WH, Colomer D, Lopez-Guillermo A, Lopez-Otin C, Puente XS, Campo E (2013) Landscape of somatic mutations and clonal evolution in mantle cell lymphoma. Proc Natl Acad Sci U S A 110(45):18250–18255. https://doi.org/10.1073/pnas.1314608110

Bellavia D, Campese AF, Alesse E, Vacca A, Felli MP, Balestri A, Stoppacciaro A, Tiveron C, Tatangelo L, Giovarelli M, Gaetano C, Ruco L, Hoffman ES, Hayday AC, Lendahl U, Frati L, Gulino A, Screpanti I (2000) Constitutive activation of NF-kappaB and T-cell leukemia/lymphoma in Notch3 transgenic mice. EMBO J 19(13):3337–3348

Bellavia D, Campese AF, Checquolo S, Balestri A, Biondi A, Cazzaniga G, Lendahl U, Fehling HJ, Hayday AC, Frati L, von Boehmer H, Gulino A, Screpanti I (2002) Combined expression of pTalpha and Notch3 in T cell leukemia identifies the requirement of preTCR for leukemogenesis. Proc Natl Acad Sci U S A 99(6):3788–3793

Bellavia D, Mecarozzi M, Campese AF, Grazioli P, Talora C, Frati L, Gulino A, Screpanti I (2007) Notch3 and the Notch3-upregulated RNA-binding protein HuD regulate Ikaros alternative splicing. EMBO J 26(6):1670–1680. https://doi.org/10.1038/sj.emboj.7601626

Bernasconi-Elias P, Hu T, Jenkins D, Firestone B, Gans S, Kurth E, Capodieci P, Deplazes-Lauber J, Petropoulos K, Thiel P, Ponsel D, Hee Choi S, LeMotte P, London A, Goetcshkes M, Nolin E, Jones MD, Slocum K, Kluk MJ, Weinstock DM, Christodoulou A, Weinberg O, Jaehrling J, Ettenberg SA, Buckler A, Blacklow SC, Aster JC, Fryer CJ (2016) Characterization of activating mutations of NOTCH3 in T-cell acute lymphoblastic leukemia and anti-leukemic activity of NOTCH3 inhibitory antibodies. Oncogene 35(47):6077–6086. https://doi.org/10.1038/onc.2016.133

Beverly LJ, Capobianco AJ (2003) Perturbation of Ikaros isoform selection by MLV integration is a cooperative event in Notch(IC)-induced T cell leukemogenesis. Cancer Cell 3(6):551–564

Beverly LJ, Felsher DW, Capobianco AJ (2005) Suppression of p53 by Notch in lymphomagenesis: implications for initiation and regression. Cancer Res 65(16):7159–7168. doi:65/16/7159[pii]. https://doi.org/10.1158/0008-5472.CAN-05-1664

Bonnet M, Loosveld M, Montpellier B, Navarro JM, Quilichini B, Picard C, Di Cristofaro J, Bagnis C, Fossat C, Hernandez L, Mamessier E, Roulland S, Morgado E, Formisano-Treziny C, Dik WA, Langerak AW, Prebet T, Vey N, Michel G, Gabert J, Soulier J, Macintyre EA, Asnafi V, Payet-Bornet D, Nadel B (2011) Posttranscriptional deregulation of MYC via PTEN constitutes a major alternative pathway of MYC activation in T-cell acute lymphoblastic leukemia. Blood 117(24):6650–6659. https://doi.org/10.1182/blood-2011-02-336842

Borggrefe T, Liefke R (2012) Fine-tuning of the intracellular canonical Notch signaling pathway. Cell Cycle 11(2):264–276. https://doi.org/10.4161/cc.11.2.18995

Bories JC, Willerford DM, Grevin D, Davidson L, Camus A, Martin P, Stehelin D, Alt FW (1995) Increased T-cell apoptosis and terminal B-cell differentiation induced by inactivation of the Ets-1 proto-oncogene. Nature 377(6550):635–638. https://doi.org/10.1038/377635a0

Bray SJ (2016) Notch signalling in context. Nat Rev Mol Cell Biol. https://doi.org/10.1038/nrm.2016.94

Bretones G, Delgado MD, Leon J (2015) Myc and cell cycle control. Biochim Biophys Acta 1849(5):506–516. https://doi.org/10.1016/j.bbagrm.2014.03.013

Buonamici S, Trimarchi T, Ruocco MG, Reavie L, Cathelin S, Mar BG, Klinakis A, Lukyanov Y, Tseng

JC, Sen F, Gehrie E, Li M, Newcomb E, Zavadil J, Meruelo D, Lipp M, Ibrahim S, Efstratiadis A, Zagzag D, Bromberg JS, Dustin ML, Aifantis I (2009) CCR7 signalling as an essential regulator of CNS infiltration in T-cell leukaemia. Nature 459(7249):1000–1004. doi:nature08020 [pii]. https://doi.org/10.1038/nature08020

Campese AF, Garbe AI, Zhang F, Grassi F, Screpanti I, von Boehmer H (2006) Notch1-dependent lymphomagenesis is assisted by but does not essentially require pre-TCR signaling. Blood 108(1):305–310. https://doi.org/10.1182/blood-2006-01-0143

Carlesso N, Aster JC, Sklar J, Scadden DT (1999) Notch1-induced delay of human hematopoietic progenitor cell differentiation is associated with altered cell cycle kinetics. Blood 93(3):838–848

Chan SM, Weng AP, Tibshirani R, Aster JC, Utz PJ (2007) Notch signals positively regulate activity of the mTOR pathway in T cell acute lymphoblastic leukemia. Blood 110(1):278–286

Chari S, Winandy S (2008) Ikaros regulates Notch target gene expression in developing thymocytes. J Immunol 181(9):6265–6274. doi:181/9/6265 [pii]

Chiang MY, Xu ML, Histen G, Shestova O, Roy M, Nam Y, Blacklow SC, Sacks DB, Pear WS, Aster JC (2006) Identification of a conserved negative regulatory sequence that influences the leukemogenic activity of NOTCH1. Mol Cell Biol 26(16):6261–6271

Chiang MY, Xu L, Shestova O, Histen G, L'Heureux S, Romany C, Childs ME, Gimotty PA, Aster JC, Pear WS (2008) Leukemia-associated NOTCH1 alleles are weak tumor initiators but accelerate K-ras-initiated leukemia. J Clin Investig 118(9):14

Chiang MY, Shestova O, Xu L, Aster JC, Pear WS (2013) Divergent effects of supraphysiologic Notch signals on leukemia stem cells and hematopoietic stem cells. Blood 121(6):905–917. https://doi.org/10.1182/blood-2012-03-416503. blood-2012-03-416503 [pii]

Chiang MY, Wang Q, Gormley AC, Stein SJ, Xu L, Shestova O, Aster JC, Pear WS (2016) High selective pressure for Notch1 mutations that induce Myc in T cell acute lymphoblastic leukemia. Blood. https://doi.org/10.1182/blood-2016-01-692855

Chiarini F, Fala F, Tazzari PL, Ricci F, Astolfi A, Pession A, Pagliaro P, McCubrey JA, Martelli AM (2009) Dual inhibition of class IA phosphatidylinositol 3-kinase and mammalian target of rapamycin as a new therapeutic option for T-cell acute lymphoblastic leukemia. Cancer Res 69(8):3520–3528. https://doi.org/10.1158/0008-5472.CAN-08-4884

Choi YJ, Li X, Hydbring P, Sanda T, Stefano J, Christie AL, Signoretti S, Look AT, Kung AL, von Boehmer H, Sicinski P (2012) The requirement for cyclin D function in tumor maintenance. Cancer Cell 22(4):438–451. https://doi.org/10.1016/j.ccr.2012.09.015

Chung J, Ebens CL, Perkey E, Radojcic V, Koch U, Scarpellino L, Tong A, Allen F, Wood S, Feng J, Friedman A, Granadier D, Tran IT, Chai Q, Onder L, Yan M, Reddy P, Blazar BR, Huang AY, Brennan TV, Bishop DK, Ludewig B, Siebel CW, Radtke F, Luther SA, Maillard I (2017) Fibroblastic niches prime T cell alloimmunity through Delta-like Notch ligands. J Clin Invest. https://doi.org/10.1172/JCI89535

Clappier E, Collette S, Grardel N, Girard S, Suarez L, Brunie G, Kaltenbach S, Yakouben K, Mazingue F, Robert A, Boutard P, Plantaz D, Rohrlich P, van Vlierberghe P, Preudhomme C, Otten J, Speleman F, Dastugue N, Suciu S, Benoit Y, Bertrand Y, Cave H (2010) NOTCH1 and FBXW7 mutations have a favorable impact on early response to treatment, but not on outcome, in children with T-cell acute lymphoblastic leukemia (T-ALL) treated on EORTC trials 58881 and 58951. Leukemia 24 (12):2023-2031. https://doi.org/10.1038/leu.2010.205. leu2010205 [pii]

Coustan-Smith E, Mullighan CG, Onciu M, Behm FG, Raimondi SC, Pei D, Cheng C, Su X, Rubnitz JE, Basso G, Biondi A, Pui CH, Downing JR, Campana D (2009) Early T-cell precursor leukaemia: a subtype of very high-risk acute lymphoblastic leukaemia. Lancet Oncol 10(2):147–156. doi:S1470-2045(08)70314-0 [pii]. https://doi.org/10.1016/S1470-2045(08)70314-0

Cullion K, Draheim KM, Hermance N, Tammam J, Sharma VM, Ware C, Nikov G, Krishnamoorthy V, Majumder PK, Kelliher MA (2009) Targeting the Notch1 and mTOR pathways in a mouse T-ALL model. Blood 113(24):6172–6181. doi:blood-2008-02-136762 [pii]. https://doi.org/10.1182/blood-2008-02-136762

Dai TY, Cao L, Yang ZC, Li YS, Tan L, Ran XZ, Shi CM (2014) P68 RNA helicase as a molecular target for cancer therapy. J Exp Clin Cancer Res 33:64. https://doi.org/10.1186/s13046-014-0064-y

Dail M, Li Q, McDaniel A, Wong J, Akagi K, Huang B, Kang HC, Kogan SC, Shokat K, Wolff L, Braun BS, Shannon K (2010) Mutant Ikzf1, KrasG12D, and Notch1 cooperate in T lineage leukemogenesis and modulate responses to targeted agents. Proc Natl Acad Sci U S A 107(11):5106–5111. https://doi.org/10.1073/pnas.1001064107

de Pooter RF, Schmitt TM, de la Pompa JL, Fujiwara Y, Orkin SH, Zuniga-Pflucker JC (2006) Notch signaling requires GATA-2 to inhibit myelopoiesis from embryonic stem cells and primary hemopoietic progenitors. J Immunol 176(9):5267–5275

Deangelo DJ, Stone RM, Silverman LB, Stock W, Attar EC, Fearen I, Dallob A, Matthews C, Stone J, Freedman SJ, Aster J (2009) A phase I clinical trial of the notch inhibitor MK-0752 in patients with T-cell acute lymphoblastic leukemia/lymphoma (T-ALL) and other leukemias. J Clin Oncol., 2006 ASCO Annual Meeting Proceedings Part I 24 (18S (June 20 Supplement)):6585

Deftos ML, He YW, Ojala EW, Bevan MJ (1998) Correlating notch signaling with thymocyte maturation. Immunity 9(6):777–786

Deftos ML, Huang E, Ojala EW, Forbush KA, Bevan MJ (2000) Notch1 signaling promotes the maturation of CD4 and CD8 SP thymocytes. Immunity 13(1):73–84

Demarest RM, Dahmane N, Capobianco AJ (2011) Notch is oncogenic dominant in T-cell acute lymphoblastic leukemia. Blood 117(10):2901–2909. doi:blood-2010-05-286351 [pii]. https://doi.org/10.1182/blood-2010-05-286351

Depew MJ, Liu JK, Long JE, Presley R, Meneses JJ, Pedersen RA, Rubenstein JL (1999) Dlx5 regulates regional development of the branchial arches and sensory capsules. Development 126(17):3831–3846

Di Ianni M, Baldoni S, Rosati E, Ciurnelli R, Cavalli L, Martelli MF, Marconi P, Screpanti I, Falzetti F (2009) A new genetic lesion in B-CLL: a NOTCH1 PEST domain mutation. Br J Haematol 146(6):689–691. https://doi.org/10.1111/j.1365-2141.2009.07816.x

Dohda T, Maljukova A, Liu L, Heyman M, Grander D, Brodin D, Sangfelt O, Lendahl U (2007) Notch signaling induces SKP2 expression and promotes reduction of p27Kip1 in T-cell acute lymphoblastic leukemia cell lines. Exp Cell Res 313(14):3141–3152. https://doi.org/10.1016/j.yexcr.2007.04.027

Dong Y, Zhang L, Zhang S, Bai Y, Chen H, Sun X, Yong W, Li W, Colvin SC, Rhodes SJ, Shou W, Zhang ZY (2012) Phosphatase of regenerating liver 2 (PRL2) is essential for placental development by down-regulating PTEN (Phosphatase and Tensin Homologue Deleted on Chromosome 10) and activating Akt protein. J Biol Chem 287(38):32172–32179. https://doi.org/10.1074/jbc.M112.393462

Dong Y, Zhang L, Bai Y, Zhou HM, Campbell AM, Chen H, Yong W, Zhang W, Zeng Q, Shou W, Zhang ZY (2014) Phosphatase of regenerating liver 2 (PRL2) deficiency impairs Kit signaling and spermatogenesis. J Biol Chem 289(6):3799–3810. https://doi.org/10.1074/jbc.M113.512079

Dudley DD, Wang HC, Sun XH (2009) Hes1 potentiates T cell lymphomagenesis by up-regulating a subset of notch target genes. PLoS One 4(8):e6678. https://doi.org/10.1371/journal.pone.0006678

Dumortier A, Durham AD, Di Piazza M, Vauclair S, Koch U, Ferrand G, Ferrero I, Demehri S, Song LL, Farr AG, Leonard WJ, Kopan R, Miele L, Hohl D, Finke D, Radtke F (2010) Atopic dermatitis-like disease and associated lethal myeloproliferative disorder arise from loss of Notch signaling in the murine skin. PLoS One 5(2):e9258. https://doi.org/10.1371/journal.pone.0009258

Egawa T, Tillman RE, Naoe Y, Taniuchi I, Littman DR (2007) The role of the Runx transcription factors in thymocyte differentiation and in homeostasis of naive T cells. J Exp Med 204(8):1945–1957. doi:jem.20070133 [pii]. https://doi.org/10.1084/jem.20070133

Eguchi-Ishimae M, Eguchi M, Kempski H, Greaves M (2008) NOTCH1 mutation can be an early, prenatal genetic event in T-ALL. Blood 111 (1):376-378

Ellisen LW, Bird J, West DC, Soreng AL, Reynolds TC, Smith SD, Sklar J (1991) TAN-1, the human homolog of the Drosophila notch gene, is broken by chromosomal translocations in T lymphoblastic neoplasms. Cell 66(4):649–661

van Es JH, van Gijn ME, Riccio O, van den Born M, Vooijs M, Begthel H, Cozijnsen M, Robine S, Winton DJ, Radtke F, Clevers H (2005) Notch/gamma-secretase inhibition turns proliferative cells in intestinal crypts and adenomas into goblet cells. Nature 435 (7044):959-963. doi:nature03659 [pii] https://doi.org/10.1038/nature03659

Espinosa L, Cathelin S, D'Altri T, Trimarchi T, Statnikov A, Guiu J, Rodilla V, Ingles-Esteve J, Nomdedeu J, Bellosillo B, Besses C, Abdel-Wahab O, Kucine N, Sun SC, Song G, Mullighan CC, Levine RL, Rajewsky K, Aifantis I, Bigas A (2010) The Notch/Hes1 pathway sustains NF-kappaB activation through CYLD repression in T cell leukemia. Cancer Cell 18(3):268–281. doi:S1535-6108(10)00302-8 [pii]. https://doi.org/10.1016/j.ccr.2010.08.006

Fabbri G, Rasi S, Rossi D, Trifonov V, Khiabanian H, Ma J, Grunn A, Fangazio M, Capello D, Monti S, Cresta S, Gargiulo E, Forconi F, Guarini A, Arcaini L, Paulli M, Laurenti L, Larocca LM, Marasca R, Gattei V, Oscier D, Bertoni F, Mullighan CG, Foa R, Pasqualucci L, Rabadan R, Dalla-Favera R, Gaidano G (2011) Analysis of the chronic lymphocytic leukemia coding genome: role of NOTCH1 mutational activation. J Exp Med 208(7):1389–1401. https://doi.org/10.1084/jem.20110921. jem.20110921 [pii]

Fabbri G, Holmes AB, Viganotti M, Scuoppo C, Belver L, Herranz D, Yan XJ, Kieso Y, Rossi D, Gaidano G, Chiorazzi N, Ferrando AA, Dalla-Favera R (2017) Common nonmutational NOTCH1 activation in chronic lymphocytic leukemia. Proc Natl Acad Sci U S A. https://doi.org/10.1073/pnas.1702564114

Fasnacht N, Huang HY, Koch U, Favre S, Auderset F, Chai Q, Onder L, Kallert S, Pinschewer DD, MacDonald HR, Tacchini-Cottier F, Ludewig B, Luther SA, Radtke F (2014) Specific fibroblastic niches in secondary lymphoid organs orchestrate distinct Notch-regulated immune responses. J Exp Med 211(11):2265–2279. https://doi.org/10.1084/jem.20132528

Felsher DW, Bishop JM (1999) Reversible tumorigenesis by MYC in hematopoietic lineages. Mol Cell 4(2):199–207

Fryer CJ, White JB, Jones KA (2004) Mastermind recruits CycC:CDK8 to phosphorylate the Notch ICD and coordinate activation with turnover. Mol Cell 16(4):509–520

Garcia-Ojeda ME, Klein Wolterink RG, Lemaitre F, Richard-Le Goff O, Hasan M, Hendriks RW, Cumano A, Di Santo JP (2013) GATA-3 promotes T-cell specification by repressing B-cell potential in pro-T cells in mice. Blood 121(10):1749–1759. https://doi.org/10.1182/blood-2012-06-440065

Geimer Le Lay AS, Oravecz A, Mastio J, Jung C, Marchal P, Ebel C, Dembele D, Jost B, Le Gras S, Thibault C, Borggrefe T, Kastner P, Chan S (2014) The tumor suppressor ikaros shapes the repertoire of notch target genes in T cells. Sci Signal 7(317). https://doi.org/10.1126/scisignal.20045457/317/ra28. [pii]

Gonzalez-Garcia S, Garcia-Peydro M, Martin-Gayo E, Ballestar E, Esteller M, Bornstein R, de la Pompa JL, Ferrando AA, Toribio ML (2009) CSL-MAML-dependent Notch1 signaling controls T lineage-specific IL-7R{alpha} gene expression in early human thymopoiesis and leukemia. J Exp Med 206 (4):779-791. doi:jem.20081922 [pii] https://doi.org/10.1084/jem.20081922

Gordon WR, Vardar-Ulu D, Histen G, Sanchez-Irizarry C, Aster JC, Blacklow SC (2007) Structural basis for autoinhibition of Notch. Nat Struct Mol Biol 14(4):295–300

Gordon WR, Roy M, Vardar-Ulu D, Garfinkel M, Mansour MR, Aster JC, Blacklow SC (2009) Structure of the Notch1-negative regulatory region: implications for normal activation and pathogenic signaling in T-ALL. Blood 113(18):4381–4390. https://doi.org/10.1182/blood-2008-08-174748

Grieselhuber NR, Klco JM, Verdoni AM, Lamprecht T, Sarkaria SM, Wartman LD, Ley TJ (2013) Notch signaling in acute promyelocytic leukemia. Leukemia 27(7):1548–1557. https://doi.org/10.1038/leu.2013.68

Gutierrez A, Sanda T, Grebliunaite R, Carracedo A, Salmena L, Ahn Y, Dahlberg S, Neuberg D, Moreau LA, Winter SS, Larson R, Zhang J, Protopopov A, Chin L, Pandolfi PP, Silverman LB, Hunger SP, Sallan SE, Look AT (2009) High frequency of PTEN, PI3K, and AKT abnormalities in T-cell acute lymphoblastic leukemia. Blood 114(3):647–650. doi:blood-2009-02-206722 [pii]. https://doi.org/10.1182/blood-2009-02-206722

Gutierrez A, Pan L, Groen RW, Baleydier F, Kentsis A, Marineau J, Grebliunaite R, Kozakewich E, Reed C, Pflumio F, Poglio S, Uzan B, Clemons P, VerPlank L, An F, Burbank J, Norton S, Tolliday N, Steen H, Weng AP, Yuan H, Bradner JE, Mitsiades C, Look AT, Aster JC (2014) Phenothiazines induce PP2A-mediated apoptosis in T cell acute lymphoblastic leukemia. J Clin Invest 124(2):644–655. https://doi.org/10.1172/JCI65093

Hattori N, Kawamoto H, Fujimoto S, Kuno K, Katsura Y (1996) Involvement of transcription factors TCF-1 and GATA-3 in the initiation of the earliest step of T cell development in the thymus. J Exp Med 184(3):1137–1147

Helbig C, Amsen D (2015) Notch Signaling: Piercing a Harness of Simplicity. Immunity 43(5):831–833. https://doi.org/10.1016/j.immuni.2015.10.022

Herranz D, Ambesi-Impiombato A, Palomero T, Schnell SA, Belver L, Wendorff AA, Xu L, Castillo-Martin M, Llobet-Navas D, Cordon-Cardo C, Clappier E, Soulier J, Ferrando AA (2014) A NOTCH1-driven MYC enhancer promotes T cell development, transformation and acute lymphoblastic leukemia. Nat Med 20(10):1130–1137. https://doi.org/10.1038/nm.3665. nm.3665 [pii]

Herranz D, Ambesi-Impiombato A, Sudderth J, Sanchez-Martin M, Belver L, Tosello V, Xu L, Wendorff AA, Castillo M, Haydu JE, Marquez J, Mates JM, Kung AL, Rayport S, Cordon-Cardo C, DeBerardinis RJ, Ferrando AA (2015) Metabolic reprogramming induces resistance to anti-NOTCH1 therapies in T cell acute lymphoblastic leukemia. Nat Med 21(10):1182–1189. https://doi.org/10.1038/nm.3955

Hodson DJ, Janas ML, Galloway A, Bell SE, Andrews S, Li CM, Pannell R, Siebel CW, MacDonald HR, De Keersmaecker K, Ferrando AA, Grutz G, Turner M (2010) Deletion of the RNA-binding proteins ZFP36L1 and ZFP36L2 leads to perturbed thymic development and T lymphoblastic leukemia. Nat Immunol 11(8):717–724. https://doi.org/10.1038/ni.1901

Hosoya T, Kuroha T, Moriguchi T, Cummings D, Maillard I, Lim KC, Engel JD (2009) GATA-3 is required for early T lineage progenitor development. J Exp Med 206(13):2987–3000. https://doi.org/10.1084/jem.20090934

Hozumi K, Mailhos C, Negishi N, Hirano K, Yahata T, Ando K, Zuklys S, Hollander GA, Shima DT, Habu S (2008) Delta-like 4 is indispensable in thymic environment specific for T cell development. J Exp Med 205(11):2507–2513. https://doi.org/10.1084/jem.20080134

Hu Y, Su H, Liu C, Wang Z, Huang L, Wang Q, Liu S, Chen S, Zhou J, Li P, Chen Z, Liu H, Qing G (2016) DEPTOR is a direct NOTCH1 target that promotes cell proliferation and survival in T-cell leukemia. Oncogene. https://doi.org/10.1038/onc.2016.275

Inuzuka H, Shaik S, Onoyama I, Gao D, Tseng A, Maser RS, Zhai B, Wan L, Gutierrez A, Lau AW, Xiao Y, Christie AL, Aster J, Settleman J, Gygi SP, Kung AL, Look T, Nakayama KI, DePinho RA, Wei W (2011) SCF(FBW7) regulates cellular apoptosis by targeting MCL1 for ubiquitylation and destruction. Nature 471(7336):104–109. https://doi.org/10.1038/nature09732. nature09732 [pii]

Jenkinson S, Koo K, Mansour MR, Goulden N, Vora A, Mitchell C, Wade R, Richards S, Hancock J, Moorman AV, Linch DC, Gale RE (2013) Impact of NOTCH1/FBXW7 mutations on outcome in pediatric T-cell acute lymphoblastic leukemia patients treated on the MRC UKALL 2003 trial. Leukemia 27(1):41–47. https://doi.org/10.1038/leu.2012.176. leu2012176 [pii]

Jeromin S, Weissmann S, Haferlach C, Dicker F, Bayer K, Grossmann V, Alpermann T, Roller A, Kohlmann A, Haferlach T, Kern W, Schnittger S (2014) SF3B1 mutations correlated to cytogenetics and mutations in NOTCH1, FBXW7, MYD88, XPO1 and TP53 in 1160 untreated CLL patients. Leukemia 28(1):108–117. https://doi.org/10.1038/leu.2013.263

Jitschin R, Braun M, Qorraj M, Saul D, Le Blanc K, Zenz T, Mougiakakos D (2015) Stromal cell-mediated glycolytic switch in CLL cells involves Notch-c-Myc signaling. Blood 125(22):3432–3436. https://doi.org/10.1182/blood-2014-10-607036

Johnson K, Hashimshony T, Sawai CM, Pongubala JM, Skok JA, Aifantis I, Singh H (2008) Regulation of immunoglobulin light-chain recombination by the transcription factor IRF-4 and the attenuation of interleukin-7 signaling. Immunity 28(3):335–345. doi:S1074-7613(08)00044-7 [pii]. https://doi.org/10.1016/j.immuni.2007.12.019

Joshi I, Minter LM, Telfer J, Demarest RM, Capobianco AJ, Aster JC, Sicinski P, Fauq A, Golde TE, Osborne

BA (2009) Notch signaling mediates G1/S cell-cycle progression in T cells via cyclin D3 and its dependent kinases. Blood 113(8):1689–1698. doi:blood-2008-03-147967 [pii]. https://doi.org/10.1182/blood-2008-03-147967

Jung C, Mittler G, Oswald F, Borggrefe T (2013) RNA helicase Ddx5 and the noncoding RNA SRA act as coactivators in the Notch signaling pathway. Biochim Biophys Acta 1833(5):1180–1189. https://doi.org/10.1016/j.bbamcr.2013.01.032

Kannan S, Sutphin RM, Hall MG, Golfman LS, Fang W, Nolo RM, Akers LJ, Hammitt RA, McMurray JS, Kornblau SM, Melnick AM, Figueroa ME, Zweidler-McKay PA (2013) Notch activation inhibits AML growth and survival: a potential therapeutic approach. J Exp Med 210(2):321–337. https://doi.org/10.1084/jem.20121527

Karube K, Martinez D, Royo C, Navarro A, Pinyol M, Cazorla M, Castillo P, Valera A, Carrio A, Costa D, Colomer D, Rosenwald A, Ott G, Esteban D, Gine E, Lopez-Guillermo A, Campo E (2014) Recurrent mutations of NOTCH genes in follicular lymphoma identify a distinctive subset of tumours. J Pathol 234(3):423–430. https://doi.org/10.1002/path.4428

Kathrein KL, Chari S, Winandy S (2008) Ikaros directly represses the notch target gene Hes1 in a leukemia T cell line: implications for CD4 regulation. J Biol Chem 283(16):10476–10484. doi:M709643200 [pii]. https://doi.org/10.1074/jbc.M709643200

Kiel MJ, Velusamy T, Betz BL, Zhao L, Weigelin HG, Chiang MY, Huebner-Chan DR, Bailey NG, Yang DT, Bhagat G, Miranda RN, Bahler DW, Medeiros LJ, Lim MS, Elenitoba-Johnson KS (2012) Whole-genome sequencing identifies recurrent somatic NOTCH2 mutations in splenic marginal zone lymphoma. J Exp Med 209(9):1553–1565. https://doi.org/10.1084/jem.20120910. jem.20120910 [pii]

King B, Trimarchi T, Reavie L, Xu L, Mullenders J, Ntziachristos P, Aranda-Orgilles B, Perez-Garcia A, Shi J, Vakoc C, Sandy P, Shen SS, Ferrando A, Aifantis I (2013) The Ubiquitin Ligase FBXW7 Modulates Leukemia-Initiating Cell Activity by Regulating MYC Stability. Cell 153(7):1552–1566. https://doi.org/10.1016/j.cell.2013.05.041. S0092-8674(13)00647-8 [pii]

Kleinmann E, Geimer Le Lay AS, Sellars M, Kastner P, Chan S (2008) Ikaros represses the transcriptional response to Notch signaling in T-cell development. Mol Cell Biol 28(24):7465–7475. https://doi.org/10.1128/MCB.00715-08

Klinakis A, Lobry C, Abdel-Wahab O, Oh P, Haeno H, Buonamici S, van De Walle I, Cathelin S, Trimarchi T, Araldi E, Liu C, Ibrahim S, Beran M, Zavadil J, Efstratiadis A, Taghon T, Michor F, Levine RL, Aifantis I (2011) A novel tumour-suppressor function for the Notch pathway in myeloid leukaemia. Nature 473(7346):230–233. doi:nature09999 [pii]. https://doi.org/10.1038/nature09999

Kluk MJ, Ashworth T, Wang H, Knoechel B, Mason EF, Morgan EA, Dorfman D, Pinkus G, Weigert O, Hornick JL, Chirieac LR, Hirsch M, Oh DJ, South AP, Leigh IM, Pourreyron C, Cassidy AJ, Deangelo DJ, Weinstock DM, Krop IE, Dillon D, Brock JE, Lazar AJ, Peto M, Cho RJ, Stoeck A, Haines BB, Sathayanrayanan S, Rodig S, Aster JC (2013) Gauging NOTCH1 Activation in Cancer Using Immunohistochemistry. PLoS One 8(6):e67306. https://doi.org/10.1371/journal.pone.0067306. PONE-D-13-13213 [pii]

Knoechel B, Roderick JE, Williamson KE, Zhu J, Lohr JG, Cotton MJ, Gillespie SM, Fernandez D, Ku M, Wang H, Piccioni F, Silver SJ, Jain M, Pearson D, Kluk MJ, Ott CJ, Shultz LD, Brehm MA, Greiner DL, Gutierrez A, Stegmaier K, Kung AL, Root DE, Bradner JE, Aster JC, Kelliher MA, Bernstein BE (2014) An epigenetic mechanism of resistance to targeted therapy in T cell acute lymphoblastic leukemia. Nat Genet 46(4):364–370. https://doi.org/10.1038/ng.2913. ng.2913 [pii]

Knoechel B, Bhattt A, DeAngelo DJ, Aster JC (2015) Complete hematologic response of early T-cell progenitor acute lymphoblastic leukemia to the γ-secretase inhibitor BMS-906024: genetic and epigenetic findings in an outlier case. Cold Spring Harb Mol Case Stud 1(1):a000539

Kobayashi M, Bai Y, Chen S, Gao R, Yao C, Cai W, Cardoso AA, Croop J, Zhang ZY, Liu Y (2016) Phosphatase PRL2 promotes oncogenic NOTCH1-induced T cell leukemia. Leukemia. https://doi.org/10.1038/leu.2016.340

Koch U, Fiorini E, Benedito R, Besseyrias V, Schuster-Gossler K, Pierres M, Manley NR, Duarte A, Macdonald HR, Radtke F (2008) Delta-like 4 is the essential, nonredundant ligand for Notch1 during thymic T cell lineage commitment. J Exp Med 205(11):2515–2523. doi:jem.20080829 [pii]. https://doi.org/10.1084/jem.20080829

Koepp DM, Schaefer LK, Ye X, Keyomarsi K, Chu C, Harper JW, Elledge SJ (2001) Phosphorylation-dependent ubiquitination of cyclin E by the SCFFbw7 ubiquitin ligase. Science 294(5540):173–177

Kox C, Zimmermann M, Stanulla M, Leible S, Schrappe M, Ludwig WD, Koehler R, Tolle G, Bandapalli OR, Breit S, Muckenthaler MU, Kulozik AE (2010) The favorable effect of activating NOTCH1 receptor mutations on long-term outcome in T-ALL patients treated on the ALL-BFM 2000 protocol can be separated from FBXW7 loss of function. Leukemia 24(12):2005–2013. https://doi.org/10.1038/leu.2010.203. leu2010203 [pii]

Kridel R, Meissner B, Rogic S, Boyle M, Telenius A, Woolcock B, Gunawardana J, Jenkins C, Cochrane C, Ben-Neriah S, Tan K, Morin RD, Opat S, Sehn LH, Connors JM, Marra MA, Weng AP, Steidl C, Gascoyne RD (2012) Whole transcriptome sequencing reveals recurrent NOTCH1 mutations in mantle cell lymphoma. Blood 119(9):1963–1971. https://doi.org/10.1182/blood-2011-11-391474. blood-2011-11-391474 [pii]

Krop I, Demuth T, Guthrie T, Wen PY, Mason WP, Chinnaiyan P, Butowski N, Groves MD, Kesari S,

Freedman SJ, Blackman S, Watters J, Loboda A, Podtelezhnikov A, Lunceford J, Chen C, Giannotti M, Hing J, Beckman R, Lorusso P (2012) Phase I pharmacologic and pharmacodynamic study of the gamma secretase (Notch) inhibitor MK-0752 in adult patients with advanced solid tumors. J Clin Oncol 30(19):2307–2313. https://doi.org/10.1200/JCO.2011.39.1540

Krysiak K, Gomez F, White BS, Matlock M, Miller CA, Trani L, Fronick CC, Fulton RS, Kreisel F, Cashen AF, Carson KR, Berrien-Elliott MM, Bartlett NL, Griffith M, Griffith OL, Fehniger TA (2016) Recurrent somatic mutations affecting B-cell receptor signaling pathway genes in follicular lymphoma. Blood. https://doi.org/10.1182/blood-2016-07-729954

La Starza R, Borga C, Barba G, Pierini V, Schwab C, Matteucci C, Lema Fernandez AG, Leszl A, Cazzaniga G, Chiaretti S, Basso G, Harrison CJ, Te Kronnie G, Mecucci C (2014) Genetic profile of T-cell acute lymphoblastic leukemias with MYC translocations. Blood 124(24):3577–3582. https://doi.org/10.1182/blood-2014-06-578856

Lange BJ, Raimondi SC, Heerema N, Nowell PC, Minowada J, Steinherz PE, Arenson EB, O'Connor R, Santoli D (1992) Pediatric leukemia/lymphoma with t(8;14)(q24;q11). Leukemia 6(7):613–618

Langenau DM, Traver D, Ferrando AA, Kutok JL, Aster JC, Kanki JP, Lin S, Prochownik E, Trede NS, Zon LI, Look AT (2003) Myc-induced T cell leukemia in transgenic zebrafish. Science 299(5608):887–890

Larrayoz M, Rose-Zerilli MJ, Kadalayil L, Parker H, Blakemore S, Forster J, Davis Z, Steele AJ, Collins A, Else M, Catovsky D, Oscier DG, Strefford JC (2016) Non-coding NOTCH1 mutations in chronic lymphocytic leukemia; their clinical impact in the UK CLL4 trial. Leukemia. https://doi.org/10.1038/leu.2016.298

Li X, Gounari F, Protopopov A, Khazaie K, von Boehmer H (2008) Oncogenesis of T-ALL and nonmalignant consequences of overexpressing intracellular NOTCH1. J Exp Med 205 (12):2851-2861. doi:jem.20081561 [pii] https://doi.org/10.1084/jem.20081561

Li N, Fassl A, Chick J, Inuzuka H, Li X, Mansour MR, Liu L, Wang H, King B, Shaik S, Gutierrez A, Ordureau A, Otto T, Kreslavsky T, Baitsch L, Bury L, Meyer CA, Ke N, Mulry KA, Kluk MJ, Roy M, Kim S, Zhang X, Geng Y, Zagozdzon A, Jenkinson S, Gale RE, Linch DC, Zhao JJ, Mullighan CG, Harper JW, Aster JC, Aifantis I, von Boehmer H, Gygi SP, Wei W, Look AT, Sicinski P (2014) Cyclin C is a haploinsufficient tumour suppressor. Nat Cell Biol 16(11):1080–1091. https://doi.org/10.1038/ncb3046

Lin S, Tian L, Shen H, Gu Y, Li JL, Chen Z, Sun X, You MJ, Wu L (2013) DDX5 is a positive regulator of oncogenic NOTCH1 signaling in T cell acute lymphoblastic leukemia. Oncogene 32(40):4845–4853. https://doi.org/10.1038/onc.2012.482

Liu H, Chi AW, Arnett KL, Chiang MY, Xu L, Shestova O, Wang H, Li YM, Bhandoola A, Aster JC, Blacklow SC, Pear WS (2010) Notch dimerization is required for leukemogenesis and T-cell development. Genes Dev 24(21):2395–2407. doi:gad.1975210 [pii]. https://doi.org/10.1101/gad.1975210

Liu Y, Easton J, Shao Y, Maciaszek J, Wang Z, Wilkinson MR, McCastlain K, Edmonson M, Pounds SB, Shi L, Zhou X, Ma X, Sioson E, Li Y, Rusch M, Gupta P, Pei D, Cheng C, Smith MA, Auvil JG, Gerhard DS, Relling MV, Winick NJ, Carroll AJ, Heerema NA, Raetz E, Devidas M, Willman CL, Harvey RC, Carroll WL, Dunsmore KP, Winter SS, Wood BL, Sorrentino BP, Downing JR, Loh ML, Hunger SP, Zhang J, Mullighan CG (2017) The genomic landscape of pediatric and young adult T-lineage acute lymphoblastic leukemia. Nat Genet 49(8):1211–1218. https://doi.org/10.1038/ng.3909

Lobry C, Ntziachristos P, Ndiaye-Lobry D, Oh P, Cimmino L, Zhu N, Araldi E, Hu W, Freund J, Abdel-Wahab O, Ibrahim S, Skokos D, Armstrong SA, Levine RL, Park CY, Aifantis I (2013) Notch pathway activation targets AML-initiating cell homeostasis and differentiation. J Exp Med 210(2):301–319. https://doi.org/10.1084/jem.20121484

Lohr JG, Stojanov P, Lawrence MS, Auclair D, Chapuy B, Sougnez C, Cruz-Gordillo P, Knoechel B, Asmann YW, Slager SL, Novak AJ, Dogan A, Ansell SM, Link BK, Zou L, Gould J, Saksena G, Stransky N, Rangel-Escareno C, Fernandez-Lopez JC, Hidalgo-Miranda A, Melendez-Zajgla J, Hernandez-Lemus E, Schwarz-Cruz y Celis A, Imaz-Rosshandler I, Ojesina AI, Jung J, Pedamallu CS, Lander ES, Habermann TM, Cerhan JR, Shipp MA, Getz G, Golub TR (2012) Discovery and prioritization of somatic mutations in diffuse large B-cell lymphoma (DLBCL) by whole-exome sequencing. Proc Natl Acad Sci U S A 109(10):3879–3884. https://doi.org/10.1073/pnas.1121343109

Loosveld M, Castellano R, Gon S, Goubard A, Crouzet T, Pouyet L, Prebet T, Vey N, Nadel B, Collette Y, Payet-Bornet D (2014) Therapeutic targeting of c-Myc in T-cell acute lymphoblastic leukemia, T-ALL. Oncotarget 5(10):3168–3172. https://doi.org/10.18632/oncotarget.1873

Malecki MJ, Sanchez-Irizarry C, Mitchell JL, Histen G, Xu ML, Aster JC, Blacklow SC (2006) Leukemia-associated mutations within the NOTCH1 heterodimerization domain fall into at least two distinct mechanistic classes. Mol Cell Biol 26(12):4642–4651

Mansour MR, Duke V, Foroni L, Patel B, Allen CG, Ancliff PJ, Gale RE, Linch DC (2007) Notch-1 mutations are secondary events in some patients with T-cell acute lymphoblastic leukemia. Clin Cancer Res 13(23):6964–6969

Mansour MR, Sulis ML, Duke V, Foroni L, Jenkinson S, Koo K, Allen CG, Gale RE, Buck G, Richards S, Paietta E, Rowe JM, Tallman MS, Goldstone AH, Ferrando AA, Linch DC (2009) Prognostic implications of NOTCH1 and FBXW7 mutations in adults with T-cell acute lymphoblastic leukemia treated on the MRC UKALLXII/ECOG E2993 protocol. J Clin Oncol 27(26):4352–4356. https://doi.org/10.1200/JCO.2009.22.0996. JCO.2009.22.0996 [pii]

Mao JH, Kim IJ, Wu D, Climent J, Kang HC, DelRosario R, Balmain A (2008) FBXW7 targets mTOR for degradation and cooperates with PTEN in tumor suppression. Science 321(5895):1499–1502. https://doi.org/10.1126/science.1162981321/5895/1499. [pii]

Marcais A, Jeannet R, Hernandez L, Soulier J, Sigaux F, Chan S, Kastner P (2010) Genetic inactivation of Ikaros is a rare event in human T-ALL. Leuk Res 34(4):426–429. doi:S0145-2126(09)00472-X [pii]. https://doi.org/10.1016/j.leukres.2009.09.012

Margolin AA, Palomero T, Sumazin P, Califano A, Ferrando AA, Stolovitzky G (2009) ChIP-on-chip significance analysis reveals large-scale binding and regulation by human transcription factor oncogenes. Proc Natl Acad Sci U S A 106(1):244–249. https://doi.org/10.1073/pnas.08064451060806445106. [pii]

Martinez D, Navarro A, Martinez-Trillos A, Molina-Urra R, Gonzalez-Farre B, Salaverria I, Nadeu F, Enjuanes A, Clot G, Costa D, Carrio A, Villamor N, Colomer D, Martinez A, Bens S, Siebert R, Wotherspoon A, Bea S, Matutes E, Campo E (2016) NOTCH1, TP53, and MAP2K1 Mutations in Splenic Diffuse Red Pulp Small B-cell Lymphoma Are Associated With Progressive Disease. Am J Surg Pathol 40(2):192–201. https://doi.org/10.1097/PAS.0000000000000523

McMillan BJ, Zimmerman B, Egan ED, Lofgren M, Xu X, Hesser A, Blacklow SC (2017) Structure of human POFUT1, its requirement in ligand-independent oncogenic Notch signaling, and functional effects of Dowling-Degos mutations. Glycobiology:1–10. https://doi.org/10.1093/glycob/cwx020

Medyouf H, Gao X, Armstrong F, Gusscott S, Liu Q, Gedman AL, Matherly LH, Schultz KR, Pflumio F, You MJ, Weng AP (2010) Acute T-cell leukemias remain dependent on Notch signaling despite PTEN and INK4A/ARF loss. Blood 115(6):1175–1184. doi:blood-2009-04-214718 [pii]. https://doi.org/10.1182/blood-2009-04-214718

Medyouf H, Gusscott S, Wang H, Tseng JC, Wai C, Nemirovsky O, Trumpp A, Pflumio F, Carboni J, Gottardis M, Pollak M, Kung AL, Aster JC, Holzenberger M, Weng AP (2011) High-level IGF1R expression is required for leukemia-initiating cell activity in T-ALL and is supported by Notch signaling. J Exp Med 208(9):1809–1822. doi:jem.20110121 [pii]. https://doi.org/10.1084/jem.20110121

Mendes RD, Sarmento LM, Cante-Barrett K, Zuurbier L, Buijs-Gladdines JG, Povoa V, Smits WK, Abecasis M, Yunes JA, Sonneveld E, Horstmann MA, Pieters R, Barata JT, Meijerink JP (2014) PTEN microdeletions in T-cell acute lymphoblastic leukemia are caused by illegitimate RAG-mediated recombination events. Blood 124(4):567–578. https://doi.org/10.1182/blood-2014-03-562751

Mendes RD, Cante-Barrett K, Pieters R, Meijerink JP (2016) The relevance of PTEN-AKT in relation to NOTCH1-directed treatment strategies in T-cell acute lymphoblastic leukemia. Haematologica 101(9):1010–1017. https://doi.org/10.3324/haematol.2016.146381

Mercher T, Raffel GD, Moore SA, Cornejo MG, Baudry-Bluteau D, Cagnard N, Jesneck JL, Pikman Y, Cullen D, Williams IR, Akashi K, Shigematsu H, Bourquin JP, Giovannini M, Vainchenker W, Levine RL, Lee BH, Bernard OA, Gilliland DG (2009) The OTT-MAL fusion oncogene activates RBPJ-mediated transcription and induces acute megakaryoblastic leukemia in a knockin mouse model. J Clin Invest 119(4):852–864. doi:35901 [pii]. https://doi.org/10.1172/JCI35901

Milano J, McKay J, Dagenais C, Foster-Brown L, Pognan F, Gadient R, Jacobs RT, Zacco A, Greenberg B, Ciaccio PJ (2004) Modulation of notch processing by gamma-secretase inhibitors causes intestinal goblet cell metaplasia and induction of genes known to specify gut secretory lineage differentiation. Toxicol Sci 82(1):341–358. https://doi.org/10.1093/toxsci/kfh254

Minervini A, Minervini CF, Anelli L, Zagaria A, Casieri P, Coccaro N, Cumbo C, Tota G, Impera L, Orsini P, Brunetti C, Giordano A, Specchia G, Albano F (2016) Droplet digital PCR analysis of NOTCH1 gene mutations in chronic lymphocytic leukemia. Oncotarget. https://doi.org/10.18632/oncotarget.13246

Minuzzo S, Agnusdei V, Pusceddu I, Pinazza M, Moserle L, Masiero M, Rossi E, Crescenzi M, Hoey T, Ponzoni M, Amadori A, Indraccolo S (2015) DLL4 regulates NOTCH signaling and growth of T acute lymphoblastic leukemia cells in NOD/SCID mice. Carcinogenesis 36(1):115–121. https://doi.org/10.1093/carcin/bgu223

Moellering RE, Cornejo M, Davis TN, Del Bianco C, Aster JC, Blacklow SC, Kung AL, Gilliland DG, Verdine GL, Bradner JE (2009) Direct inhibition of the NOTCH transcription factor complex. Nature 462 (7270):182-188. doi:nature08543 [pii] https://doi.org/10.1038/nature08543

Muthusamy N, Barton K, Leiden JM (1995) Defective activation and survival of T cells lacking the Ets-1 transcription factor. Nature 377(6550):639–642. https://doi.org/10.1038/377639a0

Nadeu F, Delgado J, Royo C, Baumann T, Stankovic T, Pinyol M, Jares P, Navarro A, Martin-Garcia D, Bea S, Salaverria I, Oldreive C, Aymerich M, Suarez-Cisneros H, Rozman M, Villamor N, Colomer D, Lopez-Guillermo A, Gonzalez M, Alcoceba M, Terol MJ, Colado E, Puente XS, Lopez-Otin C, Enjuanes A, Campo E (2016) Clinical impact of clonal and subclonal TP53, SF3B1, BIRC3, NOTCH1, and ATM mutations in chronic lymphocytic leukemia. Blood 127(17):2122–2130. https://doi.org/10.1182/blood-2015-07-659144

Nam Y, Sliz P, Pear WS, Aster JC, Blacklow SC (2007) Cooperative assembly of higher-order Notch complexes functions as a switch to induce transcription. Proc Natl Acad Sci U S A 104(7):2103–2108. doi:0611092104 [pii]. https://doi.org/10.1073/pnas.0611092104

Neumann M, Heesch S, Schlee C, Schwartz S, Gokbuget N, Hoelzer D, Konstandin NP, Ksienzyk B, Vosberg S, Graf A, Krebs S, Blum H, Raff T, Bruggemann M, Hofmann WK, Hecht J, Bohlander SK, Greif PA, Baldus CD (2013) Whole-exome sequencing in adult

ETP-ALL reveals a high rate of DNMT3A mutations. Blood 121(23):4749–4752. https://doi.org/10.1182/blood-2012-11-465138

Nie L, Xu M, Vladimirova A, Sun XH (2003) Notch-induced E2A ubiquitination and degradation are controlled by MAP kinase activities. EMBO J 22(21):5780–5792

Nowell CS, Radtke F (2017) Notch as a tumour suppressor. Nat Rev Cancer 17(3):145–159. https://doi.org/10.1038/nrc.2016.145

Ohtani Y, Xu S, Petrovic J, Xu L, Aster JC, Pear WS (2015) Modular domains within a super enhancer determine drug resistance in leukemia. Blood (ASH Annual Meeting Abstracts) 126(23):44

Onaindia A, Gomez S, Piris-Villaespesa M, Martinez-Laperche C, Cereceda L, Montes-Moreno S, Batlle A, de Villambrosia SG, Pollan M, Martin-Acosta P, Gonzalez-Rincon J, Menarguez J, Alves J, Rodriguez-Pinilla SM, Garcia JF, Mollejo M, Fraga M, Garcia-Marco JA, Piris MA, Sanchez-Beato M (2015) Chronic lymphocytic leukemia cells in lymph nodes show frequent NOTCH1 activation. Haematologica 100(5):e200–e203. https://doi.org/10.3324/haematol.2014.117705

O'Neil J, Calvo J, McKenna K, Krishnamoorthy V, Aster JC, Bassing CH, Alt FW, Kelliher M, Look AT (2006) Activating Notch1 mutations in mouse models of T-ALL. Blood 107(2):781–785

O'Neil J, Grim J, Strack P, Rao S, Tibbitts D, Winter C, Hardwick J, Welcker M, Meijerink JP, Pieters R, Draetta G, Sears R, Clurman BE, Look AT (2007a) FBW7 mutations in leukemic cells mediate NOTCH pathway activation and resistance to {gamma}-secretase inhibitors. J Exp Med 204(8):1813–1824

O'Neil J, Grim J, Strack P, Rao S, Tibbitts D, Winter C, Hardwick J, Welcker M, Meijerink JP, Pieters R, Draetta G, Sears R, Clurman BE, Look AT (2007b) FBW7 mutations in leukemic cells mediate NOTCH pathway activation and resistance to gamma-secretase inhibitors. J Exp Med 204(8):1813–1824

Palii CG, Perez-Iratxeta C, Yao Z, Cao Y, Dai F, Davison J, Atkins H, Allan D, Dilworth FJ, Gentleman R, Tapscott SJ, Brand M (2011) Differential genomic targeting of the transcription factor TAL1 in alternate haematopoietic lineages. EMBO J 30(3):494–509. https://doi.org/10.1038/emboj.2010.342

Palomero T, Barnes KC, Real PJ, Glade Bender JL, Sulis ML, Murty VV, Colovai AI, Balbin M, Ferrando AA (2006a) CUTLL1, a novel human T-cell lymphoma cell line with t(7;9) rearrangement, aberrant NOTCH1 activation and high sensitivity to gamma-secretase inhibitors. Leukemia 20(7):1279–1287. https://doi.org/10.1038/sj.leu.2404258

Palomero T, Lim WK, Odom DT, Sulis ML, Real PJ, Margolin A, Barnes KC, O'Neil J, Neuberg D, Weng AP, Aster JC, Sigaux F, Soulier J, Look AT, Young RA, Califano A, Ferrando AA (2006b) NOTCH1 directly regulates c-MYC and activates a feed-forward-loop transcriptional network promoting leukemic cell growth. Proc Natl Acad Sci U S A 103(48):18261–18266

Palomero T, Sulis ML, Cortina M, Real PJ, Barnes K, Ciofani M, Caparros E, Buteau J, Brown K, Perkins SL, Bhagat G, Agarwal AM, Basso G, Castillo M, Nagase S, Cordon-Cardo C, Parsons R, Zuniga-Pflucker JC, Dominguez M, Ferrando AA (2007) Mutational loss of PTEN induces resistance to NOTCH1 inhibition in T-cell leukemia. Nat Med 13(10):1203–1210

Pancewicz J, Taylor JM, Datta A, Baydoun HH, Waldmann TA, Hermine O, Nicot C (2010) Notch signaling contributes to proliferation and tumor formation of human T-cell leukemia virus type 1-associated adult T-cell leukemia. Proc Natl Acad Sci U S A 107(38):16619–16624. https://doi.org/10.1073/pnas.1010722107

Papayannidis C, DeAngelo DJ, Stock W, Huang B, Shaik MN, Cesari R, Zheng X, Reynolds JM, English PA, Ozeck M, Aster JC, Kuo F, Huang D, Lira PD, McLachlan KR, Kern KA, Garcia-Manero G, Martinelli G (2015) A Phase 1 study of the novel gamma-secretase inhibitor PF-03084014 in patients with T-cell acute lymphoblastic leukemia and T-cell lymphoblastic lymphoma. Blood Cancer J 5:e350. https://doi.org/10.1038/bcj.2015.80

Pear WS, Aster JC, Scott ML, Hasserjian RP, Soffer B, Sklar J, Baltimore D (1996) Exclusive development of T cell neoplasms in mice transplanted with bone marrow expressing activated Notch alleles. J Exp Med 183(5):2283–2291

Pikman Y, Alexe G, Roti G, Saur Conway A, Furman A, Lee ES, Place AE, Kim S, Saran C, Modiste R, Weinstock DM, Harris M, Kung AL, Silverman LB, Stegmaier K (2016) Synergistic Drug Combinations with a CDK4/6 Inhibitor in T-cell Acute Lymphoblastic Leukemia. Clin Cancer Res. https://doi.org/10.1158/1078-0432.CCR-15-2869

Pillai S, Cariappa A (2009) The follicular versus marginal zone B lymphocyte cell fate decision. Nat Rev Immunol 9(11):767–777. https://doi.org/10.1038/nri2656

Pinnell N, Yan R, Cho HJ, Keeley T, Murai MJ, Liu Y, Alarcon AS, Qin J, Wang Q, Kuick R, Elenitoba-Johnson KS, Maillard I, Samuelson LC, Cierpicki T, Chiang MY (2015) The PIAS-like Coactivator Zmiz1 Is a Direct and Selective Cofactor of Notch1 in T Cell Development and Leukemia. Immunity 43(5):870–883. https://doi.org/10.1016/j.immuni.2015.10.007

Piovan E, Yu J, Tosello V, Herranz D, Ambesi-Impiombato A, Da Silva AC, Sanchez-Martin M, Perez-Garcia A, Rigo I, Castillo M, Indraccolo S, Cross JR, de Stanchina E, Paietta E, Racevskis J, Rowe JM, Tallman MS, Basso G, Meijerink JP, Cordon-Cardo C, Califano A, Ferrando AA (2013) Direct reversal of glucocorticoid resistance by AKT inhibition in acute lymphoblastic leukemia. Cancer Cell 24(6):766–776. https://doi.org/10.1016/j.ccr.2013.10.022

Puente XS, Pinyol M, Quesada V, Conde L, Ordonez GR, Villamor N, Escaramis G, Jares P, Bea S, Gonzalez-Diaz M, Bassaganyas L, Baumann T, Juan M, Lopez-Guerra M, Colomer D, Tubio JM, Lopez C, Navarro A, Tornador C, Aymerich M, Rozman M, Hernandez JM, Puente DA, Freije JM, Velasco G, Gutierrez-Fernandez A, Costa D, Carrio A, Guijarro S, Enjuanes

A, Hernandez L, Yague J, Nicolas P, Romeo-Casabona CM, Himmelbauer H, Castillo E, Dohm JC, de Sanjose S, Piris MA, de Alava E, San Miguel J, Royo R, Gelpi JL, Torrents D, Orozco M, Pisano DG, Valencia A, Guigo R, Bayes M, Heath S, Gut M, Klatt P, Marshall J, Raine K, Stebbings LA, Futreal PA, Stratton MR, Campbell PJ, Gut I, Lopez-Guillermo A, Estivill X, Montserrat E, Lopez-Otin C, Campo E (2011) Whole-genome sequencing identifies recurrent mutations in chronic lymphocytic leukaemia. Nature 475(7354):101–105. https://doi.org/10.1038/nature10113

Puente XS, Bea S, Valdes-Mas R, Villamor N, Gutierrez-Abril J, Martin-Subero JI, Munar M, Rubio-Perez C, Jares P, Aymerich M, Baumann T, Beekman R, Belver L, Carrio A, Castellano G, Clot G, Colado E, Colomer D, Costa D, Delgado J, Enjuanes A, Estivill X, Ferrando AA, Gelpi JL, Gonzalez B, Gonzalez S, Gonzalez M, Gut M, Hernandez-Rivas JM, Lopez-Guerra M, Martin-Garcia D, Navarro A, Nicolas P, Orozco M, Payer AR, Pinyol M, Pisano DG, Puente DA, Queiros AC, Quesada V, Romeo-Casabona CM, Royo C, Royo R, Rozman M, Russinol N, Salaverria I, Stamatopoulos K, Stunnenberg HG, Tamborero D, Terol MJ, Valencia A, Lopez-Bigas N, Torrents D, Gut I, Lopez-Guillermo A, Lopez-Otin C, Campo E (2015) Non-coding recurrent mutations in chronic lymphocytic leukaemia. Nature 526(7574):519–524. https://doi.org/10.1038/nature14666

Qian L, Zhang W, Lei B, He A, Ye L, Li X, Dong X (2016) MicroRNA-101 regulates T-cell acute lymphoblastic leukemia progression and chemotherapeutic sensitivity by targeting Notch1. Oncol Rep. https://doi.org/10.3892/or.2016.5117

Rakowski LA, Lehotzky EA, Chiang MY (2011) Transient Responses to NOTCH and TLX1/HOX11 Inhibition in T-Cell Acute Lymphoblastic Leukemia/Lymphoma. PLoS One 6(2):e16761. https://doi.org/10.1371/journal.pone.0016761

Ramasamy SK, Kusumbe AP, Wang L, Adams RH (2014) Endothelial Notch activity promotes angiogenesis and osteogenesis in bone. Nature 507(7492):376–380. https://doi.org/10.1038/nature13146

Rao SS, O'Neil J, Liberator CD, Hardwick JS, Dai X, Zhang T, Tyminski E, Yuan J, Kohl NE, Richon VM, Van der Ploeg LH, Carroll PM, Draetta GF, Look AT, Strack PR, Winter CG (2009) Inhibition of NOTCH signaling by gamma secretase inhibitor engages the RB pathway and elicits cell cycle exit in T-cell acute lymphoblastic leukemia cells. Cancer Res 69(7):3060–3068. https://doi.org/10.1158/0008-5472.CAN-08-4295

Rasi S, Khiabanian H, Ciardullo C, Terzi-di-Bergamo L, Monti S, Spina V, Bruscaggin A, Cerri M, Deambrogi C, Martuscelli L, Biasi A, Spaccarotella E, De Paoli L, Gattei V, Foa R, Rabadan R, Gaidano G, Rossi D (2016) Clinical impact of small subclones harboring NOTCH1, SF3B1 or BIRC3 mutations in chronic lymphocytic leukemia. Haematologica 101(4):e135–e138. https://doi.org/10.3324/haematol.2015.136051

Real PJ, Tosello V, Palomero T, Castillo M, Hernando E, de Stanchina E, Sulis ML, Barnes K, Sawai C, Homminga I, Meijerink J, Aifantis I, Basso G, Cordon-Cardo C, Ai W, Ferrando A (2009) Gamma-secretase inhibitors reverse glucocorticoid resistance in T cell acute lymphoblastic leukemia. Nat Med 15 (1):50-58. doi:nm.1900 [pii] https://doi.org/10.1038/nm.1900

Reizis B, Leder P (2002) Direct induction of T lymphocyte-specific gene expression by the mammalian Notch signaling pathway. Genes Dev 16(3):295–300

Riccio O, van Gijn ME, Bezdek AC, Pellegrinet L, van Es JH, Zimber-Strobl U, Strobl LJ, Honjo T, Clevers H, Radtke F (2008) Loss of intestinal crypt progenitor cells owing to inactivation of both Notch1 and Notch2 is accompanied by derepression of CDK inhibitors p27Kip1 and p57Kip2. EMBO Rep 9 (4):377-383. doi:embor20087 [pii] https://doi.org/10.1038/embor.2008.7

Rich BE, Campos-Torres J, Tepper RI, Moreadith RW, Leder P (1993) Cutaneous lymphoproliferation and lymphomas in interleukin 7 transgenic mice. J Exp Med 177(2):305–316

Roderick JE, Tesell J, Shultz LD, Brehm MA, Greiner DL, Harris MH, Silverman LB, Sallan SE, Gutierrez A, Look AT, Qi J, Bradner JE, Kelliher MA (2014) c-Myc inhibition prevents leukemia initiation in mice and impairs the growth of relapsed and induction failure pediatric T-ALL cells. Blood 123(7):1040–1050. https://doi.org/10.1182/blood-2013-08-522698

Rossi D, Rasi S, Fabbri G, Spina V, Fangazio M, Forconi F, Marasca R, Laurenti L, Bruscaggin A, Cerri M, Monti S, Cresta S, Fama R, De Paoli L, Bulian P, Gattei V, Guarini A, Deaglio S, Capello D, Rabadan R, Pasqualucci L, Dalla-Favera R, Foa R, Gaidano G (2012a) Mutations of NOTCH1 are an independent predictor of survival in chronic lymphocytic leukemia. Blood 119(2):521–529. https://doi.org/10.1182/blood-2011-09-379966

Rossi D, Trifonov V, Fangazio M, Bruscaggin A, Rasi S, Spina V, Monti S, Vaisitti T, Arruga F, Fama R, Ciardullo C, Greco M, Cresta S, Piranda D, Holmes A, Fabbri G, Messina M, Rinaldi A, Wang J, Agostinelli C, Piccaluga PP, Lucioni M, Tabbo F, Serra R, Franceschetti S, Deambrogi C, Daniele G, Gattei V, Marasca R, Facchetti F, Arcaini L, Inghirami G, Bertoni F, Pileri SA, Deaglio S, Foa R, Dalla-Favera R, Pasqualucci L, Rabadan R, Gaidano G (2012b) The coding genome of splenic marginal zone lymphoma: activation of NOTCH2 and other pathways regulating marginal zone development. J Exp Med 209(9):1537–1551. https://doi.org/10.1084/jem.20120904. jem.20120904 [pii]

Rossi D, Spina V, Bomben R, Rasi S, Dal-Bo M, Bruscaggin A, Rossi FM, Monti S, Degan M, Ciardullo C, Serra R, Zucchetto A, Nomdedeu J, Bulian P, Grossi A, Zaja F, Pozzato G, Laurenti L, Efremov DG, Di-Raimondo F, Marasca R, Forconi F, Del Poeta G, Gaidano G, Gattei V (2013) Association between molecular lesions and specific B-cell receptor subsets in chronic lymphocytic leukemia. Blood 121(24):4902–4905. https://doi.org/10.1182/blood-2013-02-486209

Roti G, Stegmaier K (2014) New Approaches to Target T-ALL. Front Oncol 4:170. https://doi.org/10.3389/fonc.2014.00170

Roti G, Carlton A, Ross KN, Markstein M, Pajcini K, Su AH, Perrimon N, Pear WS, Kung AL, Blacklow SC, Aster JC, Stegmaier K (2013) Complementary genomic screens identify SERCA as a therapeutic target in NOTCH1 mutated cancer. Cancer Cell 23(3):390–405. https://doi.org/10.1016/j.ccr.2013.01.015S1535-6108(13)00037-8. [pii]

Ryan RJ, Drier Y, Whitton H, Cotton MJ, Kaur J, Issner R, Gillespie S, Epstein CB, Nardi V, Sohani AR, Hochberg EP, Bernstein BE (2015) Detection of Enhancer-Associated Rearrangements Reveals Mechanisms of Oncogene Dysregulation in B-cell Lymphoma. Cancer Discov 5(10):1058–1071. https://doi.org/10.1158/2159-8290.CD-15-0370

Ryan R, Petrovic J, Rausch D, Lareau C, Lee W, Donohue L, Christie A, Gillespie S, Kluk M, Nardi V, Faryabi R, Hochberg E, Weinstock D, Bernstein B, Aster J, Pear W (2016) Notch-regulated enhancers in B-cell lymphoma activate MYC and potentiate B-cell receptor signaling. Blood – ASH Annual Meeting Abstracts 128(22):457

Sade H, Krishna S, Sarin A (2004) The anti-apoptotic effect of Notch-1 requires p56lck-dependent, Akt/PKB-mediated signaling in T cells. J Biol Chem 279(4):2937–2944

Salat D, Liefke R, Wiedenmann J, Borggrefe T, Oswald F (2008) ETO, but not leukemogenic fusion protein AML1/ETO, augments RBP-Jkappa/SHARP-mediated repression of notch target genes. Mol Cell Biol 28(10):3502–3512. https://doi.org/10.1128/MCB.01966-07

Sanchez-Irizarry C, Carpenter AC, Weng AP, Pear WS, Aster JC, Blacklow SC (2004) Notch subunit heterodimerization and prevention of ligand-independent proteolytic activation depend, respectively, on a novel domain and the LNR repeats. Mol Cell Biol 24(21):9265–9273

Sanchez-Martin M, Ambesi-Impiombato A, Qin Y, Herranz D, Bansal M, Girardi T, Paietta E, Tallman MS, Rowe JM, De Keersmaecker K, Califano A, Ferrando AA (2017) Synergistic antileukemic therapies in NOTCH1-induced T-ALL. Proc Natl Acad Sci U S A. https://doi.org/10.1073/pnas.1611831114

Sanda T, Lawton LN, Barrasa MI, Fan ZP, Kohlhammer H, Gutierrez A, Ma W, Tatarek J, Ahn Y, Kelliher MA, Jamieson CH, Staudt LM, Young RA, Look AT (2012) Core transcriptional regulatory circuit controlled by the TAL1 complex in human T cell acute lymphoblastic leukemia. Cancer Cell 22(2):209–221. https://doi.org/10.1016/j.ccr.2012.06.007.S1535-6108(12)00256-5 [pii]

Sawai CM, Freund J, Oh P, Ndiaye-Lobry D, Bretz JC, Strikoudis A, Genesca L, Trimarchi T, Kelliher MA, Clark M, Soulier J, Chen-Kiang S, Aifantis I (2012) Therapeutic targeting of the cyclin D3:CDK4/6 complex in T cell leukemia. Cancer Cell 22(4):452–465. https://doi.org/10.1016/j.ccr.2012.09.016

Schnell SA, Ambesi-Impiombato A, Sanchez-Martin M, Belver L, Xu L, Qin Y, Kageyama R, Ferrando AA (2015) Therapeutic targeting of HES1 transcriptional programs in T-ALL. Blood 125(18):2806–2814. https://doi.org/10.1182/blood-2014-10-608448

Scripture-Adams DD, Damle SS, Li L, Elihu KJ, Qin S, Arias AM, Butler RR, 3rd, Champhekar A, Zhang JA, Rothenberg EV (2014) GATA-3 dose-dependent checkpoints in early T cell commitment. J Immunol 193 (7):3470-3491. https://doi.org/10.4049/jimmunol.1301663

Sharma VM, Calvo JA, Draheim KM, Cunningham LA, Hermance N, Beverly L, Krishnamoorthy V, Bhasin M, Capobianco AJ, Kelliher MA (2006) Notch1 contributes to mouse T-cell leukemia by directly inducing the expression of c-myc. Mol Cell Biol 26(21):8022–8031

Shimizu D, Taki T, Utsunomiya A, Nakagawa H, Nomura K, Matsumoto Y, Nishida K, Horiike S, Taniwaki M (2007) Detection of NOTCH1 mutations in adult T-cell leukemia/lymphoma and peripheral T-cell lymphoma. Int J Hematol 85(3):212–218. https://doi.org/10.1532/IJH97.06165

Shin HM, Minter LM, Cho OH, Gottipati S, Fauq AH, Golde TE, Sonenshein GE, Osborne BA (2006) Notch1 augments NF-kappaB activity by facilitating its nuclear retention. EMBO J 25(1):129–138. https://doi.org/10.1038/sj.emboj.7600902

Shochat C, Tal N, Bandapalli OR, Palmi C, Ganmore I, te Kronnie G, Cario G, Cazzaniga G, Kulozik AE, Stanulla M, Schrappe M, Biondi A, Basso G, Bercovich D, Muckenthaler MU, Izraeli S (2011) Gain-of-function mutations in interleukin-7 receptor-alpha (IL7R) in childhood acute lymphoblastic leukemias. J Exp Med 208(5):901–908. doi:jem.20110580 [pii]. https://doi.org/10.1084/jem.20110580

Shukla V, Shukla A, Joshi SS, Lu R (2016) Interferon regulatory factor 4 attenuates notch signaling to suppress the development of chronic lymphocytic leukemia. Oncotarget. https://doi.org/10.18632/oncotarget.9596

Sicinska E, Aifantis I, Le Cam L, Swat W, Borowski C, Yu Q, Ferrando AA, Levin SD, Geng Y, von Boehmer H, Sicinski P (2003) Requirement for cyclin D3 in lymphocyte development and T cell leukemias. Cancer Cell 4(6):451–461

Song C, Gowda C, Pan X, Ding Y, Tong Y, Tan BH, Wang H, Muthusami S, Ge Z, Sachdev M, Amin SG, Desai D, Gowda K, Gowda R, Robertson GP, Schjerven H, Muschen M, Payne KJ, Dovat S (2015a) Targeting casein kinase II restores Ikaros tumor suppressor activity and demonstrates therapeutic efficacy in high-risk leukemia. Blood 126(15):1813–1822. https://doi.org/10.1182/blood-2015-06-651505

Song C, Pan X, Ge Z, Gowda C, Ding Y, Li H, Li Z, Yochum G, Muschen M, Li Q, Payne KJ, Dovat S (2015b) Epigenetic regulation of gene expression by Ikaros, HDAC1 and Casein Kinase II in leukemia. Leukemia. https://doi.org/10.1038/leu.2015.331

Sportoletti P, Baldoni S, Cavalli L, Del Papa B, Bonifacio E, Ciurnelli R, Bell AS, Di Tommaso A, Rosati E, Crescenzi B, Mecucci C, Screpanti I, Marconi P,

Martelli MF, Di Ianni M, Falzetti F (2010) NOTCH1 PEST domain mutation is an adverse prognostic factor in B-CLL. Br J Haematol 151(4):404–406. https://doi.org/10.1111/j.1365-2141.2010.08368.x

Stier S, Cheng T, Dombkowski D, Carlesso N, Scadden DT (2002) Notch1 activation increases hematopoietic stem cell self-renewal in vivo and favors lymphoid over myeloid lineage outcome. Blood 99(7):2369–2378

Stoeck A, Lejnine S, Truong A, Pan L, Wang H, Zang C, Yuan J, Ware C, MacLean J, Garrett-Engele PW, Kluk M, Laskey J, Haines BB, Moskaluk C, Zawel L, Fawell S, Gilliland G, Zhang T, Kremer BE, Knoechel B, Bernstein BE, Pear WS, Liu XS, Aster JC, Sathyanarayanan S (2014) Discovery of biomarkers predictive of GSI response in triple-negative breast cancer and adenoid cystic carcinoma. Cancer Discov 4(10):1154–1167. https://doi.org/10.1158/2159-8290.CD-13-0830

Strohmaier H, Spruck CH, Kaiser P, Won KA, Sangfelt O, Reed SI (2001) Human F-box protein hCdc4 targets cyclin E for proteolysis and is mutated in a breast cancer cell line. Nature 413(6853):316–322. https://doi.org/10.1038/35095076

Sulis ML, Williams O, Palomero T, Tosello V, Pallikuppam S, Real PJ, Barnes K, Zuurbier L, Meijerink JP, Ferrando AA (2008) NOTCH1 extracellular juxtamembrane expansion mutations in T-ALL. Blood 112(3):733–740. https://doi.org/10.1182/blood-2007-12-130096

Tan Y, Sementino E, Xu J, Pei J, Liu Z, Ito TK, Cai KQ, Peri S, Klein-Szanto AJ, Wiest DL, Testa JR (2017) The homeoprotein Dlx5 drives murine T-cell lymphomagenesis by directly transactivating Notch and upregulating Akt signaling. Oncotarget. https://doi.org/10.18632/oncotarget.14784

Tatarek J, Cullion K, Ashworth T, Gerstein R, Aster JC, Kelliher MA (2011) Notch1 inhibition targets the leukemia-initiating cells in a Tal1/Lmo2 mouse model of T-ALL. Blood 118(6):1579–1590. https://doi.org/10.1182/blood-2010-08-300343. blood-2010-08-300343 [pii]

Thiel VN, Giaimo BD, Schwarz P, Soller K, Vas V, Bartkuhn M, Blatte TJ, Dohner K, Bullinger L, Borggrefe T, Geiger H, Oswald F (2017) Heterodimerization of AML1/ETO with CBFbeta is required for leukemogenesis but not for myeloproliferation. Leukemia. https://doi.org/10.1038/leu.2017.105

Thomas M, Calamito M, Srivastava B, Maillard I, Pear WS, Allman D (2007) Notch activity synergizes with B-cell-receptor and CD40 signaling to enhance B-cell activation. Blood 109(8):3342–3350

Thompson BJ, Buonamici S, Sulis ML, Palomero T, Vilimas T, Basso G, Ferrando A, Aifantis I (2007) The SCFFBW7 ubiquitin ligase complex as a tumor suppressor in T cell leukemia. J Exp Med 204(8):1825–1835

Tosello V, Ferrando AA (2013) The NOTCH signaling pathway: role in the pathogenesis of T-cell acute lymphoblastic leukemia and implication for therapy. Ther Adv Hematol 4(3):199–210. https://doi.org/10.1177/2040620712471368. 10.1177_2040620712471368 [pii]

Trimarchi T, Bilal E, Ntziachristos P, Fabbri G, Dalla-Favera R, Tsirigos A, Aifantis I (2014) Genome-wide mapping and characterization of Notch-regulated long noncoding RNAs in acute leukemia. Cell 158(3):593–606. https://doi.org/10.1016/j.cell.2014.05.049

Trinquand A, Tanguy-Schmidt A, Ben Abdelali R, Lambert J, Beldjord K, Lengline E, De Gunzburg N, Payet-Bornet D, Lhermitte L, Mossafa H, Lheritier V, Bond J, Huguet F, Buzyn A, Leguay T, Cahn JY, Thomas X, Chalandon Y, Delannoy A, Bonmati C, Maury S, Nadel B, Macintyre E, Ifrah N, Dombret H, Asnafi V (2013) Toward a NOTCH1/FBXW7/RAS/PTEN-based oncogenetic risk classification of adult T-cell acute lymphoblastic leukemia: a Group for Research in Adult Acute Lymphoblastic Leukemia study. J Clin Oncol 31(34):4333–4342. https://doi.org/10.1200/JCO.2012.48.5292

Troen G, Wlodarska I, Warsame A, Hernandez Llodra S, De Wolf-Peeters C, Delabie J (2008) NOTCH2 mutations in marginal zone lymphoma. Haematologica 93(7):1107–1109. https://doi.org/10.3324/haematol.11635

Uren AG, Kool J, Matentzoglu K, de Ridder J, Mattison J, van Uitert M, Lagcher W, Sie D, Tanger E, Cox T, Reinders M, Hubbard TJ, Rogers J, Jonkers J, Wessels L, Adams DJ, van Lohuizen M, Berns A (2008) Large-scale mutagenesis in p19(ARF)- and p53-deficient mice identifies cancer genes and their collaborative networks. Cell 133 (4):727-741. doi:S0092-8674(08)00436-4 [pii] https://doi.org/10.1016/j.cell.2008.03.021

Valls E, Lobry C, Geng H, Wang L, Cardenas M, Rivas M, Cerchietti L, Oh P, Yang SN, Oswald E, Graham CW, Jiang Y, Hatzi K, Agirre X, Perkey E, Li Z, Tam W, Bhatt K, Leonard JP, Zweidler-McKay PA, Maillard I, Elemento O, Ci W, Aifantis I, Melnick A (2017) BCL6 Antagonizes NOTCH2 to Maintain Survival of Human Follicular Lymphoma Cells. Cancer Discov. https://doi.org/10.1158/2159-8290.CD-16-1189

Van Vlierberghe P, Ambesi-Impiombato A, Perez-Garcia A, Haydu JE, Rigo I, Hadler M, Tosello V, Della Gatta G, Paietta E, Racevskis J, Wiernik PH, Luger SM, Rowe JM, Rue M, Ferrando AA (2011) ETV6 mutations in early immature human T cell leukemias. J Exp Med 208(13):2571–2579. https://doi.org/10.1084/jem.20112239

VanDussen KL, Carulli AJ, Keeley TM, Patel SR, Puthoff BJ, Magness ST, Tran IT, Maillard I, Siebel C, Kolterud A, Grosse AS, Gumucio DL, Ernst SA, Tsai YH, Dempsey PJ, Samuelson LC (2012) Notch signaling modulates proliferation and differentiation of intestinal crypt base columnar stem cells. Development 139(3):488–497. https://doi.org/10.1242/dev.070763. dev.070763 [pii]

Vilimas T, Mascarenhas J, Palomero T, Mandal M, Buonamici S, Meng F, Thompson B, Spaulding C, Macaroun S, Alegre ML, Kee BL, Ferrando A, Miele L, Aifantis I (2007) Targeting the NF-kappaB signaling pathway in Notch1-induced T-cell leukemia. Nat Med 13(1):70–77. doi:nm1524 [pii]. https://doi.org/10.1038/nm1524

Wang JH, Nichogiannopoulou A, Wu L, Sun L, Sharpe AH, Bigby M, Georgopoulos K (1996) Selective defects in the development of the fetal and adult lymphoid system in mice with an Ikaros null mutation. Immunity 5(6):537–549

Wang H, Zou J, Zhao B, Johannsen E, Ashworth T, Wong H, Pear WS, Schug J, Blacklow SC, Arnett KL, Bernstein BE, Kieff E, Aster JC (2011a) Genome-wide analysis reveals conserved and divergent features of Notch1/RBPJ binding in human and murine T-lymphoblastic leukemia cells. Proc Natl Acad Sci U S A 108(36):14908–14913. https://doi.org/10.1073/pnas.1109023108. 1109023108 [pii]

Wang L, Lawrence MS, Wan Y, Stojanov P, Sougnez C, Stevenson K, Werner L, Sivachenko A, DeLuca DS, Zhang L, Zhang W, Vartanov AR, Fernandes SM, Goldstein NR, Folco EG, Cibulskis K, Tesar B, Sievers QL, Shefler E, Gabriel S, Hacohen N, Reed R, Meyerson M, Golub TR, Lander ES, Neuberg D, Brown JR, Getz G, Wu CJ (2011b) SF3B1 and other novel cancer genes in chronic lymphocytic leukemia. N Engl J Med 365(26):2497–2506. https://doi.org/10.1056/NEJMoa1109016

Wang H, Zang C, Taing L, Arnett KL, Wong YJ, Pear WS, Blacklow SC, Liu XS, Aster JC (2014) NOTCH1-RBPJ complexes drive target gene expression through dynamic interactions with superenhancers. Proc Natl Acad Sci U S A 111(2):705–710. https://doi.org/10.1073/pnas.1315023111 1315023111. [pii]

Wang Z, Hu Y, Xiao D, Wang J, Liu C, Xu Y, Shi X, Jiang P, Huang L, Li P, Liu H, Qing G (2017) Stabilization of Notch1 by the Hsp90 Chaperon is Crucial for T Cell Leukemogenesis. Clin Cancer Res. https://doi.org/10.1158/1078-0432.CCR-16-2880

Wei W, Jin J, Schlisio S, Harper JW, Kaelin WG, Jr. (2005) The v-Jun point mutation allows c-Jun to escape GSK3-dependent recognition and destruction by the Fbw7 ubiquitin ligase. Cancer Cell 8 (1):25-33. https://doi.org/10.1016/j.ccr.2005.06.005

Wei G, Abraham BJ, Yagi R, Jothi R, Cui K, Sharma S, Narlikar L, Northrup DL, Tang Q, Paul WE, Zhu J, Zhao K (2011) Genome-wide analyses of transcription factor GATA3-mediated gene regulation in distinct T cell types. Immunity 35(2):299–311. https://doi.org/10.1016/j.immuni.2011.08.007

Welcker M, Orian A, Jin J, Grim JA, Harper JW, Eisenman RN, Clurman BE (2004) The Fbw7 tumor suppressor regulates glycogen synthase kinase 3 phosphorylation-dependent c-Myc protein degradation. Proc Natl Acad Sci U S A 101(24):9085–9090

Wendorff AA, Koch U, Wunderlich FT, Wirth S, Dubey C, Bruning JC, MacDonald HR, Radtke F (2010) Hes1 is a critical but context-dependent mediator of canonical Notch signaling in lymphocyte development and transformation. Immunity 33(5):671–684. doi:S1074-7613(10)00420-6 [pii]. https://doi.org/10.1016/j.immuni.2010.11.014

Weng AP, Nam Y, Wolfe MS, Pear WS, Griffin JD, Blacklow SC, Aster JC (2003) Growth suppression of pre-T acute lymphoblastic leukemia cells by inhibition of notch signaling. Mol Cell Biol 23(2):655–664

Weng AP, Ferrando AA, Lee W, JPt M, Silverman LB, Sanchez-Irizarry C, Blacklow SC, Look AT, Aster JC (2004) Activating mutations of NOTCH1 in human T cell acute lymphoblastic leukemia. Science 306(5694):269–271

Weng AP, Millholland JM, Yashiro-Ohtani Y, Arcangeli ML, Lau A, Wai C, Del Bianco C, Rodriguez CG, Sai H, Tobias J, Li Y, Wolfe MS, Shachaf C, Felsher D, Blacklow SC, Pear WS, Aster JC (2006) c-Myc is an important direct target of Notch1 in T-cell acute lymphoblastic leukemia/lymphoma. Genes Dev 20(15):2096–2109

Wertz IE, Kusam S, Lam C, Okamoto T, Sandoval W, Anderson DJ, Helgason E, Ernst JA, Eby M, Liu J, Belmont LD, Kaminker JS, O'Rourke KM, Pujara K, Kohli PB, Johnson AR, Chiu ML, Lill JR, Jackson PK, Fairbrother WJ, Seshagiri S, Ludlam MJ, Leong KG, Dueber EC, Maecker H, Huang DC, Dixit VM (2011) Sensitivity to antitubulin chemotherapeutics is regulated by MCL1 and FBW7. Nature 471(7336):110–114. https://doi.org/10.1038/nature09779. nature09779 [pii]

Witkowski MT, Cimmino L, Hu Y, Trimarchi T, Tagoh H, McKenzie MD, Best SA, Tuohey L, Willson TA, Nutt SL, Busslinger M, Aifantis I, Smyth GK, Dickins RA (2015) Activated Notch counteracts Ikaros tumor suppression in mouse and human T-cell acute lymphoblastic leukemia. Leukemia 29(6):1301–1311. https://doi.org/10.1038/leu.2015.27

Wouters BJ, Alberich Jorda M, Keeshan K, Louwers I, Erpelinck-Verschueren CA, Tielemans D, Langerak AW, He Y, Yashiro-Ohtani Y, Zhang P, Hetherington CJ, Verhaak RG, Valk PJ, Lowenberg B, Tenen DG, Pear WS, Delwel R (2007) Distinct gene expression profiles of acute myeloid/T-lymphoid leukemia with silenced CEBPA and mutations in NOTCH1. Blood 110(10):3706–3714. https://doi.org/10.1182/blood-2007-02-073486

Wu Y, Cain-Hom C, Choy L, Hagenbeek TJ, de Leon GP, Chen Y, Finkle D, Venook R, Wu X, Ridgway J, Schahin-Reed D, Dow GJ, Shelton A, Stawicki S, Watts RJ, Zhang J, Choy R, Howard P, Kadyk L, Yan M, Zha J, Callahan CA, Hymowitz SG, Siebel CW (2010) Therapeutic antibody targeting of individual Notch receptors. Nature 464 (7291):1052-1057. https://doi.org/10.1038/nature08878. nature08878 [pii]

Xu W, Carr T, Ramirez K, McGregor S, Sigvardsson M, Kee BL (2013) E2A transcription factors limit expression of Gata3 to facilitate T lymphocyte lineage commitment. Blood 121(9):1534–1542. https://doi.org/10.1182/blood-2012-08-449447

Yada M, Hatakeyama S, Kamura T, Nishiyama M, Tsunematsu R, Imaki H, Ishida N, Okumura F, Nakayama K, Nakayama KI (2004) Phosphorylation-dependent degradation of c-Myc is mediated by the F-box protein Fbw7. EMBO J 23(10):2116–2125

Yao D, Huang Y, Huang X, Wang W, Yan Q, Wei L, Xin W, Gerson S, Stanley P, Lowe JB, Zhou L (2011) Protein O-fucosyltransferase 1 (Pofut1) regulates lymphoid and myeloid homeostasis through modulation of Notch receptor ligand interactions. Blood 117(21):5652–5662. https://doi.org/10.1182/blood-2010-12-326074

Yashiro-Ohtani Y, He Y, Ohtani T, Jones ME, Shestova O, Xu L, Fang TC, Chiang MY, Intlekofer AM, Blacklow SC, Zhuang Y, Pear WS (2009) Pre-TCR signaling inactivates Notch1 transcription by antagonizing E2A. Genes Dev 23(14):1665–1676. doi:23/14/1665 [pii]. https://doi.org/10.1101/gad.1793709

Yashiro-Ohtani Y, Wang H, Zang C, Arnett KL, Bailis W, Ho Y, Knoechel B, Lanauze C, Louis L, Forsyth KS, Chen S, Chung Y, Schug J, Blobel GA, Liebhaber SA, Bernstein BE, Blacklow SC, Liu XS, Aster JC, Pear WS (2014) Long-range enhancer activity determines Myc sensitivity to Notch inhibitors in T cell leukemia. Proc Natl Acad Sci U S A 111(46):E4946–E4953. https://doi.org/10.1073/pnas.1407079111

Yatim A, Benne C, Sobhian B, Laurent-Chabalier S, Deas O, Judde JG, Lelievre JD, Levy Y, Benkirane M (2012) NOTCH1 nuclear interactome reveals key regulators of its transcriptional activity and oncogenic function. Mol Cell 48(3):445–458. https://doi.org/10.1016/j.molcel.2012.08.022S1097-2765(12)00740-X. [pii]

Yoda M, Kimura T, Tohmonda T, Uchikawa S, Koba T, Takito J, Morioka H, Matsumoto M, Link DC, Chiba K, Okada Y, Toyama Y, Horiuchi K (2011) Dual functions of cell-autonomous and non-cell-autonomous ADAM10 activity in granulopoiesis. Blood 118(26):6939–6942. https://doi.org/10.1182/blood-2011-06-357210

Yu S, Zhao DM, Jothi R, Xue HH (2010) Critical requirement of GABPalpha for normal T cell development. J Biol Chem 285(14):10179–10188. https://doi.org/10.1074/jbc.M109.088740

Yu VW, Saez B, Cook C, Lotinun S, Pardo-Saganta A, Wang YH, Lymperi S, Ferraro F, Raaijmakers MH, Wu JY, Zhou L, Rajagopal J, Kronenberg HM, Baron R, Scadden DT (2015) Specific bone cells produce DLL4 to generate thymus-seeding progenitors from bone marrow. J Exp Med 212(5):759–774. https://doi.org/10.1084/jem.20141843

Yuan T, Yang Y, Chen J, Li W, Li W, Zhang Q, Mi Y, Goswami RS, You JQ, Lin D, Qian MD, Calin S, Liang Y, Miranda RN, Calin GA, Zhou X, Ma L, Zweidler-Mc PA, Liu B, Weng AP, Medeiros LJ, Zhang Y, You MJ (2017) Regulation of PI3K signaling in T cell acute lymphoblastic leukemia: A novel PTEN/Ikaros/miR-26b mechanism reveals a critical targetable role for PIK3CD. Leukemia. https://doi.org/10.1038/leu.2017.80

Zenatti PP, Ribeiro D, Li W, Zuurbier L, Silva MC, Paganin M, Tritapoe J, Hixon JA, Silveira AB, Cardoso BA, Sarmento LM, Correia N, Toribio ML, Kobarg J, Horstmann M, Pieters R, Brandalise SR, Ferrando AA, Meijerink JP, Durum SK, Yunes JA, Barata JT (2011) Oncogenic IL7R gain-of-function mutations in childhood T-cell acute lymphoblastic leukemia. Nat Genet. https://doi.org/10.1038/ng.924ng.924. [pii]

Zhang J, Ding L, Holmfeldt L, Wu G, Heatley SL, Payne-Turner D, Easton J, Chen X, Wang J, Rusch M, Lu C, Chen SC, Wei L, Collins-Underwood JR, Ma J, Roberts KG, Pounds SB, Ulyanov A, Becksfort J, Gupta P, Huether R, Kriwacki RW, Parker M, McGoldrick DJ, Zhao D, Alford D, Espy S, Bobba KC, Song G, Pei D, Cheng C, Roberts S, Barbato MI, Campana D, Coustan-Smith E, Shurtleff SA, Raimondi SC, Kleppe M, Cools J, Shimano KA, Hermiston ML, Doulatov S, Eppert K, Laurenti E, Notta F, Dick JE, Basso G, Hunger SP, Loh ML, Devidas M, Wood B, Winter S, Dunsmore KP, Fulton RS, Fulton LL, Hong X, Harris CC, Dooling DJ, Ochoa K, Johnson KJ, Obenauer JC, Evans WE, Pui CH, Naeve CW, Ley TJ, Mardis ER, Wilson RK, Downing JR, Mullighan CG (2012) The genetic basis of early T-cell precursor acute lymphoblastic leukaemia. Nature 481(7380):157–163. https://doi.org/10.1038/nature10725nature10725. [pii]

Zhao B, Zou J, Wang H, Johannsen E, Peng CW, Quackenbush J, Mar JC, Morton CC, Freedman ML, Blacklow SC, Aster JC, Bernstein BE, Kieff E (2011) Epstein-Barr virus exploits intrinsic B-lymphocyte transcription programs to achieve immortal cell growth. Proc Natl Acad Sci U S A 108(36):14902–14907. https://doi.org/10.1073/pnas.1108892108. 1108892108 [pii]

Zhou L, Li LW, Yan Q, Petryniak B, Man Y, Su C, Shim J, Chervin S, Lowe JB (2008) Notch-dependent control of myelopoiesis is regulated by fucosylation. Blood 112(2):308–319. https://doi.org/10.1182/blood-2007-11-115204

Zuurbier L, Petricoin EF, 3rd, Vuerhard MJ, Calvert V, Kooi C, Buijs-Gladdines JG, Smits WK, Sonneveld E, Veerman AJ, Kamps WA, Horstmann M, Pieters R, Meijerink JP (2012) The significance of PTEN and AKT aberrations in pediatric T-cell acute lymphoblastic leukemia. Haematologica 97 (9):1405-1413. https://doi.org/10.3324/haematol.2011.059030

Zweidler-McKay PA, Deangelo DJ, Douer D, Dombret H, Ottoman OG, Vey N, Thomas DA, Zhu L, Huang F, Bajaj G, Fischer BS (2014) The Safety and Activity of BMS-906024, a Gamma Secretase Inhibitor (GSI) with Anti-Notch Activity, in Patients with Relapsed T-Cell Acute Lymphoblastic Leukemia (T-ALL): Initial Results of a Phase 1 Trial Blood (ASH Annual Meeting Abstracts) 124 (21)

Index

T. Borggrefe, B. D. Giaimo (eds.), *Molecular Mechanisms of Notch Signaling*, Advances in Experimental Medicine and Biology 1066, https://doi.org/10.1007/978-3-319-89512-3

MIX
Papier aus verantwortungsvollen Quellen
Paper from responsible sources
FSC® C105338
FSC
www.fsc.org

If you have any concerns about our products,
you can contact us on
ProductSafety@springernature.com

In case Publisher is established outside the EU,
the EU authorized representative is:
Springer Nature Customer Service Center GmbH
Europaplatz 3, 69115 Heidelberg, Germany

Printed by Libri Plureos GmbH
in Hamburg, Germany